ELASTIC BEAMS AND FRAMES

In architecture, as in all other operative arts, the end must direct the operation. The end is to build well. Well building hath three conditions. Commodity, firmness and delight.

Henry Wotton: *Elements of Architecture* (1624)

Axisymmetric Engineering Problems With Finite Element Programs
C.T.F. Ross, Professor of Structural Dynamics, University of Portsmouth

Signal Processing In Electronic Communications
M.J. Chapman et al, University of Coventry

Microelectronics: Microprocessor-Based Systems
D.M. Boniface, University of Portsmouth

Missile Guidance And Pursuit: Kinematics, Dynamics & Control
N.A. Shneydor, Technion Institute of Technology, Haifa, Israel

Landfill Waste Pollution And Control
K. Westlake, Loughborough University of Technology

Advance Applied Finite Element Methods
C.T.F. Ross, Professor of Structural Dynamics, University of Portsmouth

Finite Element Programs In Structural Engineering And Continuum Mechanics
C.T.F. Ross, Professor of Structural Dynamics, University of Portsmouth

Pressure Vessels: External Pressure Technology
C.T.F. Ross, Professor of Structural Dynamics, University of Portsmouth

Engineering Simulation And Control For Automotive Systems
T. Gordon & H. Williams. Loughborough University & Jaguar Cars Research

Analysis Of Engineering Structures
B. Bedenik & C. Besant, University of Maribor, Slovenia & Imperial College, London

Finite Elements Techniques In Structural Mechanics
C.T.F. Ross, Professor of Structural Dynamics. University of Portsmouth

Macro-Engineering: Mit Brunel Lectures On Global Infrastructure
F.P. Davidson et al, Massachusetts Institute of Technology

Macro-Engineering & The Earth: World Projects For Year 2000
U. Kitzinger & E.G. Frankel, Templeton College, Oxford & Massachusetts Institute of Technology

Dynamics Of Mechanical Systems
C.T.F. Ross, Professor of Structural Dynamics, University of Portsmouth

Mechanics Of Solids
C.T.F. Ross, Professor of Structural Dynamics, University of Portsmouth

Engineering Thermodynamics: Converting Gas & Steam Cycles To Work
G. Cole, Dept. of Engineering Design & Manufacture, Hull University

ELASTIC BEAMS AND FRAMES

Second Edition

JOHN D. RENTON
Department of Engineering Science
University of Oxford
Oxford

WOODHEAD PUBLISHING

Oxford Cambridge Philadelphia New Delhi

Published by Woodhead Publishing Limited,
80 High Street, Sawston, Cambridge CB22 3HJ
www.woodheadpublishing.com

Woodhead Publishing, 1518 Walnut Street, Suite 1100, Philadelphia,
PA 19102-3406, USA

Woodhead Publishing India Private Limited, G-2, Vardaan House, 7/28 Ansari Road,
Daryaganj, New Delhi – 110002, India
www.woodheadpublishingindia.com

First published by Horwood Publishing Limited, 2002
Reprinted by Woodhead Publishing Limited, 2011

British Library Cataloguing in Publication Data
A catalogue record for this book is available from the British Library

ISBN 978-1-898563-86-0

Table of Contents

Preface

I cannot doubt but that these things, which now seem to us so mysterious, will be no mysteries at all; that the scales will fall from our eyes; that we shall learn to look on things in a different way - when that which is now a difficulty will be the only common-sense and intelligible way of looking at the subject.

(Lord Kelvin)

Early Developments in the Theory of Elasticity

1638 Galileo Galilei

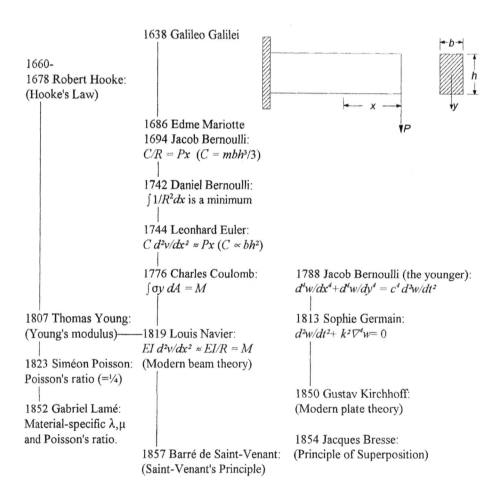

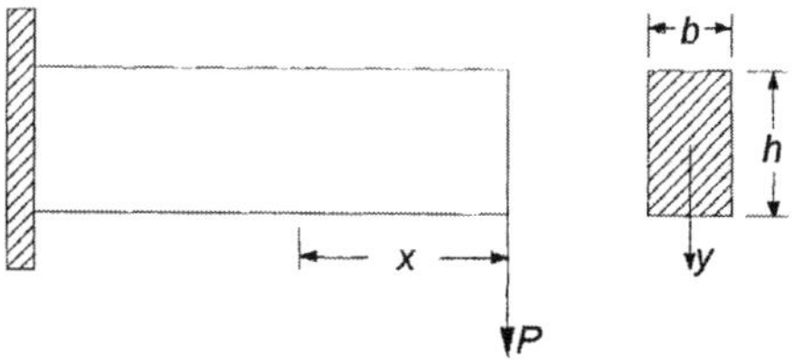

1660-
1678 Robert Hooke:
(Hooke's Law)

1686 Edme Mariotte
1694 Jacob Bernoulli:
$C/R = Px$ $(C = mbh^3/3)$

1742 Daniel Bernoulli:
$\int 1/R^2 dx$ is a minimum

1744 Leonhard Euler:
$C\, d^2v/dx^2 \approx Px$ $(C \propto bh^2)$

1776 Charles Coulomb:
$\int \sigma y\, dA = M$

1788 Jacob Bernoulli (the younger):
$d^4w/dx^4 + d^4w/dy^4 = c^4\, d^2w/dt^2$

1807 Thomas Young:
(Young's modulus)——1819 Louis Navier:
$EI\, d^2v/dx^2 \approx EI/R = M$
(Modern beam theory)

1813 Sophie Germain:
$d^2w/dt^2 + k^2\nabla^4 w = 0$

1823 Siméon Poisson:
Poisson's ratio (=¼)

1850 Gustav Kirchhoff:
(Modern plate theory)

1852 Gabriel Lamé:
Material-specific λ, μ
and Poisson's ratio.

1854 Jacques Bresse:
(Principle of Superposition)

1857 Barré de Saint-Venant:
(Saint-Venant's Principle)

Preface

The history of science is one of observation, measurement, and the postulation of theories to explain the observed phenomena. As Karl Popper said, any theory worthy of being called scientific should be capable of making predictions which can be tested. With the passage of time, these processes become more precise and sophisticated. Theories which had earlier seemed to be satisfactory turn out to be inadequate, either because more precise measurements and analysis show them to be false or because they lack the scope to explain new phenomena. Also, they may be replaced by new methods and ideas which are more elegant or useful. Thus Young's modulus, E as used today, should really be attributed to Navier. Young's own definition was "The modulus of elasticity of any substance is a column of the same substance, capable of producing a pressure on its base which is to the weight causing a certain degree of compression as the length is to the diminution of its length". Poisson was originally convinced that his ratio, ν, had a fixed value of ¼ until the body of experimental evidence showed that it must take different values for different materials. Lamé (1852) was probably the first to accept that this must be so.

Galileo first examined the resistance of a cantilever beam to failure. He had no concept of elasticity, and implicitly assumed a constant stress distribution at the support with failure occurring by rotation about the bottom edge. Although this lead to poor predictions of the failure of a cantilever, it came to be known as 'Galileo's problem'. Working independently from Hooke, Mariotte found the elastic stress distribution in bending. This was the correct linear response, with the neutral axis at the centroid. However, owing to a mathematical error, he concluded that this gave the same result for the resistance of the beam as taking a tensile linear variation in stress from a fulcrum at the bottom edge. The name 'Bernoulli' is often used generically for the whole family from Jacob[1] (the elder) to Daniel. The former is usually credited with the invention of beam theory, having determined that beam curvature was proportional to the local bending moment. However, he incorporated Mariotte's mistake in his work. Even Euler, who first applied differential calculus to beam theory, wrongly estimated the stiffness of a rectangular beam to be proportional to the square of its depth. Again, it is Navier who should take the credit for beam theory as it is normally used today[2].

Perhaps surprisingly, the earliest work on the theory of plates was concerned with their vibration. Jacob Bernoulli treated plates as if they were square grids of beams (subject to bending only) and so derived an incorrect differential equation for their lateral displacement w. Sophie Germain found the correct form for the equation, but was at a loss to give an expression for the plate's bending stiffness. Poisson (1814) deduced the same equation as Sophie, but claimed that it was too complicated to be solved. He also insisted that three boundary conditions could be imposed on each edge, a debate which continued until Kirchhoff's definitive work on the subject in 1850. The Bernoulli-Euler hypothesis for beams, that plane sections remain plane, has an equivalent for plates and shells known as the Kirchhoff-Love hypothesis. This hypothesis has been challenged from the earliest times, most notably by Barré de Saint-Venant. It will be seen that his name and that of Bresse have been added to the above chart. In Chapter 8, it will be shown that their principles unify the engineering theories of flexural, torsional and axial response. Together with the concept of strain energy, the shear response of beams can be determined from exactly the same principles. They also lead to the determination of the characteristic responses of other

---

[1]Also known as James or Jacques.

[2]However, it is commonly known as the Bernoulli-Euler theory. In October 1742, Daniel Bernoulli wrote to Leonhard Euler proposing his minimum principle. From this, Euler deduced his differential equation (see chart) using the calculus of variations that he had invented. It then still required Coulomb to discover the correct stress distribution and Young to devise his modulus before all the elements of bending theory as used today could be combined by Navier.

linearly-elastic structures. This is made possible by discarding the constraint of assuming that 'plane sections remain plane'. Here and elsewhere in the book it has been necessary to re-examine approaches which have been hallowed by convention.

The general reader may be more interested in practical applications rather than the more abstract aspects of the theory. The intention is to provide an introduction to more advanced ideas on the subject than is commonly available in a student text book. There are, of course, many excellent books on the basic theory of structures. The aim here will be to present the basic theory in a different light and add to it material which is useful but is not readily available. Efforts have been made to maintain an overall consistency of notation. As one form is more familiar for two-dimensional problems and another for three-dimensional problems, some compromises have been made. As far as possible, the more advanced sections which may not be required at a first reading will be enclosed in curly brackets: **{ }** , and will normally be located after the elementary theory, even at the risk of some duplication, and specialist formulae and data consigned to the appendices.

JOHN D. RENTON

Acknowledgments

I would like to thank Prof. D.L. Dean, Prof. W. Gutkowski, Prof. K. Heki, Prof. W.S. Hemp, Prof. V. Kolár, Prof. Z.S. Makowski, Prof. M.V. Soare and Staff Scientist Thein Wah for providing me with their research material, too little of which has been incorporated in a book which attempts to cover such a wide field. I am particularly grateful to Dr. I. Ecsedi for enlightening me on the subject of the shear centre.

Chapter 1 Introduction

1.1 Loads, Deflexions, Joints and Supports

The most common notation will be found in Appendix 1 and is also defined where it is first used. In addition, it is worthwhile making clear some of the distinctions in terminology which will be used in this book. A *force* is an action or influence on a body which tends to cause it to move in a particular direction. A *moment* is an action or influence on a body which has a turning effect on it, about a particular axis. The word *load* will be taken as a general term, referring to forces, moments, distributed forces (such as pressure) and distributed moments. Likewise, *deflexion* may be taken to refer to a linear displacement, a rotation or a general movement of a structure or component. Loads can have *corresponding deflexions* which are the deflexions through which they do work. Thus the corresponding deflexion to a force is its displacement along its line of action, and the corresponding deflexion to a moment acting about some axis is its rotation about that axis. The work done by a load is then given by the integral of that load with incremental changes in its corresponding deflexion.

Forces, displacements, moments and rotations are, in general, vector quantities possessing both magnitude and direction. Figure 1.1 shows the convention that will be used in two dimensions. $\boldsymbol{F}$ is a force vector and $\boldsymbol{u}$ is the deflexion vector corresponding to it. The horizontal and vertical components of $\boldsymbol{F}$ are H and V in the x and y directions respectively. If the magnitude of $\boldsymbol{F}$ is F, then the values of H and V are $F \cos \alpha$ and $F \sin \alpha$ respectively. The horizontal and vertical components of $\boldsymbol{u}$, u and v, are similarly related to the magnitude of $\boldsymbol{u}$. In a right-handed coordinate system, the z axis would come out of the paper. The moment M and the corresponding rotation θ would be clockwise about this axis, as viewed looking along it.

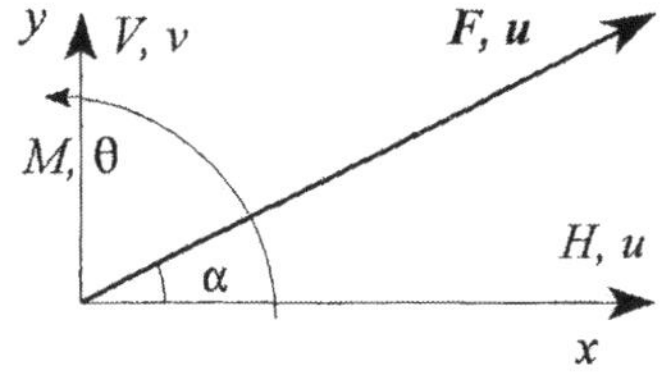

Figure 1.1 Loads and deflexions in two dimensions.

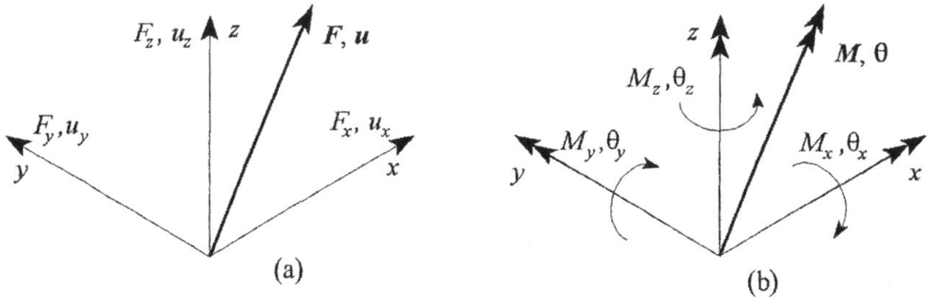

Figure 1.2 Load and deflexion vectors in three dimensions.

In three dimensions, the force $\boldsymbol{F}$ and the corresponding deflexion $\boldsymbol{u}$ have components (F_x, F_y, F_z) and (u_x, u_y, u_z) respectively in the x, y and z directions, as shown in Figure 1.2a. The moment $\boldsymbol{M}$ and the corresponding rotation $\boldsymbol{\theta}$ can be regarded as vectors too, with components (M_x, M_y, M_z) and (θ_x, θ_y, θ_z) in the x, y and z directions, as shown in Figure 1.2b. These components are taken to be positive in the clockwise sense, as viewed along the axis about which they act. The double-headed arrow convention is used to indicate that they are *rotation vectors*.

Most joints and supports are taken to be workless. That is, they neither absorb nor give out work (or energy). The most common exceptions are elastic supports or joints and frictional supports. Here, we will consider only the workless variety. Most frameworks are classified into either pin-jointed frames or rigid-jointed frames. In practice, most frames are not entirely one or the other, but from the analytic point of view, the classification is useful. Figure 1.3 shows the two types of joint. Neither permits relative displacement of the ends of the structural members joined to it, and both can sustain forces applied to them. However, the pin joint permits relative rotation of the ends of members joined to it. As it is a workless joint, it follows that it can exert no moments on these ends and also it cannot sustain an external moment applied to it. Usually, the members of a pin-jointed framework are straight and are not subject to lateral loads between their ends. In this case, they only sustain axial forces. In three dimensions, the pin joint permits relative rotation of the ends about all three axes, and is commonly referred to as a *universal joint*. A rigid joint does not permit any relative rotation of the ends attached to it, and can exert moments as well as forces on the member ends.

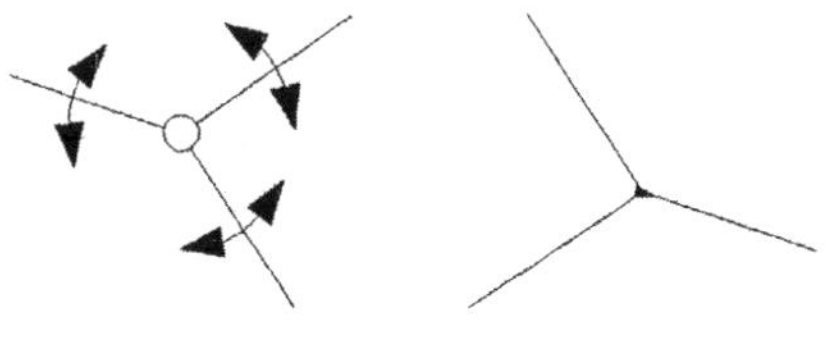

(a) A pin joint (b) A rigid joint

Figure 1.3 Common types of joint.

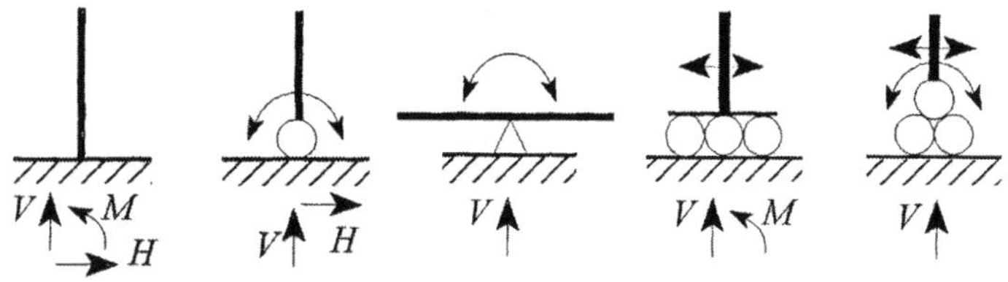

(a) (b) (c) (d) (e)

Figure 1.4 Common types of workless support (in two dimensions)

The common types of workless support are shown in Figure 1.4. These can exert either a reaction (force or moment), in which case the corresponding deflexion of the frame at that point is zero, or a particular reaction does not exist, in which case the corresponding deflexion of the frame at that point is free to take place. (The extreme example of the latter is a free end, which is subject to no reactions at all.) Other workless supports could exist, such that both reactions and deflexions took place at the support, but they would be interrelated in such a way that no net work was ever done. However, these can be simulated by rigid mechanisms attached to supports of the above kind. Figure 1.4a shows a *fixed* or *encastré* end. All rotation and displacement of the end are prevented, requiring the three reactions shown in two dimensions (or six reactions in three dimensions). Figure 1.4b shows a *pin support* which permits end rotation but no end displacement. In three dimensions, a pin support will be taken to permit rotation about all three axes, and a *rocker* support taken as one which permits rotation about a particular axis (the rock axis). Figure 1.4c shows a *knife-edge* support which permits rotation of the structure about it but no motion normal to it (in either direction). It can then exert a normal reaction but no moment reaction. This kind of support is postulated in examining the flexure of beams, when it is assumed that the axial beam loading in the beam is insignificant, so that the question of any lateral reaction provided by the knife edge does not arise. Figure 1.4d shows a *slider* support which permits transverse motion of the end, but prevents any rotation or normal motion of it with the aid of normal and moment reactions, M and V, if necessary. Finally, 1.4e shows a pinned slider support which allows rotation and lateral displacement of the end but prevents any normal motion with a normal reaction V. More possible workless support conditions exist in three dimensions, but no commonly-accepted symbols have been devised for them.

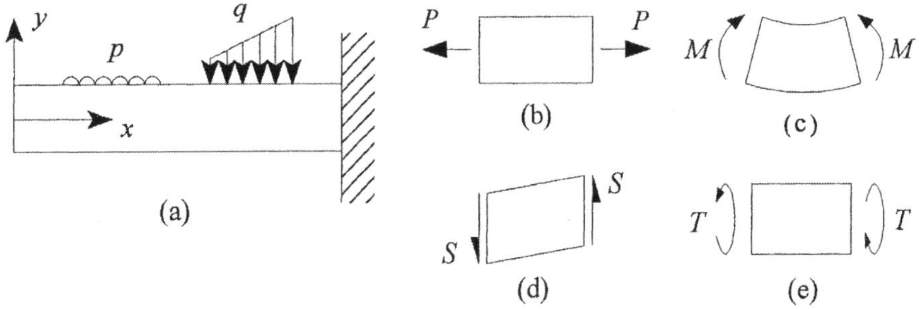

Figure 1.5 Loading on a structural member.

The above loading descriptions are related to some overall set of coordinates and not to a particular structural member. Figure 1.5a shows distributed loads p and q acting on a beam. These are usually forces per unit length, although sometimes they are used to indicate forces per unit area. The symbol used for p will be used to indicate that it is of constant intensity, whereas q, as shown, indicates a variable intensity. Figures 1.5b to 1.5e show resultant internal load pairs acting on short sections of the beam. Each pair is acting in the positive sense used in this book, which is that the corresponding deflexion tends to increase along the axis (here the x axis) of the beam. The axial force P in Figure 1.5b is tensile positive, producing an increase in the axial displacement u along the beam. In Figure 1.5c, the moment M produces 'sagging' of the beam, corresponding to a positive increase in the rotation θ along the beam. The shear force S in Figure 1.5d acts in such a way as to increase the lateral displacement v in the y direction. Lastly, the torque T is such that the beam's rotation about its longitudinal (x) axis increases along its length.

1.2 Small Deflexion Theory

In order to illustrate the mode of deflexion of a structure, this mode is usually exaggerated. The actual deflexions are usually too small to be detected with the naked eye. Typically, the maximum displacement of a beam is no more than 1/300th of its length and the induced rotations are likewise of the order of a fraction of a degree. For many purposes, the change in geometry of a structure under load can be ignored. For example, its equilibrium under load is usually examined relative to the position of the loading in the undeformed state. Also, the changes in orientation of parts of the structure, produced by rotations under load, are often (implicitly) taken as negligible. If the rotations were large, then the effect of a sequence of rotations about different axes, which themselves are affected by these rotations, would depend on the order in which the rotations took place. (The method of analysing the effects of large rotations will be discussed in setting up the stiffness matrices for beams with arbitrary orientations in three dimensions, see §11.4.) However, if the rotations are small, the effect of a sequence of rotations is given by their vector sum.

Unless specifically stated otherwise, radian measure will be used for such rotations. Then there is no need to distinguish between these rotations and their sines and tangents, as they are the same, if only the first term of each of the power expansions for these functions is taken to be significant. In particular, this means that the slope (or gradient) of a beam produced by flexure, and the flexural rotation of the beam at the same location are taken as the same.

Figure 1.6 shows the displacement of the end C of a rigid bar AC of length l resulting from a rotation θ about the end A. These are given by the horizontal and vertical components, δ and ϵ, where

$$\delta = l\sin\theta \quad ; \quad \epsilon = l(1-\cos\theta).$$

As θ is very small, δ can be taken as $l\theta$ and ϵ as zero (using the first terms in the expansions of $\sin\theta$ and $\cos\theta$ only). However, for stability analyses it will be seen later that it is necessary to use the more accurate expression $\frac{1}{2}l\theta^2$ for ϵ, given by including the second term in the expansion for $\cos\theta$.

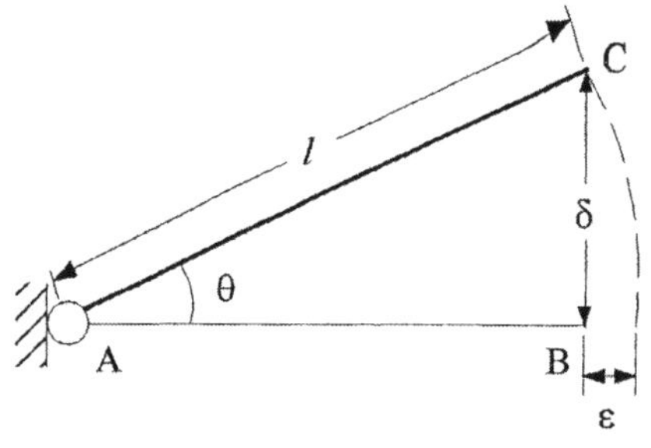

Figure 1.6 A rigid-body rotation.

1.3 Energy, Equilibrium and Stability

The laws of thermodynamics form an important part of the theoretical basis of structural analysis. They can be expressed in various ways, but put simply, they are as follows. The first law states that in any closed system, energy is conserved. That is, if we can encapsulate the structural system within a real or imaginary boundary, so that energy passes neither in nor out, the total energy within the system will remain constant. (Workless supports can form a useful part of that boundary.) However, the forms of energy within the system can change. The second law is related to what (natural) changes in these forms can take place. They tend to be such that the potential for further change to take place is reduced. The system tends to become less organised and more random (the 'entropy' increases), so that the amount of 'useful' energy is reduced. To put it another way, you cannot have your cake and eat it, or there are no free lunches.

A loaded structure is generally designed to reach a state of balance of loads (both external and internal) such that no further movement occurs. This state holds not only for the structure as a whole but for every part of it. It is then said to be in a state of *static equilibrium*. If the loaded structure tends to move away from a given deflected state, then it is not in equilibrium. The energy of movement (*kinetic energy*) may be provided from two sources. The applied loads can do work, thus reducing their *potential energy*, or the deformed structure may release *strain energy* in moving back towards its undeformed state. A clock driven by falling weights and one driven by a spring are examples of kinetic energy derived from these two sources. Then if a loaded structure is in equilibrium, it has no tendency to move resulting in the release of energy from these two sources. This can be analysed by examining the net energy released from these two sources during any imaginary small deflexion (*virtual deflexion*) from the equilibrium state. The conditions of static equilibrium of the loads on a body are that these loads have no resultant force or moment. This means that during any *rigid-body motion* of the body, in which it displaces or rotates without any change in its deformation, no net work is done and hence no kinetic energy can be generated.

A loaded structure is not always safe, even when it is in equilibrium. Although in theory it could remain in that state if undisturbed, some small imperfection or external perturbation may

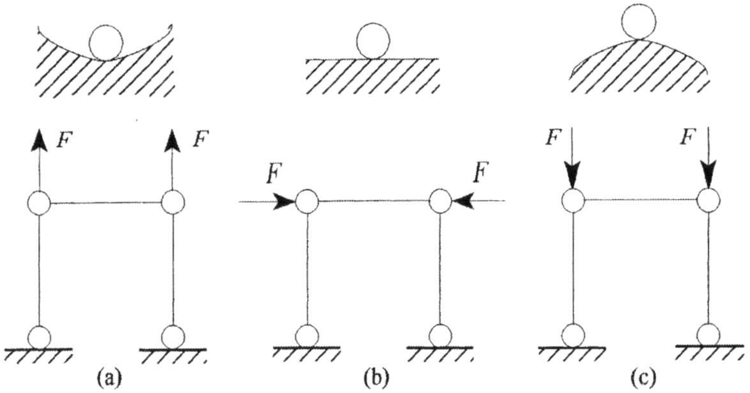

Figure 1.7 Basic types of equilibrium.

induce movements away from the assumed state. Figure 1.7 illustrates three different kinds of equilibrium. The upper diagrams show a heavy ball free to roll over a surface and the lower ones a mechanism consisting of three rigid bars linked by pin joints. The pair of diagrams shown in Figure 1.7a illustrate *stable equilibrium.* If the ball is moved slightly, it will tend to move back to its original position. Likewise, the pair of forces F will tend to pull the mechanism back to its initial state if it is perturbed slightly. In Figure 1.7b, if either the ball or the mechanism is displaced from its initial state, neither gravity acting on the ball nor the forces F have any influence on the perturbation and both systems tend to remain in the displaced position. This is known as *neutral equilibrium.* Any tendency of either the ball or the mechanism in Figure 1.7c to move from their equilibrium states is amplified by the forces acting on them. This is therefore known as *unstable equilibrium.* In the cases shown, the nature of the equilibrium state is determined from the potential of the loading in an immediately adjacent state. If this potential then permits the loading to do work in either restoring the system to its original state or enhancing the perturbation, then the state is either stable or unstable respectively. More generally, the strain energy of the system has to be considered as well. Structural systems are usually stable, because during any further growth of a small perturbation, more energy would be absorbed in straining the structure than would be released from the loss of potential energy of the applied loads in doing work. However, under certain loading conditions the structure grows unstable, as its ability to absorb strain energy becomes less than that of the loading to do work. The transition is usually marked by a state of neutral equilibrium.

{1.4 Linear Response}

{Most structural engineering analyses involve the solution of linear simultaneous equations of the form

$$\begin{aligned} a_{11}x_1 + a_{12}x_2 + \dots + a_{1n}x_n &= y_1 \\ a_{21}x_1 + a_{22}x_2 + \dots + a_{2n}x_n &= y_2 \\ \dots\dots\dots\dots\dots\dots\dots\dots\dots \\ a_{m1}x_1 + a_{m2}x_2 + \dots + a_{mn}x_n &= y_m \end{aligned} \tag{1.1}$$

or

$$\sum_{j=1}^{n} a_{ij} x_j = y_i \qquad (i = 1 \text{ to } m) \tag{1.2}$$

These equations can be written in matrix form as

$$Ax = y \tag{1.3}$$

where A is the matrix of coefficients a_{ij}, and x and y are the column vectors of the parameters x_j and y_i. Such equations arise in writing the equations of equilibrium, the relationships between (small) deflections, and the linear-elastic relationships between stresses and strains, for example. If a third column vector z of parameters z_k ($k = 1$ to p) is related to y by

$$By = z \tag{1.4}$$

where B is the matrix of coefficients b_{ki}, then

$$Cx = BAx = By = z \tag{1.5}$$

where

$$c_{kj} = \sum_{i=1}^{m} b_{ki} a_{ij} \tag{1.6}$$

so that the relationship between x and z is also linear. In any linear relationship, the following results apply.

Let x_1, x_2, x_3 be particular values of the column vector x and y_1, y_2, y_3 particular values of the column vector y such that

$$\begin{aligned} Ax_1 = y_1 \;, \quad Ax_2 = y_2 \;, \\ x_3 = x_1 + x_2 \;, \quad y_3 = y_1 + y_2 \;. \end{aligned} \tag{1.7}$$

Then

$$Ax_3 = A(x_1 + x_2) = Ax_1 + Ax_2 = y_1 + y_2 = y_3 \tag{1.8}$$

Suppose that x is the column vector of deflexions of a structure in response to a loading given by y. Then (1.8) shows that a response to a combination of loadings y_1 and y_2 is the sum of the individual responses to these loadings applied separately. This is known as the *principle of superposition* attributed to Bresse, as noted in the preface. It also follows that if y_2 is a scalar multiple, K, of y_1 then Kx_1 is a solution, for

$$A(Kx_1) = KAx_1 = Ky_1 = y_2 \tag{1.9}$$

so that this response increases in proportion to the loading.

There is not necessarily a unique solution to linear simultaneous equations such as (1.3). If there are more equations than unknowns ($m>n$), then there may be no solution. If there are fewer equations than unknowns ($m<n$), then the parameters x_j cannot be completely determined. If the structure is insufficiently constrained so that it forms a mechanism (such as that shown in Figure 1.7), the above equations will be insufficient to find the deflexions. Likewise, if too many constraints are applied to the structure, it may not be possible to find its internal loading from the equations of equilibrium alone. It is then called a *statically-indeterminate* or *redundant* structure. (This will be discussed further in Chapter 2.) Even if the number of equations is equal to the number of unknowns ($m=n$), there may still not be a unique solution. This would be because some

of the equations contain no information which could not be deduced from the other equations. In this case, some of the rows of A would be linear combinations of other rows of A, so that the determinant of A would be zero. This happens when the structure just reaches its buckling load and passes from a state of stable equilibrium to a state of neutral equilibrium where its deflexions are no longer determinate. As will be seen in §11.5, this zero determinant is a useful indicator of the loss of stability.

In engineering analyses, most problems have *unique* solutions. If a structure is constrained so that it neither forms a mechanism nor is free to move bodily in space (*rigid-body motion*), then its deflexions can be determined from the applied loading. If, in addition, there are no redundant constraints, so that the internal loading can be determined from the applied loading by means of the equations of equilibrium only, then the structure is said to be *just-stiff*. It will be assumed that the initial (unloaded) state is one of zero stress and strain, and so one of zero strain energy. As will be seen in Chapter 3, it can then be shown that the internal stresses and strains in a structure are uniquely determined for given loads and displacements applied to it. Suppose that in (1.3), x_1 and x_2 are two different responses of a structure to a particular loading y_1. Then

$$A x_1 = y_1 \quad , \quad A x_2 = y_1 \; . \tag{1.10}$$

and so

$$A(x_1 - x_2) = y_1 - y_1 = 0 \tag{1.11}$$

Thus, if a unique solution exists, so that x_1 and x_2 must be equal, then the initial (unloaded) state must be one of zero response. Uniqueness implies that if *a* solution has been found which satisfies all the necessary conditions imposed on it, then it is *the* solution. The importance of uniqueness will be seen, for example, in proving Betti's reciprocal theorem in Chapter 3.}

{1.5 Symmetry and Antisymmetry}

{A structure is said to be symmetrical when, under some transformation of the coordinate system from which it is viewed, it appears unchanged. That is to say it is *invariant* with respect to the transformation. The most common type of symmetry is *mirror-symmetry*. The two-dimensional structure shown in Figure 1.8a is symmetrical about its centre line. That is, if a mirror M-M is placed on this centre line as in Figure 1.8b, the right-hand half of the structure would appear as the mirror image of the left-hand half. The system of reference shown by the (x,y) axes and the anticlockwise sense of rotation also has its mirror image, (x',y') and the clockwise rotation.

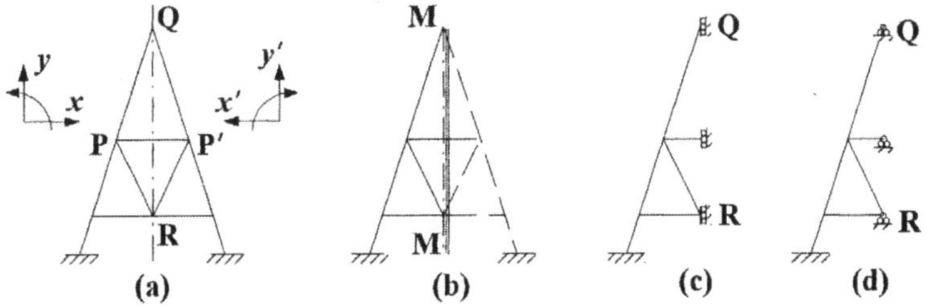

Figure 1.8 Behaviour of a symmetric framework.

The structure appears identical when viewed from either system of reference, except for changes in the joint labels. The typical joint labelled P seen with respect to the (x,y) system is the same as the joint P′seen with respect to the (x',y') system. Points on the plane of the mirror such as Q and R are their own images.

Let the loads and deflexions at P, defined in terms of the (x,y) system of reference, be denoted by the column vectors $\boldsymbol{f}_P$ and $\boldsymbol{d}_P$ respectively, and those at P′ relative to (x',y') be $\boldsymbol{f}_P'$ and $\boldsymbol{d}_P'$ respectively. Suppose that it is possible to write a pair of matrix equations of the form

$$\begin{aligned} \boldsymbol{A}\boldsymbol{d}_P + \boldsymbol{B}\boldsymbol{d}_P' &= \boldsymbol{f}_P \\ \boldsymbol{C}\boldsymbol{d}_P + \boldsymbol{D}\boldsymbol{d}_P' &= \boldsymbol{f}_P' \end{aligned} \qquad \textbf{(1.12)}$$

which have unique solutions for the deflexions. As the structure is symmetrical, interchanging the primed and unprimed terms must yield the same equations, as the physical problem is identical. This means that in the above equations the matrices $\boldsymbol{A}$ and $\boldsymbol{D}$ are identical, and so are $\boldsymbol{B}$ and $\boldsymbol{C}$. Consider the loadings and deflexions given by

$$\begin{aligned} \boldsymbol{f}_P^+ &= \frac{1}{2}(\boldsymbol{f}_P + \boldsymbol{f}_P') \ ; \ \boldsymbol{f}_P^- = \frac{1}{2}(\boldsymbol{f}_P - \boldsymbol{f}_P') \ ; \\ \boldsymbol{d}_P^+ &= \frac{1}{2}(\boldsymbol{d}_P + \boldsymbol{d}_P') \ ; \ \boldsymbol{d}_P^- = \frac{1}{2}(\boldsymbol{d}_P - \boldsymbol{d}_P') \ . \end{aligned} \qquad \textbf{(1.13)}$$

Given the above relationships, it follows that if both $\boldsymbol{f}_P$ and $\boldsymbol{f}_P'$ are $\boldsymbol{f}_P^+$, then the equations are satisfied if both $\boldsymbol{d}_P$ and $\boldsymbol{d}_P'$ are $\boldsymbol{d}_P^+$. That is to say, a symmetrical loading induces a symmetrical response. If $\boldsymbol{f}_P$ and $\boldsymbol{f}_P'$ are equal and opposite, the loading is said to be *antisymmetrical* (i.e. $\boldsymbol{d}_P$ and $\boldsymbol{d}_P'$ will be equal and opposite). From the above relationships, it then follows that the deflexions will also be antisymmetrical. Equations (1.13) can be generalized by taking the vectors not to refer to the joints P and P′ only, but to all the joints on the left- and right-hand sides of the framework respectively.

Joints lying on the centreline can be considered as half-joint pairs lying on either side of it. Because these pairs are coincident points, their loading and motion under conditions of symmetry are restricted . Figure 1.9 shows the effects on joint R of Figure 1.8, split into two half-joints. In the case of mirror-symmetrical behaviour about the line M-M, vertical motion of each half-joint is the same, but their rotations and horizontal displacements would be equal and opposite. As the two half-joints are coincident, no rotation or horizontal displacement can take place, so that the joint is free only to move vertically, as shown in Figure 1.9a. Interactions between the two halves prevent rotation and horizontal displacement. This is like replacing the connection to the right-hand half with a vertical slider as shown in Figure 1.8c (cf. Figure 1.4d). There is no vertical interaction between the two halves, but if an external vertical force of $2V$ is applied to the joint, it is shared it is shared as equal forces V by the two half-joints as shown in Figure 1.9b. Antisymmetry about the same line, marked as A-A in Figures 1.9c and 1.9d, means that the image behaves in exactly the opposite way to a mirror image. Rotation and horizontal motion of the joint R now becomes possible, as both half-joints will move in the same sense and by the same amount. However, vertical motion of the joint is ruled out, as the as the

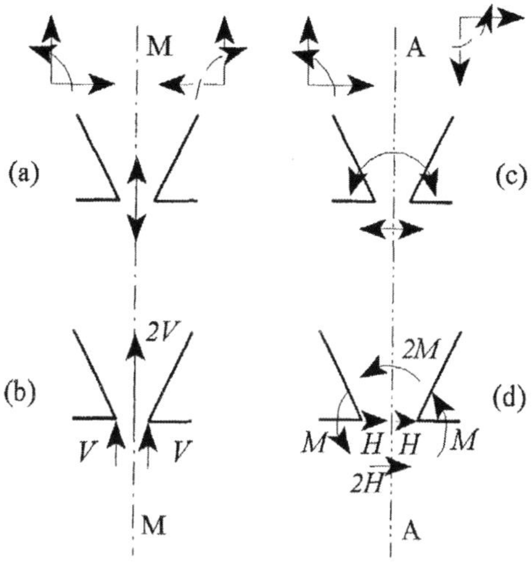

Figure 1.9 Behaviour of a joint on the centreline.

two halves would respond in opposite senses. Vertical interactions between the two halves will prevent this. In this case, the connection to the right-hand half can be replaced by a pinned slider support, as shown in Figure 1.8d (cf. Figure 1.4e).

Any loading on a symmetrical structure can be divided into symmetrical components of the kind f_P^+ applied to both halves and antisymmetrical components of the kind f_P^- applied to the left-hand side and $-f_P^-$ applied to the right-hand side. The analysis is then much simplified by analysing the half-structures with the appropriate support conditions for the two cases inserted on the centre line, as described above. The two solutions can be combined to give the full solution, using the principle of superposition. The same process can be used to analyse three-dimensional structures. In this case, the mirror images of linear vectors (of the kind shown in Figure 1.2a) parallel to the plane of symmetry are in the same sense and of those normal to the plane of symmetry are in the opposite sense. The mirror images of rotation vectors (of the kind shown in Figure 1.2b) parallel to the plane of symmetry are in the *opposite* sense and of those normal to the plane of symmetry are in the *same* sense. Again, the rules for antisymmetrical images are exactly opposite to those for symmetrical images. Symmetry about more than one plane means that it may be possible to simplify the analysis even further, using the same principles. These possibilities are built into the structural analysis software described later.

It is possible to give a more general definition of symmetry. An object is symmetrical if it is invariant (i.e. does not change its appearance) when the coordinate system undergoes an affine transformation in which orthogonality is retained. That is to say, if two vectors were at right-angles when viewed with respect to the original coordinate system, they remain at right-angles when viewed from the transformed coordinate system, so that their scalar product remains zero. In two dimensions, the affine transformation from $\{x,y\}$ to $\{x',y'\}$ is given by

$$\begin{bmatrix} x' \\ y' \end{bmatrix} = \begin{bmatrix} a_{11} & a_{12} \\ a_{21} & a_{22} \end{bmatrix} \begin{bmatrix} x \\ y \end{bmatrix} + \begin{bmatrix} x_o \\ y_o \end{bmatrix} ,$$

$$(a_{11}a_{22} - a_{12}a_{21} \neq 0) .$$

and to retain orthogonality,

$$a_{11}a_{12} + a_{22}a_{21} = 0 ,$$

$$a_{11}^2 + a_{21}^2 = a_{12}^2 + a_{22}^2 .$$

Various kinds of symmetry are special cases of this.

a) Mirror symmetry (about y axis)

$$a_{11} = -1 , \quad a_{22} = 1 , \quad a_{12} = a_{21} = 0 , \quad x_o = y_o = 0 .$$

b) Rotational symmetry (through a rotation α)

$$a_{11} = a_{22} = \cos\alpha , \quad a_{12} = -a_{21} = \pm\sin\alpha , \quad x_o = y_o = 0 .$$

c) Amplification symmetry (by a factor k)

$$a_{11} = a_{22} = k \quad , \quad a_{12} = a_{21} = x_o = y_o = 0 \ .$$

d) Translation symmetry or regularity (under shifts s and t in the x and y directions)

$$a_{11} = a_{22} = 1 \quad , \quad a_{12} = a_{21} = 0 \quad , \quad x_o = s \quad , \quad y_o = t \ .$$

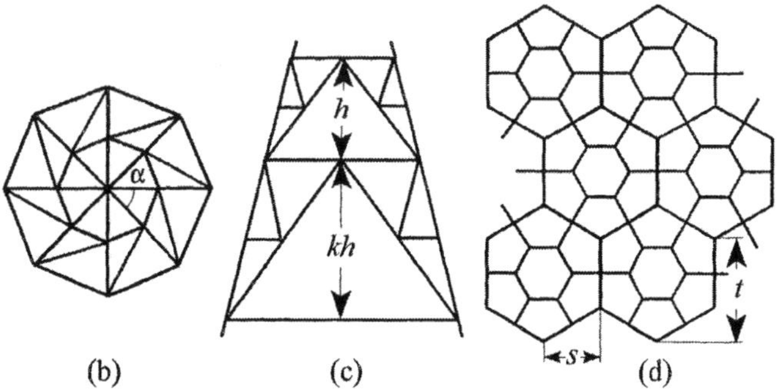

Figure 1.10 Other forms of symmetrical structure.

Mirror symmetry has already been examined. The rotational symmetry of any finite structure must be with respect to an angle α which is rational fraction of 2π radians. (It is $\pi/4$ in Figure 1.10b.) The amplification symmetry shown in Figure 1.10c is for a value of k equal to 1.5. Such structures are self-similar fractals. For this case and that of translation symmetry illustrated by the regular grid shown in Figure 1.10d, the structure is necessarily infinite. However, this does not mean that simplifications in the analysis cannot be found for substructures of these forms. Other symmetries are given by combinations of the above. In fact, the grid shown in Figure 1.10d also exhibits mirror and rotational symmetry. The forms of symmetry have been discussed by Hermann Weyl (1952) who classifies them algebraically rather than geometrically. Further consideration will be given to symmetrical behaviour later, in particular regular structures are examined in Chapter 15.}

Chapter 2 Statics

2.1 Work, Energy and Static Equilibrium

Loads have a potential for doing work. In doing work, they expend potential energy which, in accordance with the first law of thermodynamics, must be turned into some other form of energy (see §1.3). For static equilibrium, this will not be turned into kinetic energy during any conceivable infinitesimal deflexion from the equilibrium state. This can be a an imaginary deflexion of the structure as a whole, or of any part of it. Statics is concerned with deflexions which do not involve any (further) deformation of the structure's material, so that the work cannot be turned into strain energy either. In fact, it is assumed that the work cannot be turned into any other form of energy, so that during such a deflexion, no net work is done by the applied and internal loads acting on the structure as a whole or any part of it. These are known as the *conditions of statics* and are often expressed as ones of balance of loading. In the course of this chapter, it will be seen that the two approaches are equivalent, although each has particular advantages.

2.2 Motion of a Rigid Body, Resultants and Equilibrium

The work done by a constant force $\boldsymbol{F}$ is the product of its magnitude, F, and the distance through which it moves in the direction in which it acts (its *corresponding deflexion*). If $\boldsymbol{F}$ is variable, then the work done is found by integrating the instantaneous magnitude F with small increments of the deflexion.

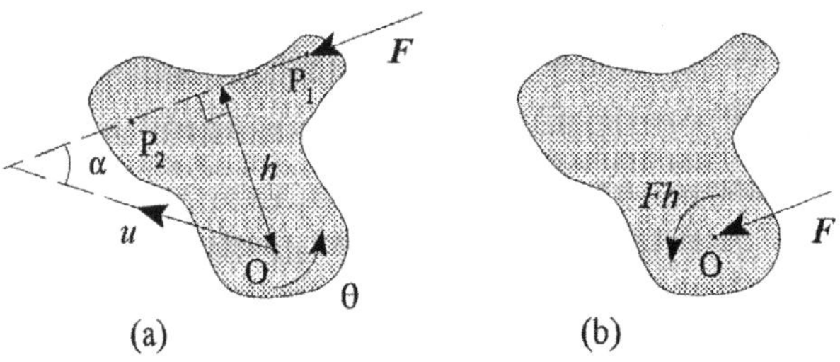

Figure 2.1 Plane rigid-body motion.

Consider the case of a force $\boldsymbol{F}$ acting at a point P_1 on the plane rigid body shown in Figure 2.1a. Any motion of the body in the plane can be expressed by the displacement $\boldsymbol{u}$ of the point O on it and its in-plane rotation θ about O. We will discuss only small movements, as discussed in §1.2. The corresponding deflexion of $\boldsymbol{F}$ resulting from the displacement $\boldsymbol{u}$ is of magnitude $u\cos\alpha$, where α is the angle between the directions of $\boldsymbol{F}$ and $\boldsymbol{u}$ and u is the magnitude of $\boldsymbol{u}$. The corresponding deflexion resulting from the rotation θ is of magnitude $h\theta$, where h is the perpendicular distance from O to the line through P_1 in the direction of $\boldsymbol{F}$ (called the *line of action* of $\boldsymbol{F}$). Thus the work done by $\boldsymbol{F}$ during any arbitrary motion of the body can be expressed by $F(u\cos\alpha + h\theta)$. The expression would be the same if $\boldsymbol{F}$ was applied at some other point P_2 on its line of action. The two cases are said to be *statically equivalent*. From the point of view of the static analysis, $\boldsymbol{F}$ can be considered to act anywhere on its line of action.

The corresponding deflexion to a moment $\boldsymbol{M}$ is a rotation θ about the same axis. If the moment is constant and of magnitude M and the rotation is of magnitude θ, then the work it does is $M\theta$. Suppose that the force $\boldsymbol{F}$ in Figure 2.1a is replaced by a parallel force $\boldsymbol{F}$ acting through O together with an anticlockwise moment Fh about O. The work done during an arbitrary motion of the body is again $F(u\cos\alpha + h\theta)$, so that the loading in Figure 2.1b is statically equivalent to that in Figure 2.1a.

In three dimensions, suppose that the force $\boldsymbol{F}$ acts on a body at a point P which is at a position relative to O given by the vector $\boldsymbol{r}$, as shown in Figure 2.2. Any arbitrary rigid motion of the body is given by the vector displacement $\boldsymbol{u}$ of O and the vector rotation θ of the body about O. The work done by $\boldsymbol{F}$ during the displacement $\boldsymbol{u}$ is again $Fu\cos\alpha$, where α is the angle between the two vectors. This can be written as the scalar product $\boldsymbol{F}.\boldsymbol{u}$. In terms of the components shown in Figure 1.2a, this can be written as

$$\boldsymbol{F}.\boldsymbol{u} = F_x u_x + F_y u_y + F_z u_z \qquad \textbf{(2.1)}$$

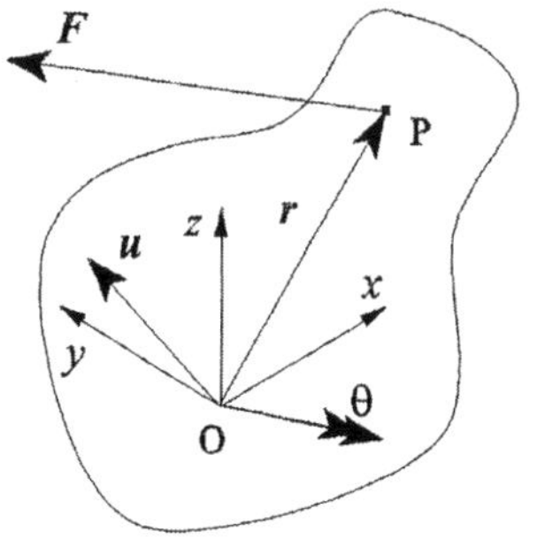

Figure 2.2 Force on a rigid body in three dimensions.

Suppose that $\boldsymbol{r}$ has the components (x,y,z) and θ the components $(\theta_x, \theta_y, \theta_z)$ as shown in Figure 1.2. Then the displacements of the point P due to the rotation θ are shown in Figure 2.3.

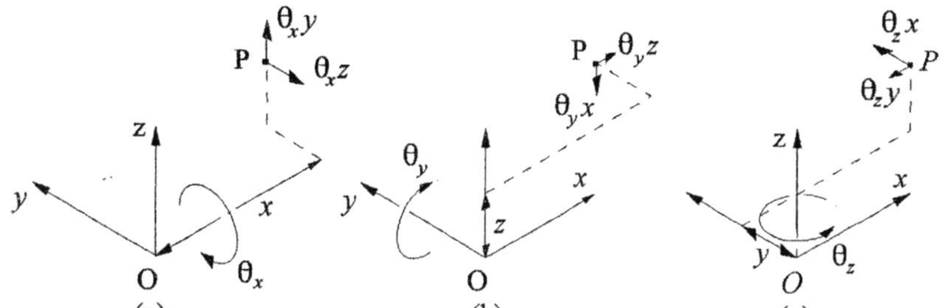

Figure 2.3 Displacements of a point P produced by the components of a small rotation θ.

The displacement in the x direction, d_x, is given by the rotation of the ordinate z through a small angle θ_y shown in Figure 2.3b and the rotation of the y ordinate through a small angle θ_z shown in Figure 2.3c, giving

$$d_x = \theta_y z - \theta_z y$$

and the displacements d_y and d_z in the y and z directions are

$$d_y = \theta_z x - \theta_x z$$
$$d_z = \theta_x y - \theta_y x$$

so that if these displacements are the components of a vector $\boldsymbol{d}$, it can be expressed as the vector product

$$\boldsymbol{d} = \theta \times \boldsymbol{r} \qquad \textbf{(2.2)}$$

The work done by $\boldsymbol{F}$ during this rotation is then

$$\boldsymbol{F}.\boldsymbol{d} = \boldsymbol{F}.(\theta \times \boldsymbol{r}) \equiv (\boldsymbol{r} \times \boldsymbol{F}).\theta \qquad \textbf{(2.3)}$$

Thus the total work done, W, by $\boldsymbol{F}$ during a small arbitrary rigid-body motion is

$$W = \boldsymbol{F}.\boldsymbol{u} + (\boldsymbol{r}\times\boldsymbol{F}).\boldsymbol{\theta} \tag{2.4}$$

If a moment $\boldsymbol{M}$ is applied about O, its corresponding deflexion is $\boldsymbol{\theta}$ so that the work it does is

$$\boldsymbol{M}.\boldsymbol{\theta} = M_x\theta_x + M_y\theta_y + M_z\theta_z \tag{2.5}$$

As before, suppose that $\boldsymbol{F}$ is shifted so that its line of action passes through O and that a moment $\boldsymbol{M}$ equal to $\boldsymbol{r}\times\boldsymbol{F}$ is applied about it, then the same work W is done during an arbitrary rigid-body motion as before. This loading is then statically-equivalent to the loading shown in Figure 2.2.

Suppose that a set of forces $\boldsymbol{F}_i$ ($i = 1$ to n) is applied to the body at radii $\boldsymbol{r}_i$ from O. The total work done is then

$$W_T = (\sum_{i=1}^{n}\boldsymbol{F}_i).\boldsymbol{u} + (\sum_{i=1}^{n}\boldsymbol{r}_i\times\boldsymbol{F}_i).\boldsymbol{\theta} \tag{2.6}$$

This is statically equivalent to a force $\boldsymbol{F}$ at a radius $\boldsymbol{r}$ where

$$\sum_{i=1}^{n}\boldsymbol{F}_i = \boldsymbol{F} \quad , \quad \sum_{i=1}^{n}(\boldsymbol{r}_i\times\boldsymbol{F}_i) = \boldsymbol{r}\times\boldsymbol{F} \tag{2.7}$$

This force $\boldsymbol{F}$ is called the *resultant* of the forces $\boldsymbol{F}_i$. Although $\boldsymbol{F}$ can be determined from the first of these equations, $\boldsymbol{r}$ is at best only partially determinate. This can be seen by adding to $\boldsymbol{r}$ a vector $\boldsymbol{r}^p$ which is parallel to $\boldsymbol{F}$. Then if $\boldsymbol{r}$ satisfies (2.7) so does $(\boldsymbol{r}+\boldsymbol{r}^p)$, for $\boldsymbol{r}^p\times\boldsymbol{F}$ is zero. This means that $\boldsymbol{r}$ could be a radius vector to any point on the line of action of $\boldsymbol{F}$. In general, we can take $\boldsymbol{r}$ to be the sum $(\boldsymbol{r}^n+\boldsymbol{r}^p)$ where $\boldsymbol{r}^n$ is a vector normal to $\boldsymbol{F}$. Then

$$\boldsymbol{F}\times(\boldsymbol{r}\times\boldsymbol{F}) = \boldsymbol{F}\times(\boldsymbol{r}^n\times\boldsymbol{F}) \equiv (\boldsymbol{F}.\boldsymbol{F})\boldsymbol{r}^n - (\boldsymbol{F}.\boldsymbol{r}^n)\boldsymbol{F} = (\boldsymbol{F}.\boldsymbol{F})\boldsymbol{r}^n$$

so that

$$\boldsymbol{r}^n = \boldsymbol{F}\times(\boldsymbol{r}\times\boldsymbol{F})/(\boldsymbol{F}.\boldsymbol{F}) \tag{2.8}$$

where $\boldsymbol{F}$ and $\boldsymbol{r}\times\boldsymbol{F}$ are given by (2.7). A particular case of interest occurs when the first of equations (2.7) implies that $\boldsymbol{F}$ is zero. From (2.8), $\boldsymbol{r}^n$ is then indeterminate. The resultant is then a pure moment $\boldsymbol{M}$. The simplest example of this is when only two forces, $\boldsymbol{F}_1$ and $\boldsymbol{F}_2$, act on the body and $\boldsymbol{F}_2$ is equal to $-\boldsymbol{F}_1$, as shown in Figure 2.4. The resultant force is zero and the resultant moment $\boldsymbol{M}$ is $(\boldsymbol{r}_1 - \boldsymbol{r}_2)\times\boldsymbol{F}_1$. The magnitude of this is equal to the magnitude F_1 of the forces multiplied by the perpendicular distance between them. Such a force pair is referred to as a *couple* of magnitude M. If the distance between the forces is allowed to tend to zero, but their magnitudes are increased so that M remains constant, the couple tends to become statically-equivalent to a point moment $\boldsymbol{M}$ at the limit.

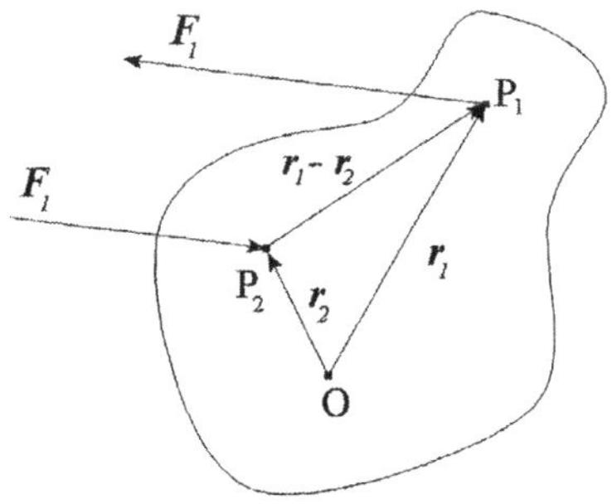

Figure 2.4 Action of a couple on a body.

Static equilibrium is achieved if no work is done by the set of forces $\boldsymbol{F}_i$ under any rigid motion of the body. In (2.6), W_T must be zero for any arbitrary small deflexions $\boldsymbol{u}$ and $\boldsymbol{\theta}$. This is then an example of a *virtual work equation*. It is true if both of the expressions in (2.7) are zero. More generally, if a set of point moments $\boldsymbol{M}_j$ (j = 1 to m) is also applied to the body, the conditions for static equilibrium become

$$\sum_{i=1}^{n} \boldsymbol{F}_i = \boldsymbol{0} \quad , \quad \sum_{i=1}^{n} (\boldsymbol{r}_i \times \boldsymbol{F}_i) + \sum_{j=1}^{m} \boldsymbol{M}_j = \boldsymbol{0} \; . \tag{2.9}$$

In terms of components of the above vectors, there are six equations of equilibrium for a single rigid body. These are given by

$$\begin{aligned} \sum_{i=1}^{n} F_{xi} = 0 \quad &, \quad \sum_{i=1}^{n} (y_i F_{zi} - z_i F_{yi}) + \sum_{j=1}^{m} M_{xi} = 0 \; , \\ \sum_{i=1}^{n} F_{yi} = 0 \quad &, \quad \sum_{i=1}^{n} (z_i F_{xi} - x_i F_{zi}) + \sum_{j=1}^{m} M_{yi} = 0 \; , \\ \sum_{i=1}^{n} F_{zi} = 0 \quad &, \quad \sum_{i=1}^{n} (x_i F_{yi} - y_i F_{xi}) + \sum_{j=1}^{m} M_{zi} = 0 \; . \end{aligned} \tag{2.9a}$$

If another origin O′, at a position relative to O given by $\boldsymbol{r}_o$, is chosen and the positions of the forces $\boldsymbol{F}_i$ with respect to O′ are given by $\boldsymbol{r}_i\,'$, then a new set of equations is given by replacing $\boldsymbol{r}_i$ with $\boldsymbol{r}_i'$ in (2.9). This of course only affects the second part of these equations. This now takes the form

$$\sum_{i=1}^{n} \boldsymbol{r}_i' \times \boldsymbol{F}_i + \sum_{j=1}^{m} \boldsymbol{M}_j = \sum_{i=1}^{n} (\boldsymbol{r}_i - \boldsymbol{r}_o) \times \boldsymbol{F}_i + \sum_{j=1}^{m} \boldsymbol{M}_j = \sum_{i=1}^{n} \boldsymbol{r}_i \times \boldsymbol{F}_i + \sum_{j=1}^{m} \boldsymbol{M}_j - \boldsymbol{r}_o \times \sum_{i=1}^{n} \boldsymbol{F}_i = \boldsymbol{0}$$

(It may help to consider P_2 at a position $\boldsymbol{r}_2$ relative to O in Figure 2.4 to be the point O′ at a position $\boldsymbol{r}_o$.) However, the sum of the forces $\boldsymbol{F}_i$ is already known to be zero and thus no independent equations arise by choosing a new position for O.

Suppose that all the forces are parallel. For example, the body might consist of a set of point masses m_i under the influence of the acceleration due to gravity, g. In vector terms, this means that each force $\boldsymbol{F}_i$ is a scalar multiple F_i of a unit vector, $\boldsymbol{U}$ say. Equations (2.7) then become

$$\begin{aligned} \sum_{i=1}^{n} \boldsymbol{F}_i = (\sum_{i=1}^{n} F_i)\boldsymbol{U} = \boldsymbol{F} \; , \; \sum_{i=1}^{n} (\boldsymbol{r}_i \times \boldsymbol{F}_i) &= (\sum_{i=1}^{n} F_i \boldsymbol{r}_i) \times \boldsymbol{U} = \boldsymbol{r} \times \boldsymbol{F} \\ &= \boldsymbol{r} \times (\sum_{i=1}^{n} F_i)\boldsymbol{U} = (\sum_{i=1}^{n} F_i)\boldsymbol{r} \times \boldsymbol{U} \end{aligned} \tag{2.10}$$

It is possible to find a point through which the resultant $\boldsymbol{F}$ acts regardless of the direction of the forces. The position of this point is given by the vector $\boldsymbol{r}$, where

$$\boldsymbol{r} = (\sum_{i=1}^{n} F_i \boldsymbol{r}_i) / \sum_{i=1}^{n} F_i$$

for the second of the conditions in (2.10) is satisfied, regardless of the value of $\boldsymbol{U}$. Suppose that the above collection of masses has a total mass m. Then

$$F = (\sum_{i=1}^{n} m_i) g = mg \quad , \quad \boldsymbol{r} = (\sum_{i=1}^{n} m_i \boldsymbol{r}_i) / m \tag{2.11}$$

and the point given by the position vector $\boldsymbol{r}$ is known as the *centre of mass*.

For problems confined to two dimensions, only three of equations (2.9a) apply. Thus for the *x-y* plane, the first equations on the first two lines of (2.9a) and the second equation on the third line are used. In the past, structural analyses were largely confined to two-dimensional problems. Most common engineering structures can be reduced to a series of plane substructures which can be analysed individually. This reduces considerably the number of equations required for a solution, which is a particularly important consideration for manual analyses. Also, graphical methods of analysis were developed for these problems, such as Bow's notation and the funicular polygon, which have been made obsolete by the arrival of the computer. However, force diagrams remain a useful visual aid to the student in understanding the subject.

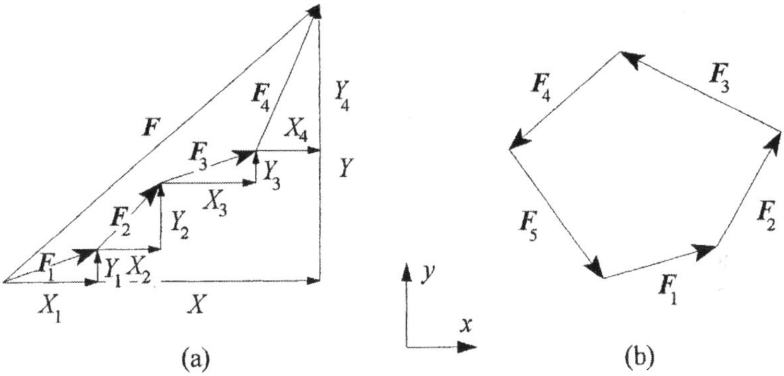

Figure 2.5 Graphic representation of the summation of force vectors.

In Figure 2.5 the magnitude and direction of a force are represented by a length and direction of a line vector, marked with an arrow head. The components X_i and Y_i of the force $\boldsymbol{F}_i$ ($i = 1$ to 4) in the x and y directions are likewise represented by lines in Figure 2.5a. These lines can be summed algebraically to give the components X and Y of the resultant force $\boldsymbol{F}$. This can be seen as equivalent to joining the vector lines for the individual forces end-to-end to give the vector line for $\boldsymbol{F}$. This is the graphical representation of the first of equations (2.7) in two dimensions. If the resultant force is zero, then the chain of vector lines has no gap between its beginning and end, so that it joins up to form a closed polygon, as in Figure 2.5b. This is the graphical representation of the first of the equilibrium equations (2.9) in two dimensions and is known as the *polygon of forces.*

2.3 Distributed Mass and Load, Force Fields

The concepts of point force and mass are useful simplifications of more realistic representations of the physical world. True point forces would exert infinite pressure on any body to which they were applied and point masses would be black holes. In practice, a force applied *externally* to a body is distributed as a pressure over a finite *area* of its surface. Likewise, mass is distributed over a body of finite volume and can be found by integrating the density, ρ, over that volume. The force of gravity then acts on every elementary volume of the body and so produces an *internal* force per unit *volume*. A force distribution in space is known as a *force field.* Sometimes the problem can be conveniently reduced to a two-dimensional analysis. In this case, the external force is considered to be distributed over a length of the boundary of the body and the internal force is taken to be distributed over an internal area of it. The summations over a finite number of forces or masses in the preceding equations can here be replaced with integrations of elementary masses and loads. Thus for a body of volume V and uniform density ρ, the finite masses m_i in (2.11) are replaced by infinitesimal masses ρdV, and it becomes

$$F = \int_V \rho g \, dV = mg \quad , \quad \boldsymbol{r}_o = (\int_V \rho \boldsymbol{r} \, dV)/m \tag{2.12}$$

where $\boldsymbol{r}_o$ is the position of the centre of mass with respect to the origin of coordinates and $\boldsymbol{r}$ is the position of the small volume dV with respect to this origin. (The second integral can be evaluated by noting that each component of the vector $\boldsymbol{r}_o$, (x_o , y_o or z_o) can be found individually by carrying out the integration with respect to the corresponding components of $\boldsymbol{r}$, (x, y or z).)

{The gravitational field is a particular case of a *conservative force field*. In general, the force acting on a particle within the field is a function of the particle's position. A conservative force field is such that the total work done by the field on the particle when it moves from one fixed point, A, to another, B, is the same, regardless of the path between A and B. Thus the total work done by the force field on a particle moving along path (i) in Figure 2.6 is the same as that done when it moves along path (ii). The work done by $\boldsymbol{F}$ in moving a small distance $\boldsymbol{ds}$ along path (i) is $\boldsymbol{F.ds}$ and the total work done in moving from A to B is

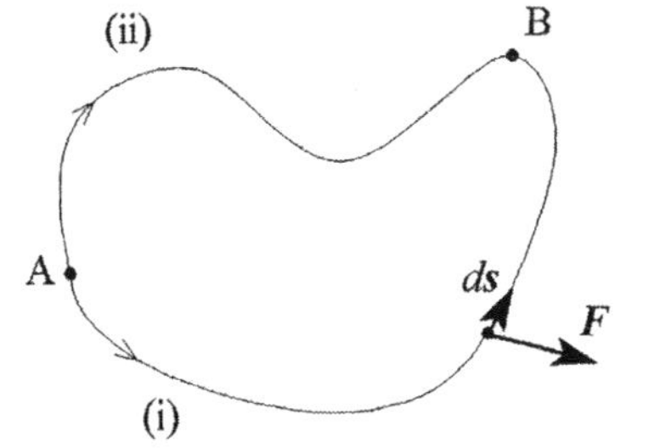

Figure 2.6 Work done on a particle in a force field.

$$W = \int_A^B \boldsymbol{F}.\boldsymbol{ds} = \int_A^B (F_x dx + F_y dy + F_z dz) \tag{2.13}$$

When the field is conservative, the value of W is independent of the path from A to B. If the path is followed in the reverse direction, from B to A, then the work done is $-W$. Thus if the particle moves along path (i) from A to B and then back to A along path (ii), the total work done is zero. No net work is done on any particle moving around a closed loop in a conservative force field. This can readily be seen in a gravitational field, where equal and opposite amounts of work are done when a particle moves 'up' or 'down' over the same interval. A *potential* ϕ can be assigned to each point in a conservative force field so that the work done in moving from A to B is equal to the *loss* of potential, $-(\phi_B - \phi_A)$, in the transition. In (2.13), $\boldsymbol{F.ds}$ is then the perfect differential $-d\phi$ or

$$\int_A^B \left(F_x dx + F_y dy + F_z dz\right) = -\int_A^B d\phi$$

$$\equiv -\int_A^B \left(\frac{\partial \phi}{\partial x} dx + \frac{\partial \phi}{\partial y} dy + \frac{\partial \phi}{\partial z} dz \right)$$

so that

$$F_x = -\frac{\partial \phi}{\partial x} \, , \quad F_y = -\frac{\partial \phi}{\partial y} \, , \quad F_z = -\frac{\partial \phi}{\partial z} \, ,$$

$$\text{or} \quad \boldsymbol{F} = -\nabla \phi$$

Note that the work done is associated with a loss of potential. The potential of a particle in a force field can be taken as a form of energy (potential energy) which is

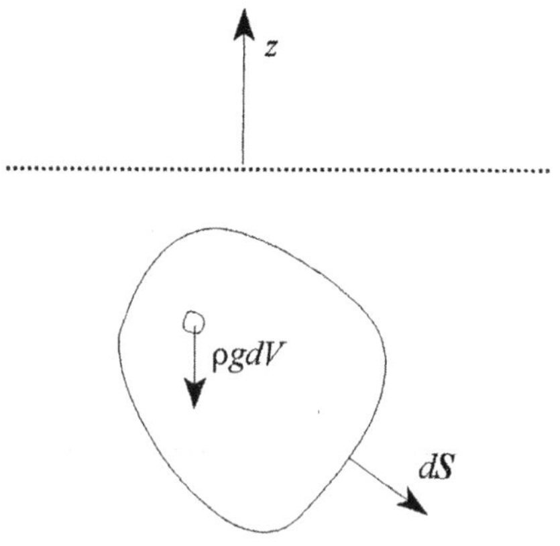

Figure 2.7 Equilibrium of a liquid volume.

lost when the field does work on it. In the case of a simple constant gravitational force field acting on a particle of mass *m*, where *z* is in the 'upwards' direction, ϕ is given by mgz. The position of the origin does not matter, as only changes in the potential are of any consequence. In the gravitational field outside a uniform spherical body of mass *M*, the potential of the mass *m* is given by $-GmM/r$ where *r* is the radius of the particle from the centre of the body and *G* is the universal gravitational constant (6.67×10^{-11} m^3 / kg sec^2).

The potential of a field can be related to the force per unit volume of a body within it. For the constant gravitational field acting on a body of density ρ the potential per unit volume ϕ is ρgz. Suppose that the body is an arbitrary volume of a dense liquid, as shown in Figure 2.7. The gravitational force on the elementary volume dV is then $-\nabla\phi\, dV$ or $-\rho g dV\, \boldsymbol{k}$ (i.e. downwards). An elementary surface area of the body of the liquid can be represented by the vector $d\boldsymbol{S}$ which has a magnitude equal to the area of the elementary surface and direction which is that of the *outwards* normal to the surface. The liquid pressure on this area of surface is $\rho gz d\boldsymbol{S}$, or $\phi d\boldsymbol{S}$. (As the body is below the surface, this force is *inwards*.)The total gravitational force on the body is given by integrating over its volume *V*, and the total pressure force on it is found by integrating over its surface *S*. The resultant force on the body is the sum of these two and must be zero for equilibrium, so that

$$-\int_V \nabla\phi\, dV + \int_S \phi\, d\boldsymbol{S} = \boldsymbol{0}$$

This is in fact a particular case of Gauss's law and so applies to any arbitrary potential ϕ.}

2.4 Particular Cases of Equilibrium

Equations (2.9) give the equations of equilibrium of a single rigid body, where the origin of the coordinate system can be chosen arbitrarily. Suppose that only two forces, $\boldsymbol{F}_1$ and $\boldsymbol{F}_2$, act on the body. The first equation indicates that these two must be equal and opposite. Let the origin be somewhere on the line of action of $\boldsymbol{F}_1$. Then $\boldsymbol{r}_1 \times \boldsymbol{F}_1$ must be zero and so that the only other term on the left-hand side of the second equation, $\boldsymbol{r}_2 \times \boldsymbol{F}_2$, must also be zero. This means that the perpendicular distance from the origin to the line of action of $\boldsymbol{F}_2$ is zero. This result can be summarised in the following statement. *If only two forces are applied to a body, then for equilibrium they must be equal and opposite and collinear.*

Suppose now that only three forces are applied to the body. The origin will again be taken to lie on the line of action of $\boldsymbol{F}_1$. From the second equation of equilibrium, it follows that $\boldsymbol{r}_2 \times \boldsymbol{F}_2$ and $\boldsymbol{r}_3 \times \boldsymbol{F}_3$ must be equal and opposite. The vector product of two vectors is normal to the plane containing them. This means that $\boldsymbol{r}_2$, $\boldsymbol{F}_2$, $\boldsymbol{r}_3$ and $\boldsymbol{F}_3$ must lie in a common plane. Unless $\boldsymbol{F}_1$ also lies in this plane, it must have a component normal to it. As this would be the only component of force normal to this plane, it must be zero for equilibrium. Then if the three forces are in equilibrium, they are coplanar. Unless they are all parallel, the lines of action of at least two of them

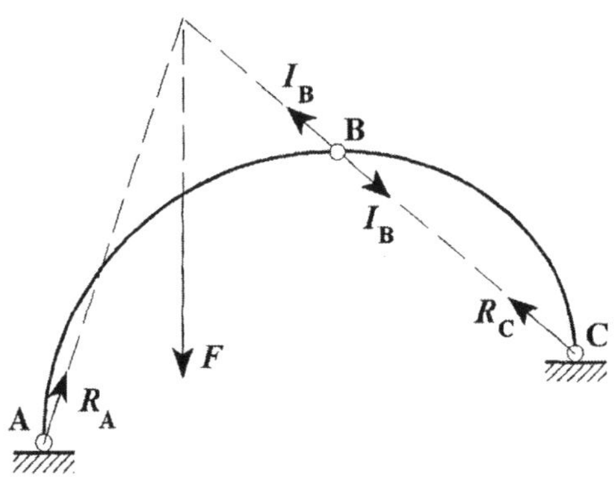

Figure 2.8 Equilibrium of a three-pin arch.

will intersect. Shifting the origin of coordinates to this intersection point, the second of the equilibrium equations is only satisfied if the line of action of the third force also passes through this point (by the same reasoning as before). This leads to the following conclusion. *If only three forces are applied to a body, then for equilibrium they must be coplanar and their lines of action must pass through a common point. (If they are all parallel, this point lies at infinity.)*

Figure 2.8 shows an application of the above two conditions to a two-dimensional problem. A curved bar AB is pinned at its ends, A and B. Pin A is fixed to a support and pin B links it to another curved bar BC which is pinned to a support at C. A force F is applied to the bar AB, but the only forces acting on BC are the actions of the pins, I_B and R_C as shown. The first condition implies that these must be equal and opposite and collinear along the broken line shown. This line is known as the *line of thrust.* On AB, the pin at B exerts an equal and opposite force I_B to that which it exerts on BC, so that its total action is zero. There is also the action R_A of the pin at A and the force F acting on AB. The second condition applies to these three forces, so that their lines of action must pass through a common point, as shown. This determines the directions of the support reactions, R_A and R_C. If F is a vertical force, the horizontal components of R_A and R_C must be equal and opposite, and the algebraic sum of their vertical components and F must be zero.

If a structure consists of a number of bodies linked together, then the equations of equilibrium found for a single body given by (2.9) can be applied to each of them, or the virtual work equations can be applied to the whole or any part of the structure. Imaginary cuts can be made in it, and the appropriate interaction loads applied to the faces of each cut. Then the net work done by the loads (including the interaction loads) during any possible rigid-body motion of the system is zero. (Rigid-body motion implies that no energy is used in straining any component of the system.) Figure 2.9 shows a simple example of the application of this approach. The structure consists of four uniform heavy bars, AB, BC, CD and DA each of weight W and length l which are pinned together to form a rhombus as shown in Figure 2.9a. A light horizontal bar between the joints B and D holds them apart. As the only loads in it are the two forces transmitted by the pinned ends, it is subject only to an axial force P. This can be found directly by making an imaginary cut in BD. This means that a small rigid-body rotation $\delta\theta$ of the bars AB and AD from their initial angle to the vertical, θ, becomes possible. The small change in the initial position (x,y) of joint B can be found by differentiating its initial position with respect to θ:

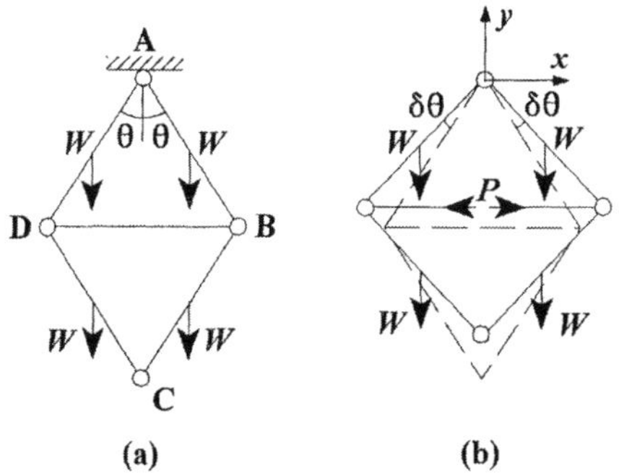

Figure 2.9 Equilibrium of a linked structure.

$$\delta x = \frac{dx}{d\theta}\,\delta\theta = \frac{d}{d\theta}(l\sin\theta)\,\delta\theta = l\cos\theta\,\delta\theta \;;\; \delta y = \frac{dy}{d\theta}\,\delta\theta = \frac{d}{d\theta}(-l\cos\theta)\,\delta\theta = l\sin\theta\,\delta\theta$$

Joint D goes through the same vertical displacement and through the opposite horizontal displacement. Joint C moves upwards towards BD by the same amount by which BD moves upwards towards A. For the purposes of static analysis, the weight W of each bar can be considered to be concentrated at its centre of gravity, which is at the centre of the bar. The motion of each centre can be found from the average of the end displacements. It is now possible to write down the total virtual work done by the loads during this virtual displacement and equate it to zero:

$$[-2W\times\tfrac{1}{2}l\sin\theta - 2W\times\tfrac{3}{2}l\sin\theta + 2P\times l\cos\theta]\delta\theta = 0$$

As $\delta\theta$ is not zero, the contents of the square brackets is zero, giving the value of P as $2W\tan\theta$. This result could have been obtained by applying equations (2.9) to each bar individually, but this would have involved calculating the joint interactions as well, rather than finding the axial force P directly.

2.5 Method of Sections

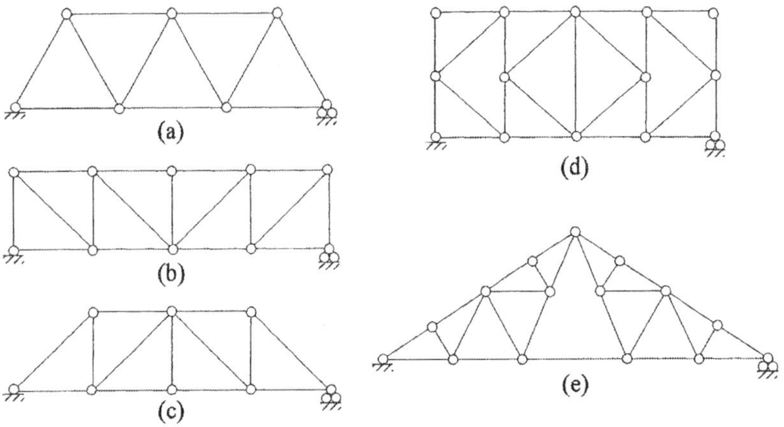

Figure 2.10 Common types of plane truss.

It was mentioned in the last section that the equilibrium of part of a structure could be examined by making an imaginary cut in it. This is often done with pin-jointed trusses when particular bar forces are of interest. Figure 2.10 shows the best-known types of plane pin-jointed truss. Those shown in Figures 2.10a to 2.10d are known as a Warren truss, a Pratt truss, a Howe truss and a K-truss respectively and are often used to support bridge decks. That shown in Figure 2.10e is known as a Fink truss and is used to support rooves. Such trusses are normally only loaded at their joints by forces acting in the plane of the truss. This means that the line of thrust in each bar is along the axis of the bar, so that they are subject only to pure tension or compression. Also the static analysis of each part, considered as a single rigid body, is governed by a subset of (2.9) which is three independent equations only, as mentioned in §2.2. These will be taken here as the conditions that the resultant horizontal and vertical forces are zero and that the moment about some point is zero.

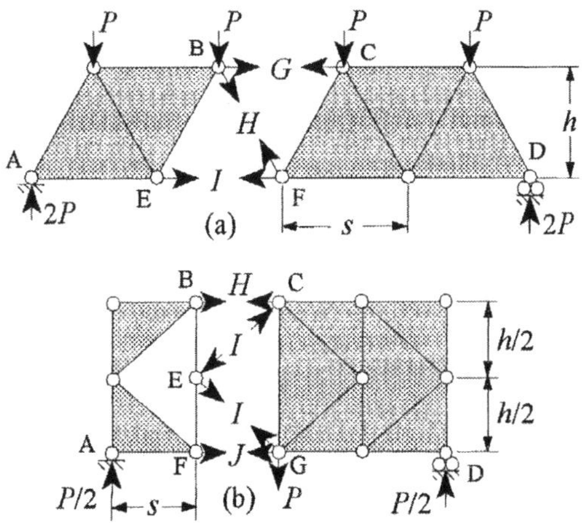

Figure 2.11 Method of sections for plane trusses.

Figure 2.11 shows how the method of sections can be applied to two such trusses. Each has bays of depth h and span s. In both cases, pin A is prevented from displacing and pin D is free to move horizontally. This means that the reaction at D must be vertical. Treating the truss as a single rigid body and taking the origin to be at A, this reaction can be determined from the equilibrium of moments about this point. Since the other external forces are vertical, the reaction at A can have no horizontal component and so it can now be found by resolving vertically. This gives the values of the reactions shown in the figure.

Figure 2.11a shows a four-bay equilateral triangular Warren truss which has been sectioned through the bars BC, BF and EF. These have unknown tensions in them, G, H and I respectively. The sectioning process cuts the frame into the two bodies shown by the shaded areas. By a judicious choice of the origin about which moments are taken and the order in which the equilibrium equations are solved, the need to solve simultaneous equations to find the above tensions can usually be avoided. Thus if moments are taken about B for the left-hand body, the clockwise moment is $2P \times 1.5s$ and the anticlockwise moment is $(P \times s + I \times h)$. Equating these to give a zero resultant moment gives the value of I as $2Ps/h$. Likewise, G can be found directly to be $-2Ps/h$ by taking moments about joint F for the equilibrium of the right-hand body. (The negative sign indicates that the bar BC is in compression.) The force H can be found by resolving vertically. However, there is no net vertical force in either body apart from the vertical component of G, so that the tension in BF must be zero.

It is not possible to section the K-truss shown in Figure 2.11b without cutting through at least four bars. This means that the three equations of plane equilibrium for a single body are not enough. However, if there is to be no net horizontal force on joint E, the horizontal components of the forces in EC and EG acting on it must be equal and opposite. Thus if the tension in EG is I, the compression in EC must also be I. Resolving vertically for the equilibrium of the right-hand shaded block gives the value of the vertical component of I as $P/4$. The actual value of I is found by multiplying this by the secant of the angle between the direction of I and the vertical, which is $\sqrt{(1 + 4s^2/h^2)}$. Resolving horizontally for either side of the cut shows that H and J must be equal and opposite, as the forces I have no horizontal resultant. Taking moments about E for the left-hand side gives J as $Ps/2h$ and H as $-Ps/2h$.

2.6 Joint Resolution

This method relies on the equilibrium equations for the individual joints of the structure to find the loads in its other components. Joints may be subject to loads from the individual components of a structure and also to external loads applied to the structure. Usually, joints are considered to have no size, so that the equilibrium equations reduce to the conditions that the vector sum of the applied forces is zero and the vector sum of the applied moments is zero. (If the size of a joint is considered to be significant, for example if large gusset plates are used, then allowance can be made for this. As will be seen in §11.7, one method of doing this is to incorporate rigid end extensions into the other components. These begin at the edge of the joint and extend to a common point at its centre. This centre can then be treated as a joint of zero size.)

Figure 2.12 shows a simple pin-jointed frame ABCD where joint A is fixed in space and joint D is free to roll horizontally. All the component bars are vertical, horizontal or at 45° to the horizontal. An external force of $2\sqrt{2}W$ is applied to joint B at 45° to the horizontal. By taking moments about A for the whole structure the vertical reaction at D is found to be W. Resolving vertically and horizontally for the whole frame gives the vertical and horizontal reactions at A as W and $2W$ respectively, as shown in the figure. The bar forces can now be found by joint resolution.

These forces are denoted by the italicised letters F to J and their actions on the joints shown by the arrows on the bars. As the reactions of the joints are equal and opposite, the bar forces F, G and I are compressive as shown and the bar forces J and H are tensile. As all the joints are pinned, no moments can act on them, so that the only equilibrium condition that can be applied to them is that the resultant force acting on each of them is zero. In graphical terms, the vectors representing these forces form a closed polygon, as in Figure 2.5b. A particular case of this which occurs frequently is the triangle of forces. The force polygon then comprises only three forces, as at joints B, C and D. If two of the forces are known in direction and the third is known in magnitude and direction, then the unknown magnitudes of first two forces can be found by construction. Thus at joint B, the external force of $2\sqrt{2}W$ is known in magnitude and direction and so can be drawn as the line vector forming one side of the triangle of forces shown adjacent to joint B in Figure 2.12. The other two forces, F and I, are represented by line vectors forming the other two sides of the triangle. As their directions are known, construction lines can be drawn for them at either end of the line vector for the external force. The length of these two lines, and hence the magnitudes of the forces they represent, is determined by their intersection. Hence I and F are both equal to $2W$. Knowing F, the triangle for joint C can now be drawn similarly. Drawing the triangle of forces for joint D now determines the only remaining unknown bar force H. The polygon of forces for the equilibrium of joint A is also shown in Figure 2.12. It will be seen that each bar force occurs in two of the force polygons. In fact, all four polygons can be superimposed on one another to form a single diagram in which each bar force is represented by a single line. In order to do this, a consistent method has to be employed in drawing the polygons of force, as each polygon can be drawn in more than one way, depending on the order in which the line vectors are joined together. This consistent method is known as Bow's notation. However, such graphical constructions are no longer used in practice.

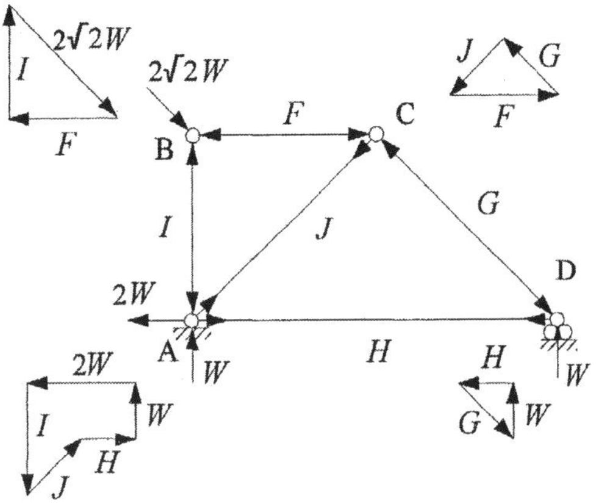

Figure 2.12 Joint resolution for a pin-jointed frame.

The above graphical approach is equivalent to using the first equations on the first two lines of (2.9a) to give the equilibrium of the joints. The explicit form of these equations is

$$\begin{aligned} 2W - F = 0 \quad &, \quad I - 2W = 0, \\ F - (J + G)/\sqrt{2} = 0 \quad &, \quad G - J = 0, \\ G/\sqrt{2} - H = 0 \quad &, \quad W - G/\sqrt{2} = 0. \end{aligned} \tag{2.14}$$

which is given by resolving horizontally and vertically at joints B, C and D respectively.

2.7 Tension Coefficients

This method is the same in principle as that of joint resolution. However, by expressing the unknown parameters as the bar forces divided by the bar lengths, the equations can be written in a much simpler form. This is because the equations are expressed in terms of the resolved parts of the bar forces in the directions of the global axes, x, y and z. These are given by multiplying the bar forces by the cosines of the angles between the bars and these axes (the *direction cosines*). Such a cosine is given by the projected length of a bar in the direction of a particular axis divided by its actual length, and is often an irrational number. However, if the parameter used is the bar force divided by the bar length, then this, multiplied by the projected length, gives the required term in the equilibrium equation. Taking the example given in the previous section, if the length of BC is taken as l, the forces can be expressed in terms of such parameters by

$$f = F/l\,,\ g = G/\sqrt{2}l\,,\ h = H/2l\,,\ i = I/l\,,\ j = J/\sqrt{2}l.$$

On dividing equations (2.14) by l, they now become

$$\begin{aligned} 2W/l - f &= 0 \quad , \quad i - 2W/l = 0, \\ f - j - g &= 0 \quad , \quad g - j = 0, \\ g - 2h &= 0 \quad , \quad W/l - g = 0. \end{aligned} \tag{2.15}$$

A slight simplification in the equations will be noticed, but the method lends itself particularly well to the analysis of three-dimensional pin-jointed frames. A more methodical approach can be made by assuming that all the bars are in tension. Then the components of the force T_{AB} in a bar AB of length l_{AB} are $t_{AB}\,x_{AB}$, $t_{AB}\,y_{AB}$ and $t_{AB}\,z_{AB}$ in the x, y and z directions, where t_{AB} is the *tension coefficient* T_{AB}/l_{AB} for the bar, x_{AB}, y_{AB} and z_{AB} being the projections of its length in the directions of the axes as shown in Figure 2.13. The correct signs for the components of this force acting on joint A are given by giving the appropriate signs to the projected lengths. This is done by defining them as the coordinates B relative to A. Conversely, when looking at the components of the bar force acting on joint B, the projected lengths are defined as the coordinates of A relative B, that is, they are equal and opposite to those used for joint A.

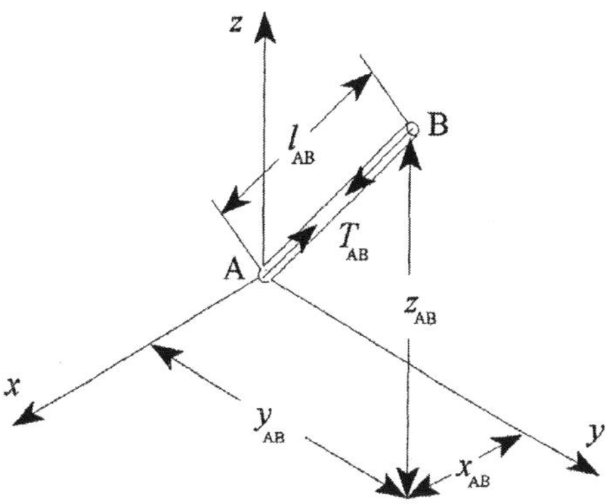

Figure 2.13 Tension components in three dimensions.

Figure 2.14 shows a simple three-dimensional frame which will be used to demonstrate this method. The coordinates of each joint are shown in Figure 2.14a. Joints C,D and E are fixed in position and a vertical load of 10 units[1] applied downwards at joint A. Figures 2.14b and 2.14c show projections of this frame on the y-z and x-y planes respectively.

[1]Here and elsewhere, units will only be specified when necessary. Provided that they are used in a consistent manner, the system of units does not matter. In most computer programs, the units are defined only for the input and output at the user interface, and not for the main body of the analysis.

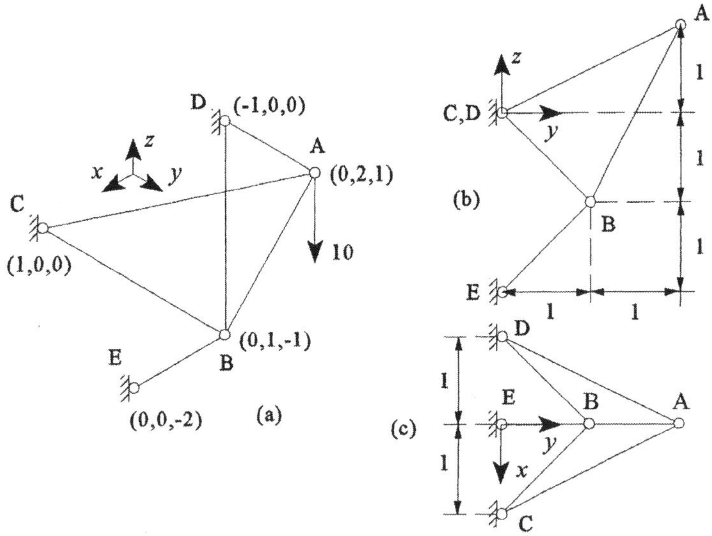

Figure 2.14 Views of a three-dimensional pin-jointed frame.

Although the frame is symmetrical about a vertical plane through joints A and B, for the purposes of the example it will not be assumed that the tensions in AC and AD are the same, nor that the tensions in BC and BD are the same. Then the equations of equilibrium for joint A are

$$\begin{aligned} t_{AC} - t_{AD} &= 0\,, \\ -t_{AB} - 2t_{AC} - 2t_{AD} &= 0\,, \\ -2t_{AB} - t_{AC} - t_{AD} - 10 &= 0\,. \end{aligned}$$

resolving in the x,y and z directions respectively. Likewise, the equations for joint B are

$$\begin{aligned} t_{BC} - t_{BD} &= 0\,, \\ t_{AB} - t_{BC} - t_{BD} - t_{BE} &= 0\,, \\ 2t_{AB} + t_{BC} + t_{BD} - t_{BE} &= 0\,. \end{aligned}$$

The values of the tension coefficients found from these equations are multiplied by the bar lengths to give the bar tensions. This can be set out in tabular form as follows

Bar	Tension coefficient	Length	Tension
AB	-20/3	$\sqrt{5}$	-14.91
AC,AD	5/3	$\sqrt{6}$	4.08
BC,BD	5/3	$\sqrt{3}$	2.89
BE	-10	$\sqrt{2}$	-14.14

2.8 Static Analysis of Beams[1]

The resultant loads in the bars of a pin-jointed frame are axial forces only, unless the bars are loaded directly, or are curved as in Figure 2.8. If the frame has rigid joints, they can carry and transmit moments as well as forces. The distribution of bending moments and shear forces in the component beams of such frames is often critical in structural analysis and can be represented diagrammatically as in Figure 2.15. Here, the common convention for two-dimensional problems, as used in Figure 1.5, has been adopted. The local axes for the beam are taken as x axis along it and the y axis normal to it. For a right-handed coordinate system, the z axis must be out of the paper, and the local rotation θ of the bar is taken as clockwise, looking along this axis. The local bending moment pair, M, is taken to be positive when it causes θ to increase with x (cf. Figure 1.5c). The local shear force pair, S, is such that it causes the shear displacement in the y direction to increase with x (cf. Figure 1.5d).

Figure 2.15 shows the shear force and bending moment diagrams for a simple cantilever of length l, loaded at its free end by a bending moment M_0 and a shear force F_0 as shown. In terms of the above sign convention, this induces a constant shear force of $-F_0$ at all points along the beam, as plotted in Figure 2.15b. (This is because the resultant force induced on the *left-hand* face of shaded area is F_0 upwards. This face corresponds to the left-hand face of the short element shown in Figure 1.5d, on which the shear force is downwards-positive.) The bending moment at a distance x from the free end of the cantilever is M_0+F_0x clockwise on the left-hand face of the shaded area. Similarly, by comparing this with Figure 1.5c, this is seen to be positive. Thus the ordinates on the bending-moment diagram shown in Figure 2.15c are positive. More generally, the ordinates can be found by looking at the total effect on the left-hand face of the shaded area of the loading to the left of it.

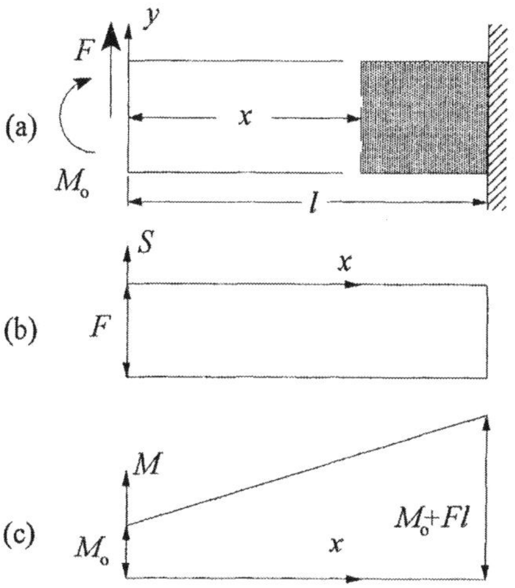

Figure 2.15 Shear force and bending moment diagrams.

Figure 2.16 shows the effects of loads distributed along the beam. Now the additional force F and the distributed load q will also contribute to the resultant loads on the section at a distance x' along the beam. The resultant force produced by q on an elementary length dx of the beam is $q dx$. If, as shown in Figure 2.16b, this is at a distance x from the left-hand end of the beam, then the moment produced by this force x' along is given by multiplying it by the lever arm $(x'-x)$. The total effect of the distributed load is given by integrating these elementary effects. Then the total shear force S' and bending moment M' at the section x' along is

[1] The Oxford English Dictionary implies that a beam was originally a bar used to support a roof or the deck of a ship, and so is subject to bending. Here it will be used to indicate a bar subject to some loading other than just an axial force.

$$S' = \int_{x_q}^{x'} q\,dx + F - F_o$$

$$M' = M_0 - \int_{x_q}^{x'} q(x'-x)dx - F(x'-x_f) + F_o x'$$

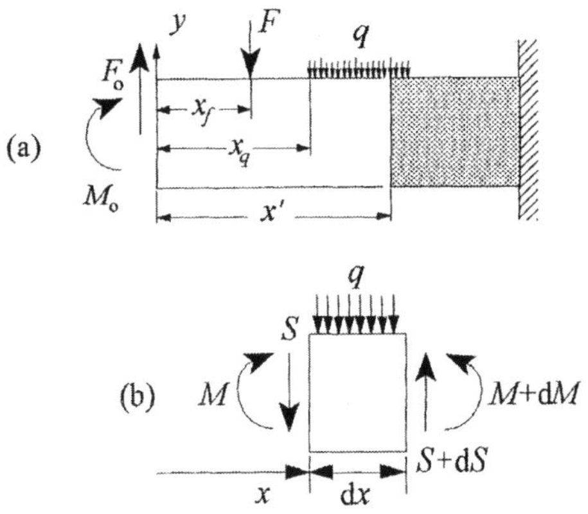

Figure 2.16 Effects of loads along a beam.

where q is in general a function of x. If x' is less than x_q then the terms in q disappear. Likewise, if x' is less than x_f then the terms in F disappear. This means that these expressions are not continuous functions but *step functions*. These will be examined more fully when discussing Macaulay's method later. Differentiating M' with respect to x' gives $-S'$. This can be seen to be true in general by examining the equilibrium of the elementary length of beam shown in Figure 2.16b. The resultant anticlockwise moment about the right-hand face of the element is zero for equilibrium, giving

$$M + \mathrm{d}M - M + S\,\mathrm{d}x + q\,\mathrm{d}x \times \tfrac{1}{2}\mathrm{d}x = 0 \text{ . Hence } \frac{\mathrm{d}M}{\mathrm{d}x} = -S \,. \quad \textbf{(2.16)}$$

(The term in q can be ignored as it is of second order of smallness.) Likewise, the resultant vertical force on the element is zero, giving

$$S + \mathrm{d}S - S - q\,\mathrm{d}x = 0 \text{ . Hence } \frac{\mathrm{d}S}{\mathrm{d}x} = q \,. \quad \textbf{(2.17)}$$

It follows from (2.16) that the bending moment is *locally* a maximum or minimum where the shear force is zero. However, it must be remembered that these expressions are discontinuous functions and apply only to the finite length of the beam. This means that if the greatest bending moment or shear force is sought, it is not enough to look at the points where their differentials are zero. In addition, their values at the discontinuities, where the loading applied to the beam suddenly changes, and also the values at the ends of the beam must be examined.

Figure 2.17 shows examples of shear-force and bending-moment diagrams for simply-supported beams. The reactions at either end can be found from the condition that the resultant moment about the other end is zero. This gives the end reactions in the three examples as

$$\text{(a): } R_A = \frac{Fb}{a+b}\,; \quad \text{(b): } R_A = -R = \frac{-m}{a+b}\,; \quad \text{(c): } R_A = \frac{pb^2}{2(a+b)}\,;$$

$$R_B = \frac{Fa}{a+b}\,. \qquad R_B = R = \frac{m}{a+b}\,. \qquad R_B = \frac{pb(2a+b)}{2(a+b)}\,.$$

where the reactions are taken as upwards-positive. In case (a) there is a step discontinuity of F in the shear-force diagram at the point where the shear force F is applied, and similarly in case (b) there is a step discontinuity of m in the bending-moment diagram where the point moment m is

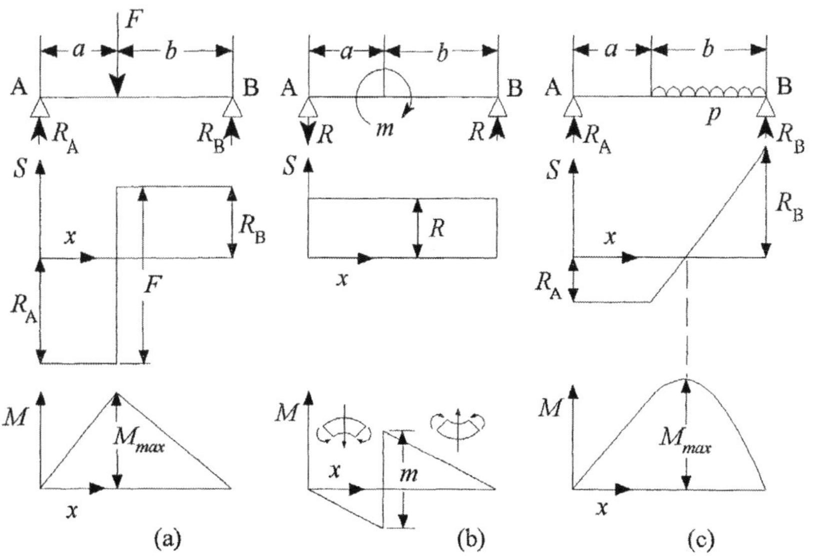

Figure 2.17 Examples of shear force and bending moment diagrams.

applied[1]. In case (c), there is no step discontinuity in either diagram at the point at which the distributed load of constant intensity p commences, but it follows from (2.17) that there is a discontinuity of slope of the shear-force diagram at this point. In case (c), the absolute maximum bending moment, $M_{\max}$, occurs at the point of zero slope of the bending-moment diagram, at which S is zero in accordance with (2.16). In the other two cases, this value occurs at a discontinuity. For the three examples, they are

$$\text{(a): } M_{\max} = \frac{Fab}{a+b} \quad \text{(b): } M_{\max} = \frac{ma}{a+b}\ (a>b) \text{ or } \frac{mb}{a+b}\ (b>a) \quad \text{(c): } M_{\max} = \frac{pb^3(4a+5b)}{8(a+b)^2}\ .$$

In case (b), the bending moments to the left of the point moment m are negative. As shown in the sketch above that part of the diagram, they induce local curvature about a centre which is below the beam. This is known as *hogging* curvature. To the right of m, the induced local curvature is about a centre above the beam. This is known as *sagging* curvature. The point where the flexural curvature changes from one to the other is known as a *point of contraflexure*. If the bending-moment diagram is locally continuous, this occurs at a point where the bending moment is zero, so that analytically the problem remains the same if a pin joint is inserted at this point. The analysis can sometimes be simplified in this way if the points of contraflexure are known.

Bending-moment diagrams are often superimposed on outlines of the structures to which they refer. Figure 2.18 shows an example of this. The frame shown is subject to a distributed side load which varies linearly in intensity from zero at the base, A, to q_o at B. (This might represent a wind load.) The resultant of any such distribution can be found by integration. This is given by the area of the distribution diagram and acts through its centroid. Thus the magnitude of its

[1]It is common practice to join the graphs on either side of point discontinuities, even though this would seem to imply that the functions have no unique values at such points. In fact, point forces and point moments are only convenient approximations to real physical loads and what happens in the immediate vicinity of their application requires more detailed examination of the true conditions.

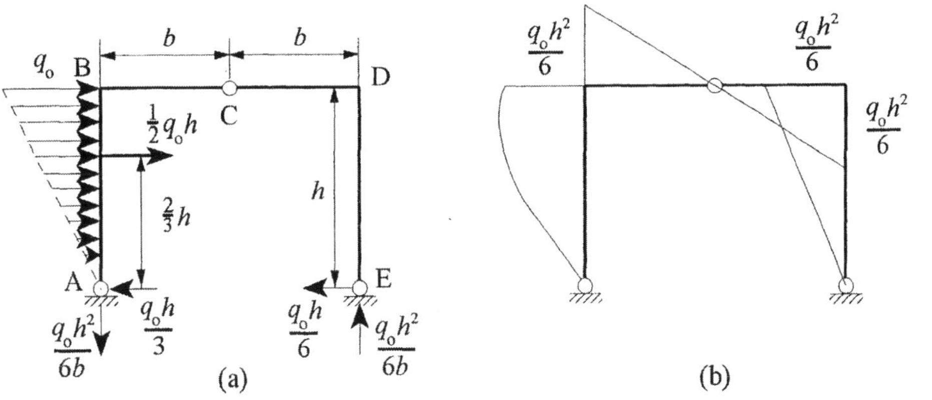

Figure 2.18 Bending moment diagram for a frame.

resultant is $\frac{1}{2}q_0h$ and acts at $\frac{2}{3}h$ above the base. The vertical reaction at E can be found by taking moments about A; that at A must be equal and opposite to this. The horizontal reactions at A and E can be found by taking moments about C for ABC and CDE. It is now possible to draw the bending-moment diagram on the outline of the frame. The convention that positive bending moments are represented by upwards ordinates cannot be used for the stanchions AB and DE. However, the convention that the ordinate is in the direction of the local centre of curvature, as shown in Figure 2.17b, can still be applied to give the diagram shown in Figure 2.18b.

2.9 Static Determinacy

The problems analysed so far illustrate cases where the resultant loads throughout the structure can be found from the conditions of static equilibrium. Such structures are said to be *statically determinate*. If in addition the structure resists loading applied to it in any manner, it said to be *just stiff*. If, instead, the structure can be induced to change its form without resistance (apart possibly from friction) a *mechanism* is said to exist. The mechanism is usually defined in terms of the type of motion possible. The structure is *statically indeterminate* if the conditions of equilibrium are insufficient to determine all the resultant loads on the component bars. The structure is also *redundant* because there are more constraints on it than are necessary to make it just stiff. Each constraint inhibits one possible deflexion or relative deflexion. The number of these additional constraints is known as the *degree of redundancy*. It is also possible to imagine structures which do not fall into any one of the above categories, as they may have characteristics of more than one type.

The three basic types are illustrated in Figure 2.19. The upper diagrams show a single beam which is classified according to its support conditions. The lower diagrams show a frame which is classified primarily by the constraints provided by its constituent bars. Figures 2.19a show simple mechanisms with one degree of freedom. The upper one is a pendulum and the lower one is a *four-bar chain*. (The fourth bar in this case is considered to be the ground.) The addition of one constraint to each of these can produce a just-stiff structural system. Thus if, as shown in Figure 2.19b, the free end of the pendulum is supported by a knife edge, it behaves as a simply-supported beam and if a diagonal bar is added to the four-bar chain, it can no longer deflect freely. If the supports at the ends of the beam are replaced by fixed ends and a extra diagonal bar is added to the frame, then the redundant structural forms shown in Figure 2.19c result.

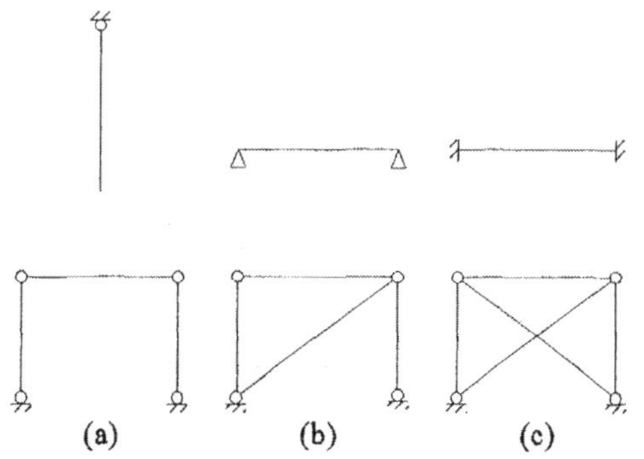

Figure 2.19 Degrees of freedom and redundancy.

A structure is statically determinate if there are (at least) as many independent equilibrium conditions as there are unknown loadings to be found. The loadings may be the support reactions or the resultant loads acting on the component bars. If a bar is supported only at it ends and the resultant loads acting at one section of a bar are known, then the resultant loads at all points can be found by treating it as a simple cantilever between the section and an end (cf. §2.8). A frame is composed of bars, each connected to two nodes (joints or free ends) and supports.

Figure 2.20a shows a plane pin-jointed frame where the nodes are pin joints. Each has two degrees of freedom (horizontal and vertical motion) as shown in Figure 2.20b. Each degree of freedom can be associated with an independent equilibrium or virtual work equation. (Joint rotation yields no equations as no moments are applied to the joints.) Each bar has an unknown axial force in it. This may be found from the equations associated with joint motion, as in the example shown in Figure 2.9. The only additional independent equations give any variation of resultant loading on sections along the bars. (Here, the loading is normally a constant axial force). Every support restraint corresponds to an additional unknown load. Thus, if there are j pin joints there will be a corresponding $2j$ independent equations. If there are b bars linking the joints and r support restraints, these equations will be needed to determine $(b+r)$ unknowns. Alternatively, as each bar constrains the relative motion of two joints, the number of degrees of freedom left is $(2j-b-r)$. This is the *degrees-of-freedom formula*. If this expression is zero, then the frame will normally be just stiff and statically determinate, as the number of unknowns is equal to the number of equations.

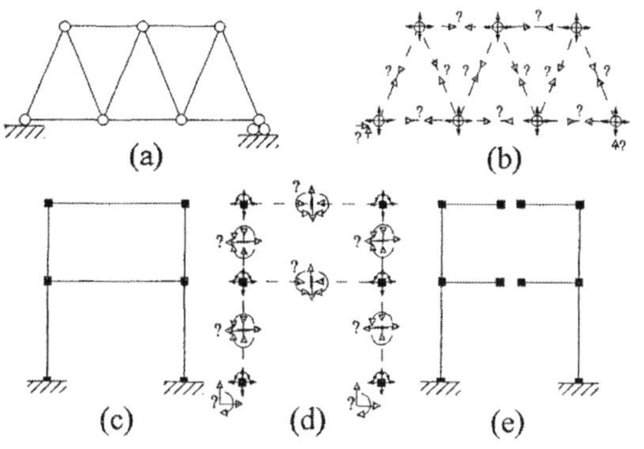

Figure 2.20 Static determinacy of plane frames.

Take for example the frame shown in Figure 2.20a. There are seven joints, eleven bars and three support restraints. Then the value of $2j-b-r$ is (14-11-3), or zero, which corresponds to a just-stiff, statically-determinate frame. The values of this expression for the frames shown in the lower parts of Figures 2.19a, b and c are one, zero and minus one, corresponding to a mechanism with one degree of freedom, a just-stiff frame and a frame with one redundancy. In three dimensions, each free joint has three degrees of freedom, and each support which completely fixes a joint in space provides three restraints. The degrees-of-freedom formula then takes the form ($3j$-b-r) and when applied to the frame shown in Figure 2.14 gives zero (15-6-9) indicating that it is also a just-stiff frame.

The nodes of the rigid plane frames shown in Figures 2.20c to 2.20e are shown as black squares. Each has three degrees of freedom, two displacements and a rotation. Each bar of the frame now has three unknown resultant loads acting on a cross-section, an axial force, a shear force and a bending moment. The degrees-of-freedom formula now becomes ($3n$-$3b$-r) where n is number of nodes. (Where all the nodes are joints, as in the case shown in Figure 2.20c, j can be used in the formula instead of n.) In this example, each support provides three restraints, so that the formula becomes (18-18-6), so that the frame has six redundancies. Cutting each horizontal bar of this frame gives the two frames shown in Figure 2.20e. These are known as *tree structures*. Each has two nodes which are not joints and the formula gives (15-12-3) for either of them, indicating that they are both statically determinate. Tree structures are composed of a trunk, branches and twigs like a tree, are always statically determinate. Another way of finding the degree of redundancy of a rigid-jointed frame is to count the number of cuts in its bars necessary to reduce it to tree structures. Each cut introduces three extra degrees of freedom, two relative displacements and one relative rotation of the two sides of the cut. The sum of the degrees of freedom introduced, in this case six, is equal to the number of redundancies of the original frame. In three dimensions, each node has six degrees of freedom, three rotations and three rotations, and each bar induces six constraints corresponding to six unknown resultants. The degrees-of-freedom formula now becomes ($6n$-$6b$-r). Other types of structure can be reduced to pin-jointed or rigid-jointed forms by introducing degrees of freedom or adding constraints and then applying the formula, taking into account the increase or decrease in the degrees of freedom introduced by the modification.

The formula is not always sufficient to ensure that a frame is statically determinate or just stiff. Figure 2.21 shows three pin-jointed frames which are just stiff according to the formula, but exhibit characteristics of mechanisms and redundant frames. This is because a rigid-body motion is possible which does not involve work being done by either the bar loads or the support reactions. This means that the corresponding equilibrium equation imposes conditions on the applied loading only, and there are not enough equations left to determine the other loads. In case (a), the right-hand rectangle can deform into a parallelogram, that is, it is a four-bar chain. The left-hand rectangle is over-braced, having two diagonals, and so has one redundancy. Cases (b) and (c) are both examples of *critical forms*. These are frames which become just-stiff with a slight change in geometry. In case (b), if the centre lines of all three legs are extended, they meet at a point O. This means that the frame can rotate slightly about O. (The point O acts as an 'instantaneous centre'.) This means that the applied loading can have no resultant moment about O. Also, the axial forces in the legs cannot be found from statics. Changing the directions of the legs so that their centre lines do not meet at a single point, even at infinity, makes the frame statically determinate. The regular hexagonal frame shown in Figure 2.21c is also capable of

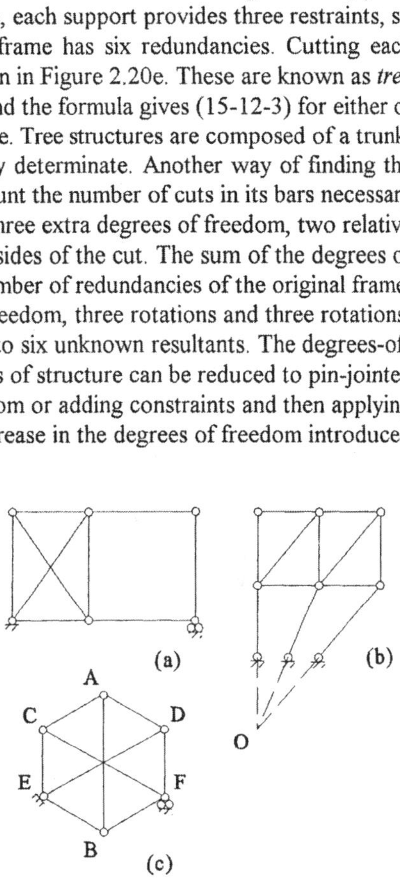

Figure 2.21 Mixed types of structure.

deforming slightly during a rigid-body motion of its components, without cutting any of the bars. Joints C and D can move apart horizontally by a small amount 2δ, while joints E and F move towards one another by the same amount, and joints A and B move down relative to the others by $\delta\sqrt{3}$. Any applied loading which does work during this motion is inadmissible if the frame is to be in equilibrium. The loss of one equilibrium equation means that the bar forces cannot be determined from statics. For example, the frame would be in equilibrium under zero applied loading if all three diagonals had some indeterminate tension T in them and all six peripheral bars were subject to compressive forces of the same magnitude T. Again, the frame becomes statically determinate if the geometry is changed, for example by moving the position of joint B upwards so that it is in line with joints E and F.

2.10 Displacement Diagrams

An equation of static equilibrium for the frame shown in Figure 2.9 was obtained by making an imaginary cut in the horizontal bar and finding the virtual-work equation for the resulting mechanism. In that particular case, it was a relatively simple matter to find the motion of the mechanism. In more complex cases, the motion can be found from a displacement diagram. This is a vector diagram of relative and absolute displacements. It is constructed on the same principles as velocity diagrams in mechanics.

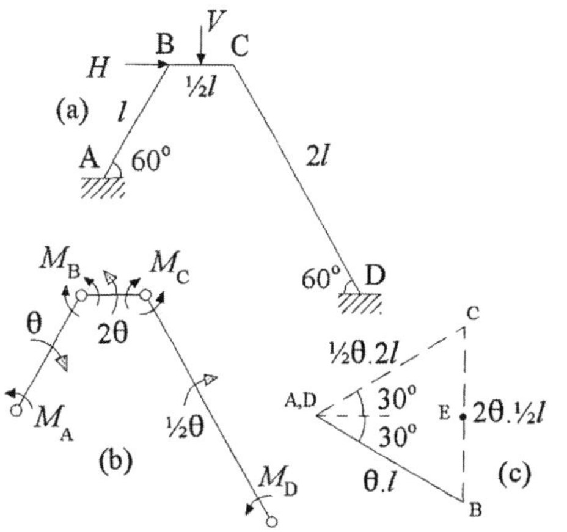

Figure 2.22 Sway of a sloping portal frame.

Figure 2.22a shows a rigid-jointed portal frame with sloping legs. A horizontal force H is applied at B and vertical force V applied at E, the midpoint of BC. A more complete analysis of this frame will be given in §4.5, but in order to carry this out, it is necessary to find a relationship between the joint moments M_A to M_D and the applied loads. By replacing the rigid joints with pins, a mechanism is created permitting a rigid-body motion of the bars in which only these loads and the joint moments do work. Suppose that the bar AB rotates about A through a small angle θ as shown in Figure 2.22b. The end B will then move through a distance θl at an angle of 30° below the horizontal. This is represented by the line AB in the vector diagram shown in Figure 2.22c. The point A is at what is called the pole of this diagram. The pole is the point from which all absolute motion is measured. As joint D does not move either, D is also at this pole. The relative motion of C with respect to B must be perpendicular to BC, i.e. vertical. On the vector diagram this is represented by a line from B to C. At this stage, its length is not known, and so is represented by a vertical broken line. Likewise, the motion of C relative to D must be along a line at 30° above the horizontal and this is represented by a broken line DC in the same direction on the vector diagram. The point C must be at the junction of these two lines, producing the equilateral triangular diagram as shown. All the sides are then of length θl. However, the line DC represents the motion of C relative to D as a result of the rotation of DC which is of length $2l$. This rotation must then be ½θ. Likewise, the diagram shows that C moves vertically upwards relative to B through a distance of θl. This means that BC must rotate anticlockwise through an angle of 2θ.

The horizontal motion of B and C is the horizontal component of their absolute motion, that is, $\frac{1}{2}\sqrt{3}\theta l$. The movement of E is the average of the movements of B and C and hence is given by the midpoint E of the line BC on the vector diagram. It follows from the diagram that E moves horizontally but not vertically, so that V does no work during this motion. The total virtual work done during this motion is zero, giving the equation

$$(-M_A + M_B)\theta + (M_B - M_C)2\theta + (-M_C - M_D)\tfrac{1}{2}\theta + H\tfrac{\sqrt{3}}{2}l\theta = 0$$
$$\text{or} \quad 2M_A - 6M_B + 5M_C + M_D = \sqrt{3}Hl \qquad \textbf{(2.18)}$$

on removing the common factor θ.

If the rigid joints of the frame shown in Figure 2.20c are replaced by pins, as shown in Figure 2.23a, then it has two degrees of freedom. (As seen earlier, the original frame had six degrees of redundancy. The pins at the top and bottom joints each permit one new relative or absolute rotation and the middle pins introduce two independent relative rotations. Thus eight degrees of freedom are introduced, six of which remove the redundancies in the frame.) These two degrees of freedom can be represented by the two independent sway modes shown in Figures 2.23b and 2.23c. Virtual work equations, relating joint moments to the applied loading, can be written for each of these modes by using the above method.

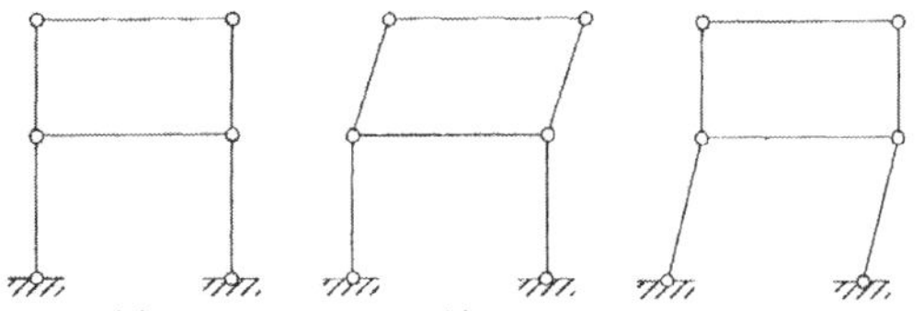

(a) (b) (c)

Figure 2.23 Sway of a two-storey portal frame.

{2.11 Full Determinacy Analysis}

{It was pointed out in §2.9 that the degrees-of-freedom formulae were not always sufficient criteria of determinacy. This is because some restraints may simply duplicate constraints which already exist. If the frame in Figure 2.21a is composed of rigid bars, removing one diagonal bar from the left-hand bay does not make it any less rigid. In the other two cases the duplication of restraint is less obvious, but can be determined from a full analysis.

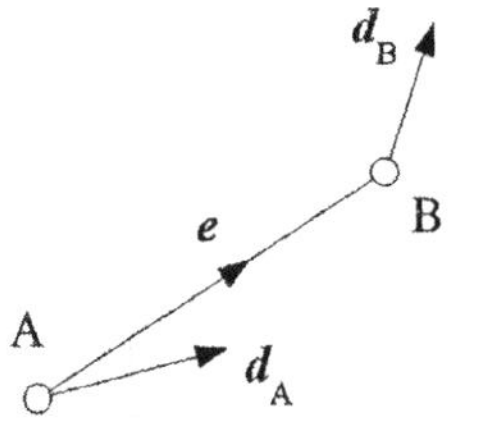

Figure 2.24 Displacement restraint of a rigid bar.

Figure 2.24 shows a rigid bar linking two pin joints, A and B. This prevents any relative displacement of the joints in the direction of the bar. Suppose that the displacement of joint A is given by the vector $\boldsymbol{d}_A$, that of joint B is given by $\boldsymbol{d}_B$ and a unit vector in the direction AB is given by $\boldsymbol{e}$. Then the displacement of joint A in the direction AB is $\boldsymbol{d}_A \cdot \boldsymbol{e}$ and that of joint B is $\boldsymbol{d}_B \cdot \boldsymbol{e}$. The constraint provided by the rigid bar means that these two displacements are equal, or

$$(\boldsymbol{d}_B - \boldsymbol{d}_A)\cdot\boldsymbol{e} = 0 \qquad \textbf{(2.19)}$$

If the length of the bar is l, then $l\boldsymbol{e}$ is the difference of the two position vectors of joints A and B, $(\boldsymbol{r}_B - \boldsymbol{r}_A)$, so that (2.19) can be written as

$$(\boldsymbol{d}_B - \boldsymbol{d}_A)\cdot(\boldsymbol{r}_B - \boldsymbol{r}_A) = 0 \qquad \textbf{(2.20)}$$

One equation of this kind can be written for each bar of a pin-jointed frame. Taken together with the support conditions[1], these equations give the full set of restraints applied to the joint displacements. If these equations are independent and equal to the number of independent joint displacements, the frame is just stiff.

[1] It is assumed here that the support conditions are such that individual joint displacements or linear combinations of them are zero.

Take for example the hexagonal frame shown in Figure 2.21c. For convenience, the sides of the hexagon will be taken to be of length 2, so that the diagonal bars are of length 4. The horizontal and vertical joint displacements of the bars will be denoted by u and v respectively. The support constraints give

$$u_E = v_E = v_F = 0 . \qquad \textbf{(2.21)}$$

Applying (2.20) to bar to the bar AD gives

$$(u_D - u_A)\sqrt{3} + (v_D - v_A)(-1) = 0 \qquad \textbf{(2.22)}$$

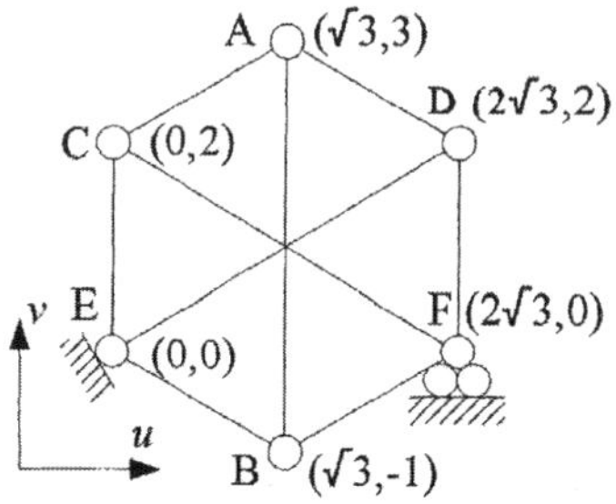

Figure 2.25 Determinacy of a hexagonal frame.

Using (2.21), the full set of bar restraint equations can be written. Starting with AD and proceeding clockwise around the peripheral bars and ending with the diagonal bars, AB, CF and DE, they are

$$\begin{bmatrix} -1 & 1 & 0 & 0 & 0 & 0 & 1 & -1 & 0 \\ 0 & 0 & 0 & 0 & 0 & 0 & 0 & 2 & 0 \\ 0 & 0 & -1 & -1 & 0 & 0 & 0 & 0 & 1 \\ 0 & 0 & 1 & -1 & 0 & 0 & 0 & 0 & 0 \\ 0 & 0 & 0 & 0 & 0 & 2 & 0 & 0 & 0 \\ 1 & 1 & 0 & 0 & -1 & -1 & 0 & 0 & 0 \\ 0 & 4 & 0 & -4 & 0 & 0 & 0 & 0 & 0 \\ 0 & 0 & 0 & 0 & -2 & 2 & 0 & 0 & 2 \\ 0 & 0 & 0 & 0 & 0 & 0 & 2 & 2 & 0 \end{bmatrix} \begin{bmatrix} u_A\sqrt{3} \\ v_A \\ u_B\sqrt{3} \\ v_B \\ u_C\sqrt{3} \\ v_C \\ u_D\sqrt{3} \\ v_D \\ u_F\sqrt{3} \end{bmatrix} = \begin{bmatrix} 0 \\ 0 \\ 0 \\ 0 \\ 0 \\ 0 \\ 0 \\ 0 \\ 0 \end{bmatrix} \qquad \textbf{(2.23)}$$

The general solution of these equations is

$$v_C = u_D = v_D = 0 \ , \ 2u_A\sqrt{3} = 2v_A = 2u_B\sqrt{3} = 2v_B = u_C\sqrt{3} = u_F\sqrt{3} \ (= k \text{ say}). \qquad \textbf{(2.24)}$$

The frame is not then just-stiff. A simple mechanism exists giving displacements for non-zero values of k. The factor √3 could be replaced throughout these equations with any other number and the result would be the same. This corresponds to stretching (or compressing) the form of the frame in the horizontal sense. Thus although deforming the frame so that B is in line with EF produces a statically-determinate form, this stretching has no such effect. In general, a frame may have more than one degree of freedom . A formal method of finding the modes associated with each of these is known as the Gauss-Jordan process and is discussed in §14.4.

Exactly the same method can be applied to three-dimensional pin-jointed structures. Figure 2.26 shows a single tier of a six-sided Schwedler dome. This has a height h, the base hexagon is of side $2k$+2 and the upper is of side $2k$. The joint coordinates are then as shown. If the

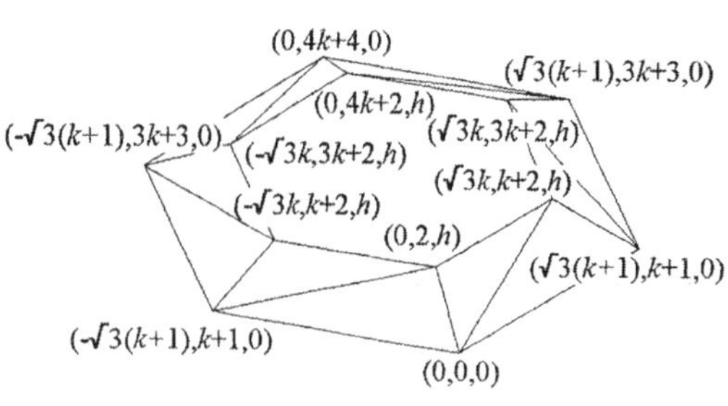

Figure 2.26 A tier of singly-braced Schwedler dome.

lower joints are fixed in position, there are 18 bar restraint equations governing the displacements of the six upper joints. Setting up these equations and solving them as before, the only solution is that all these displacements must be zero for any non-zero values of h and k. Again, stretching the form in any direction does not affect this result. The upper joints are then fixed in position so that the frame is just-stiff. If a further similar tier is added, it follows from the same arguments that it remains just-stiff. The Schwedler dome is discussed again in §15.9 and its general form is shown in Figure 15.26b.

The matrix in (2.23) is the kinematic matrix $\boldsymbol{B}$ defined by Pellegrino and Calladine[1]. By virtual work methods, they show it to be the transpose of an equilibrium matrix $\boldsymbol{A}$ relating bar tensions to joint forces.}

There remain some structures which, despite having mechanisms according to the above analyses, are essentially stable. This is because the mechanisms are for infinitesimal motion, so that after a very small amount of 'give', the structure resists further deflexion. The best known examples are some of the 'tensegrity' structures popularised by Buckminster Fuller. Definitions of tensegrity structures vary[2], but they are usually composed of isolated struts within a web of tensile components. This may be a set of pretensioned cables which induce compression in the struts. The Skylon at the 1951 Festival of Britain exhibition was a simple example of a tensegrity structure, and they are usually used as architectural features.

---

[1]Pellegrino, S. and Calladine, C.R. (1986) Matrix Analysis of Statically and Kinematically Indeterminate Frameworks. *Int. J. Solids Structures* V.22 No.4 p.409.

[2]See for example Motro, R. and Raducanu, V. (2001) Tensegrity Systems and Tensile Structures, *Proc. IASS Conference, Nagoya* Paper TP140.

Chapter 3 Elasticity

3.1 Stress and Equilibrium

In considering the detailed behaviour of a body under load, the response to forces acting per unit area of surface will be examined. This surface can be an external surface, in which case these forces per unit area are referred to as *surface tractions* or they may be on the surface of any imaginary section through the body, and are then called *stresses.*

Using cartesian coordinates, it is possible to define an elementary rectangular block of the formed by imaginary plane 'x', 'y' and 'z' sections through the body, which are perpendicular to the x, y and z axes respectively. This block is shown in Figure 3.1. Each pair of surfaces perpendicular to an axis is comprised of one 'negative' face and one 'positive' face. The positive face is the one which is the further of the two along the corresponding axis. Thus the block in Figure 3.1 only shows its positive faces. The stresses acting on these faces can be resolved into components acting in the directions of the axes. On each face, one component will be acting in a direction normal to the face, known as a *normal stress*, and two components acting parallel to the face, known as *shear stresses*. The symbol σ is used to indicate a normal stress and the symbol τ indicates a shear stress. The first subscript attached to these symbols indicates the face on which the stress acts and the second indicates the direction in which it acts. (As it is possible to tell from the subscripts alone whether a stress is a normal stress or a shear stress, a single symbol, σ, is sometimes used for both.) The sign convention is that stresses acting on positive faces are assumed to act in the positive directions of the axes, and those on negative faces taken to act in the opposite directions. (Each negative face is attached to the positive face of an adjacent block, so that the negative actions on the negative face are matched by positive reactions on the adjacent positive face.)

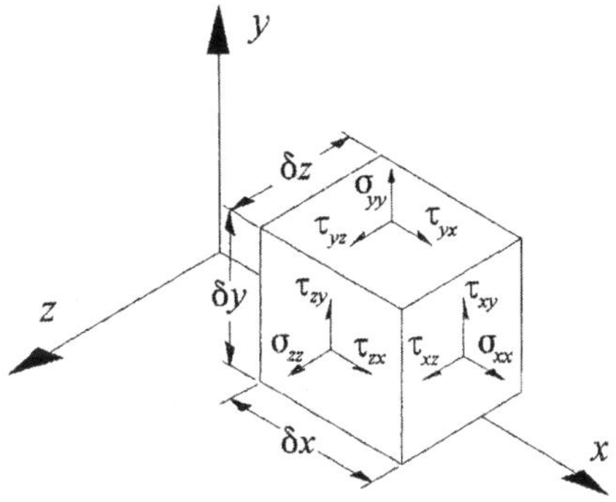

Figure 3.1 Conventions for stresses.

In a state of equilibrium, the body as a whole and every part of it is in equilibrium. It is then useful to examine the equilibrium of the elementary block shown in Figure 3.1. In addition to the stresses on its surface, there may also be body forces acting on the material of the block. These might arise from the effects of gravity for example. These forces are defined in terms of their magnitudes X, Y and Z per unit volume of material, in the x,y and z directions respectively. In general, there will be a slight variation in the magnitudes of the stresses between the negative and positive faces. The incremental changes in the stresses acting parallel to the x axis are shown in Figure 3.2. The small lengths δx, δy and δz of the sides of the block are also shown. Then the incremental changes of stress shown can be related to the rates of change by

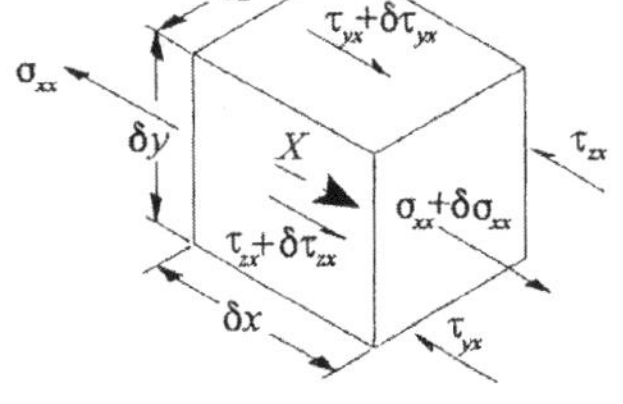

Figure 3.2 Force equilibrium.

$$\delta\sigma_{xx} = \frac{\partial\sigma_{xx}}{\partial x}\,\delta x \ ; \quad \delta\tau_{yx} = \frac{\partial\tau_{yx}}{\partial y}\,\delta y \ ; \quad \delta\tau_{zx} = \frac{\partial\tau_{zx}}{\partial z}\,\delta z \ . \qquad \textbf{(3.1)}$$

The equal and opposite components of stress on each pair of faces produce no resultant forces. Then for equilibrium the above increments, multiplied by the areas on which they act, together with the body force per unit volume multiplied by the volume on which it acts must produce no resultant force. This gives the equation

$$\delta\sigma_{xx}\,\delta y\,\delta z + \delta\tau_{yx}\,\delta z\,\delta x + \delta\tau_{zx}\,\delta x\,\delta y + X\delta x\,\delta y\,\delta z = 0 \tag{3.2}$$

for equilibrium of the block in the x direction. On using (3.1) and eliminating the common factor $\delta x\delta y\delta z$ throughout, this becomes

$$\frac{\partial\sigma_{xx}}{\partial x} + \frac{\partial\tau_{yx}}{\partial y} + \frac{\partial\tau_{zx}}{\partial z} + X = 0. \tag{3.3}$$

Likewise, resolving in the y and z directions,

$$\frac{\partial\tau_{xy}}{\partial x} + \frac{\partial\sigma_{yy}}{\partial y} + \frac{\partial\tau_{zy}}{\partial z} + Y = 0. \tag{3.4}$$

$$\frac{\partial\tau_{xz}}{\partial x} + \frac{\partial\tau_{yz}}{\partial y} + \frac{\partial\sigma_{zz}}{\partial z} + Z = 0. \tag{3.5}$$

The resultant moment about the z axis produced by the shear stresses on the block can be seen by reference to Figure 3.3. (The increments in shear stress produce moments of a smaller order of magnitude and the resultants of the axial stresses can be taken to act through the centroid of the block.) The pair of stresses τ_{yx} acting on an area δz by δx produce a clockwise couple with a lever arm δy, and the stresses τ_{xy} acting on an area δz by δy produce an anticlockwise couple with a lever arm δx. For equilibrium, the net moment on the block must be zero, giving

$$\tau_{yx}\delta z\delta x\;\delta y = \tau_{xy}\delta z\delta y\;\delta x \quad\text{or}\quad \tau_{yx} = \tau_{xy} \tag{3.6}$$

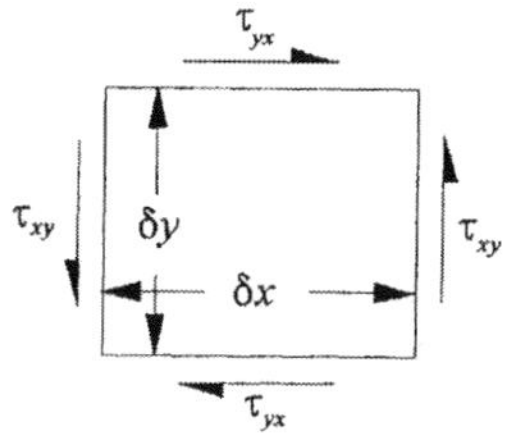

Figure 3.3 Moment equilibrium.

Likewise, on taking moments about the x and y axes,

$$\tau_{yz} = \tau_{zy} \quad ; \quad \tau_{xz} = \tau_{zx} \tag{3.7}$$

The results of (3.6) and (3.7) can be generalised by saying that a shear stress acting in a given direction on a given face is matched by a *complementary shear stress*. This acts on a face normal to the direction of the first shear stress and in a direction normal to that of the first face.

3.2 Strain and Compatibility

Strains are defined in terms of the rates of change of displacements within a body. The displacements in the x, y and z directions of a point P in a body will be denoted by u, v and w respectively. These displacements will be taken as functions of the initial coordinates of the point P and their rates of change determined with respect to the partial derivatives with respect to these coordinates.

Figure 3.4a shows a section of the elementary block in Figure 3.1 parallel to the x-y plane. The point P is chosen to be at the bottom left-hand corner of this element, and the other corners are labelled as shown. The displacements of P_1 can be taken to differ from those of P by the rates of change of u, v and w with respect to x multiplied by the distance between them, δx.[1] Likewise, those of P_2 differ from those of P by their rates of change in the y direction multiplied by δy. Those of P_3 relative to P are the sum of the above two variations. Some of these variations can arise from the rigid-body rotation of the element, but the existence of strains implies that some deformation has taken place.

Normal strain may be defined as the rate of extension of a body in a given direction and is usually denoted by the symbol ϵ. If this is sensibly constant over a fibre of a given length, then the strain of the fibre is

$$\epsilon = \frac{\text{final length} - \text{original length}}{\text{original length}} = \frac{\text{extension}}{\text{original length}} \qquad (3.8)$$

Thus the fibre between P and P_1 extends from its original length of δx by an amount equal to the distance through which P_1 moves further in the x direction than P, or

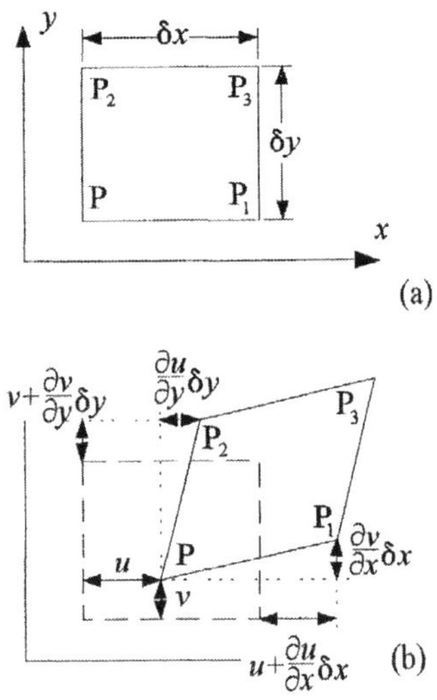

Figure 3.4 Strains and displacements.

$$\epsilon_{xx} = \frac{(\delta x + \frac{\partial u}{\partial x}\delta x) - \delta x}{\delta x} = \frac{\partial u}{\partial x} \qquad (3.9)$$

Strains of fibres in the y and z directions can be defined in the same way, giving

$$\epsilon_{yy} = \frac{\partial v}{\partial y} \ ; \ \epsilon_{zz} = \frac{\partial w}{\partial z} \qquad (3.10)$$

The element shown in Figure 3.4a may also distort from a rectangular form to a parallelogram, without necessarily extending the fibres in the x or y directions. The point P_1 can move further in the y direction than P by an amount $\partial v/\partial x \; \delta x$ corresponding to a small rotation of the line PP_1 of $\partial v/\partial x$ anticlockwise. Likewise, there can be a small rotation of the line PP_2 of magnitude $\partial u/\partial y$. The sum of these two rotations gives the amount by which the angle P_1PP_2 changes from a right-angle, and hence does not correspond to a rigid-body rotation of the element as a whole. Such a distortion is called a *shear strain* and is usually denoted by γ. This particular shear strain is in the x-y plane and is given by

$$\gamma_{xy} = \frac{\partial u}{\partial y} + \frac{\partial v}{\partial x} \qquad (3.11)$$

Similarly, the shear strains in the y-z and z-x planes are given by

$$\gamma_{yz} = \frac{\partial v}{\partial z} + \frac{\partial w}{\partial y} \qquad (3.12)$$

$$\gamma_{zx} = \frac{\partial w}{\partial x} + \frac{\partial u}{\partial z} \qquad (3.13)$$

Since the six strains given by (3.9) to (3.13) are functions of only three variables, u, v and w, they

---

[1] This might be seen by using a Taylor's series for the displacement functions. The terms in δx^2 and higher powers of δx can be ignored as being of too small an order.

are not completely independent of one another. For example, from (3.9), (3.10) and (3.11),

$$\frac{\partial^2\gamma_{xy}}{\partial x\partial y} = \frac{\partial^3 u}{\partial x\partial y^2} + \frac{\partial^3 v}{\partial x^2\partial y} = \frac{\partial^2\epsilon_{xx}}{\partial y^2} + \frac{\partial^2\epsilon_{yy}}{\partial x^2} \qquad \textbf{(3.14)}$$

{Likewise, from (3.9) to (3.13),

$$\frac{\partial^2\gamma_{yz}}{\partial y\partial z} = \frac{\partial^2\epsilon_{yy}}{\partial z^2} + \frac{\partial^2\epsilon_{zz}}{\partial y^2} \qquad \textbf{(3.15)}$$

$$\frac{\partial^2\gamma_{zx}}{\partial z\partial x} = \frac{\partial^2\epsilon_{zz}}{\partial x^2} + \frac{\partial^2\epsilon_{xx}}{\partial z^2} \qquad \textbf{(3.16)}$$

and

$$2\frac{\partial^2\epsilon_{xx}}{\partial y\partial z} = \frac{\partial}{\partial x}\left[-\frac{\partial\gamma_{yz}}{\partial x} + \frac{\partial\gamma_{zx}}{\partial y} + \frac{\partial\gamma_{xy}}{\partial z}\right] \qquad \textbf{(3.17)}$$

$$2\frac{\partial^2\epsilon_{yy}}{\partial z\partial x} = \frac{\partial}{\partial y}\left[\frac{\partial\gamma_{yz}}{\partial x} - \frac{\partial\gamma_{zx}}{\partial y} + \frac{\partial\gamma_{xy}}{\partial z}\right] \qquad \textbf{(3.18)}$$

$$2\frac{\partial^2\epsilon_{zz}}{\partial x\partial y} = \frac{\partial}{\partial z}\left[\frac{\partial\gamma_{yz}}{\partial x} + \frac{\partial\gamma_{zx}}{\partial y} - \frac{\partial\gamma_{xy}}{\partial z}\right] \qquad \textbf{(3.19)}$$

These equations are not entirely independent of one another. They are related by *Bianchi's identities*, see Renton (1987) §1.5.5.}

3.3 Linear Elastic Behaviour of Isotropic Materials

A material is said to be *isotropic* if it exhibits the same properties in any direction within it. Most engineering materials are taken to be isotropic, although the overall properties of fibre-reinforced materials must be related to the direction of the fibres. The consideration of such *anisotropic* materials will be left until later. A material is *elastic* if the strain response to loading is exactly reversed on unloading. If this response is such that the strains induced are exactly in proportion to the stresses applied, then the material is said to be linearly elastic. Usually, engineering materials are treated as linearly elastic in the working range of their loading. In an elastic design analysis, the maximum stresses are restricted to some fraction of the yield stress, and typically this corresponds to a normal strain of the order of 0.001[1]. At such small deformations, small deflexion theory is usually applicable (see §1.2) and for linear elastic behaviour, the properties of a linear response also normally apply (see §1.4).

Hooke's law merely stated that there was a linear relationship between load and deflexion, and marked the beginning of the theory of elasticity (see the Preface). Young's modulus, E, in the generally accepted form defined by Navier, relates a normal stress, σ, to the normal strain which it induces in the same direction, ϵ, in the equation

$$\sigma = E\epsilon \qquad \textbf{(3.20)}$$

---

[1] Note that strains are dimensionless. This same strain could have been written as a percentage (0.1%) or as 1000 microstrains (a microstrain being 10^{-6}).

In a direction perpendicular to that in which the stress acts, it will usually induce a contraction in the material, giving a lateral strain of $-\varepsilon$. Poisson's ratio, ν, is the ratio of the magnitudes of these two, or

$$\nu = \frac{\varepsilon}{\epsilon} \tag{3.21}$$

Here, the parameters E and ν will be restricted to describing the properties of isotropic materials. In some texts, they are also used for anisotropic materials, but this can be misleading. Typical values for ν are between 0.25 and 0.35, but as will be seen in §3.5, its maximum possible value is 0.5 and there is a lower limit of -1.0 (implying that a tensile stress can produce a lateral expansion).

In the general three-dimensional case, there will be three normal stresses acting which produce the three normal strains given by

$$\begin{aligned}
\epsilon_{xx} &= \frac{1}{E}[\sigma_{xx} - \nu(\sigma_{yy} + \sigma_{zz})] \\
\epsilon_{yy} &= \frac{1}{E}[\sigma_{yy} - \nu(\sigma_{zz} + \sigma_{xx})] \\
\epsilon_{zz} &= \frac{1}{E}[\sigma_{zz} - \nu(\sigma_{xx} + \sigma_{yy})]
\end{aligned} \tag{3.22}$$

Note that from the definition of ν, it is assumed that a normal tensile stress produces *contractions* in the directions perpendicular to it. However, the three strains above are defined as tensile positive, so that in these equations the terms involving Poisson's ratio are negative.

Many problems involve only the stresses acting in a plane or the strains occurring in a plane. These are known as plane stress and plane strain problems respectively. Normally, the *x-y* plane is chosen. Then for plane stress, σ_{zz}, τ_{zx} and τ_{zy} are zero and for plane strain ϵ_{zz}, γ_{zx} and γ_{zy} are zero. Then the form of the first two of the above equations for plane stress is given by dropping the terms in σ_{zz}. For plane strain, σ_{zz} can be replaced by $\nu(\sigma_{xx} + \sigma_{yy})$ in these two equations, a result which follows from taking ϵ_{zz} as zero in the third equation. Then these two equations can be rearranged in the form

$$\begin{aligned}
\epsilon_{xx} &= \frac{1-\nu^2}{E}\left(\sigma_{xx} - \frac{\nu}{1-\nu}\sigma_{yy}\right) \\
\epsilon_{yy} &= \frac{1-\nu^2}{E}\left(\sigma_{yy} - \frac{\nu}{1-\nu}\sigma_{xx}\right)
\end{aligned} \tag{3.23}$$

The form of these equations is the same as that for plane stress. In fact if E^* is used for Young's modulus and ν^* used for Poisson's ratio in the plane-stress equations, where

$$E^* = \frac{E}{1-\nu^2}, \quad \nu^* = \frac{\nu}{1-\nu} \tag{3.24}$$

then they become identical to the plane strain equations. It will be seen later that this substitution means that solutions to problems in plane stress can be used as solutions to problems in plane strain. (It would also be possible to derive solutions for plane stress problems from solutions for plane strain problems by a similar process.) The conditions of plane stress are taken to hold with a sufficient degree of accuracy to analyse thin flat plates loaded in their planes only. The conditions of plane strain are approximately met on the cross-sections of long, prismatic bodies loaded identically at each cross-section and fixed to flat supports at their ends. Also, the resultant loading applied to each cross-section must be zero. The problems to which plane strain solutions are applied include tunnels, straight dams, pipes under pressure and spinning shafts.

{The equations of thermoelasticity include the apparent strain induced by a temperature rise T. The first of equations (3.22) then becomes

$$\epsilon_{xx} = \frac{1}{E}[\sigma_{xx} - \nu(\sigma_{yy} + \sigma_{zz})] + \alpha T \qquad \textbf{(3.25)}$$

where α is the *linear coefficient of expansion.* The term αT is also added to the right-hand side of the other two equations for the strains. The conditions for plane stress and plane strain are as before. Again, solutions for plane stress problems can be converted into solutions for plane strain problems, using the substitute parameters given by (3.24) and α^* (which is equal to $\alpha(1+\nu)$).}

Just as a normal stress produces a normal strain, a shear stress produces a shear strain. For isotropic materials, the two are related by the equation

$$\tau = G\gamma \qquad \textbf{(3.26)}$$

where G is called the *shear modulus.* As will be seen later, it is not independent of E and ν, which are the only two parameters needed to specify the linear elastic properties of an isotropic material. In terms of the shear stresses shown in Figure 3.1 are related to the shear strains defined by (3.11) to (3.13) by

$$\tau_{xy} = G\gamma_{xy} \ , \quad \tau_{yz} = G\gamma_{yz} \ , \quad \tau_{zx} = G\gamma_{zx} \ . \qquad \textbf{(3.27)}$$

From (3.22) and (3.27), the complete set of relationships between the stresses and strains can be written in matrix form as

$$\begin{bmatrix} \epsilon_{xx} \\ \epsilon_{yy} \\ \epsilon_{zz} \\ \gamma_{yz} \\ \gamma_{zx} \\ \gamma_{xy} \end{bmatrix} = \begin{bmatrix} 1/E & -\nu/E & -\nu/E & 0 & 0 & 0 \\ -\nu/E & 1/E & -\nu/E & 0 & 0 & 0 \\ -\nu/E & -\nu/E & 1/E & 0 & 0 & 0 \\ 0 & 0 & 0 & 1/G & 0 & 0 \\ 0 & 0 & 0 & 0 & 1/G & 0 \\ 0 & 0 & 0 & 0 & 0 & 1/G \end{bmatrix} \begin{bmatrix} \sigma_{xx} \\ \sigma_{yy} \\ \sigma_{zz} \\ \tau_{yz} \\ \tau_{zx} \\ \tau_{xy} \end{bmatrix} \quad \text{or} \quad \epsilon = S\sigma \qquad \textbf{(3.28)}$$

where ϵ and σ are the column vectors of strain and stress respectively and S is the *compliance matrix* relating them.

It is sometimes useful to consider the *bulk strain*, ϵ_V, of the material. (This is also referred to as the *volumetric strain.*) The definition is similar to that given by (3.8) for normal strain, only in terms of an elementary volume of the material rather than an elementary fibre:

$$\epsilon_V = \frac{\text{final volume} - \text{original volume}}{\text{original volume}} = \frac{\text{volume increase}}{\text{original volume}} \qquad \textbf{(3.29)}$$

This can be expressed in terms of the extensions of the sides of the elementary block[1] shown in Figure 3.1 as given by the normal strains:

$$\epsilon_V = \frac{[(1 + \epsilon_{xx})\delta x(1 + \epsilon_{yy})\delta y(1 + \epsilon_{zz})\delta z] - \delta x \delta y \delta z]}{\delta x \delta y \delta z} \approx \epsilon_{xx} + \epsilon_{yy} + \epsilon_{zz} \qquad \textbf{(3.30)}$$

ignoring, as before, terms of smaller magnitude. If a uniform normal stress σ ($= \sigma_{xx} = \sigma_{yy} = \sigma_{zz}$) is applied to the material in all directions[2] then from (3.22)

---

[1] The effect of the shear strains on the volume of the block are of a smaller order of magnitude.

[2] In practice, a uniform *pressure* p ($= -\sigma$) is more common. This is known as a hydrostatic pressure.

$$\epsilon_V = \frac{(1 - 2\nu)}{E}(\sigma_{xx} + \sigma_{yy} + \sigma_{zz}) = \frac{\sigma}{K} \tag{3.31}$$

where K, the *bulk modulus*, is given by

$$K = \frac{E}{3(1 - 2\nu)} \tag{3.32}$$

Note that if ν was greater than 0.5, the material would expand under a uniform pressure. This would mean that a supposedly inert material could be persuaded to do work, and so flouting the laws of thermodynamics. Thus 0.5 imposes an upper limit to Poisson's ratio. At this limit, the bulk modulus becomes infinite. Such materials are said to be *incompressible* and are much beloved by applied mathematicians, probably because the equations governing their behaviour are relatively simple. Rubber is almost incompressible, having a Poisson's ratio of around 0.49.

3.4 Strain Energy of a Body

The initial unstrained state of the material of an elastic body will be defined as being one in which it is not capable of doing work. The body forces acting on it may then do work, losing potential energy in the process, and the stresses acting on it may do work in deforming the material. As the body is elastic, the process can be exactly reversed by unloading it, with work being given out by the body as it returns to its unstrained state. The initial work done by the stresses is said to be stored up by the body as strain energy, which is released to do work on unloading.

Figure 3.5a shows a linearly-elastic body under the action of a set of loads, L_1 to L_n, whose corresponding deflexions are D_1 to D_n respectively[1]. The loads have been applied gradually and in proportion to one another up to their final values. Provided that an arbitrary rigid-body motion of the body is prevented, the deflexions D_1 to D_n will increase in proportion to the increase in the loading[2]. A plot of the growth of a typical load L_i against that of its corresponding deflexion D_i is shown in Figure 3.5c. The work done by applying the load L_i is given by integrating the current value of the load with each increment of the corresponding deflexion, from zero up to D_i. This integral is given by the triangular diagonally-hatched area in the figure and has the value $\frac{1}{2}L_iD_i$.

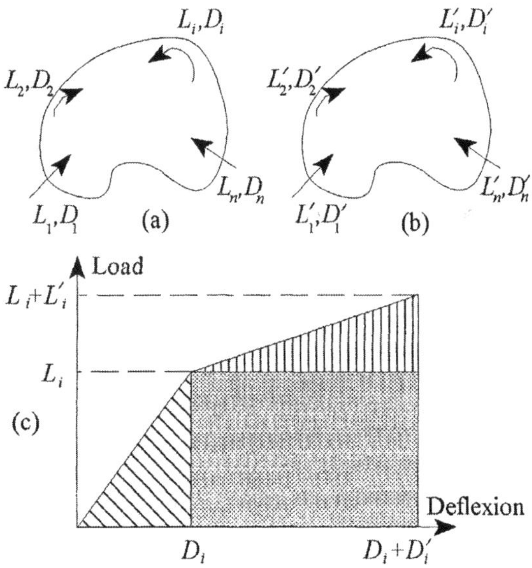

Figure 3.5 Work done by two load systems.

Consider now that a second set of loads, L_1' to L_n' are applied to the body. These are of

---

[1] See §1.1 for the general definition of a load and a corresponding deflexion.

[2] This is a linear elastic response within the constraints of small-deflexion theory.

the same types as L_1 to L_n and applied in the same localities, but their magnitudes can be different and not in the same ratio to one another. The corresponding deflexions to these loads are D_1' to D_n' respectively. Superimposing these additional loads on the first set gives the final loads L_1+L_1' to L_n+L_n' and final corresponding deflexions D_1+D_1' to D_n+D_n' respectively (using the principle of superposition). The work done by the increments of load up to the additional load L_i' is shown by the vertically-hatched triangular area in Figure 3.5c and has the value $\frac{1}{2}L_i'D_i'$. However, the original loading does additional work during the additional deflexions. This is shown by the shaded area and is equal to L_iD_i'. As the loading was applied gradually, the work done is stored as strain energy. (A sudden change of loading could produce dynamic effects, so that some of the work done would be transformed into kinetic energy.) The total strain energy is then

$$U_T = \sum_{i=1}^{n} (\tfrac{1}{2}L_iD_i + \tfrac{1}{2}L'_iD'_i + L_iD'_i) \tag{3.33}$$

Had the loading been applied in the reverse order (i.e. the L_i' loads before the L_i loads) the same final state would have been reached (see the discussion of uniqueness in Chapter 1). This means that the total strain energy would also be the same. However, the expression for it as given by (3.33) would be different, the last term in the summation being $L_i'D_i$ instead of L_iD_i'. It follows that

$$\sum_{i=1}^{n} L_iD_i' = \sum_{i=1}^{n} L_i'D_i \tag{3.34}$$

This is the general form of *Betti's reciprocal theorem.* Various applications of this result will be discussed in context later. One which is of importance in this chapter concerns to the nature of the equations relating loads to their corresponding deflexions. As a linear response is being examined, the equations can be written in matrix form (cf. §1.4).

$$\begin{aligned} \sum_{j=1}^{n} k_{ij}D_j &= L_i \quad (i=1 \text{ to } n) \quad \text{or} \quad \boldsymbol{KD} = \boldsymbol{L} \\ \sum_{j=1}^{n} f_{ij}L_j &= D_i \quad (i=1 \text{ to } n) \quad \text{or} \quad \boldsymbol{FL} = \boldsymbol{D} \end{aligned} \tag{3.35}$$

where $\boldsymbol{L}$ and $\boldsymbol{D}$ are the column vectors of the loads L_i and the corresponding deflexions D_i and $\boldsymbol{K}$ and $\boldsymbol{F}$ are the matrices of the coefficients k_{ij} and f_{ij}, known as a *stiffness matrix* and a *flexibility matrix* respectively.

{If $\boldsymbol{L}$ is the column vector of all the loads on a body in equilibrium, not all its elements will be independent of one another (unless each load is a self-equilibrating system), as the conditions of statics must be applied to them. In this case, there will then be no unique set of corresponding deflexions, $\boldsymbol{D}$, as an arbitrary rigid-body motion can be added to them. Then although it may be possible to find $\boldsymbol{L}$ for a given $\boldsymbol{D}$ from the first equation, the second equation cannot be written as $\boldsymbol{D}$ is not unique. The matrix $\boldsymbol{K}$ is then said to be *singular*, and its inverse, $\boldsymbol{F}$, does not exist.}

Equations (3.35) also relate the set of loads $\boldsymbol{L}'$ to their corresponding deflexions $\boldsymbol{D}'$. Then from (3.34) and the first of equations (3.35),

$$\boldsymbol{L}^T\boldsymbol{D}' = \boldsymbol{L}'^T\boldsymbol{D} \quad \text{or} \quad \boldsymbol{D}^T\boldsymbol{K}^T\boldsymbol{D}' = \boldsymbol{D}'^T\boldsymbol{K}^T\boldsymbol{D} = \boldsymbol{D}^T\boldsymbol{K}\boldsymbol{D}' \tag{3.36}$$

where the superscript T indicates a transpose. (In the above equations, the last equality is true because the product is a scalar, and hence equal to its transpose.) Now $\boldsymbol{D}$ and $\boldsymbol{D}'$ are arbitrary sets of deflexions. Suppose that all the terms in $\boldsymbol{D}$ are zero except for D_j and all the terms in $\boldsymbol{D}'$ are zero except for D_i' and that both of these are unit deflexions. Substituting this case into (3.36) shows that k_{ij} and k_{ji} are equal. As this is true for any i and j, $\boldsymbol{K}$ is a symmetric matrix. Similarly, if $\boldsymbol{F}$ exists, it can be shown to be a symmetric matrix.

The strain energy stored when a set of loads $\boldsymbol{L}$ is applied is given by (3.33) and (3.35) as $\frac{1}{2}\boldsymbol{D}^T\boldsymbol{K}\boldsymbol{D}$ or $\frac{1}{2}\boldsymbol{L}^T\boldsymbol{F}\boldsymbol{L}$. These products, known as *quadratic forms*, must be positive (or zero) for any $\boldsymbol{D}$ or $\boldsymbol{L}$. (Otherwise, it would be possible to get work out of a body with zero strain energy in it.) These forms are then said to be *positive-definite*. It can be shown that this implies that the principal minors of $\boldsymbol{K}$ and $\boldsymbol{F}$ must be positive. In particular, the terms on their leading diagonals and all the determinants formed from their first r rows and columns ($r = 1$ to n) are positive.

3.5 Strain energy density

The *strain energy density*, U_V, will be defined here as the strain energy per unit (unstrained) volume of the material[1]. This can be found by examining the work done by the stresses in straining the elementary block shown in Figure 3.1. The pair of stresses σ_{xx} produce equal and opposite forces $\sigma_{xx}\delta y\delta z$ on the x faces. These surfaces move apart by an amount δu or $\epsilon_{xx}\delta x$ in the x direction (from (3.9)). Again assuming that the loading is applied gradually, the work done on the block by this pair of stresses is then $\frac{1}{2}\sigma_{xx}\epsilon_{xx}\,\delta x\delta y\delta z$. Similar expressions give the work done on the block by the stresses σ_{yy} and σ_{zz}.

The pair of shear stresses τ_{xy} produce equal and opposite forces $\tau_{xy}\delta y\delta z$ on the x faces in the y direction. They do no net work during the overall motion v of the block but only as a result of the relative motion of the x faces in the y direction, $\partial v/\partial x\ \delta x$ (cf. Figure 3.4). Again assuming that these stresses grow in proportion to the growth of this motion, the work done by them is $\frac{1}{2}\tau_{xy}\ \partial v/\partial x\ \delta x\delta y\delta z$. Likewise, the work done by the pair of stresses τ_{yx} during a relative motion $\partial u/\partial y\ \delta y$ of the y faces in the x direction is $\frac{1}{2}\tau_{yx}\,\partial u/\partial y\ \delta x\delta y\delta z$. From (3.6) and (3.11) the sum of these two can be written as $\frac{1}{2}\tau_{xy}\gamma_{xy}\,\delta x\delta y\delta z$. Similar expressions can be written for the work done by the other shear stresses. As $\delta x\delta y\delta z$ is the (unstrained) volume of the elementary block, its strain energy density can be written as

$$\begin{aligned} U_V &= \tfrac{1}{2}[\sigma_{xx}\epsilon_{xx} + \sigma_{yy}\epsilon_{yy} + \sigma_{zz}\epsilon_{zz} + \tau_{xy}\gamma_{xy} + \tau_{yz}\gamma_{yz} + \tau_{zx}\gamma_{zx}] \\ &= \tfrac{1}{2}\sigma^T\epsilon = \tfrac{1}{2}\sigma^T \boldsymbol{S}\sigma \\ &= \frac{1}{2E}[\sigma_{xx}^2 + \sigma_{yy}^2 + \sigma_{zz}^2 - 2\nu(\sigma_{xx}\sigma_{yy} + \sigma_{yy}\sigma_{zz} + \sigma_{zz}\sigma_{xx})] + \frac{1}{2G}[\tau_{xy}^2 + \tau_{yz}^2 + \tau_{zx}^2] \end{aligned} \qquad \textbf{(3.37)}$$

from (3.28). The strain energy density can only be positive (or zero) so that from the previous section, the principal minors of the compliance matrix $\boldsymbol{S}$ must be positive (or zero). In particular,

$$\frac{1}{E} \geq 0 \ , \quad \frac{1}{E^2}(1 - \nu^2) \geq 0 \ , \quad \frac{1}{E^3}(1 - 3\nu^2 - 2\nu^3) \geq 0 \qquad \textbf{(3.38)}$$

from the first three principal minors, giving the range of possible values of ν as between -1.0 and 0.5. (Some plastic foams appear to exhibit a negative Poisson's ratio[2].)

The strain energy density is a measure of the absolute physical state of the material at a given location and so is independent of the orientation of the coordinate axes used to describe it. The expression for U_V in (3.37) is then said to be *invariant*.

3.6 Saint-Venant's Principle

This principle states that if a system of loads is applied to a small part of the surface of an elastic body, then the stresses and strains induced at large distances from the part (in comparison

---

[1] Elsewhere, it is defined as the strain energy per unit mass.

[2] See Lakes, R.S. (1987) Negative Poisson's ratio materials. *Science*, V.238, p.551.

with the linear dimensions of the part) depend almost entirely on the resultant force and moment of the system of loads. That is to say, if two different systems of loads with the same resultants are applied (separately) to the same part, then the difference in stress and strain in the two cases is likely to be significant only in the vicinity of the part. As will be seen later, the normal engineering theory of beams and frames relies on this being true, and this is often implicitly assumed rather than quoted as Saint-Venant's principle.

{From the principle of superposition, it is sufficient to show that the difference between the two sets of loads produces stresses and strains which are only significant near the part to which it is applied. This difference is statically equivalent to zero loading, that is, it has zero resultant force or moment. Let the part to which this difference in loads is applied have linear dimensions of a small order a. Then its area will be no greater than the order of a^2. Let the magnitude of the surface tractions on the part from the difference loading be of order σ. Then the surface forces will be of order σa^2 (or less). The strains at the surface will be of order σ/E. The maximum relative displacements of the part, other than those due to rigid-body motion, are then of order $\sigma a/E$. The difference loading, having no resultant, does no work during rigid-body motion. The work it does is then given by the surface forces moving through the above relative displacements and so is of order $\sigma^2 a^3/E$. The strain energy density in the region of the part is of order σ^2/E. Then the work done can be absorbed by the strain energy of this density in a volume of order a^3 in the region of the part. Stresses of order σ are then unlikely to extend beyond a linear dimension of order a from this part.

There are special cases where it may only apply weakly, if at all. In the torsion of thin-walled beams, local stresses may give rise to *bimoments* which have no resultants. The strain energy may decay slowly away from the section to which such bimoments are applied. The resulting stress distribution can then be significant at a considerable distance from the section. Some structures may behave like partial mechanisms under the action of certain kinds of local loading. These partial mechanisms do not absorb the work done by the local loading, but pass it on to more remote parts of the structure. Such special cases will be discussed in context later.[1]}

3.7 Stress Transformation and Principal Stresses

Within a material under stress, local planes can be found on which no shear stresses act. These are called *principal planes* and the stresses acting normal to these planes are called *principal stresses*. The simpler case of plane stress will be considered first. A plane triangular element of unit thickness is shown in Figure 3.6. The sloping face is at an angle θ to the y axis and its length is δs. In considering the equilibrium of the element under the stresses shown, body forces can be ignored as their effects are of a smaller order of magnitude. Then resolving perpendicular and parallel to the sloping face,

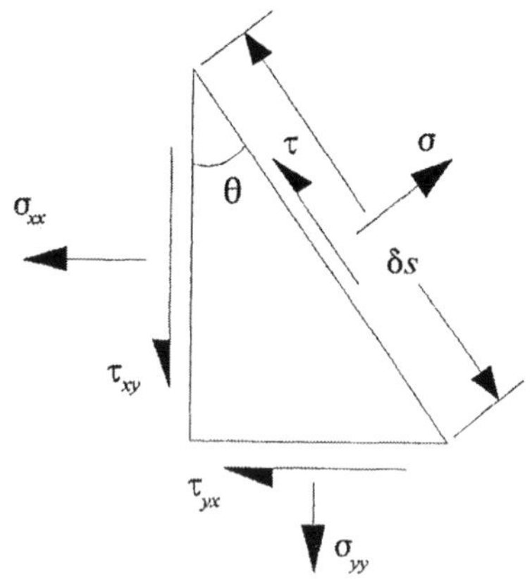

Figure 3.6 Equilibrium of a plane triangular element.

[1] See Boresi, A.P. and Chong, K.P. (1987) §4-15 for a further discussion.

$$\sigma\,\delta s = \sigma_{xx}\delta s\cos^2\theta + \sigma_{yy}\delta s\sin^2\theta + \tau_{xy}\delta s\cos\theta\sin\theta + \tau_{yx}\delta s\sin\theta\cos\theta$$
$$\tau\,\delta s = -\sigma_{xx}\delta s\cos\theta\sin\theta + \sigma_{yy}\delta s\sin\theta\cos\theta + \tau_{xy}\delta s\cos^2\theta - \tau_{yx}\delta s\sin^2\theta$$

Using (3.6) and simplifying these equations, reduces them to the form

$$\begin{aligned}\sigma &= \tfrac{1}{2}(\sigma_{xx} + \sigma_{yy}) + \tfrac{1}{2}(\sigma_{xx} - \sigma_{yy})\cos 2\theta + \tau_{xy}\sin 2\theta \\ \tau &= -\tfrac{1}{2}(\sigma_{xx} - \sigma_{yy})\sin 2\theta + \tau_{xy}\cos 2\theta\end{aligned} \qquad \textbf{(3.39)}$$

It follows from the second of these equations that τ is zero if

$$\tan 2\theta = \frac{\tau_{xy}}{\tfrac{1}{2}(\sigma_{xx} - \sigma_{yy})} \qquad \textbf{(3.40)}$$

Then there must be two mutually perpendicular planes on which the shear stress is zero, because if a particular value of θ is found which satisfies (3.40) then so does the angle $(\theta+\tfrac{1}{2}\pi)$. The values of σ on these two planes, σ_1 and σ_2 say, are the *principal stresses*. Figure 3.7 shows a graphical construction which relates these principal stresses to σ_{xx}, σ_{yy} and τ_{xy}. The principal stresses have fixed values at a given location, but the other stresses are dependent on the orientation of the coordinates x and y. All possible values of the latter then lie on a circle of radius r known as *Mohr's circle for stress.*

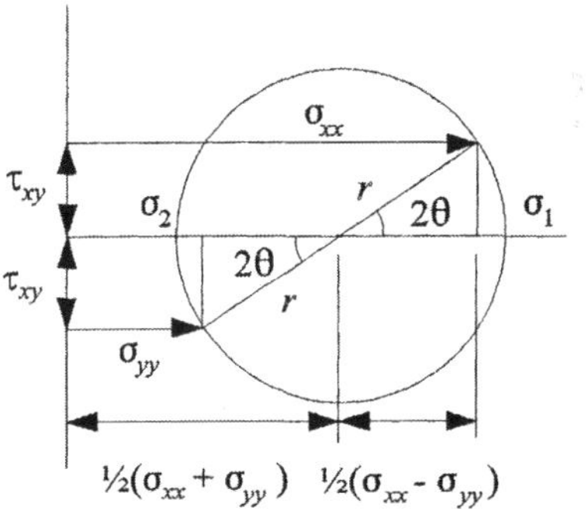

Figure 3.7 Mohr's circle for stress.

{Figure 3.8 shows a tetrahedral element formed by the intersections of planes perpendicular to the x,y and z axes and some arbitrary plane. The orientation of this plane is given by a unit normal vector $\boldsymbol{n}$ which has components (direction cosines) given by $\{l,m,n\}$. It can be thought of as a plane which slices off the corner nearest the origin of the block shown in Figure 3.1, so that all the other planes are 'negative faces' of the block. The stresses on these faces all act in negative directions, as shown for the negative 'x' face. The inward unit normals to these faces are the unit vectors in the x,y and z directions, $\boldsymbol{i}$, $\boldsymbol{j}$ and $\boldsymbol{k}$ respectively. The cosines of the angles between these normals and the normal $\boldsymbol{n}$ are given by $\boldsymbol{i}.\boldsymbol{n}$, $\boldsymbol{j}.\boldsymbol{n}$ and $\boldsymbol{k}.\boldsymbol{n}$, that is, l, m and n respectively. These will also be cosines of the angles between the fourth face and the x,y and z faces. The x,y and z faces of the tetrahedron are projections of the fourth face. Then if its area is δA, the areas of the x,y and z faces are $l\delta A$, $m\delta A$ and $n\delta A$ respectively. The stresses acting on these faces must be multiplied by these areas to give the corresponding forces.

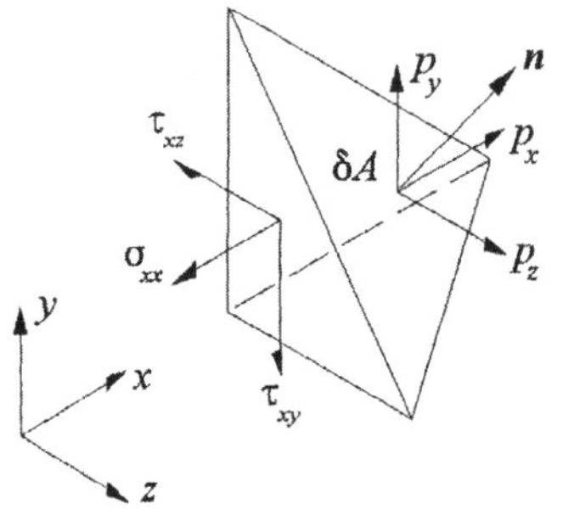

Figure 3.8 Equilibrium of a tetrahedral element.

Suppose that the fourth face forms part of the surface of the body. If this has a resultant force per unit area $\boldsymbol{p}$ acting on it, where, in terms of the components shown in Figure 3.8,

$$p = p_x\boldsymbol{i} + p_y\boldsymbol{j} + p_z\boldsymbol{k}$$

The vector equation of force equilibrium for the tetrahedron can be written as

$$\boldsymbol{p}\delta A = (\sigma_{xx}\boldsymbol{i}+\tau_{xy}\boldsymbol{j}+\tau_{xz}\boldsymbol{k})l\delta A + (\tau_{yx}\boldsymbol{i}+\sigma_{yy}\boldsymbol{j}+\tau_{yz}\boldsymbol{k})m\delta A + (\tau_{zx}\boldsymbol{i}+\tau_{zy}\boldsymbol{j}+\sigma_{zz}\boldsymbol{k})n\delta A$$

Taking the components of force in the direction of each axis separately and dividing by δA gives the three equations

$$\begin{aligned} p_x &= \sigma_{xx}l + \tau_{yx}m + \tau_{zx}m \\ p_y &= \tau_{xy}l + \sigma_{yy}m + \tau_{zy}n \\ p_z &= \tau_{xz}l + \tau_{yz}m + \sigma_{zz}n \end{aligned} \qquad \textbf{(3.41)}$$

Suppose that only a normal stress σ acts on the fourth face. This will produce a resultant force $\boldsymbol{p}\delta A$ equal to $\sigma\boldsymbol{n}\delta A$. These equations can be written in matrix form as

$$\begin{bmatrix} \sigma_{xx}-\sigma & \tau_{yx} & \tau_{zx} \\ \tau_{xy} & \sigma_{yy}-\sigma & \tau_{zy} \\ \tau_{xz} & \tau_{yz} & \sigma_{zz}-\sigma \end{bmatrix} \begin{bmatrix} l \\ m \\ n \end{bmatrix} = \begin{bmatrix} 0 \\ 0 \\ 0 \end{bmatrix} \qquad \textbf{(3.41a)}$$

The column vector $\{l, m, n\}$ represents the unit vector $\boldsymbol{n}$ and so is not zero. This means that the determinant of the matrix of stresses must be zero, giving the cubic in σ:

$$\sigma^3 - I_1\sigma^2 + I_2\sigma - I_3 = 0 \qquad \textbf{(3.42)}$$

This equation always has three real roots[1] which are the principal stresses. As these will be the same, regardless of the orientation of the coordinate system, I_1 to I_3 are said to be *stress invariants*. They are

$$\begin{aligned} I_1 &= \sigma_{xx} + \sigma_{yy} + \sigma_{zz} \\ I_2 &= \sigma_{xx}\sigma_{yy} + \sigma_{yy}\sigma_{zz} + \sigma_{zz}\sigma_{xx} - \tau_{xy}^2 - \tau_{yz}^2 - \tau_{zx}^2 \\ I_3 &= \sigma_{xx}\sigma_{yy}\sigma_{zz} + 2\tau_{xy}\tau_{yz}\tau_{zx} - \sigma_{xx}\tau_{yz}^2 - \sigma_{yy}\tau_{zx}^2 - \sigma_{zz}\tau_{xy}^2 \end{aligned} \qquad \textbf{(3.43)}$$

Any other function of the stresses which takes the same value regardless of the orientation of the coordinate system can be expressed in terms of the above three invariants. In particular, if the direction of the *z* axis is fixed, then the normal stress, σ_{zz} and the resultant shear stress, $\sqrt{(\tau^2_{yz}+\tau^2_{zx})}$ on the *z* faces does not change. Then from the expressions for I_1 and I_2 in (3.43) it follows that the expressions in terms of σ_{xx}, σ_{yy} and τ_{xy} for the radius of Mohr's circle and the distance of its centre from the baseline are invariants.}

3.8 Mohr's Circle for Strain

It is possible to derive a similar construction for strain to Mohr's circle for stress using the stress-strain relationships for isotropic materials. However, this disguises the fact that these constructions apply to any material, even if it is not elastic or isotropic.

---

[1] Cardan's solution can be applied. See Renton (1987) §1.6.6 for example.

In two dimensions, the strains are given in terms of the differentials of displacement measured with respect to a coordinate system x, y. These must be related to the differentials of displacement measured with respect to a coordinate system x', y' at some arbitrary angle θ to the original coordinate system, as shown in Figure 3.9. The partial differentials of a function f in of position in the two coordinate systems are related by

$$\begin{aligned} \frac{\partial f}{\partial x'} &= \frac{\partial f}{\partial x}\frac{\partial x}{\partial x'} + \frac{\partial f}{\partial y}\frac{\partial y}{\partial x'} = \frac{\partial f}{\partial x}\cos\theta + \frac{\partial f}{\partial y}\sin\theta \\ \frac{\partial f}{\partial y'} &= \frac{\partial f}{\partial x}\frac{\partial x}{\partial y'} + \frac{\partial f}{\partial y}\frac{\partial y}{\partial y'} = -\frac{\partial f}{\partial x}\sin\theta + \frac{\partial f}{\partial y}\cos\theta \end{aligned} \tag{3.44}$$

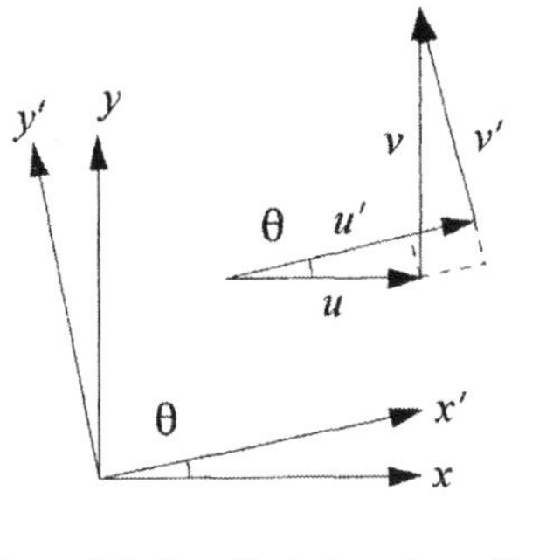

Figure 3.9 Coordinate transformation.

Here, the function f will be the displacements u or v. These displacements are related to the displacements u' and v' measured with respect to the new coordinate system by

$$\begin{aligned} u' &= u\cos\theta + v\sin\theta \\ v' &= -u\sin\theta + v\cos\theta \end{aligned} \tag{3.45}$$

as can be seen from Figure 3.9. Using (3.9) to (3.11), (3.44) and (3.45), the strains in terms of the new coordinate system can now be related to those in the old coordinate system:

$$\begin{aligned} \epsilon'_{xx} = \frac{\partial u'}{\partial x'} &= \frac{\partial}{\partial x}(u\cos\theta + v\sin\theta)\cos\theta + \frac{\partial}{\partial y}(u\cos\theta + v\sin\theta)\sin\theta \\ &= \frac{\partial u}{\partial x}\cos^2\theta + \frac{\partial v}{\partial y}\sin^2\theta + \left(\frac{\partial v}{\partial x} + \frac{\partial u}{\partial y}\right)\sin\theta\cos\theta \\ &= \epsilon_{xx}\cos^2\theta + \epsilon_{yy}\sin^2\theta + \gamma_{xy}\sin\theta\cos\theta \\ &= \frac{1}{2}(\epsilon_{xx} + \epsilon_{yy}) + \frac{1}{2}(\epsilon_{xx} - \epsilon_{yy})\cos 2\theta + \frac{1}{2}\gamma_{xy}\sin 2\theta \end{aligned} \tag{3.46}$$

Similarly

$$\begin{aligned} \epsilon'_{yy} &= \frac{1}{2}(\epsilon_{xx} + \epsilon_{yy}) - \frac{1}{2}(\epsilon_{xx} - \epsilon_{yy})\cos 2\theta - \frac{1}{2}\gamma_{xy}\sin 2\theta \\ \frac{1}{2}\gamma'_{xy} &= -\frac{1}{2}(\epsilon_{xx} - \epsilon_{yy})\sin 2\theta + \frac{1}{2}\gamma_{xy}\cos 2\theta \end{aligned} \tag{3.47}$$

The directions of the axes x',y' can be chosen so that γ'_{xy} is zero. This is such that

$$\tan 2\theta = \frac{\frac{1}{2}\gamma_{xy}}{\frac{1}{2}(\epsilon_{xx} - \epsilon_{yy})} \tag{3.48}$$

Then ϵ'_{xx} and ϵ'_{yy} are the *principal strains.* The form of (3.46) to (3.48) can be compared to (3.39) and (3.40). It leads to a construction which is similar to Mohr's circle for stress, in which σ_{xx}, σ_{yy} and τ_{xy} are replaced by ϵ_{xx}, ϵ_{yy} and $\frac{1}{2}\gamma_{xy}$. This is *Mohr's circle for strain*[1]. For an isotropic elastic medium, the principal strains, being the greatest and least normal strains, are in the same directions as the principal stresses. Suppose that the material is in a state of plane stress where the two principal stresses are equal and opposite, σ and $-\sigma$ say. From (3.22), the principal strains are then $(1+\nu)\sigma/E$ and $-(1+\nu)\sigma/E$. At an angle of 2θ on Mohr's circle equal to 90° to the principal

---

[1] {In cartesian tensor notation, the stress and strain components are the same apart from the shear strain, which is defined as $\frac{1}{2}\gamma_{xy}$. The comparison between the two constructions is then more direct.}

stresses and strains, the normal stresses and strains will be zero. At this orientation, the material is seen to be in a state of pure shear. From the form of the Mohr's circles, it follows that the shear stresses and strains at this orientation are σ and $2(1+\nu)\sigma/E$ respectively. However, these are related by (3.26), which implies that

$$G = \frac{E}{2(1 + \nu)} \tag{3.49}$$

It is again possible to infer that the lower limit of Poisson's ratio must be -1, when G becomes infinite if E is finite. The shear modulus is unchanged when expressed in terms of the dummy coefficients E^* and ν^* given by (3.24). This is because pure shear is both a plane stress and a plane strain phenomenon.

3.9 Failure Criteria for Ductile Materials

The theoretical failure stresses for metals in pure tension are about one tenth to one fortieth of their Young's moduli. However, these values are for pure metals with perfect crystalline structures. Near-perfect whiskers of metal have been grown with failure stresses approaching these values, but normal industrial metals fail at very much lower stresses. For example, whiskers of pure iron have been found to fail at 13 GPa. This about thirty times the failure stress of structural steel and forty times that of malleable cast iron. Because of the large effect of imperfections and impurities, semi-empirical formulae for failure are used. For isotropic materials with similar properties in tension and compression, these formulae predict failure under any combination of stresses, knowing only the failure stress, σ_f, in a simple tension test.

Experimental evidence indicates that some distortion of ductile materials must take place for failure to occur. From considerations of symmetry, a sphere of isotropic material under hydrostatic pressure should only become a smaller sphere. Such a sphere would not break in the usual sense. Distortion implies that there must be shearing in some direction and *Tresca's criterion* of failure is based on this. The criterion states that there is a maximum shearing stress that the material will take without failure occurring. The maximum shearing stress given by a particular Mohr's stress circle diagram is the radius of the circle. Then the maximum shear stress, τ_f, is equal to half the difference of the principal stresses. In the case of a simple tension test there is only one non-zero principal stress. From the conditions at failure it follows that τ_f is $\frac{1}{2}\sigma_f$. In the case of plane stress, the third principal stress, σ_3 say, is zero. If the other two principal stresses, σ_1 and σ_2, are of the same sign, then the difference between them will be less than the difference between the greater of the two and zero. The failure condition will then be that $\frac{1}{2}\sigma_1$ or $\frac{1}{2}\sigma_2$ will be equal to $\pm\tau_f$. (Only the magnitude and not the sign of the shear is significant.) If the first two principal stresses are of opposite sign, then half their difference gives the worst shear stress. The above conditions for failure in plane stress can be written as

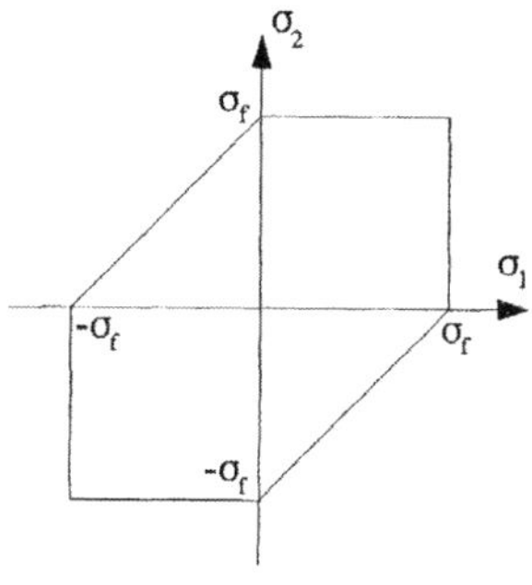

Figure 3.10 Tresca's criterion for plane stress.

$$\sigma_1 = \pm\sigma_f \;, \quad \sigma_2 = \pm\sigma_f \quad \text{or} \quad \sigma_1 - \sigma_2 = \pm\sigma_f \tag{3.50}$$

These six limiting conditions are plotted as the failure surface shown in Figure 3.10. All points with coordinates (σ_1, σ_2) within the hexagon correspond to states of stress where failure has not occurred.

Beltrami and Haigh suggested that failure occurred when a maximum strain energy density was reached. This criterion has not been found to work well in practice. However, if the component of the strain energy associated with distortion is used as the limiting condition, the resulting predictions of failure agree well with the experimental evidence. Under pure hydrostatic pressure, an isotropic material undergoes no distortion. The overall state of stress is then broken down into two components. One produces a change in volume without distortion, and the other produces distortion without a change in volume. In terms of the principal stresses, these two components are given by

$$\sigma_1 = \sigma + s_1 \;,\quad \sigma_2 = \sigma + s_2 \;,\quad \sigma_3 = \sigma + s_3 \tag{3.51}$$

where the *stress deviations* s_1 to s_3 produce no change in volume. The hydrostatic stress σ must then produce the same change of volume, or volumetric strain, as the original principal stresses. Then from (3.31),

$$\frac{(1-2\nu)}{E}(\sigma_1 + \sigma_2 + \sigma_3) = \frac{(1-2\nu)}{E}(3\sigma) \quad \text{or} \quad \sigma = \tfrac{1}{3}(\sigma_1 + \sigma_2 + \sigma_3) \tag{3.52}$$

In terms of the principal stresses, the expression for the strain energy density given by (3.37) becomes[1]

$$U_V = \frac{1}{2E}[\sigma_1^2 + \sigma_2^2 + \sigma_3^2 - 2\nu(\sigma_1\sigma_2 + \sigma_2\sigma_3 + \sigma_3\sigma_1)] \tag{3.53}$$

The distortion strain energy density, U_s, is given by the same expression, except that σ_1 to σ_3 are replaced by s_1 to s_3. From (3.51) to (3.53) this can be written in terms of the principal stresses as

$$U_s = \frac{(1+\nu)}{6E}[(\sigma_1 - \sigma_2)^2 + (\sigma_2 - \sigma_3)^2 + (\sigma_3 - \sigma_1)^2] \tag{3.54}$$

At failure in a simple tension test, one of these principal stresses is σ_f and the other two are zero. If the value of U_s in this case is the same as that under all other conditions of failure, then

$$(\sigma_1 - \sigma_2)^2 + (\sigma_2 - \sigma_3)^2 + (\sigma_3 - \sigma_1)^2 = 2\sigma_f^2 \tag{3.55}$$

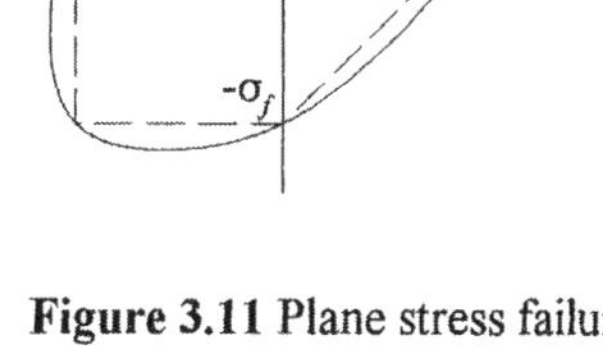

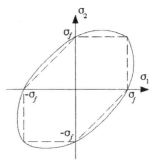

Figure 3.11 Plane stress failure.

This is known as *von Mises' criterion* although the names of others are also associated with it. The form for plane stress is given by setting σ_3 to zero and the failure surface the resulting equation generates is plotted in Figure 3.11. The failure surface according to Tresca's criterion is shown by the broken lines. The two surfaces coincide at the apices of the hexagon. Most experimental results for ductile materials indicate that von Mises' criterion is slightly the more accurate of the two. However, Tresca's criterion is safer, being more conservative, and is sometimes easier to apply. The failure surfaces for the general three-dimensional case are plotted in Figure 3.12. The failure surface according to von Mises' criterion then forms a circular cylinder of radius $\sqrt{(2/3)}\sigma_f$ with its axis on the line $\sigma_1 = \sigma_2 = \sigma_3$. Tresca's criterion gives a failure surface which forms a regular hexagonal prism

[1] This is invariant and can be expressed in terms of the stress invariants by $U_V = [I_1^2 - 2(1+\nu)I_2]/2E$.

which just fits inside this cylinder.

Other failure surfaces have been proposed for combined stress and also for other combinations of loading. Drucker has shown, from quite general considerations, that all such surfaces must be convex. That is, if a straight line is drawn between any two 'safe' points within the surface, then all points on the line are also 'safe'. This can be used to give simple but safe criteria for a wide variety of problems.

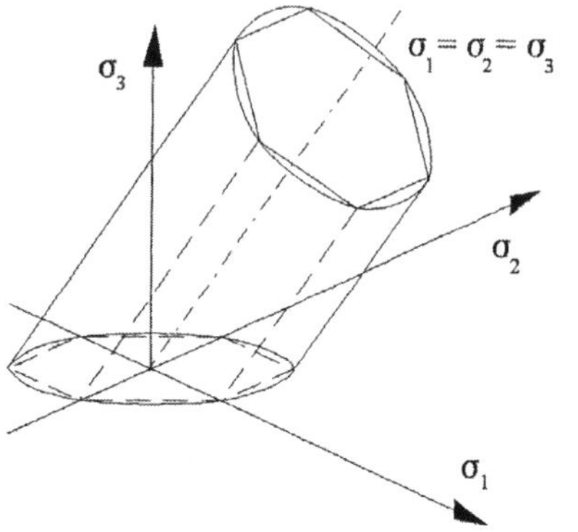

Figure 3.12 Failure in three dimensions.

3.10 Cylindrical Polar Coordinates

In some three-dimensional analyses, it is more convenient to find solutions in terms of cylindrical polar coordinates (r, θ, z). In two dimensions these become the polar coordinates (r,θ). Figure 3.13 shows an elementary block of material with its faces normal to the directions of varying r,θ and z. The conventions for defining stresses on the faces of this block are the same as those used in §3.1. The block shows its positive r and z faces and its negative θ face. The first subscript of each stress indicates the face on which it acts and the second indicates the direction in which its acts. On the positive faces, the direction is of increasing r,θ or z, and the stresses act in the opposite direction on negative faces. The angle θ is clockwise positive about the z axis. The body forces per unit volume within the block in the r, θ and z directions will be denoted by R, Θ and Z respectively. In writing the equations of equilibrium for the block, it is necessary to allow for the difference of width of the r faces and the difference in orientation of the θ faces. The components of stress and body force which need to be taken into account when resolving for force equilibrium in the r direction are shown in Figure 3.14. The resulting equation is then

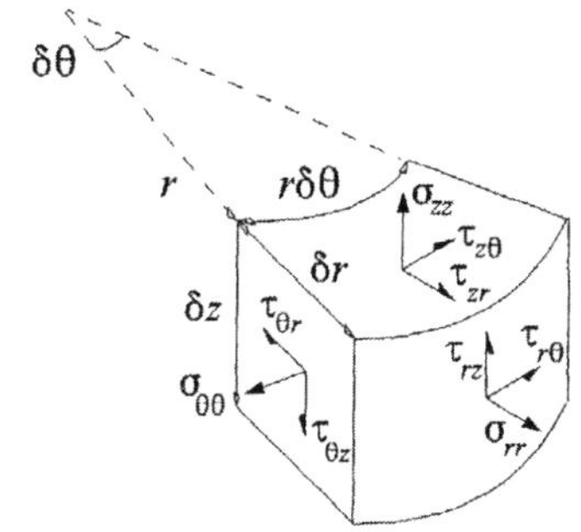

Figure 3.13 A polar elementary block.

$$\delta(\sigma_{rr} r)\delta\theta\,\delta z - \sigma_{\theta\theta}\,\delta r\,\delta z\,\delta\theta + \delta\tau_{\theta r}\,\delta r\,\delta z + \delta\tau_{zr}\, r\,\delta\theta\,\delta r + R r\,\delta\theta\,\delta r\,\delta z = 0 \qquad (3.56)$$

This includes all the net terms no smaller than the third order of smallness. At the limit, this equation becomes

$$\frac{\partial\sigma_{rr}}{\partial r} + \frac{\sigma_{rr} - \sigma_{\theta\theta}}{r} + \frac{1}{r}\frac{\partial\tau_{\theta r}}{\partial\theta} + \frac{\partial\tau_{zr}}{\partial z} + R = 0 \qquad (3.57)$$

Likewise, for force equilibrium in the directions of increasing θ and z,

$$\begin{aligned} &\frac{\partial\tau_{r\theta}}{\partial r} + \frac{2}{r}\tau_{r\theta} + \frac{1}{r}\frac{\partial\sigma_{\theta\theta}}{\partial\theta} + \frac{\partial\tau_{z\theta}}{\partial z} + \Theta = 0 \\ &\frac{\partial\tau_{rz}}{\partial r} + \frac{1}{r}\tau_{rz} + \frac{1}{r}\frac{\partial\tau_{\theta z}}{\partial\theta} + \frac{\partial\sigma_{zz}}{\partial z} + Z = 0 \end{aligned} \qquad (3.58)$$

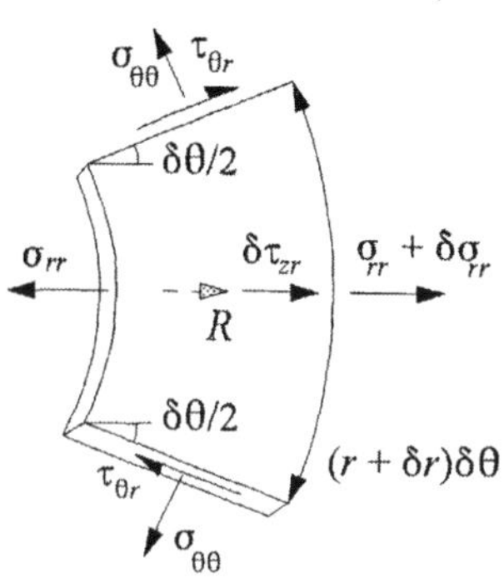

Figure 3.14 Radial force components.

These stresses are identical to the stresses in a cuboidal block of the type shown in Figure 3.1 located in the same position and oriented in the same way. The local strains in polar coordinates are likewise identical to those of the cuboid. This means that the stress-strain relationships given by (3.22) and (3.27) can immediately be rewritten in terms of their polar coordinate equivalents as

$$\begin{aligned} \epsilon_{rr} &= \frac{1}{E}[\sigma_{rr} - \nu(\sigma_{\theta\theta} + \sigma_{zz})] \\ \epsilon_{\theta\theta} &= \frac{1}{E}[\sigma_{\theta\theta} - \nu(\sigma_{zz} + \sigma_{rr})] \\ \epsilon_{zz} &= \frac{1}{E}[\sigma_{zz} - \nu(\sigma_{rr} + \sigma_{\theta\theta})] \\ \tau_{r\theta} &= G\gamma_{r\theta}\,,\ \tau_{\theta z} = G\gamma_{\theta z}\,,\ \tau_{zr} = G\gamma_{zr}\,. \end{aligned} \qquad \textbf{(3.59)}$$

In expressing the strains in terms of the radial, tangential and axial displacements, u^P, v^P and w^P, allowance has to be made for the slight rotation $\delta\theta$ of the reference axes when moving a small distance in the circumferential direction $r\delta\theta$ from P to Q in Figure 3.15. This is sensibly the same as moving through a distance δy with respect to the coincident cartesian coordinate system. The increments in displacement in this direction are denoted by δ', and those in the radial sense are denoted by δ. In moving a small distance δr (or δy) in the radial direction from P to R, the two sets of coordinate directions remain coincident. At P, the definitions of the displacements are the same in both coordinate systems. Likewise, the definitions of the $\delta(\ )$ increments at R are the same for both systems. Thus

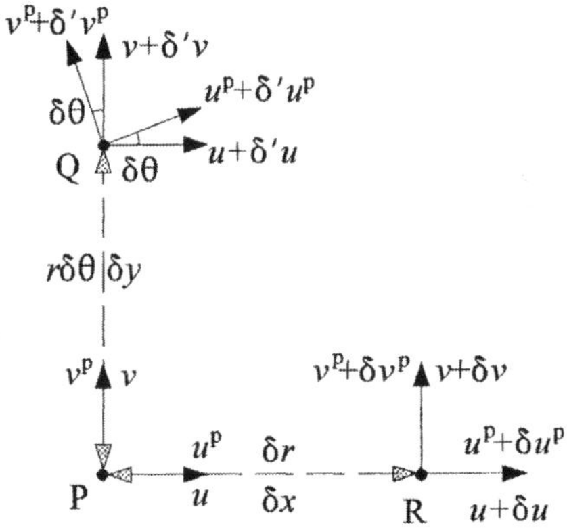

Figure 3.15 Displacements in rectangular and polar coordinates.

$$\delta u = \frac{\partial u}{\partial x}\delta x = \delta u^P = \frac{\partial u^P}{\partial r}\delta r \quad \delta v = \frac{\partial v}{\partial x}\delta x = \delta v^P = \frac{\partial v^P}{\partial r}\delta r \quad (\delta x = \delta r) \qquad \textbf{(3.60)}$$

However, allowing for the slight rotation of the polar coordinate directions in the tangential sense,

$$\begin{aligned} u + \delta' u &= u + \frac{\partial u}{\partial y}\delta y = u^P + \delta' u^P - v^P\delta\theta = u^P + \frac{\partial u^P}{\partial \theta}\delta\theta - v^P\delta\theta \\ v + \delta' v &= v + \frac{\partial v}{\partial y}\delta y = v^P + \delta' v^P + u^P\delta\theta = v^P + \frac{\partial v^P}{\partial \theta}\delta\theta + u^P\delta\theta \quad (\delta y \approx r\delta\theta) \end{aligned} \qquad \textbf{(3.61)}$$

ignoring terms of less than first order of smallness. The local z direction is identical in both systems, as are w and w^P. From (3.60) and (3.61), the strains can now be expressed in terms of the material displacements as

$$\epsilon_{rr} = \frac{\partial u}{\partial x} = \frac{\partial u^P}{\partial r}, \quad \epsilon_{\theta\theta} = \frac{\partial v}{\partial y} = \frac{1}{r}\left(\frac{\partial v^P}{\partial \theta} + u^P\right), \quad \epsilon_{zz} = \frac{\partial w}{\partial z} = \frac{\partial w^P}{\partial z},$$
$$\gamma_{r\theta} = \frac{\partial u}{\partial y} + \frac{\partial v}{\partial x} = \frac{1}{r}\left(\frac{\partial u^P}{\partial \theta} - v^P\right) + \frac{\partial v^P}{\partial r}, \tag{3.62}$$
$$\gamma_{\theta z} = \frac{\partial v}{\partial z} + \frac{\partial w}{\partial y} = \frac{\partial v^P}{\partial z} + \frac{1}{r}\frac{\partial w^P}{\partial \theta}, \quad \gamma_{zr} = \frac{\partial w}{\partial x} + \frac{\partial u}{\partial z} = \frac{\partial w^P}{\partial r} + \frac{\partial u^P}{\partial z}.$$

{3.11 Anisotropic Elasticity}

{A linear elastic relationship between stresses and strains for isotropic materials was given in matrix form by (3.28). This can be generalised to give a relationship which holds for any linearly-elastic material. This takes the form

$$\begin{bmatrix}\epsilon_{xx}\\ \epsilon_{yy}\\ \epsilon_{zz}\\ \gamma_{yz}\\ \gamma_{zx}\\ \gamma_{xy}\end{bmatrix} = \begin{bmatrix} s_{11} & s_{12} & s_{13} & s_{14} & s_{15} & s_{16}\\ s_{21} & s_{22} & s_{23} & s_{24} & s_{25} & s_{26}\\ s_{31} & s_{32} & s_{33} & s_{34} & s_{35} & s_{36}\\ s_{41} & s_{42} & s_{43} & s_{44} & s_{45} & s_{46}\\ s_{51} & s_{52} & s_{53} & s_{54} & s_{55} & s_{56}\\ s_{61} & s_{62} & s_{63} & s_{64} & s_{65} & s_{66}\end{bmatrix} \begin{bmatrix}\sigma_{xx}\\ \sigma_{yy}\\ \sigma_{zz}\\ \tau_{yz}\\ \tau_{zx}\\ \tau_{xy}\end{bmatrix} \quad \text{or} \quad \epsilon = S\sigma \tag{3.63}$$

If these stresses act on a unit elementary cube, then the stress column vector also represents a column vector of self-equilibrating forces. Then the column vector of strains also represents the corresponding deflexions through which these forces do work. The compliance matrix S relating the two is then a flexibility matrix. It then follows from §3.4 that S is symmetric (i.e. $s_{ij} = s_{ji}$). Its inverse is a symmetric stiffness matrix C (with elements c_{ij} such that $c_{ij} = c_{ji}$).

Coordinate transformations for stresses and strains in the x-y plane were given in §3.7 and §3.8. These were for a rotation θ of the coordinate axes about the z axis. During such a rotation, the perceived normal stress and strain in the z direction do not change. The shear stresses τ_{yz} and τ_{zx} transform like ordinary vector forces, because either the directions in which they act rotate or the surfaces on which they act rotate, but not both. Then for a clockwise rotation θ about the z axis, the stresses with respect to the new coordinate system x', y' are given by

$$\begin{bmatrix}\sigma'_{xx}\\ \sigma'_{yy}\\ \sigma'_{zz}\\ \tau'_{yz}\\ \tau'_{zx}\\ \tau'_{xy}\end{bmatrix} = \begin{bmatrix} \cos^2\theta & \sin^2\theta & 0 & 0 & 0 & \sin 2\theta\\ \sin^2\theta & \cos^2\theta & 0 & 0 & 0 & -\sin 2\theta\\ 0 & 0 & 1 & 0 & 0 & 0\\ 0 & 0 & 0 & \cos\theta & -\sin\theta & 0\\ 0 & 0 & 0 & \sin\theta & \cos\theta & 0\\ -\frac{1}{2}\sin 2\theta & \frac{1}{2}\sin 2\theta & 0 & 0 & 0 & \cos 2\theta\end{bmatrix} \begin{bmatrix}\sigma_{xx}\\ \sigma_{yy}\\ \sigma_{zz}\\ \tau_{yz}\\ \tau_{zx}\\ \tau_{xy}\end{bmatrix} \quad \text{or} \quad \sigma' = T\sigma \tag{3.64}$$

The inverse of T (i.e T^{-1}) is of course given by replacing θ by $-\theta$ in the above matrix. The transformation matrix corresponding to inversion of the z axis is given by equating θ to π in the above matrix. The work done on the elementary cube is equal to the strain energy density and so is the same in all coordinate systems. Then if ϵ' is the strain column vector in the new coordinate system,

$$\frac{1}{2}\sigma^T\epsilon = \frac{1}{2}\sigma'^T\epsilon' = \frac{1}{2}\sigma^T T^T\epsilon' \qquad \textbf{(3.65)}$$

This is true for any σ^T so that

$$\epsilon = T^T\epsilon' \quad \text{or} \quad \epsilon' = (T^T)^{-1}\epsilon \qquad \textbf{(3.66)}$$

It follows from (3.63) to (3.66) that if S' is the compliance matrix in the new coordinate system, then

$$\epsilon' = (T^T)^{-1}\epsilon = (T^T)^{-1}S\sigma = (T^T)^{-1}ST^{-1}\sigma' = S'\sigma' \quad \text{or} \quad S' = (T^T)^{-1}ST^{-1} \qquad \textbf{(3.67)}$$

The material may have a degree of symmetry, so that its properties are the same for different orientations of the coordinates. This means that S and S' are identical under certain transformations. Comparing the individual elements of S with those of S' formed from S using (3.67) gives the conditions which must apply to the compliances. Here are the conditions for some materials which are invariant under certain coordinate rotations or axis inversions:

(i) **Ternary symmetry (Trigonal system)**
Invariance under rotations of 120° about the z axis.
(7 independent constants). $s_{11} = s_{22} = s_{12} + \frac{1}{2}s_{66}$;
$s_{13} = s_{23}$; $s_{14} = -s_{24} = \frac{1}{2}s_{56}$; $s_{15} = -s_{25} = -\frac{1}{2}s_{46}$;
$s_{44} = s_{55}$; $s_{16} = s_{26} = s_{34} = s_{35} = s_{36} = s_{45} = 0$.

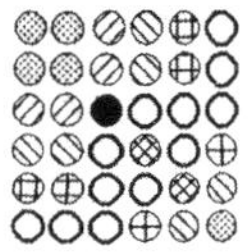

(ii) **Quadric symmetry (Tetragonal system)**
Invariance under rotations of 90° about the z axis.
(7 independent constants).
$s_{11} = s_{22}$; $s_{13} = s_{23}$; $s_{16} = -s_{26}$; $s_{44} = s_{55}$;
$s_{14} = s_{15} = s_{24} = s_{25} = s_{34} = s_{35} = s_{36} = s_{45} = s_{46} = s_{56} = 0$.

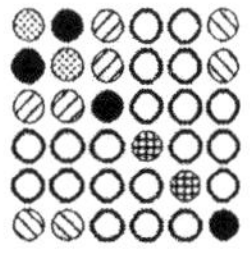

(iii) **Plane isotropy (and Hexagonal system)**
Invariance under rotations of 60° (and all other angles) about the z axis. (5 independent constants).
$s_{11} = s_{22} = s_{12} + \frac{1}{2}s_{66}$; $s_{13} = s_{23}$; $s_{44} = s_{55}$;
$s_{14} = s_{15} = s_{16} = s_{24} = s_{25} = s_{26} = s_{34} = s_{35} = s_{36} = s_{45} = s_{46} = s_{56} = 0$.

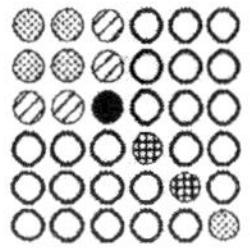

(iv) **Cubic system**
Invariance under rotations of 90° about the x,y and z axes.
(3 independent constants). $s_{11} = s_{22} = s_{33}$; $s_{12} = s_{13} = s_{23}$; $s_{44} = s_{55} = s_{66}$;
$s_{14} = s_{15} = s_{16} = s_{24} = s_{25} = s_{26} = s_{34} = s_{35} = s_{36} = s_{45} = s_{46} = s_{56} = 0$.

(v) **Monoclinic system**
Invariance under inversion of the z axis. (13 independent constants).
(This is the same as invariance under rotations of 180° about the z axis.)
$s_{14} = s_{15} = s_{24} = s_{25} = s_{34} = s_{35} = s_{46} = s_{56} = 0$.

(vi) **Orthotropy (Orthorhombic system)**
Invariance under the inversion of any two axes. (Invariance under inversion of the third axis is then a necessary consequence.
(9 independent constants).
$s_{14} = s_{15} = s_{16} = s_{24} = s_{25} = s_{26} = s_{34} = s_{35} = s_{36} = s_{45} = s_{46} = s_{56} = 0$.

The dot patterns show the corresponding arrays of elements. The black dots show unrelated elements, the patterned dots related elements and the empty circles are zeros.

Similarly it can be shown that the stiffness matrix C transforms to the matrix C'in the new coordinate system, where

$$C' = TCT^T \tag{3.68}$$

The constraints on the elements c_{ij} of the stiffness matrix for various kinds of material symmetry are similar to those for the elements s_{ij} above (see for example Renton (1987) §1.8.3). Other conditions are imposed by the positive-definiteness of these matrices (see §3.5).}

{3.12 Stress and Strain Tensors}

{Many of the equations given earlier in this chapter can often be represented more conveniently in tensor form. A fuller treatment of this will be found elsewhere (Renton (1987) for example). Here, only cartesian tensors will be discussed. These are tensors whose elements (or resolved parts) are expressed relative to a cartesian coordinate system. First-order tensors are the same as ordinary vectors. In three dimensions, they have three elements distinguished from one another by the subscripts 1 to 3.Expressed as column vectors, the first-order tensors which have already been used in this chapter are

$$\boldsymbol{u} \sim \begin{bmatrix} u_1 \\ u_2 \\ u_3 \end{bmatrix} \equiv \begin{bmatrix} u \\ v \\ w \end{bmatrix} , \quad \boldsymbol{n} \sim \begin{bmatrix} n_1 \\ n_2 \\ n_3 \end{bmatrix} \equiv \begin{bmatrix} l \\ m \\ n \end{bmatrix} , \quad \boldsymbol{X} \sim \begin{bmatrix} X_1 \\ X_2 \\ X_3 \end{bmatrix} \equiv \begin{bmatrix} X \\ Y \\ Z \end{bmatrix} , \quad \boldsymbol{p} \sim \begin{bmatrix} p_1 \\ p_2 \\ p_3 \end{bmatrix} \equiv \begin{bmatrix} p_x \\ p_y \\ p_z \end{bmatrix} . \tag{3.69}$$

where $\boldsymbol{u}$ is the vector of the displacements of a typical point P of an elastic medium, $\boldsymbol{n}$ is the unit (outwards) vector normal to a particular point on the surface of the medium, $\boldsymbol{X}$ is the vector of body forces at a typical point in the medium and $\boldsymbol{p}$ is the traction vector of forces per unit area of surface. The state of stress and strain at a point in the medium can be expressed by the stress and strain tensors with the elements σ_{ij} and e_{ij}. Arranging these elements in matrix form, they are related to the stresses and strains defined earlier by

$$\begin{bmatrix} \sigma_{11} & \sigma_{12} & \sigma_{13} \\ \sigma_{21} & \sigma_{22} & \sigma_{23} \\ \sigma_{31} & \sigma_{32} & \sigma_{33} \end{bmatrix} \equiv \begin{bmatrix} \sigma_{xx} & \tau_{yx} & \tau_{zx} \\ \tau_{xy} & \sigma_{yy} & \tau_{zy} \\ \tau_{xz} & \tau_{yz} & \sigma_{zz} \end{bmatrix} , \quad \begin{bmatrix} e_{11} & e_{12} & e_{13} \\ e_{21} & e_{22} & e_{23} \\ e_{31} & e_{32} & e_{33} \end{bmatrix} \equiv \begin{bmatrix} \epsilon_{xx} & \frac{1}{2}\gamma_{yx} & \frac{1}{2}\gamma_{zx} \\ \frac{1}{2}\gamma_{xy} & \epsilon_{yy} & \frac{1}{2}\gamma_{zy} \\ \frac{1}{2}\gamma_{xz} & \frac{1}{2}\gamma_{yz} & \epsilon_{zz} \end{bmatrix} \tag{3.70}$$

respectively. The definitions of γ_{yx}, γ_{zy} and γ_{xz} are exactly the same as those given by (3.11) to (3.13) respectively. Then from (3.6) and (3.7), all the above matrices are symmetrical.

Differentiation of a tensor element by x,y or z is denoted by adding a comma after the element followed by the subscript 1, 2 or 3 respectively. Thus equation (3.3) can be written as

$$\sigma_{11,1} + \sigma_{12,2} + \sigma_{13,3} + X_1 = 0$$

All three of equations (3.3) to (3.5) can be represented by the single expression

$$\sigma_{ij,j} + X_i = 0 \tag{3.71}$$

Here it is understood that the subscript i can take any value in its permitted range (1 to 3) and that where a subscript is repeated within a term (or a product of terms) summation over the range of that subscript is intended. Thus in the above case, j is repeated within the first term, which thus represents the sum of $\sigma_{i1,1}$ $\sigma_{i2,2}$ and $\sigma_{i3,3}$. This is known as the *Einstein summation convention* and j is called a *dummy subscript* because it does not refer to a particular value of the subscript only. Similarly, equations (3.9) to (3.13) become

$$e_{ij} = \frac{1}{2}(u_{i,j} + u_{j,i}) \tag{3.72}$$

it being understood that this relationship is true for all possible values of i and j. The compatibility equations (3.14) to (3.19) can now be written as

$$e_{ij,kl} + e_{kl,ij} = e_{ik,jl} + e_{jl,ik} = \frac{1}{2}(u_{i,jkl} + u_{j,ikl} + u_{k,ijl} + u_{l,ijk}) \tag{3.73}$$

where every subscript following a comma indicates a differentiation of the term preceding the comma.

The relationships between the surface tractions p_i on a surface with direction cosines n_j and the stresses σ_{ij} follow from (3.41) which now can be written as

$$p_i = \sigma_{ij} n_j \tag{3.74}$$

The stress-strain relationships for a linear anisotropic material take the form

$$\sigma_{ij} = c_{ijkl} e_{kl} \tag{3.75}$$

where, according to the summation convention, the right-hand side indicates a summation of terms for all values of k and l in the range 1 to 3. As it is not normally necessary to distinguish between σ_{ij} and σ_{ji} or e_{kl} and e_{lk}, without loss of generality c_{ijkl}, c_{jikl}, c_{ijlk} and c_{jilk} can be taken as having the same value. The stress-strain relationships can be written in a more compact form using the compliance matrix $\boldsymbol{S}$ as in (3.63) or in terms its inverse, the stiffness matrix $\boldsymbol{C}$. This is because these distinctions are not made in writing the stress and strain column vectors, resulting in only thirty six stiffness coefficients, as opposed to eighty one in (3.75). Also, as $\boldsymbol{C}$ is symmetric, it follows that c_{ijkl} and c_{klij} have the same value. This reduces the number of independent coefficients to twenty one at most. For an isotropic material, (3.75) becomes

$$\sigma_{ij} = \lambda \delta_{ij} e_{kk} + 2\mu e_{ij} \tag{3.76}$$

where the *Lamé constants*, λ and μ, are given by

$$\lambda = \frac{2G\nu}{1 - 2\nu} \quad , \quad \mu = G \; . \tag{3.77}$$

and the *Kronecker delta* δ_{ij} is defined as

$$\delta_{ij} = 0 \quad (i \neq j) \quad , \quad \delta_{ij} = 1 \quad (i = j) \; . \tag{3.78}$$

The strain energy per unit volume of a linearly-elastic material was given by the first line of (3.37). From (3.75) this becomes, in terms of tensor notation,

$$U_V = \frac{1}{2}\sigma_{ij} e_{ij} = \frac{1}{2}[\sigma_{11} e_{11} + \sigma_{12}(e_{12} + e_{21}) + \ldots] = \frac{1}{2} c_{ijkl} e_{ij} e_{kl} \tag{3.79}$$

(Note that e_{12} and e_{21} are the same and both equal to one half of γ_{xy} , so that the resulting expression is the same as that in (3.37). Using (3.76), this expression takes the form

$$U_V = \frac{1}{2}[\lambda \delta_{ij} e_{kk} + 2\mu e_{ij}] e_{ij} = \frac{1}{2}[\lambda (e_{kk})^2 + 2\mu e_{ij} e_{ij}] \tag{3.80}$$

for an isotropic material.

The inverse of (3.75) can be written as

$$e_{ij} = s_{ijkl} \sigma_{kl} \tag{3.81}$$

where the compliances s_{ijkl} have the same symmetry properties as the stiffnesses c_{ijkl}. For isotropic materials, (3.81) becomes

$$e_{ij} = \frac{1}{2G}\sigma_{ij} - \frac{\nu}{E}\delta_{ij}\sigma_{kk} \qquad \textbf{(3.82)}$$

Further information on linear elasticity in terms of cartesian tensors will be found in Ford and Alexander (1977), Renton (1987) and Sokolnikoff (1956) for example.}

Chapter 4 Beams with Axial Stresses

4.1 Introduction

In the following three chapters, the normal theory of beams will be explained. It should be made clear at the outset that this is just a particular case of a more general theory with much wider applications. Here, a beam will be taken to be a prismatic continuum, made from isotropic, homogeneous, linearly-elastic material. A prismatic beam is one which has a straight axis and every cross-section normal to this axis is the same. In analyses which are essentially two-dimensional, this axis will be taken as the x axis of a local cartesian coordinate system. In fully three-dimensional analyses, the axis of the beam will be taken as the z axis of the local coordinate system. The x and y axes then lie in the plane of a normal cross-section of the beam. A homogeneous material is one which has the same properties at all points within it.

In the engineering analysis of beams subject to axial load and bending, it is usually assumed that plane cross-sections remain plane and normal to the centroidal axis of the beam. It will be shown later in this section that such assumptions are not necessary. In fact, unnecessary errors can result if these assumptions are applied to beams other than the kind described in the previous paragraph. There is, of course, no unique stress distribution over the cross-section which has a given resultant bending moment, axial force or torque. However, there is a unique *characteristic response* to each of these. If the resultant is applied in the form of any other stress pattern over the cross-section, this pattern will decay towards this characteristic response further along the beam, in accordance with Saint-Venant's principle. The characteristic response to a bending moment, axial force or torque is necessarily one of constant stress and strain along the beam. A response must then be found which satisfies this condition, the conditions of equilibrium and compatibility within the beam and the condition that the lateral surfaces of the beam are stress-free. Since the characteristic response is unique, any pattern of stress and strain satisfying these conditions is the characteristic response. These arguments will be discussed more fully in Chapter 8. The results used in the later sections of this chapter are those which also result from the standard assumptions. For this reason, the first-time reader may wish to skip the proofs given below.

{What will be employed here is known as *Saint-Venant's semi-inverse method*. That is, a pattern of stress is assumed and then shown to satisfy the necessary criteria. The most general state of stress which will be considered in this chapter is

$$\sigma_{zz} = a + bx + cy \tag{4.1}$$

In the absence of any body forces, this automatically satisfies the equations of equilibrium given by (3.3) to (3.5). From (3.22), the corresponding strains are

$$\epsilon_{xx} = \epsilon_{yy} = -\nu(a + bx + cy)/E \quad , \quad \epsilon_{zz} = (a + bx + cy)/E \; . \tag{4.2}$$

These satisfy the equations of compatibility given by (3.14) to (3.16). The transverse stresses σ_{xx} σ_{yy} and τ_{xy} are all zero so that it follows that any transverse stresses acting on a plane parallel to the z axis will be zero. Such stresses will be of the kind shown by σ and τ acting on the triangular element in Figure 3.6. This plane will be taken to form part of the surface of the continuum. A three-dimensional picture of the element is shown in Figure 4.1. This shows the other possible stresses on such an element, which act in the z direction. For equilibrium of the element in this direction,

$$\tau_s \delta s \delta z = \tau_{xz} \delta s \cos\theta \, \delta z + \tau_{yz} \delta s \sin\theta \, \delta z$$
$$\text{or} \quad \tau_s = \tau_{xz} \cos\theta + \tau_{yz} \sin\theta \qquad (4.3)$$

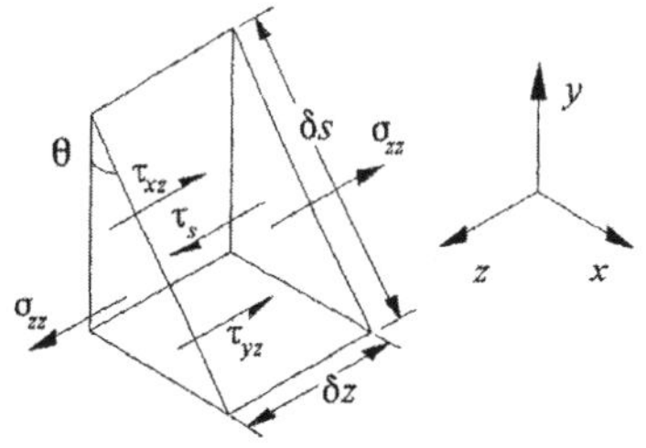

Figure 4.1 Equilibrium of a triangular element.

Note that the end stresses σ_{zz} balance one another out and so do not enter the equation. Even if σ_{zz} varied longitudinally, the increment would not appear in the equation, as it is of a smaller magnitude than the other terms. In the present case, τ_{xz} and τ_{yz} are zero and so τ_s is zero. As the other two surface tractions, σ and τ, are also zero, the surface is free from stress. This surface will be taken to form a closed boundary around the continuum. Then (4.1) describes a prismatic beam with its axis parallel to the z direction. The lateral surfaces of this beam are free from stress, so that it is loaded at its ends only.

The resultant pairs of end loads produced by this stress distribution are the moments M_x and M_y and the tensile forces P shown in Figure 4.2. If the area of the cross-section is A and a small part of it is denoted by dA,

$$M_x = \int_A y \sigma_{zz} dA = aA\bar{y} + bI_{xy} + cI_{xx}$$
$$M_y = -\int_A x \sigma_{zz} dA = -(aA\bar{x} + bI_{yy} + cI_{xy})$$
$$P = \int_A \sigma_{zz} dA = aA + bA\bar{x} + cA\bar{y} \qquad (4.4)$$

where

$$I_{xx} = \int_A y^2 dA \,, \quad I_{xy} = \int_A xy dA \,, \quad I_{yy} = \int_A x^2 dA$$
$$\bar{x} = \frac{1}{A}\int_A x dA \,, \quad \bar{y} = \frac{1}{A}\int_A y dA \,. \qquad (4.5)$$

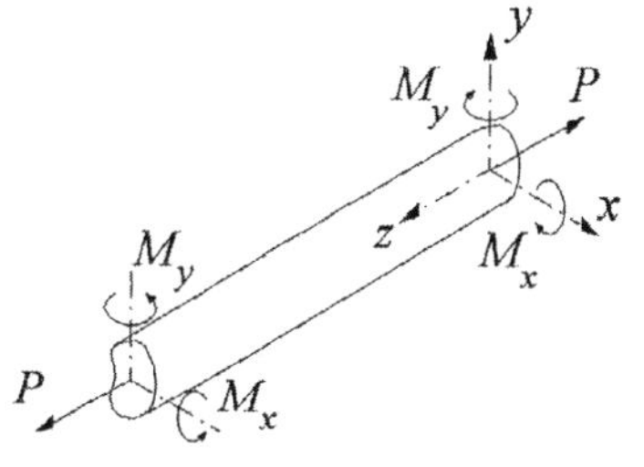

Figure 4.2 Resultant loads on a beam.

Equations (4.4) can be uncoupled by shifting and rotating the coordinate axes. If the origin is shifted to the centroid of the section at $(\bar{x}, \bar{y})$, the terms in a in the first two equations and the terms in b and c in the third equation disappear. (This is because the new coordinates of the centroid are (0,0) .) If the coordinate axes are rotated through a clockwise angle θ about the z axis, as shown in Figure 4.3, then the new x and y coordinates, (x', y'), are related to the old coordinates by

$$x' = x\cos\theta + y\sin\theta$$
$$y' = -x\sin\theta + y\cos\theta \qquad (4.6)$$

In terms of the new coordinate system, the product of inertia is given by

$$I_{xy}' = \int_A x'y' dA = I_{xy}\cos 2\theta + \tfrac{1}{2}(I_{xx} - I_{yy})\sin 2\theta \qquad (4.7)$$

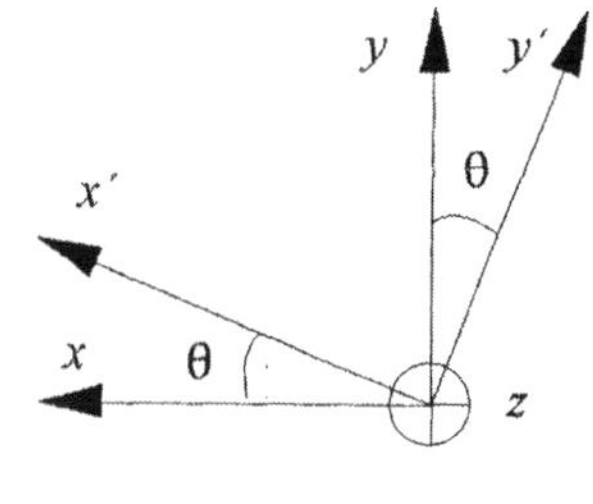

Figure 7.2 **Figure 4.3** Rotation of coordinates.

Then if a new set of coordinates is chosen at an angle θ

such that

$$\tan 2\theta = \frac{I_{xy}}{\frac{1}{2}(I_{yy} - I_{xx})} \tag{4.8}$$

the product of inertia disappears, so that the three equations (4.4) are in terms of c, b and a respectively. The new axes x and y are called the *principal axes* of the cross-section[1].

In terms of these new axes, (4.4) can now be inverted to give

$$a = \frac{P}{A} \quad , \quad b = -\frac{M_y}{I_{yy}} \quad , \quad c = \frac{M_x}{I_{xx}} \ . \tag{4.9}$$

so that the state of stress is given by

$$\sigma_{zz} = \frac{P}{A} - \frac{M_y x}{I_{yy}} + \frac{M_x y}{I_{xx}} \tag{4.10}$$

The stresses and strains given by (4.1) and (4.2) do not vary with z. Suppose that the beam shown in Figure 4.2 is very long. According to Saint-Venant's principle, any resultant end loads P, M_x and M_y must produce stresses and strains very similar to these well away from the ends. If the resultants arise from different end stress distributions, these must then decay towards the steady-state pattern given by (4.1) away from the ends. It follows that this is the *characteristic response* of a beam to these three resultants. This forms the basis of the engineering theory of the bending and axial loading of beams. It is assumed that if the local values of these resultants are known, then the local behaviour of the beam is given with sufficient accuracy by the characteristic response.

The state of strain is given completely by (4.2), i.e. the shear strains γ_{xy}, γ_{yz} and γ_{zx} are zero. From the definitions of strain given by (3.9) to (3.13) it is now possible to find the displacements u, v and w by integration. Leaving out arbitrary rigid-body displacements and rotations of the beam,

$$\begin{aligned} u &= -\frac{\nu}{E}(ax + \tfrac{1}{2}bx^2 + cxy) + \frac{\nu b y^2}{2E} - \frac{bz^2}{2E} \\ v &= -\frac{\nu}{E}(ay + bxy + \tfrac{1}{2}cy^2) + \frac{\nu c x^2}{2E} - \frac{cz^2}{2E} \\ w &= \frac{1}{E}(az + bxz + cyz) \end{aligned} \tag{4.11}$$

The deflections of any normal cross-section of the beam are given by the above equations for a constant value of z. In particular, w is a linear function of x and y. This corresponds to the usual assumption that plane sections normal to the axis of the beam remain plane during bending (although they may rotate and displace). As these sections remain plane, there is no problem in defining the angles θ_x and θ_y through which they rotate about

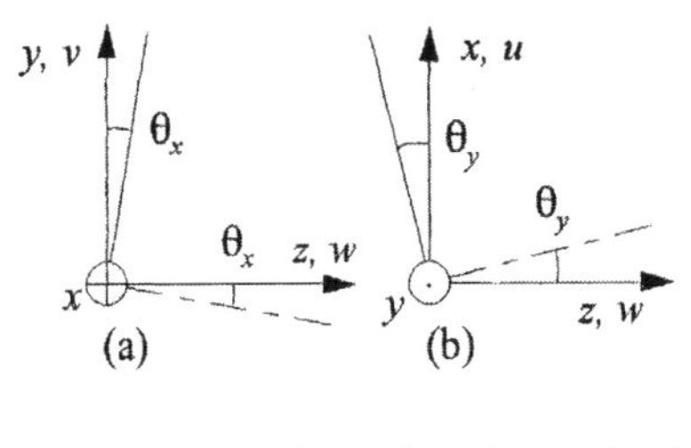

Figure 4.4 Rotations of section and axis.

[1] Equation (4.8) may be compared with (3.40) and (3.48) of the previous chapter. In fact, a Mohr's circle for inertia can be drawn too. This is because stress, strain and inertia can be represented by symmetrical second-order tensors, all of which transform in the same way.

the x and y axes. Using the assumptions of small deflexion theory given in §1.2, these are

$$\theta_x = \frac{\partial w}{\partial y} \quad \left(= \frac{cz}{E} = -\frac{\partial v}{\partial z} \right)$$
$$\theta_y = -\frac{\partial w}{\partial x} \quad \left(= -\frac{bz}{E} = \frac{\partial u}{\partial z} \right) \tag{4.12}$$

where the equalities in brackets are derived from (4.11). These equalities mean that the centroidal axis also rotates through the same angles, as shown in Figure 4.4, so that plane sections normal to the centroidal axis remain normal to it, as is also assumed in the usual derivation of the engineering theory of bending.

From (4.9) and (4.11), the deflections of the centroidal axis are given by

$$u_c = \frac{M_y z^2}{2EI_{yy}} \quad , \quad v_c = -\frac{M_x z^2}{2EI_{xx}} \quad , \quad w_c = \frac{Pz}{EA} \quad . \tag{4.13}$$

By differentiating these equations with respect to z, the local resultants can be related to the local curvature and axial deformation of the centroidal axis, giving

$$M_x = -EI_{xx}\frac{d^2 v_c}{dz^2} \quad , \quad M_y = EI_{yy}\frac{d^2 u_c}{dz^2} \quad , \quad P = EA\frac{dw_c}{dz} \quad . \tag{4.14}$$

In solving differential equations of this kind, the subscript c will be dropped, it being understood that the deflexion of the centroidal axis characterises that of the beam as a whole. However, as will be seen later, when more accurate analyses are needed, this assumption can be misleading.}

4.2 The Differential Equations of Flexure

The simple engineering theory of bending is based on the characteristic response of a beam to a resultant bending moment, as outlined in §4.1. Account is taken of local loads on the side of the beam and resultant shear forces on the cross-section only in so far as they affect the variation of the bending moment. This was considered in the two-dimensional static analysis of beams given in §2.8. For simplicity of notation, the conventions used there will be adopted for the rest of this chapter. Also, the axial *compressive* stress will denoted by σ_b and the second moment of area of the cross-section about the axis normal to the plane of bending denoted by I. It will be assumed that this axis is a principal axis of the cross-section. Then from (4.10), (4.14), (2.16) and (2.17) respectively, taking EI as constant,

$$\sigma_b = \frac{My}{I} \quad , \quad M = EI\frac{d^2 v}{dx^2}$$
$$S = -\frac{dM}{dx} = -EI\frac{d^3 v}{dx^3} \quad , \quad q = \frac{dS}{dx} = -EI\frac{d^4 v}{dx^4} \quad . \tag{4.15}$$

where v is taken to typify the lateral displacement of the beam in the y direction, as before. The first of equations (4.15) shows that the greatest value of the bending stress, σ_{max}, at any given section occurs at the greatest distance, y_{max}, from the neutral axis. The *section modulus*[1], Z, is defined as I/y_{max} so that σ_{max} is given by M/Z.

In the case of statically-determinate beams, the value of M will be known at all points along the beam, so that it is possible to start from the above second-order differential equation

[1] This is sometimes referred to as the *elastic* section modulus, Z_e, to distinguish it from the plastic section modulus, Z_p. However, problems of plasticity will not be dealt with in this book.

relating v to M. More generally, it is necessary to start from the fourth order differential equation relating v to q. Integrating this four times gives four constants of integration. These constants can be determined from two conditions at each end of the beam[1]. The most common end conditions are related to the *workless supports* discussed in §1.1. Here, the end reactions can be a bending moment M_e or a shear force S_e. The deflexions corresponding to these are an end rotation θ_e and an end displacement v_e. The two end conditions at a normal workless support are that either M_e or θ_e is zero and that either S_e or v_e is zero. At a simple support (pin or knife-edge) M_e and v_e are zero and at a fixed (or encastré) support θ_e and v_e are zero.

The end moment and rotation will be taken as *anticlockwise*[2] positive and the end shear force and displacement as positive in the y direction. In terms of this convention, any rotation of the beam, θ, is equal to the local slope of the axis, $\partial v/\partial x$ (to the degree of accuracy of small deflexion theory discussed in §1.2). It must be noted that these are *vector conventions*. The way in which M and S are defined above are in terms of *deformation conventions*. That is, M is positive if it produces sagging curvature in the beam and S is positive if it tends to produce a shear which increases v along the beam (see Figure 2.17). The conventions for M and M_e are the same, as are those for S and S_e, at the right-hand end of the beam. Conversely, the two conventions are opposite in sign at the left-hand end of the beam. These conventions are related to the beam AB shown in Figure 4.5.

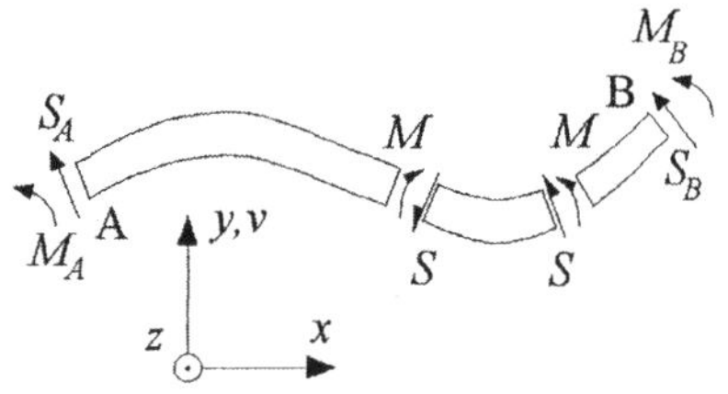

Figure 4.5 Loading conventions.

In order to solve equations (4.15), it is often necessary to integrate discontinuous functions. Figure 4.6a shows a discontinuous function which has no values for x less than a. It is convenient to denote this function as $f\{x\text{-}a\}$ or $f\{X\}$. Usually, this is just an nth power function of the argument, $\{x\text{-}a\}^n$ or $\{X\}^n$. The curly brackets are used to indicate that the function only has non-zero values when its argument is positive. Such brackets are known as *Macaulay brackets*. The integral of this function is shown by the shaded area. This too has no values when X is negative. In Figure 2.17a, the shear

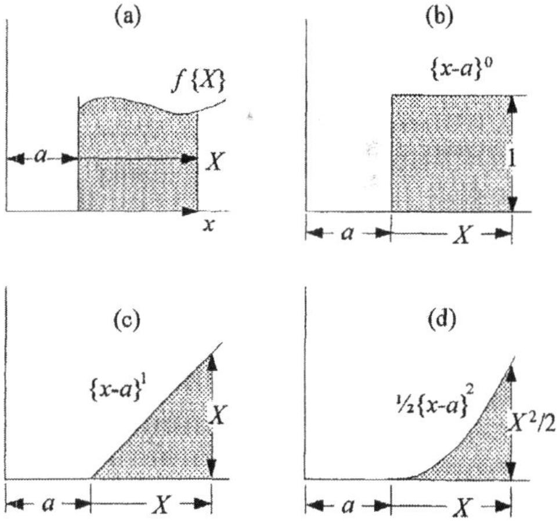

Figure 4.6 Step functions.

[1] In some analyses, it is possible to determine these constants from the conditions at one end only, but two of these conditions will always depend on the state of affairs at the other end of the beam.

[2] They will be *clockwise* about the z axis of a right-handed coordinate system, as in Figure 1.1. This also conforms to the convention used for polar coordinates.

force is discontinuous because a point force F is applied to the beam[1]. This produces a step of magnitude F in the shear force diagram. There is a discontinuity in the slope of the bending moment diagram because the rate of change of bending moment is equal to the shear force. Figure 4.6b shows a unit step function (also known as the Heaviside step function). The integral of this is given by the shaded area underneath the step and is shown in Figure 4.6c. This is known as the unit ramp function. In Figure 2.17a, the discontinuity in the shear force diagram is given by superimposing a step function of magnitude F onto the shear force produced by the left-hand reaction, giving

$$S = -R_A + F\{x-a\}^0 \tag{4.16}$$

Likewise, the bending moment diagram is given by superimposing a ramp function of magnitude $-F$ onto the bending moment produced by the left-hand reaction, giving

$$EI\frac{d^2v}{dx^2} = M = R_A x - F\{x-a\}^1 \tag{4.17}$$

making use of (4.15). Integrating the ramp function shown in Figure 4.6c gives the discontinuous parabolic function shown in Figure 4.6d and so on. Likewise, the point moment m in Figure 2.17b is represented by a step function in Figure 2.17b and the uniform distributed load p in Figure 2.17c produces a ramp function in the shear force diagram and a discontinuous parabolic function in the bending moment diagram. Again, these are statically-determinate problems so that it is possible to write down the expressions for M in cases 2.17b and 2.17c respectively:

$$\begin{aligned} EI\frac{d^2v}{dx^2} &= M = -\frac{mx}{a+b} + m\{x-a\}^0 \\ EI\frac{d^2v}{dx^2} &= M = R_A x - \tfrac{1}{2}p\{x-a\}^2 \end{aligned} \tag{4.18}$$

These differential equations can be integrated twice to give the deflexion curves for the beams. The two constants of integration which result in each case are determined from the conditions that v is zero at each end. (The other two end conditions are that M is zero at each end. These have already been taken into account in setting up the second-order differential equations.) The resulting expressions are:

$$\begin{aligned} v &= \frac{F}{6EI}\left[\frac{bx(x^2-2ab-a^2)}{a+b} - \{x-a\}^3\right] \\ v &= \frac{m}{2EI}\left[\frac{x(a^2+2ab-2b^2-x^2)}{3(a+b)} + \{x-a\}^2\right] \\ v &= \frac{p}{24EI}\left[\frac{b^2x(2x^2-2a^2-4ab-b^2)}{a+b} - \{x-a\}^4\right] \end{aligned} \tag{4.19}$$

for cases 2.17a to 2.17c respectively, using the expressions for R_A in §2.8.

[1] Starting from the fourth order differential equation relating v to q, the force F could be represented using a Dirac delta function. In practice, it is only necessary to know that the integral of this delta function is a unit step function.

Figure 4.7 shows how the method can be applied to a number of related problems. If several loads are applied at different locations along the beam, then step functions beginning at each location will be needed. Thus in Figure 4.7a the bending moment is given by

$$M = R_A x - F_1\{X_1\} - F_2\{X_2\} \quad \textbf{(4.20)}$$

where, if F_1 and F_2 are at distances a and b from the left-hand end of the beam, X_1 and X_2 can be replaced by $(x-a)$ and $(x-b)$ respectively. If a continuous load does not extend to the right-hand end of the beam, it can be terminated earlier by superimposing a continuous load of the opposite magnitude to cancel it out. Thus in Figure 4.7b, the real load of intensity p extends from the origin of the ordinate X_1 to the origin of the ordinate X_2. This can be simulated by superimposing the load of intensity $-p$ which starts at $X_2 = 0$. Then the bending moment is given by

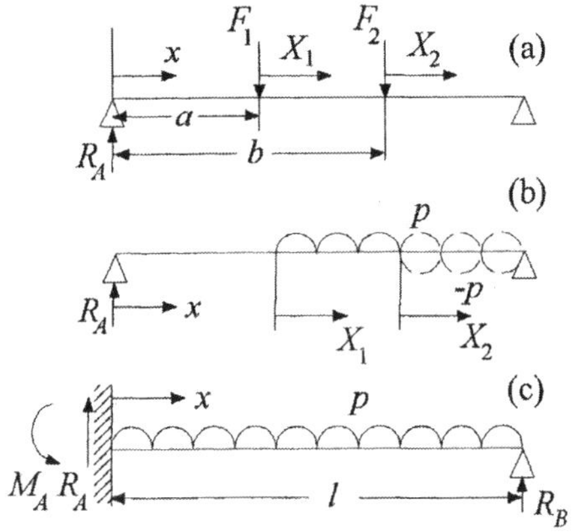

Figure 4.7 Various beam problems.

$$M = R_A x - \tfrac{1}{2}p\{X_1\}^2 + \tfrac{1}{2}p\{X_2\}^2 \quad \textbf{(4.21)}$$

When the beam is not statically determinate, it is still possible to start from the second-order differential equation related to M. The left-hand end reactions are included as unknown quantities which are evaluated from the end conditions after integration. Thus the expression for M for the beam shown in Figure 4.7c could be written as

$$M = -M_A + R_A x - \tfrac{1}{2}px^2 = EI\frac{d^2v}{dx^2} \quad \textbf{(4.22)}$$

However, in this case there is only one redundant reaction; the other is related to it from the condition that M is zero at the right-hand end, giving

$$M_A = R_A l - \tfrac{1}{2}pl^2 \quad \textbf{(4.23)}$$

On integrating once, the condition that the slope dv/dx is zero at the left-hand end can be used to eliminate the constant of integration and on integrating a second time, the conditions of zero displacement at the left- and right-hand ends can be used to eliminate the second constant of integration and to determine the unknown redundancy. The deflexion curve is then given by

$$v = \frac{p}{48EI}(2x^4 + 3x^2l^2 - 5x^3l) \quad \textbf{(4.24)}$$

Results for a number of simple beam problems are given in §A6.1.

Beams which are continuous over several supports can be analysed in a similar fashion. For each unknown redundant reaction there will be an extra equation expressing the condition that the corresponding deflexion is zero[1]. Often a pin is inserted into a continuous beam to give a zero bending moment locally. This means that although the bending moment and displacement have no local discontinuity, there will usually be a local discontinuity in the slope of the beam. This can

[1]This is true for the common type of workless reaction. Elastic supports are considered a little later in this section.

be catered for by introducing a step function of unknown magnitude as a constant of integration when the expression for the bending moment is integrated. Consider the case shown in Figure 4.8. This has fixed ends, A and E, at a distance $4a$ apart. It is supported by a knife-edge support at its midpoint, C. There is a pin at D, midway between C and E, which links the two parts of the beam, AD and DE. The structure is loaded by a downwards force P at B, midway between A and C. If the reactions at A are a force R_A upwards and a moment M_A anticlockwise and the reaction at C is a force R_C upwards, then the bending moment at any distance x from A is given by

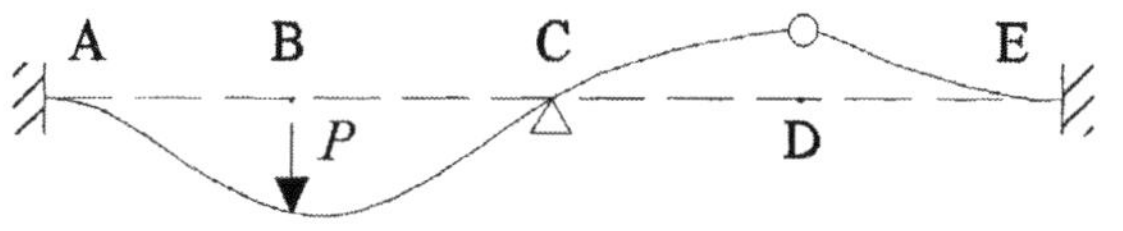

Figure 4.8 Deflexion of a continuous beam.

$$EI\frac{d^2v}{dx^2} = M = -M_A + R_Ax - P\{x-a\} + R_C\{x-2a\} \qquad \textbf{(4.25)}$$

The bending moment is zero at x=$3a$, where the pin is located, so that

$$0 = -M_A + 3R_Aa - 2Pa + R_Ca \qquad \textbf{(4.26)}$$

Integrating (4.25) once and allowing for a slope discontinuity at D of unknown magnitude G,

$$EI\frac{dv}{dx} = -M_Ax + \tfrac{1}{2}R_Ax^2 - \tfrac{1}{2}P\{x-a\}^2 + \tfrac{1}{2}R_C\{x-2a\}^2 + G\{x-3a\}^0 + A \qquad \textbf{(4.27)}$$

The slope is zero at the fixed ends, giving A as zero from the condition at x=0, and at x=$4a$,

$$0 = -4M_Aa + 8R_Aa^2 - 4.5Pa^2 + 2R_Ca^2 + G \qquad \textbf{(4.28)}$$

as $\{4a\text{-}3a\}^0$ is unity. Integrating once more gives

$$EIv = -\tfrac{1}{2}M_Ax^2 + \tfrac{1}{6}R_Ax^3 - \tfrac{1}{6}P\{x-a\}^3 + \tfrac{1}{6}R_C\{x-2a\}^3 + G\{x-3a\} + B \qquad \textbf{(4.29)}$$

The conditions of zero displacement at x= 0, $2a$ and $4a$ give B as zero and

$$\begin{aligned} 0 &= -2M_Aa^2 + \tfrac{4}{3}R_Aa^3 - \tfrac{1}{6}Pa^3 \\ 0 &= -8M_Aa^2 + \tfrac{32}{3}R_Aa^3 - \tfrac{9}{2}Pa^3 + \tfrac{4}{3}R_Ca^3 + Ga \end{aligned} \qquad \textbf{(4.30)}$$

respectively. The four unknowns, M_A, R_A, R_C and G, can be determined from (4.26), (4.28) and (4.30) giving

$$M_A = \frac{9Pa}{28}\ , \quad R_A = \frac{17P}{28}\ , \quad R_C = \tfrac{1}{2}P\ , \quad G = -\frac{Pa^2}{14}\ . \qquad \textbf{(4.31)}$$

and hence the deflexion curve shown in Figure 4.8. In general, there is one deflexion condition corresponding to each unknown load and one load condition corresponding to an unknown deflexion. Here, the unknown deflexion is the relative rotation at D and the corresponding load condition is that the bending moment is zero at this point. In the case of *elastic supports*, the reaction will be in the opposite sense to the corresponding deflexion and equal to some multiple k of it, where k is known as the support stiffness. This condition replaces the condition at a simple workless support, that a reaction or its corresponding deflexion is zero.

An example is shown in Figure 4.9. The beam of length l has elastic supports at its ends which resist vertical displacement only, with springs of stiffness k. Along its length there is an elastic foundation which resists a vertical displacement v by exerting a vertical force per unit length kv/l in the opposite sense. A uniform load of intensity p per unit length is applied to the beam. Because the beam is symmetrical, it is convenient to take the origin of the coordinate system at its midpoint. The bending moment can no longer be expressed as a simple function of x and so it becomes necessary to start from the equation in (4.15) which relates the fourth differential of v to the distributed load on the beam. In addition to p there is the foundation reaction, so that

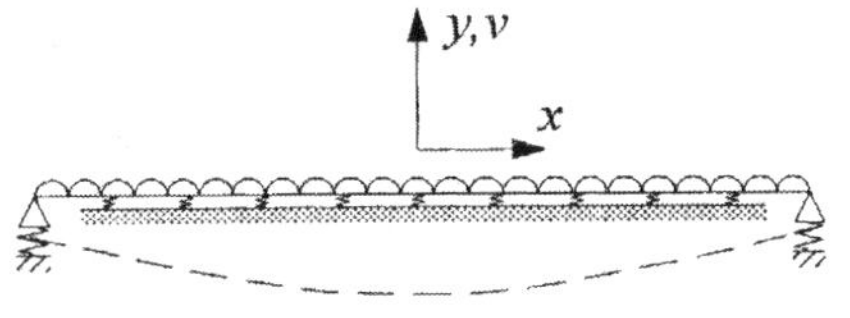

Figure 4.9 Elastically supported beam.

$$p + \frac{kv}{l} = -EI\frac{d^4v}{dx^4} \tag{4.32}$$

The solution of this equation takes the general form

$$v = \cosh cx(A\cos cx + B\sin cx) + \sinh cx(C\cos cx + D\sin cx) - \frac{pl}{k} \quad \text{where } c^4 = \frac{k}{4EIl} \tag{4.33}$$

This must be symmetrical about the origin, so that the coefficients of the antisymmetrical functions, B and C, must be zero. Then it is only necessary to satisfy the conditions at the right-hand end, $x=\frac{1}{2}l$, and the same conditions will automatically be satisfied at the left-hand end. The first condition is that the bending moment, and hence the beam curvature, is zero at the ends, so that

$$0 = \frac{d^2v}{dx^2} = -2Ac^2\sinh d\sin d + 2Dc^2\cosh d\cos d \quad \text{where } d = \frac{1}{2}cl \tag{4.34}$$

The other condition is that the shear force given by (4.15) is equal to the reaction $-kv$, produced by the elastic support, or

$$-EI\frac{d^3v}{dx^3} = 2c^3[-A(\cosh d\sin d + \sinh d\cos d) + D(\sinh d\cos d - \cosh d\sin d)]$$
$$= -kv = -k\left(A\cosh d\cos d + D\sinh d\sin d - \frac{pl}{k}\right) \tag{4.35}$$

The constants A and D can be determined from the above two equations, yielding the final expression for the deflected form,

$$v = \frac{pl}{k}(A^*\cosh cx\cos cx + D^*\sinh cx\sin cx - 1)$$
$$\text{where } A^* = 4\frac{d}{e}\cosh d\cos d\,,\ D^* = 4\frac{d}{e}\sinh d\sin d\,, \tag{4.36}$$
$$e = 4d(\cos^2 d + \sinh^2 d) + \sin d\cos d + \sinh d\cosh d\,.$$

This curve is shown by the broken line in Figure 4.9.

{4.3 Non-Prismatic Beams and Other Exceptional Cases}

{The theory of beams is often applied to non-prismatic beams. Some idea of the accuracy with which the theory still applies can be gathered from an examination of known solutions to some two-dimensional problems. Consider the plane wedge shown in Figure 4.10. The stresses induced by an axial force P and a bending moment M applied at the tip as given by Michell[1] and Carothers[2] are

$$\sigma_{rr} = \frac{2M\sin 2\theta}{r^2(\sin\alpha - \alpha\cos\alpha)} - \frac{2P\cos\theta}{r(\alpha + \sin\alpha)}$$
$$\tau_{r\theta} = \frac{M(\cos\alpha - \cos 2\theta)}{r^2(\sin\alpha - \alpha\cos\alpha)} \qquad \textbf{(4.37)}$$

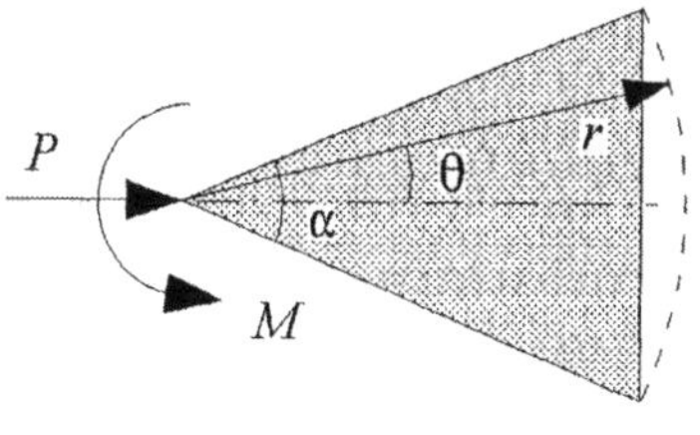

Figure 4.10 A wedge-shaped beam.

where θ is the anticlockwise angle from the centreline of the wedge. The radial stress produced by M at the surface of the wedge can be compared with the axial stress at the same point, as predicted by the engineering theory of bending. This shows that the engineering theory overestimates the local stress in this case. The predicted axial stress exceeds the radial stress by 10% when α is a little over 45°, as shown in Figure 4.10. If the axial stress produced by P at the centreline given by the two analyses is compared, the engineering theory gives the lower value. It is 10% less than that given by (4.37) when α is approximately 95°. The three-dimensional problem of the bending of a cone is discussed in Appendix 7.

Figure 4.11 shows a curved beam with an inner radius a and an outer radius b. It is subject to a constant moment M which acts in the sense shown. The solution of the two-dimensional problem was first found by Golovin[3]. The bending moment induces both radial stresses σ_{rr} and tangential stresses $\sigma_{\theta\theta}$. The latter can be compared with those predicted by the engineering theory of bending. They are given by

$$\sigma_{rr} = \frac{4M}{N}\left(-\frac{a^2b^2}{r^2}\ln\frac{b}{a} + a^2\ln\frac{r}{a} + b^2\ln\frac{b}{r} \right)$$
$$\sigma_{\theta\theta} = \frac{4M}{N}\left(\frac{a^2b^2}{r^2}\ln\frac{b}{a} + a^2\ln\frac{r}{a} + b^2\ln\frac{b}{r} + a^2 - b^2 \right)$$
$$\text{where } N = (b^2 - a^2)^2 - \left(2ab\ln\frac{b}{a} \right)^2 \qquad \textbf{(4.38)}$$

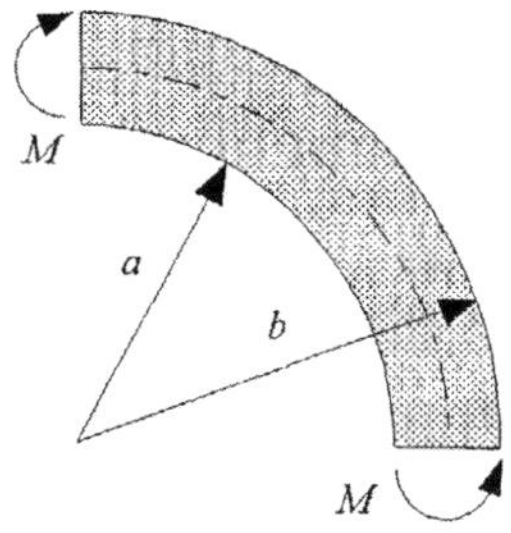

Figure 4.11 A curved beam.

The value of $\sigma_{\theta\theta}$ is greatest at the inner surface of the beam ($r=a$) and exceeds that predicted by the engineering theory of bending. This prediction is 10% less than that given by (4.38) when b/a is approximately 1.35, which is the curvature depicted in Figure 4.11.

The exact solution of the three-dimensional problem of the bending of curved beams has been found by Sadowsky and Sternberg[4] for the case of a beam with a circular cross-section. An approximation for curved beams of arbitrary cross-section is to assume that plane radial cross-sections through the centre of curvature remain plane. The state of stress is then taken to be given by the hoop stress $\sigma_{\theta\theta}$ arising from the extension and compression of the circumferential fibres.

---

[1] Michell, J.H. (1900) *London Math. Soc. Proc.* **32**, 29.

[2] Carothers, S.D. (1912) *Proc. Roy. Soc. Edinburgh* **23**, 292.

[3] Golovin, H. (1881) *Trans. Inst. Tech.*, St. Petersburg.

[4] Sadowsky, M.A. and Sternberg, E. (1953) *J. Appl. Mech.* **20**, 215.

The radial stress σ_{rr} which would necessarily arise for internal equilibrium (cf. (3.57)) is ignored. This is known as the Winkler[1] theory of curved beams and proves to be sufficiently accurate for many purposes. Those fibres between radial planes which are closer to the centre of curvature are shorter than the fibres further out. This means that the same relative displacement of the radial planes induces a higher strain (and hence a higher stress) on the inner fibres than it does on the outer fibres. Then the neutral axis in pure bending moves inwards by a distance n from the centroid in order that the state of stress gives rise to no resultant axial force. If R is the radius of curvature to the centroidal axis, y is the distance radially outwards of an elementary area dA from this axis and the cross-sectional area is A, the zero axial force condition is given by

$$\int_A \left(\frac{y + n}{y + R} \right) dA = 0 \tag{4.39}$$

which determines the value of n. The bending stress is then

$$\sigma_{\theta\theta} = \frac{M}{R(B-A)} \left[\frac{R}{R+y} - \frac{B}{A} \right] \quad \text{where} \quad B = \int_A \left(\frac{R}{R+y} \right) dA \tag{4.40}$$

For example, for a beam with a circular section of radius a curved to a radius R of $5a$, the value of B is $3.1737a^2$. This gives a bending stress of $-1.10M/a^2$ at the outer edge of the section and $1.49M/a^2$ the inner edge. Taking Poisson's ratio as 0.3, the correspond values given by Sadowsky and Sternberg are $-1.13M/a^2$ and $1.62M/a^2$ respectively. A further discussion of curved beams will be found in §8.6.

The engineering theory of bending is also applied to prismatic beams made from more than one (isotropic) material. A simple solution can be found if the materials have a common Poisson's ratio (or if the junction of the materials forms the neutral axis of the beam). Under these conditions, a solution can be found which corresponds to axial stresses only, which remain constant along the beam. For compatibility, the axial strains at the junctions of the materials will be the same. The transverse strains at the junctions will then also be the same, because the materials have the same Poisson's ratio. Muskhelishvili (1963) discusses this problem in detail in Chapter 24, but the result of his analysis is the same as that normally used in engineering analyses. Consider the particular case of the displacement field defined in (4.11) given by

$$u = -\nu Cxy \ , \quad v = \tfrac{1}{2}C[\nu(x^2 - y^2) - z^2] \ , \quad w = Cyz \ . \tag{4.41}$$

where a and b are zero and c/E has been replaced by C, and ν is the same for all materials in the beam. The value of C is equal to the flexural curvature $-\partial^2 v/\partial x^2$ (cf. (4.12)). The strains and stresses resulting from this are

$$\begin{aligned} \epsilon_{xx} = \epsilon_{yy} = -\nu Cy \ , \quad \epsilon_{zz} = Cy \ , \quad \gamma_{xy} = \gamma_{yz} = \gamma_{zx} = 0 \ . \\ \sigma_{xx} = \sigma_{yy} = 0 \ , \quad \sigma_{zz} = E_L Cy \ , \quad \tau_{xy} = \tau_{yz} = \tau_{zx} = 0 \ . \end{aligned} \tag{4.42}$$

where E_L is the local value of Young's modulus. These strains and stresses satisfy the necessary internal conditions of compatibility and equilibrium and the condition that the lateral surfaces of the beam are free from surface tractions. The cross-section will be taken as divided into zones of constant E_L. The resultant loads on the cross-section are then

---

[1] A fuller exposition will be found in Oden (1967) Chapters 4 and 5. See also the extensive tables based on the Winkler theory in Young (1989) Chapter 8 and Pilkey (1994) Chapter 16.

$$
\begin{aligned}
M_x &= \int_A \sigma_{zz} y\,dA = \int_A E_L C y^2 dA = C\Sigma E_L I_{xxL} \\
M_y &= -\int_A \sigma_{zz} x\,dA = -\int_A E_L Cxy\,dA = -C\Sigma E_L I_{xyL} \\
P &= \int_A \sigma_{zz} dA = \int_A E_L Cy\,dA = C\Sigma E_L (A\bar{y})_L
\end{aligned}
\tag{4.43}
$$

where the summation is of all the zones and the subscript L indicates a value related to a particular zone. Pure bending (P=0) occurs when the third summation is zero. Here, $(A\bar{y})_L$ is the first moment of area of a zone about the x axis. If the origin is shifted a distance y_0 in the y direction, this expression becomes

$$\Sigma E_L (A\bar{y})_L = \Sigma E_L [A(\bar{y}' + y_0)]_L = \Sigma E_L (A\bar{y}')_L + \Sigma E_L A_L y_0 \tag{4.44}$$

where $(A\bar{y}')_L$ is the first moment of area about the new x axis. Then on taking the displacement field given by (4.41) to refer to the new origin, P is zero if

$$y_0 = \frac{\Sigma E_L (A\bar{y})_L}{\Sigma E_L A_L} \tag{4.45}$$

Likewise, for pure flexure about the y axis, the neutral axis needs to be shifted a distance x_0 in the x direction, where

$$x_0 = \frac{\Sigma E_L (A\bar{x})_L}{\Sigma E_L A_L} \tag{4.46}$$

The 'equivalent centroid' of the section is then at (x_0, y_0) relative to the original origin. It is possible to define principal axes (x', y') through this point such that flexure about an axis only requires a bending moment about the same axis. For this to be true for flexure about the x' axis, the second summation in (4.43), taken relative to the new axes, must be zero. Suppose that the new axes are at an anticlockwise angle θ to the old axes. Then the relationship between the two sets of coordinates is given by (4.6) and the above condition becomes

$$
\begin{aligned}
0 &= \Sigma E_L I_{xyL}{}' = \int_A E_L (x\cos\theta + y\sin\theta)(-x\sin\theta + y\cos\theta)\,dA \\
&= \Sigma E_L [(I_{xxL} - I_{yyL})\sin\theta\cos\theta + I_{xyL}(\cos^2\theta - \sin^2\theta)] \\
\text{or} \quad \tan 2\theta &= \frac{\Sigma E_L I_{xyL}}{\frac{1}{2}(\Sigma E_L I_{yyL} - \Sigma E_L I_{xxL})}
\end{aligned}
\tag{4.47}
$$

This can be compared with (4.8). A geometrical approach to the analysis is often made using an equivalent cross-section in which local areas are magnified in proportion to the local values of Young's modulus. Although this can be useful, it can lead to misconceptions.

When Poisson's ratio is not the same for all the materials used in a compound beam, a more complex stress system is required to ensure compatibility at the material junctions. This has been analysed by Muskhelishvili (1963) in Chapter 25. A solution is given for a compound circular beam consisting of an inner core of radius R_1 encased in a tube of a different material of outer radius R_2. The core material has a Young's modulus E_1 and Poisson's ratio ν_1. The corresponding values for the tube material are E_2 and ν_2 respectively. For a uniform Poisson's ratio, the value of the bending stiffness about any axis through the centre of the beam is then given by the above analysis as

$$I_E = \Sigma E_L I_L = \frac{\pi}{4}[E_1 R_1^4 + E_2(R_2^4 - R_1^4)] \qquad \textbf{(4.48)}$$

where I_E is the notation used by Muskhelishvili for the bending stiffness. The total bending stiffness is given by adding a term K_{11} to this, where

$$K_{11} = \frac{\pi(\nu_1 - \nu_2)^2(R_2^4 - R_1^4)R_1^4}{\alpha_1(R_2^4 - R_1^4) + \alpha_2 R_1^4 + \beta_2 R_2^4} \qquad \textbf{(4.49)}$$

where $\alpha_1 = (3-4\nu_1)(1+\nu_1)/E_1$, $\alpha_2 = (3-4\nu_2)(1+\nu_2)/E_2$, $\beta_2 = (1+\nu_2)/E_2$.

However, the significance of this additional term is usually quite small. Its greatest value in proportion to I_E is usually when ν_1 is small and ν_2 is large, the outer radius is little more than the inner radius and E_2 is much larger than E_1 . For example, on taking ν_1 and ν_2 as 0.2 and 0.5 respectively, K_{11} is less than a thirtieth of I_E for all practical cases.

The final example in Figure 4.12 illustrates the limitations of simple engineering beam theory. Consider a cantilever of length l loaded by a uniform distributed moment m per unit length as shown in Figure 4.12a. The bending moment then increases linearly from zero at the free end to ml at the fixed end, as shown in Figure 4.12b. However, this is exactly the same bending moment diagram that would be produced by a shear force of magnitude m acting at the free end, as shown in Figure 4.12c. (Note that the distributed moment m has units of force.) Similarly, if the distributed moment increased linearly along the beam from zero intensity at the free end to an intensity $m_o l$ at the fixed end, the bending moment diagram would be identical to that for a cantilever with a uniform load of intensity m_o acting upwards on it. A similar phenomenon is encountered in the description of the edge shear conditions in the analysis of plates. The total distributed shear is taken to be that found from static equilibrium considerations plus the equivalent shear found from the rate of change of the distributed edge torque. The shear behaviour of a beam will distinguish between the two types of loading. This is covered in Chapter 6.}

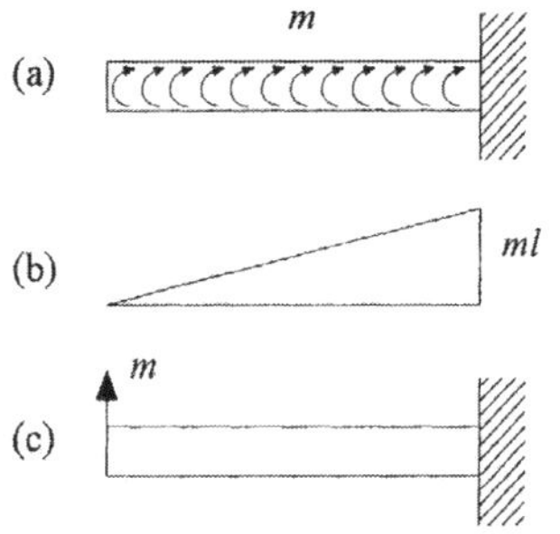

Figure 4.12 A distributed moment.

4.4 Moment-Area Methods

In §4.3 it was seen that ordinary beam theory can be used to predict the behaviour of non-uniform beams, provided that EI does not vary too rapidly. From §4.2, the deflexion curve is given by integrating the expression for the bending moment divided by the bending stiffness (M/EI). Sometimes this curve can be found conveniently by using the geometric properties of such integrals. Figure 4.13 shows the plot of M/EI for a segment of a beam. (This will be proportional to the bending moment diagram if EI is constant.) Properties of the deflexion curve can be deduced from the area and the first moment of area of this diagram. Integrating the second of equations (4.15) once gives

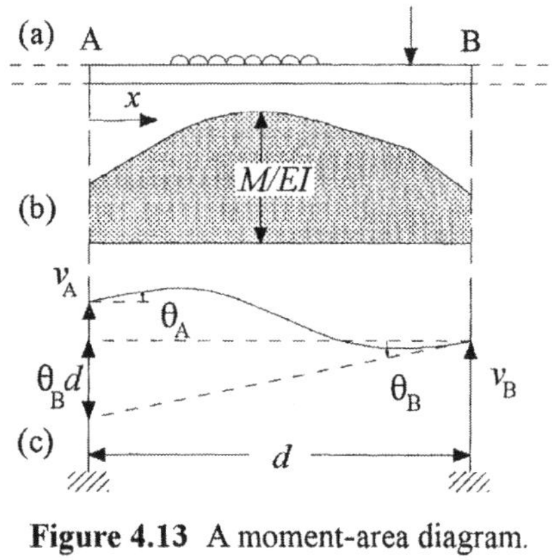

Figure 4.13 A moment-area diagram.

$$\int_A^B \frac{M}{EI}\,dx = \int_A^B \frac{d^2v}{dx^2}\,dx = \left(\frac{dv}{dx}\right)_B - \left(\frac{dv}{dx}\right)_A = \theta_B - \theta_A \qquad \textbf{(4.50)}$$

where θ_A and θ_B are the local (anticlockwise) rotations of the beam. (Note that A and B are not necessarily at the end of the beam but can be any two points along it, a distance d apart.) The difference of the rotations at these two points is then equal to the area of the M/EI diagram between them. This result is known as *Mohr's First Moment-Area Theorem*.

Suppose that both sides of the equation had been multiplied by x before integrating (where x is measured from A). Then

$$\int_A^B \frac{M}{EI}\,x\,dx = \int_A^B \frac{d^2v}{dx^2}\,x\,dx = \left|\frac{dv}{dx}\,x\right|_A^B - \int_A^B \frac{dv}{dx}\,dx = \theta_B d - v_B + v_A \qquad \textbf{(4.51)}$$

Thus from Figure 4.13c it can be seen that the first moment of the portion A to B of the M/EI diagram about A is equal to the displacement of A above the tangent to the beam at B. This is known as *Mohr's Second Moment-Area Theorem*.

An application of the first theorem is illustrated in Figure 4.14. If the beam shown supporting a uniform distributed load p was simply supported on knife edges at its ends, the bending moment would vary parabolically with a maximum value of $pl^2/8$ at the centre. This can be determined from statics. As the ends are fixed, there will be equal and opposite moments M_o at the ends of the beam which prevent end rotation. The fixed-ended bending moment diagram can then be found from the pin-ended diagram by superimposing the bending-moment diagram for the end moments alone. The M/EI diagram can be found similarly, as shown in Figure 4.14. The area of the parabola is ⅔ of its base length times its height, so that the simply-supported diagram has an area of $pl^3/12EI$. From this must be subtracted the diagram for the fixed-end moments, shown by the shaded rectangle. From the first theorem, the net area must be zero, so that M_o must be $pl^2/12$.

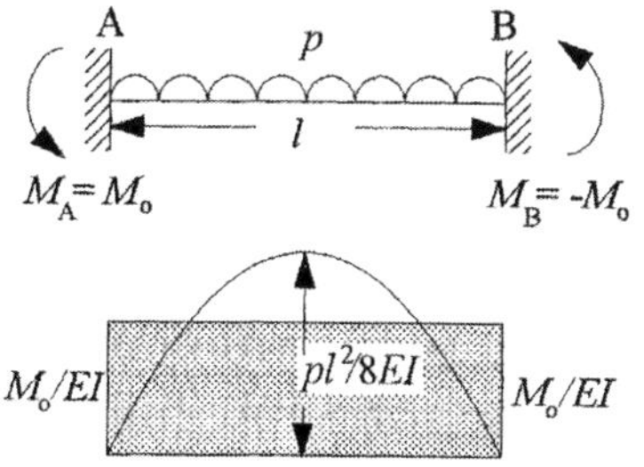

Figure 4.14 A fixed-ended beam.

Figure 4.15 shows a cantilever with a bending stiffness which varies linearly from zero at the free end to EI_o at the fixed end. The uniform load p produces a bending-moment diagram with a quadratic variation from zero at the free end to $pl^2/2$ at the fixed end. The M/EI diagram then varies linearly from zero to $pl^2/2EI_o$ at the fixed end. The first moment of area of this diagram about the free end is the distance to its centroid, ⅔l, multiplied by its area, $-pl^3/4EI_o$. It then follows from the second moment-area theorem that the downwards deflexion of the free end is $pl^4/6EI_o$.

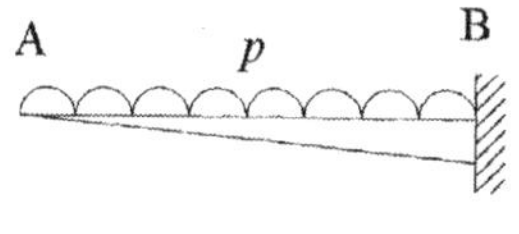

Figure 4.15 A cantilever of varying bending stiffness.

One way of applying the moment-area theorems is known as the *conjugate beam method*. This draws an analogy between these theorems and the static equations for a distributed load q^* applied between A and B. Let M_A^*, S_A^*, M_B^* and S_B^* be the local bending moment and shear associated with this loading (in terms of the *deformation conventions* discussed in §4.2). Then

$$\int_A^B q^*\,dx = S_B^* - S_A^* \quad , \quad \int_A^B q^* x\,dx = S_B^* d + M_B^* - M_A^* \qquad \textbf{(4.52)}$$

This can be compared with (4.50) and (4.51). If q^* corresponds to M/EI, then S^* corresponds to θ and M^* corresponds to $-v$. These correspondences can also be deduced from (4.15).

Real Beam			**Conjugate Beam**		
Condition	v	θ	*Condition*	M^*	S^*
Free end	?	?	Fixed end	?	?
Knife-edge end	0	?	Knife-edge end	0	?
Fixed end	0	0	Free end	0	0
Intermediate knife edge	0	=	Intermediate pin	0	=
Intermediate pin	=	?	Intermediate knife edge	=	?

Geometrical conditions on the real beam can now be reinterpreted as load conditions on a 'conjugate beam'. These are shown for workless supports in the preceding table. The symbols used in the second, third, fifth and sixth columns of the table, '0','?' and '=', indicate whether the quantity listed is zero, undetermined by the support condition alone, or equal on either side of an intermediate condition. From the table, it will be seen that if a beam is fixed at both ends, its conjugate beam has no supports at all. This means that the conjugate distributed loading q^* must be self-balancing. Such a loading is given by the M/EI diagram in Figure 4.14. The net diagram, which is the difference between the parabola and the shaded rectangle, has no net area or first moment of area. Then q^* has no net resultant force or moment.

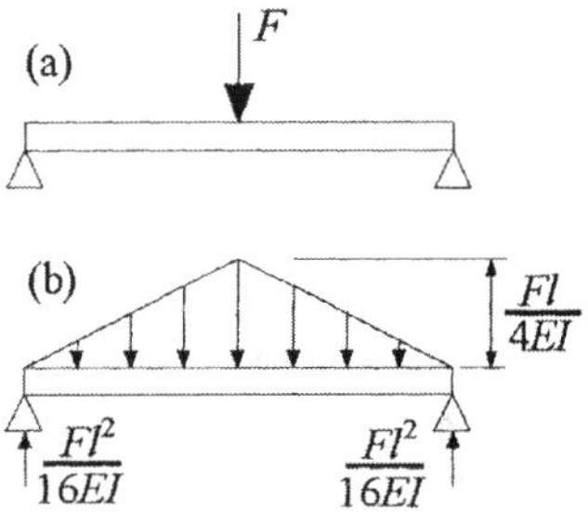

Figure 4.16 A conjugate beam.

Consider the simply-supported beam of length l shown in Figure 4.16a. This is a particular case of the beam shown in Figure 2.17a, and the central bending moment is $Fl/4$. From the above table, the conjugate beam is also simply supported and has a linearly-varying distributed load of maximum intensity $Fl/4EI$ as shown in Figure 4.16b. By resolving vertically, the conjugate reactions are then found to be $Fl^2/16EI$ as shown. The central moment on the conjugate beam is then given by this reaction multiplied by ½l, less the resultant load to the left of centre (which is of the same magnitude as the reaction) multiplied by $l/6$ (the distance from the centre to the line of action of this resultant). This conjugate moment is then $Fl^3/48EI$ and is equal to the downwards displacement ($-v$) of the centre of the real beam.

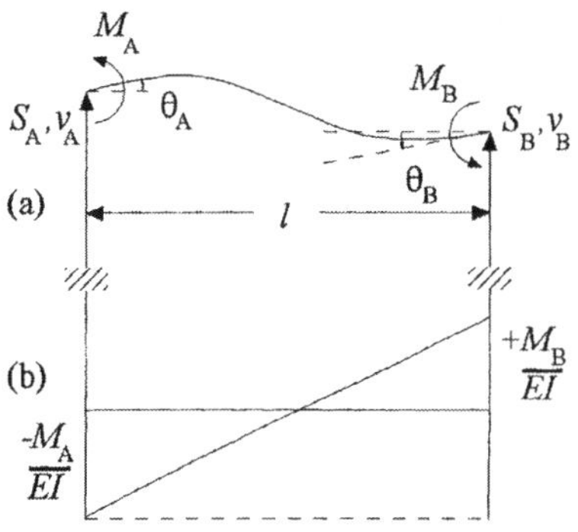

Figure 4.17 An end-loaded beam.

4.5 The Slope-Deflexion Equations

If only end moments and shear forces are applied to a beam, as shown in Figure 4.17a, the resulting M/EI diagram is as shown in Figure 4.17b.

This can be thought of as a traingular diagram which increases from zero at the left-hand end to $(M_A+M_B)/EI$ at the right-hand end, from which is subtracted a rectangular diagram of height M_A/EI. Applying (4.50) to this gives

$$\frac{1}{EI}\left(\tfrac{1}{2}l(M_A+M_B) - M_A l\right) = \theta_B - \theta_A \tag{4.53}$$

and using the second moment-area theorem given by (4.51) yields

$$\frac{1}{EI}\left(\tfrac{1}{2}l(M_A+M_B)\times\tfrac{2}{3}l - M_A l\times\tfrac{1}{2}l\right) = \theta_B l - v_B + v_A \tag{4.54}$$

These equations can be rearranged to give

$$\begin{aligned} M_A &= \frac{2EI}{l}\left(2\theta_A + \theta_B - \tfrac{3}{l}(v_B - v_A)\right) \\ M_B &= \frac{2EI}{l}\left(2\theta_B + \theta_A - \tfrac{3}{l}(v_B - v_A)\right) \end{aligned} \tag{4.55}$$

As there are no loads along the beam, the end shear forces are equal and opposite and for there to be no net moment on the beam,

$$S_A l - M_A - M_B = 0$$

so that from (4.55),

$$S_A = \frac{6EI}{l^2}\left(\theta_A + \theta_B - \tfrac{2}{l}(v_B - v_A)\right) = -S_B \tag{4.56}$$

These equations allow the end moments and shear forces to be determined from the end rotations and displacements of a beam with no loads along its length. However, these equations cannot be inverted to give the end deflexions in terms of the end loads. This is because, in the absence of any further information about the beam, an arbitrary rigid-body displacement and rotation of the beam could be added to any values satisfying the above equations. (The equations of equilibrium ensure that the end loads do no work during such motion.)

To analyse the general case where there are loads along the length of the beam, a second analysis has to be carried out to find the fixed-end reactions, M_A^f, S_A^f, M_b^f and S_b^f. Using the principle of superposition, these can then be added to components of the end reactions found in terms of the end deflexions alone to give the total end reactions. One way of determining the fixed-end reactions is to make use of the results for Case 6a of the table in §A6.1. These give the reactions for a force P at some arbitrary point along the beam. For a set of point forces and distributed loads, these results can be summed and integrated along the beam to give the total fixed-end reactions.

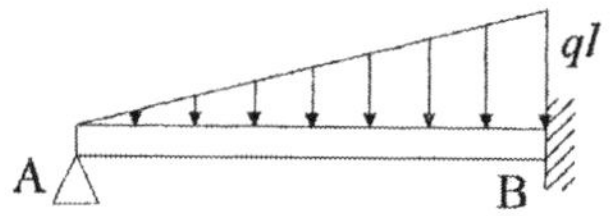

Figure 4.18 An application of slope-deflexion equations.

Consider the beam shown in Figure 4.18. The intensity of the distributed loading at any distance x from the left-hand end is qx. From the above table, the fixed-end reactions are

$$\begin{aligned} M_A^f &= \int_A^B \frac{qx.x(l-x)^2}{l^2}\,dx = \frac{ql^3}{30}, \quad M_B^f = \int_A^B -\frac{qx.x^2(l-x)}{l^2}\,dx = -\frac{ql^3}{20}, \\ S_A^f &= \int_A^B \frac{qx(l-x)^2(l-2x)}{l^3}\,dx = \frac{3ql^2}{20}, \quad S_B^f = \int_A^B \frac{qx.x(3l-2x)^2}{l^3}\,dx = \frac{7ql^2}{20}. \end{aligned} \tag{4.57}$$

The total end reactions are the sum of the above and those given by the slope-deflexion equations (4.55) and (4.56). The only end deflexion is the rotation θ_A at A. Then

$$M_A = \frac{4EI}{l}\theta_A + \frac{ql^3}{30}\ ;\ M_B = \frac{2EI}{l}\theta_A - \frac{ql^3}{20}\ ;\ S_A = \frac{6EI}{l^2}\theta_A + \frac{3ql^2}{20}\ ;\ S_B = -\frac{6EI}{l^2}\theta_A + \frac{7ql^2}{20}.$$

As end A is simply-supported, M_A is zero giving θ_A as equal to $-ql^4/120EI$ and so

$$M_B = -\frac{ql^3}{15}\ ;\ S_A = \frac{ql^2}{10}\ ;\ S_B = \frac{2ql^2}{5}.$$

The analysis of complete frameworks will be dealt with more fully in Chapter 11. The slope-deflexion equations are often used to analyse the behaviour of plane rigid-jointed frames using the simplifying assumption that the axial compression or extension of the component bars is negligible. For example, consider a square beam of side a and length l used as a cantilever to support a force at its free end. The ratio of the end axial deflexion, when the force is applied axially, to the end sideways deflexion, when the force is applied transversely, is $a^2/4l^2$. If this ratio is only 1%, a is $0.2l$ and for such short, fat beams the engineering theory of bending can be rather inaccurate anyway.

Figure 4.19 shows the frame which was partially analysed in Chapter 2 (cf. Figure 2.22). Here, the horizontal and vertical forces will both be taken to be of magnitude P. The analysis of the portal as a mechanism can again be used to find the relative displacements of the ends of the beams during sway. In terms of the sway angle ϕ, the relative displacements, as measured normal to each of the beams AB, BC and CD respectively, are

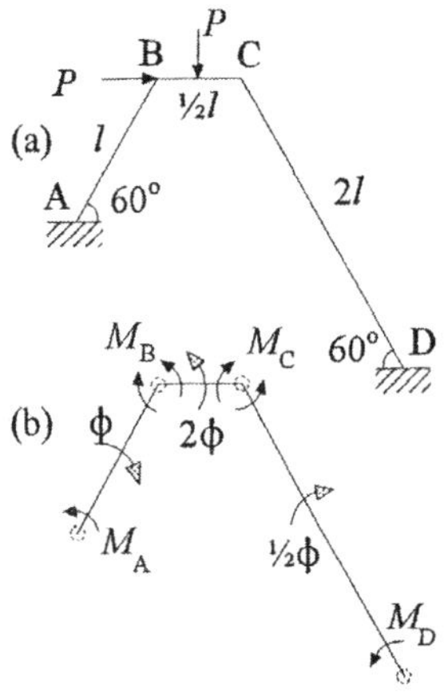

Figure 4.19 The sloping portal frame again.

$$\frac{1}{l}(v_B - v_A) = -\phi\ ,\ \frac{2}{l}(v_C - v_B) = 2\phi\ ,\ \frac{1}{2l}(v_D - v_C) = -\frac{1}{2}\phi.$$

bearing in mind that the lengths of these beams are l, $\frac{1}{2}l$ and $2l$ respectively. In terms of the notation used in Figure 4.19, equations (4.55) and (2.18) then give

$$\begin{aligned}
M_A &= \frac{2EI}{l}(\theta_B + 3\phi)\ ,\quad -M_B = \frac{2EI}{l}(2\theta_B + 3\phi)\ ,\\
M_B &= \frac{4EI}{l}(2\theta_B + \theta_C - 6\phi) + \frac{Pl}{16}\ ,\quad -M_C = \frac{4EI}{l}(2\theta_C + \theta_B - 6\phi) - \frac{Pl}{16}\ ,\\
M_C &= \frac{EI}{l}(2\theta_C + \tfrac{3}{2}\phi)\ ,\quad M_D = \frac{EI}{l}(\theta_C + \tfrac{3}{2}\phi)\ ,\\
2M_A &- 6M_B + 5M_C + M_D = \sqrt{3}Pl\ .
\end{aligned} \tag{4.58}$$

These can be reduced to three equations in θ_B, θ_C and ϕ and solved to give

$$\theta_B = 0.00736\frac{Pl^2}{EI}\ ,\quad \theta_C = 0.04431\frac{Pl^2}{EI}\ ,\quad \phi = 0.01822\frac{Pl^2}{EI}.$$

and hence the end moments found by substituting these values back in the original equations. Note that for every unknown joint rotation there is a corresponding joint moment equilibrium equation.

(Here it is that the sum of the moments exerted by a free joint on the bars connected to it is zero. This is because no external moments are exerted on the free joints.) Also, for every independent sway mode, there is an independent equation of equilibrium resulting from the corresponding virtual work equation.

4.6 Strain Energy of Bending and Axial Loading

The strain energy per unit volume (strain energy density), U_V, of an isotropic elastic body is given by (3.37). From this, the strain energy per unit length, U_l, of a beam can be found. Applications of this will be found in Chapter 7; only two simple examples will be given here. The general state of stress considered in this chapter is given by σ_{zz} as expressed by (4.10). The strain energy per unit length is then given by integrating σ_{zz} over the cross-section of the beam, or

$$U_l = \int_A \frac{\sigma_{zz}^2}{2E}\,dA = \int_A \frac{1}{2E}\left(\frac{P}{A} - \frac{M_y x}{I_{yy}} + \frac{M_x y}{I_{xx}}\right)^2 dA = \frac{P^2}{2EA} + \frac{M_y^2}{2EI_{yy}} + \frac{M_x^2}{2EI_{xx}} \qquad \textbf{(4.59)}$$

making use of the fact that the origin is at the centroid of the section and the axes are the principal axes of the section.

The strain energy stored in an elastic body is equal to the work done on it. The pin-jointed frame shown in Figure 4.20 consists of a horizontal bar BC of length l and a bar AC at 45° to the horizontal of length $\sqrt{2}l$. The vertical force F at joint C induces a tensile force $\sqrt{2}F$ in AC and a compressive force F in BC. Both bars have a cross-sectional area A and are made from a material with Young's modulus E. The strain energy in the bars is found from the first term on the right-hand side of (4.59). As the strain energy per unit length in these bars is constant along each bar, this term can be multiplied by the bar length to give the total bar strain energy. This can be equated to the work done by F in moving through the corresponding deflexion, δ, giving

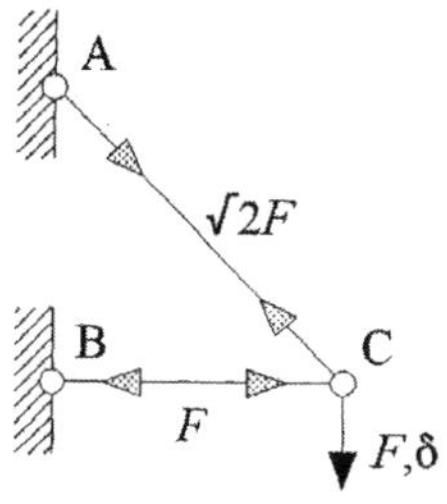

Figure 4.20 Deflexion of a simple pin-jointed frame.

$$\frac{F^2 l}{2EA} + \frac{(\sqrt{2}F)^2\sqrt{2}l}{2EA} = \frac{1}{2}F\delta \quad \text{so that} \quad \delta = \frac{Fl}{EA}(1 + 2\sqrt{2})\,. \qquad \textbf{(4.60)}$$

Figure 4.21 shows a semi-circular three-pin arch of radius r with a vertical load F applied at its apex. (Such arches were examined in §2.4.) As the ends of the two members AB and BC are pinned, the reactions at A and C are forces of magnitude $F/\sqrt{2}$ whose lines of action (shown by broken lines) pass through B. Then each member is symmetrically loaded about its midpoint and in the same way. The strain energy for the whole arch can then be found by calculating the strain energy in the upper half of AB and multiplying this by four. At a typical point in this upper half, at an angle θ from the midpoint as measured from the centre, the axial force P and bending moment M are

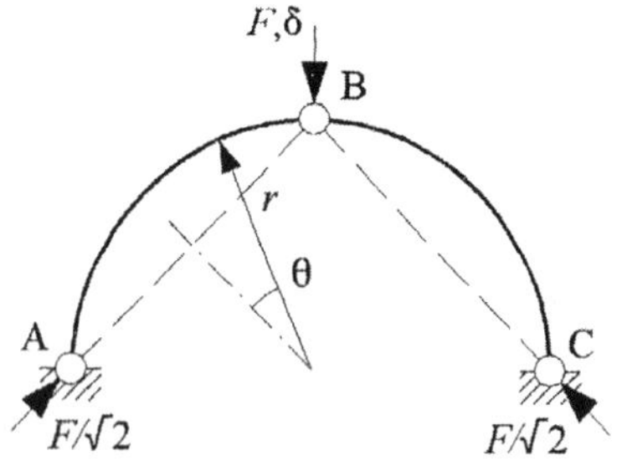

Figure 4.21 Deflexion of a three-pin arch.

$$P = \frac{F}{\sqrt{2}}\cos\theta \quad , \quad M = \frac{F}{\sqrt{2}}\left(r\cos\theta - \frac{r}{\sqrt{2}} \right) .$$

If δ is again the displacement through which F does work, then

$$\frac{1}{2}F\delta = 4\int_0^{\pi/4}\left(\frac{P^2}{2EA} + \frac{M^2}{2EI} \right) r d\theta$$

$$\text{giving} \quad \delta = \frac{Fr}{4}\left(\frac{1}{EA}(\pi + 2) + \frac{r^2}{EI}(2\pi - 6) \right) \tag{4.61}$$

Then if the arch is made from a beam with a square cross-section of side length equal to one tenth of the radius r, the component of δ resulting from axial compression is 1.5% of that resulting from flexure.

{4.7 Anisotropic Beams Subject to Axial Stresses}

{In analysing homogeneous anisotropic beams, Lekhnitiskii (1981) starts from the state of stress given by (4.10). The axes chosen are again the principal axes of the cross-section, taken through its centroid. Because the state of stress is as before, the equations of equilibrium and the conditions of zero stress on the lateral surfaces are again satisfied. The state of strain can now be found from σ_{zz} using (3.63) giving

$$\frac{\epsilon_{xx}}{s_{13}} = \frac{\epsilon_{yy}}{s_{23}} = \frac{\epsilon_{zz}}{s_{33}} = \frac{\gamma_{yz}}{s_{43}} = \frac{\gamma_{zx}}{s_{53}} = \frac{\gamma_{xy}}{s_{63}} = \sigma_{zz} = \frac{P}{A} - \frac{M_y x}{I_{yy}} + \frac{M_x y}{I_{xx}} \tag{4.62}$$

which must satisfy the compatibility conditions. This means that the strains can be deduced from some deflected form of the beam, although this form will have arbitrary rigid-body displacements associated with it. Omitting these arbitrary terms, the form associated with an axial tensile force P is

$$\begin{aligned} u &= \frac{P}{A}(s_{13}x + \tfrac{1}{2}s_{63}y) \\ v &= \frac{P}{A}(\tfrac{1}{2}s_{63}x + s_{23}y) \\ w &= \frac{P}{A}(s_{53}x + s_{43}y + s_{33}z) \end{aligned} \tag{4.63}$$

the form associated with a bending moment M_x is

$$\begin{aligned} u &= \frac{M_x}{2I_{xx}}(2s_{13}xy + s_{63}y^2 + s_{53}yz) \\ v &= \frac{M_x}{2I_{xx}}(-s_{13}x^2 - s_{53}xz + s_{23}y^2 - s_{33}z^2) \\ w &= \frac{M_x}{2I_{xx}}(s_{53}xy + s_{43}y^2 + 2s_{33}yz) \end{aligned} \tag{4.64}$$

and the form associated a bending moment M_y is

$$
\begin{aligned}
u &= -\frac{M_y}{2I_{yy}}(s_{13}x^2 - s_{23}y^2 - s_{43}yz - s_{33}z^2) \\
v &= -\frac{M_y}{2I_{yy}}(s_{63}x^2 + 2s_{23}xy + s_{43}xz) \\
w &= -\frac{M_y}{2I_{yy}}(s_{53}x^2 + s_{43}xy + 2s_{33}xz)
\end{aligned} \tag{4.65}
$$

From the expressions for w in (4.64) and (4.65), it can be seen that a plane of constant z may warp in pure flexure. Thus the usual assumption that plane sections remain plane is not in general the appropriate starting point for an engineering theory of beams. It also creates a problem in defining the absolute rotation of a cross-section. As will be seen in Chapter 6, this problem causes particular difficulties when considering the deformation of a beam under shear. The concept of 'corresponding deflexions' was introduced in Chapter 1. It can be used to define the rotation of a cross-section as being that through which the corresponding moment on it does work. Thus if M_x is the moment acting on a cross-section, the rotation of the cross-section, θ_x, is such that the work done by the stresses acting on the cross-section produced by M_x is $\frac{1}{2}M_x\theta_x$. The work done by these stresses can be determined from the component of σ_{zz} produced by M_x in (4.62) and and the expression for w in (4.64). Then on $z = 0$,

$$
\frac{1}{2}M_x\theta_{x0} = \frac{1}{2}\int_A \sigma_{zz} w \, dA = \frac{1}{2}\int_A \frac{M_x y}{I_{xx}} \frac{M_x}{2I_{xx}}(s_{53}xy + s_{43}y^2) \tag{4.66}
$$

This gives

$$
\theta_{x0} = M_x(K_{5x}s_{53} + K_{4x}s_{43}) \quad \text{where} \quad K_{5x} = \frac{1}{2I_{xx}^2}\int_A xy^2 dA \, , \; K_{4x} = \frac{1}{2I_{xx}^2}\int_A y^3 dA \, . \tag{4.67}
$$

The rotation θ_{y0} of the section at $z = 0$ related to the moment M_y can similarly be found from (4.62) and (4.65) giving

$$
\theta_{y0} = M_y(K_{5y}s_{53} + K_{4y}s_{43}) \quad \text{where} \quad K_{5y} = \frac{1}{2I_{yy}^2}\int_A x^3 dA \, , \; K_{4y} = \frac{1}{2I_{yy}^2}\int_A x^2 y \, dA \, . \tag{4.68}
$$

To simulate a fixed end at $z = 0$ when M_x is applied, the rigid-body rotation θ_{x0} must be subtracted from the solution given by (4.64). This means *adding* the term $\theta_{x0}z$ to the expression for v and subtracting $\theta_{x0}y$ from the expression for w in (4.64). Likewise, the rigid-body rotation θ_{y0} is subtracted from (4.65) to simulate such a fixed end for bending about the y axis. This means subtracting the term $\theta_{y0}z$ from the expression for u and adding the term $\theta_{y0}x$ to the expression for w in (4.65). If the section is symmetrical about the x axis, then K_{4x} and K_{4y} are zero. If the section is symmetrical about the y axis, then K_{5x} and K_{5y} are zero. Then for bisymmetrical sections, such as rectangular, circular, ellipsoidal and I-sections, these corrective terms are never necessary.

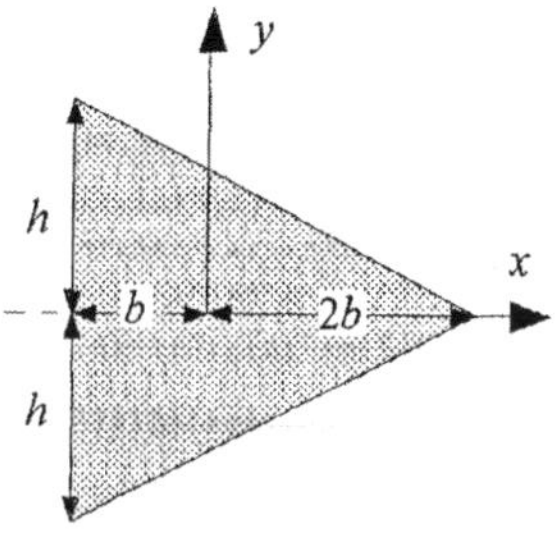

Figure 4.22 An isosceles triangular section.

An example of a section for which these terms must be calculated is shown in Figure 4.22. This is in the form of an isosceles triangle which is symmetrical about the x axis so that K_{4x} and K_{4y} are zero. The required integrations are of functions which are symmetrical in y and so can be carried out using

$$\int_A f(x,y)\,dA = 2\int_0^h \int_{-b}^{2b-3by/h} f(x,y)\,dx\,dy$$

This gives

$$I_{xx} = \frac{1}{2}bh^3\,,\; I_{yy} = \frac{3}{2}b^3h\,,\; \int_A xy^2 dA = -\frac{1}{5}b^2h^3\,,\; \int_A x^3 dA = \frac{3}{5}b^4h$$

and so

$$K_{5x} = -\frac{2}{5h^3}\quad,\quad K_{5y} = \frac{2}{15b^2h}\quad.$$

so that for an equilateral triangular section, ($h = \sqrt{3}b$), these two terms are equal and opposite.

Apart from this correction to the expression for local rotation, which is only needed in very special circumstances, the *flexural* behaviour of anisotropic beams can be analysed using the methods outlined earlier in this chapter and taking I/s_{33} as the flexural stiffness instead of EI. Likewise, the axial stiffness can be taken as A/s_{33} instead of EA.

In the simple cases of uniform bending given by (4.64) and (4.65), it is possible to infer the rates of flexural and torsional rotation induced. This is found from the general solution for a state of strain which is uniform along the axis of the beam, given in §8.2. The shear behaviour will be considered in the Chapter 6.}

Chapter 5 Torsion of Beams

5.1 Introduction

In the last chapter, the engineering theory of beams was based on a characteristic responses to a moment or axial force. These were given by states of stress and strain which do not vary along the beam. Likewise, the characteristic response to a torque does not vary along the beam. If the resultant of the loading on the end of a beam is just a torque, any other pattern of stress and strain will decay along the beam towards this unique characteristic response. However, in some cases this decay can be quite slow, and so it can be necessary to take it into account. This non-uniform torsion is examined later in this chapter.

The analysis of isotropic beams in pure torsion starts from the premise that plane cross-sections of the beam do not deform in their own planes, but rotate at a constant rate of twist ϑ along the axis of the beam. This produces constant shear strains and shear stresses acting in longitudinal planes of the beam, but no other stresses or strains. The torsional behaviour of circular beams is particularly important and relatively simple to analyse and will be dealt with first.

5.2 Isotropic Beams with Circular Sections

Figure 5.1 shows a short section of a circular beam. It has two flat faces, normal to the axis of the beam and δl apart. The point P_1 on one face is at a radius r from the centre of the beam, as is the corresponding point P_2 on the other face. After twisting, these two faces are assumed to remain flat, but the former has rotated through an angle θ and the latter through an angle $\theta+\delta\theta$ about the axis of the beam, as shown in Figure 5.1b. This means that the point P_2 moves a small distance $r\delta\theta$ further around the circle of radius r than P_1 does. The line P_1P_2 of length δl, which was originally parallel to the axis of the beam, is now at an angle γ to it, as shown in Figure 5.1b. This means that

$$\gamma\,\delta l = r\,\delta\theta \quad \text{or} \quad \gamma = r\vartheta \tag{5.1}$$

where ϑ is the rate of twist along the beam. As the end faces are assumed to remain flat and normal to the axis of the beam, γ is the angle through which the fibre connecting P_1 to P_2 shears, as shown in Figure 5.2a. From (3.26), the shear stress τ on the fibre is equal to $G\gamma$. From (5.1), all fibres at the same radius will have the same shear stress τ. The ends of these fibres form a ring of thickness δr and the resultant of the stresses τ on this ring produces a torque δT as shown in Figure 5.2b where

$$\delta T = 2\pi r\,\delta r \times \tau \times r$$

The total torque T on the far face of the short section is given by integrating all the elementary torques δT over the area of the face. If this is a solid circle of radius a,

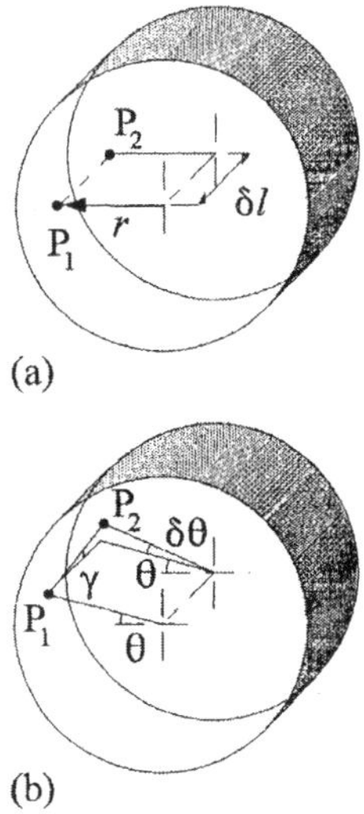

Figure 5.1 Torsion of a circular section.

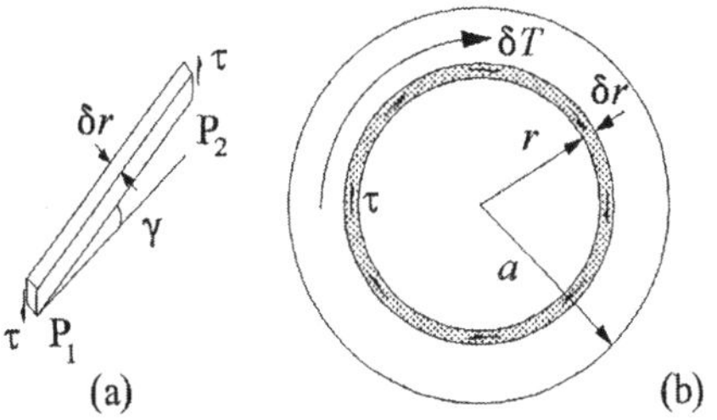

Figure 5.2 Shear stresses on a circular section.

$$T = \int_0^a 2\pi r^2 \tau \, dr = \int_0^a 2\pi r^3 G\vartheta \, dr = \frac{\pi a^4}{2} G\vartheta = GJ\vartheta \qquad \textbf{(5.2)}$$

using (3.26) and (5.1). Here J, known as the (uniform) torsional constant of the section, is also the second polar moment of area of the cross-section. For a hollow section of inner radius b and outer radius a, the above relationship becomes

$$T = \int_b^a 2\pi r^2 \tau \, dr = \int_b^a 2\pi r^3 G\vartheta \, dr = \frac{\pi}{2}(a^4 - b^4)G\vartheta = GJ\vartheta \qquad \textbf{(5.3)}$$

Again, J is the second polar moment of area of the section. If the hollow section is in the form of a thin tube of thickness t ($t \ll b$) and mean radius c, so that a is $c+\frac{1}{2}t$ and b is $c-\frac{1}{2}t$, then

$$a^4 - b^4 = 4c^3t + ct^3 , \quad \therefore \quad J \approx 2\pi c^3 t \qquad \textbf{(5.4)}$$

The above relationships can be rearranged in the form

$$\frac{T}{J} = \frac{\tau}{r} = \frac{G\gamma}{r} = G\vartheta \qquad \textbf{(5.5)}$$

It must be emphasised that these relationships apply to circular sections only.

5.3 Thin Tubes and the Approximate Analysis of Non-Circular Sections

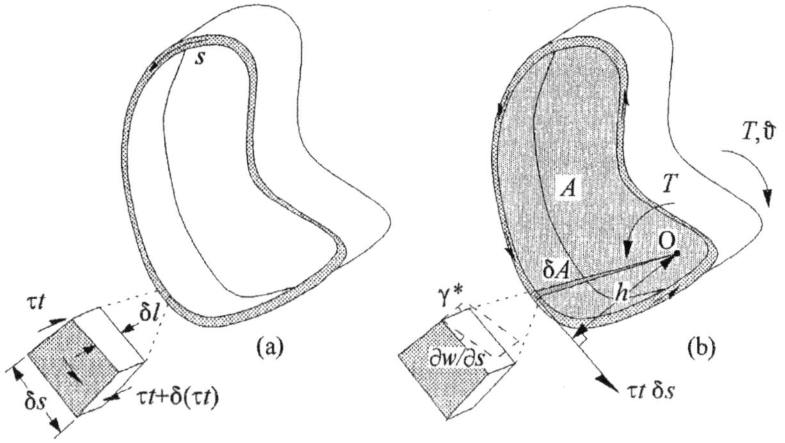

Figure 5.3 Torsion of a thin tube.

The torsion of a thin circular tube was considered above. An approximate analysis of non-circular tubes is based on assuming that the shear stress τ, induced by torsion, is uniform across the thickness, t, of the wall of the tube. This thickness need not be constant. It is no longer satisfactory to assume that plane cross-sections of the tube remain plane after torsion, because then the resulting shear stresses would not normally be in equilibrium. The out-of-plane warping of the cross-section will be denoted by a displacement w in the axial direction. Distances along the middle line of the wall will be expressed in terms of the parameter s as shown in Figure 5.3a.

Consider a small length δs of the wall of the tube which is δl long in the axial direction, as shown in the detail of Figure 5.3a. The inner and outer faces of this element are stress-free, as they are parts of the free surfaces of the beam. The element is assumed to be in pure shear under the action of the shear stress τ on the faces of the cross-section (the nearer one shown shaded) and the complementary shear stresses on the upper and lower faces of the element. As there are no other longitudinal stresses, these complementary stresses must be in balance with one another. Both the shear stress τ and the thickness t can change between the two. The change in the product of the shear stress and the thickness is denoted by $\delta(\tau t)$. The length of the faces on which these complementary shear stresses act is δl, so that equating the forces which they produce,

$$\tau t\,\delta l = [\tau t + \delta(\tau t)]\delta l \quad \text{giving} \quad \delta(\tau t) = 0\,.$$

The product τt is thus constant around the wall of the tube. This product is known as the *shear flow* and is often denoted by q. The shear stress on the 'near' face of the element produces a resultant force $\tau t \delta s$ as shown in Figure 5.3b. The line of action of this force is at a perpendicular distance h from some point O in the same plane. This force then produces a small torque about O, given by the product of this force and h. The total torque T is given by integrating all these elementary torques around the section, or

$$T = \oint \tau t \times h\,ds = \oint q \times 2dA = 2qA \tag{5.6}$$

because q is constant and the area of the elementary triangle δA is the product of half the base length, δs, and its height, h.

Suppose that O lies on the axis about which the beam twists, and that the rate of twist is again ϑ. The far end of the element will then move clockwise through a distance $h\vartheta\delta l$ in the direction of the middle line relative to the near end. By the same arguments used in the previous section, this means that the shear angle γ^* shown in Figure 5.3b is equal to $h\vartheta$. However, this is no longer the actual shear strain, because the other side of the corner of the element rotates in the same sense through an angle $\partial w/\partial s$ The distortion of the beam is proportional to the rate of twist ϑ, so that w can be written as $\vartheta\omega$, where ω is a warping function of position on the cross-section. The true shear strain in the plane of the wall of the tube is then given by the difference of these two rotations, or

$$\frac{\tau}{G} = \gamma = \gamma^* - \frac{\partial w}{\partial s} = \vartheta\left(h - \frac{\partial \omega}{\partial s}\right) \tag{5.7}$$

using (3.26) again. This can be integrated around the wall of the section to give

$$\frac{q}{G}\oint \frac{1}{t}\,ds = \frac{1}{G}\oint \tau\,ds = \oint \vartheta\left(h - \frac{\partial \omega}{\partial s}\right)ds = \vartheta \oint h\,ds = 2A\vartheta \tag{5.8}$$

(cf. (5.6)). Note that the second term of the integrand disappears because on integration this would give the difference between the value of ω at the end of the path and its value at the start of the path. As these are at the same point, the difference is zero. Then from (5.6) and (5.8),

$$T = 2qA = G\left(\frac{4A^2}{\oint \frac{1}{t}ds}\right)\vartheta = GJ\vartheta \tag{5.9}$$

where the torsional constant J is given by the contents of the brackets. If the wall thickness t is constant around the perimeter, then the line integral becomes p/t where p is the perimeter length. For example, for a thin circular tube of mean radius c,

$$A = \pi c^2 \ , \quad p = 2\pi c \quad \therefore \quad T = 2\pi c^2 t\tau = G(2\pi c^3 t)\vartheta = GJ\vartheta\ . \tag{5.10}$$

This is the same result as that given by the approximate analysis of a thin circular tube in the previous section (cf. 5.4)).

Multicell tubes can be analysed in a similar fashion. Figure 5.4 shows a tube with a thin interior wall which divides the tube into two cells. The detail at the bottom of this figure shows the junction of this interior wall with the outer wall at P. This is of length δl and has longitudinal shear stresses acting on the cut lengths of wall. These produce the forces shown by the product of the shear flows and δl. Then for longitudinal equilibrium the shear flows must balance, or

$$q_1 = q_2 + q_3 \tag{5.11}$$

(eliminating the common factor δl). The same equation holds at the junction R. In terms of the complementary shear flows on the face of the cross-section, it can be seen that the sum of the shear flows into a junction is matched by the sum of the shear flows out of it. The torque T produced by the shear stresses can be found by the same method used to derive (5.6) giving

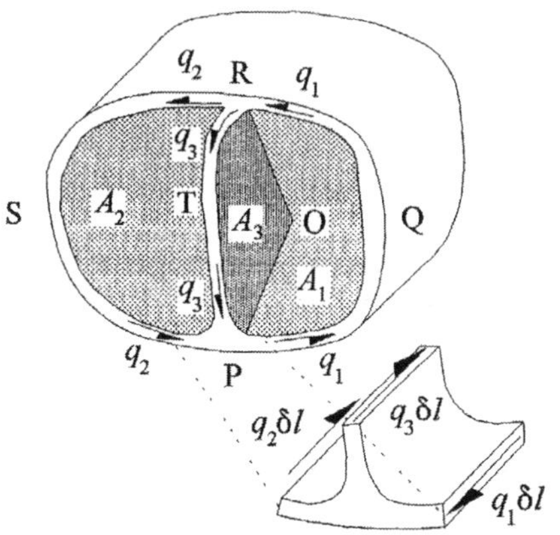

Figure 5.4 Equilibrium of a multicell tube.

$$\begin{aligned} T &= q_1 \int_{PQR} 2dA + q_2 \int_{RSP} 2dA + q_3 \int_{RTP} 2dA \\ &= 2[q_1A_1 + q_2(A_2+A_3) + q_3A_3] = 2[q_1(A_1+A_3) + q_2A_2] \end{aligned} \tag{5.12}$$

making use of (5.11) to give the final equality. The areas A_1 to A_3 are shown shaded in Figure 5.4. Following the same process that gave rise to (5.8), the shear stress can be integrated around any closed loop formed by the walls of the section to give an expression for the rate of twist. (This is sometimes referred to as the *shear circulation theorem.*) In the present case,

$$\frac{1}{G}\oint_{RSPTR} \frac{q}{t} ds = 2A_2\vartheta \quad , \quad \frac{1}{G}\oint_{RTPQR} \frac{q}{t} ds = 2(A_1+A_3)\vartheta \tag{5.13}$$

Note that (A_1+A_3), and hence (5.12) and (5.13), are independent of the choice of the pole O.

Equations of the above type are sufficient to analyse multicell tubes. The example shown in Figure 5.5 has a constant wall thickness t. From (5.12),

$$T = 2[q_1(2c^2) + q_2c^2]$$

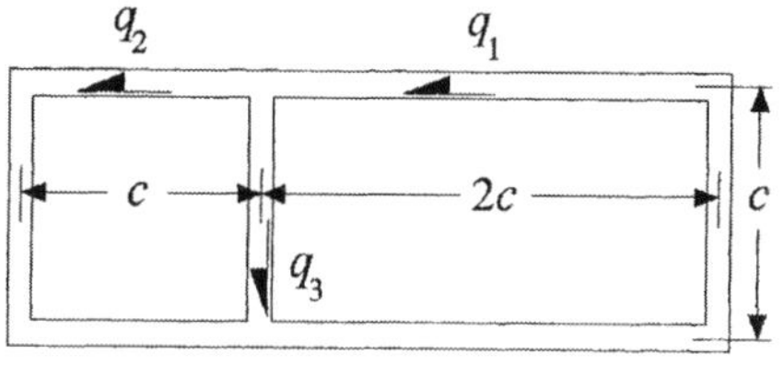

Figure 5.5 An example of a multicell tube.

and from (5.13),

$$\frac{1}{G}\left(\frac{q_2}{t}(3c) - \frac{q_3}{t}c\right) = 2c^2\vartheta \ ,$$

$$\frac{1}{G}\left(\frac{q_3}{t}c + \frac{q_1}{t}(5c)\right) = 2(2c^2)\vartheta \ .$$

Together with (5.11), this gives

$$q_1 = \frac{9T}{26c^2} \ , \quad q_2 = \frac{8T}{26c^2} \ , \quad q_3 = \frac{T}{26c^2} \ , \quad T = G(4\tfrac{12}{23}c^3t)\vartheta \ . \tag{5.14}$$

If the internal wall is omitted, the earlier analysis for a simple tube give the value of the torsional constant J as $4\frac{1}{2}c^3t$, so that the inclusion of this wall does not increase the torsional stiffness of the tube significantly.

The analysis of the uniform torsion of solid non-circular sections also starts from the assumption that only shear stresses act on the plane of the cross-section. The resultant shear stress at the boundary of the cross-section must be parallel to it. (If the resultant had a component normal to the boundary, there would be a complementary shear stress acting longitudinally on the free surface of the beam.) Thus the outermost layer of the section can be thought of as a thin tube with a constant shear flow around it during torsion. Within this lies another tube for which the same will be true, and so on. The whole cross-section then resembles a sliced onion, the local resultant shear stress at any point being tangential to the path of the local layer. These paths are known as *shear stress trajectories*.

Figure 5.6 shows a rectangular section in which the shear stress trajectories are assumed to follow rectangular paths. The section then behaves like a series of rectangular tubes, one inside the other. Each tube will be taken to have a uniform wall thickness all the way round. A typical tube is shown shaded. This has a breadth of $2x$, a height of $2x+b-a$ and a wall thickness of δx (taking b to be greater than a as shown). If the shear flow in this tube produces a torque δT, then from (5.9)

$$\delta T = G\left(\frac{4[2x(2x+b-a)]^2}{8x+2b-2a}\,\delta x\right)\vartheta$$

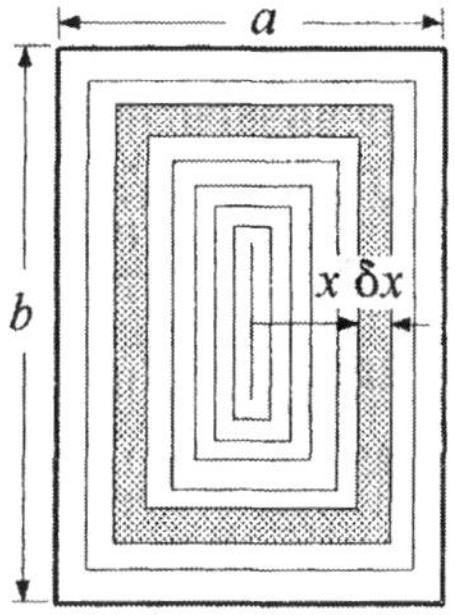

Figure 5.6 Torsion of a rectangular section.

Integrating this for all such elementary tubes (from $x=0$ to $x=\frac{1}{2}a$) gives the total torque as

$$T = G\frac{1}{32}\left[8a^2b^2 - 2ab(a^2+b^2) + (b-a)^4\ln\left(\frac{b+a}{b-a}\right)\right]\vartheta$$

For a square section of side a, this gives the torsional constant as

$$J = 0.125a^4 \quad (\text{versus the correct result of } J = 0.1406a^4).$$

For a thin strip of thickness t and depth b, the above expression gives

$$J = \frac{1}{3}bt^3\{-\frac{1}{3}t^4...\} \ (\text{versus the correct result of } J = \frac{1}{3}bt^3 - 0.21t^4...). \qquad \textbf{(5.15)}$$

(It is necessary to expand the logarithmic expression to obtain this result.) The exact expressions will be found in Timoshenko and Goodier (1970) for example. They are obtained by using Saint-Venant's semi-inverse method, which is the subject of the next section. Many common beams have open thin-walled sections (that is, the walls of the section do not join up to form closed loops). The value for J used for such a section is taken to be the sum of the approximate J values ($\frac{1}{3}bt^3$) for each strip of constant thickness composing the section.

5.4 Saint-Venant Torsion

In §4.1 a pattern of stress was assumed, and the semi-inverse method used to show that it met the necessary criteria for a theory of the bending of beams. Here, the method will be used again, only starting from a pattern of deformation. As with the theory of thin tubes, it will be assumed that the cross-section of a beam undergoes rigid-body rotation during torsion, but can warp out of its plane. This warping will be taken as proportional to the rate of twist, ϑ. This rate of twist is assumed to be constant, and the axial stress, σ_{zz}, taken as zero. Cases where these assumptions do not hold will be examined later.

Figure 5.7 shows the cross-section of a beam which has a rigid-body rotation ϑz about the point O. From the above, the displacements of a point on the cross-section with coordinates (x,y) are

$$u = -\vartheta yz \quad , \quad v = \vartheta xz \quad , \quad w = \vartheta g(x,y) \ . \tag{5.16}$$

(Note that the z axis is out of the paper and that ϑz is clockwise about it.) The shear stresses on the cross-section are

$$\begin{aligned} \tau_{zx} &= G\gamma_{zx} = G\left(\frac{\partial w}{\partial x} + \frac{\partial u}{\partial z}\right) = G\vartheta\left(\frac{\partial g}{\partial x} - y\right) \\ \tau_{zy} &= G\gamma_{zy} = G\left(\frac{\partial w}{\partial y} + \frac{\partial v}{\partial z}\right) = G\vartheta\left(\frac{\partial g}{\partial y} + x\right) \end{aligned} \tag{5.17}$$

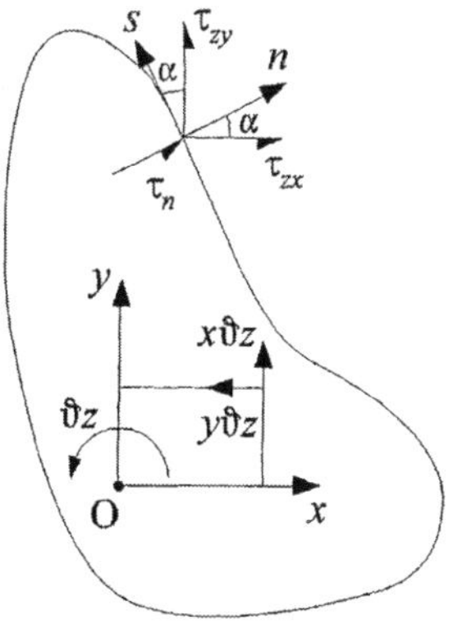

Figure 5.7 Torsion of a section.

In the absence of body forces and of any longitudinal stress σ_{zz}, (3.5) becomes

$$\frac{\partial \tau_{zx}}{\partial x} + \frac{\partial \tau_{zy}}{\partial y} = 0 \tag{5.18}$$

so that from (5.17),

$$G\vartheta\left(\frac{\partial^2 g}{\partial x^2} + \frac{\partial^2 g}{\partial y^2}\right) = 0 \quad \text{or} \quad \nabla^2 g = 0 \tag{5.19}$$

as $G\vartheta$ is not zero. At the boundary of the cross-section, the shear stress τ_n normal to the boundary must be zero. This is because there would otherwise be a complementary shear stress τ_n on the free surface of the beam. This shear stress can be found by simple resolution of the shear stresses τ_{zx} and τ_{zy} which act on the same surface. Then from Figure 5.7,

$$\tau_n = \tau_{zx}\cos\alpha + \tau_{zy}\sin\alpha = 0 \tag{5.20}$$

If s is the parameter of position around the boundary, measured anticlockwise from some point on it and n is the distance outwards from the boundary along the local normal, then

$$\cos\alpha = \frac{dy}{ds} = \frac{dx}{dn} \quad , \quad \sin\alpha = -\frac{dx}{ds} = \frac{dy}{dn} \ . \tag{5.21}$$

Then from (5.20) and (5.21),

$$\tau_n = -\tau_{zy}\frac{dx}{ds} + \tau_{zx}\frac{dy}{ds} \tag{5.22}$$

The shear stresses are now defined in terms of the gradients of a function ϕ known as *Prandtl's stress function*:

$$\frac{\partial \phi}{\partial x} = -\tau_{zy} \quad , \quad \frac{\partial \phi}{\partial y} = \tau_{zx} \ . \tag{5.23}$$

This definition satisfies (5.18) immediately, and from (5.17),

$$\nabla^2\phi = \frac{\partial^2 \phi}{\partial x^2} + \frac{\partial^2 \phi}{\partial y^2} = -\frac{\partial \tau_{zy}}{\partial x} + \frac{\partial \tau_{zx}}{\partial y} = -2G\vartheta \tag{5.24}$$

Using (5.23) and the first relationships for cosα and sinα in (5.21), the boundary condition given by (5.20) becomes

$$\tau_n = \frac{\partial \phi}{\partial x}\frac{dx}{ds} + \frac{\partial \phi}{\partial y}\frac{dy}{ds} = \frac{d\phi}{ds} \tag{5.25}$$

so that lines on which τ_n is zero, such as the boundary of the section, are lines of constant ϕ (i.e. ϕ contours). These lines are such that the resultant shear stress is tangential to the line. That is to say, they are *shear stress trajectories*. Changing ϕ by a constant value does not affect the relationships given by (5.23). Then for a simply-connected[1] cross-section, ϕ can be taken as zero on the boundary without loss of generality. Similarly, the shear stress parallel to the boundary is

$$\tau_s = \tau_{zy}\cos\alpha - \tau_{zx}\sin\alpha = -\frac{\partial \phi}{\partial x}\frac{dx}{dn} - \frac{\partial \phi}{\partial y}\frac{dy}{dn} = -\frac{d\phi}{dn} \tag{5.26}$$

More generally, the magnitude of the shear stress along any shear stress trajectory is equal to the gradient of ϕ in the direction normal to it. The shear flow (the product of shear stress and width) between two adjacent ϕ contours is then constant. Thus these contours can be thought of as dividing the section into a set of thin rings, in the manner used in §5.3.

From Figure 5.8, the total torque on the section is

$$\begin{aligned} T &= \int_A (\tau_{zy}x - \tau_{zx}y)\,dA \\ &= \iint -\left(\frac{\partial \phi}{\partial x}x + \frac{\partial \phi}{\partial y}y\right) dx\,dy \end{aligned} \tag{5.27}$$

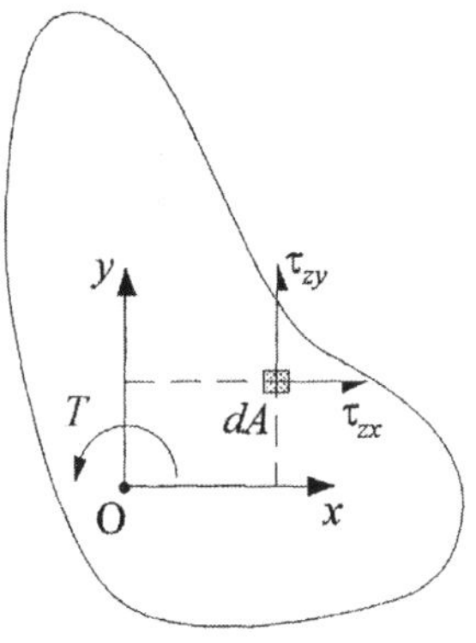

Figure 5.8 Torque produced by shear on a section.

from (5.23). On integrating by parts,

$$\begin{aligned} T &= \int\left(-\left|\phi x\right| + \int \phi\, dx\right)dy + \int\left(-\left|\phi y\right| + \int \phi\, dy\right)dx \\ &= 2\int\int \phi\, dx\,dy \end{aligned} \tag{5.28}$$

(The first terms of the bracketed integrands vanish because ϕ is zero on the boundary of the section.) Thus if the ordinates of ϕ are plotted normal to the section, the torque is given by twice the volume underneath the ϕ surface.

The solution of the torsion problem for a particular cross-section then requires a function ϕ to be found which satisfies (5.24) and is zero on the boundary of the cross-section. The shear stresses associated with a particular rate of twist ϑ are then given by (5.23) and the torque by (5.28). For example, for a solid elliptic section with major and minor semi-axes of lengths a and b aligned with the x and y axes respectively,

$$\phi = -\frac{a^2 b^2 G\vartheta}{a^2 + b^2}\left(\frac{x^2}{a^2} + \frac{y^2}{b^2} - 1\right) \tag{5.29}$$

from which it follows that the shear stresses and the torque are given by

---

[1]A simply-connected body is such that any path in the body between two points can be deformed into any other path in the body linking these two points without crossing any boundary of the body. What is important here is that the boundary of a simply-connected section is described by a single (closed) continuous line.

$$\tau_{zy} = \frac{2b^2 x G\vartheta}{a^2+b^2} \ , \quad \tau_{zx} = -\frac{2a^2 y G\vartheta}{a^2+b^2} \ , \quad T = \left(\frac{\pi a^3 b^3}{a^2+b^2}\right) G\vartheta \ . \tag{5.30}$$

where the bracketed term in the last expression is the torsional constant J. In the limiting case when the ellipse becomes a circle ($b=a$), this result is the same as that given in §5.2. However, some differences from the earlier analysis should be noted. The torsional constant J is not in general the second polar moment of area. Also, the greatest shear stress is not at the furthest distance from the centroid of the section but at the end of the semi-minor axis and is given by

$$\tau_{max} = \frac{2a^2 b G\vartheta}{a^2+b^2} = \frac{2T}{\pi a b^2} \tag{5.31}$$

{Polar coordinates

The above analysis can be expressed in terms of cylindrical polar coordinates (cf. §3.10). If the relationship between the polar and cartesian coordinate systems is as shown in Figure 5.9, then the increments in position δx and δy are related to the increments δr and $\delta\theta$ shown in Figure 5.9b by

$$\begin{aligned}\delta x &= \cos\theta\,\delta r - r\sin\theta\,\delta\theta \\ &= \frac{\partial x}{\partial r}\delta r + \frac{\partial x}{\partial \theta}\delta\theta \\ \delta y &= \sin\theta\,\delta r + r\cos\theta\,\delta\theta \\ &= \frac{\partial y}{\partial r}\delta r + \frac{\partial y}{\partial \theta}\delta\theta\end{aligned} \tag{5.32}$$

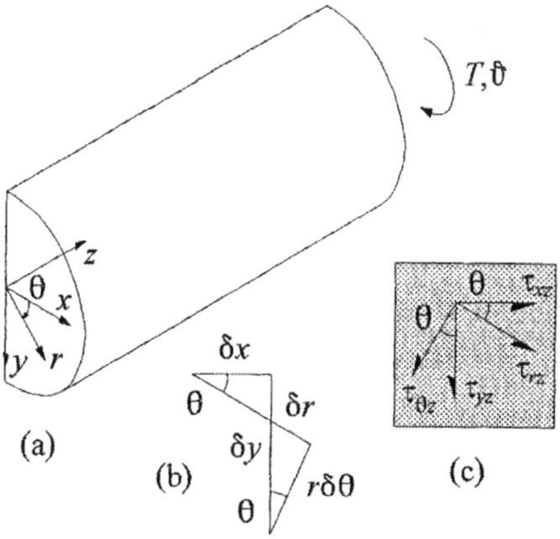

Figure 5.9 Torsion of a semicircular section.

The shear stresses acting on the positive z face shown in Figure 5.9c can be resolved like ordinary vectors to give

$$\begin{aligned}\tau_{rz} &= \tau_{xz}\cos\theta + \tau_{yz}\sin\theta = \frac{\partial\phi}{\partial y}\frac{1}{r}\frac{\partial y}{\partial\theta} + \frac{\partial\phi}{\partial x}\frac{1}{r}\frac{\partial x}{\partial\theta} = \frac{1}{r}\frac{\partial\phi}{\partial\theta} \\ \tau_{\theta z} &= -\tau_{xz}\sin\theta + \tau_{yz}\cos\theta = -\frac{\partial\phi}{\partial y}\frac{\partial y}{\partial r} - \frac{\partial\phi}{\partial x}\frac{\partial x}{\partial r} = -\frac{\partial\phi}{\partial r}\end{aligned} \tag{5.33}$$

using (5.23) where the harmonic equation for ϕ given by (5.24) becomes

$$\nabla^2\phi = \frac{1}{r^2}\left[r\frac{\partial}{\partial r}\left(r\frac{\partial\phi}{\partial r}\right) + \frac{\partial^2\phi}{\partial\theta^2}\right] = -2G\vartheta \tag{5.34}$$

where ϕ is zero on the boundary of the section. For the semicircular section of radius a shown in Figure 5.9,

$$\phi = G\vartheta\left[-r^2\cos^2\theta + \frac{8a^2}{\pi}\sum_{m=0}^{\infty}(-1)^{m+1}\left(\frac{r}{a}\right)^{2m+1}\frac{\cos(2m+1)\theta}{(2m+1)[(2m+1)^2-4]}\right] \tag{5.35}$$

giving

$$\left.\begin{aligned}\gamma_{rz} &= \frac{1}{G}\tau_{rz} = \vartheta a\left[\frac{r}{a}\sin 2\theta + \frac{8}{\pi}\sum_{m=0}^{\infty}(-1)^m\left(\frac{r}{a}\right)^{2m}\frac{\sin(2m+1)\theta}{(2m+1)[(2m+1)^2-4]}\right]\\ \gamma_{\theta z} &= \frac{1}{G}\tau_{\theta z} = \vartheta a\left[2\frac{r}{a}\cos^2\theta + \frac{8}{\pi}\sum_{m=0}^{\infty}(-1)^m\left(\frac{r}{a}\right)^{2m}\frac{\cos(2m+1)\theta}{(2m+1)[(2m+1)^2-4]}\right]\end{aligned}\right\} \quad \textbf{(5.36)}$$

The contours of ϕ given by (5.35) are the shear stress trajectories shown in Figure 5.10. (The sense of the resultant shear stress shown by the arrows is for a positive z face.) The maximum shear stress, τ_{max} occurs at the origin of the coordinates. This and the torsional stiffness are given by

$$\tau_{max} = 0.849G\vartheta \quad , \quad GJ = 0.298Ga^4. \qquad \textbf{(5.37)}\}$$

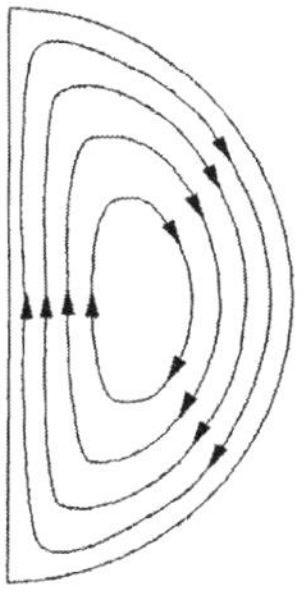

Figure 5.10 Shear stress trajectories for a semi-circular section.

5.5 The Membrane Analogy

An insight into the form of the shear stress distribution during torsion can be obtained from the *membrane analogy*. A thin membrane, or soap film, is imagined to cover a slot in the shape of the cross-section. This has a uniform tension per unit length S in it and a uniform pressure p is applied under the membrane, as shown in Figure 5.11. The membrane deflects a small amount w in the z direction, normal to the membrane. Consider the equilibrium of a small portion of the membrane, δx long by δy broad. Figure 5.11b shows this element as viewed in the x-z plane. The tension on the edge of length δy has a resultant $S\delta y$ which has a component in the z direction which is given by multiplying this resultant by the slope of the membrane in the x direction. Owing to the curvature of the membrane, this component changes slightly between the edges of the element. There are similar vertical components of the tensile forces $S\delta x$ on the other pair of edges. The pressure p produces a resultant force $p\delta x\delta y$ on the element. Then for equilibrium in the z direction,

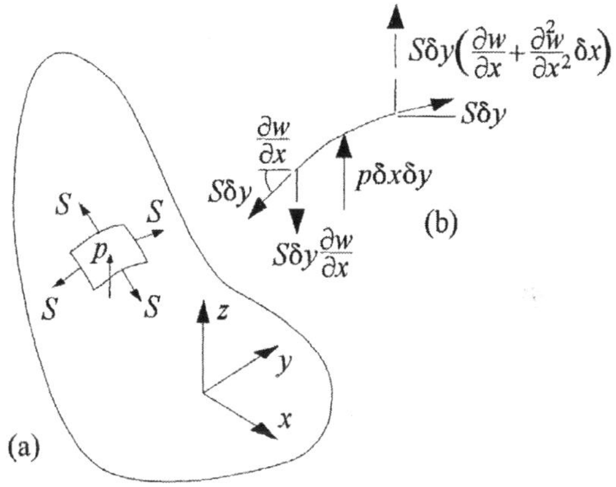

Figure 5.11 Membrane analogy for torsion.

$$S\left(\frac{\partial^2 w}{\partial x^2}\delta x\right)\delta y + S\left(\frac{\partial^2 w}{\partial y^2}\delta y\right)\delta x + p\delta x\delta y = 0 \quad \text{or} \quad \nabla^2 w = -\frac{p}{S} \qquad \textbf{(5.38)}$$

The displacement w is zero around the boundary of the section, just as ϕ is. Comparing (5.38) with (5.24), if w is analogous to ϕ then p/S is analogous to $2G\vartheta$. Also, from (5.28), the volume enclosed between the deformed membrane and the plane of the slot is analogous to half the applied torque. Experimental apparatus based on this analogy was used in the past to determine the torsional properties of different cross-sections, but this has been superseded by computer methods. However, based on the common everyday experience of soap films, it is possible to draw useful conclusions about the nature of the shear stress distribution during torsion. Figure 5.12 shows an imaginary equilateral triangular slot which is covered with a soap film. This film bulges under lateral pressure, and the contours of the deformed shape are sketched in the figure. This sketch shows that the contours are most tightly packed half way along the sides of the triangle. This must be where the slope of the film is at its greatest, and hence where the shear stress during torsion of a cross-section with the same profile as the slot is greatest. Equation (5.24) and the condition that ϕ is zero on the boundary are satisfied by

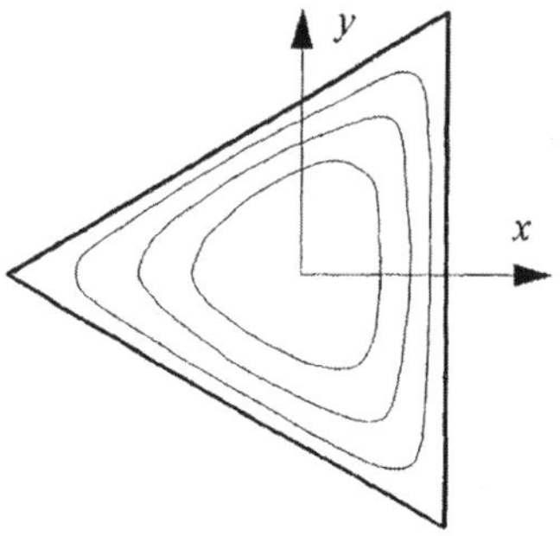

Figure 5.12 Membrane analogy applied to a triangular section.

$$\phi = -\frac{G\vartheta}{54a}(3x - a)(3x - 3\sqrt{3}y + 2a)(3x + 3\sqrt{3}y + 2a) \qquad \textbf{(5.39)}$$

where a is the height of the triangle from base to apex. The first bracket is zero on the right-hand edge of the triangle ($x = \frac{1}{3}a$), the second bracket is zero on the upper edge and the third bracket is zero on the lower edge. The shear stress τ_{zy} parallel to the boundary ($x=\frac{1}{3}a$) is given by (5.23) as $-\partial\phi/\partial x$. At its midpoint ($y=0$) from (5.39),

$$\tau = -\frac{\partial\phi}{\partial x} = \frac{G\vartheta}{54a}(81x^2 - 81y^2 + 54xa) = \frac{1}{2}G\vartheta a \quad (x=\tfrac{1}{3}a\ ,\ y=0) \qquad \textbf{(5.40)}$$

and this is the maximum shear stress.

The membrane analogy can be used to justify the value of J used for thin-walled open sections, mentioned at the end of §5.3. For a given pressure, the total volume under the analogous soap film will be proportional to J. For a set of strips, this volume will be almost the same, whether the strips are joined together or not (unless the strips form a closed loop). Hence the total torsional stiffness of an open thin-walled section can be taken as the sum of the stiffnesses of each strip. If the strips form a closed loop, the corresponding slots isolate an area of plate which is then free to lift up under pressure. The total volume then has to include the volume under this isolated area. The torsional stiffness then increases dramatically with the closure of a loop of thin-walled strips and the thin-walled tube analysis has to be used. The analogous soap film also does not change greatly if the strip is slightly curved. Thus (5.15) is also used to give the value of J for slightly curved strips.

Using the membrane analogy, it is possible to see that high shear stresses are likely to occur at reentrant (or concave) parts of the section profile. This is where the contours on the corresponding deformed membrane tend to 'pile up', producing a high gradient. Thus the worst torsional shear stress in a thin-walled open section usually occurs where the flange meets the web. A generalisation of such a junction is shown in Figure 5.13. The fillet radius is r and the tangent to the fillet rotates through an angle ψ in going around it. The diameter of the greatest inscribed circle which touches the fillet is D and the total area of the cross-section is A. The maximum shear stress induced is then given by

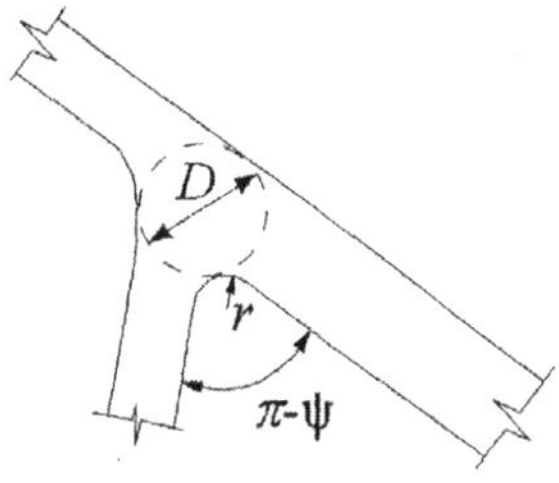

Figure 5.13 Stress concentration.

$$\tau = G\vartheta c \quad \text{where} \quad c = \frac{D}{1 + \dfrac{\pi^2 D^4}{16A^2}}\left\{1 + \left[0.118 \ln\left(1 + \frac{D}{2r}\right) + 0.238\frac{D}{2r}\right]\tanh\frac{2\psi}{\pi}\right\} \tag{5.41}$$

This formula and others for the torsional properties of sections will be found in Table 20 of Young (1989). See also Appendix 4 and the program TORCONS.EXE listed in Appendix 3.

5.6 Strain Energy of Torsion

In §4.6, the strain energy per unit length, U_l, produced by the axial stresses associated with axial forces and bending moments was found by integrating the strain energy per unit volume over the cross-sectional area of the beam. Here, using (3.37) and (5.23), it is given by

$$U_l = \frac{1}{2G}\int_A (\tau_{xz}^2 + \tau_{yz}^2)\,dA = \frac{1}{2G}\int_A \left(\frac{\partial\phi}{\partial x}\right)^2 + \left(\frac{\partial\phi}{\partial y}\right)^2 dA \tag{5.42}$$

where A is the area of the cross-section. From Stokes' theorem,

$$\iint_A \left[\frac{\partial}{\partial x}\left(\phi\frac{\partial\phi}{\partial x}\right) + \frac{\partial}{\partial y}\left(\phi\frac{\partial\phi}{\partial y}\right)\right]dA = \int_C \phi\left[-\frac{\partial\phi}{\partial y}dx + \frac{\partial\phi}{\partial x}dy\right] = 0 \tag{5.43}$$

where the second integral is around the contour of the section and is zero because ϕ is zero on the boundary. It then follows from (5.24) and (5.28) that

$$\int_A \left(\frac{\partial\phi}{\partial x}\right)^2 + \left(\frac{\partial\phi}{\partial y}\right)^2 dA = -\int_A \phi\left(\frac{\partial^2\phi}{\partial x^2} + \frac{\partial^2\phi}{\partial y^2}\right)dA = -\int_A \phi(-2G\vartheta)\,dA = TG\vartheta \tag{5.44}$$

so that from (5.42),

$$U_l = \frac{1}{2G}TG\vartheta = \frac{1}{2}GJ\vartheta^2 = \frac{T^2}{2GJ} \quad \text{where } T = GJ\vartheta \tag{5.45}$$

as is also the case in §4.6, this result could also be obtained by equating the strain energy per unit length to the work done on that length by the resultant loading.

Strain-energy methods can be used to find the deflexion of an open-coil helical spring, such as the one shown in Figure 5.14. In this example, the spring is subject either to an axial tensile force P or to a torque T. The coil will be taken to have a mean radius R, as seen in plan. The pitch angle of the coil to a plane normal to its axis is α, as shown in the detail of Figure 5.14. The axial force P produces a moment PR on every section of the coil which can be

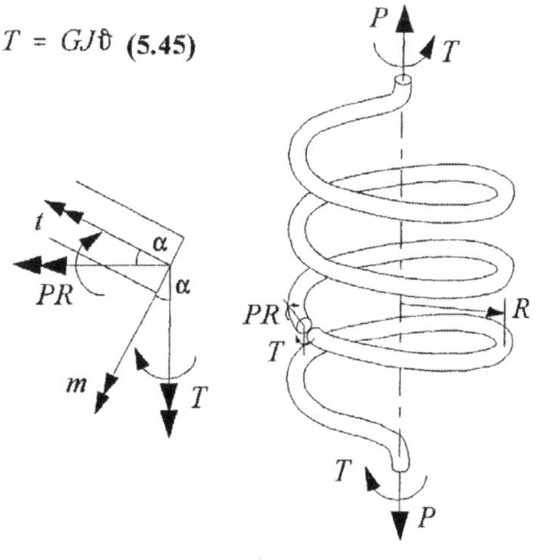

Figure 5.14 An open-coil spring.

resolved into a local bending moment m and a local torque t, where

$$m = PR\sin\alpha \quad , \quad t = PR\cos\alpha \tag{5.46}$$

Likewise, the torque T can be resolved to give

$$m = T\cos\alpha \quad , \quad t = -T\sin\alpha \tag{5.47}$$

The coil will be taken to have n complete turns. This means that the total length of the coil wire is $2\pi Rn\sec\alpha$. The total strain energy of its bending and torsion is then

$$U = 2\pi Rn\sec\alpha\left(\frac{m^2}{2EI} + \frac{t^2}{2GJ}\right) \tag{5.48}$$

using (4.59) and (5.45). This strain energy can be equated either to the work done ($\frac{1}{2}P\delta$) by the axial force in extending the spring by an amount δ, or to the work done ($\frac{1}{2}T\theta$) by the torque T in twisting the ends of the spring through a relative rotation θ. This gives

$$\delta = 2\pi PR^3 n\sec\alpha\left(\frac{\sin^2\alpha}{EI} + \frac{\cos^2\alpha}{GJ}\right) \quad , \quad \theta = 2\pi TR\sec\alpha\left(\frac{\cos^2\alpha}{EI} + \frac{\sin^2\alpha}{GJ}\right) . \tag{5.49}$$

The axial force P also produces a twist of the spring and the torque T a relative axial displacement of its ends. How these are found will be discussed in Chapter 7.

5.7 Non-Prismatic Bars and Other Exceptional Cases

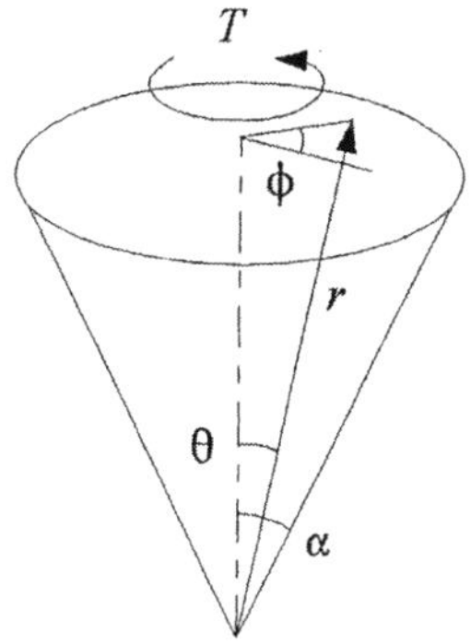

Figure 5.15 Torsion of a cone.

The theory of torsion outlined above is often applied to cases where, strictly speaking, it does not apply. For example, if the cross-section of the bar varies along its length, it is taken that the relationship between torque and twist still applies locally, using the local torsional stiffness. The general form of the solution for the torsion of bars with a circular cross-section of varying diameter is due to Michell[1]. The solution for a conical bar was found by Föppl[2] (see also Timoshenko and Goodier (1970) §119). This solution can be used as an indication of the accuracy of the above approximate analysis. It is best expressed in spherical polar coordinates (see Appendix 7). The cone has a half-angle α as shown in Figure 5.15. It is taken to consist of spherical surfaces of constant radius r as measured from the tip of the cone. These surfaces are taken to remain undistorted during torsion by a torque T, but to rotate with respect to one another about the axis of the cone. This gives only a tangential displacement w of the cone material which is taken to be

$$w = \frac{C}{r^2}\sin\theta, \quad \text{so that (A7.1.1)} \quad \gamma_{r\phi} = -\frac{3C}{r^3}\sin\theta . \tag{5.50}$$

No other strains are induced, and so the only stress, $\tau_{r\phi}$, is G times this shear strain. This shear stress, acting on a small area δA of the spherical surface produces a torque of $\tau_{r\phi}\delta A.r\sin\theta$ about

[1] Michell, J.H. (1899) *Proc. London Math. Soc.* V.31 p.141.

[2] Föppl, A. (1905) *Sitzungsberichte Bayer. Akademie Wissenschaften* V.35 p.249.

the axis of the cone. The total torque T is then given by

$$T = \int_A \tau_{r\phi} r \sin\theta \, dA = \int_0^\alpha -\frac{3GC}{r^3} \sin\theta \, r\sin\theta \, 2\pi r \sin\theta \, r d\theta$$
$$= 2\pi GC(3\cos\alpha - \cos^3\alpha - 2) \quad \textbf{(5.51)}$$

After some manipulation, this gives the maximum shear stress at a radius r as

$$\tau_{max} = \frac{3T}{2\pi r^3(1 - \cos\alpha)^2(2 + \cos\alpha)} \quad \textbf{(5.52)}$$

The elementary theory, extended to analyse varying cross-sections, gives

$$\tau_{max} = \frac{2T}{\pi r^3 \sin^3\alpha} \quad \textbf{(5.53)}$$

The ratio of the two tends to one as α tends to zero and the elementary theory overestimates the maximum stress by 10% when α is 30.4°. The ratio of the strain energies per unit length predicted by the two methods is inversely proportional to the predicted torsional stiffnesses. As will be seen in Chapter 7, the approximate stiffness found this way is always greater than the true stiffness. The ratio, J_R, of the former to the latter is given by

$$J_R = \frac{1}{3(1 + 2\cos^2\alpha)} \left(\frac{4(2 + \cos\alpha)}{(1 + \cos\alpha)^2} \right)^2 \quad \textbf{(5.54)}$$

This ratio increases with α and shows that the approximate torsional stiffness exceeds the true torsional stiffness by 10% when α is 15.3°.

The shear modulus G may vary over the cross-section of a bar in torsion. Under certain circumstances, the previous solutions may still be applicable. If the material is uniform between the stress trajectories found for a homogeneous bar, then the section behaves like a set of nesting tubes, each with its own shear modulus.

The torsion of a circular bar of radius r_1 made from a material with shear modulus G_1 with a second circular bar of radius r_2 and modulus G_2 was analysed by Vekua and Rukhadze[1]. Suppose that the centre of the second bar is offset from the centre of the first by a distance e as shown in Figure 5.16. The overall torsional stiffness of the section, GJ_o say, is then given by

$$GJ_o = G_1 J_1 + \frac{2G_1(G_2 - G_1)}{G_1 + G_2} I_{p2} \quad \textbf{(5.55)}$$

approximately. Here, J_1 is the torsional constant for a solid circular section of radius r_1 and I_{p2} is the second polar moment of area of the second circle about the centre of the first, that is

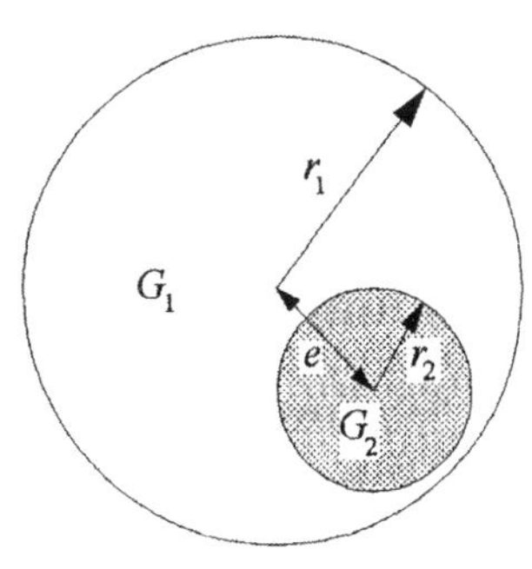

Figure 5.16 Torsion of a compound cross-section.

[1] Vekua, I.N. and Rukhadze, A.K. (1933) *Izv. A. N. S.S.S.R.* No. 3 p.373.

$$J_1 = \frac{1}{2}\pi r_1^4 \quad , \quad I_{p2} = \frac{1}{2}\pi r_2^4 + \pi r_2^2 e^2 \tag{5.56}$$

(see Muskhelishvili (1963) §140a for example). This result is particularly useful for the analysis of fibre-reinforced rods, when (5.55) can be extended by adding further terms of the second kind on the right-hand side for each fibre. Muskhelishvili also analyses the problem of the torsion of a rectangular strip of two materials, as shown in Figure 5.17. Provided that the width of each part is at least 2.5 times the thickness of the strip, the overall torsional stiffness of the compound bar is given approximately as

$$GJ_o = \frac{1}{3}(G_1 b_1 + G_2 b_2)t^3 - 0.21\left(\frac{G_1^2 + G_2^2}{G_1 + G_2}\right)t^4 \tag{5.57}$$

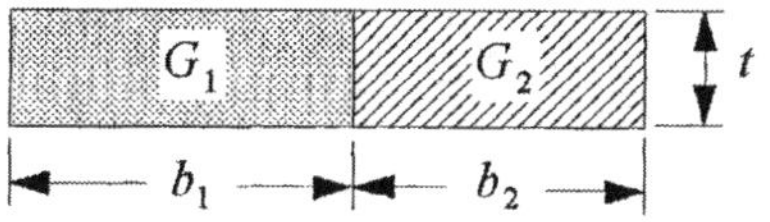

Figure 5.17 A compound strip.

This reduces to the form given by (5.15) when the two materials are the same.

{5.8 Anisotropic Beams in Torsion}

{This problem has been examined by Leknitskii (1981) and Sokolnikoff (1956). Some of the material is complex and specialised, so that only some of the simpler and more commonly applicable results will be discussed here. It will be seen in §8.2 that a bending moment applied to an anisotropic beam can produce torsion and that a torque can produce bending when the compliances s_{43} and s_{53} are not zero. If a torque T is applied in the sense used earlier, taking the axis of the beam as the z axis, flexure is produced about the principal axes of the section which is given by the rates of rotation

$$\vartheta_x = -\frac{Ts_{53}}{2I_{xx}} \quad , \quad \vartheta_y = -\frac{Ts_{43}}{2I_{yy}} \quad . \tag{5.58}$$

(cf. Lekhnitskii, ibid. (50.5)). However, the general derivation of the torque/twist relationship is not easy.

The problem becomes much simpler if the material is symmetric with regard to inversion of the z axis. This is the monoclinic system given in §3.11. The compliances in (5.58) are then zero so that there is no coupling between bending and torsion. Also, it then becomes possible to use a generalisation of Saint-Venant's analysis given in §5.4. The material displacements are again those given by (5.16). This means that the only non-zero strains are γ_{yz} and γ_{zx}. Then the only non-zero stresses are τ_{yz} and τ_{zx}. As in (5.23), these will be given by the gradients of a stress function ϕ where

$$\frac{\partial\phi}{\partial x} = -\tau_{yz} \quad , \quad \frac{\partial\phi}{\partial y} = \tau_{zx} \ .$$

The equilibrium equation (5.18) is automatically satisfied. The stress-strain relationships previously given by (5.17) now become

$$s_{45}\tau_{yz} + s_{55}\tau_{zx} = \gamma_{zx} = \left(\frac{\partial g}{\partial x} - y\right)\vartheta$$
$$s_{44}\tau_{yz} + s_{45}\tau_{zx} = \gamma_{yz} = \left(\frac{\partial g}{\partial y} + x\right)\vartheta \tag{5.59}$$

so that

$$s_{45}\frac{\partial \tau_{yz}}{\partial y} + s_{55}\frac{\partial \tau_{zx}}{\partial y} - s_{44}\frac{\partial \tau_{yz}}{\partial x} - s_{45}\frac{\partial \tau_{zx}}{\partial x} = -2\vartheta \tag{5.60}$$

or in terms of ϕ this becomes

$$s_{44}\frac{\partial^2 \phi}{\partial x^2} - 2s_{45}\frac{\partial^2 \phi}{\partial x \partial y} + s_{55}\frac{\partial^2 \phi}{\partial y^2} = -2\vartheta \tag{5.61}$$

The condition of zero shear stress on the sides of the beam is again satisfied by taking ϕ as constant around the boundary of the section (or zero if the section is simply-connected). Solving (5.61) with this boundary condition then determines the state of stress of the beam for a given rate of twist. This can be integrated, as in (5.27), to give the necessary applied torque.

To simplify the solution, (5.61) can be expressed in terms of a new set of coordinates $(\bar{x},\bar{y})$ at an angle α to (x,y) such that

$$\bar{x} = x\cos\alpha + y\sin\alpha \;,\; \bar{y} = -x\sin\alpha + y\cos\alpha \qquad \text{where} \quad \tan 2\alpha = \frac{2s_{45}}{s_{55} - s_{44}} \tag{5.62}$$

It then takes the form

$$S_{44}\frac{\partial^2 \phi}{\partial \bar{x}^2} + S_{55}\frac{\partial^2 \phi}{\partial \bar{y}^2} = -2\vartheta \qquad \text{where}$$
$$S_{44} = s_{44}\cos^2\alpha + s_{55}\sin^2\alpha - s_{45}\sin 2\alpha \;,\quad S_{55} = s_{44}\sin^2\alpha + s_{55}\cos^2\alpha + s_{45}\sin 2\alpha \tag{5.63}$$

From the positive-definiteness of the compliance matrix, it can readily be shown that S_{44} and S_{55} are both positive. A further coordinate transformation reduces this to the harmonic form

$$\frac{\partial^2 \phi}{\partial X^2} + \frac{\partial^2 \phi}{\partial Y^2} = -2G^*\vartheta \qquad \text{where}$$
$$X = gx \;,\quad gY = y \;,\quad g = \left(\frac{S_{55}}{S_{44}}\right)^{1/4} \;,\quad G^* = \frac{1}{\sqrt{(S_{44}S_{55})}} \tag{5.64}$$

A function ϕ is now sought which satisfies (5.64) and is constant on the boundary of the section, as expressed in terms of X and Y. For example, what was a square boundary may now become a parallelogram. A circular boundary becomes an ellipse. In this case, if the bar is of radius a,

$$\phi = -\frac{\vartheta}{S_{44} + S_{55}}\left(\frac{X^2}{g^2} + Y^2 g^2 - a^2\right) = -\frac{\vartheta}{s_{44} + s_{55}}(x^2 + y^2 - a^2) \tag{5.65}$$

giving a shear stress distribution like that of an isotropic circular bar,

$$\tau_{yz} = \frac{2x\vartheta}{s_{44} + s_{55}} \;,\quad \tau_{zx} = -\frac{2y\vartheta}{s_{44} + s_{55}} \tag{5.66}$$

(although the cross-section may warp in a complex manner). Integrating over the cross-section to find the torque gives

$$T = \int_A \frac{2\vartheta}{s_{44} + s_{55}}(x^2 + y^2)dA = \frac{\pi a^4}{s_{44} + s_{55}}\vartheta \tag{5.67}$$

Thus a circular bar made from monoclinic material in torsion behaves very much like an isotropic bar, if the inverse of the shear modulus ($1/G$) is taken as the average of s_{44} and s_{55}. Other cases are examined in the references given at the beginning of this section. For example, Lekhnitskii (1981) analyses the torsion of orthotropic bars with elliptic and rectangular cross-sections and gives approximate methods for other problems. See also Sokolniknoff (1956) §51, Hearmon (1961) §4.4, and Love (1952) §226. }

{5.9 Non-Uniform Torsion of Thin-Walled Open Sections}

{The value of J for a thin circular tube of radius c and thickness t is given by (5.10) as approximately $2\pi c^3 t$ (when $c >> t$). If a longitudinal slit is cut in the tube, J drops to the value given by (5.15) of approximately $\tfrac{2}{3}\pi c t^3$. The reason for this sudden drop is that the slit tube can warp much more readily than the closed tube, the two sides of the slit being free to move longitudinally with respect to one another. However, if the ends of this tube are constrained from warping, its overall torsional stiffness can increase significantly. The problem of torsion with constrained warping been analysed by Vlasov (1961) and discussed by Zbirohowski-Koscia (1967).

In his analysis, Vlasov makes two assumptions. One is that the form of the cross-section does not change, so that its in-plane motion is as a rigid body. (This was also the basic assumption of Saint-Venant torsion described in §5.4.) Also, it is assumed that the overall longitudinal shear strain of any part of the section wall (denoted by γ in (5.7)) is zero. At first sight, this is a surprising assumption. However, it is analogous to allowing that shear forces (and shear stresses) can arise as a result of a varying bending moment, but ignoring the shear displacements which must be produced. This is implicit in Chapter 4, and can be justified on the grounds that these shear displacements are usually relatively small[1]. The constraint of warping gives rise to axial stresses which produce localised bending effects. These effects are sometimes referred to as *flexural torsion.* In order to uncouple the effects of these stresses from those of other kinds of loading, such stress systems can be taken to give rise to no resultant axial force or bending moments, without loss of generality. Such stress systems are referred to as *bimoments* and are denoted by the letter B.

The two assumptions mentioned above are now used to find the warping of the cross-section. As in (5.7), the warping of the cross-section is given by $\vartheta\omega$ where ϑ is the rate of twist and ω is a warping function. In the present case though, the shear strain is taken as zero, so that

$$0 = \gamma = \vartheta\left(h - \frac{\partial\omega}{\partial s}\right) \quad \text{or} \quad \frac{\partial\omega}{\partial s} = h \tag{5.68}$$

[1] In §5.3, the resultant shear force per unit length of section wall was denoted by the shear flow q. From equilibrium considerations, this was shown not to vary along any simple strip. However, at the end of an open strip, q must be zero because the complementary shear stresses would be on a stress-free surface. Then in *uniform* torsion, an open strip has no shear flow.

The warping function at distance s along the middle line, as measured from the start, a, is given by

$$\omega = \omega_a + \int_a^s h\,ds = \omega_a + 2A_s \tag{5.69}$$

where A_s is the sectorial area shown lightly shaded in Figure 5.18 and ω_a is the warping displacement at a. The mean warping ω_m of the cross-section , allowing for a possible variation in the wall thickness[1], t, is

$$\omega_m = \frac{\int_a^b (\omega_a + 2A_s)\,t\,ds}{\int_a^b t\,ds} \tag{5.70}$$

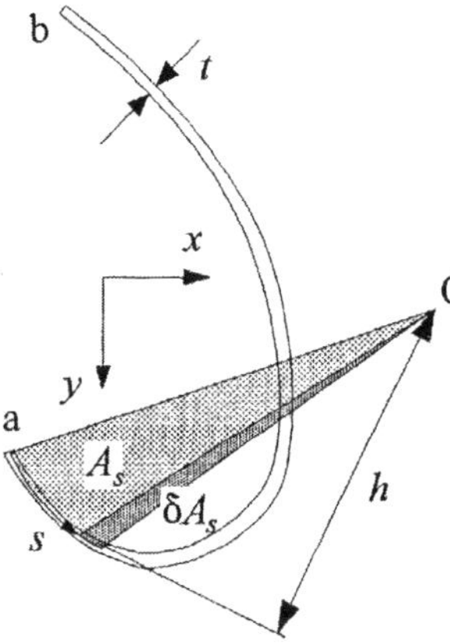

Figure 5.18 Sectorial areas.

where the end-point of the middle line is at b.

The position of O is determined from the condition that flexural warping is uncoupled from normal flexure. A longitudinal warping strain ϵ_{zz} results if there is a change in warping between adjacent cross-sections. This change will be produced by a change in the rate of twist $\dot{\theta}$. Using primes to indicate the rate of change in the longitudinal direction, z, the associated warping stress σ_{zz} is given by

$$\sigma_{zz} = E\epsilon_{zz} = Ew' = E\dot{\theta}'\omega = E\theta''\omega \tag{5.71}$$

If this is to produce no axial force,

$$\begin{gathered} 0 = \int_a^b \sigma_{zz} t\,ds = E\theta'' \int_a^b \omega t\,ds = E\theta'' \int_a^b (\omega_a + 2A_s)\,t\,ds \\ \text{giving} \quad \omega_a \int_a^b t\,ds = -2\int_a^b A_s t\,ds \end{gathered} \tag{5.72}$$

In other words, the mean warping, ω_m, must be zero. For this stress system to produce no resultant bending moments about the y and z axes,

$$\begin{gathered} 0 = \int_a^b \sigma_{zz} x t\,ds = E\theta'' \int_a^b (\omega_a + 2A_s)\,x t\,ds \quad \text{giving} \quad \omega_a \int_a^b x t\,ds = -2\int_a^b A_s x t\,ds \\ 0 = \int_a^b \sigma_{zz} y t\,ds = E\theta'' \int_a^b (\omega_a + 2A_s)\,y t\,ds \quad \text{giving} \quad \omega_a \int_a^b y t\,ds = -2\int_a^b A_s y t\,ds \end{gathered} \tag{5.73}$$

The position of O and the value of ω_a must be found which satisfy (5.72) and (5.73). The change in A_s given by measuring it with respect to some point C instead of with respect to O can be determined from Figure 5.19.

[1] For a discussion of the nature of ω_m, see Chilver, A.H. (1955) *Journal of the Mechanics and Physics of Solids,* V.3 p. 267 (1955).

A short length δs of the middle line of the section subtends the triangle OP_1P_2 of area δA_s at O. With respect to C, the triangle CP_1P_2 has an area δA_s^c. Then

$$\begin{aligned}\delta A_s^c &= CP_1Q + CQP_2 - P_1P_2Q \\ &= \tfrac{1}{2}(\delta y.QR + \delta x.OR) - P_1P_2Q, \\ \delta A_s &= OP_1Q + OQP_2 - P_1P_2Q \\ &= \tfrac{1}{2}[\delta y.(QR + x^\circ) + \delta x.(OR - y^\circ)] - P_1P_2Q.\end{aligned} \tag{5.74}$$

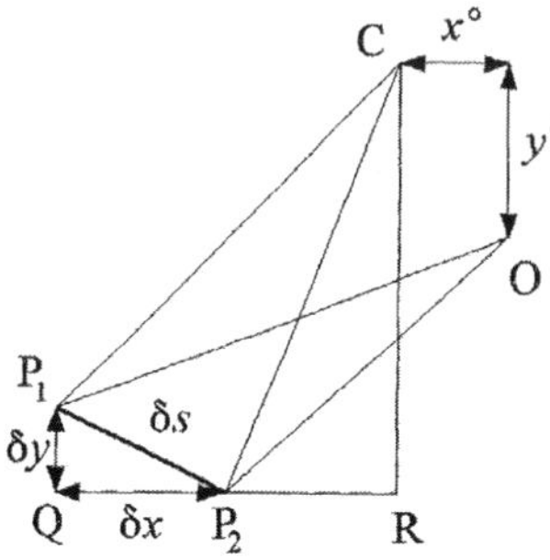

Figure 5.19 Elementary areas.

Integrating the difference in these expressions to give the difference between A_s and A_s^c gives

$$2(A_s - A_s^c) = \int_a^s 2(dA_s - dA_s^c) = \int_a^s (x^\circ dy - y^\circ dx) = x^\circ(y - y_a) - y^\circ(x - x_a) \tag{5.75}$$

The value of A_s^c can be determined for some arbitrary point C and the value of A_s then expressed in terms of x° and y°. There are now three unknowns x°, y° and ω_a which can be determined from the three equations given by (5.72) and (5.73).

If the origin of coordinates is chosen to be at the centroid of the section, then the left-hand sides of the final forms of equations (5.73) are zero. It is also convenient to choose C to be at the centroid and to take the axes to be the principal axes of the section. The coordinates of O are then (x°, y°) and certain terms in the required integration disappear. For the slit tube shown in Figure 5.20, from (5.75)

$$2A_s = c^2\alpha + x^\circ(c\sin\alpha) - y^\circ(-c\cos\alpha + c)$$

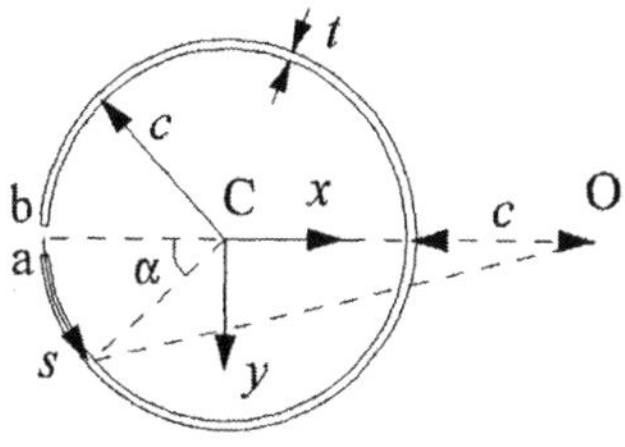

Figure 5.20 Constrained warping of a slit circular tube.

Substituting this in (5.72) and (5.73) gives

$$\omega_a(2\pi ct) = -2\pi^2c^3t - 2\pi y^\circ c^2 t$$

$$0 = \pi c^3 t y^\circ \quad , \quad 0 = 2\pi c^4 t - \pi c^3 t x^\circ .$$

This means that the position of O is as shown in the figure, the value of ω_a is $-\pi c^2$ and

$$\omega = \omega_a + 2A_s = c^2(\alpha - \pi) + 2c^2\sin\alpha \tag{5.76}$$

The centre of rotation O is also known as the *shear centre* and will be examined again in the chapter on shear.

The equilibrium of a longitudinal strip of the section wall is illustrated in Figure 5.21. The axial stress σ_{zz} and the shear flow τt must produce no net axial force on the strip, giving

$$\delta(\tau t)\delta z = \delta\sigma_{zz} t\delta s$$

or, from (5.71),

$$\frac{\partial \tau t}{\partial s} = t\frac{\partial \sigma_{zz}}{\partial z} = Et\theta'''\omega \tag{5.77}$$

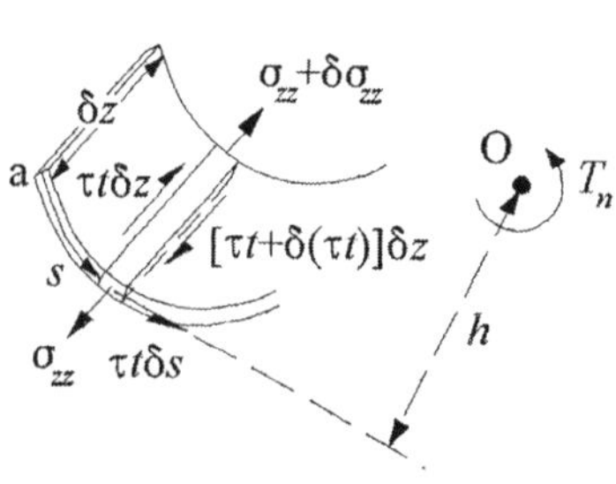

Figure 5.21 Non-uniform torsional stresses.

so that from (5.69),

$$\tau t = E\theta''' \int_a^s \omega t ds = E\theta''' \int_a^s (\omega_a + 2A_s) t ds \tag{5.78}$$

Note that there is no additional constant because τt must be zero at a, as explained earlier. The torque T_n about O produced by the shear flow is found by the same process as in (5.6) and is given by

$$T_n = \int_a^b \tau t h ds = E\theta''' \int_a^b \left(\int_a^s \omega t ds \right) h ds \tag{5.79}$$

so that, integrating by parts,

$$T_n = E\theta''' \int_a^b \left(\int_a^s \omega t ds \right) \frac{d\omega}{ds} ds = E\theta''' \left\{ \left| \left(\int_a^s \omega t ds \right) \omega \right|_a^b - \int_a^b \omega^2 t ds \right\} \tag{5.80}$$

Evaluating the first term on the right-hand side of (5.80), the integral is obviously zero when the upper limit is also at a, and from (5.72) it is also zero when the upper limit is at b. Then

$$T_n = -EI_\omega \theta''' \qquad \text{where } I_\omega = \int_a^b \omega^2 t ds \tag{5.81}$$

The warping constant I_ω is sometimes denoted by Γ or K. For the split circular tube,

$$I_\omega = t \int_0^{2\pi} [c^2(\alpha - \pi) + 2c^2 \sin\alpha]^2 c d\alpha = \pi c^5 t(\tfrac{2}{3}\pi^2 - 4) \tag{5.82}$$

The total torque T is now the sum of that given by the uniform torsion theory and the non-uniform torque given by (5.81):

$$T = GJ\theta' - EI_\omega \theta''' \tag{5.83}$$

If the torque is constant along the beam, the general form of the solution of this equation is

$$\theta = \frac{Tz}{GJ} + A + C\sinh kz + D\cosh kz \qquad \left(k^2 = \frac{GJ}{EI_\omega} \right) \tag{5.84}$$

and A, C and D are constants. Consider a bar of length l with end fixtures which prevent warping. This means that $\dot\theta$ (i.e. θ') is zero at the ends. The relative rotation of the two ends produced by the torque T is then given by

$$\theta_l - \theta_0 = \frac{Tl}{GJ}\left(1 - \frac{2}{kl}\tanh\frac{kl}{2} \right) \tag{5.85}$$

as kl becomes very small, this relative rotation is given by

$$\theta_l - \theta_0 \approx \frac{Tl^3}{12EI_\omega} \tag{5.86}$$

In the case of the slit tube, this overall torsional stiffness is not the same as that for the closed tube. If Poisson's ratio is a third and the tube thickness t is a tenth of its radius c, the end-constrained slit tube has the same overall torsional stiffness as the closed tube when the length of the tube is about $0.16c$. However, ordinary beam theory is not appropriate for such short beams.

Because it has no resultant force or moment, a bimoment does no work during a rigid-body motion of the beam cross-section. The work done by a bimoment results from the warping $\vartheta\omega$ of the cross-section produced by the twist ϑ. This twist is the corresponding deflexion to the bimoment, B, so that the work done by it is $\frac{1}{2}B\vartheta$ (cf. §3.4). This means that a bimoment is defined by the warping characteristics of the section on which it acts; the same set of loads applied to a different section will not necessarily constitute the same bimoment. Take for example the I-section shown in Figure 5.22a, which has a flange width of b and a web depth of h. A rate of twist ϑ induces warping of the flanges which increases or decreases linearly from zero at the web junction to $\pm\frac{1}{4}hb\vartheta$ at the flange ends. Point forces P, which increase in proportion to ϑ, are applied to these ends in the sense shown, so that they each do positive work during warping. The total work done by these forces is $\frac{1}{2}Phb\vartheta$, so that this loading represents a pure bimoment of magnitude Phb. (Note that this system of forces has no resultant force or moment.) This loading is now applied to the I-section of flange width h and web depth b shown in Figure 5.22b. However, the pattern of warping now produces exactly the *opposite* motions of the forces P when the same rate of twist occurs. Thus the same set of forces now represents a bimoment of $-Phb$ in this case.

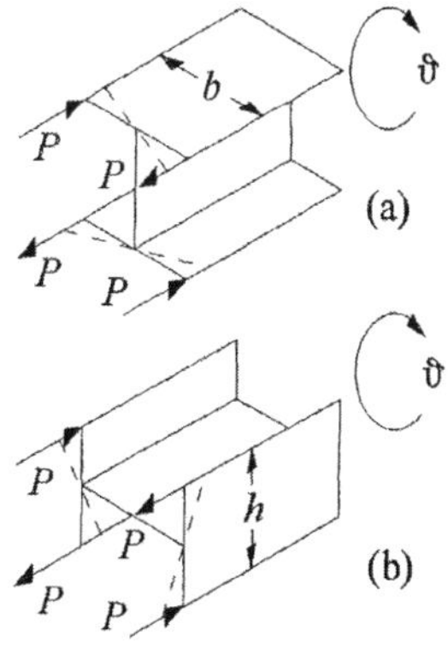

Figure 5.22 Bimoments.

The system of axial stresses σ_{zz} given by (5.71) represents the bimoment pair. In the case of the I-beam shown in Figure 5.23, the rate of change of twist θ'' (or ϑ') induces bending of the flanges in their own planes, and the warping stresses σ_{zz} are the associated bending stresses. In accordance with the convention used for force and moment pairs acting on elementary lengths of beam (cf. Figure 1.5), a positive bimoment pair will be taken to produce a positive rate of change of the corresponding deflexion, ϑ, along the beam. This means that the component of a positive pair, shown acting on the element face further along the beam, is also positive in the sense used in the last paragraph. The work done by these warping stresses is then given by integrating over the area A of the cross-section (or from a to b in the case of a simple strip):

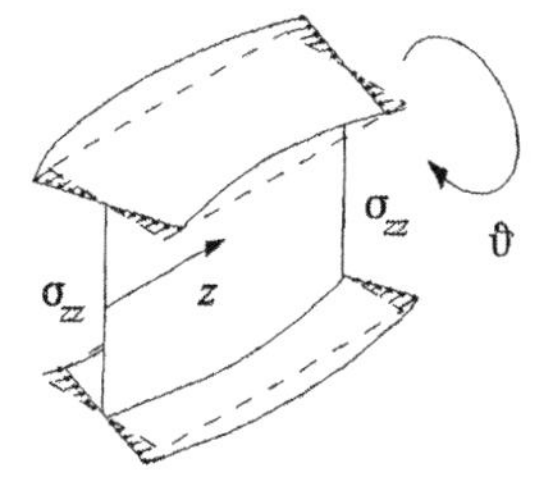

Figure 5.23 A bimoment pair.

$$\frac{1}{2}B\vartheta = \frac{1}{2}\int_A \sigma_{zz}\vartheta\omega t\,ds = \frac{1}{2}\int_A E\theta''\vartheta\omega^2 t\,ds \quad \text{or} \quad B = E\theta''\int_A \omega^2 t\,ds = EI_\omega\theta'' \qquad \textbf{(5.87)}$$

using (5.71) and (5.81). It should be noted that this definition of a positive bimoment pair is the opposite of that given by Vlasov (1961) but is consistent with the way in which load pairs are defined throughout this book.

In the case of uniform torsion, the constants C and D in (5.84) are zero, allowing two boundary conditions to be specified. These are an end torque and an end rotation. In the case of non-uniform torsion, four end conditions can be specified, the extra two relating to the end warping and end bimoments. For a workless reaction, either the end torque or the end rotation is zero, and either the end bimoment or the end warping is zero. A workless joint subject to warping will be such that the total work done by the bimoments acting on the ends of the beams meeting at the joint is zero. As a bimoment is only defined in terms of the beam on which it acts, it cannot be said that if two beams only meet at a joint, the bimoments acting on their ends will be equal and opposite[1]. Elastic joints on which bimoments can do work have also been analysed[2]. The webs between rigid gusset plates at the joint are treated as a short beam in torsion, and the relative rotation of the plates is related to the warping of the beams meeting at the joint.

For non-uniform torsion, the equivalent of the slope-deflexion equations given in §4.5 take the form

$$\begin{bmatrix} T_o \\ B_o/l \end{bmatrix} = \frac{GJ}{l}\begin{bmatrix} 1+2\alpha & \alpha \\ \alpha & \alpha-\beta \end{bmatrix}\begin{bmatrix} \theta_o \\ \theta_o' l \end{bmatrix} + \frac{GJ}{l}\begin{bmatrix} -(1+2\alpha) & \alpha \\ -\alpha & \beta \end{bmatrix}\begin{bmatrix} \theta_l \\ \theta_l' l \end{bmatrix} \tag{5.88}$$

and

$$\begin{bmatrix} T_l \\ B_l/l \end{bmatrix} = \frac{GJ}{l}\begin{bmatrix} -(1+2\alpha) & -\alpha \\ \alpha & \beta \end{bmatrix}\begin{bmatrix} \theta_o \\ \theta_o' l \end{bmatrix} + \frac{GJ}{l}\begin{bmatrix} 1+2\alpha & -\alpha \\ -\alpha & \alpha-\beta \end{bmatrix}\begin{bmatrix} \theta_l \\ \theta_l' l \end{bmatrix} \tag{5.89}$$

where the subscripts o and l refer to the 'near' and 'far' ends of the beam ($z=0$ and $z=l$ respectively), T is a clockwise torque applied to the beam, θ is the clockwise rotation of the beam, B is the bimoment which does work of $\tfrac{1}{2}B\theta'$ during warping of the section on which it acts (as described earlier) and

$$\alpha = \frac{\cosh kl - 1}{kl\sinh kl - 2(\cosh kl - 1)} \ , \quad \beta = \frac{\sinh kl - kl}{kl[kl\sinh kl - 2(\cosh kl - 1)]} \tag{5.90}$$

and so

$$\begin{aligned} \alpha &\approx \frac{6}{(kl)^2} + \frac{1}{10} \quad (kl \le 1) \ , \qquad \alpha \approx \frac{1}{kl-2} \quad (kl \ge 10) \ , \\ \beta &\approx \frac{2}{(kl)^2} - \frac{1}{30} \quad (kl \le 1) \ , \qquad \beta \approx \frac{1}{kl(kl-2)} \quad (kl \ge 10) \ . \end{aligned} \tag{5.91}$$

For example, it readily follows from (5.88) and (5.89) that if the 'near' end of a beam is prevented from rotating and a bimoment B_o is applied to it, the rotation θ_l of the free 'far' end is B_o/GJ, whatever the length l of the beam may be.

The strain energy associated with non-uniform torsion can be determined from the work done by a pair of equal and opposite bimoments B acting on a short length dz of a beam. As already mentioned, the work done by a bimoment B on a cross-section when it warps is given by $\tfrac{1}{2}B\vartheta$, where ϑ is the rate of twist, θ', and B is given by (5.87). On the other face of the short length dz, the rate of twist will have changed to $\theta'+\theta''dz$. The other bimoment of the pair which acts on this face will then do work $\tfrac{1}{2}B(\theta'+\theta''dz)$. This will be in the opposite sense to the former

[1] A counter example to this occurs when two similar I-beams are joined together at right-angles. If their flanges are coplanar, the bimoments acting on each at the junction are equal and opposite. However, if their webs are coplanar, these bimoments are (almost) equal. (See Renton, J.D. (1974) On the transmission of non-uniform torsion through joints. *Oxford University Engineering Laboratory Report No. 1086.*)

[2] Gorbunov, B.H. & Strel'bitskaya, A.I. (1955) The analysis of the rigidity of thin-walled bar systems. (In Russian). *Analysis of Space Structures* V.2, Moscow.

work, so that the net work is the difference of the two. This is equal to the strain energy absorbed by the length *dz*. From (5.87), the strain energy per unit length is then

$$U_l = \frac{1}{2}B\theta'' = \frac{1}{2}EI_\omega(\theta'')^2 = \frac{B^2}{2EI_\omega} \qquad \textbf{(5.92)}$$

(cf. (5.45)). The same result can be obtained by integrating the strain energy per unit volume generated by the warping stress σ_{zz} (given by (5.71)) over a unit length of the beam.}

Chapter 6 Shear of Beams

6.1 Introduction

In Chapter 4, shear forces acting on beams were seen to produce varying flexural curvature and hence varying bending stresses. More complete analyses include the shear deformation and shear stresses produced by such loading. In thin-walled beams, it is usually sufficient to determine the response from the mean shear flow (cf. §5.3 and §5.9) but as with torsion, there is a more exact theory due to Saint-Venant. In general, a shear force acting on a beam will produce torsion as well as bending. The sense of this torsion will depend on the position of the line of action of the shear force relative to the cross-section of the beam. If this line of action passes through the *shear centre* of the beam, there is no tendency for the force to twist it either way. Conversely, if a pure torque is applied to the beam, its cross-section tends to twist about this shear centre. Were this not so, a shear force acting through the shear centre would do work during torsion, which would mean that it had a torsional effect on the beam. The deformation of a beam caused by a shear force means that plane cross-sections no longer remain plane. This has implications on defining what is meant by the rotation of a beam's cross-section and the nature of support conditions.

Because the beam's cross-section can no longer be considered rigid, it is necessary to consider what meaning should be attached to its overall rotation and displacement. To choose the rotation and displacement at the centroid, for example, is arbitrary. Also, the use of a simple average rotation and displacement can lead to inconsistencies. Later in this chapter and in the general theory expounded in Chapter 8, work and energy principles will be employed.

6.2 The Engineering Theory of Shear of Thin-Walled Sections

From §4.2 it follows that the resultant shear force S produces a rate of change of bending stress σ. Thus from (4.15) the rate of variation of *tensile* stress along a longitudinal fibre of a prismatic beam is given by

$$\frac{\partial \sigma}{\partial x} = -\frac{y}{I}\frac{dM}{dx} = \frac{Sy}{I} \qquad \textbf{(6.1)}$$

Consider the elementary longitudinal fibre of cross-sectional area δA and length δx shown at the right of Figure 6.1. This area occupies the full width t of the bar's section and a short length δs along its middle line. The resultant force produced by the longitudinal stresses on the end areas is $\delta\sigma\delta A$. This must be balanced by the shear stresses τ acting on the other two internal faces. These stresses are expressed in terms of the *shear flow* q (= τt). The resultant longitudinal force produced by these shear stresses is $\delta q \delta x$, so that for equilibrium,

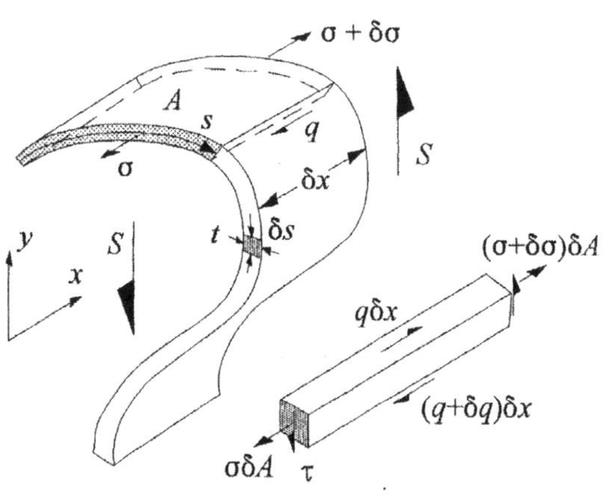

Figure 6.1 Shear of a thin-walled section.

$$\delta\sigma\,\delta A \;=\; \delta\sigma\, t\,\delta s \;=\; \delta q\,\delta x$$

$$\text{or}\quad \frac{\partial\sigma}{\partial x}\delta x\, t\,\delta s = \frac{\partial q}{\partial s}\delta s\,\delta x \quad \text{so that} \quad t\frac{\partial\sigma}{\partial x} = \frac{\partial q}{\partial s} \qquad \textbf{(6.2)}$$

This equation can be integrated with respect to s to give the change in q between two points on the middle line. Suppose that s is measured from a free end of the section wall, as shown in Figure 6.1. At this origin, the complementary shear stress, which acts on a free surface, must be zero, so that the shear flow q is zero. Then integrating (6.2) gives the shear flow at any point along the section wall:

$$q \;=\; \int_0^s t\frac{\partial\sigma}{\partial x}ds \;=\; \int_0^A \frac{\partial\sigma}{\partial x}dA \;=\; \frac{S}{I}\int_0^A y\,dA \;=\; \frac{S}{I}A\bar{y} \qquad \textbf{(6.3)}$$

using (6.1) and taking $\bar{y}$ as the ordinate of the centroid of the area A shown shaded in Figure 6.1. The complementary shear stress, τ, to that used in (6.2) acts along the middle line on the area δA as shown by the shear arrows shown in Figure 6.1. The shear force S is the resultant of the elementary shear forces $\tau\delta A$ integrated over the cross-section.

Example 6.1 A rectangular section

The section shown in Figure 6.2 has a breadth b which may not be small in comparison with its depth d. However, it will be assumed that the shear stress τ is constant across the breadth of the section. (A more accurate analysis, which avoids this simplifying approximation, will be given later.) The variation of bending moment along the beam produces the variation of bending stress, $\partial\sigma/\partial x\;\delta x$, which varies linearly with y as shown in the figure. The shear flow q at a distance y above the neutral axis can be determined from (6.3) where in this case

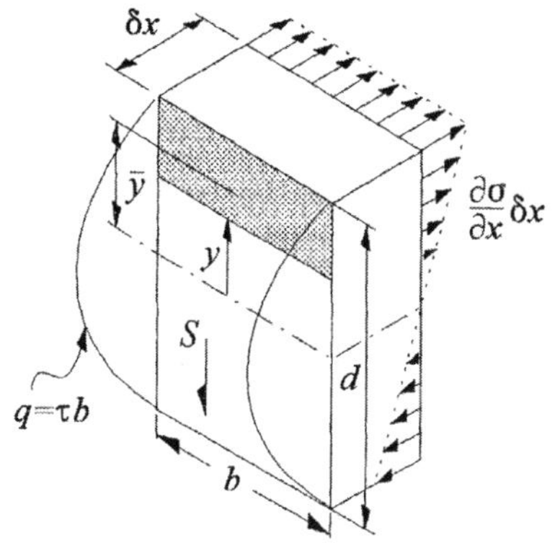

$$I = \frac{bd^3}{12}\;,\; A = b(\frac{d}{2}-y)\;,\; \bar{y} = \tfrac{1}{2}(\frac{d}{2}+y).$$

Figure 6.2 Shear of a rectangular section.

$$\text{giving} \qquad q \;=\; \tau b \;=\; \frac{3S}{2d}\left(1 - \frac{4y^2}{d^2}\right)$$

This gives the parabolic distribution of shear flow shown. Integrating this over the cross-section shows that the resultant of the shear flow is the shear force:

$$\int_{-\frac{d}{2}}^{\frac{d}{2}} q\,dy \;=\; \int_{-\frac{d}{2}}^{\frac{d}{2}} \frac{3S}{2d}\left(1 - \frac{4y^2}{d^2}\right)dy \;=\; S$$

From the derivation of (6.2) it follows that the shear flow is equal to the integral of the increment in bending stress over the area A. Thus a linear variation of bending stress gives rise to a parabolic distribution of shear stress as shown in Figure 6.2. This relationship can readily be

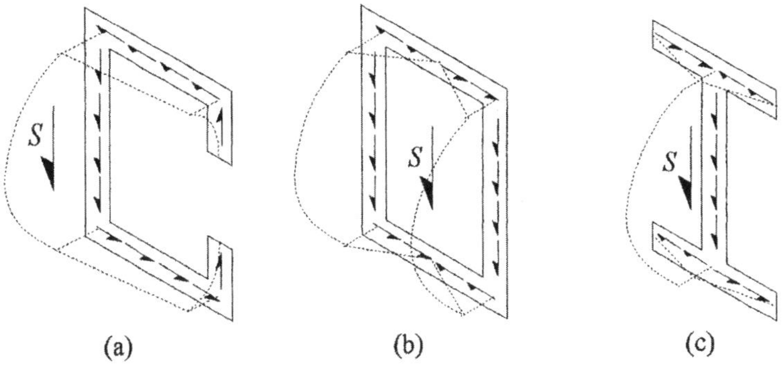

Figure 6.3 Shear flow around various sections.

used to sketch the variation of shear stress in the sections of constant wall thickness shown in Figure 6.3. The bending stress varies linearly in the lips of the channel shown in Figure 6.3a and in the webs of all three sections. Thus the shear stress varies parabolically in these zones. In the flanges of all three sections the bending stress is constant, so that the shear stress varies linearly. From symmetry, it follows that the shear stress is zero at the centres of the flanges of the box section shown in Figure 6.3b. For the I-section shown in Figure 6.3c, longitudinal equilibrium of the complementary stresses at the junction of the web and flanges requires that the shear flow into the junction is equal to the shear flow out of it (cf. (5.11)).

Example 6.2 Shear centre of a channel section

Suppose that the channel section shown in Figure 6.3 has a second moment of area of I about its horizontal centroidal axis through C. The rate of change of bending stress in the upper flange induced by a shear force S is $-Sd/2I$. From (6.2) this means that the shear stress in the flange must increase linearly from zero at A to $Sbd/2I$ at B, producing a shear force in the flange of $Sb^2td/4I$. An equal and opposite force is produced in the lower flange, so that shear stresses produce a couple about the web of $Sb^2td^2/4I$. This means that the line of action of the shear force must be offset from the web by a distance e where $\quad e = b^2td^2/4I \quad .$

If the shear force is offset by any other distance, it will produce torsion as well as bending of the section. For bending only about the vertical y axis, it follows from symmetry that the line of action of the shear force must be along the horizontal axis through C. Then a shear force acting in any direction which only produces bending must have vertical and horizontal components along the above two lines of action. The line of action of the resultant of these components must then pass through the point O where these two lines of action meet. This point is called the *shear centre*.

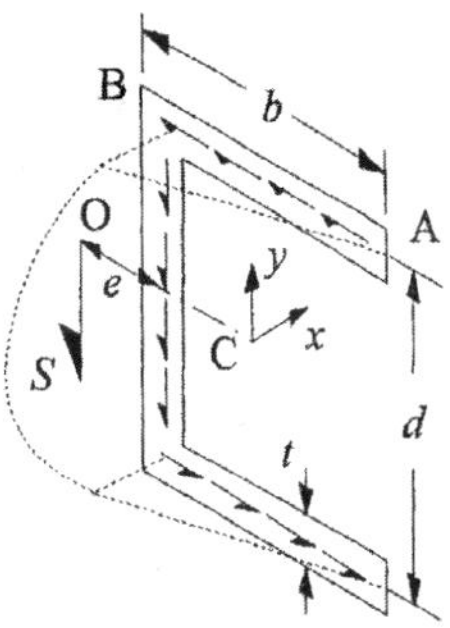

Figure 6.4 Shear centre of a channel section.

For bisymmetric sections, such as I-sections and box sections, the shear centre and the centroid are coincident. When the section consists of straight ribs radiating from a single junction, the shear centre is at this junction. This is because all the resultant rib shear forces must pass through this junction. The most common examples of this angle sections, tee sections and cruciform sections. The shear centre of a split circular tube of radius c is at a distance $2c$ from the centroid of the tube in the opposite direction to the slit. This can be shown by the same method as above, by finding the torque produced by the bending shear stresses about the centroid, or by finding the centre of rotation as in §5.9. The shear centre of a Z-section is also at its centroid. For such a section, the principal axes are not parallel to either the web or flanges. The shear stress distribution in such a case is most readily found by examining non-uniform *flexure* (rather than a non-uniform bending moment) about axes through the centroid, parallel and perpendicular to the web. In the first case, this produces a constant shear stress in the web which parabolically decreases to zero in the flanges. In the second case, the shear distribution in the web and flanges is similar to that in Figure 6.3. Linear combinations of these two distributions will give the desired resultant shear force.

6.3 Shear Strain Energy and the Shear Stiffness of Thin-Walled Sections

When a shear force is applied to a beam without twisting it, the force will induce linearly-varying bending stresses together with shear stresses of the kind described above. In finding the total strain energy of the bar, the strain energy per unit length related to the normal stresses is given by the second and third terms on the right-hand side of (4.59). The remaining strain energy arises from the shear terms in (3.37). This can be expressed in terms of the shear strain energy per unit length as $S^2/2GkA$ where G and A are the shear modulus and area of the cross-section and GkA is the shear stiffness of the section. This energy is produced from the work done by the local shear force S in moving through a shear displacement γ per unit length so that

$$\tfrac{1}{2}S\gamma = \frac{S^2}{2GkA} \qquad \text{or} \qquad \gamma = \frac{S}{GkA} \tag{6.4}$$

Example 6.3 Shear of circular tubes

From symmetry, the shear flow in the solid circular tube shown in Figure 6.5a must be zero at the top and bottom of the section. Also, at symmetrical points on either side if the y axis the shear flow must be equal and in opposite senses. In (6.3) then, the origin for s is taken to lie on the y axis, where the shear flow must be zero. The shear force S is the resultant of the shear stresses acting on the face of the section. The shear flow is then given by

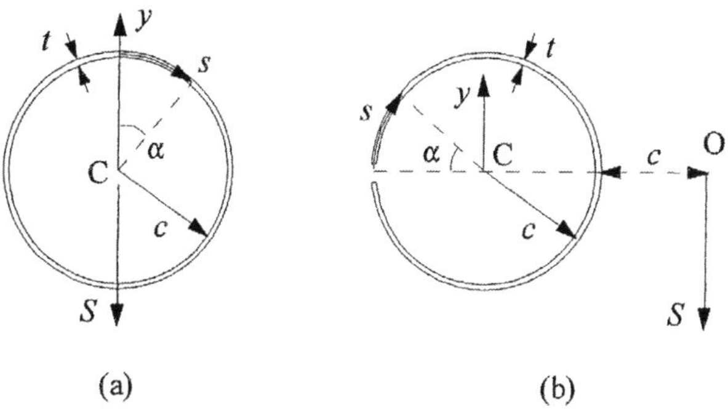

Figure 6.5 Closed and a slit circular tubes.

$$\tau t = q = \frac{S}{I}\int_0^s yt\,ds = \frac{S}{\pi c}\sin\alpha \qquad \text{where} \quad I = \pi c^3 t \tag{6.5}$$

From (3.37), the shear strain energy per unit length is then given by

$$U_l = \int_0^{2\pi} \frac{\tau^2}{2G} tc\,d\alpha = \frac{S^2}{2G\pi ct} = \frac{S^2}{2GkA} \qquad \textbf{(6.6)}$$

so that in this case k is 0.5.

For the slit tube shown in Figure 6.5b, (6.3) gives

$$\tau t = q = \frac{S}{I}\int_0^s yt\,ds = \frac{S}{\pi c}(1 - \cos\alpha) \qquad \textbf{(6.7)}$$

Integrating $\tau t ds$ multiplied by the distance c from the centroid gives a resultant moment of $2Sc$ about C, indicating that the shear force acts through O as shown. This confirms the position of the shear centre found previously (cf. Figure 5.20). Integrating as before in (6.6) to find the shear strain energy gives a value of 1/6 for k.

{6.4 A Closer Examination of Deflexion and Support Conditions}

{Some of the problems related to the analysis of shear can be seen from the examination of the plane-stress analysis of a simple cantilever given by Timoshenko and Goodier (1970) and others, Dugdale and Ruiz (1971) and Ford and Alexander (1977) for example. This cantilever is shown in Figure 6.6. It is of unit thickness, length l, depth $2c$, and a parabolic shear stress distribution is applied to the free end giving a downwards resultant force denoted by S. The state of stress given by

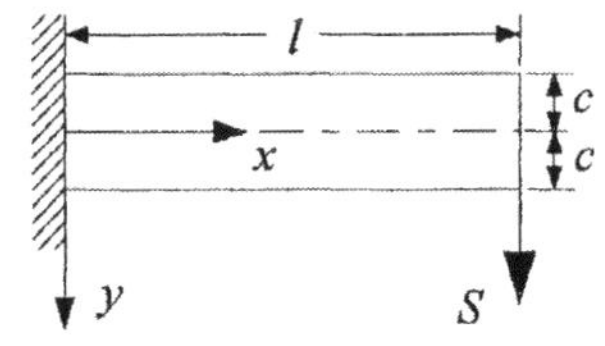

Figure 6.6 A simple cantilever.

$$\sigma_{xx} = -\frac{Sy}{I}(l - x) \ , \quad \sigma_{yy} = 0 \ , \quad \tau_{xy} = \frac{S}{2I}(c^2 - y^2) \qquad \textbf{(6.8)}$$

satisfies all the internal and boundary conditions except those at the fixed end. The x axis lies along the middle line of the beam, and the origin will be taken at the fixed end. The expressions for the deflections found from the resulting strains then take the form

$$\begin{aligned} u &= \frac{Sy}{2EI}(l - x)^2 - \frac{Sy^3}{6EI}(2 + \nu) - ey + g \\ v &= \frac{\nu S y^2}{2EI}(l - x) + \frac{S}{6EI}(l - x)^3 + d(l - x) + h \end{aligned} \qquad \textbf{(6.9)}$$

where

$$d + e = -\frac{Sc^2}{2GI} \ , \quad I = \tfrac{2}{3}c^3 \qquad \textbf{(6.10)}$$

and G is the shear modulus. Cross-sections no longer remain plane after deformation, as u is a cubic function of y. This poses problems in defining the fixed-end boundary conditions. Figure 6.7 shows the deformed profile of the cross-section described by (6.9) at the fixed end The constants d to h can be chosen to rotate

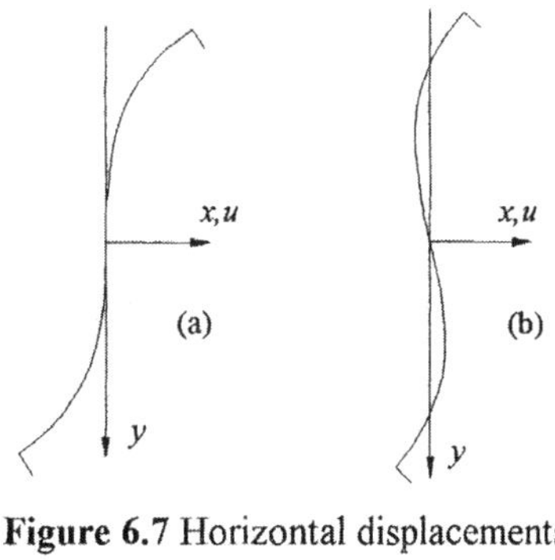

Figure 6.7 Horizontal displacements at a fixed end.

and displace this profile in order to give the best match to the support conditions. In the above references, these are taken to be that the centroid does not move at the fixed end, and locally the vertical face at the centroidal axis does not rotate. This implies that the fixed-end horizontal displacement is

$$u_o = -\frac{Sy^3}{6EI}(2+\nu) \tag{6.11}$$

as shown in Fig. 6.7a, and the vertical displacement of the middle line is given by

$$v_c = \frac{S}{EI}[\tfrac{1}{6}(-x^3 + 3x^2 l) + c^2 x(1 + \nu)] \tag{6.12}$$

However, this gives undue weight to the conditions at one particular point on the cross-section. Instead, it is possible to define the rotation of the cross-section as that through which the moment on it does work, and the displacement of the cross-section as that through which the force on it does work. (This is a simple extension of the ideas about 'corresponding deflexions' used earlier.) These rotations and displacements may have no relationship to those of a particular zone on the cross-section. Taking the fixed end to provide a workless reaction, the appropriate conditions to use on the end section are

$$\int_{-c}^{c} \sigma_{xx} u\, dy = \int_{-c}^{c} \tau_{xy} v\, dy = 0 \tag{6.13}$$

giving in particular

$$u_o = \frac{Sy}{EI}(2 + \nu)(\frac{c^2}{10} - \frac{y^2}{6}) \tag{6.14}$$

as shown in Fig. 6.7b and

$$v_c = \frac{S}{EI}[\tfrac{1}{6}(-x^3 + 3x^2 l) + \frac{c^2 x}{10}(8 + 9\nu) - \frac{\nu c^2 l}{10}] \tag{6.15}$$

By measuring x from the fixed end, only the cubic and quadratic terms in (6.12) and (6.15) are associated with normal flexure, and are the same in both cases. The remaining terms are related to shear and rigid-body motion.

This can be investigated using finite-element analysis. This has been done using a cantilever 1m long, 10cm deep and 1cm wide, divided into a thousand square plane-stress finite elements. A shear force of 1kN is applied to the free end, using a parabolic distribution of shear forces on the elements. All displacements of the elements are prevented at the fixed end. Using a best-fit cubic polynomial, the expression for v_c[1] is

$$v_c = [-967.0x^3 + 2{,}901x^2 + 11.81x - 0.565]\mu m \tag{6.16}$$

where x is in metres, Young's modulus is 206.8 GPa, and Poisson's ratio is 0.29. The

[1] Renton, J.D. (1991) Generalized beam theory applied to shear stiffness. *Int. J. Solids Structures* V.27, p.1955.

corresponding result given by (6.12) is

$$v_c = [-967.1x^3 + 2{,}901x^2 + 18.71x]\,\mu m \tag{6.17}$$

and that given by (6.15) is

$$v_c = [-967.1x^3 + 2{,}901x^2 + 15.39x - 0.421]\,\mu m \tag{6.18}$$

It will be seen that the above expression for workless reactions corresponds more closely than (6.17) to the finite-element analysis. In turn, the finite-element analysis indicates that truly fixed ends provide even more constraint against deflexion.

The above arguments concerned the meaning which should be attached to the 'rotation' and 'displacement' of a cross-section when the cross-section is itself deformed. Rotation is taken to be zero when the total work done by the bending stresses is zero and displacement is taken to be zero when the resultant shear force on the section does no work. It can then be erroneous to relate these deflexions to those of any particular point on the cross-section. The shear displacements of a beam (as defined above) can now be determined by integrating the shear slope γ given by (6.4) along the beam, or by using energy methods. The common types of support, such as those shown in Figure 1.4, can be taken as having workless reactions in the above sense. In §4.2, the general differential equation was a fourth-order one relating v and q, which meant that two boundary conditions could be specified at each end of the beam. Here, the relationship between v and q is given by a second order differential equation,

$$S = GkA\gamma = GkA\frac{dv}{dx} \qquad \therefore \quad q = \frac{dS}{dx} = GkA\frac{d^2v}{dx^2} \tag{6.19}$$

using (4.15). This means that in relation to the shear deflexion, only one boundary condition can be specified at each end, usually either the end shear force or the corresponding shear displacement. The difference between the shear deflexion curve and the bending deflexion curve is particularly apparent in the case of a cantilever supporting a uniform distributed load. The slope of the flexural curve is zero at the fixed end and a maximum at the free end. However, because the shear force is a maximum at the support and zero at the free end, so is the shear slope. This effect will be seen again later, in the response of multi-storey buildings to wind loading, as in Figure 15.8.

The slope-deflexion equations modified for shear

The slope-deflexion equations derived in §4.5 take no account of the shear deflexion of the beam. In most common problems, both the shear deflexion and the axial deflexion of a beam are negligible in comparison with the bending deflexion. The effects of shear deflexion can be incorporated in the slope-deflexion equations, using the definitions of displacement and rotation given above. Suppose that the beam shown in Figure 4.17 is rigidly held at its right-hand end so that it can neither rotate nor displace ($\theta_B = v_B = 0$). Integrating the strain energy per unit length given by (4.59) and (6.6) along the beam, the total strain energy in the beam is given by

$$U = \int_0^l \frac{M^2}{2EI} + \frac{S^2}{2GkA}\,dx = \int_0^l \frac{1}{2EI}(M_A - S_A x)^2 + \frac{S_A^2}{2GkA}\,dx \tag{6.20}$$

where x is the distance along the beam from A. The rotation and displacement of the left-hand end, $\theta_A{}^o$ and $v_A{}^o$ say, are the corresponding deflexions to M_A and S_A so that[1] from (6.20),

$$\theta_A^o = \frac{\partial U}{\partial M_A} = \frac{M_A l}{EI} - \frac{S_A l^2}{2EI}$$

$$v_A^o = \frac{\partial U}{\partial S_A} = -\frac{M_A l^2}{2EI} + \frac{S_A l^3}{3EI}(1 + s/4) \tag{6.21}$$

$$\text{where} \quad s = \frac{12EI}{GkAl^2}$$

Inverting these equations to find the end loads in terms of the corresponding deflexions,

$$(1 + s)M_A = \frac{EI}{l}\left[(4 + s)\theta_A^o + \frac{6}{l}v_A^o\right]$$

$$(1 + s)S_A = \frac{EI}{l}\left[\frac{6}{l}\theta_A^o + \frac{12}{l^2}v_A^o\right] \tag{6.22}$$

Suppose that a rigid-body movement is given to the beam and the fixed end as a whole, so that the right-hand end now has an anticlockwise rotation θ_B and an upwards displacement v_B. The rotation θ_A and displacement v_A of the left-hand end are now

$$\theta_A = \theta_A^o + \theta_B \;, \quad v_A = v_A^o + v_B - \theta_B l \tag{6.23}$$

Using the resulting expressions for $\theta_A{}^o$ and $v_A{}^o$ in (6.22) gives

$$M_A = \frac{EI}{l(1 + s)}\left[(4 + s)\theta_A + (2 - s)\theta_B + \frac{6}{l}(v_A - v_B)\right]$$

$$S_A = \frac{EI}{l(1 + s)}\left[\frac{6}{l}(\theta_A + \theta_B) + \frac{12}{l^2}(v_A - v_B)\right] \tag{6.24}$$

From the equations of statics, the end loads at B are given by

$$M_B = \frac{EI}{l(1 + s)}\left[(2 - s)\theta_A + (4 + s)\theta_B + \frac{6}{l}(v_A - v_B)\right]$$

$$S_B = -\frac{EI}{l(1 + s)}\left[\frac{6}{l}(\theta_A + \theta_B) + \frac{12}{l^2}(v_A - v_B)\right] \tag{6.25}$$

These equations may be used in the same way as (4.55) and (4.56).}

{6.5 The Exact Analysis of Flexural Shear}

{The above analysis of the shearing stresses associated with non-uniform bending applies to thin-walled sections where taking the shear stress as parallel to the wall and constant across its thickness is a reasonable approximation. The general analysis required for isotropic prismatic beams was developed by Saint-Venant in 1856 and is given by Love (1952) for example. This method is rather cumbersome and the simpler approach given by Timoshenko and Goodier (1970) will be examined in detail here.

---

[1] See Castigliano's theorems, discussed in Chapter 7.

A shear force S is applied in the $-y$ direction to the free end of the prismatic beam shown in Figure 6.8. It will be assumed that only a bending stress σ_{zz} of the kind given by (4.15) and the shear stresses τ_{xz} and τ_{yz} which are not functions of z are induced by S. The bending stress is then

$$\sigma_{zz} = \frac{Syz}{I} \tag{6.26}$$

In the absence of body forces, (3.3) and (3.4) are then automatically satisfied and (3.5) becomes

$$\frac{\partial \tau_{xz}}{\partial x} + \frac{\partial \tau_{yz}}{\partial y} + \frac{Sy}{I} = 0 \tag{6.27}$$

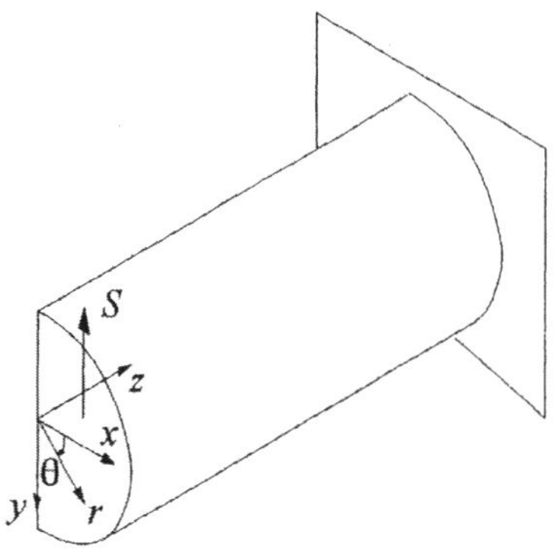

Figure 6.8 Shear of a cantilever.

The strains are given in terms of the above stresses by

$$\epsilon_{xx} = \epsilon_{yy} = -\frac{\nu\sigma_{zz}}{E}, \ \epsilon_{zz} = \frac{\sigma_{zz}}{E}, \ \gamma_{xy} = 0, \ \gamma_{yz} = \frac{\tau_{yz}}{G}, \ \gamma_{zx} = \frac{\tau_{zx}}{G}. \tag{6.28}$$

From the assumed form of these stresses and the relationship between E and G given by (3.49), the compatibility conditions given by (3.14) to (3.16) and (3.19) are automatically satisfied and (3.17) and (3.18) become

$$\frac{\nu S}{I(1+\nu)} = \frac{\partial}{\partial x}\left(\frac{\partial \tau_{yz}}{\partial x} - \frac{\partial \tau_{zx}}{\partial y}\right), \quad 0 = \frac{\partial}{\partial y}\left(\frac{\partial \tau_{yz}}{\partial x} - \frac{\partial \tau_{zx}}{\partial y}\right) \tag{6.29}$$

giving

$$\frac{\partial \tau_{yz}}{\partial x} - \frac{\partial \tau_{zx}}{\partial y} = \frac{\nu S x}{I(1+\nu)} + K \tag{6.30}$$

where K is a constant. If the line of action of S is changed, this is equivalent to adding a pure torque to the loading. The change of displacements in the plane of the cross-section produced by a change of twist ϑ_{ch} is given by (5.16). Then the change to the left-hand side of (6.30) is

$$\left(\frac{\partial \tau_{yz}}{\partial x} - \frac{\partial \tau_{zx}}{\partial y}\right)_{ch} = G\left(\frac{\partial \gamma_{yz}}{\partial x} - \frac{\partial \gamma_{zx}}{\partial y}\right)_{ch}$$
$$= G\left[\frac{\partial}{\partial x}\left(\frac{\partial v}{\partial z} + \frac{\partial w}{\partial y}\right) - \frac{\partial}{\partial y}\left(\frac{\partial u}{\partial z} + \frac{\partial w}{\partial x}\right)\right]_{ch} = G\left[\frac{\partial}{\partial z}\left(\frac{\partial v}{\partial x} - \frac{\partial u}{\partial y}\right)\right]_{ch} = 2G\vartheta_{ch} \tag{6.31}$$

Shifting the line of action of S then changes the value of the constant K by $2G\vartheta_{ch}$. Timoshenko and Goodier (1970) argue that by taking the line of action of S to be such that K is zero, the mean rotation of points on the cross-section is also zero. However, under the action of a shear force, this mean rotation is not necessarily the appropriate indicator of the torsion of the section. If the section is symmetrical about the y axis, setting K zero is sufficient to ensure that the shear force produces no torsion. Here, K will be taken as zero as a first step and the line of action of S (and implicitly the value of K) is adjusted to produce zero torsion later.

A modified form of the torsional stress function ϕ will be used here, taking the general solution of (6.27) for these shear stresses as

$$\tau_{yz} = -\frac{\partial\phi}{\partial x} - \frac{Sy^2}{2I} + f(x) \quad , \quad \tau_{zx} = \frac{\partial\phi}{\partial y} \quad . \tag{6.32}$$

(cf. (5.23)). The only remaining internal condition to be satisfied is (6.30) which now becomes

$$\frac{\partial^2\phi}{\partial x^2} + \frac{\partial^2\phi}{\partial y^2} = \frac{df}{dx} - \frac{\nu Sx}{I(1+\nu)} \tag{6.33}$$

As in the torsion problem, the shear stress normal to the boundary of the section, τ_n, is zero so that if s is the parameter of position around the boundary, (5.22) becomes

$$0 = \tau_n = \tau_{zx}\frac{dy}{ds} - \tau_{zy}\frac{dx}{ds} = \frac{\partial\phi}{\partial y}\frac{dy}{ds} + \left(\frac{\partial\phi}{\partial x} + \frac{Sy^2}{2I} - f(x)\right)\frac{dx}{ds}$$
$$\text{or} \quad \frac{\partial\phi}{\partial s} = \left(f(x) - \frac{Sy^2}{2I}\right)\frac{dx}{ds} \tag{6.34}$$

The initial problem is now to find functions $\phi(x,y)$ and $f(x)$ which satisfy (6.33) within the boundary of the section and (6.34) on the boundary. The function $f(x)$ will be chosen so that the right-hand side of (6.34) is zero at all points on the boundary[1]. The function $\phi(x,y)$ is then constant around the boundary and for a simply-connected cross-section it can be taken as zero without loss of generality, just as in the case of Saint-Venant torsion.

Polar coordinates

By the same methods that were used to derive (5.33) and (5.34), the expressions in (6.32) and (6.33) can be rewritten in polar coordinates as

$$\tau_{rz} = \frac{1}{r}\frac{\partial\phi}{\partial\theta} - \sin\theta\left[\frac{Sy^2}{2I} - f(x)\right] \quad , \quad \tau_{\theta z} = -\frac{\partial\phi}{\partial r} - \cos\theta\left[\frac{Sy^2}{2I} - f(x)\right] . \tag{6.35}$$

$$\nabla^2\phi = \frac{1}{r^2}\left[r\frac{\partial}{\partial r}\left(r\frac{\partial\phi}{\partial r}\right) + \frac{\partial^2\phi}{\partial\theta^2}\right] = \frac{df}{dx} - \frac{\nu Sx}{I(1+\nu)} \tag{6.36}$$

This hybrid form, retaining cartesian terms on the right-hand sides of the equations, seems more convenient than a full conversion to polar coordinates.

Example 6.4 Shear of Circular and Semicircular Sections

For a circular section of radius a, the right-hand side of (6.34) is zero if

$$f(x) = \frac{S}{2I}(a^2 - x^2) \tag{6.37}$$

Then (6.36) becomes

---

[1] This means that where the boundary is perpendicular to the x axis, the value of $f(x)$ is arbitrary.

$$\frac{1}{r^2}\left[r\frac{\partial}{\partial r}\left(r\frac{\partial\phi}{\partial r}\right) + \frac{\partial^2\phi}{\partial\theta^2}\right] = -\frac{S(1+2\nu)}{I(1+\nu)}r\cos\theta \tag{6.38}$$

and for zero ϕ on the boundary $r = a$, the solution is

$$\phi = -\frac{S(1+2\nu)}{8I(1+\nu)}(r^2 - a^2)r\cos\theta \tag{6.39}$$

giving the shear stresses

$$\begin{aligned}\tau_{rz} &= \frac{S(3+2\nu)}{8I(1+\nu)}(a^2 - r^2)\sin\theta \\ \tau_{\theta z} &= \frac{S}{8I(1+\nu)}[a^2(3+2\nu) - r^2(1-2\nu)]\cos\theta\end{aligned} \tag{6.40}$$

where I is $\pi a^4/4$. As $\tau_{\theta z}$ is zero on the y axis ($\theta = \pm\pi/2$), this boundary is also a free surface. This solution then also gives the shear stresses in a semicircular section either to the right or to the left of the y axis, each carrying a shear force of $S/2$.

From the symmetry of the solution, the shear force must act along the y axis and there is no tendency of the bar to twist in either direction. This is not true when the same solution is applied to the semi-circular bar shown in Figure 6.8. Suppose that the line of action of the force S is offset by a distance e along the x axis Then the moment it produces is

$$Se = \int_A \tau_{\theta z} r\,dA \tag{6.41}$$

$$= \int_{-\pi/2}^{\pi/2}\int_0^a \frac{S}{\pi a^4(1+\nu)}[a^2(3+2\nu) - r^2(1-2\nu)]\cos\theta\, r^2\,dr\,d\theta = \frac{8Sa(3+4\nu)}{15\pi(1+\nu)}$$

(bearing in mind the sense of S and $\tau_{\theta z}$ on a positive z face as shown in Figure 6.9).

Here, the position of the shear centre will be found by applying Betti's reciprocal theorem. The two loading systems on the beam consist of a shear force only ($L_1 = S$) applied through the shear centre so that it has a shear displacement of γ per unit length but produces no twist ($D_2 = 0$) and a pure torque only ($L_2' = T$) which produces a rate of twist ($D_2' = \vartheta'$). Then from (3.34)

$$L_1D_1' + L_2D_2' = S\gamma' + 0\times\vartheta' = L_1'D_1 + L_2'D_2 = 0\times\gamma + T\times 0 \quad .$$

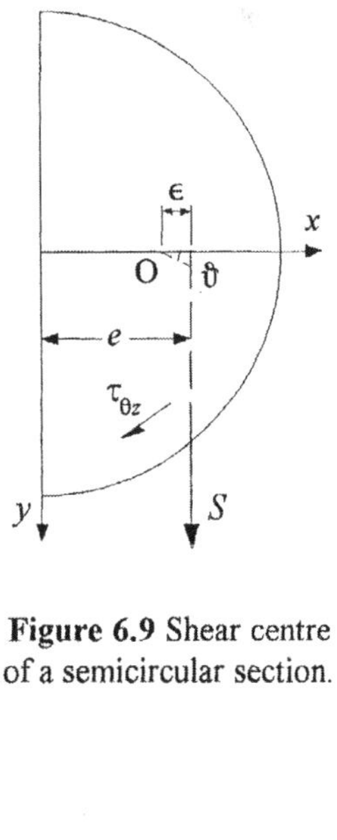

Figure 6.9 Shear centre of a semicircular section.

This means that the shear displacement γ' of the shear centre under the action of a pure torque is zero. Suppose then that the resultant shear force S of the stress system found above acts along a line offset by ϵ from the shear centre O as shown in Figure 6.9. During pure torsion of the section, it will then do work $S\epsilon\vartheta$ per unit length of the bar. This work can be found from the work done by the shear stresses associated with the shear force in moving through the shear strains associated with the torque. Then from (5.36) and (6.40)

$$S\epsilon\vartheta = \int_A (\tau^S_{rz}\gamma^T_{rz} + \tau^S_{\theta z}\gamma^T_{\theta z})dA$$

$$= \frac{S\vartheta a}{\pi(1+\nu)}\int_{\theta=-\pi/2}^{\pi/2}\int_{\rho=0}^{1}(3+2\nu)(1-\rho^2)\sin\theta\left[\rho\sin 2\theta + \frac{8}{\pi}\sum_{m=0}^{\infty}(-\rho^2)^m\frac{\sin(2m+1)\theta}{[(2m+1)^2-4]}\right]$$

$$+[(3+2\nu)-(1-2\nu)\rho^2]\cos\theta\left[2\rho\cos^2\theta + \frac{8}{\pi}\sum_{m=0}^{\infty}(-\rho^2)^m\frac{\cos(2m+1)\theta}{[(2m+1)^2-4]}\right]\rho\, d\rho\, d\theta$$

$$= \frac{8S\vartheta a\nu}{15\pi(1+\nu)} \tag{6.42}$$

where the shear stress distribution produced by the shear force is denoted by τ^S, the shear strain distribution resulting from pure torsion is denoted by γ^T and ρ is r/a. From this, the value of ϵ can be deduced and hence the coordinate of O, x_o say, is given by

$$x_o = e - \epsilon = \frac{8a(3+4\nu)}{15\pi(1+\nu)} - \frac{8a\nu}{15\pi(1+\nu)} = \frac{8a}{5\pi} \tag{6.43}$$

Solutions for the shear of other sections will be found in Appendix 4.

The Trefftz expression for the position of the shear centre

The position of the shear centre depends on the exact definition which is used. Osgood[1] for example lists four possible definitions. On taking Poisson's ratio as 0.3, these give values of x_o in the above case ranging from $0.424a$ to $0.548a$. The value of x_o given by e is attributed to Timoshenko and is quoted by Young (1989). However, in terms of the general beam theory discussed in later chapters, the only meaningful definition is that which uncouples the work done in bending shear from that which is done during pure torsion. It will be seen that in the above case, the position of this shear centre is independent of Poisson's ratio. Trefftz[2] obtained a general expression for the shear centre which shows this to be true in general.

Suppose that the shear force S is acting through the shear centre. The moment it produces about the origin of the coordinate system is then Sx_o. This is the same as the moment produced by its shear stress distribution, or

$$Sx_0 = \int\int(\tau^S_{zy}x - \tau^S_{zx}y)\,dx\,dy \tag{6.44}$$

(cf. (5.27) for example). During the shear strains associated with torsion, the shear stresses associated with the shear force are taken to do no net work. These shear strains are given by (5.17), so that this condition is given by

$$\vartheta\int\int\left\{\tau^S_{zx}\left(\frac{\partial g}{\partial x} - y\right) + \tau^S_{zy}\left(\frac{\partial g}{\partial y} + x\right)\right\}dx\,dy = 0 \tag{6.45}$$

from which it follows that

---

[1] Osgood, W.R. (1943) The Center of Shear Again. *J. Appl. Mech.* Trans. A.S.M.E. V.65 p.A-62.

[2] Trefftz, E. (1935) Über den Schubmittelpunkt in einem durch eine Einzellast gebogenen Balken. *Ztschr. f. angew. Math. und Mech.* V.15 p.220.

$$\iint (x\tau_{zy}^S - y\tau_{zx}^S)\,dx\,dy = -\iint\left(\tau_{zx}^S\frac{\partial g}{\partial x} + \tau_{zy}^S\frac{\partial g}{\partial y}\right)dx\,dy \tag{6.46}$$

From the plane form of Green's theorem or partial integration, the right hand side of (6.46) can be written as

$$\iint\left(\tau_{zx}^S\frac{\partial g}{\partial x} + \tau_{zy}^S\frac{\partial g}{\partial y}\right)dx\,dy = -\iint g\left(\frac{\partial \tau_{zx}^S}{\partial x} + \frac{\partial \tau_{zy}^S}{\partial y}\right)dx\,dy + \oint g(\tau_{zx}^S dy - \tau_{zy}^S dx) \tag{6.47}$$

However, as τ_n^S is zero, the second integrand on the right-hand side of (6.47) is zero (cf. (6.34)). From (6.27), the first bracketed term on the right-hand side is given by $-Sy/I$. It then follows from the above equations that

$$Sx_o = -\frac{S}{I}\iint gy\,dx\,dy \quad \text{or} \quad x_o = -\frac{1}{I}\int_A gy\,dA \tag{6.48}$$

In the particular case of the semicircular section analysed above,

$$g = a^2\left[\frac{1}{2}\rho^2\sin 2\theta + \frac{8}{\pi}\sum_{m=0}^{\infty}\frac{(-1)^m\rho^{2m+1}}{2m+1}\frac{\sin(2m+1)\theta}{(2m+1)^2-4}\right] \tag{6.49}$$

which gives the same position of the shear centre as before.

The general expression for the shear strain energy and the shear stiffness

In the following analysis[1], only simply-connected sections will be considered, so that ϕ can be taken as zero on the contour of the section. However, the form of the shear stiffness applies more generally, as will be seen from the example of a circular tube. From (6.32), the shear strain energy per unit length is given by

$$\begin{aligned} U_S &= \frac{1}{2G}\int_A (\tau_{xz}^2 + \tau_{yz}^2)\,dA \\ &= \frac{1}{2G}\int_A \left[\left(\frac{\partial\phi}{\partial x}\right)^2 + \left(\frac{\partial\phi}{\partial y}\right)^2\right] + \left[\frac{Sy^2}{2I} - f\right]\left[2\frac{\partial\phi}{\partial x} + \frac{Sy^2}{2I} - f\right]dA \end{aligned} \tag{6.50}$$

where A is the area of the cross-section. From Stokes' theorem,

$$\int_A\left[\frac{\partial}{\partial x}\left(\phi\frac{\partial\phi}{\partial x}\right) + \frac{\partial}{\partial y}\left(\phi\frac{\partial\phi}{\partial y}\right)\right]dA = \int_C \phi\left[-\frac{\partial\phi}{\partial y}dx + \frac{\partial\phi}{\partial x}dy\right] = 0 \tag{6.51}$$

where the second integral is around the contour of the section and is zero because ϕ is zero on the boundary. It then follows from (6.33) and (6.51) that

$$\int_A\left[\left(\frac{\partial\phi}{\partial x}\right)^2 + \left(\frac{\partial\phi}{\partial y}\right)^2\right]dA = -\int_A \phi\left(\frac{\partial^2\phi}{\partial x^2} + \frac{\partial^2\phi}{\partial y^2}\right)dA = -\int_A \phi\left(\frac{df}{dx} - \frac{\nu}{1+\nu}\frac{Sx}{I}\right)dA \tag{6.52}$$

[1] Renton, J.D. (1997) A note on the form of the shear coefficient. *Int. J. Solids Structures* V.34, p.1681.

Also, it follows from Stokes' theorem that

$$\int_A 2\frac{\partial}{\partial x}\left[\phi\left(\frac{Sy^2}{2I} - f\right)\right] dA = \int_C 2\phi\left(\frac{Sy^2}{2I} - f\right) dy = 0$$

$$\text{or} \quad \int_A 2\frac{\partial\phi}{\partial x}\left(\frac{Sy^2}{2I} - f\right) dA = \int_A 2\phi\frac{df}{dx} dA \tag{6.53}$$

From (6.52) and (6.53), (6.50) now becomes

$$U_S = \frac{1}{2G}\int_A \left(\frac{Sy^2}{2I} - f\right)^2 + \phi\left(\frac{df}{dx} + \frac{\nu}{1+\nu}\frac{Sx}{I}\right) dA \tag{6.54}$$

Suppose that

$$\phi = \phi_0 + \frac{\nu}{1+\nu}\phi_1 \tag{6.55}$$

where ϕ_0 is the solution for the case when Poisson's ratio is zero. As ϕ_0 and ϕ are zero on the boundary of the section, ϕ_1 will also be zero. From (6.33), the conditions on these new functions within the cross-sectional area are

$$\frac{\partial^2\phi_0}{\partial x^2} + \frac{\partial^2\phi_0}{\partial y^2} = \frac{df}{dx} \quad , \quad \frac{\partial^2\phi_1}{\partial x^2} + \frac{\partial^2\phi_1}{\partial y^2} = -\frac{Sx}{I} . \tag{6.56}$$

Then neither of these are functions of Poisson's ratio. Now

$$\int_A \phi\left(\frac{df}{dx} + \frac{\nu}{1+\nu}\frac{Sx}{I}\right) dA$$

$$= \int_A \left(\phi_0 + \frac{\nu}{1+\nu}\phi_1\right)\left[\frac{\partial^2\phi_0}{\partial x^2} + \frac{\partial^2\phi_0}{\partial y^2} - \frac{\nu}{1+\nu}\left(\frac{\partial^2\phi_1}{\partial x^2} + \frac{\partial^2\phi_1}{\partial y^2}\right)\right] dA \tag{6.57}$$

But from the plane form of Green's theorem, if n is the local outward normal to the boundary of the section and s is measured anticlockwise around this boundary,

$$\int_A \left[-\phi_0\left(\frac{\partial^2\phi_1}{\partial x^2} + \frac{\partial^2\phi_1}{\partial y^2}\right) + \phi_1\left(\frac{\partial^2\phi_0}{\partial x^2} + \frac{\partial^2\phi_0}{\partial y^2}\right)\right] dA$$

$$= \int_C \left(-\phi_0\frac{\partial\phi_1}{\partial n} + \phi_1\frac{\partial\phi_0}{\partial n}\right) ds = 0 \tag{6.58}$$

The last equality holds because ϕ_0 and ϕ_1 are zero on the boundary. Then from (6.57) and (6.58), (6.54) becomes

$$U_S = \frac{1}{2G}\int_A \left[\frac{Sy^2}{2I} - f\right]^2 + \phi_0\left[\frac{\partial^2\phi_0}{\partial x^2} + \frac{\partial^2\phi_0}{\partial y^2}\right] - \left(\frac{\nu}{1+\nu}\right)^2 \phi_1\left[\frac{\partial^2\phi_1}{\partial x^2} + \frac{\partial^2\phi_1}{\partial y^2}\right] dA \qquad \textbf{(6.59)}$$

It follows from (6.34) and (6.56) that f, ϕ_0 and ϕ_1 will be directly proportional to S/I and that none of them is a function of Poisson's ratio. Then if

$$f = F\frac{S}{I}\ ,\quad \phi_0 = \Phi_0\frac{S}{I}\ ,\quad \phi_1 = \Phi_1\frac{S}{I} \qquad \textbf{(6.60)}$$

(6.59) can be written as

$$U_S = \frac{S^2}{2GA}\left[B + C\left(\frac{\nu}{1+\nu}\right)^2\right] = \frac{S^2}{2GkA} \quad \text{or} \quad k = \frac{1}{B + C\left(\dfrac{\nu}{1+\nu}\right)^2} \qquad \textbf{(6.61)}$$

where

$$B = \frac{A}{I^2}\int_A \left[\left(\frac{1}{2}y^2 - F\right)^2 + \Phi_0\left(\frac{\partial^2\Phi_0}{\partial x^2} + \frac{\partial^2\Phi_0}{\partial y^2}\right)\right] dA$$

$$C = -\frac{A}{I^2}\int_A \Phi_1\left[\frac{\partial^2\Phi_1}{\partial x^2} + \frac{\partial^2\Phi_1}{\partial y^2}\right] dA \qquad \textbf{(6.62)}$$

and GkA is the shear stiffness of the section (cf. (6.4)). The conditions governing Φ_0, Φ_1 and F are

$$\frac{\partial^2\Phi_0}{\partial x^2} + \frac{\partial^2\Phi_0}{\partial y^2} = \frac{dF}{dx}\ ,\quad \frac{\partial^2\Phi_1}{\partial x^2} + \frac{\partial^2\Phi_1}{\partial y^2} = -x \qquad \textbf{(6.63)}$$

within the cross-section and

$$\Phi_0 = \Phi_1 = (F - \tfrac{1}{2}y^2)\frac{dx}{ds} = 0 \qquad \textbf{(6.64)}$$

on the boundary. For example, for a circular section of radius R,

$$F = \tfrac{1}{2}(R^2 - x^2)\ ,\quad -\Phi_0 = \Phi_1 = \tfrac{1}{8}(R^2 - x^2 - y^2)x\ ,\quad A = \pi R^2\ ,\quad I = \frac{\pi R^4}{4} \qquad \textbf{(6.65)}$$

giving $B = 7/6$, $C = 1/6$.

The expression for k is valid for the full range of admissible values of ν, which lie between the limits of minus one and plus one half. An upper limit to the stiffness can be found using minimum potential energy principles, as detailed in §7.2 and §7.4. A kinematically-admissible displacement field is applied to the beam and the work done by the end loading is compared with the internal strain energy stored. This yields a beam stiffness which is greater than (or equal to) the true beam stiffness. In the present case, the work done in shearing is compared with the shear strain energy. Take the shear deformation to be a constant value γ along the beam and over the cross-section. Then for a beam of length ℓ the work done by the shear force S during shear is $\frac{1}{2}S\gamma\ell$ and the shear strain energy stored is $\frac{1}{2}G\gamma^2A\ell$. On comparing the two, an upper limit for

GkA is given by S/γ which is equal to GA. Examining the expression for GkA in (6.61) for the particular case when ν is zero then shows that B cannot be less than unity. As ν tends towards minus one, the term in C becomes dominant in the denominator of this expression. As k must be positive, it follows that C cannot be negative.

Shear of a circular tube

A circular tube does not have a simply-connected cross-section, so that the above analysis, based on Timoshenko's approach, can no longer be applied. Love (1952) §231 describes the older Saint-Venant solution for prismatic beams subject to bending and shear in which the state of stress is deduced from a function χ. For a circular tube of inner radius b and outer radius a, χ is given in polar coordinates by

$$\chi = -(\tfrac{3}{4} + \tfrac{1}{2}\nu)\left[(a^2 + b^2)r + \frac{a^2b^2}{r}\right]\cos\theta + \tfrac{1}{4}r^3\cos 3\theta + \text{const.} \qquad \textbf{(6.66)}$$

where r is the radius from the centre and θ is measured anticlockwise from the inwards direction of the shear force. After some manipulation, this yields the shear stresses in polar coordinates

$$\begin{aligned} \tau_{rz} &= \frac{S\cos\theta}{2(1+\nu)I}(\tfrac{3}{4} + \tfrac{1}{2}\nu)\left[r^2 - a^2 - b^2 + \frac{a^2b^2}{r^2}\right] \\ \tau_{\theta z} &= \frac{S\sin\theta}{2(1+\nu)I}\left[(\tfrac{3}{4} + \tfrac{1}{2}\nu)\left(a^2 + b^2 + \frac{a^2b^2}{r^2}\right) + (\tfrac{1}{2}\nu - \tfrac{1}{4})r^2\right] \end{aligned} \qquad \textbf{(6.67)}$$

As expected, the average tangential shear stress is that predicted from the simple engineering theory. The shear stiffness can be deduced from the shear strain energy and is given by

$$GkA = \frac{6GA(a^2 + b^2)^2}{(7a^4 + 34a^2b^2 + 7b^4) + (a^2 - b^2)^2\left(\dfrac{\nu}{1+\nu}\right)^2} \qquad \textbf{(6.68)}$$

This reduces to the solution for a solid circular section as b tends to zero, but see also §A4.1c. It also corresponds to the general form for GkA found previously.}

{6.6 Non-Prismatic and Inhomogeneous Beams}

{Consider again the problem of a plane wedge of unit thickness examined in §4.3. This time, a shear force S acts at the apex of the wedge cantilever shown in Figure 6.10 and normal to its axis. This can be carried by radial stresses σ_{rr} only, where

$$\sigma_{rr} = \frac{2S\sin\theta}{r(\alpha - \sin\alpha)} \qquad \textbf{(6.69)}$$

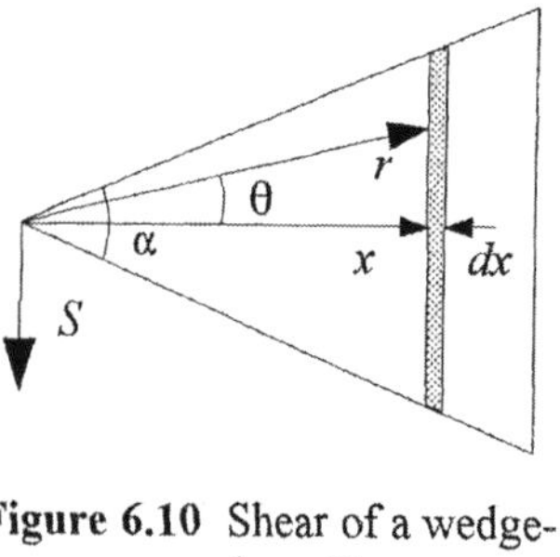

Figure 6.10 Shear of a wedge-shaped beam.

(Both the response to an axial force and to a shear force are given in the paper by Michell cited in §4.3.)

The work done by S is equal to half its final value multiplied by the displacement through which it moves. It is also equal to the strain energy stored in the cantilever. Consider the strain energy stored in a strip of width dx perpendicular to the axis and at a distance x along it from the apex. The approximate strain energy given by ordinary beam theory, U_a say, is the bending strain energy given by the second or third term in (4.59) and the shear strain energy given by (6.4). This gives

$$\begin{aligned} U_a &= \left(\frac{M^2}{2EI} + \frac{S^2}{2GkA} \right) dx \\ &= \left(\frac{12S^2x^2}{2E(2x\tan\alpha/2)^3} + \frac{1.2S^2}{2G(2x\tan\alpha/2)} \right) dx \end{aligned} \quad \textbf{(6.70)}$$

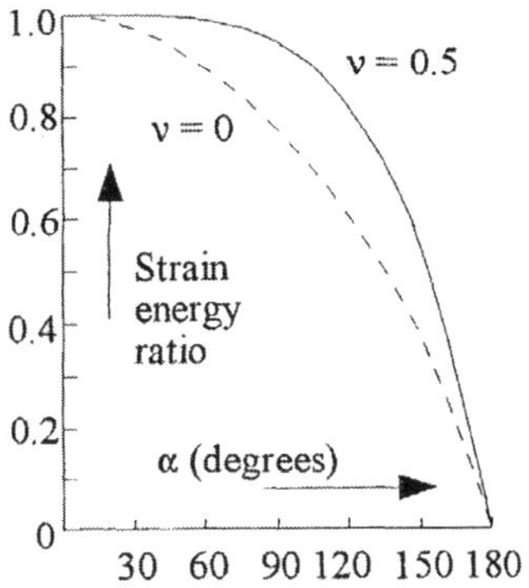

Figure 6.11 Ratio of approximate strain energy to exact strain energy.

It follows from (6.69) that the exact strain energy in the strip is given by

$$U_e = \int_A \frac{\sigma_{rr}^2}{2E} dA = \frac{1}{2E} \int_{-\alpha/2}^{\alpha/2} \left(\frac{2S\sin\theta}{r(\alpha - \sin\alpha)} \right)^2 \{x\sec^2\theta\, d\theta\} dx = \frac{S^2 dx}{xE(\alpha - \sin\alpha)} \quad \textbf{(6.71)}$$

where the term in curly brackets is the length of the strip which subtends an angle $d\theta$. The ratio of U_a/U_e is shown in Figure 6.11. As will be seen in Chapter 7, this ratio is less than or equal to one. This indicates that the approximate solution predicts a stiffer response of the wedge than the correct result, because the work done by the shear force S must also be less than the true value. It is not possible to assign an exact simple shear stiffness to a typical section of the wedge. This is because there is coupling between bending and shear effects. This coupling will be discussed further in Chapter 8.

The effect of a shear force on a plane, curved beam was determined by Golovin (op. cit. §4.3). The stresses produced in the beam at an angle θ from the plane on which S acts, as measured from the centre of curvature, and at a radius r from it are

$$\begin{aligned} \sigma_{rr} &= \frac{S}{Nr^3}[r^4 + a^2b^2 - r^2(a^2+b^2)]\sin\theta \\ \sigma_{\theta\theta} &= \frac{S}{Nr^3}[3r^4 - a^2b^2 - r^2(a^2+b^2)]\sin\theta \\ \tau_{r\theta} &= -\frac{S}{Nr^3}[r^4 + a^2b^2 - r^2(a^2+b^2)]\cos\theta \end{aligned}$$

$$\text{where} \quad N = a^2 - b^2 + (a^2+b^2)\ln\frac{b}{a} \quad \textbf{(6.72)}$$

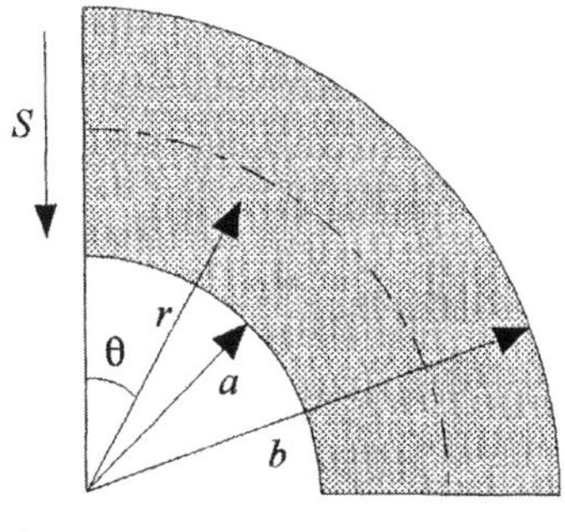

Figure 6.12 A curved beam (b/a=2).

The axial and shear stresses, $\sigma_{\theta\theta}$ and $\tau_{r\theta}$, predicted by the elementary theory of beams vary with θ in the same way. They are given by

$$\sigma_{\theta\theta} = \frac{S}{(b-a)^3}[6r(a+b) - 4(a^2+ab+b^2)]\sin\theta$$
$$\tau_{r\theta} = \frac{6S}{(b-a)^3}[r(a+b) - r^2 - ab]\cos\theta \quad (6.73)$$

The stresses predicted by the two analyses are compared in Figure 6.13 for the case when b/a is two. In each case the local stress for some value of r/a is expressed as a proportion of the mean stress. The ratios found from (6.72) are shown by solid curves and those found from (6.73) are shown by dotted curves. The curve for σ_{rr} as given by (6.72) is not shown, but it follows from this equation that it is the same in form as the $\tau_{r\theta}$ curve. It will be seen that the difference between the two analyses is quite small, considering the degree of curvature implied by this b/a ratio. (A beam with this curvature is shown in Figure 6.12).)

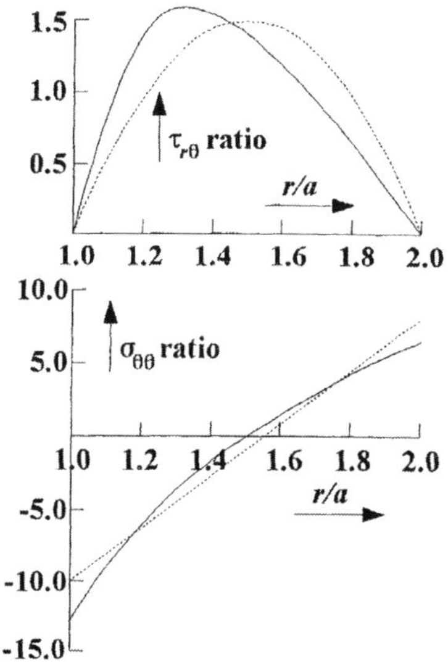

Figure 6.13 Stress ratios for b/a=2.

Muskhelisvili (1963) examines the problem of shear force applied to a compound bar composed of different isotropic materials with a common Poisson's ratio in §144. The solution for the specific case of a hollow circular bar composed of two concentric tubes is given, and the solutions for the cases where the circular boundaries are not concentric, for confocal ellipses and for epitrochoids (clover-leaf forms) are listed. The solution of the problem when the materials do not have a common Poisson's ratio is examined in §150. It is noted that the bending stiffness is again given by the bending stiffness for the case when the materials have a common Poisson's ratio plus a term K_{11} of the kind given by (4.49).}

{6.7 Anisotropic Beams}

{The general form of this problem was first stated by Voigt in 1928 and examined in more detail by Lekhnitskii (1981). The analysis will be summarised here, with some additional comments and corrections. It was seen in §4.7 that a suitable starting point for the flexure of anisotropic beams was to take the same linear variation of the axial stress σ_{zz} across the section as used for isotropic beams. It will be seen later that the characteristic response to a shear force S is given by this response to the moment produced by S locally combined with a response which is constant along the beam. In the example shown in Figure 6.14, this means that all the stresses are functions of x and y only, apart from σ_{zz} which will have the additional term Szy/I_{xx} . The equilibrium equations (3.3) to (3.5) then become

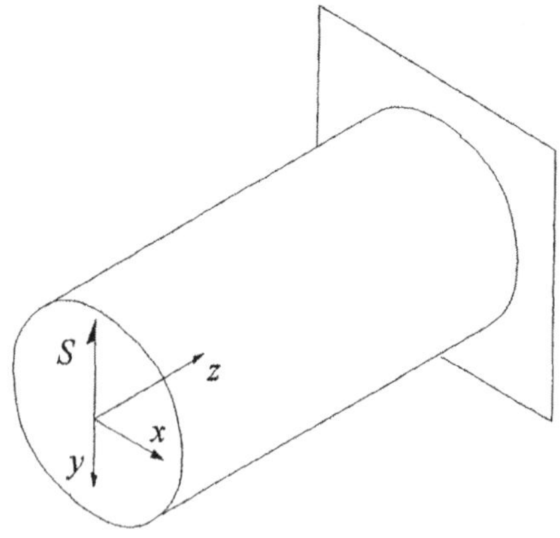

Figure 6.14 An anisotropic cantilever.

$$\frac{\partial \sigma_{xx}}{\partial x} + \frac{\partial \tau_{xy}}{\partial y} = 0 \;, \quad \frac{\partial \tau_{xy}}{\partial x} + \frac{\partial \sigma_{yy}}{\partial y} = 0 \;, \quad \frac{\partial \tau_{xz}}{\partial x} + \frac{\partial \tau_{yz}}{\partial y} + \frac{Sy}{I_{xx}} = 0. \tag{6.74}$$

From (3.63), the stress-strain relationships are given by

$$\begin{bmatrix} \epsilon_{xx} \\ \epsilon_{yy} \\ \epsilon_{zz} \\ \gamma_{yz} \\ \gamma_{zx} \\ \gamma_{xy} \end{bmatrix} = \begin{bmatrix} s_{11} & s_{12} & s_{13} & s_{14} & s_{15} & s_{16} \\ s_{21} & s_{22} & s_{23} & s_{24} & s_{25} & s_{26} \\ s_{31} & s_{32} & s_{33} & s_{34} & s_{35} & s_{36} \\ s_{41} & s_{42} & s_{43} & s_{44} & s_{45} & s_{46} \\ s_{51} & s_{52} & s_{53} & s_{54} & s_{55} & s_{56} \\ s_{61} & s_{62} & s_{63} & s_{64} & s_{65} & s_{66} \end{bmatrix} \begin{bmatrix} \sigma_{xx} \\ \sigma_{yy} \\ \sigma^o_{zz} + Szy/I_{xx} \\ \tau_{yz} \\ \tau_{zx} \\ \tau_{xy} \end{bmatrix} \tag{6.75}$$

where σ^o_{zz} is not a function of z and the matrix is symmetrical. The general solution is given by

$$\begin{aligned} u &= \frac{Sz}{2I_{xx}}(2s_{13}xy + s_{63}y^2 + \tfrac{1}{2}s_{53}yz) - \tfrac{1}{2}Pz^2 - \vartheta_z yz + U(x,y) \\ v &= \frac{Sz}{2I_{xx}}(-s_{13}x^2 - \tfrac{1}{2}s_{53}xz + s_{23}y^2 - \tfrac{1}{3}s_{33}z^2) - \tfrac{1}{2}Qz^2 + \vartheta_z xz + V(x,y) \\ w &= \frac{Sz}{2I_{xx}}(s_{53}xy + s_{43}y^2 + s_{33}yz) + (Px + Qy + R)z + W(x,y) \end{aligned} \tag{6.76}$$

excluding arbitrary rigid-body displacements. (The differential of these forms with respect to z may be compared with (4.64). This is because the result is then the linear response to a stress system which is the differential of the above σ_{zz}, i.e. Sy/I_{xx}.) The conditions that the resultant force on the cross-section is the shear force S and the resultant bending moment on it is Sz are expressed by

$$\int_A \tau_{xz}\,dA = 0 \;, \quad \int_A \tau_{yz}\,dA = S \;, \quad \int_A \sigma_{zz}\,dA = 0 \;, \quad \int_A \sigma_{zz}y\,dA = Sz \;, \quad \int_A \sigma_{zz}x\,dA = 0. \tag{6.77}$$

The first two conditions are satisfied automatically if the above equilibrium equations are satisfied and the third to fifth equations imply that

$$P = Q = 0 \;, \quad R = \frac{Ss_{34}}{2A}. \tag{6.78}$$

If the shear centre lies on the y axis, as would be the case if the beam were fully symmetrical about this axis, then no torque will be produced about the centroid by a bending shear force. This gives rise to an equation from which the twist ϑ_z can be determined:

$$\int_A (\tau_{yz}x - \tau_{xz}y)\,dA = 0 \tag{6.79}$$

Then on writing the strains in terms of u, v and w and using the stress-strain relationships (6.75), U, V and W must satisfy the conditions

$$
\begin{aligned}
\frac{\partial U}{\partial x} &= b_{11}\sigma_{xx} + b_{12}\sigma_{yy} + b_{14}\tau_{yz} + b_{15}\tau_{zx} + b_{16}\tau_{xy} + s_{13}D \\
\frac{\partial V}{\partial y} &= b_{21}\sigma_{xx} + b_{22}\sigma_{yy} + b_{24}\tau_{yz} + b_{25}\tau_{zx} + b_{26}\tau_{xy} + s_{23}D \\
\frac{\partial W}{\partial y} &= b_{41}\sigma_{xx} + b_{42}\sigma_{yy} + b_{44}\tau_{yz} + b_{45}\tau_{zx} + b_{46}\tau_{xy} + s_{43}D + \frac{S}{2I_{xx}}(s_{13}x^2 - s_{23}y^2) - \vartheta_z x \\
\frac{\partial W}{\partial x} &= b_{51}\sigma_{xx} + b_{52}\sigma_{yy} + b_{54}\tau_{yz} + b_{55}\tau_{zx} + b_{56}\tau_{xy} + s_{53}D - \frac{S}{2I_{xx}}(2s_{13}xy + s_{63}y^2) + \vartheta_z y \\
\frac{\partial U}{\partial y} + \frac{\partial V}{\partial x} &= b_{61}\sigma_{xx} + b_{62}\sigma_{yy} + b_{64}\tau_{yz} + b_{65}\tau_{zx} + b_{66}\tau_{xy} + s_{63}D
\end{aligned}
\qquad \textbf{(6.80)}
$$

where

$$
b_{ij} = s_{ij} - \frac{s_{i3}s_{j3}}{s_{33}} \quad , \quad D = \frac{S}{2s_{33}}\left[\frac{s_{34}}{A} + \frac{1}{I_{xx}}(s_{35}xy + s_{34}y^2)\right] \qquad \textbf{(6.81)}
$$

and

$$
\sigma_{zz} = \frac{Szy}{I_{xx}} + D - \frac{1}{s_{33}}(s_{13}\sigma_{xx} + s_{23}\sigma_{yy} + s_{34}\tau_{yz} + s_{35}\tau_{zx} + s_{36}\tau_{xy}) \qquad \textbf{(6.82)}
$$

(The proof that the integrals of certain expressions over the cross-section disappear is left to the reader. It will also be found in §19 of Lekhnitskii's book.)

The equilibrium equations (6.74) can be satisfied by the use of stress functions Ψ and Φ similar to those used by Airy and Prandtl, where

$$
\sigma_{xx} = \frac{\partial^2\Psi}{\partial y^2} \; , \; \sigma_{yy} = \frac{\partial^2\Psi}{\partial x^2} \; , \; \tau_{xy} = -\frac{\partial^2\Psi}{\partial x \partial y} \; , \; \tau_{xz} = \frac{\partial\Phi}{\partial y} + \tau_1 \; , \; \tau_{yz} = -\frac{\partial\Phi}{\partial x} + \tau_2 \qquad \textbf{(6.83)}
$$

and τ_1 and τ_2 are arbitrary particular solutions of

$$
\frac{\partial\tau_1}{\partial x} + \frac{\partial\tau_2}{\partial y} + \frac{Sy}{I_{xx}} = 0 \qquad \textbf{(6.84)}
$$

By double differentiation of the first, second and fifth equations of (6.80) with respect to x and/or y and then combining them, U and V can be eliminated, giving an equation in terms of the above stress functions. Likewise, differentiating the third equation with respect to x and subtracting the fourth differentiated with respect to y yields an equation independent of W. These are

$$
\begin{aligned}
\nabla_4\Psi + \nabla_3\Phi &= \frac{S}{2I_{xx}a_{33}}(a_{35}a_{36} - 2a_{13}a_{34}) - \frac{\partial^2}{\partial y^2}(b_{14}\tau_2 + b_{15}\tau_1) - \frac{\partial^2}{\partial x^2}(b_{24}\tau_2 + b_{25}\tau_1) \\
&\quad + \frac{\partial^2}{\partial x \partial y}(b_{64}\tau_2 + b_{65}\tau_1) \\
\nabla_3\Psi + \nabla_2\Phi &= \frac{S}{2I_{xx}a_{33}}[(4a_{13}a_{33} - a_{35}^2)x + (2a_{33}a_{36} - a_{34}a_{35})y] - 2\vartheta_z \\
&\quad + \frac{\partial}{\partial x}(b_{44}\tau_2 + b_{45}\tau_1) - \frac{\partial}{\partial y}(b_{45}\tau_2 + b_{55}\tau_1)
\end{aligned}
\qquad \textbf{(6.85)}
$$

where

$$\nabla_2 = b_{44}\frac{\partial^2}{\partial x^2} - 2b_{45}\frac{\partial^2}{\partial x \partial y} + b_{55}\frac{\partial^2}{\partial y^2}$$

$$\nabla_3 = -b_{24}\frac{\partial^3}{\partial x^3} + (b_{25}+b_{46})\frac{\partial^3}{\partial x^2 \partial y} - (b_{14}+b_{56})\frac{\partial^3}{\partial x \partial y^2} + b_{15}\frac{\partial^3}{\partial y^3} \tag{6.86}$$

$$\nabla_4 = b_{22}\frac{\partial^4}{\partial x^4} - 2b_{26}\frac{\partial^4}{\partial x^3 \partial y} + (2b_{12}+b_{66})\frac{\partial^4}{\partial x^2 \partial y^2} - 2b_{16}\frac{\partial^4}{\partial x \partial y^3} + b_{11}\frac{\partial^4}{\partial y^4}$$

For the lateral surfaces of the beam to be free from stress,

$$\frac{\partial \Psi}{\partial x} = C_1 \ , \quad \frac{\partial \Psi}{\partial y} = C_2 \ , \quad \Phi = \int_0^s (\tau_2 dx - \tau_1 dy) + C_3 \tag{6.87}$$

on the boundary of the section, where the constants C_1 to C_3 can be taken as zero for a simply-connected section without loss of generality. Also, it may be possible to choose the ratio of τ_1 to τ_2 so that the integral of $(\tau_2\, dx - \tau_1\, dy)$ around the boundary is zero. Then both Ψ and Φ can be taken as zero on the boundary of the section.

Example 6.5 Shear of a homogeneous cantilever with an elliptic cross-section

Suppose that the cantilever shown in Figure 6.14 has an elliptic cross-section with semi-axes of lengths a and b in the x and y directions respectively. The values of I_{xx} is then $\pi a b^3/4$ and by taking

$$\tau_1 = 0 \ , \quad \tau_2 = \frac{2S}{\pi ab}\left(1 - \frac{x^2}{a^2} - \frac{y^2}{b^2}\right) \tag{6.88}$$

(6.84) is satisfied and τ_1 and τ_2 are zero on the boundary, so that Ψ and Φ can be taken as zero on the boundary too. To satisfy (6.87), they take the form

$$\Psi = p\left(\frac{x^2}{a^2} + \frac{y^2}{b^2} - 1\right)^2 \ , \quad \Phi = (qx + ry)\left(\frac{x^2}{a^2} + \frac{y^2}{b^2} - 1\right) \tag{6.89}$$

Substituting these expressions into (6.85) gives three equations for p, q and r. (This is because the coefficients of x and y in the second equation must be compared separately.) If there is material as well as geometric symmetry about the y axis, it follows from (6.76) that ϑ_z will be zero. The particular case of an orthotropic[1] beam was solved by Saint-Venant[2]. In this case, p and r are zero and

$$q = \frac{2S(b^2 s_{44} - 2a^2 s_{13})}{\pi ab(3b^2 s_{44} + a^2 s_{55})} \tag{6.90}$$

---

[1] The properties of an orthotropic material are invariant under inversion of all three axes. This means that $b_{ij} = s_{ij} = 0$, $(i \neq j,\ 4 \leq j \leq 6)$, leaving nine independent elastic constants.

[2] Saint-Venant, B. (1856) Mémoire sur la flexion des prisms *J. math. pures et appl. (Liouville)* Sér II, V.1, p.89.

The shear stresses resulting from this analysis are

$$\tau_{xz} = -\frac{4S(b^2 s_{44} - 2a^2 s_{13})}{\pi a b^3 (3b^2 s_{44} + a^2 s_{55})} xy$$

$$\tau_{yz} = -\frac{2S}{\pi a b^3 (3b^2 s_{44} + a^2 s_{55})} \{b^2 x^2 (s_{55} + 6s_{13}) + (y^2 - b^2)[2b^2 s_{44} + a^2 (s_{55} + 2s_{13})]\} \tag{6.91}$$

Chapter 7 Energy Methods

7.1 Introduction

As the theory of elastic beams and frames rests on the laws of thermodynamics, the concepts of energy and potential are fundamental to structural analysis. That is why energy, equilibrium and stability were introduced as early as §1.3 and the conditions of static equilibrium were expressed in terms of work and energy early in Chapter 2. The reciprocal theorems of Betti *et al.* (§3.4) and Saint-Venant's principle (§3.6), which forms the real basis of beam theory, also follow from the use of energy methods.

This chapter is concerned with exact and approximate methods of obtaining solutions to problems in elasticity. Firstly, the previous expressions for strain energy will be summarised. The strain energy per unit volume, U_V, of a linearly-elastic body can be expressed in matrix form as $\frac{1}{2}\sigma^T S \sigma$ (cf. (3.63) and (3.65)) which for an isotropic material becomes

$$\begin{aligned} U_V &= \tfrac{1}{2}[\sigma_{xx}\epsilon_{xx} + \sigma_{yy}\epsilon_{yy} + \sigma_{zz}\epsilon_{zz} + \tau_{xy}\gamma_{xy} + \tau_{yz}\gamma_{yz} + \tau_{zx}\gamma_{zx}] \\ &= \frac{1}{2E}[\sigma_{xx}^2 + \sigma_{yy}^2 + \sigma_{zz}^2 - 2\nu(\sigma_{xx}\sigma_{yy} + \sigma_{yy}\sigma_{zz} + \sigma_{zz}\sigma_{xx})] + \frac{1}{2G}[\tau_{xy}^2 + \tau_{yz}^2 + \tau_{zx}^2] \end{aligned} \quad (7.1)$$

as in (3.37). In terms of strains, this becomes $\frac{1}{2}\epsilon^T C \epsilon$ (cf. §3.11) or, for an isotropic body,

$$U_V = \frac{G\nu}{1-2\nu}(\epsilon_{xx} + \epsilon_{yy} + \epsilon_{zz})^2 + G(\epsilon_{xx}^2 + \epsilon_{yy}^2 + \epsilon_{zz}^2) + \frac{1}{2}G(\gamma_{xy}^2 + \gamma_{yz}^2 + \gamma_{zx}^2) \quad (7.2)$$

For the case of plane stress, ($\tau_{xz} = \tau_{yz} = \sigma_{zz} = 0$), this becomes

$$U_V = \frac{G}{1-\nu}[\epsilon_{xx}^2 + \epsilon_{yy}^2 + 2\nu\epsilon_{xx}\epsilon_{yy}] + \frac{1}{2}G\gamma_{xy}^2 \quad (7.3)$$

and for plane strain ($\gamma_{xz} = \gamma_{yz} = \epsilon_{zz} = 0$), (3.37) becomes

$$U_V = \frac{1}{4G}[(1-\nu)(\sigma_{xx}^2 + \sigma_{yy}^2) - 2\nu\sigma_{xx}\sigma_{yy}] + \frac{1}{2G}\tau_{xy}^2 \quad (7.4)$$

The strain energy per unit length, U_l, of an homogeneous isotropic beam has been expressed in terms of the resultant loads acting on it locally. For axial loading and bending moments about the principal axes it is given by (4.59), for torsion by (5.45) and for shear by (6.6). Allowing for shear in the x and y directions, the general expression is then

$$U_l = \frac{M_x^2}{2EI_{xx}} + \frac{M_y^2}{2EI_{yy}} + \frac{T^2}{2GJ} + \frac{S_x^2}{2Gk_xA} + \frac{S_y^2}{2Gk_yA} + \frac{P^2}{2EA} \quad (7.5)$$

or, in terms of the corresponding rates of rotation ϑ_x, ϑ_y and ϑ_z about the x,y and z axes of the beam and the rates displacement γ_x, γ_y and ε_z caused by the deformation of the beam,

$$U_l = \frac{1}{2}EI_{xx}\vartheta_x^2 + \frac{1}{2}EI_{yy}\vartheta_y^2 + \frac{1}{2}GJ\vartheta_z^2 + \frac{1}{2}Gk_xA\gamma_x^2 + \frac{1}{2}Gk_yA\gamma_y^2 + \frac{1}{2}EA\varepsilon_z^2 \quad (7.6)$$

The total strain energy of a body, U, is given by integrating the strain energy per unit volume[1], U_V, over the volume of the body. In the case of a beam, it is given by integrating the strain energy per unit length, U_l, along the beam.

---

[1]For large deformations, it is necessary to specify that the strain energy density is measured per unit unstrained volume or use the strain energy per unit mass instead.

The strain energy per unit volume, U_V, was found from the work done by the stresses in moving through the corresponding strains of a unit cube of the material.

$$U_V = \int_0^{\text{final state}} \sigma_{xx} d\epsilon_{xx} + \sigma_{yy} d\epsilon_{yy} \ldots = \int_0^{\epsilon_f} \sigma^T d\epsilon \tag{7.7}$$

using the column-vector notation of §3.11, where the subscript f indicates the final state. The *complementary* work done is given by integrating the deflexion with change in the corresponding load. It is then possible to define the complementary strain energy per unit volume, $\bar{U}_V$, as

$$\bar{U}_V = \int_0^{\text{final state}} \epsilon_{xx} d\sigma_{xx} + \epsilon_{yy} d\sigma_{yy} \ldots = \int_0^{\sigma_f} \epsilon^T d\sigma \tag{7.8}$$

For linearly-elastic materials, the set of stresses increase in direct proportion to the set of strains, so that (7.7) gives U_V as $\frac{1}{2}\sigma_f^T\epsilon_f$ and (7.8) gives $\bar{U}_V$ as $\frac{1}{2}\epsilon_f^T\sigma_f$. As these expressions are the same, there is no quantitive distinction between them in this case. However, the strain energy is derived by treating the stresses as functions of the strains, yielding expressions of the kind given by (7.2), (7.3) and (7.6). Likewise, the complementary strain energy (sometimes referred to as the stress energy) is expressed as a function of the stresses or loads, as in (7.1), (7.4) and (7.5). For other elastic materials, the two are not the same, but related by

$$\begin{aligned} U_V + \bar{U}_V &= \int_0^{\text{final state}} \sigma_{xx} d\epsilon_{xx} + \epsilon_{xx} d\sigma_{xx} + \sigma_{yy} d\epsilon_{yy} + \epsilon_{yy} d\sigma_{yy} \ldots \\ &= \int_0^{\text{final state}} d(\sigma_{xx}\epsilon_{xx} + \sigma_{yy}\epsilon_{yy} \ldots) = \int_0^{\text{final state}} d(\sigma^T\epsilon) = \sigma_f^T\epsilon_f \;\; (\equiv \epsilon_f^T\sigma_f) \end{aligned} \tag{7.9}$$

For the purposes of the analyses in this book, the loading applied to the body (body forces and surface tractions) will normally be taken as constant. The work done by this loading is equal to loss in potential energy. The necessary expression for the potential energy, Φ, (cf. §2.3) can then be determined from the changes of position of the loads from some arbitrary initial state[1]. The initial state (of zero potential) will be taken as being when the loads are at the locations on the *undeformed* body to which they will be applied. The subsequent loss of potential is given by the work done, W, by these (constant) loads in moving to their positions in the deformed state of the body, or in terms of the notation of §3.12,

$$-\Phi = W = \int_V X_i u_i \, dV + \int_S p_i u_i \, dS \tag{7.10}$$

where the integrals are over the volume, V, and the surface, S, of the body and summation over $i = 1$ to 3 is understood. (Point loads or resultants could be thought of as special cases of the surface tractions p_i in (7.10). However, these will be dealt with separately later.) The total potential energy[2] of the body and its loads, Π, includes both the strain energy, U, and the potential energy, Φ, or

---

[1] Here, the *absolute* value of the potential energy is not significant. A different assumption about the initial positions of the loads will change the absolute value of the potential energy, but not its incremental change.

[2] More generally, the thermal energy of the body would be considered. It is possible to distinguish between isothermal and adiabatic changes, but for most purposes the distinction is negligible.

$$\Pi = U + \Phi = \int_V U_V dV - \int_V X_i u_i \, dV - \int_S p_i u_i \, dS \tag{7.11}$$

The strain energy of any unstrained part of an elastic body is taken to be zero. Its strain energy in any other state must be positive. If this were not so, it would follow from the first law of thermodynamics that the part could be made to give out energy by straining it[1]. The above expressions for the strain energy are then positive-definite (cf. §3.5). This will also be true of the corresponding expressions for anisotropic linear elasticity and non-linear elasticity. The total complementary energy, $\bar{\Pi}$, can be defined as[2]

$$\bar{\Pi} = \bar{U} + \bar{\Phi} = \int_V \bar{U}_V dV - \int_S p_i u_i \, dS \tag{7.12}$$

Variational energy methods examine the effects of small perturbations from the loaded state. The most common problem is to find the stresses and deformations in a body and the reactions at its supports induced by prescribed loads. These may be body forces and surface tractions. Surface displacements may also be prescribed, but not their corresponding tractions at the same location. *Displacement methods* relate to small perturbations of the displacements of material points within the body, the compatible perturbations of strain implied and the resulting variation in the total energy, Π. *Force methods* relate to small perturbations of the loading on the body, the internal stress in equilibrium with this loading and the resulting variation in the total complementary energy, $\bar{\Pi}$. In the next three sections, minimum principles for Π and $\bar{\Pi}$ will be established for linearly-elastic materials. Weaker, stationary principles exist for non-linear elastic materials (see Ogden (1984) for example).

{7.2 The Principle of Minimum Potential Energy}

{Consider a linearly-elastic body in equilibrium with a prescribed set of body forces, X_i, and surface tractions, p_i. A given set of surface displacements may also be prescribed. Corresponding to the induced state of strain, ϵ say, the induced strain energy will be $U(\epsilon)$. In terms of the tensor notation of §3.12, this is given by

$$U(\epsilon) = \frac{1}{2}\int_V \sigma_{ij} e_{ij} \, dV = \frac{1}{2}\int_V c_{ijkl} e_{ij} e_{kl} \, dV \tag{7.13}$$

where summation in the range of the repeated indices (1 to 3) is understood. Suppose that the body in a different geometrically-possible state of strain, ϵ^*, also has the same prescribed surface displacements, but is not necessarily in equilibrium with the prescribed loads. The state of strain, ϵ^*-ϵ, will then correspond to a geometrically-possible state in which the prescribed surface displacements are zero. As the strain energy is always positive, $U(\epsilon)$, $U(\epsilon^*)$ and $U(\epsilon^*-\epsilon)$ will all be positive. Then

---

[1] This would imply that the part was unstable under certain conditions.

[2] Tauchert (1974) includes a term for the complementary potential of the body forces. These are permitted to vary with the stresses rather than being prescribed. The above definition follows those of Prager (1961), Sokolnikoff (1956) and Washizu (1968).

$$0 < U(\epsilon^* - \epsilon) = \frac{1}{2}\int_V c_{ijkl}(e_{ij}^* - e_{ij})(e_{kl}^* - e_{kl})dV$$

$$= U(\epsilon^*) - U(\epsilon) + \int_V c_{ijkl}(e_{ij}e_{kl} - \frac{1}{2}e_{ij}^* e_{kl} - \frac{1}{2}e_{ij}e_{kl}^*)dV \tag{7.14}$$

$$= U(\epsilon^*) - U(\epsilon) + \int_V (\sigma_{ij}e_{ij} - \sigma_{ij}e_{ij}^*)dV \quad (c_{ijkl} = c_{klij})$$

giving

$$\int_V \sigma_{ij}(e_{ij}^* - e_{ij})dV < U(\epsilon^*) - U(\epsilon) \tag{7.15}$$

Let the material displacements corresponding to ϵ be u_i and those corresponding to ϵ^* be u_i^*. Then as the original state of strain was in equilibrium, from (3.71) it follows that

$$\int_V (\sigma_{ij},_j + X_i)(u_i^* - u_i)\,dV = 0 \tag{7.16}$$

From Gauss' divergence theorem,

$$\int_V \sigma_{ij},_j(u_i^* - u_i)dV + \int_V \sigma_{ij}(u_i^* - u_i),_j dV = \int_V [\sigma_{ij}(u_i^* - u_i)],_j dV$$

$$= \int_S \sigma_{ij}(u_i^* - u_i)n_j\,dS \tag{7.17}$$

Note that the integrand of the last integral is zero when the surface displacements are prescribed, because u_i^* and u_i are the same. It will be taken here that where the surface displacements are not prescribed, the surface tractions are prescribed[1]. These are related to the actual state of stress by (3.74) so that

$$\int_S \sigma_{ij}(u_i^* - u_i)n_j\,dS = \int_{S_p} p_i(u_i^* - u_i)\,dS \tag{7.18}$$

where S_p is the surface on which the tractions are prescribed. Then from (7.16) to (7.18) and (3.72),

$$\int_V X_i(u_i^* - u_i)dV + \int_{S_p} p_i(u_i^* - u_i)dS = \int_V \sigma_{ij}(u_i^* - u_i),_j dV$$

$$= \int_V \frac{1}{2}\sigma_{ij}(u_i^* - u_i),_j + \frac{1}{2}\sigma_{ji}(u_j^* - u_j),_i dV = \int_V \sigma_{ij}(e_{ij}^* - e_{ij})dV \quad (\sigma_{ij} = \sigma_{ji}) \tag{7.19}$$

Consider the variation of deformation of the body from the equilibrium state of strain of the body, ϵ, to the state ϵ^* in which the prescribed loads, X_i and p_i, are kept constant and the prescribed surface displacements[2] are still satisfied. The variation $\Delta\Pi$ of the total potential energy in moving from the equilibrium state is then

$$\Delta\Pi = \int_V (U_V^* - U_V)dV - \int_V X_i(u_i^* - u_i)dV - \int_{S_p} p_i(u_i^* - u_i)dS$$

$$= U(\epsilon^*) - U(\epsilon) - \int_V \sigma_{ij}(e_{ij}^* - e_{ij})dV > 0 \tag{7.20}$$

---

[1] A surface traction or surface displacement may be prescribed as zero, and often are.

[2] Also called 'kinematic boundary conditions'.

using (7.15) and (7.19). Then *in the equilibrium state, Π is an absolute minimum under any such variations of deformation*[1]. In particular, when the variations of deformation are infinitesimal, the value of Π remains stationary.}

{7.3 The Principle of Minimum Complementary Energy}

{The total complementary energy is allowed to vary with any stress field within the body which satisfies the equations of equilibrium for the prescribed body forces and surface tractions. The proof that the actual state of stress then corresponds to an absolute minimum of $\bar{\Pi}$ follows similar lines to the proof in the last section.

From (3.81), (7.8) and (7.12), the complementary strain energy for an induced state of stress σ in the body is given by

$$\bar{U}(\sigma) = \frac{1}{2}\int_V s_{ijkl}\sigma_{ij}\sigma_{kl}\,dV \tag{7.21}$$

Suppose that the body is in a different state of stress, σ*, which is in equilibrium with the same prescribed surface loads p_i and body forces X_i. As the strain energy is positive, the complementary strain energy is necessarily positive for linearly-elastic materials. Then $\bar{U}(\sigma)$ and $\bar{U}(\sigma^*)$ are positive, as will be $\bar{U}(\sigma^*-\sigma)$, which is the complementary energy for the stress field given by the difference of σ* and σ. Then corresponding to (7.14),

$$\begin{aligned} 0 < \bar{U}(\sigma^* - \sigma) &= \frac{1}{2}\int_V s_{ijkl}(\sigma^*_{ij} - \sigma_{ij})(\sigma^*_{kl} - \sigma_{kl})\,dV \\ &= \bar{U}(\sigma^*) - \bar{U}(\sigma) + \int_V e_{ij}(\sigma_{ij} - \sigma^*_{ij})\,dV \end{aligned} \tag{7.22}$$

Both states of stress, σ and σ*, satisfy the equations of equilibrium with the specified body forces. Then from (3.71),

$$\sigma_{ij},_j + X_j = 0 \quad , \quad \sigma^*_{ij},_j + X_j = 0 \quad \text{so that} \quad (\sigma^*_{ij} - \sigma_{ij}),_j = 0 \tag{7.23}$$

within the body. On the surface S_p where the tractions are specified, from (3.74),

$$\sigma_{ij}n_j = p_i \quad , \quad \sigma^*_{ij}n_j = p_i \quad \text{so that} \quad (\sigma^*_{ij} - \sigma_{ij})n_j = 0 \tag{7.24}$$

From (3.72), the symmetry of the stress tensor, (7.23), Gauss' divergence theorem and (7.24),

$$\begin{aligned} \int_V e_{ij}(\sigma_{ij} - \sigma^*_{ij})\,dV &= \int_V u_{i},_j(\sigma_{ij} - \sigma^*_{ij})\,dV \\ &= \int_V [u_i(\sigma_{ij} - \sigma^*_{ij})],_j\,dV - \int_V u_i(\sigma_{ij} - \sigma^*_{ij}),_j\,dV \\ &= \int_S u_i(\sigma_{ij} - \sigma^*_{ij})n_j\,dS = \int_{S_u} u_i(p_i - p^*_i)\,dS \end{aligned} \tag{7.25}$$

where S_u is the portion of the surface on which the displacements are prescribed . (As in §7.2, the remaining portion of the surface is taken to be S_p , on which the tractions are prescribed.) Then from (7.22) and (7.25),

---

[1] Note that only the surface integral on S_p appears in the expression for ΔΠ. This is because on the rest of the surface the displacements are prescribed so that $u_i = u_i$*. On S_p, the tractions p_i are prescribed.

$$\bar{U}(\sigma) - \int_{S_u} p_i u_i \, dS < \bar{U}(\sigma^*) - \int_{S_u} p_i^* u_i \, dS \tag{7.26}$$

It then follows from (7.12) that *the total complementary energy of the actual stressed state is the absolute minimum of all possible stress fields which satisfy the equations of equilibrium for the prescribed loading (body forces and surface tractions)*.[1] In particular, for infinitesimal variations of the stress field, the value of $\bar{\Pi}$ remains stationary.}

7.4 Prescribed Resultants, Corresponding Deflexions and Work

In engineering problems, it is usually convenient to prescribe the resultant loads (forces and moments) acting on an elastic system. The corresponding deflexion, D_j, to a resultant load, L_j, is that through which the load does work. In terms of such resultants, the expression for the loss of potential given by (7.10) becomes

$$-\Phi = W = \int_V X_i u_i \, dV + \sum_{j=1}^{n} L_j D_j \tag{7.27}$$

where it is taken that the work done by the n resultants L_j accounts for all the work done by the surface tractions p_i. In some cases, this will be exactly true. If the prescribed deflexions on a portion of the surface S_u are zero, the work done by the surface tractions will be zero. Then taking the deflexions D_j as zero will give exactly the right expression. Such is the case at the built-in end of an encastré beam. Likewise, if the prescribed tractions on a portion of the surface S_p are zero, then the resultants on that surface can be taken as zero. Such is the case with any free surface. Workless supports, discussed in §1.1, can be taken to fall into the same categories as the above, a resultant reaction or its corresponding deflexion being prescribed as zero.

The 'point' loads often used in engineering analyses are a convenient fiction without which most common problems would not have been resolved. A true point force applied to the surface of a body would induce infinite stresses locally, thus forcing its way through the material and emerging on the other side of the body. In reality, a system of tractions is applied to a small portion of the surface. Then by invoking Saint-Venant's principle, a sufficiently accurate result can be obtained by using the resultants L_j on that portion instead of the real and possibly unknown tractions. The corresponding deflexions D_j cannot necessarily be determined from purely geometric principles. What is required are values such that the summation in (7.27) best represents the work done. Thus, for example, for the resultant forces

$$L_i = \int_{S_p} p_i \, dS \quad (i = 1 \text{ to } 3) \tag{7.28}$$

the mean displacements

$$D_i = \frac{1}{S_p} \int_{S_p} u_i \, dS \quad (i = 1 \text{ to } 3) \tag{7.29}$$

might be used. This will give the exact expression needed in (7.27) if p_i is constant on S_p but not necessarily otherwise. More generally, if the exact expression is needed,

---

[1] Note that only the surface integrals on S_u appear in (7.26). This is because the stresses are considered to vary without changing the surface displacements, the equations of compatibility not necessarily continuing to be satisfied.

$$D_i = \frac{1}{L_i}\int_{S_p} p_i u_i \, dS \quad (i = 1 \text{ to } 3, \ \text{no summation}) \tag{7.30}$$

should be used.

When only the resultant loads acting on certain parts of the body are specified, the question of the uniqueness of the solution arises. However, in engineering analyses, constraints are placed on the admissible stress (and strain) distributions which ensure that the solutions are unique for stable linearly-elastic problems.

Example 7.1 Torsion of a beam with a square cross-section

In the last chapter, the theory of Saint-Venant torsion was derived by imposing geometrical constraints on the deformation and equilibrium constraints on the state of stress. Exact solutions were found by imposing both sets of constraints and, in addition, the distribution of stress and strain was considered to be constant along the beam. (It will be seen in Chapter 8 that this is a requirement of generalized beam theory.)

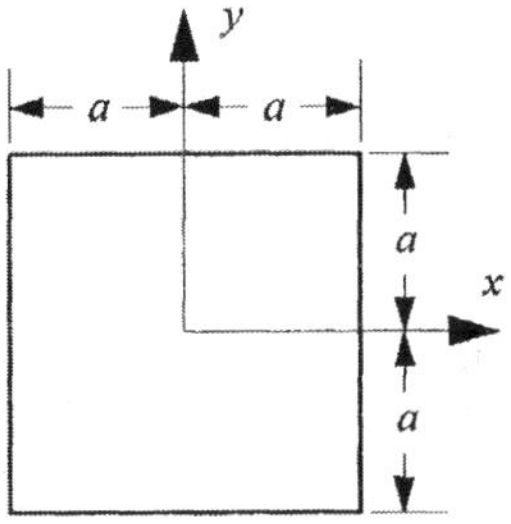

Figure 7.1 A square cross-section.

The minimum energy criteria established in §7.2 and §7.3 permit the deformation conditions and the stress conditions to be examined separately to determine upper and lower bounds to the stiffness of a section. These methods will be used to estimate the torsional stiffness of the square cross-section of side $2a$ shown in Figure 7.1[1]. A torque T will be applied to the end faces of the beam, l apart, inducing a rate of twist, ϑ.

Applying the theory of §7.2, an appropriate displacement field is assumed, from which an upper bound to the total potential energy is determined. This is given by assuming that the cross-sections remain undeformed, but rotate at a constant rate ϑ along the axis of the beam, z. This gives the displacement field

$$u = -\vartheta y z \ , \quad v = \vartheta x z \ , \quad w = 0 \tag{7.31}$$

(cf. (5.16)). The only non-zero strains are then

$$\gamma_{xz} = -\vartheta y \ , \quad \gamma_{yz} = \vartheta x \tag{7.32}$$

From (7.6) and (7.11), the real total potential energy is given by

$$\Pi = \frac{1}{2} G J \vartheta^2 l - T \vartheta l \tag{7.33}$$

Using the above approximation to the deformed state, the torque T is prescribed on the end faces and the rate of twist ϑ is chosen to be common to both states. Then from (7.2),

$$\begin{aligned} \Pi^* &= \frac{1}{2} G \int_V (\gamma_{yz}^2 + \gamma_{zx}^2)\, dV - T\vartheta l = \frac{1}{2} G \vartheta^2 l \int_A (x^2 + y^2)\, dA - T\vartheta l \\ &= \frac{1}{2} G I_p \vartheta^2 l - T \vartheta l \end{aligned} \tag{7.34}$$

where I_p is the polar second moment of area. Then from (7.20), Π is less than Π^*, provided that

---

[1] More accurate bounds to this problem will be found in Prager (1961) and Washizu (1968) where similar methods are used. Here, accuracy has been sacrificed for the sake of simplicity. See also the example in §5.3.

the two are not identical states,[1] so that

$$GJ < GI_p = 2\tfrac{2}{3}Ga^4 \tag{7.35}$$

The arguments used to establish that the shear stiffness is not more than GA in §6.5 follow similar lines.

To use the minimum complementary energy principle established in §7.3, a state of stress must be found which satisfies the equilibrium conditions. In §5.4, a stress function ϕ was used in order to satisfy these conditions. The only condition of internal equilibrium, (5.18), was satisfied by the relationship between ϕ and the shear stresses given by (5.23). Equilibrium with the end torque is satisfied by (5.27) and equilibrium with the surface tractions, specified as zero on the lateral surfaces, are satisfied by taking ϕ as zero on the boundary of the section. Then a suitable stress function for the section shown in Figure 7.1 is given by

$$\phi = C(x^2 - a^2)(y^2 - a^2) \tag{7.36}$$

Equilibrium with the end torque is satisfied by

$$T = 2\int_A C(x^2 - a^2)(y^2 - a^2)\,dA = \tfrac{32}{9}Ca^6 \tag{7.37}$$

hence giving the value of C. Then the stresses given by (5.23) are

$$\tau_{yz} = -\frac{9T}{16a^6}x(y^2 - a^2) \quad , \quad \tau_{zx} = \frac{9T}{16a^6}y(x^2 - a^2) \tag{7.38}$$

From (7.1), the complementary strain energy is given by

$$\bar{U}^* = \frac{1}{2G}\int_V (\tau_{yz}^2 + \tau_{zx}^2)\,dV = \frac{l}{2G}\frac{81T^2}{128a^{12}}\int_A x^2(y^2 - a^2)^2 + y^2(x^2 - a^2)^2\,dA = \frac{9T^2 l}{40Ga^4} \tag{7.39}$$

From (7.5) and (7.12), the real total complementary energy is given by

$$\bar{\Pi} = \frac{T^2 l}{2GJ} - T\vartheta l \tag{7.40}$$

and from (7.39) the approximation to the total complementary energy is

$$\bar{\Pi}^* = \frac{9T^2 l}{40Ga^4} - T\vartheta l \tag{7.41}$$

Then from the inequality given by (7.26),

$$\frac{T^2 l}{2GJ} < \frac{9T^2 l}{40Ga^4} \quad \text{or} \quad \frac{20}{9}Ga^4 < GJ \tag{7.42}$$

so that from (7.35) and (7.42), the torsional stiffness of the section lies in the range

$$2.222Ga^4 < GJ < 2.667Ga^4 \tag{7.43}$$

From more accurate analyses, GJ is found to be $2.250Ga^4$.

In addition to resultant forces and moments, other parameters can be used to represent loading. Most important amongst these is the *load pair*, introduced in §1.1. If two loads acting on a body are always of the same magnitude, they can both be labelled L. If one has a corresponding deflexion D_1 and the other has a corresponding deflexion D_2 then the loss of

---

[1] For circular cross-sections, Π and Π^* represent identical states, so that $J = I_p$.

potential for the pair is $(D_1 + D_2)$. The load pair can then be thought of as a generalized load L with a corresponding generalized deflexion[1] D equal to the sum of D_1 and D_2. Usually, the load pair consists of equal and opposite moments or collinear forces, so that the pair has no resultant. Under such conditions, the pairs do no work during any overall rigid-body motion of the body on which they act but only as a result of their relative motion with respect to each other in the sense in which they act. For example, the bimoments discussed in §5.9 can be thought of as a pair of equal and opposite couples. If an imaginary cut is made in a structure, the interaction between the two faces of the cut can be represented by equal-load pairs. In Figure 7.2a, such an imaginary cut is made in a diagonal bar. If the bar is only loaded through its pinned ends, the only resultant interaction on the cut is by the pair of forces R. The upper one does positive work by moving down along the mutual line of action and the lower one by moving up along the same line. The total loss of potential is then given by multiplying R by e, the amount by which the two forces approach one another. If the bar fits perfectly, then e is, and remains, zero. Consider the case when the bar is initially too short. Starting from the unstrained state of the structure, the faces of the imaginary cut would initially be e apart. Applying the load pair R would bring the faces together, giving a term Re in the expression for the loss of potential. Once this gap has been closed, this expression for the loss of potential by R remains the same for all subsequent loading and deformation of the frame. In Figure 7.2b, an imaginary pin is inserted in a continuous beam. The interaction moments between the faces of the real beam is simulated by the moment pair m on either side of the pin. This generalized loading has a corresponding deflexion θ. This is the magnitude of the relative rotation of the component moments. In the real beam, which has no discontinuity at the pin, this would be zero. As will be seen in §7.6, these concepts of load pairs can be used to analyse redundant structures.

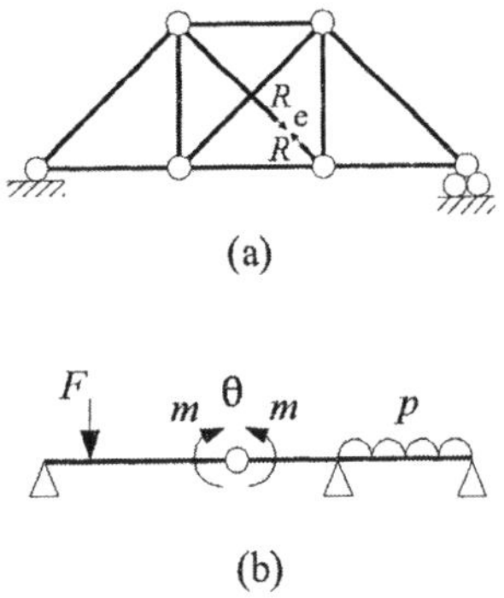

Figure 7.2 Load pairs.

7.5 Castigliano's Strain Energy Theorem

This theorem is often referred to as "Castigliano's first theorem", but this is also taken[2] to refer the theorem in the next section as well. This is because they were published as Parts I and II of the same theorem[3]. Here, it will be considered as a special case of the Principle of Minimum Potential Energy, derived in §7.2 and described by (7.20). The elastic system will be taken as subject to a set of loads L_i (i = 1 to n) only, which produce corresponding deflexions D_i. (These may be generalized loads and deflexions, as described in the last section.) The strains in the elastic system are taken to be continuous functions of these deflexions only, so that (7.11) becomes

$$\Pi = U(D_1, D_2, \ldots D_n) - \sum_{i=1}^{n} L_i D_i \qquad \textbf{(7.44)}$$

---

[1] Exactly the same concept is used in mechanics, where generalized loads and deflexions are usually denoted by Q_i and q_i respectively.

[2] See Southwell (1936) for example.

[3] Castigliano, A. (1879) *Théorie de l'équilibre des systèmes élastiques et ses applications.* A.F. Negro, Turin.

As Π is a minimum for all admissible states of strain,

$$0 = \frac{\partial \Pi}{\partial D_j} = \frac{\partial U}{\partial D_j} - L_j \qquad (j = 1 \text{ to } n) \tag{7.45}$$

The theorem may be expressed as follows. *If the strain energy of an elastic system can be expressed as a function of the deflexions corresponding to a set of applied loads, the partial differential of this strain energy with respect to one of these deflexions gives the corresponding load.*

This theorem is not often used. The reason is that the strain energy often has to be expressed in terms of deflexions where the corresponding load is zero. Usually, this means that many more equations have to be written to find the desired characteristics of the loaded state than are required by the method used in the next section. It is most usefully employed for highly redundant structures.

Example 7.2 Deflexions of a pin-jointed structure

The joints A, C, E and F of the pin-jointed frame shown in Figure 7.3 are rigidly fixed to the ground. They are a apart and joints B and D are a distance a above joints C and E respectively. Then the state of strain of the frame can be expressed in terms of the horizontal displacements, u_B and u_D, of joints B and D and their vertical displacements, v_B and v_D respectively. The strains in the bars AB to DF are

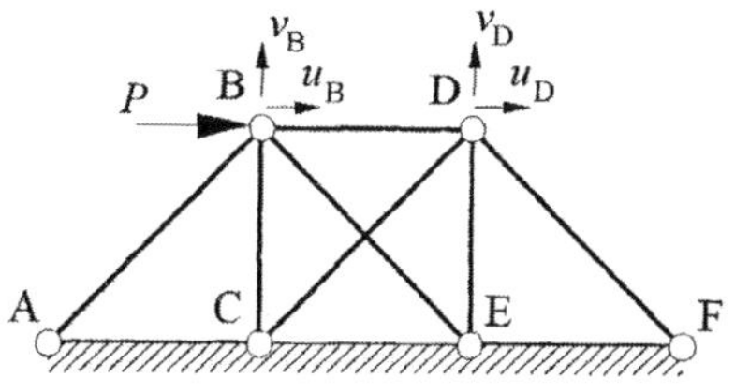

Figure 7.3 A plane, pin-jointed frame.

$$\epsilon_{AB} = \frac{1}{2a}(u_B + v_B)\ ,\ \epsilon_{BC} = \frac{v_B}{a}\ ,\ \epsilon_{BE} = \frac{1}{2a}(v_B - u_B)\ ,\ \epsilon_{BD} = \frac{1}{a}(u_D - u_B)$$
$$\epsilon_{DC} = \frac{1}{2a}(u_D + v_D)\ ,\ \epsilon_{DE} = \frac{v_D}{a}\ ,\ \epsilon_{DF} = \frac{1}{2a}(v_D - u_D)\ . \tag{7.46}$$

Each diagonal bar will be taken to have a cross-sectional area of $\sqrt{2}A$ and the rest to have a cross-sectional area of A. The material used has a Young's modulus of E. Then from (7.6) the strain energy of the frame is given by

$$U = \frac{EA}{a}(u_B^2 - u_B u_D + u_D^2 + v_B^2 + v_D^2) \tag{7.47}$$

Then if the only load applied is a horizontal force P at joint B,

$$P = \frac{\partial U}{\partial u_B} = \frac{EA}{a}(2u_B - u_D)\ ,\ 0 = \frac{\partial U}{\partial v_B} = \frac{2EA}{a}v_B\ ,$$
$$0 = \frac{\partial U}{\partial u_D} = \frac{EA}{a}(2u_D - u_B)\ ,\ 0 = \frac{\partial U}{\partial v_D} = \frac{2EA}{a}v_D\ . \tag{7.48}$$

giving

$$u_B = \frac{2aP}{3EA}\ ,\ u_D = \frac{aP}{3EA}\ ,\ v_B = v_D = 0\ . \tag{7.49}$$

7.6 Castigliano's and Crotti's Complementary Energy Theorems

This theorem was first proposed for linear elastic systems by Castigliano in 1876 and modified by Crotti[1] for non-linear elastic systems. It is often ascribed to Engesser[2] who derived it independently later. Here, it will be derived from the Principle of Minimum Complementary Energy. The stationary complementary energy principle follows immediately from this[3]. As in the last section, an elastic system will be examined which is subject to a set of generalized loads L_i only. The system will be taken as statically determinate, so that its state of stress can be found in terms of these loads. The complementary strain energy can then be expressed as a function of these loads, so that from (7.12), the total complementary energy can be written as

$$\bar{\Pi} = \bar{U}(L_1, L_2, \ldots L_n) - \sum_{i=1}^{n} L_i D_i \tag{7.50}$$

As this is a minimum for all statically-admissible stress states, it must be stationary for infinitesimal changes of stress, giving

$$0 = \frac{\partial \bar{\Pi}}{\partial L_j} = \frac{\partial \bar{U}}{\partial L_j} - D_j \qquad (j = 1 \text{ to } n) \tag{7.51}$$

This equation can be expressed as the following theorem. *If the complementary strain energy of an elastic system can be expressed as a function of a set of applied loads, the partial differential of this complementary strain energy with respect to one of these loads gives the corresponding deflexion.*

Example 7.3 A redundant frame

The horizontal beam AD of the frame shown in Figure 7.4 is continuous, the pin at C being attached to the top of it. This means that the frame has one degree of redundancy. To make it statically-determinate, an imaginary cut can be made in the vertical bar BC. This bar will be taken as having a perfect fit. Then if the force pair acting on the faces of the cut is X, the corresponding deflexion to X is zero. The state of stress in the frame can now be found in terms of the vertical load W at C, the uniform distributed load of intensity p on AD and X. The frame will be taken as symmetrical, with the bars AB and BD sloping at 30° to the horizontal. The axial forces in the frame can then be found in terms of X as shown in Figure

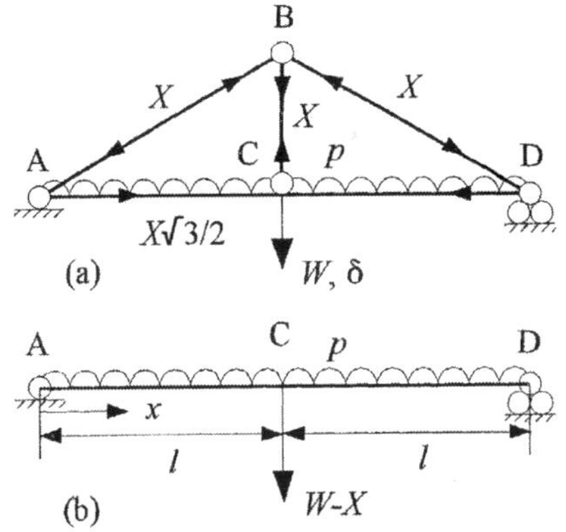

Figure 7.4 A redundant frame.

---

[1] Crotti, F. (1878) Esposizione del Teorema Castigliano e suo raccordo colla teoria dell' elasticà. *Atti del Collegio degli Ingegneri ed Architetti in Milano*, V.11 p. 224.

[2] Engesser, F. (1889) Über statisch unbestimmte Träger bei beliebigem Formänderungs - Gesetze und über den Satz von der klienster Ergänzungsarbeit. *Ztschr. des Arch.- und Ing.-Vereins zu Hannover* V.35 p.733.

[3] For non-linear elastic materials, the stationary principles will be found in Ogden (1984) §5.4.

7.4a. In addition, there will be a variable bending moment M in AD. This can be determined from Figure 7.4b. At a distance x from A ($x < l$) it is given by

$$M = \left[\tfrac{1}{2}(W - X) + pl\right]x - \tfrac{1}{2}px^2 \quad \textbf{(7.52)}$$

For the simplicity of the resulting expression, it will be taken that the cross-sectional areas of BC, AB, BD and AD are $A/\sqrt{3}$, $A/\sqrt{3}$, $A/\sqrt{3}$ and $A/2$ respectively. The bending stiffness of AD will be taken as EI. Then the complementary strain energy is found from (7.4) to be

$$\bar{U} = 2\int_0^l \frac{M^2}{2EI}dx + \frac{X^2}{2EA}(l + 2l + 2l + 3l) \quad \textbf{(7.53)}$$

Then if the deflexion corresponding to W is δ, from (7.51)

$$\delta = \frac{\partial \bar{U}}{\partial W} = 2\int_0^l \frac{M}{EI}\frac{\partial M}{\partial W}dx = \frac{l^3}{24EI}[4(W - X) + 5pl]$$

$$0 = \frac{\partial \bar{U}}{\partial X} = 2\int_0^l \frac{M}{EI}\frac{\partial M}{\partial X}dx + \frac{8Xl}{EA} = \frac{l^3}{24EI}[4(X - W) - 5pl] + \frac{8Xl}{EA} \quad \textbf{(7.54)}$$

from which X and δ can be found.

The problem of the beam BD on its own loaded by W and p can be simulated by taking X as zero in the first equation. This then gives

$$\delta = \frac{Wl^3}{6EI} + \frac{5pl^4}{24EI} \quad \textbf{(7.55)}$$

Of course, the deflexion at C caused by p alone is given by taking W as zero in (7.55). Note that this result could not have been found without including W in the first place, and only setting it to zero after the process of partial differentiation. This is known as the *dummy load method.*

{As $\bar{U}$ is a function of all the loads including p, it is instructive to ask what its partial differential with respect to p signifies. From (7.51) and (7.53), it is

$$D_p = \frac{\partial \bar{U}}{\partial p} = \frac{l^4}{120EI}[25(W - X) + 32pl] \quad \textbf{(7.56)}$$

On inspection, D_p will be seen to have units of area, and it is in fact the area swept out by the distributed loading during deflexion. Similarly, in three dimensions, partially differentiating the complementary strain energy of a pressure vessel with respect to its internal pressure gives the increase in volume of the vessel under load.}

As an example of the use of both equations (7.54), the value of I will be taken as $Al^2/48$. These equations then give

$$0 = 4(X - W) - 5pl + 4X \quad \text{or} \quad 8X = 4W + 5pl$$

$$\delta = \frac{2l}{EA}[4(W - X) + 5pl] = \frac{l}{EA}[4W + 5pl] \quad \textbf{(7.57)}$$

so that the pinned bars AB, BC and BD reduce the central deflexion by one half in this case.

Example 7.4 A non-linearly elastic structure

Consider a non-linear elastic material governed by the stress-strain relationship

$$\epsilon = S\sigma^n \quad \textbf{(7.58)}$$

If a bar of cross-sectional area A and length l is loaded so that it has a final uniform axial strain ϵ_f, the strain energy in the bar is given by

$$U = \int_V \left(\int_0^{\epsilon_f} \sigma d\epsilon \right) dV = \frac{nSP^{n+1}l}{(n+1)A^n} \qquad \textbf{(7.59)}$$

where P is the resultant axial tensile force, $\sigma_f A$. The complementary strain energy is given by

$$\bar{U} = \int_V \left(\int_0^{\sigma_f} \epsilon d\sigma \right) dV = \frac{SP^{n+1}l}{(n+1)A^n} \qquad \textbf{(7.60)}$$

From (7.59) and (7.60),

$$U + \bar{U} = S\sigma_f^{n+1}Al = V\sigma_f \epsilon_f \qquad \textbf{(7.61)}$$

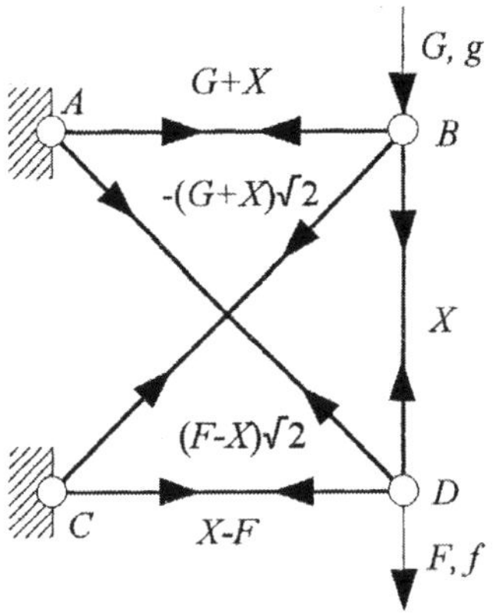

Figure 7.5 A non-linearly elastic redundant truss.

as predicted by (7.9). The frame shown in Figure 7.5 has one degree of redundancy. The joints B and D are loaded by vertical forces G and F as shown. The corresponding deflexions are g and f. By assuming that the force in BD is X, the axial forces in the bars, as shown in the figure, can be found from statics. The lengths of the diagonal bars will be taken as $l\sqrt{2}$ and the remainder are l long. The cross-sectional areas of all but the diagonal bars are A, and AD and BC have cross-sectional areas $A\sqrt{2}$ and $A/\sqrt{2}$ respectively. The complementary strain energy of the whole frame is then

$$\bar{U} = \frac{Sl}{(n+1)A^n}\{(G+X)^{n+1}[1+(-2)^{n+1}] + X^{n+1} + (X-F)^{n+1}[1+2(-1)^{n+1}]\} \qquad \textbf{(7.62)}$$

It will be seen from (7.60) and (7.62) that the complementary strain energy, for a bar and for the whole frame, is positive-definite provided that S is positive and n is odd. Taking n as odd and BC to be a perfect fit,

$$f = \frac{\partial \bar{U}}{\partial F} = \frac{3Sl}{A^n}(F-X)^n \ , \quad g = \frac{\partial \bar{U}}{\partial G} = \frac{Sl}{A^n}(G+X)^n(1+2^{n+1}) \ ,$$
$$0 = \frac{\partial \bar{U}}{\partial X} = \frac{Sl}{A^n}[(G+X)^n(1+2^{n+1}) + X^n + 3(X-F)^n] \ . \qquad \textbf{(7.63)}$$

Excluding complex roots for X, this gives the following results:

	$F = 1\ ,\ G = 0\ .$			$F = 0\ ,\ G = 1\ .$		
n	X	fA^n/Sl	gA^n/Sl	X	fA^n/Sl	gA^n/Sl
1	0.3333	2.0000	1.6667	-0.5556	1.6667	2.2222
3	0.3350	0.8050	0.7606	-0.6183	0.7091	0.9454
5	0.3502	0.3476	0.3424	-0.6359	0.3209	0.4159

By comparing the fourth and sixth columns, it will be seen that the reciprocal theorem only applies when n is one. (If n is even, there are no real solutions for X when only a unit F is applied, and if only G is applied, X must be equal and opposite to it. However, such values of n are not physically admissible, as they can give rise to negative strain energy.)

7.7 The Rayleigh-Ritz Method

The basic method was outlined by Lord Rayleigh[1] and later generalized by Ritz[2]. The method makes use of the minimum energy principles of §7.2 and §7.3. Approximate solutions to problems are found by expressing the loading or deflexion as a linear combination of functions. The coefficients multiplying these functions are allowed to vary in order to minimise the total potential. (These coefficients themselves could be thought of as generalized loads and deflexions in the sense discussed in §7.4. The processes of finding stationary values then follow those given in §7.5 and §7.6.) The method is best illustrated with two examples. Both relate to the simply-supported beam of length l and distributed load q shown in Figure 7.6. The first example finds an approximation to the deflected form for a given loading by minimising the potential energy of the system. The second example finds an approximation to the loading required to produce a particular deflected form by minimising the complementary potential of the system.

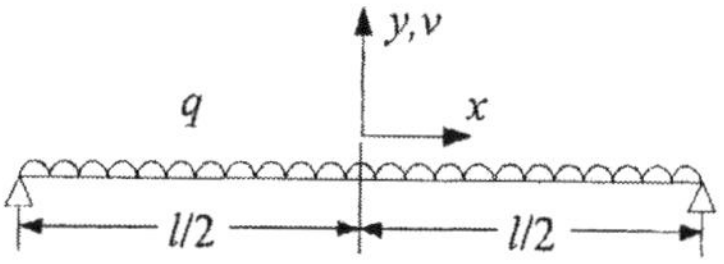

Figure 7.6 A simply-supported beam with a distributed load of variable intensity q.

Example 7.5 Deflexion under a sinusoidally-varying distributed load

Suppose that the beam is subjected to a distributed loading

$$q = Q\cos\frac{\pi x}{l} \tag{7.64}$$

A kinematically-admissible deflexion mode for the beam (giving zero v at each end) is

$$v = v_2\left[x^2 - \left(\frac{l}{2}\right)^2\right] + v_4\left[x^4 - \left(\frac{l}{2}\right)^4\right] \tag{7.65}$$

The flexural curvature is given by the second derivative of v so that from (7.6), (7.11), (7.64) and (7.65), the total potential can be written as

$$\begin{aligned}\Pi^* &= 2\int_0^{l/2}\frac{EI}{2}\left(\frac{d^2v}{dx^2}\right)^2 dx + 2\int_0^{l/2} vQ\cos\frac{\pi x}{l}\,dx \\ &= EI(2lv_2^2 + 2l^3v_2v_4 + 0.9l^5v_4^2) - 2Q\left[2v_2\left(\frac{l}{\pi}\right)^3 + 3v_4(\pi^2-8)\left(\frac{l}{\pi}\right)^5\right]\end{aligned} \tag{7.66}$$

(note that v and q are positive in opposite senses). The total potential is minimised by varying v_2 and v_4. This gives the stationary conditions

---

[1] Strutt, J.W., *Baron Rayleigh* (1877) *Device of the Variable Parameter.* The Theory of Sound V.1§182.

[2] Ritz, W. (1908) *Journal für der reine und angewandte Mathematik* V.135 pp.1-61.

$$0 = \frac{\partial \Pi^*}{\partial v_2} = EI(4lv_2 + 2l^3 v_4) - 4Q\left(\frac{l}{\pi}\right)^3$$
$$0 = \frac{\partial \Pi^*}{\partial v_4} = EI(2l^3 v_2 + 1.8 l^5 v_4) - 6Q(\pi^2 - 8)\left(\frac{l}{\pi}\right)^5 \qquad (7.67)$$

From which the parameters v_2 and v_4 are found to be

$$v_2 = \frac{Ql^2}{\pi^5 EI}(30 - 1.5\pi^2) \;,\quad v_4 = -\frac{Q}{\pi^5 EI}(60 - 5\pi^2) \;. \qquad (7.68)$$

giving the central deflexion, v_0, potential and strain energy to be

$$v_0 = -0.9973\frac{Ql^4}{\pi^4 EI} \;,\quad -\Pi^* = U^* = \frac{12Q^2 l^5}{\pi^{10} EI}(120 - 20\pi^2 + \pi^4) = 0.0025650\frac{Q^2 l^5}{EI} \qquad (7.69)$$

The exact deflexion mode is

$$v = -\frac{Ql^4}{\pi^4 EI}\cos\frac{\pi x}{l} \qquad (7.70)$$

giving

$$v_0 = -\frac{Ql^4}{\pi^4 EI} \;,\quad -\Pi = U = \frac{Q^2 l^5}{4\pi^4 EI} = 0.0025665\frac{Q^2 l^5}{EI} \qquad (7.71)$$

Thus the actual central deflexion is 0.3% larger, the actual strain energy 0.06% greater and the total potential 0.06% less (more negative) than given by the approximation. Normally, the method would be used for problems where no exact solution can be found, for example, when the bending stiffness varies along the beam.

Example 7.6 Distributed load required to produce a sinusoidal deflexion

This is the inverse of the previous problem. The desired deflexion mode is given by

$$v = -V\cos\frac{\pi x}{l} \qquad (7.72)$$

An approximation to the required distributed load will be taken of the form

$$q = Q_0 + Q_2 x^2 \qquad (7.73)$$

From statics, the bending moment in the beam is given by

$$M = \frac{Q_0 l^2}{8} + \frac{Q_2 l^4}{192} - \frac{Q_0 x^2}{2} - \frac{Q_2 x^4}{12} \qquad (7.74)$$

The complementary potential given by (7.12) is then

$$\bar{\Pi}^* = 2\int_0^{l/2}\frac{M^2}{2EI}dx + 2\int_0^{l/2} qv\,dx$$
$$= \frac{l^5}{240EI}\left(Q_0^2 + \frac{2Q_0Q_2 l^2}{21} + \frac{Q_2^2 l^4}{432}\right) - 2a\left[\frac{Q_0 l}{\pi} + \frac{Q_2 l^3}{4\pi^3}(\pi^2 - 8)\right] \qquad (7.75)$$

The least value of $\bar{\Pi}^*$ can be found by minimising it with respect to Q_0 and Q_2. Its stationary point found by partial differentiation gives the values of these parameters as

$$Q_0 = 96.524765\frac{EIa}{l^4} \quad , \quad Q_2 = -422.73824\frac{EIa}{l^6} \tag{7.76}$$

This gives the complementary potential and complementary strain energy as

$$-\bar{\Pi}^* = \bar{U}^* = 24.3522605\frac{EIa^2}{l^3} \tag{7.77}$$

The exact solution is given by

$$q = EIa\frac{\pi^4}{l^4}\cos\frac{\pi x}{l} \tag{7.78}$$

which gives the correct values as

$$-\bar{\Pi} = \bar{U} = 24.3522728\frac{EIa^2}{l^3} \tag{7.79}$$

so that the correct solution gives a slightly lower (more negative) complementary potential. The exact intensity of loading required at the origin is 0.9% greater than that predicted by the approximate analysis.

7.8 The Calculus of Variations

In some problems, the differential equations governing the deformed state of the elastic system may not be fully known. For a beam, it might be possible to write down the strain energy of the elastic curve and the work done during deflexion by the loads applied to it as integrals over its length. The correct total potential Π thus found would be a minimum for all possible curves. Finding the necessary conditions for stationary values of such integrals is the problem addressed by the calculus of variations.

Suppose that the curve for which the desired integral is stationary is $v(x)$. The expression for the integral might be in terms of an integrand which is a function of x, v, v' and v'', where the primes indicate differentiation with respect to x. This would be of the form

$$I = \int_0^l F(x, v, v', v'')\, dx \tag{7.80}$$

Let there be a small variation of the curve from this, given by $\epsilon w(x)$, where ϵ is an amplitude which does not vary with x. The new curve, $v(x)+\epsilon w(x)$, must still satisfy the end conditions. The integral now takes the form

$$I(\epsilon) = \int_0^l F(x, v+\epsilon w, v'+\epsilon w', v''+\epsilon w'')\, dx \tag{7.81}$$

Treating this as a function of ϵ as indicated, it has a stationary value when ϵ is zero so that

$$\begin{aligned} \left.\frac{\partial I(\epsilon)}{\partial \epsilon}\right|_{\epsilon=0} &= \int_0^l \left.\frac{\partial}{\partial \epsilon}\right|_{\epsilon=0} F(x, v+\epsilon w, v'+\epsilon w', v''+\epsilon w'')\, dx \\ &= \int_0^l \left.\frac{\partial}{\partial \epsilon}\right|_{\epsilon=0} F(x, V, V', V'')\, dx \qquad (V = v+\epsilon w\,,\ V' = v'+\epsilon w'\,,\ V'' = v''+\epsilon w'') \end{aligned} \tag{7.82}$$

is zero. Examining the integrand,

$$\frac{\partial}{\partial\epsilon}_{\epsilon=0} F(x,V,V',V'')\,dx = \frac{\partial F}{\partial V}\frac{\partial V}{\partial\epsilon} + \frac{\partial F}{\partial V'}\frac{\partial V'}{\partial\epsilon} + \frac{\partial F}{\partial V''}\frac{\partial V''}{\partial\epsilon} \quad (\epsilon = 0)$$
$$= \frac{\partial F}{\partial v}w(x) + \frac{\partial F}{\partial v'}w'(x) + \frac{\partial F}{\partial v''}w''(x) \tag{7.83}$$

Examining the second term on the lower right-hand side of this,

$$\int_0^l \frac{\partial F}{\partial v'}w'(x)\,dx = \left|\frac{\partial F}{\partial v'}w(x)\right|_0^l - \int_0^l \frac{d}{dx}\left(\frac{\partial F}{\partial v'}\right)w(x)\,dx \tag{7.84}$$

If $w(x)$ is zero at the ends, then V satisfies the same end conditions as v and the first term on the right-hand side of (7.84) vanishes. Similarly, imposing the condition that $w'(x)$ vanishes at the ends, gives

$$\int_0^l \frac{\partial F}{\partial v''}w''(x)\,dx = \int_0^l \frac{d^2}{dx^2}\left(\frac{\partial F}{\partial v''}\right)w(x)\,dx \tag{7.85}$$

Then from (7.83) to (7.85), (7.82) becomes

$$\frac{\partial I(\epsilon)}{\partial\epsilon}_{\epsilon=0} = \int_0^l \left[\frac{\partial F}{\partial v} - \frac{d}{dx}\left(\frac{\partial F}{\partial v'}\right) + \frac{d^2}{dx^2}\left(\frac{\partial F}{\partial v''}\right)\right]w(x)\,dx \tag{7.86}$$

As this must be zero for any admissible $w(x)$, the contents of the square brackets must be zero, or

$$\frac{\partial F}{\partial v} - \frac{d}{dx}\left(\frac{\partial F}{\partial v'}\right) + \frac{d^2}{dx^2}\left(\frac{\partial F}{\partial v''}\right) = 0 \tag{7.87}$$

This is an extended form of *Euler's equation.*

Example 7.7 Equations governing an axially-loaded beam

A beam subject to a lateral distributed load of intensity $q(x)$ and a large axial force P is shown in Figure 7.7. The potential associated with the distributed load is given by integrating qv along the beam, the deflexion v being in the opposite sense to q. As P is large, the amount by which the ends of the beam approach one another during this deflexion has to be taken into account. A small length δx of the beam is taken to rotate through an angle ϕ as a result of this deflexion. Its projected length, $\delta x \cos\phi$, is less than its original length, δx. Taking ϕ to be a very small angle, the first two terms in the expansion of $\cos\phi$ can be used to show that this apparent shortening is $\frac{1}{2}\phi^2\delta x$. Also, ϕ is approximately equal to $\sin\phi$ which is given by $-dv/dx$, as shown in Figure 7.7. It is then possible to integrate this increment in shortening along the beam to find the loss of potential of P. The strain energy per unit length of the beam due to bending is given by terms like the first and second on the right-hand side of (7.6), where the curvature ϑ is given by v''. Taking the bending stiffness of the beam to be EI, the

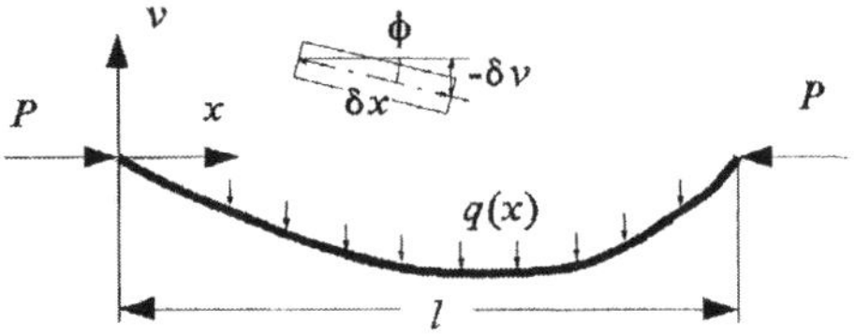

Figure 7.7 An axially-loaded beam.

potential given by (7.11) is then

$$\Pi = \int_0^l \left[\frac{1}{2} EI(v'')^2 - \frac{1}{2} P(v')^2 + qv \right] dx \tag{7.88}$$

Then for Π to be stationary, it follows from (7.87) that

$$EIv'''' + Pv'' + q = 0 \tag{7.89}$$

{Example 7.8 Equations governing a Timoshenko beam}

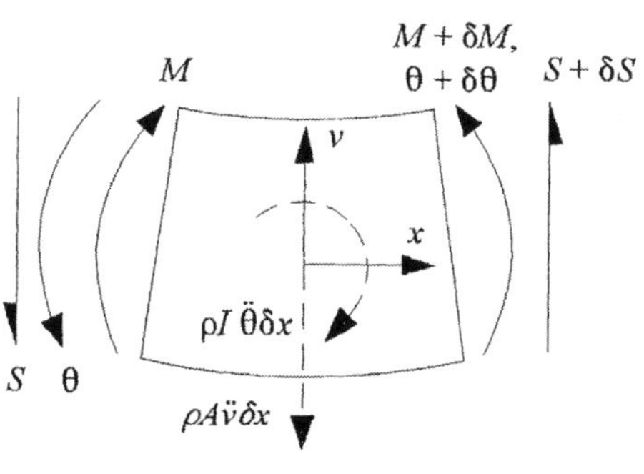

Figure 7.8 Element of a vibrating beam.

{A vibrating beam for which the shear behaviour and rotational inertia of an element, as well as its bending behaviour and linear inertia are taken into account is known as a Timoshenko beam. Three parameters will be used to define its deflexion. The effective mass displacement of a beam element of length δx will denoted by v. The angle through which an anticlockwise moment M acting on the element does work will be denoted by θ. This will also be taken as the effective angle of rotation of the mass of the element. The effective shear through which a shear force S acting on the element does work in straining it will be denoted by γ. As plane sections cannot be assumed to remain plane, an attempt to establish a direct geometrical relationship between these three parameters could be erroneous. Taking the bending stiffness of the beam to be EI and its shear stiffness to be GkA, where A is the cross-sectional area of the beam, the local bending moment and shear force acting on the beam are

$$M = EI\theta' \quad , \quad S = GkA\gamma \; . \tag{7.90}$$

where the prime indicates differentiation with respect to x. Applying d'Alembert's principle, the equations of dynamic equilibrium for the inertia loads acting on the element shown in Figure 7.8 are

$$I\rho\ddot{\theta} - M' - S = 0 \quad , \quad \rho A\ddot{v} - S' = 0 \; . \tag{7.91}$$

where a dot indicates differentiation with respect to time t. Then from (7.90),

$$I\rho\ddot{\theta} - EI\theta'' - GkA\gamma = 0 \quad , \quad \rho A\ddot{v} - GkA\gamma' = 0 \; . \tag{7.92}$$

Applying Hamilton's principle[1], stationary values of the time integral of the Lagrangian function L are sought. This is given by the difference between the kinetic energy T and the potential. In the present case, this potential is given completely by the strain energy. For a beam, these can be expressed as functions per unit length and integrated along the length l of the beam. That is,

$$\delta \int_0^\tau \int_0^l L_l \, dx \, dt = 0 \quad , \quad \text{where} \quad L_l = T_l - U_l \tag{7.93}$$

---

[1] See for example Flügge (1962) *Handbook of Engineering Mechanics* §23.5.

where the time interval is τ and the subscript l refers to the values per unit length, given by

$$T_l = \frac{1}{2}\rho A\dot{v}^2 + \frac{1}{2}\rho I\dot{\theta}^2 \quad , \quad U_l = \frac{1}{2}EI(\theta')^2 + \frac{1}{2}GkA\gamma^2 \tag{7.94}$$

(cf. (7.6)). The stationary principle is thus expressed here in terms of v, θ and γ which can be represented by the generalized coordinates q_1 to q_3 respectively. These are not independent, as they are subject to the *nonholonomic constraints* given by (7.92). (Such constraints are non-integrable relationships between derivatives of these coordinates.) These conditions can be taken into account by including Lagrange multipliers[1] $\lambda_j(x,t)$ in the expression for L_l. It then becomes

$$\begin{aligned}\bar{L}_l = {} & \frac{1}{2}\rho A\dot{v}^2 + \frac{1}{2}\rho I\dot{\theta}^2 - \frac{1}{2}EI(\theta')^2 - \frac{1}{2}GkA\gamma^2 \\ & + \lambda_1(x,t)[\rho I\ddot{\theta} - EI\theta'' - GkA\gamma] + \lambda_2(x,t)[\rho A\ddot{v} - GkA\gamma']\end{aligned} \tag{7.95}$$

The extended form of Euler's equation appropriate to this problem is given by

$$\frac{\partial \bar{L}}{\partial q_i} - \frac{\partial}{\partial x}\left(\frac{\partial \bar{L}}{\partial q_i'}\right) - \frac{\partial}{\partial t}\left(\frac{\partial \bar{L}}{\partial \dot{q}_i}\right) + \frac{\partial^2}{\partial x^2}\left(\frac{\partial \bar{L}}{\partial q_i''}\right) + \frac{\partial^2}{\partial t^2}\left(\frac{\partial \bar{L}}{\partial \ddot{q}_i}\right) = 0 \quad , \quad (i = 1 \text{ to } 3) \tag{7.96}$$

This gives rise to the three equations

$$\begin{aligned} -\rho A\ddot{v} + \rho A\ddot{\lambda}_2 &= 0 \\ EI\theta'' - \rho I\ddot{\theta} - EI\lambda_1'' + \rho I\ddot{\lambda}_1 &= 0 \\ -GkA\gamma - GkA\lambda_1 + GkA\lambda_2' &= 0 \end{aligned} \tag{7.97}$$

The first two are satisfied by taking λ_2 as v and λ_1 as θ. The third equation then becomes

$$\gamma = v' - \theta \tag{7.98}$$

Substituting this expression for γ in (7.92) then gives a pair of simultaneous equations in v and θ. Taking the beam to be vibrating with simple harmonic motion , so that

$$v = V(x)\cos(\omega t + \phi) \quad , \quad \theta = \Theta(x)\cos(\omega t + \phi) \tag{7.99}$$

these can be solved to give

$$\begin{aligned} V(x) &= C_1\cosh\frac{\alpha x}{l} + C_2\sinh\frac{\alpha x}{l} + C_3\cos\frac{\beta x}{l} + C_4\sin\frac{\beta x}{l} \\ \Theta(x) &= \frac{p^2/r + \alpha^2}{\alpha l}\left(C_1\sinh\frac{\alpha x}{l} + C_2\cosh\frac{\alpha x}{l}\right) + \frac{p^2/r - \beta^2}{\alpha l}\left(C_3\sin\frac{\beta x}{l} - C_4\cos\frac{\beta x}{l}\right) \end{aligned} \tag{7.100}$$

where

$$\begin{aligned} \alpha^2, \beta^2 &= \frac{p^2}{2}\left[\mp(1+r) + \sqrt{(1-r)^2 + s^2}\right] \\ p^2 &= \frac{\rho\omega^2 l^2}{E} \quad , \quad r = \frac{E}{kG} \quad , \quad s^2 = \frac{4EA}{\rho\omega^2 I} \end{aligned} \tag{7.101}$$

the negative alternative in the first expression giving α^2 and the positive alternative giving β^2.

Fixed end conditions are given by taking v and θ as zero, pinned end conditions by taking v and M as zero and free end conditions by taking M and S as zero. From the following equations, the natural frequency of vibration, ω, can be found for various beams of length l. For a cantilever,

---

[1] See for example Gelfland, I.M. and Fomin, S.V. (1963) *Calculus of Variations* §12.2.

$$2 + \frac{\alpha^2 - \beta^2}{\alpha\beta}\sinh\alpha\sin\beta + \left(\frac{\beta^2 - p^2}{\alpha^2 + p^2} + \frac{\alpha^2 + p^2}{\beta^2 - p^2}\right)\cosh\alpha\cos\beta = 0 \tag{7.102}$$

For a fixed-ended beam,

$$\tan\frac{\beta}{2} + \frac{\beta}{\alpha}\frac{r\alpha^2 + p^2}{r\beta^2 - p^2}\tanh\frac{\alpha}{2} = 0 \quad \text{or} \quad \tan\frac{\beta}{2} - \frac{\alpha}{\beta}\frac{r\beta^2 - p^2}{r\alpha^2 + p^2}\tanh\frac{\alpha}{2} = 0 \tag{7.103}$$

where the first equation applies to symmetric modes and the second to antisymmetric modes. The corresponding results for a beam with free ends follow on substituting $r = 1$ in (7.103). For a beam with pinned ends,

$$\omega^2 = \frac{n^2\pi^2 kG}{2l^2\rho}\left[1 + r + \frac{l^2 A}{n^2\pi^2 I} \pm \sqrt{\left(1 + r + \frac{l^2 A}{n^2\pi^2 I}\right)^2 - 4r}\,\right] \tag{7.104}$$

where n is any integer. (See Flügge (1962) §61.5 for further details.)

The accuracy of Timoshenko beam theory can be tested[1] by comparing the results for an infinite thin strip of unit thickness and depth $2d$ in a state of plane stress. Then A is $2d$, I is $2d^3/3$ and k is 5/6. Let λ be $n\pi/l$ and γ be the non-dimensional parameter $\rho\omega^2/G\lambda^2$. Simple beam theory gives the relationship

$$\gamma = \tfrac{2}{3}(1 + \nu)\lambda^2 d^2 \tag{7.105}$$

(see §10.3) and (7.104) becomes

$$\gamma_1, \gamma_2 = (1 + \nu) + \tfrac{5}{12} + \frac{5}{4\lambda^2 d^2} \pm \sqrt{\left[(1 + \nu) + \tfrac{5}{12} + \frac{5}{4\lambda^2 d^2}\right]^2 - \tfrac{5}{3}(1 + \nu)} \tag{7.106}$$

where γ_1 will be taken as the lower of the two roots.

An exact plane stress solution in the x-y plane can be found, taking the x axis to lie along the middle line of the strip. Assuming simple harmonic motion, the equations of dynamic equilibrium are

$$\frac{\partial\sigma_{xx}}{\partial x} + \frac{\partial\tau_{xy}}{\partial y} + \rho\omega^2 = 0 \;, \quad \frac{\partial\tau_{xy}}{\partial x} + \frac{\partial\sigma_{yy}}{\partial y} + \rho\omega^2 = 0 \tag{7.107}$$

where the stresses are given in terms of the displacements by

$$\sigma_{xx} = \frac{2G}{1 - \nu}\left(\frac{\partial u}{\partial x} + \nu\frac{\partial v}{\partial y}\right) \;, \quad \sigma_{yy} = \frac{2G}{1 - \nu}\left(\frac{\partial v}{\partial y} + \nu\frac{\partial u}{\partial x}\right) \;, \quad \tau_{xy} = G\left(\frac{\partial u}{\partial y} + \frac{\partial v}{\partial x}\right) . \tag{7.108}$$

The displacements are taken to be of the form

$$u = \sum_1^2 u_i \sinh m_i y \sin(\lambda x + \alpha)\sin(\omega t + \epsilon) \;, \quad v = \sum_1^2 u_i \cosh m_i y \cos(\lambda x + \alpha)\sin(\omega t + \epsilon) . \tag{7.109}$$

Substituting equations (7.109) and (7.108) into (7.107) yields relationships between u_i and v_i given by

[1] Cowper, G.R. (1968), On the Accuracy of Timoshenko's Beam Theory. *Proc. ASCE Journal of Engineering Mechanics Division* V.245 p.1447.

$$\begin{bmatrix} (1-\nu)(\mu_i^2+\gamma) - 2 & -(1+\nu)\mu_i \\ (1+\nu)\mu_i & (1-\nu)(\gamma-1) + 2\mu_i^2 \end{bmatrix} \begin{bmatrix} u_i \\ v_i \end{bmatrix} = \begin{bmatrix} 0 \\ 0 \end{bmatrix} \tag{7.110}$$

where μ_i is m_i/λ. For non-zero u_i and v_i, the determinant of the above matrix must be zero. This gives the following two roots for μ_i^2

$$\mu_1^2 = 1 - \gamma \quad \text{and} \quad \mu_2^2 = 1 - \tfrac{1}{2}\gamma(1-\nu) \tag{7.111}$$

The two solutions which these roots generate can be used to satisfy the free surface conditions that σ_{xx} and τ_{xy} are zero on $y=d$. (The form of (7.109) ensures that these conditions are then also satisfied on $y=-d$.) Using the relationships between u_i and v_i given by (7.110), these conditions become

$$\begin{bmatrix} (2-\gamma)\cosh\mu_1\lambda d & 2\cosh\mu_2\lambda d \\ \dfrac{2-2\gamma}{\mu_1\lambda d}\sinh\mu_1\lambda d & \dfrac{2-\gamma}{\mu_2\lambda d}\sinh\mu_2\lambda d \end{bmatrix} \begin{bmatrix} v_1 \\ v_2 \end{bmatrix} = \begin{bmatrix} 0 \\ 0 \end{bmatrix} \tag{7.112}$$

As

$$\cosh x = 1 + \frac{x^2}{2!} + \frac{x^4}{4!} \ldots + \frac{x^{2n}}{2n!} \quad \text{and} \quad \frac{1}{x}\sinh x = 1 + \frac{x^2}{3!} + \frac{x^4}{5!} \ldots + \frac{x^{2n}}{(2n+1)!}$$

the terms in (7.112) remain real, even for imaginary μ_i. The determinant of the matrix in (7.112) must be zero for non-zero v_1 and v_2. This yields a relationship between λd and γ, i.e. between wavelength and frequency.

Figure 7.9 shows the lowest three values of γ found by this method for a range of wavelength/beam depth ($\pi/\lambda d$) ratios, taking ν as 0.3. These curves are marked 1 to 3. The simple beam solution given by (7.105) is marked B and the two roots of Timoshenko's solution are marked T1 and T2. The former remains accurate for short wavelengths, underestimating the exact frequency by only 2% when the wavelength is equal to the beam depth. However, the simple beam analysis overestimates it by 22% at a five times greater wavelength.

The equivalent solution for travelling waves is given by replacing (7.109) with

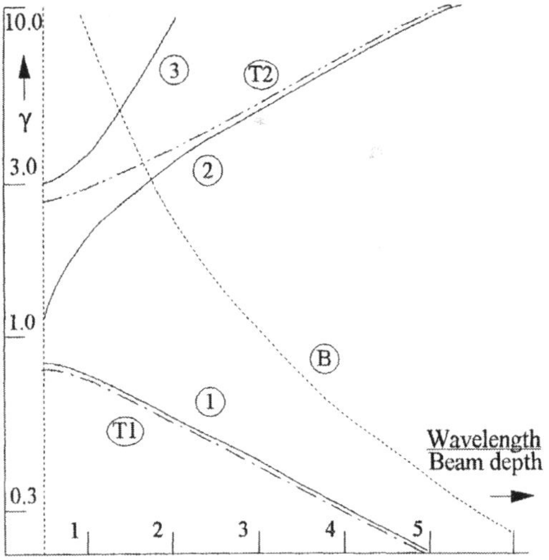

Figure 7.9 Predictions of frequencies at a wavelength.

$$u = \sum_1^2 u_i \sinh m_i y \sin[\lambda(x-ct)+\alpha] \;,\; v = \sum_1^2 u_i \cosh m_i y \cos[\lambda(x-ct)+\alpha] \tag{7.114}$$

The solution is then exactly the same, with $-\lambda c$ replacing ω, so that γ in Figure 7.9 is now $\rho c^2/G$ where c is the wave velocity.}

Chapter 8 The General Theory of Beams

8.1 Introduction

Exact stress distributions for some beams loaded along their lengths can be found. A simple example is that of a thin rectangular strip of depth h subject to an exponentially-decaying distributed load of intensity

$$q = q_0 e^{-ax/h} \tag{8.1}$$

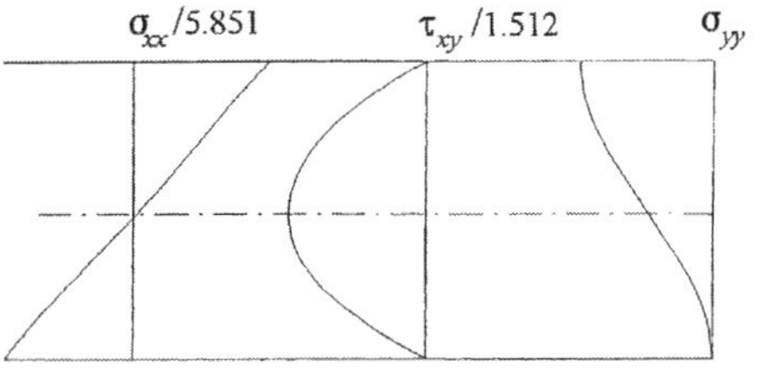

Figure 8.1 Stress distributions for $a = 1$.

applied to its upper surface $y=h$ where x is in the direction of the axis of the beam. It can be taken as an infinitely long cantilever, fixed at its left-hand end and free at its right-hand end. Treating this as a problem in plane stress, the resulting distribution is

$$\begin{aligned} \sigma_{xx} &= [A(\sin ky + ky\cos ky) + B(2\cos ky - ky\sin ky)]e^{-kx} \\ \tau_{xy} &= [Aky\sin ky + B(\sin ky + ky\cos ky)]e^{-kx} \\ \sigma_{yy} &= [A(\sin ky - ky\cos ky) + Bky\sin ky]e^{-kx} \end{aligned} \tag{8.2}$$

where

$$k = \frac{a}{h} \quad , \quad A = \frac{q_0(\sin a + a\cos a)}{a^2 - \sin^2 a} \quad , \quad B = -\frac{q_0 a\sin a}{a^2 - \sin^2 a} \tag{8.3}$$

The stress distributions for unit a is shown in Figure 8.1. The diagrams have been normalised by the factors shown so that each has a maximum amplitude of q. The longitudinal stress σ_{xx} is almost linear and the shear stress τ_{xy} is almost parabolic, as predicted by elementary beam theory. The theory does not predict the stress distribution σ_{yy}, which is almost cubic. However, for the much more rapid decay of the distributed load q given by taking a as ten, the stress distribution given by the elementary theory of bending bears no resemblance to the actual distribution given by Figure 8.2. In some sense, the beam ceases to cope with the rapid change in the loading by means of its characteristic response. This is a response to the resultant loading (moment and force) acting on a section of a beam.

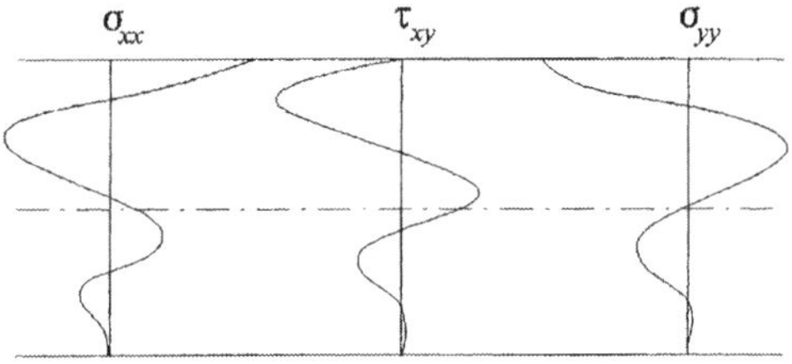

Figure 8.2 Stress distributions for $a = 10$.

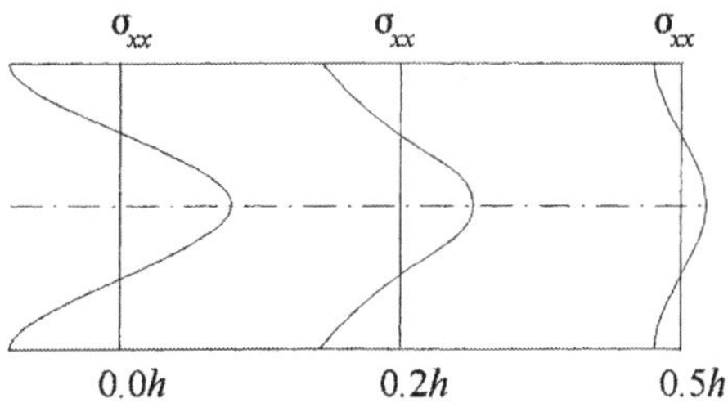

Figure 8.3 Decay of an axial stress system.

Where a stress system acting on the cross section of a beam has no resultant, it follows from Saint-Venant's principle that it tends to decay further along the beam. Figure 8.3 shows a rectangular strip subject to an axial stress distribution σ_{xx} which is sinusoidal in form and has no resultant. The maximum amplitude of this distribution decays to less than a quarter of its initial value at half the depth of the beam, $0.5h$, along the beam.

What seems to be fundamental to beam theory is neither Bernoulli's hypothesis that plane sections remain plane, nor that the stress distribution varies linearly across the section[1]. It is that there is a characteristic response to a resultant loading towards which the beam will tend, in the absence of further perturbation along it. In this sense, the basis of beam theory was laid by Saint-Venant [2].

{8.2 The Constant Response

Southwell (1936) argues in §93 as follows. If moments, axial forces or torques are applied to the ends of a beam, then the resultant load at any section will be the same. There must be a unique stress (or strain) distribution which minimises the strain energy per unit length for any given resultant. If there were two such states with the same minimum strain energy, then by the principle of superposition, a non-zero state of stress could exist which gave rise to zero strain energy. Southwell refers to this as the 'standard' response; here the term 'characteristic' response will be used. For the whole beam, there will be zones near the ends in which the stress distribution will be dominated by the way in which these loads are applied. However, in zones distant from the ends, the stress distribution is governed only by the resultant itself, in accordance with Saint-Venant's principle. The beam as a whole seeks to minimise the strain energy for the given end loading. Thus, in the zones away from the ends, the state of stress approximates to the 'characteristic' response. The longer the beam, the more closely it can approach this response over substantial portions of its length. The characteristic response can then be thought of as a constant response infinitely remote from the ends.

Consider the z axis to be along the axis of a prismatic beam. Then from (3.9) to (3.13) the state of constant strain is given by

$$\frac{\partial^2 u}{\partial x \partial z} = \frac{\partial^2 v}{\partial y \partial z} = \frac{\partial^2 w}{\partial z^2} = 0 \quad , \quad \frac{\partial^2 u}{\partial y \partial z} + \frac{\partial^2 v}{\partial x \partial z} = \frac{\partial^2 v}{\partial z^2} + \frac{\partial^2 w}{\partial y \partial z} = \frac{\partial^2 w}{\partial x \partial z} + \frac{\partial^2 u}{\partial z^2} = 0 \,. \tag{8.4}$$

The most general form of solution of these equations, excluding rigid-body rotations about the x and y axes, is

$$\begin{aligned} u &= -yz\vartheta_z + \tfrac{1}{2}\vartheta_y z^2 + U(x,y) \\ v &= xz\vartheta_z - \tfrac{1}{2}\vartheta_x z^2 + V(x,y) \\ w &= z(\varepsilon_z - \vartheta_y x + \vartheta_x y) + W(x,y) \end{aligned} \tag{8.5}$$

where ϑ_x, ϑ_y and ϑ_z are the (overall) rates of rotation about the x, y and z axes and ε_z is the (overall) tensile axial strain, as in (7.6).

Comparing (4.63) and (8.5), it can be seen that P produces an overall tensile strain of Ps_{33}/A. Likewise, comparing (4.64) and (8.5),

---

[1] This latter assumption works for homogenous anisotropic sections, see Leknitskii (1981) for example.

[2] Adhémar Jean Claude Barré, Comte de Saint-Venant 1797-1886. Arguably the most significant elastician ever. He applied unsuccessfully for membership of the Paris Académie des Sciences several times. On the first occasion, he came fourth out of six candidates and the place was won by Morin, best known for his experiments on the cohesion of bricks and mortar. He was eventually elected in 1868.

$$\vartheta_x = \frac{M_x s_{33}}{I_{xx}} \quad , \quad \vartheta_z = -\frac{M_x s_{53}}{2I_{xx}} \ . \tag{8.6}$$

and comparing (4.65) and (8.5),

$$\vartheta_y = \frac{M_y s_{33}}{I_{yy}} \quad , \quad \vartheta_z = -\frac{M_y s_{43}}{2I_{yy}} \ . \tag{8.7}$$

Thus the overall tensile and flexural responses of the beam are similar to those for an isotropic beam except that $1/E$ is replaced by s_{33}, but there is in addition a torsional response to bending moments given by the second equalities in (8.6) and (8.7). From the reciprocal theorem, it follows that if a unit bending moment produces twist, a unit torque must produce flexural rotation rate of the same magnitude. This implies the torque-flexure relationships given by (5.58):

$$\vartheta_x = -\frac{T s_{53}}{2I_{xx}} \quad , \quad \vartheta_y = -\frac{T s_{43}}{2I_{yy}} \ . \tag{8.8}$$

The theory will also be applied to modular beams. Although these are not the same at each cross-section, they are made from a number of identical modules joined end to end. Trusses and castellated beams are examples of modular beams.

A modular beam is shown in Figure 8.4. At one end, it is acted upon by an axial force P, a bending moment M and a torque T. The Nth module is a long way from the end, and so according to Saint-Venant's principle[1], it responds to these resultants rather than the way in which they are applied. If the beam is sectioned at the Zth module, the resultants acting on it are the same, as shown in the lower diagram. The response of the (N+Z)th module must then be same as that of the Nth module. If this were not so, then the lower diagram could be shifted Z modules to the left and a difference in response between the upper and lower beam found in the Nth module. This would be the response to zero resultant end loading, which should be zero. Thus the characteristic response to the end loading for each module is constant[2] and is unique.

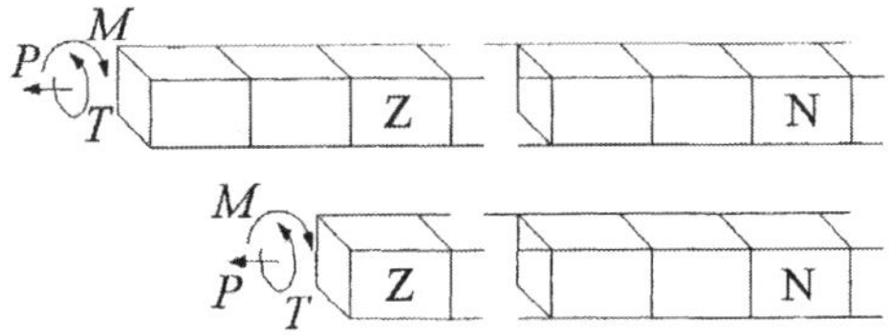

Figure 8.4 Regular response of a modular beam.

The above argument relies on the invariance of the beam under a longitudinal shift of origin. This corresponds to the translation symmetry discussed in §1.5. This is the basis of Saint-Venant's semi-inverse method, in which certain simplifying assumptions are made about the state of stress and strain. For a prismatic system, the condition can be stated as follows. *If a constant response to a resultant moment, torque or axial force can be found which satisfies the internal conditions and stress-free boundary conditions locally, it is the characteristic response.*

---

[1] A caveat must be entered here. Each module must be capable of resisting the loading applied to it, so that any loading which has no resultant is dissipated by the strain energy it induces in a module. An often quoted example of where the modules form mechanisms and so transmit rather than absorb such loading is given by Hoff, N.J. (1945) The Applicability of Saint-Venant's Principle to Airplane Structures. *Journal of Aeronautical Science* V.12, p.455, (see §15.4).

[2] There may be variations *within* a module, but these will be the same for each module.

Equations (8.5) can also be written in terms of cylindrical polar coordinates (see §3.10) again taking z to be directed along the axis of the beam:

$$\begin{aligned} u^P &= \frac{1}{2}\vartheta_\theta z^2 + U^P(r,\theta) \\ v^P &= rz\vartheta_z - \frac{1}{2}\vartheta_r z^2 + V^P(r,\theta) \\ w^P &= z(\varepsilon_z - r\vartheta_\theta) + W^P(r,\theta) \end{aligned} \tag{8.9}$$

where ϑ_r and ϑ_θ are the flexural rotation rates about the local radial and tangential axes (in the directions of increasing r and θ). Solutions of the form given by (8.9) can be found for the behaviour of beams with certain kinds of *curvilinear anisotropy*. These are material properties which are specified in terms of the local directions of the radial, tangential and longitudinal axes. They relate the stresses and strains expressed in polar coordinates, as given by (3.59) and (3.62). (See Lekhnitskii (1981) §§10, 17, 42, 43 and 60, and Renton (1987) Appendix A1.3).

8.3 The Linear Response

Here, it will be assumed that the stresses and strains may vary linearly along the beam. The differentials of the strains in (8.4) will again be zero after differentiating once more with respect to z. The general form of the solution is now that given by (8.5), integrated with respect to z:

$$\begin{aligned} u &= -\frac{1}{2}yz^2\vartheta_z^* + \frac{1}{6}\vartheta_y^* z^3 + zU^*(x,y) + U(x,y) \\ v &= \frac{1}{2}xz^2\vartheta_z^* - \frac{1}{6}\vartheta_x^* z^3 + zV^*(x,y) + V(x,y) \\ w &= \frac{1}{2}z^2(\varepsilon_z^* - \vartheta_y^* x + \vartheta_x^* y) + zW^*(x,y) + W(x,y) \end{aligned} \tag{8.10}$$

Similarly, (8.9) can be integrated with respect to z. The only end resultant which could produce such a response is a shear force. For example, on comparing the terms in (8.10) with those in (6.76),

$$\begin{aligned} \vartheta_x^* &= \frac{Ss_{33}}{I_{xx}} \quad , \quad U^*(x,y) = \frac{S}{2I_{xx}}(2s_{13}xy + s_{63}y^2) - \vartheta_z y \\ \vartheta_y^* &= 0 \quad , \quad V^*(x,y) = \frac{S}{2I_{xx}}(-s_{13}x^2 + s_{23}y^2) + \vartheta_z x \\ \vartheta_z^* &= -\frac{Ss_{53}}{2I_{xx}} \quad , \quad W^*(x,y) = \frac{S}{2I_{xx}}(s_{53}xy + s_{43}y^2) + R \quad , \quad \varepsilon_z^* = P = Q = 0 \end{aligned} \tag{8.11}$$

Note that this comparison establishes that P and Q must be zero by other means than those invoked for (6.78). Also, it can be seen that the linearly-varying flexural curvature given by ϑ_x* is of the same kind as that for pure bending, given by ϑ_x in (8.6), where M_x is Sz.

The characteristic response of a continuous beam can be seen as a particular case of the response of a modular beam. In Figure 8.5a, a modular beam is loaded at one end by a bending moment M and a shear force S acting in a direction normal to the axis of M. The Nth module is remote from the end and so its response is to the

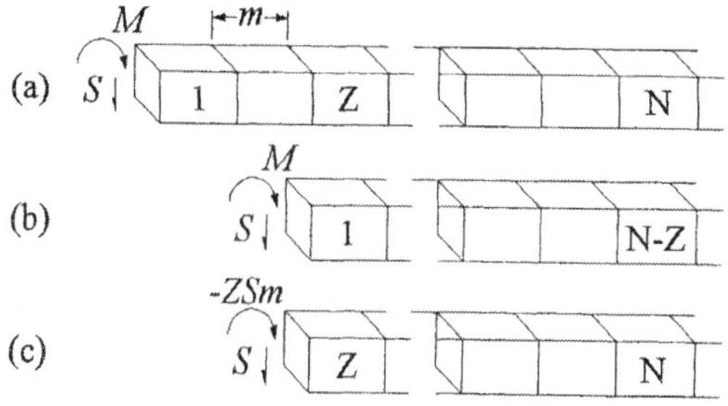

Figure 8.5 Shear of a modular beam.

resultants only, regardless of the way in which they are applied. In Figure 8.5b, the beam has been shifted Z modules to the right, (Z<<N) so that the (N-Z)th module also responds only to the end resultants. Figure 8.5c shows the beam shown in Figure 8.5a cut off at the Zth module when no moment M is applied to the first module. The resultants acting on the Zth module are then as shown. If M in Figure 8.5b is $-Zsm$, then the response of the (N-Z)th module must be indistinguishable from that of the Nth module shown in Figure 8.5c. If this were not so, the difference in loadings on the identical frames shown in Figures 8.5b and 8.5c, which has no resultant, would have a non-zero response at distances far from the end. This leads to the following theorem:

If a shear force is applied to the end of a modular beam, the patterns of stress and strain in modules far from the end can be divided into a linearly-varying component which is the characteristic response to a moment and a constant component which is the characteristic response to a shear force. By the same arguments as before, if a linearly-varying pattern of stress and strain can be found which has only a resultant shear force and bending moment, this must be a combination of the characteristic responses to each.

8.4 The Deformation Matrix

If no assumptions are made about the way in which a unit length of a linearly-elastic beam distorts under the action of resultant loads on its ends, the response can be written as

$$\begin{bmatrix} \vartheta_x \\ \vartheta_y \\ \vartheta_z \\ \gamma_x \\ \gamma_y \\ \varepsilon_z \end{bmatrix} = \begin{bmatrix} f_{11} & f_{12} & f_{13} & f_{14} & f_{15} & f_{16} \\ f_{21} & f_{22} & f_{23} & f_{24} & f_{25} & f_{26} \\ f_{31} & f_{32} & f_{33} & f_{34} & f_{35} & f_{36} \\ f_{41} & f_{42} & f_{43} & f_{44} & f_{45} & f_{46} \\ f_{51} & f_{52} & f_{53} & f_{54} & f_{55} & f_{56} \\ f_{61} & f_{62} & f_{63} & f_{64} & f_{65} & f_{66} \end{bmatrix} \begin{bmatrix} M_x \\ M_y \\ T \\ S_x \\ S_y \\ P \end{bmatrix} \quad \text{or} \quad \boldsymbol{d} = \boldsymbol{FP} \tag{8.12}$$

where $\boldsymbol{F}$ will be referred to as the *deformation matrix*. The strain energy per unit length is

$$U_l = \frac{1}{2}\boldsymbol{P}^T\boldsymbol{d} = \frac{1}{2}\boldsymbol{P}^T\boldsymbol{FP} = \frac{1}{2}\sum_{i=1}^{6}\sum_{j=1}^{6} P_i f_{ij} P_j \tag{8.13}$$

where P_i and P_j are the ith and jth elements of $\boldsymbol{P}$, as listed in (8.12). As the strain energy is positive definite, the final product is then a positive definite quadratic form. This implies constraints on the elements of $\boldsymbol{F}$ similar to those imposed on the stress and strain matrices in §3.5 (see also §3.11). Just as it follows from the reciprocal theorem that the stress and strain matrices are symmetrical, so also is $\boldsymbol{F}$. That is f_{ij} and f_{ji} are the same (i,j = 1 to 6).

For a homogeneous anisotropic beam, it follows from (8.8) that f_{13} and f_{23} are given by $-s_{53}/2I_{xx}$ and $-s_{43}/2I_{yy}$ respectively. More generally, the coefficients of the deformation matrix can be found by energy methods. The characteristic responses to resultant loads, discussed in Chapters 4 to 6, enable the stresses and strains to be written as linear functions of these loads. Suppose that all these resultants are acting simultaneously on a typical cross-section. The strain energy per unit length, U_l, can be found by integrating the strain energy per unit volume over the cross-section (cf. §7.1). This will result in a quadratic function of the resultant loads. (Note that when using the expressions for the stresses and strains produced by the shear forces at the tip of a cantilever, the terms in $S_y z$ and $-S_x z$ should be attributed to the local moments M_x and M_y respectively. The resulting quadratic is then not a function of z.) Then from (8.12) and (8.13),

$$\frac{\partial^2 U_l}{\partial P_i \partial P_j} = \frac{1}{2}(f_{ij} + f_{ji}) = f_{ij} = f_{ji} \tag{8.14}$$

The formal process of partial differentiation is not necessary. The elements f_{ij} can be found by direct comparison of the coefficients of the load products $P_i P_j$.

The beam and its material may be symmetrical. This will impose further conditions on the form of the deformation matrix. Figure 8.6 shows a beam which is symmetrical both in form and its material properties[1] under inversion of the x axis. The resultant loads acting on a section and the corresponding deformations are shown in Figure 8.6a. The mirror image of this is, resulting from inversion of the x axis, is shown in Figure 8.6b. The section itself is unchanged by this inversion. However, the senses of some of the loads and their corresponding deformations have been reversed. Thus the same M_x, S_y and P now produce opposite ϑ_y, ϑ_z and γ_x. As the section is identical to that before inversion, the two states (a) and (b) can be combined, using the principle of superposition. It now becomes clear that M_x, S_y and P produce zero ϑ_y, ϑ_z and γ_x. This implies that in the deformation matrix,

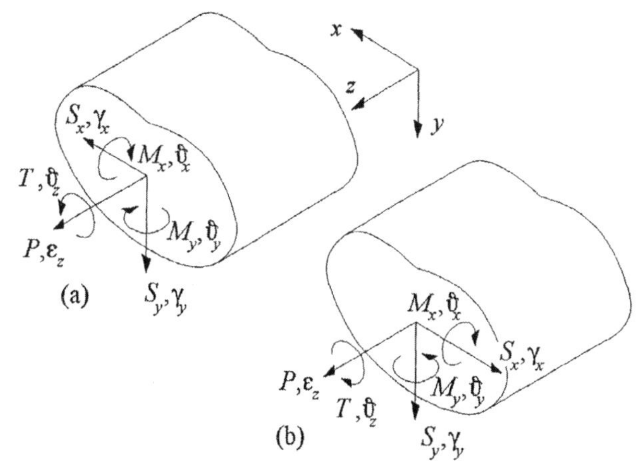

Figure 8.6 Symmetry under inversion of the x axis.

$$f_{21} = f_{25} = f_{26} = f_{31} = f_{35} = f_{36} = f_{41} = f_{45} = f_{46} = 0$$

Bearing in mind the symmetry of this matrix, this produces the pattern of zero elements shown in Figure 8.7a. These are shown by white circles, the black circles being the non-zero elements.

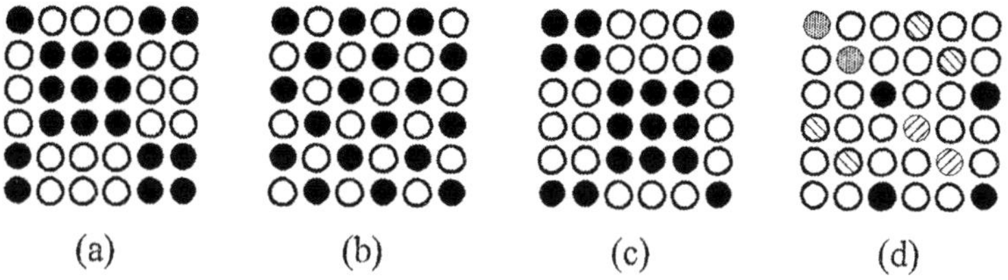

Figure 8.7 Zero-element patterns for symmetrical beams.

Figures 8.7b and 8.7c show the patterns for invariance under inversion of the y and z axes respectively. Note that f_{14}, f_{25} and f_{36} are zero in all three cases. Figure 8.7d shows the pattern for

---

[1] That is to say, the material is at least monoclinic under inversion of the x axis (see §3.11).

invariance under a rotation ($\neq 180°$) about the z axis. (Elements with the same pattern are equal). This is for a typical beam of isotropic material and one axis of symmetry. If the section is homogeneous, the coupling indicated by the f_{16} term can be eliminated by defining the axial force as acting through the centroid of the section. Likewise, the coupling between torsion and the shear force S_x can be eliminated by defining it as acting through the shear centre. Then f_{34} is zero and the matrix becomes diagonal. The expression for the (complementary) strain energy per unit length given by (8.13) then takes the form given by (7.5). However, it will be shown that beam theory can be applied to a far wider range of problems. For example, for an homogeneous anisotropic beam, it has already been seen from (8.8) that the coefficients f_{13} and f_{23} may exist.

If there is only symmetry with respect to z axis inversion, as in Figure 8.7c, full uncoupling of the deformation equations can still be obtained, as can partial uncouplings for all other cases. The line of action of P will be defined as being through the 'centroid', C, of the section. Its position, given by the coordinates (x_c, y_c) will be defined[1] as being such that P produces no flexural effects (i.e. f_{16} and f_{26} are zero). Likewise, the resultant shear forces S_x and S_y will act through the shear centre by definition. Its position, (x_o, y_o), will be such that these forces produce no torsional effects (i.e. f_{34} and f_{35} are zero). The principal flexural axes of the section will be chosen in such a way that there is no coupling between the moment M_x and the rate of flexural rotation ϑ_y (or conversely between M_y and ϑ_x) so that f_{21} (and so necessarily f_{12}) are zero. It may be necessary to define the orientation of a separate set of principle axes[2] so that there is no coupling between the shear force S_x and the shear displacement γ_y (or between S_y and γ_x) so that f_{54} (and f_{45}) are zero.

Figure 8.8 shows the resultant loads and corresponding deflexions defined with respect to a coordinate system which is arbitrary, apart from the direction of the z axis which is parallel to the axis of the beam. The same resultant loading can be defined with respect to the 'centroid' and shear centre. The magnitudes of the forces remain unchanged but the moments must take account of the shift of origin. Likewise, the shifts of origin do not change the definitions of the rotation rates, but do affect the definitions of the overall axial strain and shear strains. The changes produced by the shift to the 'centroid' are given by

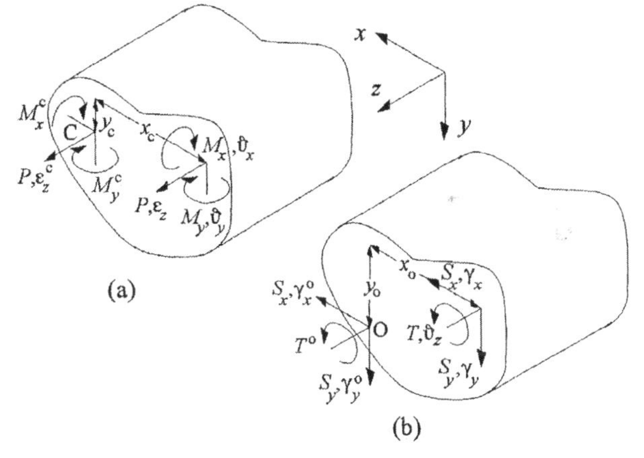

Figure 8.8 The 'centroid' and shear centre.

$$M_x = M_x^c + Py_c \;,\quad M_y = M_y^c - Px_c \;,\quad \varepsilon_z^c = \vartheta_x y_c - \vartheta_y x_c + \varepsilon_z \,. \qquad \textbf{(8.16)}$$

and the shift to the shear centre are given by

[1] In the general case examined here, it is not necessarily the geometric centroid.

[2] These can be significantly different. See Schramm, U., Kitis, L., Kang, W. and Pilkey, W.D (1994) On the shear deformation coefficient in beam theory. *Finite Elements in Design* V.16 p.141.

$$T = T^o - S_x y_o + S_y x_o \quad , \quad \gamma_x^o = -\vartheta_z y_o + \gamma_x \quad , \quad \gamma_y^o = \vartheta_z x_o + \gamma_y \; . \tag{8.17}$$

In matrix terms, these relationships become

$$\begin{bmatrix} M_x \\ M_y \\ T \\ S_x \\ S_y \\ P \end{bmatrix} = \begin{bmatrix} 1 & 0 & 0 & 0 & 0 & y_c \\ 0 & 1 & 0 & 0 & 0 & -x_c \\ 0 & 0 & 1 & -y_o & x_o & 0 \\ 0 & 0 & 0 & 1 & 0 & 0 \\ 0 & 0 & 0 & 0 & 1 & 0 \\ 0 & 0 & 0 & 0 & 0 & 1 \end{bmatrix} \begin{bmatrix} M_x^c \\ M_y^c \\ T^o \\ S_x \\ S_y \\ P \end{bmatrix} \quad , \quad \begin{bmatrix} \vartheta_x \\ \vartheta_y \\ \vartheta_z \\ \gamma_x^o \\ \gamma_y^o \\ \varepsilon_z^c \end{bmatrix} = \begin{bmatrix} 1 & 0 & 0 & 0 & 0 & 0 \\ 0 & 1 & 0 & 0 & 0 & 0 \\ 0 & 0 & 1 & 0 & 0 & 0 \\ 0 & 0 & -y_o & 1 & 0 & 0 \\ 0 & 0 & x_o & 0 & 1 & 0 \\ y_c & -x_c & 0 & 0 & 0 & 1 \end{bmatrix} \begin{bmatrix} \vartheta_x \\ \vartheta_y \\ \vartheta_z \\ \gamma_x \\ \gamma_y \\ \varepsilon_z \end{bmatrix} \tag{8.18}$$

These equations can be written as

$$\boldsymbol{P} = \boldsymbol{T}\boldsymbol{P}' \quad , \quad \boldsymbol{d}' = \boldsymbol{T}^T\boldsymbol{d} \tag{8.19}$$

so that the deformation equation now becomes

$$\boldsymbol{d}' = \boldsymbol{F}'\boldsymbol{P}' \qquad \text{where} \qquad \boldsymbol{F}' = \boldsymbol{T}^T\boldsymbol{F}\boldsymbol{T} \tag{8.20}$$

Defining the positions of the 'centroid' and shear centre so that the coupling terms mentioned earlier are eliminated,

$$\boldsymbol{F}' = \begin{bmatrix} f_{11} & f_{12} & f_{13} & f_{14}-y_o f_{13} & f_{15}+x_o f_{13} & 0 \\ f_{21} & f_{22} & f_{23} & f_{24}-y_o f_{23} & f_{25}+x_o f_{23} & 0 \\ f_{31} & f_{32} & f_{33} & 0 & 0 & f'_{36} \\ f_{41}-y_o f_{31} & f_{42}-y_o f_{32} & 0 & f_{44}-y_o f_{34} & f_{45}+x_o f_{43} & f_{46}-y_o f'_{36}+y_c f_{41}-x_c f_{42} \\ f_{51}+x_o f_{31} & f_{52}+x_o f_{32} & 0 & f_{54}+x_o f_{34} & f_{55}+x_o f_{35} & f_{56}+x_o f'_{36}+y_c f_{51}-x_c f_{52} \\ 0 & 0 & f'_{63} & f_{64}-y_o f'_{63}+y_c f_{14}-x_c f_{24} & f_{65}+x_o f'_{63}+y_c f_{15}-x_c f_{25} & f_{66}+y_c f_{61}-x_c f_{62} \end{bmatrix} \tag{8.21}$$

where

$$x_o = -\frac{f_{35}}{f_{33}} \quad , \quad y_o = \frac{f_{34}}{f_{33}} \quad , \quad x_c = \frac{f_{11}f_{26}-f_{16}f_{21}}{f_{11}f_{22}-f_{12}f_{21}} \quad , \quad y_c = \frac{f_{12}f_{26}-f_{16}f_{22}}{f_{11}f_{22}-f_{12}f_{21}} \; ,$$
$$f'_{36} = f'_{63} = f_{63}+y_c f_{13}-x_c f_{23} \; . \tag{8.22}$$

It will be seen that there may be coupling between bending and shear implied by the coefficients f_{15} and f_{24} in the deformation matrix. These are both zero if the beam is invariant under inversion of the z axis (cf. Figure 8.7c). These terms imply that a shear force induces a flexural curvature other than that resulting from the bending moment it produces (and that the bending moment produces a shear deformation). For homogeneous isotropic beams, the bending strain energy is a function of the axial stress σ_{zz} which is produced by the bending moment only. Likewise, the shear strain energy is produced by the shear stresses which are functions of the shear force only. In the expression for U_l given by (8.13) there are then no terms involving both the moment M and the shear force S, so that the above coupling coefficients do not exist. However, for the shear of the anisotropic beam analysed in §6.7 coupling could occur. In (6.75), the axial stress due to the bending moment is Szy/I_{xx} but there is also an axial stress σ^o_{zz} resulting directly from the shear force. In evaluating the strain energy resulting from the axial stress, coupling terms involving the local moment and the shear force may then occur. Uncoupling can be achieved by defining a *flexural centre* which is offset axially at a fixed distance z_f along the beam from the section being examined. The reference bending moment used in (8.12) and (8.13) is then not that at the section, but that z_f further along. Taking a shear force S only to act at the end of a beam the strain energy per unit length produced at a distance z along the beam can be written as

$$U_l = \frac{1}{2}[(Sz)^2 f_{mm} + 2(Sz)Sf_{ms} + S^2 f_{ss}] \equiv \frac{1}{2}[(Sz + Sz_f)^2 f'_{mm} + S^2 f'_{ss}]$$
$$\text{where} \quad S = S_y \;,\; f_{mm} = f_{11} \;,\; f_{ms} = f_{15} \;,\; f_{ss} = f_{55} \qquad \textbf{(8.23)}$$
$$\text{or} \quad S = S_x \;,\; f_{mm} = f_{22} \;,\; f_{ms} = -f_{24} \;,\; f_{ss} = f_{44}$$

The expressions for U_l must be the same for all values of z so that on comparing coefficients,

$$f'_{mm} = f_{mm} \;,\; f'_{ss} = f_{ss} - f_{ms}^2/f_{mm} \;,\; z_f = f_{ms}/f_{mm} \,. \qquad \textbf{(8.24)}$$

Redefining the moments as those acting at the flexural centres has the effect of changing the columns of the deformation matrix, so that

$$f'_{i4} = f_{i4} - f_{i2}f_{24}/f_{22} \;,\; f'_{i5} = f_{i5} - f_{i1}f_{15}/f_{11} \quad (i = 1 \text{ to } 6) \qquad \textbf{(8.25)}$$

Redefining the shears γ_x and γ_y as those relative to the extrapolated rotations at the flexural centres has a similar effect on the fourth and fifth columns of the matrix, without further changes to f_{44} and f_{55} (as the modified forms of f_{24} and f_{15} are now zero[1]). The modified matrix is now symmetrical again. The flexural centre is of particular importance in analysing modular structures, as will be seen in the next section.

8.5 The Slope-Deflexion Equations for Modular Beams

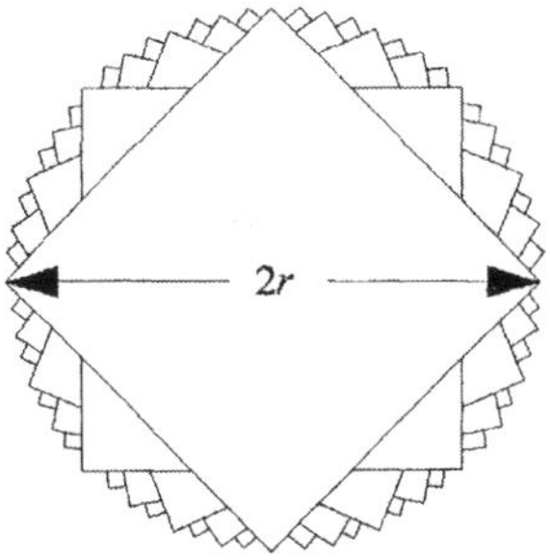

Figure 8.9 'Squaring the circle'.

Modular beams are 'discrete' or 'lumped' systems formed by a set of identical units or 'modules' joined end-to-end. A truss is the most common example of modular beam, but any prismatic structure with a regularly repeated form, such as a castellated beam, is also a modular beam. Efforts to analyse them in terms of equivalent ordinary beams have met with partial success[2]. Using the appropriate bending stiffness for an equivalent uniform beam seems to provide a good first approximation to the behaviour of a modular beam. However, the choice of an appropriate shear stiffness seems to vary with the end conditions. The problems of representing a piecewise-continuous system by a continuous one are illustrated in Figure 8.9. A square of diagonal length $2r$ is turned through 45° and placed on its original image. This forms an eight-pointed star which is turned through 22.5° and the process repeated *ad infinitum*. The circle-like outline of this object contains an area of πr^2, but the outline is $2\sqrt{2}\pi r$ long. Also, the slope of the outline is indeterminate at all points on it.

In such problems, difference calculus can be a more useful analytic tool than differential

[1] This process of uncoupling by defining the flexural centre is counter-productive if either f_{13} or f_{23} are non-zero. This is because uncoupling achieved by defining the shear centre is lost.

[2] See for example Saka, T. and Heki, K. (1981) On the effective rigidities of bar-like plane trusses. *Memoirs of the Faculty of Engng, Osaka City Univ.* V.22, p.167.

calculus. This will be discussed in more detail in Chapter 15. Operations are carried out with integer parameters rather than real variables. The most common operators are

$$\begin{aligned} \mathrm{E}f(\mathrm{X}) &= f(\mathrm{X}+1) \\ \Delta f(\mathrm{X}) &= f(\mathrm{X}+1) - f(\mathrm{X}) = (\mathrm{E}-1)f(\mathrm{X}) \\ \nabla f(\mathrm{X}) &= f(\mathrm{X}) - f(\mathrm{X}-1) = (1-\mathrm{E}^{-1})f(\mathrm{X}) \\ \Delta\!\!\!\nabla f(\mathrm{X}) &= \Delta[\nabla f(\mathrm{X})] = (\mathrm{E}+\mathrm{E}^{-1}-2)f(\mathrm{X}) \end{aligned} \tag{8.26}$$

where X is usually an integer. This notation follows that in Brown (1965) §13.2. Here, X denotes the Xth module from the left-hand end, as in Figure 8.10. Here the notation used has been chosen to correspond to that in §4.5 and §6.4 so that comparisons can be made. The length of each module is taken as m. The flexural centre is usually in the centre of the module. In general, it will be taken to be me to the right of it, and the bending moment at that position denoted by M_x', as shown in Figure 8.10b. The other loads and deflexions specific to a module are related to its right-hand face as shown. Thus the loads, M_B and S_B and the deflexions θ_B and v_B at end B of the beam are those specific to module N. (However, more thought is required to determine the loads and deflexions at end A.) The force P_x is applied downwards at the junction of modules X and X+1. The from statics,

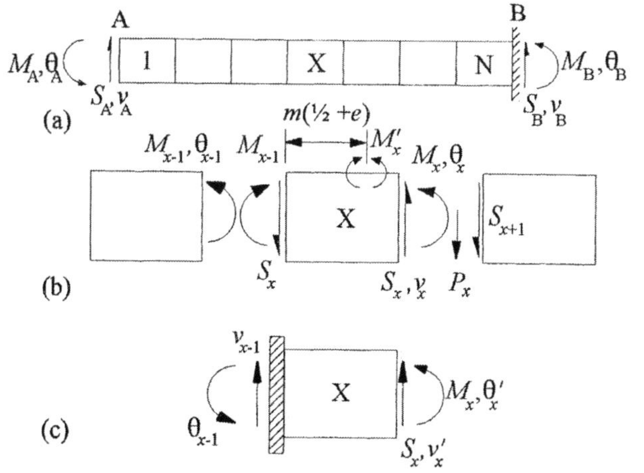

Figure 8.10 Loads and deflexions of a modular beam.

$$P_x = S_{x+1} - S_x = \Delta S_x \quad , \quad S_x m = M_{x-1} - M_x = -\nabla M_x \quad , \quad M_x' = M_x + S_x m(\tfrac{1}{2} - e) \tag{8.27}$$

It will be assumed that the only resultant loads acting on any module are the bending moments and shear forces shown. Using the notation of (8.23) for the flexibility coefficients, the strain energy in module X can be written as

$$U_m = mU_l = \tfrac{1}{2}m(M_x'^{\,2} f_{mm} + S_x^2 f_{ss}') \tag{8.28}$$

Suppose that a rigid-body movement is given to this module, so that the rotation and displacement of its left-hand side are both zero for the moment. This now acts like a cantilever, as shown in Figure 8.10c. Applying Castigliano's complementary energy theorem (cf. §7.6) to the module,

$$\begin{aligned} \theta_x' &= \frac{\partial U_m}{\partial M_x} = mf_{mm}[M_x + mS_x(\tfrac{1}{2}-e)] \\ v_x' &= \frac{\partial U_m}{\partial S_x} = m^2 f_{mm}(\tfrac{1}{2}-e)[M_x + mS_x(\tfrac{1}{2}-e)] + mf_{ss}'S_x \,. \end{aligned} \tag{8.29}$$

Rotating and displacing the left-hand side of the module so that it now has the true rotation θ_{x-1} and v_{x-1}, the rotation and displacement of the right-hand side are given by

$$\theta_x = \theta_x' + \theta_{x-1} \ , \quad v_x = v_x' + v_{x-1} + m\theta_{x-1} \ . \qquad \textbf{(8.30)}$$

Then from (8.29) and (8.30),

$$\nabla\theta_x = mf_{mm}[M_x + mS_x(\tfrac{1}{2} - e)]$$
$$\nabla v_x^m = m^2 f_{mm}(\tfrac{1}{2} - e)[M_x + mS_x(\tfrac{1}{2} - e)] + m\theta_{x-1} \ , \quad \nabla v_x^s = mf_{ss}' S_x \ . \qquad \textbf{(8.31)}$$

where v_x is the sum of v_x^m, associated with the flexure, and v_x^s, associated with shear. These equations can either be solved directly, or further use made of (8.27) to derive the forms

$$\Delta\nabla v_x^s = mf_{ss}' P_x$$
$$(\Delta\nabla)^2 v_x^m = -m^3 f_{mm}[1 + \Delta\nabla(\tfrac{1}{4} - e^2)]P_x \qquad \textbf{(8.32)}$$
$$\Delta^2\nabla\theta_x = -m^2 f_{mm}[1 + \Delta(\tfrac{1}{2} + e)]P_x$$

Taking the ordinate x as equivalent to mX, the distributed load p_x as equivalent to P_x/m and using the Taylor expansion

$$\mathrm{E}f(x) = f(x+m) = f(x) + m\frac{df}{dx} + \frac{m^2}{2}\frac{d^2 f}{dx^2} + \ldots \qquad \textbf{(8.33)}$$

and then allowing m to tend to zero in the above equations, gives at the limit,

$$\frac{d^2 v_x^s}{dx^2} = f_{ss}' p_x \ , \quad \frac{d^4 v_x^m}{dx^4} = -f_{mm} p_x \ , \quad \frac{d^3\theta_x}{dx^3} = -f_{mm} p_x \ . \qquad \textbf{(8.34)}$$

This corresponds to the governing differential equations for ordinary beams on taking f_{ss}' as $1/GkA$, f_{mm} as $1/EI$ and θ_x as the slope of v_x^m. However, applying these differential equations to a modular beam overlooks terms in the difference equations related to the flexural behaviour which are as significant as those related to the shear behaviour.

At end A of the beam, the deflexions θ_A and v_A are those specific to an imaginary 0th module, θ_0 and v_0. Likewise, the end moment M_A is $-M_0$ and the end shear force S_A is $-S_0$ (assuming that the end force P_0 can be defined separately). The case where there is no distributed load P_x will now be examined. From the conditions of equilibrium,

$$-S_B Nm = S_A Nm = M_A + M_B \ , \quad S_x = -S_A \ , \quad M_x = S_A \mathrm{X} m - M_A \ . \qquad \textbf{(8.35)}$$

From the effect of the operators in (8.26) on polynomials in X, the following inverse operations can be established:

$$\nabla^{-1}(1) = \mathrm{X} + C \ , \quad \nabla^{-1}(\mathrm{X}) = \tfrac{1}{2}(\mathrm{X}^2 + \mathrm{X}) + C \ , \quad \nabla^{-1}(\mathrm{X}^2) = \tfrac{1}{6}(2\mathrm{X}^3 + 3\mathrm{X}^2 + \mathrm{X}) + C \ . \qquad \textbf{(8.36)}$$

where C is a constant. Using (8.35) and (8.36), the first of equations (8.31) can be solved and then the remaining two equations. The constants can be found in terms of the deflexions specific to the 0th module, θ_A and v_A, giving

$$\theta_x = mf_{mm}\left[S_A m(\tfrac{1}{2}\mathrm{X}^2 + e\mathrm{X}) - M_A \mathrm{X}\right] + \theta_A$$
$$v_x = m^2 f_{mm}\left\{mS_A\left[\tfrac{1}{6}\mathrm{X}^3 + \mathrm{X}\left(\tfrac{1}{12} - e^2\right)\right] + M_A\left[e\mathrm{X} - \tfrac{1}{2}\mathrm{X}^2\right]\right\} - mf_{ss}' S_A \mathrm{X} + m\theta_A \mathrm{X} + v_A \qquad \textbf{(8.37)}$$

In particular, the deflexions specific to the Nth module, θ_B and v_B, can be found. Using (8.35) and rearranging the equations,

$$M_A = \frac{1}{f_{mm} m^2 N(N^2+s)}\{m\theta_A[4N^2+s+12e(e+N)] + m\theta_B[2N^2-s-12e^2] + 6\delta(N+2e)\}$$

$$M_B = \frac{1}{f_{mm} m^2 N(N^2+s)}\{m\theta_A[2N^2-s-12e^2] + m\theta_B[4N^2+s+12e(e-N)] + 6\delta(N-2e)\}$$

$$S_A = -S_B = \frac{6}{f_{mm} m^3 N(N^2+s)}[m\theta_A(N+2e) + m\theta_B(N-2e) + 2\delta] \tag{8.38}$$

where δ is $v_A - v_B$, $1/f_{mm}$ can be replaced by EI for the purposes of comparison with (6.24) and (6.25) and

$$s = \frac{12 f'_{ss}}{m^2 f_{mm}} - 1 \tag{8.39}$$

(cf. (6.21)). These equations are much simpler when e is zero, which is often the case. Further examples will be found in §A5.1. The coefficients f_{ij} for various modular beams formed by trusses will also be found in Appendix 5.

8.6 The Characteristic Response of Circular Beams

Saint-Venant's principle can also be applied to a circular beam to find a characteristic stress or strain response. At first sight, this appears improbable, particularly if its radius of curvature of the beam is not much greater than the dimensions of its cross-section. Any real beam must then be 'short', in that its length will be of the order of 2π times this radius. Mathematically speaking however, the beam need not interfere with itself after one revolution, unless this physical condition is imposed in addition to the other equations. A regular response to the loading at one end can then be sought. This is the response towards which the beam would tend at very large values of θ (the angle through which the undeformed beam curves from its free end). As it is a function of the resultant end loading only, it can also be thought of as a function of the local resultants produced by the end loading only. It follows immediately that any such response must be either constant or have a period of 2π.

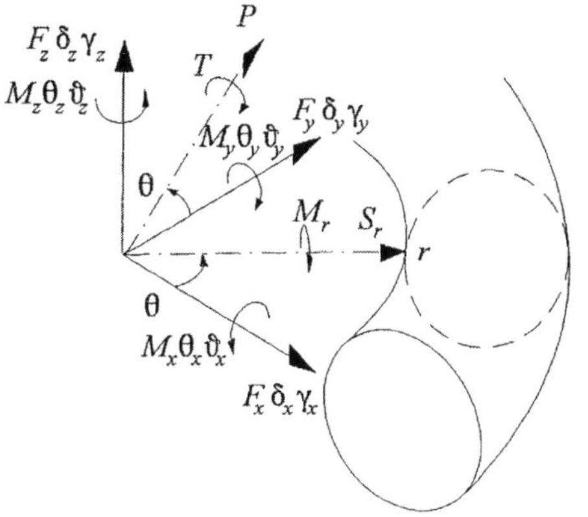

Figure 8.11 Loads and deflexions for a circular beam.

Figure 8.11 shows the notation which will be used in this case. Cylindrical polar coordinates r, θ, z will be used (cf. §3.10) with the origin situated at the centre of curvature of the beam. The angle θ is measured from the x axis of a cartesian coordinate system x,y,z. The end moments M_x, M_y and M_z and forces F_x, F_y and F_z are expressed relative to this coordinate system. The local rates of rotation ϑ_x, ϑ_y and ϑ_z and strain displacement γ_x, γ_y and γ_z measure the rates of change with θ of the global rotations θ_x, θ_y and θ_z and displacements δ_x, δ_y and δ_z respectively. The local displacements in polar coordinates will be denoted by u^p, v^p and w^p as in §3.10. The local bending moment, M_r, and torque, T, are related to the global end moments M_x and M_y by

$$M_r = M_x \cos\theta + M_y \sin\theta$$
$$T = -M_x \sin\theta + M_y \cos\theta \tag{8.40}$$

The local shear force S_r and axial force P are related to the global end forces F_x and F_y in the same way. The local moment M_z shear force F_z are the same as the corresponding global end loads. As these are constant, these loads give rise to constant strains, which on integration produce displacement fields of the form

$$\begin{aligned} u^p &= U_1(r,z) \\ v^p &= V_1(r,z) + \vartheta_z r\theta \\ w^p &= W_1(r,z) + \gamma_z \theta \end{aligned} \tag{8.41}$$

ignoring rigid-body displacements (cf. (8.5)). Note that ϑ_z and γ_z are constants. As mentioned in §4.3, solutions exist for a curved beam of circular cross-section subject to the moment M_z. Michell[1] was the first to find solutions to the problem of the response to an axial force F_z for rectangular and near-circular cross-sections. Freiberger[2] produces a solution for a circular cross-section in terms of toroidal coordinates. This is based on taking U_1, W_1 and ϑ_z as zero, which means that for an isotropic beam, only $\tau_{r\theta}$ and $\tau_{\theta z}$ are non-zero. Results for the shear deflexion γ_z can be compared with the approximate analysis of a spring (taking $\alpha = 0$) given in §5.6. The true deflexion is only 2.5% greater than the approximate one when the mean spring radius to the section radius is three. This diminishes to 0.1% when this ratio is ten. However, the maximum shear stresses can be significantly different from those predicted by the elementary theory for straight rods. When the above ratio of radii is ten, the true stress is about 14% greater and when the ratio is three, it is about 49% greater.

It follows from (8.40) that the characteristic response to the other end loads will be sinusoidal. This response falls in to two groups. If the normal strains and γ_{rz} vary with sin θ,

$$\begin{aligned} u^p &= U_2(r,z)\sin\theta + (\vartheta_y z + \gamma_x)\theta\cos\theta \\ v^p &= V_2(r,z)\cos\theta - (\vartheta_y z + \gamma_x)\theta\sin\theta \\ w^p &= W_2(r,z)\sin\theta - \vartheta_y r\theta\cos\theta \end{aligned} \tag{8.42}$$

where ϑ_y and γ_x are constant. If the normal strains and γ_{rz} vary with cos θ,

$$\begin{aligned} u^p &= U_3(r,z)\cos\theta - (\vartheta_x z - \gamma_y)\theta\sin\theta \\ v^p &= V_3(r,z)\sin\theta - (\vartheta_x z - \gamma_y)\theta\cos\theta \\ w^p &= W_3(r,z)\cos\theta + \vartheta_x r\theta\sin\theta \end{aligned} \tag{8.43}$$

where ϑ_x and γ_y are constant. In these two cases, no simplifying assumptions based on any analogy with ordinary beam theory seem appropriate. However, the above forms can be used as a basis for the approximate energy methods discussed in Chapter 7. Corrections for both curvature and pitch of helical springs have been given by Ancker and Goodier[3]. See also Young (1989) Chapter 8 and Pilkey (1994) Chapter 16.

---

[1] Michell, J.H. (1899) The Uniform Torsion and Flexure of Incomplete Tores, With Application to Helical Springs. *Proc. London Mathematical Soc.* V.31 p.130.

[2] Freiberger, W. and Smith, R.C.T. (1949) The Uniform Torsion of an Incomplete Tore. *Australian Journal of Scientific Research* ser. A, V.2 p.354.

[3] Ancker, C.J. and Goodier, J.N. (1958) Pitch and Curvature Corrections for Helical Springs. *Journal of Applied Mechanics* V.25 p. 466.

Example 8.1 Bending and torsion of an open ring

Suppose that a circular beam of mean radius R has a circular cross-section of radius a. It is convenient to use local coordinates ρ, the radius of a point from the centre of the cross-section, and ϕ, the angle of this radius to the r direction. These coordinates are related to the cylindrical coordinates as shown in Figure 8.12, giving

$$r = R + \rho\cos\phi \ , \ z = \rho\sin\phi . \qquad \textbf{(8.44)}$$

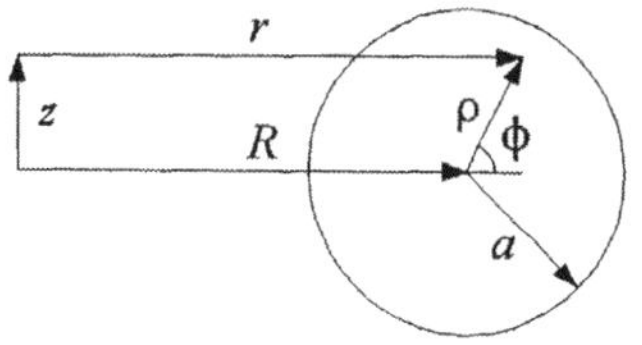

Figure 8.12 The coordinate system.

and so

$$\frac{\partial}{\partial r} = \cos\phi\frac{\partial}{\partial\rho} - \sin\phi\frac{1}{\rho}\frac{\partial}{\partial\phi} \quad , \quad \frac{\partial}{\partial z} = \sin\phi\frac{\partial}{\partial\rho} + \cos\phi\frac{1}{\rho}\frac{\partial}{\partial\phi} \ . \qquad \textbf{(8.45)}$$

Suppose that the beam forms a complete open circle and that a moment M_y is applied to one end. A characteristic response will be sought, producing a rate of rotation ϑ_y. As an approximation to the true deflexion mode given by (8.42), the form

$$\begin{aligned} u^P &= A\rho\sin\phi\sin\theta + \vartheta_y z\theta\cos\theta \\ v^P &= Cz\cos\theta - \vartheta_y z\theta\sin\theta \\ w^P &= -A\rho\cos\phi\sin\theta - \vartheta_y r\theta\cos\theta \end{aligned} \qquad \textbf{(8.46)}$$

will be used. Here A corresponds to a rotation of the cross-section in its own plane about its centre and C corresponds to its rotation about the r axis. From (3.62), the non-zero strains are

$$\epsilon_{\theta\theta} = \frac{z}{r}(B - \vartheta_y)\sin\theta \ , \ \gamma_{r\theta} = \frac{z}{r}(B + \vartheta_y)\cos\theta \ , \ \gamma_{\theta z} = -(B + \vartheta_y - \frac{1}{r}RA)\cos\theta \qquad \textbf{(8.47)}$$

where B is A-C. (If $A = C$ these terms correspond to a rigid-body rotation -A about the x axis and displacement -AR in the z direction of the whole beam, resulting in no strains.) From (7.1), the strain energy per unit volume is given by

$$U_V = \frac{G(1-\nu)}{1-2\nu}(B-\vartheta_y)^2\frac{z^2}{r^2}\sin^2\theta + \tfrac{1}{2}G\left[(B+\vartheta_y)^2\frac{z^2}{r^2}\cos^2\theta + \left(B+\vartheta_y - \frac{1}{r}RA\right)^2\cos^2\theta\right] \qquad \textbf{(8.48)}$$

The total strain energy is then given by integrating this over the volume of one turn, giving

$$U = 2\pi^2 GR^3\left[\frac{(1-\nu)}{(1-2\nu)}S(B-\vartheta_y)^2 + \tfrac{1}{2}[(B+\vartheta_y)^2(S+\tfrac{1}{2}\alpha) - A\alpha(B+\vartheta_y) + A^2T\right]$$

$$\text{where} \qquad \alpha = \frac{a^2}{R^2} \ , \ S = \tfrac{1}{2}\alpha - \tfrac{1}{3} + \tfrac{1}{3}(1-\alpha)^{3/2} \ , \ T = 1 - (1-\alpha)^{1/2} \qquad \textbf{(8.49)}$$

The total potential energy Π given by (7.11) can now be calculated, where the loss of potential energy Φ is given by the work done by M_y, which is $2\pi M_y \vartheta_y$. The total potential can be minimised by equating its partial derivatives with respect to ϑ_y, A and B to zero. This gives

$$M_y = \pi GR^3\frac{8S(1-\nu)(4ST + 2T\alpha - \alpha^2)}{(12-16\nu)ST + (1-2\nu)(2T\alpha - \alpha^2)}\vartheta_y \qquad \textbf{(8.50)}$$

This gives an upper bound to the stiffness of the curved beam.

A lower bound to the stiffness can be found from ordinary engineering beam theory. The local bending moment, M, and torque, T, acting on the beam cross-section are given by

$$M = M_y \sin\theta \quad , \quad T = M_y \cos\theta \,. \tag{8.51}$$

so that for thin rings, the complementary strain energy is given by

$$\bar{U} = \int_0^{2\pi} M_y^2 \left[\frac{\cos^2\theta}{2GJ} + \frac{\sin^2\theta}{2EI} \right] R\,d\theta = \frac{M_y^2 R(2 + \nu)}{Ga^4(1 + \nu)} \tag{8.52}$$

Minimising the total complementary potential with respect to M_y then gives

$$M_y = G\pi R^3 \frac{\alpha^2(1 + \nu)}{2 + \nu} \vartheta_y \tag{8.53}$$

The ratios of the stiffness given by (8.50) to that given by (8.53) are shown in Figure (8.13). When the curvature of the beam is very small in comparison with its cross-sectional dimensions, it tends to behave like an ordinary beam. If in addition Poisson's ratio is zero, the assumed displacement field matches that induced by the assumed stress field. Both methods then give the same stiffness, as shown in Figure 8.13. The difference between the two analyses increases with increasing Poisson's ratio. For a value of 0.5, the ratio varies between 1.6667 for zero a/R to 2.7778 for unit a/R.}

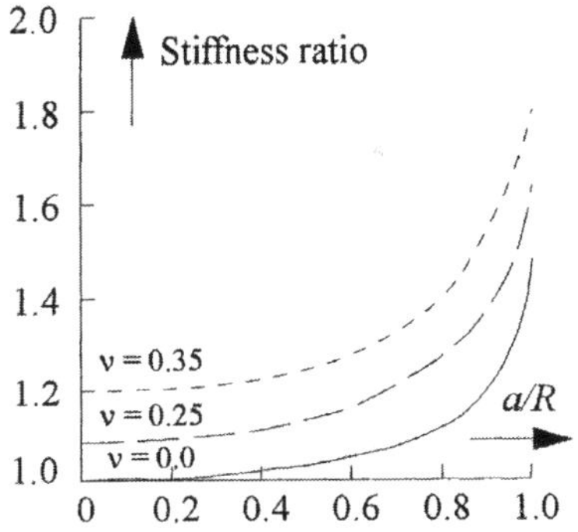

Figure 8.13 Ratio of upper bound to lower bound stiffness.

Chapter 9 The Stability of Beams

9.1 Introduction

To cover this subject satisfactorily, a book is required rather than a chapter. Such books exist, and some of them are listed in the bibliography at the end of this book. In particular, Bleich (1952), Timoshenko and Gere. (1961), Horne and Merchant (1965) and Kirby and Nethercot (1979) are recommended to the general reader. Those books relating to more specialised topics will be mentioned where these topics are considered.

The general concept of stability has already been examined in §1.3. Instability usually arises when the work done by the loading, during a small perturbation from the equilibrium state, is on the point of exceeding the strain energy absorbed by the elastic system supporting it. Any additional work is then turned into kinetic energy which can lead to the catastrophic collapse of the system. To analyse stability problems properly, large deflexion theory needs to be used. An approximation applicable to many engineering problems is to say that prior to unstable behaviour, only small deflexions have taken place and that the incipient unstable behaviour is infinitesimal. Even then, it turns out that a consistent analysis involves using higher-order elastic constants which relate the stresses to quadratic functions of the strains (see Renton (1987) Chapter 4 for example). These third-order constants are known for some materials, but they can have a stabilising effect. A conservative approximation to the loads causing instability (buckling loads) can then often be found using the usual linear elastic constants.

When the stress-strain curve for a material is severely non-linear in the region of loading where instability occurs, the local *tangent modulus*[1] is sometimes used instead of the usual elastic modulus in buckling formulae obtained for linearly-elastic materials. This is often applied even when the material is no longer elastic. A theoretical justification for this was first proposed by Shanley[2]. This shows that prior to the onset of unstable behaviour, a precursor of failure may occur with buckling deformations arising with an increasing loading. The prediction of this precursor, using the tangent moduli, acts as a safe estimate of the failure load.

Sometimes it is necessary to take account of the change of geometry of the elastic system during loading. Figure 9.1 shows two rigid bars of length l pinned together. At their other ends, one of the bars is pinned to a rigid support and the other to rollers which are restrained from moving horizontally by a spring of stiffness k. A vertical load P is applied to the common joint. If this joint is initially at a height h_o above the supports and under the load P descends to a height h, then the load and deflexion are related by

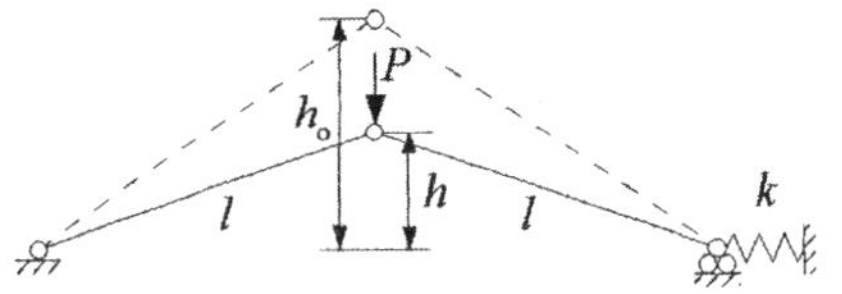

Figure 9.1 Snap through of a two-bar system.

[1] For a strictly linear material, the plot of stress against strain is linear, and the slope of the line the appropriate modulus (the shear, bulk or Young's moduli). For a non-linear curve, the slopes of the local tangents to these curves give the tangent moduli.

[2] Shanley, F.R. (1947) Inelastic column theory. *Journal of Aeronautical Sciences* V.14 No.5 p.261.

$$p = 4H\left(1 - \sqrt{\frac{L^2 - 1}{L^2 - H^2}}\right) \tag{9.1}$$

where $p = \dfrac{P}{kh_o}$, $H = \dfrac{h}{h_o}$, $L = \dfrac{l}{h_o}$.

The plot of p, a non-dimensional measure of the load, versus 1-H, a non-dimensional measure of its deflexion, is shown in Figure 9.2. This is shown for L equal to 10, but curves for greater values of L are almost identical. The load increases with deflexion until p is approximately 0.0077, after which the load must be reduced to maintain equilibrium. If not, the system then snaps through to another state of equilibrium, as indicated by the broken arrow. To reach this new state, it has passed through the mirror image below the horizontal of the unloaded state (1-H equal to 2.0). Such snap-through failures have been responsible for the collapse of shallow dome structures, such as those of the 93.5m diameter Bucharest dome in 1961 and the 51m C.W. Post Dome (Long Island University) in 1970.

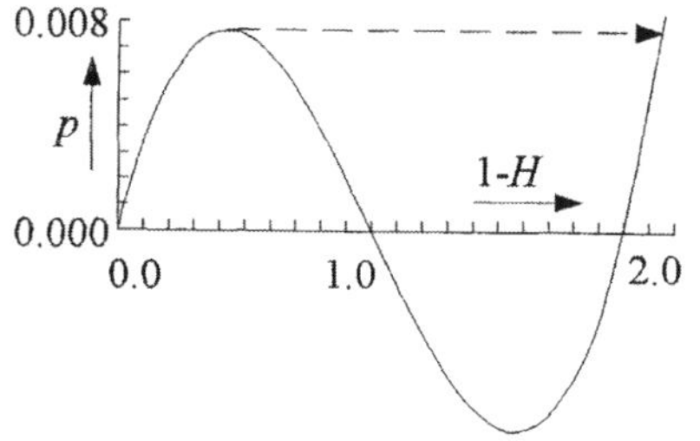

Figure 9.2 Load-deflexion curve for snap-through.

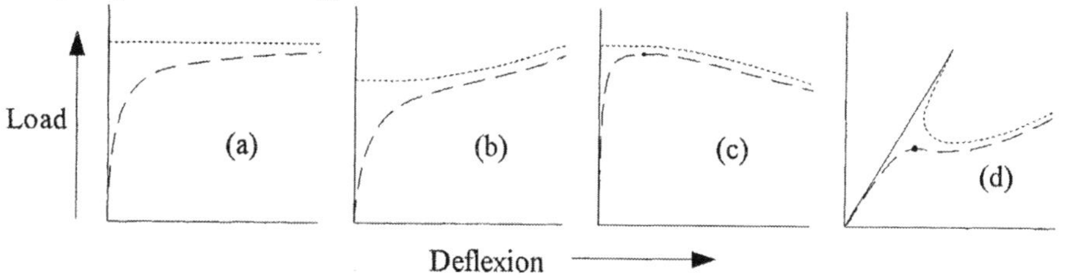

Figure 9.3 The influence of imperfections on buckling.

In theory, perfect structures can often be treated as if they behaved according to the predictions of small deflexion theory up to some critical loading state. Above this loading, they may continue to respond in the same way, or behave in an entirely different way. This split in the load-deflexion path is known as *bifurcation*. Figure 9.3 shows various load-deflexion paths. The dotted lines show possible paths following bifurcation. The broken lines show the load-deflexion paths when small imperfections of the structure, or eccentricities of the loading are present. Case (a) shows the behaviour predicted by standard small-deflexion analysis. This has a buckling mode after bifurcation which is of indeterminate magnitude, as represented by the horizontal line. However, if the strut is not perfectly straight, or the axial loading has a slight eccentricity, this analysis predicts a load-deflexion path shown by the broken line. Lateral motion begins immediately, and there is no bifurcation. The path becomes asymptotic to the buckling path. Such analyses in the presence of imperfections will be discussed again later. More accurate analyses show that the post-buckling path is often stable, as shown in (b). That is, the load has to be increased above the buckling load for further deflexion to take place. Again, if there is a slight imperfection, the path follows the broken line which becomes asymptotic to the perfect path. In such cases, the structure is said to be imperfection insensitive, because the imperfect system has no maximum load less than the critical load. Case (c) shows an unstable post-buckling path, where the equilibrium states correspond to a reduction in loading below the critical load. The system is now said to be imperfection sensitive, because there is a maximum load for the imperfect system (shown by the black dot) which is less than the critical load. Plates and columns are usually symmetric in their post-buckling behaviour, in that a lateral buckling mode could equally well be in either a positive or negative sense. Koiter (1945) showed that the reduction in the maximum load in such unstable symmetric cases is proportional to the ⅔ power of the (small) eccentricity. Typically for columns, this represents a drop of only a few percent.

Case (d) could represent the inwards deflexion of a point on the surface of a shell subject to an external pressure load. Buckling takes place by the dimpling of the surface, rather like the snap-through problem discussed above. However, this is a bifurcation problem and the post-buckling path is more like that shown by the dotted line. This mode is not symmetric, and Koiter shows that in such cases that the influence of imperfections on reducing the maximum load obeys a half-power law. The broken line again shows a typical path when imperfections are present, where the black dot shows the reduced maximum load. Because the imperfection measure is small in comparison with one, the power law in this case shows that the maximum load is reduced by imperfections considerably more than in the symmetric case (c). For thin shells, this can be very severe, the maximum load being as little as a quarter of the critical load at bifurcation, for typical imperfections. More discussion of these problems and post-buckling behaviour will be found in Thompson and Hunt (1984) for example.

In this chapter, instability will be considered to arise from conservative loading only. The work done by such loads can be expressed as a loss of a potential which has a unique value for any point in space. This means that the work done by conservative loads in moving around any closed path is zero, as there is no difference between the potential at the beginning and at the end of the path. Gravitational and elastic loads are conservative. Examples of non-conservative loads are the aerodynamic effects associated with flutter and the reaction on the end of a hosepipe emitting a jet of water. In such cases, the stability problems are seldom purely static but are associated with dynamic effects. Such problems have been considered by Bolotin (1964) and others. The interaction of stability and vibrations will be considered in the next chapter.

The standard theory of flexural instability is considered in §9.2 and §9.3. Modifications of this theory to allow for shear and torsion are given in §9.4 and §9.5 respectively. All of these analyses assume that the instability arises from large initial axial forces. However, lateral buckling may be induced by large initial moments and shear forces. This topic is covered in §9.6. In addition to overall buckling, local buckling of the section may also take place. Thin flat flanges of a section may buckle like plates subject to large axial stresses. Also, thin webs of a section may be susceptible to shear buckling. The analysis of these problems is given in §9.7. Where exact analyses are not known, or are too lengthy, approximate methods may be used to find the critical buckling conditions. Rayleigh's method, described in §9.8, is another example of the application of the principal of minimum potential energy used in Chapter 7. Finally, some general theories of stability are considered in §9.9. Further discussion of stability analyses will be found in the material covering matrix and computer methods. A program for analysing the stability of rigid-jointed three-dimensional frameworks is included, and the details of running it are given in §A2.1.2.

9.2 The Classical Problems of Flexural Stability

These are the problems first analysed by Leonhard Euler in 1744. The equation governing an axially loaded beam in simple bending was derived in Example 7.7, and is given by (7.89). Here, it will be assumed that the distributed load q is zero. The shear force S will be taken being in the v direction, that is, strictly perpendicular to the axial force P. As q is zero, it follows that δS in Figure 9.4 is zero for equilibrium. The axial force is taken to be an order of magnitude greater than the other loads. This means that in considering the moment equilibrium of the beam element of length δx shown in Figure 9.4, the moment $Pv'\delta x$ must be taken into account. The equation of equilibrium then takes the form

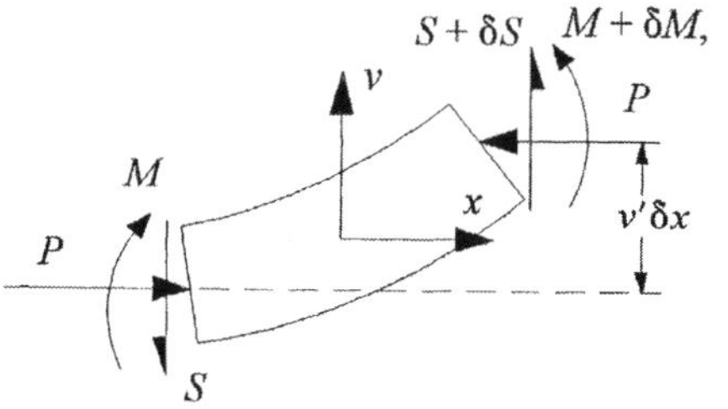

Figure 9.4 An axially-loaded beam element.

$$\delta M + Pv'\delta x + S\delta x = 0 \qquad \text{or} \qquad EIv''' + Pv' + S = 0 \tag{9.2}$$

using (4.15). The general solution of this equation is then given by

$$v = A - \frac{S}{P}x + B\cos\mu x + C\sin\mu x \qquad \text{where} \qquad \mu^2 = \frac{P}{EI} \tag{9.3}$$

This equation must satisfy two conditions at each end of the beam. As in §4.2, the conditions for a fixed end are that v and v' are zero, at a pinned end, v and v'' are zero and at a free end, the bending moment and shear force are zero so that v'' and S are zero[1].

The most common problems relating to a single beam in compression are shown in Figure 9.5a to 9.5d. The cantilevered column shown in Figure 9.5a is clamped at its lower end, x=0, and an axial force P is applied to its free end, x=l. The end conditions, expressed in terms of (9.3), are

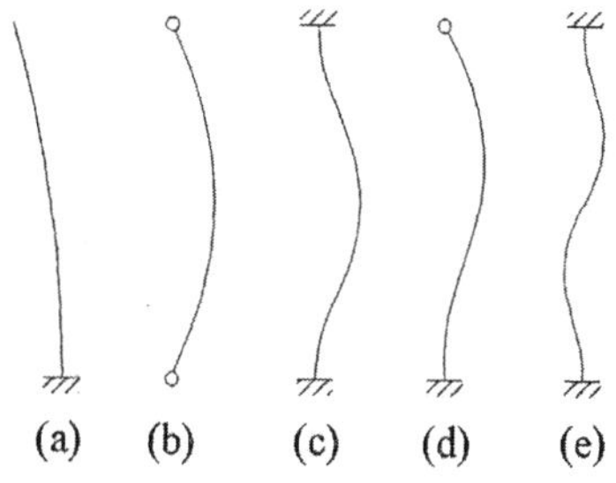

Figure 9.5 Standard buckling problems.

$$A + B = 0 \;, \quad -\frac{S}{P} + C\mu = 0 \;, \quad B\cos\mu l + C\sin\mu l = 0 \;, \quad S = 0\,. \tag{9.4}$$

Then either the constants A to D are zero and the beam does not deflect, or $\cos \mu l$ is zero and B is indeterminate. This buckling condition is satisfied when μl is $\pi/2$ (+ $n\pi$) where n is an integer. It follows from (9.3) that the lowest value of P at buckling is given by taking n as zero. Symmetrical buckling of a pin-ended beam and a fixed-ended beam, shown in Figures 9.5b and 9.5c, can be analysed similarly. The shear force S can be taken as zero in both cases, using

[1] It must be assumed implicitly that the supports permit relative axial movement of the ends of the beam, so that the axial force can be applied.

symmetry and equilibrium. This is not true in the case of the clamped and pin ended beam shown in Figure 9.5d. Applying the end conditions given earlier, the buckling criteria for these three cases are

$$\text{(b)}: \sin\mu l = 0\ , \quad \text{(c)}: \cos\mu l = 1\ , \quad \text{(d)}: \tan\mu l = \mu l\ . \tag{9.5}$$

so that the lowest Euler buckling loads, P_e, in the above four cases are

$$\text{(a)}: P_e = \frac{\pi^2 EI}{4l^2}\ , \quad \text{(b)}: P_e = \frac{\pi^2 EI}{l^2}\ , \quad \text{(c)}: P_e = \frac{4\pi^2 EI}{l^2}\ , \quad \text{(d)}: P_e = 20.191\frac{EI}{l^2}\ . \tag{9.6}$$

Figure 9.5e shows antisymmetric buckling of a fixed-ended beam. As there is a point of contraflexure in the centre, it behaves like two beams of type (d) joined end-to-end. The Euler buckling load for this case is then given by substituting $l/2$ for l in case (d) above, giving a buckling load a little over twice as large as that given by case (c).

9.3 The Slope-Deflexion Equations Allowing for Large Axial Loads

Relationships between the end loads and deflexions of an axially-loaded beam can be found using (9.3) and the end conditions

$$M_A = -EIv_A'' \ , \ M_B = EIv_B''\ , \qquad S_B = -S_A = S\ , \ \theta_A = v_A'\ , \ \theta_B = v_B' \tag{9.7}$$

where the end loads and deflexions are as shown in Figure 9.6 (and previously in Figure 4.17). The slope-deflexion equations given by (4.55) and (4.56) now become

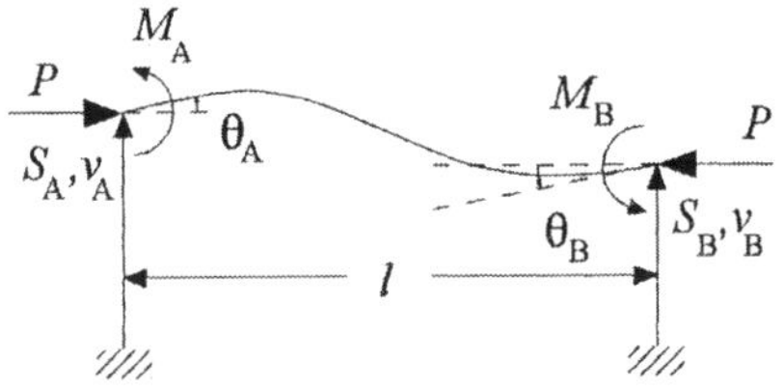

Figure 9.6 End loads and deflexions.

$$\begin{aligned} M_A &= \frac{EI}{l}[4\phi_3\theta_A + 2\phi_4\theta_B - \frac{6\phi_2}{l}(v_B - v_A)] \\ M_B &= \frac{EI}{l}[2\phi_4\theta_A + 4\phi_3\theta_B - \frac{6\phi_2}{l}(v_B - v_A)] \\ S_A = -S_B &= \frac{EI}{l}[6\phi_2\theta_A + 6\phi_2\theta_B - \frac{12\phi_1}{l}(v_B - v_A)] \end{aligned} \tag{9.8}$$

where ϕ_1 to ϕ_4 are Livesley's[1] stability functions. These can be expressed as

$$\phi_0 = \frac{\mu l}{2}\cot\frac{\mu l}{2}\ , \quad 6\phi_2 = \frac{\mu^2 l^2}{2(1-\phi_0)}$$
$$\phi_1 = \phi_0\phi_2\ , \quad 4\phi_3 = 3\phi_2 + \phi_0\ , \quad 2\phi_4 = 3\phi_2 - \phi_0 \tag{9.9}$$

Values of these functions are tabulated in §A6.2 in terms of the parameter λ (equal to $\mu^2 l^2$) for both compressive and tensile axial forces. When the axial force is zero, all the stability functions take unit values so that the slope-deflexion equations revert to their usual forms. For small values of λ, they are given approximately by

$$\phi_0 = 1 - \frac{\lambda}{12}\ , \quad \phi_1 = 1 - \frac{\lambda}{10}\ , \quad \phi_2 = 1 - \frac{\lambda}{60}\ , \quad \phi_3 = 1 - \frac{\lambda}{30}\ , \quad \phi_4 = 1 + \frac{\lambda}{60}\ . \tag{9.10}$$

---

[1] In the original (1956) version by Livesley, R.K. and Chandler, D.B. (*Stability functions for structural frameworks*. Manchester University Press) ϕ_1 was denoted by ϕ_5.

These, and the fuller expansions given in §A6.2, are equally valid for tension or compression.

Example 9.1 In-plane buckling of a portal frame

Equations (9.8) can be applied in the same way as ordinary slope-deflexion equations. As in Chapter 4, the bars are normally taken as axially rigid for manual analyses. For the rigid-jointed portal frame shown in Figure 9.7, the compatibility conditions are then that joints B and D do not move vertically and the horizontal displacements v_B and v_D are the same (v say). Also, the joint rotation θ_B is common to AB and BD and θ_D is common to BD and CD. For joint equilibrium, the sum of M_{BA} and M_{BD} is zero and the sum of M_{DB} and M_{DC} is zero. Also, the sum of the shear forces S_B and S_D on the columns is zero, as there no net sideways force. If all the members have the same length l bending stiffness EI, these three equilibrium conditions can be expressed in terms of equations (9.8) as

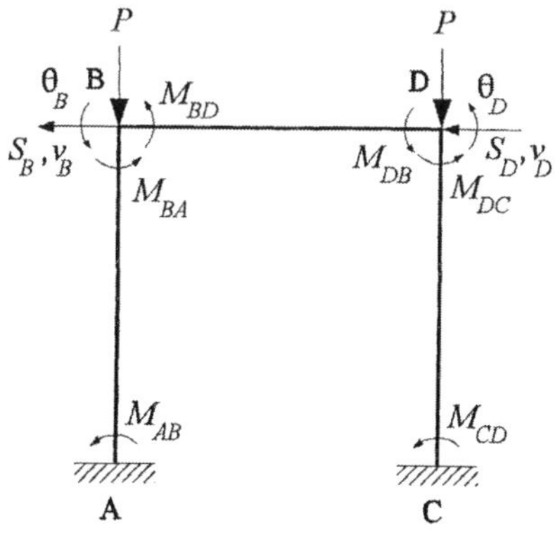

Figure 9.7 Buckling of a portal frame.

$$\begin{aligned}
\frac{4EI}{l}(\phi_3+1)\theta_B + \frac{2EI}{l}\theta_D - \frac{6EI}{l^2}\phi_2 v &= 0\\
\frac{2EI}{l}\theta_B + \frac{4EI}{l}(\phi_3+1)\theta_D - \frac{6EI}{l^2}\phi_2 v &= 0\\
-\frac{6EI}{l^2}\phi_2\theta_B - \frac{6EI}{l^2}\phi_2\theta_D + \frac{24EI}{l^3}\phi_1 v &= 0
\end{aligned} \qquad \textbf{(9.11)}$$

For non-zero deflexions, the determinant of the coefficients of θ_B, θ_D and v must be zero, i.e.

$$\begin{vmatrix}
\frac{4EI}{l}(\phi_3+1) & \frac{2EI}{l} & -\frac{6EI}{l^2}\phi_2\\
\frac{2EI}{l} & \frac{4EI}{l}(\phi_3+1) & -\frac{6EI}{l^2}\phi_2\\
-\frac{6EI}{l^2}\phi_2\theta_B & -\frac{6EI}{l^2}\phi_2 & \frac{24EI}{l^3}\phi_1
\end{vmatrix} = 0 \qquad \textbf{(9.12)}$$

This determinant factorises, giving two equations

$$\begin{aligned}
2\phi_3 + 1 &= 0\\
4\phi_1\phi_3 + 6\phi_1 - 3\phi_2^2 &= 0
\end{aligned} \qquad \textbf{(9.13)}$$

The lowest values of P satisfying these equations are

$$P_{cl} = 25.2\frac{EI}{l^2}, \quad P_{c2} = 7.38\frac{EI}{l^2} \qquad \textbf{(9.14)}$$

respectively, giving the symmetric and antisymmetric buckling modes shown in Figure 9.8.

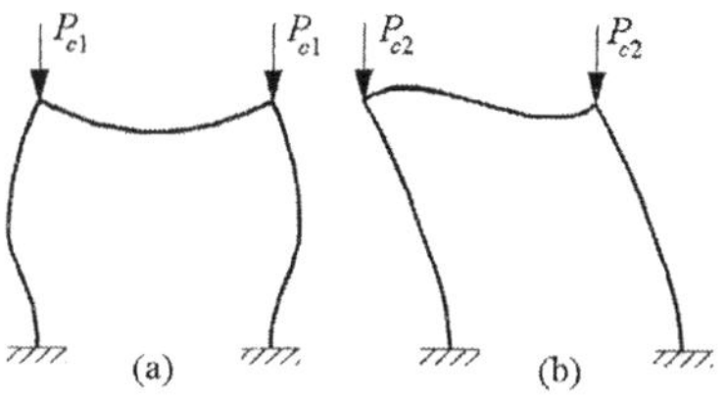

Figure 9.8 Symmetric and antisymmetric buckling.

{9.4 Flexure and Shear}

{In principle, instability of an axially-loaded beam could arise purely as a result of its shear flexibility. This is not readily seen in terms of the ordinary equations of equilibrium, but is easily analysed using work and energy principles.

Figure 9.9 shows a short length δx of a beam subject to a uniformly-distributed axial load with a resultant P. This undergoes a shear displacement of the kind discussed in Chapter 6. The rate of change of this displacement is given by γ, so that the shear strain energy stored in the length δx is $\frac{1}{2}GkA\gamma^2\delta x$. The angle γ is related to the motion of the shear force, but a reasonable approximation is to take all the fibres to rotate through it as shown in Figure 9.9b. This gives rise to an apparent shortening of the fibres in the axial direction of $\delta x(1\text{-}\cos\gamma)$, or approximately $\frac{1}{2}\delta x\gamma^2$. If P is sufficiently large to cause this motion to take place, then the work done by it during shortening matches the shear strain energy, or

$$P = P_s = GkA \tag{9.15}$$

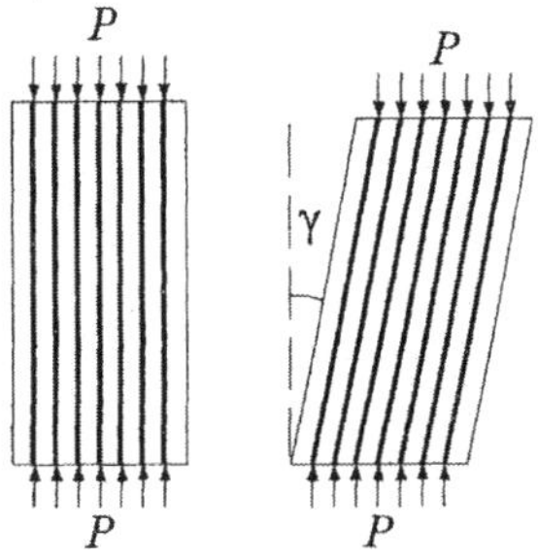

Figure 9.9 Shear instability.

where P_s is the shear failure load. This is normally so high that plastic failure of the beam would occur first. However, the concept of a shear failure load comes in useful in the following analysis.

In considering bending combined with shear, some of the ideas used in the analysis of a Timoshenko beam (Example 7.8) will be employed. The displacement through which the shear force S does work will be denoted by v, the rate of this displacement arising from beam shear denoted by γ, and the rotation through which the moment does work denoted by θ. It will be taken that the gradient v' is the sum of θ and γ, as shown in (7.98). It will be taken that the apparent axial shortening through which the axial force P does work can be determined from v'. Then in the absence of lateral loading, the total potential of a beam (as measured from the datum of the equilibrium state immediately prior to buckling) is given by

$$\Pi = \int_0^l \left[\frac{1}{2}EI(\theta')^2 + \frac{1}{2}GkA(w-\theta)^2 - \frac{1}{2}Pw^2 - Sw\right]dx \tag{9.16}$$

(cf. (7.6)) where w is the gradient v'. Euler's equations found from minimising Π with respect to small variations in w [1] and θ can then be written as

$$\begin{aligned} GkA(v' - \theta) - Pv' - S &= 0 \\ -EI\theta'' - GkA(v' - \theta) &= 0 \end{aligned} \tag{9.17}$$

The general solution of this for v is similar to (9.3) and is given by

$$v = A - \frac{S}{P}x + B\cos\upsilon x + C\sin\upsilon x \quad \text{where} \quad \frac{1}{\upsilon^2} = \frac{EI}{P}\left(1 - \frac{P}{P_s}\right) \tag{9.18}$$

and θ can then be found from the first of equations (9.17). Here, S is a constant. If it is zero, then

---

[1] Applying the calculus of variations to the parameter w implies no end constraint on the displacement v itself. The term Sw then has to be included as part of the integrand to give an expression for the work done by S.

$$\theta = \left(1 - \frac{P}{P_s}\right) v' \tag{9.19}$$

On substituting υ for μ in (9.18) it becomes (9.3). Also, the end conditions used for the classical buckling problems are the same except for fixed-end conditions in the presence of a shear force. Then for cases (a) to (c) in Figure 9.5, the critical conditions previously expressed in terms of μ are now the same conditions expressed in terms of υ. The critical axial loads allowing for shear are then given by

$$\text{(a)}: \frac{1}{P_c} = \frac{4l^2}{\pi^2 EI} + \frac{1}{P_s}, \quad \text{(b)}: \frac{1}{P_c} = \frac{l^2}{\pi^2 EI} + \frac{1}{P_s}, \quad \text{(c)}: \frac{1}{P_c} = \frac{l^2}{4\pi^2 EI} + \frac{1}{P_s} \tag{9.20}$$

All of these results can be expressed by

$$\frac{1}{P_c} = \frac{1}{P_e} + \frac{1}{P_s} \tag{9.21}$$

This equation is plotted as the solid line in the graph shown in Figure 9.10. The broken line shows the more exact solution for a pin-ended column found from solving the governing continuum equations in the presence of a large axial stress (cf. Renton (1987) Example 4.3[1]).

Unlike the three cases above, a shear force is present in the fixed-pinned beam shown in Figure 9.5d. The end conditions are that there is no displacement or bending rotation at the fixed end ($v = \theta = 0$) and that there is no displacement or bending curvature at the pinned end ($v = \theta' = 0$). From the first of equations (9.17) and (9.18), the condition for a non-zero deflexion is then

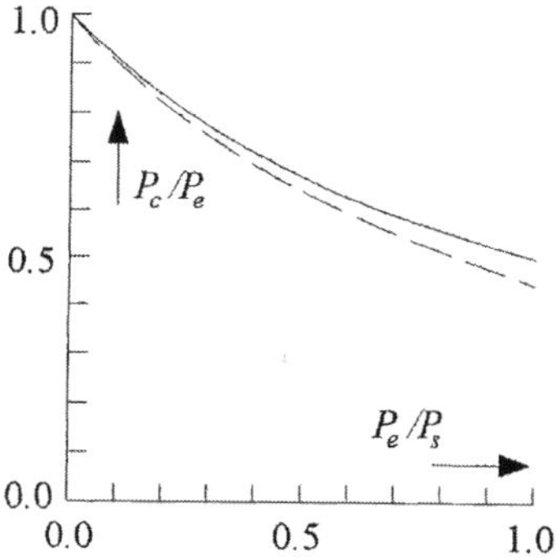

Figure 9.10 Reduction in critical load due to shear.

$$\tan \upsilon l = \upsilon l \left(1 - \frac{P}{P_s}\right) \tag{9.22}$$

This should be compared with case (d) of (9.6). In the present case, the lowest value of υl satisfying this buckling criterion is less than the lowest value of μl satisfying case (d). This is because the additional factor on the right-hand side of (9.22) reduces the required value of $\tan \upsilon l$. In this case, substituting υ for μ in (9.5d) *overestimates* the buckling load. The significance of this will be seen in §9.9.

The solution of the problem of antisymmetric buckling of a fixed-ended beam shown in Figure 9.5e is found by treating it as two fixed-pinned struts connected at the central point of contraflexure, as in §9.2. The solution given by (9.22) again applies, where l is half the overall length of the fixed-ended beam. However, the lowest buckling load is still given by the symmetric case shown in Figure 9.5c.

[1] The rod has a solid circular cross-section and Poisson's ratio is taken as 0.25.

{9.5 Flexure and Torsion}

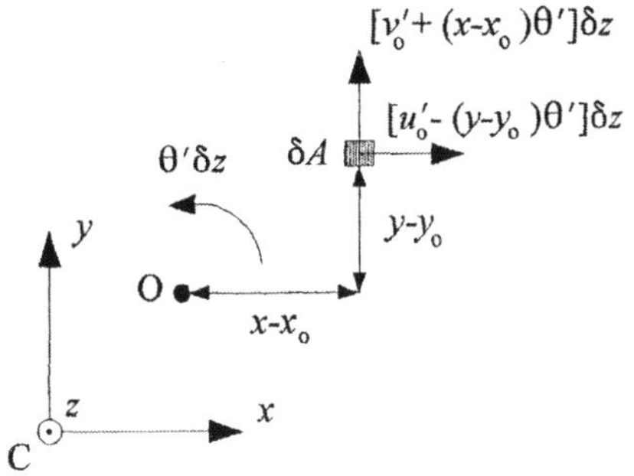

Figure 9.11 Relative motion along a fibre.

{When the beam flexes about its principal axes x and y and twists about its shear centre O, the relative motion of two sections of a longitudinal fibre of the beam, δz apart, is shown in Figure 9.11. Here u_0 and v_0 are the displacements of the shear centre in the x and y directions respectively, θ is the rotation about the z axis and δA is the cross-sectional area of the fibre. The square of the gradient of the fibre is then given to the required degree of accuracy by the sum of the squares of the relative displacements in the x and y directions divided by δz^2. The apparent shortening of the fibre in the longitudinal direction is then found in the same way as in Example 7.7. However, in this case each fibre has an individual shortening. This means that in determining the work done by the axial force P it must be treated as a uniformly distributed stress σ. The expression for the work done, δW by P on a short length δz is then

$$\begin{aligned}\delta W &= \frac{1}{2}\int_A \sigma\{[u_0' - (y-y_0)\theta']^2 + [v_0' + (x-x_0)\theta']^2\}\,dA\,\delta z \\ &= \frac{1}{2}P(u_0'^2 + v_0'^2 + 2y_0u_0'\theta' - 2x_0v_0'\theta' + r_0^2\theta'^2)\,\delta z\end{aligned} \tag{9.23}$$

where r_0 is the polar radius of gyration of the section about the shear centre. (Note that several terms disappear on integration because principal axes through the centroid of the section are chosen.) The axial and shear flexibility of the beam will not be taken into account. Using this expression for the loss of potential of the axial force, the expression for the total potential of a beam (relative to the state immediately prior to buckling) becomes

$$\begin{aligned}\Pi = \int_0^l \{&\tfrac{1}{2}[EI_{yy}u_0''^2 + EI_{xx}v_0''^2 + GJ\theta'^2 + EI_\omega\theta''^2 \\ &- P(u_0'^2 + v_0'^2 + 2y_0u_0'\theta' - 2x_0v_0'\theta' + r_0^2\theta'^2)] - S_xu_0' - S_yv_0' - T\theta'\}\,dz\end{aligned} \tag{9.24}$$

where the expressions for the strain energy can be deduced from (7.6) and (5.92). Note that the work done by the shear forces in the x and y directions, S_x and S_y, and by the torque T have been included. This is because Π will be minimised with respect to u_0', v_0' and θ', on the same basis as in §9.4. This gives the three equations

$$\begin{aligned}EI_{yy}u_0''' + Pu_0' + Py_0\theta' &= -S_x \\ EI_{xx}v_0''' + Pv_0' - Px_0\theta' &= -S_y \\ GJ\theta' - EI_\omega\theta''' - Py_0u_0' + Px_0v_0' - Pr_0^2\theta' &= T\end{aligned} \tag{9.25}$$

Differentiating each of these equations once with respect to z gives the form first obtained by Kappus[1]. (As the shear forces and torque are constant along the beam, the right-hand sides are then zero.)

---

[1] Kappus, R. (1937) Drillknicken zentrisch gedrückter stäbe mit offenem profil im elastischen bereich. *Luftfahrt-Forschung* V.19 p.444.

When the shear centre and the centroid coincide, the above three equations uncouple. The first two then relate to flexural buckling already considered in §9.2 and §9.3. The third equation governs torsional buckling, and has the general solution for Pr_0^2 less than GJ given by

$$\theta = A + \frac{T}{GJ - Pr_0^2} z + B\cosh\mu z + C\sinh\mu z \qquad \text{where} \quad \mu^2 = \frac{GJ - Pr_0^2}{EI_\omega} \tag{9.26}$$

(cf. (5.84)). When Pr_0^2 is equal to GJ, θ is given by a cubic polynomial in z. Cases of elastic buckling have been found experimentally for even greater values of P when the expression for θ becomes

$$\theta = A + \frac{T}{GJ - Pr_0^2} z + B\cos\mu z + C\sin\mu z \qquad \text{where} \quad \mu^2 = \frac{Pr_0^2 - GJ}{EI_\omega} \tag{9.27}$$

It will be seen that the form of this equation is similar to that of (9.3) and admits analogous solutions. The equivalent of a fixed end is now one which prevents twisting and warping, a 'pinned' end prevents twisting, but applies no bimoment to the column, and a 'free' end applies no torque or bimoment. The solutions to the classical flexural buckling problems of §9.2 can now be applied to torsional buckling, where μ is now defined by (9.27).

Sinusoidal solutions of Kappus' equations were found by Timoshenko[1]. These require finding the lowest value of P_c satisfying the cubic equation

$$\begin{gathered} (x_0^2 + y_0^2 - r_0^2)P_c^3 + [(P_1 + P_2 + P_3)r_0^2 - x_0^2 P_1 - y_0^2 P_2]P_c^2 \\ - r_0^2(P_1P_2 + P_2P_3 + P_3P_1)P_c + P_1P_2P_3r_0^2 = 0 \\ \text{where} \quad P_1 = \frac{n^2\pi^2 EI_{yy}}{l^2} \;, \quad P_2 = \frac{n^2\pi^2 EI_{xx}}{l^2} \;, \quad P_3 = \frac{1}{r_0^2}\left(GJ + EI_\omega \frac{n^2\pi^2}{l^2} \right) \end{gathered} \tag{9.28}$$

When n is one, this corresponds to the solution for a column when the ends are restrained from displacing or twisting, but there are no bending or warping end restraints. When n is two, it corresponds to the solution for rigidly clamped ends.

The general solution[2] is given by

$$\begin{aligned} u_0 &= \sum_{n=1}^{3} \left[U_{an} e^{q_n z} + U_{bn} e^{-q_n z} \right] + U_1 z + U_0 \\ v_0 &= \sum_{n=1}^{3} \left[V_{an} e^{q_n z} + V_{bn} e^{-q_n z} \right] + V_1 z + V_0 \\ \theta &= \sum_{n=1}^{3} \left[\Theta_{an} e^{q_n z} + \Theta_{bn} e^{-q_n z} \right] + \Theta_1 z + \Theta_0 \end{aligned} \tag{9.29}$$

where

---

[1]Timoshenko, S. (1945) Theory of bending, torsion and buckling of thin-walled members of open cross-section. *Journal of the Franklin Institute.* V.239 Nos. 3,4 &5.

[2]Renton, J.D. (1960) A Direct Solution of the Torsional-Flexural Buckling of Axially Loaded Thin-Walled Bars. *The Structural Engineer.* V.38 No.9 p.273.

$$U_{(a\text{ or }b)n} = \frac{-Py_0\Theta_{(a\text{ or }b)n}}{EI_{yy}q_n^2 + P}\ ,\quad V_{(a\text{ or }b)n} = \frac{Px_0\Theta_{(a\text{ or }b)n}}{EI_{xx}q_n^2 + P}\ ,$$

$$P(r_0^2 - x_0^2 - y_0^2)\Theta_1 = S_x y_0 - S_y x_0 - T\ ,\quad U_1 = -y_0\Theta_1 - \frac{S_x}{P}\ ,\quad V_1 = x_0\Theta_1 - \frac{S_y}{P}\ . \tag{9.30}$$

and q_n is found from the equation

$$\begin{vmatrix} q_n^2 + \dfrac{P}{EI_{yy}} & 0 & \dfrac{Py_0}{E\sqrt{I_{yy}I_\omega}} \\ 0 & q_n^2 + \dfrac{P}{EI_{xx}} & \dfrac{Px_0}{E\sqrt{I_{xx}I_\omega}} \\ \dfrac{Py_0}{E\sqrt{I_{yy}I_\omega}} & \dfrac{Px_0}{E\sqrt{I_{xx}I_\omega}} & q_n^2 - \dfrac{GJ - Pr_0^2}{EI_\omega} \end{vmatrix} = 0 \tag{9.31}$$

It will be seen that finding the roots of q_n^2 is the same as finding the eigenvalues of a real symmetric matrix. Such roots are all real, and the cubic equation in q_n^2 is known as a discriminating cubic. Further, if P is compressive it can be shown that two of the roots are negative and that the third one is such that

$$EI_\omega q_3^2 \leq GJ - P(r_0^2 - x_0^2 - y_0^2) \tag{9.32}$$

so that the third root may be positive for small P. In such circumstances, the exponential terms in (9.29) give rise to circular functions related to two of the roots and hyperbolic functions related to the third root. The other coefficients in these three equations are determined from the end conditions.

If the end fittings on a column permit its ends to rotate but not to warp, Timoshenko's solution cannot be used. The above general solution gives

$$\begin{aligned}(q_2^2-q_3^2)(q_1^2+\mu_x^2)(q_1^2+\mu_y^2)\frac{1}{q_1}\tanh\tfrac{1}{2}q_1 l + (q_3^2-q_1^2)(q_2^2+\mu_x^2)(q_2^2+\mu_y^2)\frac{1}{q_2}\tanh\tfrac{1}{2}q_2 l \\ + (q_1^2-q_2^2)(q_3^2+\mu_x^2)(q_3^2+\mu_y^2)\frac{1}{q_3}\tanh\tfrac{1}{2}q_3 l = 0\end{aligned} \tag{9.33}$$

where μ_x^2 is P/EI_{xx} and μ_y^2 is P/EI_{yy}, or

$$\begin{aligned}(q_2^2-q_3^2)(q_1^2+\mu_x^2)(q_1^2+\mu_y^2)\frac{1}{q_1}\coth\tfrac{1}{2}q_1 l + (q_3^2-q_1^2)(q_2^2+\mu_x^2)(q_2^2+\mu_y^2)\frac{1}{q_2}\coth\tfrac{1}{2}q_2 l \\ + (q_1^2-q_2^2)(q_3^2+\mu_x^2)(q_3^2+\mu_y^2)\frac{1}{q_3}\coth\tfrac{1}{2}q_3 l = 0\end{aligned} \tag{9.34}$$

The first solution is symmetric in θ about the midpoint of the column and the second is antisymmetric in θ. As two of the roots of q_n^2 will be negative, it should be noted that

$$\frac{1}{ix}\tanh ix = \frac{1}{x}\tan x\ ,\quad \frac{1}{ix}\coth ix = -\frac{1}{x}\cot x \tag{9.35}$$

More precise solutions can be found which correspond to the experimental results of

Baker and Roderick[1]. These tests were carried out using rigid ball joint fittings, of radius R say, on the ends of the struts. These were angle tee and channel sections with one axis of symmetry. Taking this as the x axis, y_0 is then zero so that the first of equations (9.25) is then uncoupled from the other two and can be solved separately for simple flexure about the y axis. The end conditions can be taken as the same as in the previous example except that there is an end bending moment given by PR times the end slope. The resulting buckling criteria are then

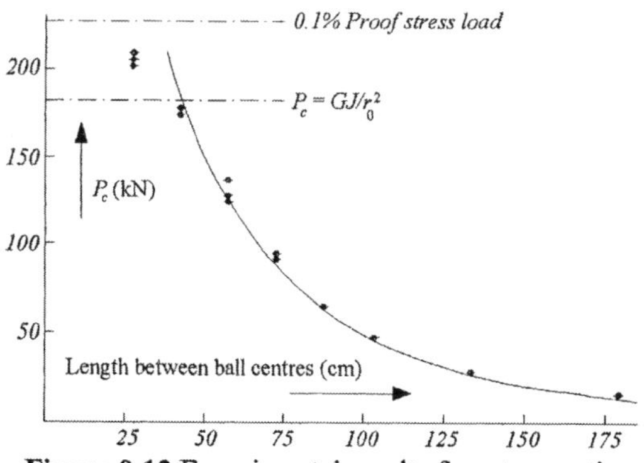

Figure 9.12 Experimental results for a tee section.

$$\cot\tfrac{1}{2}\mu_y l = R\mu_y \quad , \quad \frac{1}{q_1^2+\mu_x^2}[q_1\coth\tfrac{1}{2}q_1 l - R\mu_x^2] = \frac{1}{q_2^2+\mu_x^2}[q_2\coth\tfrac{1}{2}q_2 l - R\mu_x^2] \qquad \textbf{(9.36)}$$

for geometrically symmetrical buckling about the midpoint of the column, and

$$\tan\tfrac{1}{2}\mu_y l = -R\mu_y \quad , \quad \frac{1}{q_1^2+\mu_x^2}[q_1\tanh\tfrac{1}{2}q_1 l - R\mu_x^2] = \frac{1}{q_2^2+\mu_x^2}[q_2\tanh\tfrac{1}{2}q_2 l - R\mu_x^2] \qquad \textbf{(9.37)}$$

for antisymmetrical buckling about the midpoint. The experimental results for a tee section tested by Baker and Roderick are shown by the black dots in Figure 9.12.[2] The lowest buckling loads for this series of tests are predicted by second of equations (9.36), and this curve is plotted in the figure. However, for very short struts, antisymmetric buckling may occur at lower loads.

{9.6 Lateral Buckling}

{In this section, beam bifurcation will be examined which arise from large initial loading other than simple axial compression. Equation (9.25) was obtained by examining the work done by a large uniform compressive axial stress σ during buckling, as given by (9.23). The resultant of this stress was a large axial force P. If in addition there are large initial moments M_x and M_y, then the expression for the axial stress becomes

$$\sigma = \frac{P}{A} - \frac{M_x}{I_{xx}}y + \frac{M_y}{I_{yy}}x \qquad \textbf{(9.38)}$$

The expression for the work done by the axial stress, previously given by (9.23), now becomes

$$\delta W = \frac{1}{2}\int_A\left(\frac{P}{A} - \frac{M_x}{I_{xx}}y + \frac{M_y}{I_{yy}}x\right)\{[u_0' - (y-y_0)\theta']^2 + [v_0' + (x-x_0)\theta']^2\}\,dA\,\delta z$$
$$= \frac{1}{2}\Big[P(u_0'^2 + v_0'^2 + 2y_0u_0'\theta' - 2x_0v_0'\theta' + r_0^2\theta'^2) - M_x(b_1\theta'^2 - 2u_0'\theta') + M_y(b_2\theta'^2 + 2v_0'\theta')\Big]\delta z \qquad \textbf{(9.39)}$$

---

[1] Baker, J.F. and Roderick, J.W. (1948) The Strength of Light Alloy Struts. *The Aluminium Development Association Research Report.* No. 3.

[2] For aluminium, there is no well-defined yield stress. The 0.1% proof stress load refers to the load which leaves a permanent residual strain of 0.001 on unloading.

where

$$b_1 = \frac{1}{I_{xx}} \int_A y(x^2 + y^2)\, dA - 2y_0 \quad , \quad b_2 = \frac{1}{I_{yy}} \int_A x(x^2 + y^2)\, dA - 2x_0 \ . \qquad \textbf{(9.40)}$$

Let the constants X_0, Y_0 and R_0 be defined by the equations

$$X_0 = x_0 - \frac{M_y}{P} \quad , \quad Y_0 = y_0 + \frac{M_x}{P} \quad , \quad R_0^2 = r_0^2 - b_1 \frac{M_x}{P} + b_2 \frac{M_y}{P} \ . \qquad \textbf{(9.41)}$$

Then (9.39) becomes the same in form as (9.23), with these constants replacing x_0, y_0 and r_0 respectively. Minimising the total potential as before then yields the equations

$$\begin{aligned} EI_{yy} u_0''' + P u_0' + P Y_0 \theta' &= -S_x \\ EI_{xx} v_0''' + P v_0' - P X_0 \theta' &= -S_y \\ GJ\theta' - EI_\omega \theta''' - P Y_0 u_0' + P X_0 v_0' - P R_0^2 \theta' &= T \end{aligned} \qquad \textbf{(9.42)}$$

Which is of course the same as (9.25) only in terms of these new constants. The previous solutions for a large axial load now hold for a large axial load combined with equal and opposite end moments. In particular, this result applies to the case when the resultant axial force does not act through the centroid, but at a point on the cross-section with coordinates (e_x , e_y) so that M_x and M_y are given by $-Pe_y$ and Pe_x respectively. It should be emphasised that the deflexions found from solving (9.42) relate to the deviation mode at bifurcation. Prior to that, the moments will induce deflexions which can be found from ordinary small-deflexion theory. Likewise, the axial force P induces simple axial compression prior to the critical conditions.

A special case of these equations occurs when these moments are applied without any axial force. Suppose also that the beam is bisymmetric. Equations (9.42) then become

$$\begin{aligned} EI_{yy} u''' + M_x \theta' &= -S_x \\ EI_{xx} v''' + M_y \theta' &= -S_y \\ GJ\theta' - EI_\omega \theta''' - M_x u' - M_y v' &= T \end{aligned} \qquad \textbf{(9.43)}$$

(the centroid and shear centre now being coincident). If M_x is zero, then the first equation uncouples from the other two. The general solution for v and θ can be found in terms of exponential and linear expressions in z as before. Again, using sinusoidal terms for the bifurcation mode yields simple solutions for certain end conditions. Let

$$v = V \sin\frac{\pi z}{l} \quad , \quad \theta = \Theta \sin\frac{\pi z}{l} \ . \qquad \textbf{(9.44)}$$

This form satisfies conditions of zero lateral displacement, torsional rotation, flexural curvature about the x axis and no bimoments at the ends (z equal to 0 and l) of the beam. Also, the mode is symmetric about the midpoint of the beam, so that the right-hand sides of equations (9.43) are zero. The bifurcation condition is then

$$\begin{bmatrix} EI_{xx}\dfrac{\pi^2}{l^2} & -M_y \\ -M_y & GJ + EI_\omega \dfrac{\pi^2}{l^2} \end{bmatrix} \begin{bmatrix} V \\ \Theta \end{bmatrix} = \begin{bmatrix} 0 \\ 0 \end{bmatrix} \qquad \textbf{(9.45)}$$

For non-zero deflexions, the determinant of this matrix must be zero, giving

$$M_y = \frac{\pi}{l}\sqrt{EI_{xx}\left(GJ + EI_\omega\frac{\pi^2}{l^2}\right)} \tag{9.46}$$

This expression gives the critical value of M_y at which lateral buckling can occur. The equivalent of fixed ends is given by taking

$$v = V\left(1 - \cos\frac{2\pi z}{l}\right) \quad , \quad \theta = \Theta\left(1 - \cos\frac{2\pi z}{l}\right) . \tag{9.47}$$

This form satisfies conditions of zero displacement, rotation about the x and z axes, and zero warping at the ends (z equal to 0 and l) of the beam. Again, the mode is symmetric about the midpoint of the beam, so that the right-hand sides of equations (9.43) are zero. The constant terms satisfy these equations automatically, and the cosine terms will as well, subject to the equivalent condition to (9.45), where π^2/l^2 is replaced by $4\pi^2/l^2$. Then the same substitution in (9.46) gives the critical value of M_y for these end conditions.

It would be possible to modify the expression for the work done by the axial bending stresses given by (9.39) to analyse the lateral buckling of a cantilever with an end shear force S shown in Figure (9.13). Suppose that the beam is bisymmetrical and that the end z=0 is rigidly clamped and S is applied to the free end at z=l as shown. The moment M_y induced by S at a point z along the beam is then $-S(l - z)$, so that (9.39) becomes

$$\delta W = -S(l-z)v'\theta'\delta z \tag{9.48}$$

Using this expression to give the loss of potential energy in (9.23) gives

$$\Pi = \int_0^l \{\tfrac{1}{2}[EI_{yy}u''^2 + EI_{xx}v''^2 + GJ\theta'^2 + EI_\omega\theta''^2] + S(l-z)v'\theta' + Su' - S_y v' - T\theta'\}\,dz \tag{9.49}$$

Minimising this expression with respect to u' gives the normal flexural behaviour in the vertical plane. Minimising Π with respect to v' and θ' gives

$$\begin{aligned} EI_{xx}v''' - S(l-z)\theta' &= -S_y \\ GJ\theta' - EI_\omega\theta''' + S(l-z)v' &= T \end{aligned} \tag{9.50}$$

where as before, S_y and T are the incremental end shear force and torque associated with the buckling mode.

These equations will be examined again later. The commonly accepted solution to this problem is based on a static analysis derived by Timoshenko[1]. This involves finding the moments and torques acting about the principal axes of the deflected cross-section of the beam at a distance z from the fixed end. This is shown in Figure 9.13c. With respect to the original directions of the axes, the end shear force produces a bending moment $S(l - z)$ as shown, and a first order of smallness torque $S(v_l - v)$ as shown. To include all first order of smallness effects, the components of this bending moment about the locally rotated axes of the beam are included. This consists of a bending moment $S(l - z)\theta$ about twisted minor axis of the beam and a torque $S(l - z)v'$ in the senses shown in Figures 9.13a and 9.13b respectively. The resulting equations for flexure and torsion of the section with respect to the locally rotated coordinate system are then

---

[1] Timoshenko, S.P. (1910) Z. *Math. und Physik*, V58. p.337. See also Timoshenko & Gere (1961) §6.3.

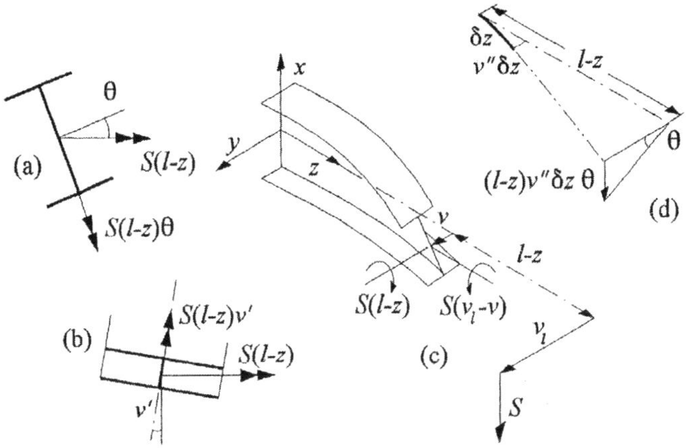

Figure 9.13 Lateral buckling of a cantilever with a large shear force S at the free end.

$$\begin{aligned} EI_{xx}v'' &= S(l-z)\theta \\ GJ\theta' - EI_{\omega}\theta''' &= S(v_l - v) - S(l-z)v' \end{aligned} \qquad \textbf{(9.51)}$$

Differentiating the second of these equations with respect to z gives

$$GJ\theta'' - EI_{\omega}\theta'''' = -S(l-z)v'' = -\frac{S^2}{EI_{xx}}(l-z)^2\theta \qquad \textbf{(9.52)}$$

This result may be derived from an expression for Π in which the strain energy terms are as before and the loss of potential is given by

$$W = \int_0^l S(l-z)v''\theta \, dz \qquad \textbf{(9.53)}$$

The first of equations (9.51) is then given by minimising Π with respect to v'' and (9.52) by minimising it with respect to θ. The physical interpretation of this is illustrated in Figure 9.13d. The position of the centroidal axis at the free end is determined from a series of corrected estimates, starting at the fixed end. At a distance z along the beam, this position is first estimated by considering that the rest of the centroidal axis remains straight. However, the next infinitesimal length δz of the axis curves through an angle $v''\delta z$ giving a correction to the estimated position of the end of $(l-z)v''\delta z$ in the direction of the local major axis. As this is inclined at an angle θ to the y axis, this gives a downwards component of $\theta(l-z)v''\delta z$. Integrated over the total length, this gives the downwards displacement of S and hence the expression for W in (9.53). In the case when EI_{ω} is zero (or negligible), (9.52) gives the condition

$$\theta'' + b^2 s^2 \theta = 0 \quad \text{where} \quad b^2 = \frac{S^2}{GJEI_{xx}} \quad \text{and} \quad s = l - z. \qquad \textbf{(9.54)}$$

and differentiation is now with respect to s. The general solution of this equation can be expressed in terms of two Bessel functions of fractional order. At the free end, there is no torque and so the rate of twist θ' is zero at s=0. This means that the Bessel function of fractional order ¼ must be zero, leaving

$$\theta = A\sqrt{s}\,J_{-1/4}\left(\frac{1}{2}bs^2\right) = B\left(1 - \frac{t}{3} + \frac{t^2}{2!(3\cdot 7)} \ldots + \frac{(-t^n)}{n!(3\cdot 7\cdot\ldots\cdot 4n-1)}\right) \quad \text{and} \quad t = \frac{1}{4}b^2 s^4 \qquad \textbf{(9.55)}$$

A and B being arbitrary constants. The rotation θ is zero at the fixed end, $s=l$, and the Bessel function is first zero when its argument is 2.0063. This gives the first critical of value of S as

$$S_c = \frac{4.013}{l^2}\sqrt{GJEI_{xx}} \tag{9.56}$$

Returning to the earlier set of governing equations given by (9.50), the incremental end loadings will be taken as zero. Again examining the case where EI_ω can be taken as zero, the governing equation in terms of v' is similarly found to be

$$(v')'' + b^2 s^2 v' = 0 \quad \text{where} \quad b^2 = \frac{S^2}{GJEI_{xx}} \quad \text{and} \quad s = l - z. \tag{9.57}$$

so that the solution is the same as the previous solution for θ. Also, the end conditions are the same as for θ, the flexural curvature $(v')'$ being zero at $s=0$ and the slope v' being zero at $s=l$. The critical value of S is then again given by (9.56). The advantage of the former approach is that it is readily generalised to deal with other problems.

Considering now the case when EI_ω is not negligible, then (9.52) yields the governing equation

$$\theta'' - al^2\theta'''' + b^2 s^2 \theta = 0 \quad \text{where} \quad a = \frac{EI_\omega}{GJl^2} \tag{9.58}$$

This has been solved by Timoshenko (ibid.) using power series expansions. The result is plotted as the broken line in Figure 9.14.The critical value of the end shear force, S_c, is expressed in terms of the critical value b_c of b. The curve is approximately given by

$$\lambda_c = b_c l^2 \approx 4.01 + 11.7\sqrt{a} \tag{9.59}$$

The solution of the alternative form of the equations given by (9.50) is found by expressing v, θ and s in terms of new parameters, V, Θ and r and introducing the parameter λ, where

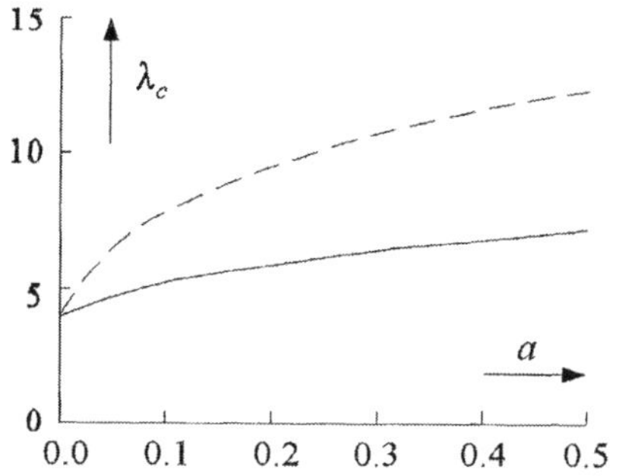

Figure 9.14 Buckling of a cantilever.

$$V(r) = v' \ , \quad \Theta(r) = \theta' l\sqrt{\frac{GJ}{EI_{xx}}} \ , \quad r = \frac{s}{l} \ , \quad \lambda = bl^2 \tag{9.60}$$

These equations then become

$$\frac{d^2V}{dr^2} - \lambda r\Theta = 0 \ , \qquad \Theta - a\frac{d^2\Theta}{dr^2} + \lambda r V = 0. \tag{9.61}$$

Taking V and Θ to be given by the series

$$V = \sum_{m=0}^{\infty} V_m r^m \ , \qquad \Theta = \sum_{m=0}^{\infty} \Theta_m r^m \tag{9.62}$$

equations (9.61) give rise to the recurrence relationships

$$m(m+1)V_{m+1} = \lambda\Theta_{m-2} \ , \qquad am(m-1)\Theta_m = \Theta_{m-2} + \lambda V_{m-3} \tag{9.63}$$

The conditions of zero bending moment and bimoment at the free end mean that V_1 and Θ_1 are zero. Two independent series for V and Θ can then be found to satisfy these conditions,

$$V = V_0\left(1 + \frac{\lambda^2 r^6}{180a} + \frac{\lambda^2 r^8}{6720a^2} + \ldots\right), \quad \Theta = \frac{\lambda V_0}{6a}\left(r^3 + \frac{r^5}{20a} + \frac{r^7}{840a^2} + \ldots\right);$$
$$V = \lambda\Theta_0\left(\frac{r^3}{6} + \frac{r^5}{40a} + \frac{r^7}{1008a^2} + \ldots\right), \quad \Theta = \Theta_0\left(1 + \frac{r^2}{2a} + \frac{r^4}{24a^2} + \frac{4\lambda^2a^2+1}{720a^3} + \ldots\right). \tag{9.64}$$

These two solutions can be used to satisfy the conditions at the fixed end (r=1). These are that the flexural slope and the warping are zero, which imply that V and Θ are zero at this end. Satisfying these conditions for non-zero deflexions gives the buckling criterion. The critical value λ_c of λ at buckling is plotted as the solid curve in Figure 9.14. This curve lies between the bounds

$$\lambda_c \approx 4.01\sqrt{(1.0 + 4.64a)} \quad \text{and} \quad \lambda_c \approx 4.01 + 4.66\sqrt{a} \tag{9.65}$$

The general process of analysing lateral buckling problems by this second method is to determine the (compressive) axial bending stress σ resulting from the initial loading in the x-z plane. This will be a function of z and x. This is then substituted in (9.39) instead of the form given by (9.38). Integrating over the cross-sectional area A then gives the work done per unit length. Then this loss of potential is substituted in the expression for the total potential Π just as (9.48) was used to obtain (9.49). The governing differential equations are then found by minimising Π with respect to v' and θ' as before. It is also possible to make allowances for the case when the shear load acts through the shear centre (and centroid) but is applied at a distance h above it. In the case of the end shear load S, this will produce an end torque T given by $Sh\theta_l$ where θ_l is the rotation of the beam where the shear load is applied. This end torque T is that given in (9.49) and the second of equations (9.50). Where there is a distributed load q per unit length along the beam applied at a distance h above the centroid, there is a term in addition to the loss of potential of the bending stresses. On the basis of the same arguments used in §9.4, Π should now also have the term $-\frac{1}{2}qh\theta^2$ in the integrand.

Formulae for the critical point loads in various lateral buckling problems are given in Table 34 of Young (1989). These relate to analyses of the type given by Timoshenko. Timoshenko and Gere (1961) also list solutions for the lateral buckling of simply-supported I-beams with a point load S or with a constant distributed load p. When S is at the centre of the beam, and acts at a distance h above the centroid, its critical value at buckling is given by

$$S_c \approx \frac{16.94\sqrt{GJEI_{xx}}}{l^2}\left(1 - \frac{1.74h}{l}\sqrt{\frac{EI_{xx}}{GJ}}\right) \tag{9.66}$$

ignoring the effects arising from non-uniform torsion. Likewise, the critical value for p applied along the centroidal axis is given by

$$p_c \approx \frac{28.3\sqrt{GJEI_{xx}}}{l^3} \tag{9.67}$$

Other solutions will be found in Young (1989), pages 680 and 681.}

{9.7 Local Buckling of Thin-Walled Sections}

{In addition to the overall buckling of a bar, local failure can occur accompanied with the deformation of the cross-section. (This is sometimes called 'crippling'.) The wall of the section acts like a plate (or shell) with membrane stresses large enough to cause buckling. A brief resumé of plate theory is given in Appendix 8. The governing equations given by (A8.3.12) were derived from the total potential of the plate. This potential will now be modified to allow for large membrane stresses[1]. As in (9.23), a compression-positive convention will be adopted. The work done by the normal membrane stresses[2] N_{xx}^- and N_{yy}^- may be determined from the apparent shortening of the membrane in the same way that the work done by the axial compressive stress σ was found in the buckling of a column in §9.4 and §9.5.

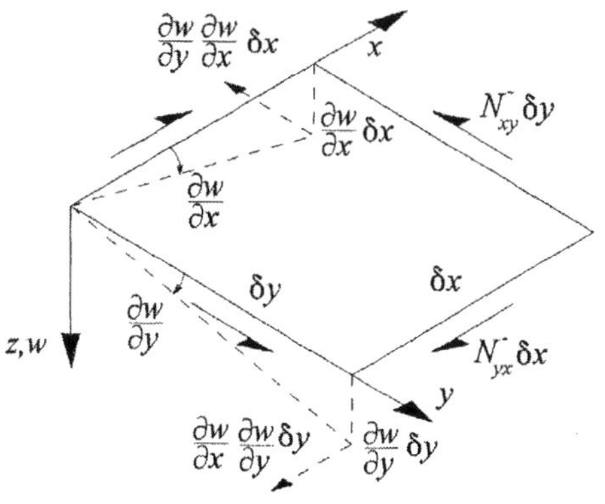

Figure 9.15 Second-order shear deformations.

The second-order effects of the lateral deflexion w of a plate element (δx by δy) on the shear membrane stresses is shown in Figure 9.15. The deflexion slope in the x direction, $\partial w/\partial x$, produces a downwards displacement $\partial w/\partial x \delta x$ of the far edge of the element. The effect of the rotation $\partial w/\partial y$ about the x axis then causes points which have undergone this displacement to move through $(\partial w/\partial y\ \partial w/\partial x\ \delta x)$ in the $-y$ direction as shown. Likewise, points on the positive y edge of the element move through $(\partial w/\partial x\ \partial w/\partial y\ \delta y)$ in the $-x$ direction. The work done by the shear membrane stresses during lateral deflexion of the plate is given by multiplying the displacements by the forces $N_{xy}^-\delta y$ and $N_{yx}^-\delta x$ respectively. Noting that N_{xy}^- and N_{yx}^- are the same, the total work done by the membrane stresses on the element is then

$$\delta W = \left[\frac{1}{2}\left(\frac{\partial w}{\partial x}\right)^2 N_{xx}^- + \frac{1}{2}\left(\frac{\partial w}{\partial y}\right)^2 N_{yy}^- + \frac{\partial w}{\partial x}\frac{\partial w}{\partial y}N_{xy}^-\right]\delta x\,\delta y \qquad \textbf{(9.68)}$$

Integrated over the area of the plate, this expression can be added to that for the work done given by (A8.3.2). The change in total potential due to a small variation δw of w can then be found as before. Taking the membrane stresses to be constant and using (A8.3.6), the relationships

$$\begin{aligned}
\delta\int_A \frac{1}{2}\left(\frac{\partial w}{\partial x}\right)^2 dA &= \int_A \frac{\partial w}{\partial x}\frac{\partial(\delta w)}{\partial x}dA \\
&= \int_A\left[\frac{\partial}{\partial x}\left(\frac{\partial w}{\partial x}\delta w\right) - \frac{\partial^2 w}{\partial x^2}\delta w\right]dA = \oint \frac{\partial w}{\partial x}\delta w\cos\alpha\, ds - \int_A \frac{\partial^2 w}{\partial x^2}\delta w\, dA \\
\delta\int_A \frac{1}{2}\left(\frac{\partial w}{\partial y}\right)^2 dA &= \int_A \frac{\partial w}{\partial y}\frac{\partial(\delta w)}{\partial y}dA \\
&= \int_A\left[\frac{\partial}{\partial y}\left(\frac{\partial w}{\partial y}\delta w\right) - \frac{\partial^2 w}{\partial y^2}\delta w\right]dA = \oint \frac{\partial w}{\partial y}\delta w\sin\alpha\, ds - \int_A \frac{\partial^2 w}{\partial y^2}\delta w\, dA
\end{aligned} \qquad \textbf{(9.69)}$$

(s and α being as shown in Figure 9.16) and

---

[1] Membrane stresses act within the plane of the membrane. They are not forces per unit area but are defined as the resultant forces per unit length of the section through the membrane on which they act.

[2] The negative superscript is used to indicate that the opposite of the usual sign convention is being used.

$$\delta \int_A \frac{\partial w}{\partial x}\frac{\partial w}{\partial y} dA = \int_A \left(\frac{\partial w}{\partial x}\frac{\partial(\delta w)}{\partial y} + \frac{\partial(\delta w)}{\partial x}\frac{\partial w}{\partial y} \right) dA$$
$$= \int_A \left[\frac{\partial}{\partial y}\left(\frac{\partial w}{\partial x}\delta w \right) + \frac{\partial}{\partial x}\left(\frac{\partial w}{\partial y}\delta w \right) - 2\frac{\partial^2 w}{\partial x \partial y}\delta w \right] dA \qquad \textbf{(9.70)}$$
$$= \oint \left(\frac{\partial w}{\partial y}\cos\alpha + \frac{\partial w}{\partial x}\sin\alpha \right) \delta w \, ds - 2\int_A \frac{\partial^2 w}{\partial x \partial y}\delta w dA$$

can be used in expressing these extra terms for the change in potential. Substituting them in (A8.3.11), equations (A8.3.12) now become

$$D\nabla^4 w + N_{xx}^- \frac{\partial^2 w}{\partial x^2} + 2N_{xy}^- \frac{\partial^2 w}{\partial x \partial y} + N_{yy}^- \frac{\partial^2 w}{\partial y^2} - p = 0$$
$$D\left((1-\nu)\frac{\partial^2 w}{\partial n^2} + \nu\nabla^2 w \right) + M_s = 0 \qquad \textbf{(9.71)}$$
$$D\left((1-\nu)\frac{\partial^3 w}{\partial n \partial s^2} + \frac{\partial}{\partial n}\nabla^2 w \right) + \left(S - \frac{\partial M_n}{\partial s} \right)$$
$$- (N_{xx}^- \cos\alpha + N_{yx}^- \sin\alpha)\frac{\partial w}{\partial x} - (N_{xy}^- \cos\alpha + N_{yy}^- \sin\alpha)\frac{\partial w}{\partial y} = 0$$

where the first equation holds *within* the boundary of the plate and the second and third equations hold on the boundary. The third equation can be expressed more simply in terms of the membrane stresses N_{nn}^- and N_{ns}^- normal and tangential to the boundary. From the equilibrium of the triangular element shown in Figure 9.16, resolving horizontally and vertically,

$$N_{xx}^- \cos\alpha + N_{yx}^- \sin\alpha = N_{nn}^- \cos\alpha - N_{ns}^- \sin\alpha$$
$$N_{xy}^- \cos\alpha + N_{yy}^- \sin\alpha = N_{nn}^- \sin\alpha + N_{ns}^- \cos\alpha \qquad \textbf{(9.72)}$$

and inverting (A8.3.7) gives

$$\frac{\partial w}{\partial n} = \cos\alpha\frac{\partial w}{\partial x} + \sin\alpha\frac{\partial w}{\partial y}$$
$$\frac{\partial w}{\partial s} = -\sin\alpha\frac{\partial w}{\partial x} + \cos\alpha\frac{\partial w}{\partial y} \qquad \textbf{(9.73)}$$

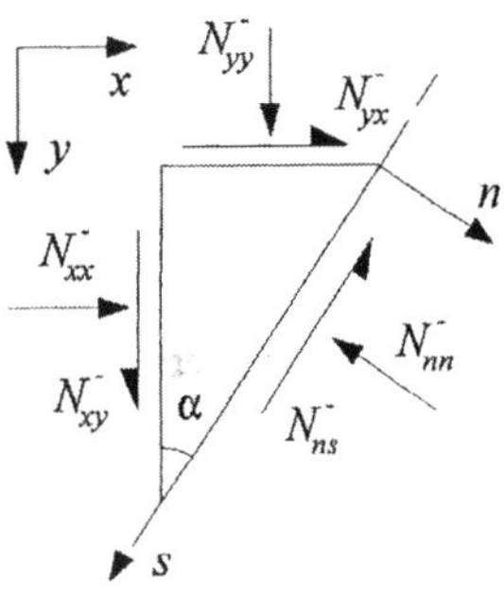

Figure 9.16 Equilibrium of membrane stresses on the boundary.

Substituting these equations into the third equations (9.71) gives

$$D\left((1-\nu)\frac{\partial^3 w}{\partial n \partial s^2} + \frac{\partial}{\partial n}\nabla^2 w \right) + \left(S - \frac{\partial M_n}{\partial s} \right) - N_{nn}^- \frac{\partial w}{\partial n} - N_{ns}^- \frac{\partial w}{\partial s} = 0 \qquad \textbf{(9.74)}$$

The problems of plate buckling most readily analysed relate to simply-supported rectangular plates subject to normal membrane stresses in the directions of the boundaries. Suppose that such a plate has sides of length a parallel to the x axis and sides of length b parallel to the y axis. If it subject to a membrane stress N_{xx}^- only, The first of equations (9.71) becomes

$$D\nabla^4 w + N_{xx}^- \frac{\partial^2 w}{\partial x^2} = 0 \qquad \textbf{(9.75)}$$

This can be solved[1] using a double Fourier series of the form

$$w = \sum_{m=1}^{\infty}\sum_{n=1}^{\infty} w_{mn} \sin\frac{m\pi x}{a} \sin\frac{n\pi y}{b} \qquad \textbf{(9.76)}$$

Each term of the series satisfies the boundary conditions of zero displacement and zero moment $M_{s,}$. Substituting this form into (9.75) gives the condition

$$\left[D\left(\frac{m^2\pi^2}{a^2} + \frac{n^2\pi^2}{b^2}\right)^2 - N_{xx}^{-}\frac{m^2\pi^2}{a^2}\right]w_{mn} = 0 \qquad \textbf{(9.77)}$$

so that w_{mn} is non-zero provided that

$$N_{xx}^{-} = \frac{\pi^2 a^2 D}{m^2}\left(\frac{m^2}{a^2} + \frac{n^2}{b^2}\right)^2 \qquad \textbf{(9.78)}$$

We seek the lowest value of N_{xx}^{-} at which a non-zero deflexion is possible. This is achieved by choosing the lowest admissible value of n in the above expression, which is unity. Minimising the resulting expression for N_{xx}^{-} with respect to m^2/a^2 gives the value of this parameter as $1/b^2$. N_{xx}^{-} can then be written as

$$N_{xx}^{-} = k\frac{\pi^2 D}{b^2} \quad (\text{minimum } k = 4) \qquad \textbf{(9.79)}$$

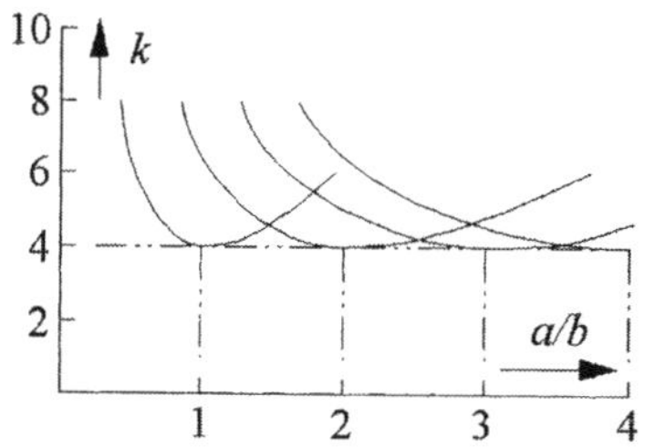

Figure 9.17 Buckling of a simply-supported rectangular plate.

The parameter m^2/a^2 can take the value $1/b^2$ if $a=b$ and m is unity or $a=2b$ and m is two and so on. More generally, a will be prescribed and it will be necessary to find the value of the integer m giving the lowest value of N_{xx}^{-}. The variation of k with a/b for integer values of m from one to four is shown by the four curves in Figure 9.17. The minimum value for m=1 is when a/b=1 and so on. The value N_{xx}^{-} given by (9.79) is then a safe lower bound to the buckling membrane stress in all cases and a reasonable estimate for large values of a/b.

Solutions for other boundary conditions on the edges of the plate parallel to the x axis can be found by taking a solution of the form

$$w = \sum_{m=0}^{\infty} \sin\frac{m\pi x}{a} f_m(y) \qquad \textbf{(9.80)}$$

Substituting this form into (9.75) gives the differential equation for $f_m(y)$

$$\frac{d^4 f_m}{dy^4} + \frac{m^2\pi^2}{a^2}\left[\left(\frac{m^2\pi^2}{a^2} - \frac{N_{xx}^{-}}{D}\right)f_m - 2\frac{d^2 f_m}{dy^2}\right] = 0 \qquad \textbf{(9.81)}$$

which can be solved by taking $f_m(y)$ as an exponential function of cy. This gives a quadratic in c^2 with the roots

$$c^2 = \alpha^2 = \frac{m^2\pi^2}{a^2} + \sqrt{\frac{N_{xx}^{-}}{D}\frac{m^2\pi^2}{a^2}}, \quad c^2 = -\beta^2 = \frac{m^2\pi^2}{a^2} - \sqrt{\frac{N_{xx}^{-}}{D}\frac{m^2\pi^2}{a^2}} \qquad \textbf{(9.82)}$$

---

[1] Bryan, G.H. (1891) *Proc. London Math. Soc.* V.22 p.54.

giving the expression[1]

$$f_m(y) = A_m \sinh \alpha y + B_m \cosh \alpha y + C_m \sin \beta y + D_m \cos \beta y \qquad \textbf{(9.83)}$$

The legs of an equal-angle section under a uniform axial load can be subject to plate buckling of the kind which can be analysed with these equations. The line where the legs join can be treated as the edge y=0 of such a plate. The legs are free to rotate about this line although the line itself does not displace during plate-like buckling of the angle. This edge can then be treated as simply-supported. If the legs are of width b, then the edge y=b can be treated as a free edge. From the second of equations (9.71) and (9.74) the boundary conditions are then

$$\begin{aligned} &\frac{\partial^2 w}{\partial y^2} + \nu \frac{\partial^2 w}{\partial x^2} = 0 \,, \qquad w = 0 \qquad \text{on } y = 0 \,. \\ &\frac{\partial^2 w}{\partial y^2} + \nu \frac{\partial^2 w}{\partial x^2} = 0 \,, \qquad \frac{\partial^3 w}{\partial y^3} + (2 - \nu) \frac{\partial^3 w}{\partial x^2 \partial y} = 0 \qquad \text{on } y = b \,. \end{aligned} \qquad \textbf{(9.84)}$$

The boundary conditions on y=0 are satisfied by taking B_m and D_m as zero in (9.83). The remaining pair of conditions on y=b yield the condition

$$\frac{1}{\alpha}\left(\alpha^2 - \nu \frac{m^2\pi^2}{a^2} \right)^2 \tanh \alpha b = \frac{1}{\beta}\left(\beta^2 + \nu \frac{m^2\pi^2}{a^2} \right)^2 \tan \beta b \qquad \textbf{(9.85)}$$

for non-zero A_m and C_m (that is, for buckling to take place). This expression can be treated as a function of m^2/a^2 and $N_{xx}^{\cdot}$. To minimise $N_{xx}^{\cdot}$, the smallest possible value of m^2/a^2 must be used. This means taking m as unity and taking a as the length of the column. The critical values of $N_{xx}^{\cdot}$ can again be expressed in terms of the parameter k in (9.79). Taking Poisson's ratio as 0.25, values of this parameter for various values of a/b are given in Table 9-2 of Timoshenko and Gere (1961). The approximation

$$k \approx \left(0.456 + \frac{b^2}{a^2} \right) \qquad \textbf{(9.86)}$$

overestimates k by 1.3% for a/b equal to 0.5 and underestimates k by 2.0% for a/b equal to 5.0.

The same general expression for w can be used for plate buckling problems when one of the lateral edges of the plate is built-in and the other is free. (This would be an approximation to the behaviour of a tee section with a thick flange and a thin web.) The first condition in (9.84) is then replaced with $\partial w/\partial y = 0$ on $y = 0$. The graph of the critical values of k is then similar to that shown in Figure 9.17, with a lower bound of 1.328. Similarly, a solution can be found for a plate with both lateral edges built-in. This could be used as an approximation for a column formed from an I-beam with thick flanges and a thin web. In this case, the lower bound to k is approximately 7.0. (In all cases, Poisson's ratio is taken as 0.25.)

Other solutions for elastically-supported lateral edges, rectangular plates subject to shear stresses, and combinations of shear stresses and linearly-varying axial stresses (as in the webs of beams) have also been found. Some of these cases were analysed by solving the differential equations and approximate energy methods used in other cases.

A reasonable approximation to the critical membrane stress associated with the buckling of a simply-supported rectangular plate in pure shear is given by

---

[1] Timoshenko, S.P. (1910) *Z. Math. Physik* V.58 p.343.

$$N_{xy}^{-} \approx \left[5.35 + 4\left(\frac{b}{a}\right)^2\right]\frac{\pi^2 D}{b^2} \tag{9.87}$$

where a and b are the lengths of the edges of the plate ($b \le a$). This result and others are given in Chapter 9 of Timoshenko and Gere (1961). Table 35 of Young (1989) gives formulae for the critical loads in other cases.}

9.8 Approximate Methods

Where exact analytic solutions are not available, approximate methods may be used. Most common amongst these is a version of the Rayleigh-Ritz method discussed in §7.7. The difference is that previously the method was used to find the conditions governing an equilibrium state. Here, it is assumed that the body is already in equilibrium but that when a critical state of loading is reached, an immediately adjacent state of equilibrium also exists. The method consists of finding an approximation to the buckling mode which satisfies the geometrical boundary conditions to the problem. The buckling load is then determined from equating the change in total potential between the two states to zero. As the buckling load is sought which is the least possible for all such hypothetical buckling modes, the buckling load found in this way will be greater than (or equal to) this critical buckling load.

An example of the application of this method is in finding the buckling load of a strut of variable cross-section. Using the notation of §9.2, consider a beam of length l with a variable bending stiffness $EI(x)$. The total potential variation is given by a particular case of (9.24),

$$\Pi = \int_0^l \frac{1}{2}[EI(x)v''^2 - Pv'^2]\,dx \tag{9.88}$$

Choosing a suitable approximation v_a to the buckling mode, an estimate of the critical load is then given by

$$P_c \le \frac{\int_0^l EI(x)v_a''^2\,dx}{\int_0^l v_a'^2\,dx} \tag{9.89}$$

This is the basic Rayleigh method. The Ritz modification is to express v_a as a linear combination of functions satisfying the geometrical boundary conditions,

$$v_a = \sum_{i=1}^{n} c_i v_i(x) \tag{9.90}$$

The resulting expression for Π given by (9.88) is then a function of the constants c_i. This can be minimised by setting the partial differentials of the expression with respect to coefficients c_i equal to zero, as in §7.7. This gives a set of n homogeneous linear equations in the unknown constants c_i. Equating the determinant of the coefficients of these constants to zero gives the buckling criterion. This reduces to the Rayleigh method when n is one.

One of the disadvantages of this approach is that if a reasonable approximation to the deflexion curve is used, the first differential of the approximation is a poorer approximation and the second differential could be very inaccurate. Suppose for example

$$v = \sin\frac{\pi x}{l} \quad , \quad v_a = \frac{3.8}{l^2}(lx - x^2) \tag{9.91}$$

where the approximation is used in the range $x=0$ to $x=l$. Then the maximum error in using v_a for v, as a proportion of the maximum amplitude of v, is 5%. However, the maximum error in using v_a' for v', as a proportion of the maximum amplitude of v', is 21% and the maximum error in using v_a'' for v'', as a proportion of the maximum amplitude of v'', is 77%. Timoshenko proposed an alternative form for the numerator of (9.89) which avoids the use of the second differential of v. It relies on being able to express the local bending moment M as Pv where v is measured from the line of action of P. This method can be used for pin-ended and cantilevered struts, but not for problems where there statically-indeterminate end reactions affect the bending moment. As M is equal to EIv'', $P_c v/EI$ can be substituted for v'' in (9.89) giving rise to the criterion

$$P_c \approx \frac{\int_0^l v_a'^2 \, dx}{\int_0^l \frac{v_a^2}{EI(x)} \, dx} \tag{9.92}$$

This is invariably a better approximation, where it can be used.

Example 9.2 A Pin-ended Non-Uniform Strut

Suppose that the expression for the bending stiffness of a non-uniform strut is given by

$$EI(x) = EI_0\left[1 - 4(1-r)\left(\frac{x}{l}\right)^2\right]^2 \tag{9.93}$$

and that it is pin-ended at the ends $x = \pm l/2$. The approximation to the buckling mode will be taken as

$$v_a = c_1 \cos\frac{\pi x}{l} + c_2 \cos\frac{3\pi x}{l} \tag{9.94}$$

Southwell (1936) §488 applies Rayleigh's method to this case in the instance when r^2 is 0.2 and uses the first term in (9.94). Then (9.89) becomes

$$P_c \le \frac{c_1^2 \frac{\pi^4}{l^4} \int_{-l/2}^{l/2} EI(x) \cos^2\frac{\pi x}{l} \, dx}{c_1^2 \frac{\pi^2}{l^2} \int_{-l/2}^{l/2} \sin^2\frac{\pi x}{l} \, dx} = \frac{\pi^2}{l^3} \int_{-l/2}^{l/2} EI(x)\left(1 + \cos\frac{2\pi x}{l}\right) dx = \lambda \frac{EI_0}{l^2} \tag{9.95}$$

giving in this case λ as 8.568[1].

Using the Ritz method and the expression for v_a given by (9.94), the expression for Π given by (9.88) becomes

---

[1] Owing to a small rounding-off error, Southwell gives $\lambda = 8.55$.

$$\Pi = \int_0^{l/2} \frac{\pi^4}{l^4} EI(x)\left(c_1^2\cos^2\frac{\pi x}{l} + 18c_1c_2\cos\frac{\pi x}{l}\cos\frac{3\pi x}{l} + 81c_2^2\cos^2\frac{3\pi x}{l}\right) - \frac{\pi^2 P}{l^2}\left(c_1^2\sin^2\frac{\pi x}{l} + 6c_1c_2\sin\frac{\pi x}{l}\sin\frac{3\pi x}{l} + 9c_2^2\sin^2\frac{3\pi x}{l}\right) dx \tag{9.96}$$

using the symmetry of the functions to integrate over only half the span of the column. Setting the partial differentials $\partial\Pi/\partial c_1$ and $\partial\Pi/\partial c_2$ to zero yields a pair of linear homogeneous equations in the terms c_1 and c_2. If these are to be non-zero, the determinant of their coefficients must be zero, giving

$$\begin{vmatrix} 8.5675 - \lambda & 12.9438 \\ 12.9438 & 563.532 - 9\lambda \end{vmatrix} = 0 \tag{9.97}$$

in the case examined previously, λ being defined by (9.95). It will be seen that if only the first term in the expression for v_a had been used, λ would again have been given as 8.568. Taking the smallest root of the quadratic represented by (9.97), it is given as 8.222.

Using the Timoshenko method,

$$P_c \approx \frac{c_1^2\dfrac{\pi^2}{l^2}\displaystyle\int_{-l/2}^{l/2}\sin^2\frac{\pi x}{l}\,dx}{c_1^2\displaystyle\int_{-l/2}^{l/2}\frac{\cos^2\frac{\pi x}{l}}{EI(x)}\,dx} = \frac{\pi^2}{l}\left[\int_{-l/2}^{l/2}\frac{1}{EI(x)}\left(1 - \cos\frac{2\pi x}{l}\right)dx\right]^{-1} = \lambda\frac{EI_0}{l^2} \tag{9.98}$$

which gives the value of λ as 8.214 in the particular case examined above. The exact solution[1] is known for this problem and is given by

$$P_c = 4\frac{EI_0}{l^2}(1 - r)\left[1 + \pi^2\left(\ln\left\{\frac{1 + \sqrt{1-r}}{1 - \sqrt{1-r}}\right\}\right)^{-2}\right] \tag{9.99}$$

so that for the value of r used above, λ is 8.1527. The Rayleigh method then overestimates the buckling load by 5.09%, the Ritz method overestimates it by 0.85% and the Timoshenko method overestimates it by 0.75%.

Another method of dealing with columns of variable stiffness is to use a stepped column approximation. The solid lines in Figure 9.18 show the variation in EI along the above column. As a rough manual approximation, this has been replaced by a stepped column with a bending stiffness of $0.2EI_0$ for $0.1l$ from either end, followed by a stiffness of $0.65EI_0$ for the next $0.2l$ from either end and a stiffness of $0.95EI_0$ for the central $0.4l$ of the column. The bending stiffness of this stepped column is represented by the shaded area in Figure 9.18. This was analysed using the software supplied, treating half the column as a cantilever. This gave a value for λ of 7.869, an underestimate of 3.48%. More accurate values could be obtained by using more steps. Also, upper and lower bounds can be

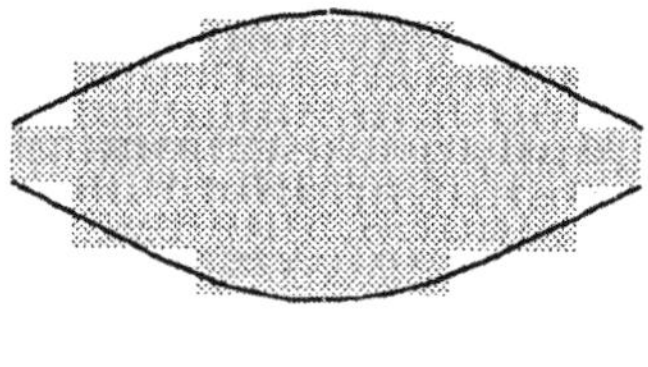

Figure 9.18 A stepped column.

[1] Cowley, W.L. and Levy, H. (1918) *Aeronautical Research Committee R. and M.* No. 484.

found by this method by ensuring that the local bending stiffness of the stepped column is either never less than or never greater than that of the actual column.

{Versions of these methods can also be used to find the buckling loads of plates. The Galerkin method is essentially a modified version of the Ritz method. Instead of the deflected form being approximated by a linear combination of functions of one ordinate, as in (9.90), it is given in terms of functions of two ordinates, x and y, as

$$w_a = \sum_{i=1}^{n} c_i w_i(x,y) \tag{9.100}$$

Also, instead of being applied directly to the original form of the expression for the total potential, it is applied to the modified form equivalent to the first integral in (A8.3.11). The variation δw is taken to be that for a variation δc_i of each of the terms c_i in (9.100) in turn, yielding simultaneous equations of the form

$$\int_A \left(D\nabla^4 w_a + \bar{N}_{xx}\frac{\partial^2 w_a}{\partial x^2} + 2\bar{N}_{xy}\frac{\partial^2 w_a}{\partial x \partial y} + \bar{N}_{yy}\frac{\partial^2 w_a}{\partial y^2} \right) w_i(x,y)\, dA = 0 \tag{9.101}$$

in the absence of lateral loading. In any given problem, the components of the membrane stress will be in a fixed ratio to one another. The amplitude of the membrane stress, λ say, at which buckling occurs is sought. This can be found from the n equations of the form given by (9.101) by setting the determinant of the coefficients of the terms c_i to zero, as in the Ritz analysis.}

9.9 General Theories of Stability

The elementary theory given in §9.2 was derived for initially-perfect struts. The initially imperfect form of a pin-ended strut can be expressed by the Fourier series

$$v_0(x) = \sum_{m=1}^{\infty} v_m \sin\frac{m\pi x}{l} \tag{9.102}$$

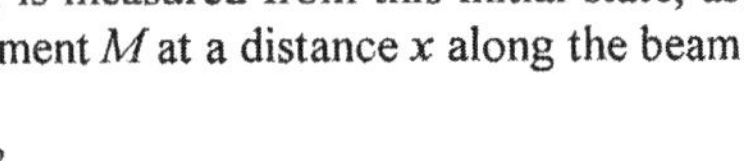

Figure 9.19 Flexure of an imperfect pin-ended column.

The additional deflexion induced by the axial load, $v(x)$, is measured from this initial state, as shown in Figure 9.19. The expression for the bending moment M at a distance x along the beam is then

$$-P(v_0 + v) = M = EI\frac{d^2v}{dx^2} \tag{9.103}$$

Using the expression for v_0 given by (9.102), the solution of this differential equation is given by

$$v = \sum_{m=1}^{\infty} \frac{v_m}{\dfrac{m^2\pi^2 EI}{Pl^2} - 1} \sin\frac{m\pi x}{l} \tag{9.104}$$

The coefficient of the mth term in this series tends to become infinite as the mth buckling load is approached. Then if the load P is sufficiently close to the first buckling load, P_e, as given by

(9.6b), the first term in the series for v dominates all the rest and we can write

$$v \approx \frac{v_1}{\frac{P_e}{P} - 1} \sin\frac{\pi x}{l} = \frac{v_e}{\frac{P_e}{P} - 1} \qquad (v_e = v_1 \sin\frac{\pi x}{l}) \qquad \textbf{(9.105)}$$

The displacement v may be measured experimentally at some position x along the beam. The buckling load can then be predicted using the linear relationships

$$v = P_e\left(\frac{v}{P}\right) - v_e \qquad \text{or} \qquad \frac{P}{v} = \frac{1}{v_e}(P_e - P) \qquad \textbf{(9.106)}$$

The first relationship gives the standard Southwell plot[1] shown in Figure 9.20a. The experimental results, shown by the circular dots, tend towards the straight line given by this equation as the load P approaches P_e. This buckling load is then predicted by measuring the slope of the line. Alternatively, from a plot of P/v against P, the experimental results lie close to a straight line when the load is close to P_e. This critical load is then predicted by the intercept of this line with the P axis, as shown in Figure 9.20b. The Southwell plot has been shown[2] to be applicable to the general flexural-torsional buckling of linearly-elastic structures.

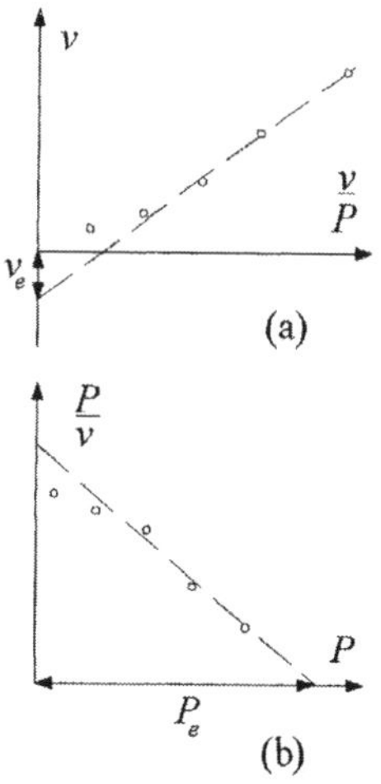

Figure 9.20 Southwell plots.

The critical behaviour of bodies which are subject to linear combinations of n different loadings can be plotted in an n-dimensional space. The boundaries between stable and unstable loading states generate *critical surfaces*. Figures 3.11 and 3.12 show critical surfaces for materials subject to two and three principal stresses. Such surfaces, as in these examples, are almost invariably convex. That is, if any two points within the stable zone are joined by a straight line, all points on that line also lie within the stable zone. This applies to both plastic failure and failure arising from elastic instability. Figure 9.21 shows the portal frame analysed in Example 9.1, except that the stanchion loads, P and Q, can be varied independently. When they are equal, the buckling load is given by the second of equations (9.14). When one of them is zero, the magnitude of the other at buckling is $14.58EI/l^2$. The values of P and Q at failure, for different ratios of one to the other, are plotted on the graph in Figure 9.21, using the non-dimensional parameters Pl^2/EI and Ql^2/EI. It can be seen that the critical surface so generated is slightly convex. If it had been a straight line joining the points on the axes, it would have predicted failure under equal loads of $7.29EI/l^2$ applied to the tops of the stanchions.

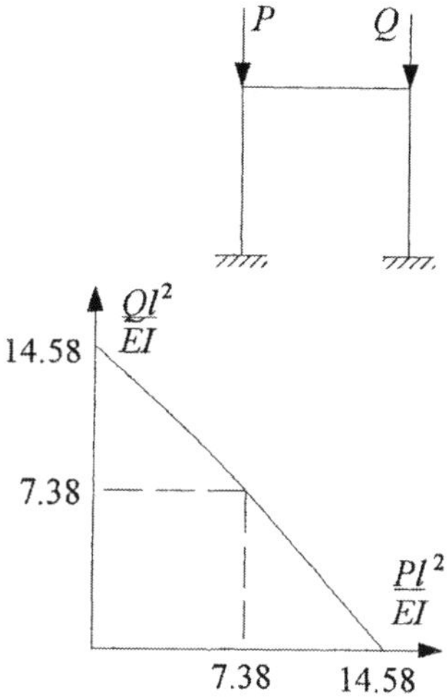

Figure 9.21 Critical loads on a portal frame.

[1] Southwell, R.V. (1932) On the Analysis of Experimental Observations in Problems of Elastic Stability. *Proc. Roy. Soc. London* Ser. A V.135 p.601.

[2] Ariaratnam, S.T. (1961) The Southwell plot for predicting critical loads of elastic structures. *Q. Jl. Mech. appl. Math.* V.14.

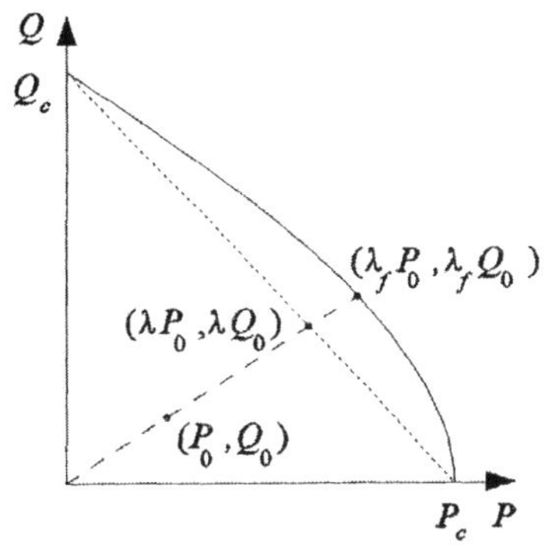

Figure 9.22 A convex critical surface.

In cases where the critical surface is convex, the following criterion may be used. Suppose that the structure is stable under the combined action of two independent loadings P_0 and Q_0, and that it becomes unstable at λ_f times this combined loading. Under each loading applied separately, it buckles at P_c which is λ_1 times P_0 and at Q_c which is λ_2 times Q_0. The dotted line shown in Figure 9.22 joins the two points $(P_c, 0)$ and $(0, Q_c)$. Its equation is given by

$$\frac{P}{P_c} + \frac{Q}{Q_c} = 1 \tag{9.107}$$

As the critical surface is convex, this line lies inside (or on) the critical surface. If the loadings P_0 and Q_0 are increased in proportion to one another up to failure, they follow the trajectory shown by the broken line. This intersects the above line at $(\lambda P_0, \lambda Q_0)$ where λ is less than (or equal to) λ_f. From (9.107) then,

$$\frac{\lambda P_0}{\lambda_1 P_0} + \frac{\lambda Q_0}{\lambda_2 Q_0} = 1 \qquad \text{or} \qquad \frac{1}{\lambda_1} + \frac{1}{\lambda_2} = \frac{1}{\lambda} \geq \frac{1}{\lambda_f} \tag{9.108}$$

Equations of this form appear in many different contexts. In general, there are n critical parameters λ_i and a convex critical surface in n dimensions. A safe lower bound to this surface is given by an $(n\text{-}1)$ dimensional plane whose equation is a generalisation of (9.107), and

$$\frac{1}{\lambda_f} \leq \sum_{i=1}^{n} \frac{1}{\lambda_i} \tag{9.109}$$

This is known as the Dunkerley theorem, although it was originally applied to vibration problems[1]. Papkovich[2] showed that it could be applied in the above sense to complete structures. Variants of this are Rankine's empirical formula

$$\frac{1}{\lambda_f} \leq \frac{1}{\lambda_P} + \frac{1}{\lambda_c} \tag{9.110}$$

where λ_f is the load parameter at which elasto-plastic buckling occurs, λ_P is the load parameter for rigid-plastic collapse and λ_c is the elastic buckling load parameter. Applied to complete frameworks, this is known as Merchant's formula. Horne[3] has shown this can be justified under certain assumptions, but is not strictly valid in all circumstances. Föppl[4] showed that it might be applied to buckling problems which had been simplified by assuming that all but one of the stiffnesses were infinite. The parameters λ_i are the critical parameters for each of these cases and

---

[1] Dunkerley, S. (1894) On the Whirling and Vibration of Shafts. *Phil. Trans. Roy. Soc. London* Ser. A V.185 p.265. See also Thomson (1988) §11.2.

[2] Papkovich, P.F. (1963) Works on the Structural Mechanics of Ships, V.4 *Gos. Soyuz. Izd. Sud. Prom. Leningrad* (In Russian).

[3] Horne, M.R. (1963) Elastic-Plastic Failure Loads for Plane Frames. *Proc. Roy. Soc. London* Ser. A V.274 p.343.

[4] Föppl, L. (1933) Über das Ausknicken von Gittermasten, insbesondere von hohen Funktürmen. *Z.A.M.M.* V.13 p.1.

λ_f is the parameter for the general case. The shear buckling relationship given by (9.21) is an example of a case where the inequality becomes an equality, but for the fixed-pinned problem given by (9.22) the formula becomes an approximation rather than a lower bound. It can also be shown that (9.109) can be applied to a set of assumed buckling modes, which approximate to the actual buckling mode (provided that certain orthogonality conditions are met). In this case, the λ_i are the critical load parameters associated with each mode and λ_f is the true buckling parameter. This may be deduced from Southwell (1936) §482.

Symmetry of a structure and its loading, described in §1.5, can be used to simplify stability and vibration analysis. Figure 1.8a shows a structure with simple symmetry about the line QR. On taking the mirror image, the zones on either side of this line 'map' onto each other, and no change is apparent. In the case of n-fold symmetry, the structure can be divided into n zones which map onto one another in n different ways (including the identity mapping, I, which does nothing). Figure 9.23 shows an equilateral triangular frame with six-fold symmetry. Rotation about a vertical axis by 120° anticlockwise can be thought of as an operation A which produces the permutation $A\{1,2,3,4,5,6\}=\{3,1,2,6,4,5\}$. This can be repeated, giving $A^2\{1,2,3,4,5,6\}=\{2,3,1,5,6,4\}$. Suppose that it is also symmetric about a horizontal plane, half way up the structure. There is then symmetry under the inversion operation $B\{1,2,3,4,5,6\} = \{4,5,6,1,2,3\}$. The operations A^3 and B^2 do nothing (that is, are equal to I), but $AB\{1,2,3,4,5,6\} = BA\{1,2,3,4,5,6\} = \{6,4,5,3,1,2\}$. Note that here the order of the operations does not matter, and the full set operations is said to be abelian. The only other distinct operation is A^2B ($=ABA = BA^2$) giving six in all. If there is n-fold abelian symmetry, the stability determinant can be divided into n sub-determinants, each $1/n$th the size of the original and each related to a particular mode of buckling[1] (or vibration). A further description will be found in Chapter 1 of Chilver (1967).

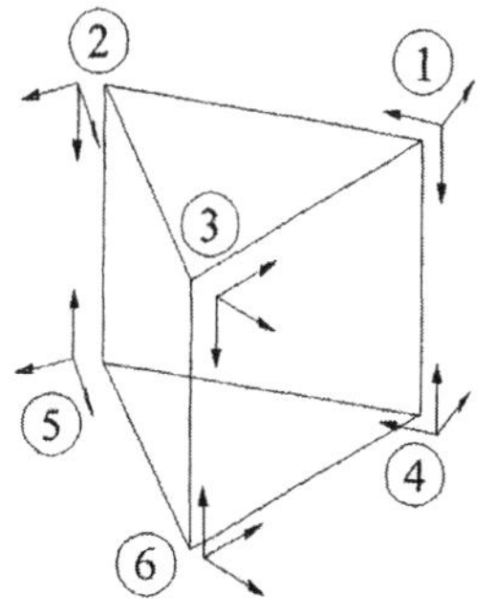

Figure 9.23 Symmetry zones.

Various theorems associated with convex critical surfaces have been proposed by Papkovich (loc. cit.). These are also discussed in Chapter 1 of Chilver (1967). There, it is asserted that the addition of constraints to a structure cannot reduce its first buckling load. This is true, provided that such constraints inhibit the corresponding buckling mode. Tarnai[2] gives counter-examples of rigid constraints which can help to induce the buckling mode. Other theories concern the location of individual critical load parameters and optimum design. These subjects will be dealt with in the appropriate chapters later in this book.

[1] Renton, J.D. (1964) On the Stability Analysis of Symmetrical Frameworks. *Quart. Journ. Mech. and Applied Math.* V.17 p.175.

[2] Tarnai, T. (1980) Destabilising Effect of Additional Restraint on Elastic Bar Structures. *Int. J. Mech. Sci.* V.22 p.379.

Chapter 10 The Dynamic Behaviour of Beams

10.1 Introduction

The main topics covered in this chapter are the vibration of elastic beams and the transmission of elastic waves through them. It is not possible to cover all the ramifications of the subject in one chapter, so that some topics will be only be summarised in this introduction. References to more detailed studies will be given where appropriate.

In the main, it will be taken that the material has the normal linear elastic response to dynamic loading. However, internal damping can be significant in some cases. It is often assumed that such material damping is like viscous damping, implying some resistance proportional to the rate of strain. For convenience, a viscous damping matrix is used which is a linear combination of the stiffness and mass matrices. In fact, what is called *hysteretic damping* is observed. The simplest model for this, known as the Voigt unit, is shown in Figure 10.1a. The material response is represented by a linearly-elastic spring of stiffness k' in parallel with a damper of dynamic stiffness k''. When vibrating sinusoidally at an angular frequency ω, this damper responds to a velocity $\dot{x}$ with a force of $k''\dot{x}/\omega$. the exciting force F is then given by $F = k'x + k''\dot{x}/\omega$. The equivalent equation relating a stress σ to a strain ϵ is $\sigma = E'\epsilon + E''\dot{\epsilon}/\omega$. The resulting plot of stress against strain is elliptical, as shown in Figure 10.1b. More realistic stress-strain plots for vibrating materials resemble the stress-strain hysteresis curve shown in Figure 10.1c. However, these require more complex models and analysis. Values of the ratio E''/E', expressed by the dimensionless parameter η_s, are tabulated in §8.2 of Lazan (1968). Typically, they are of the order of 10^{-3} for aluminium alloys and steel. Lazan (Chapter 6) also examines the stress distribution resulting from the damped vibration of various beams. Further discussion will be found in Thomson (1988) §3.9.

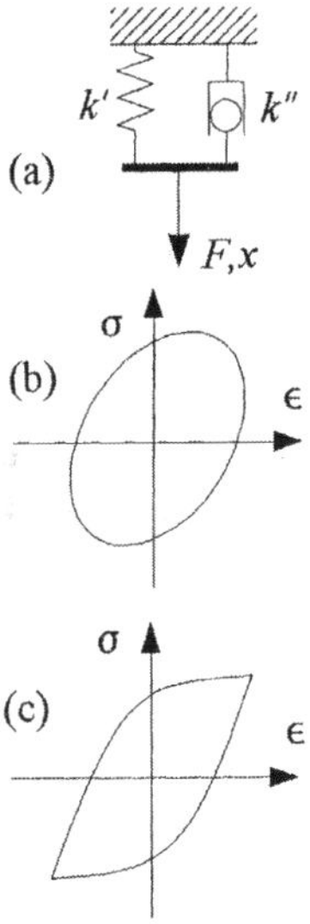

Figure 10.1
Material damping.

The dynamic interaction of fluid flow and elastic structures can be subdivided into several different phenomena. These are buffeting, vortex shedding, galloping and flutter. These effects need to be taken into account in addition to the steady-state effects of wind pressure on tall buildings, for example. Wind-induced oscillations can be devastating, particularly when they occur near the natural frequencies of the structure. Examples are the failure of the original Tacoma Narrows suspension bridge[1] in 1940 and the collapse of three cooling towers at Ferry Bridge in 1965. Even when the effects are not catastrophic, excess vibration can cause serious problems. For example, in designing skyscrapers, it is necessary to avoid any distortion of the lift shafts sufficient to cause the lifts to jam. Also, staff may suffer from motion sickness, particularly in the range of frequencies 1-8 Hz which includes some of the natural frequencies of parts of the body. One way of negating the effects of oscillation at the lowest natural frequency of a building is to install a tuned mass damper. This is a mass-spring system with a natural frequency which is the same as the frequency to be suppressed in the body to which it is attached. The Citicorp Tower in New York has a 400 tonne mass-damper installed near the top.

Buffeting results from the random turbulence usually generated by the eddies on the downward side of a structure. Its effect can be divided into a non-resonant response to a broad

[1] Analyses of this failure will be found in Bolotin (1964) §85 and Vlasov (1961) Chapter 9 §5.

band of turbulence frequencies and the resonant response at the natural frequencies of the structure. Usually, only the response at the lowest natural frequency of the structure is significant, and the influence of natural structural damping can be important. Vortex shedding is associated with oscillations induced by the regular periodic generation of vortices alternatively on either side of the slipstream behind the body. The vortex shedding frequency, ω_v , has been found experimentally to be related to the Strouhal number S. If D is a typical cross-sectional diameter of the body and V is the wind speed, the Strouhal number is given by $S = \omega_v D/V$. The Strouhal number is usually around 0.15 for rectangular shapes and around 0.2 for circular shapes (such as transmission lines). For Reynolds numbers around 10^6 and above, the Strouhal number can be considerably higher and vortex shedding may become random. The effect of vortex shedding is particularly significant when ω_v is within ±20% of the natural frequency of the body. Spiral fins are fitted to tall welded metal chimney stacks to act as spoilers to reduce vortex shedding. Rivetted structures have a significant degree of internal Coulomb (frictional) damping and are less prone to such problems.

Transmission lines may suffer from vortex shedding at their higher natural frequencies (around 100 Hz) and galloping at their fundamental frequencies (around 1 Hz). Galloping is induced by aerodynamic lift followed by stall and can induce large amplitude oscillations. Instability can be predicted from a relationship between lift and drag known as Den Hartog's criterion (see Den Hartog (1956)). Flutter can occur in laminar flow without the generation of vortices or when the flow becomes separated over part of the body (stall flutter). It is usually associated with the flexural-torsional behaviour of aircraft wings. Further details of these topics will be found in Scanlan and Rosenbaum, (1951), Den Hartog (ibid.), Flügge (1962) §63 & §80, Warburton (1976) and Smith (1988), which also contains a great deal on the practical aspects of the subject.

There are other important but specialised topics which cannot be covered within the limitations of this chapter, so that the reader will be referred to other texts. Earthquakes affect the design of buildings in Japan, the Pacific Rim, California and Mexico. Rosenblueth (1980) considers the design spectra (amplitude vs. frequency) of earthquakes used, specific structural detailing and the problems associated with masonry and concrete structures as well as steel structures. The increasing use of offshore oil platforms has given more prominence to the study of wave/structure interaction. Dean and Dalrymple (1991) examine this problem, with particular emphasis on fluid mechanics. Bachman (1995) contains a useful survey of practical guidelines.

10.2 The Flexural Vibration of Beams

The equations of motion governing the vibration of a Timoshenko beam were derived in Example 7.8. This allowed for the shear stiffness and the rotational inertia of the beam. These factors become important for relatively deep beams, but a simpler formulation can be made when they do not need to be taken into consideration. In this case, the beam behaves like an ordinary beam in flexure as governed by equations (4.15), where the distributed load q is the inertia load $\rho A\ddot{v}$ [1], as shown in Figure 10.2, where ρ is the mass density per unit volume and A is the cross-sectional area. The governing differential equation is then

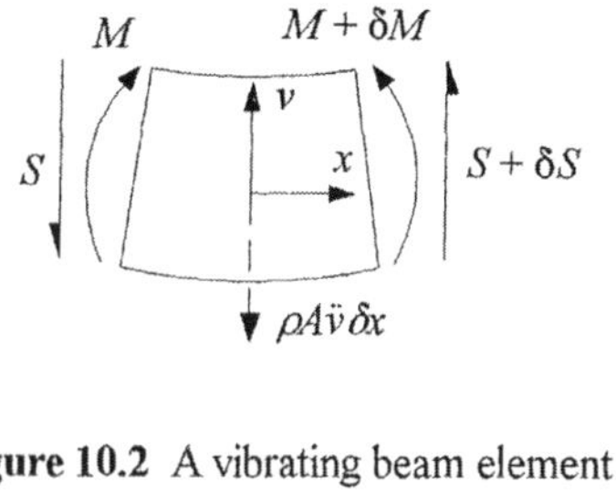

Figure 10.2 A vibrating beam element.

[1] The double dot indicates double partial differentiation with respect to time, $\partial^2/\partial t^2$.

$$\frac{\partial^2}{\partial x^2}\left(EI\frac{\partial^2 v}{\partial x^2}\right) = \frac{\partial^2 M}{\partial x^2} = -\frac{\partial S}{\partial x} = -\rho A\frac{\partial^2 v}{\partial t^2} \tag{10.1}$$

allowing for the possibility that the bending stiffness EI may vary along the beam. If the beam vibrates with an angular frequency ω, the displacement v can be expressed as

$$v = V(x)\cos(\omega t + \phi) \tag{10.2}$$

Suppose that EI is constant along the beam. Substituting the above expression in (10.1) and omitting the common sinusoidal function of time gives

$$EI\frac{d^4V}{dx^4} - \rho A\omega^2 V = 0 \tag{10.3}$$

The general solution of this equation takes the form

$$V(x) = P\cos\frac{\mu x}{l} + Q\sin\frac{\mu x}{l} + R\cosh\frac{\mu x}{l} + T\sinh\frac{\mu x}{l} \qquad \text{where} \quad \mu = \left(\frac{\rho A\omega^2}{EI}\right)^{1/4} l \tag{10.4}$$

The relationships between the end loads and end deflexions are, as before,

$$\begin{aligned} M_A &= -EIv_A^{\prime\prime}\ , \quad M_B = EIv_B^{\prime\prime} \\ S_A &= EIv_A^{\prime\prime\prime}\ , \quad S_B = -EIv_B^{\prime\prime\prime} \\ \theta_A &= v_A^{\prime}\ , \quad \theta_B = v_B^{\prime}\ . \end{aligned} \tag{10.5}$$

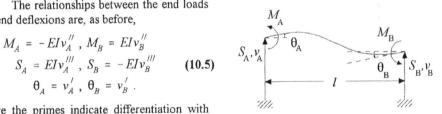

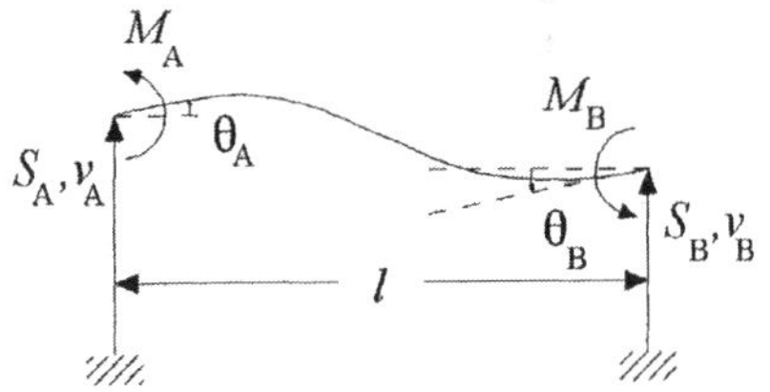

Figure 10.3 End loads and deflexions.

where the primes indicate differentiation with respect to x. Then from (10.2) and (10.4),

$$\begin{bmatrix} \theta_A \frac{l}{\mu} \\ \theta_B \frac{l}{\mu} \\ v_A \\ v_B \end{bmatrix} = \begin{bmatrix} 0 & 1 & 0 & 1 \\ -\sin\mu & \cos\mu & \sinh\mu & \cosh\mu \\ 1 & 0 & 1 & 0 \\ \cos\mu & \sin\mu & \cosh\mu & \sinh\mu \end{bmatrix} \begin{bmatrix} P \\ Q \\ R \\ T \end{bmatrix} \cos(\omega t + \phi) \tag{10.6}$$

or

$$\boldsymbol{d} = \boldsymbol{A}\,\boldsymbol{k}\,\cos(\omega t + \phi) \tag{10.6a}$$

and

$$\begin{bmatrix} M_A \\ M_B \\ S_A \frac{l}{\mu} \\ S_B \frac{l}{\mu} \end{bmatrix} = \begin{bmatrix} 1 & 0 & -1 & 0 \\ -\cos\mu & -\sin\mu & \cosh\mu & \sinh\mu \\ 0 & -1 & 0 & 1 \\ -\sin\mu & \cos\mu & -\sinh\mu & -\cosh\mu \end{bmatrix} \begin{bmatrix} P \\ Q \\ R \\ T \end{bmatrix} \frac{EI\mu^2}{l^2}\cos(\omega t + \phi) \tag{10.7}$$

or

$$\boldsymbol{p} = \boldsymbol{B}\,\boldsymbol{k}\,\frac{EI\mu^2}{l^2}\cos(\omega t + \phi) \tag{10.7a}$$

Then from (10.6a) and (10.7a),

$$\boldsymbol{p} = \frac{EI\mu^2}{l^2}\boldsymbol{B}\boldsymbol{A}^{-1}\boldsymbol{d} \tag{10.8a}$$

giving

$$\begin{bmatrix} M_A \\ M_B \\ S_A \\ S_B \end{bmatrix} = \frac{EI}{l^3} \begin{bmatrix} 4c_1 l^2 & 2c_2 l^2 & 6c_3 l & -6c_4 l \\ 2c_2 l^2 & 4c_1 l^2 & 6c_4 l & -6c_3 l \\ 6c_3 l & 6c_4 l & 12c_5 & -12c_6 \\ -6c_4 l & -6c_3 l & -12c_6 & 12c_5 \end{bmatrix} \begin{bmatrix} \theta_A \\ \theta_B \\ v_A \\ v_B \end{bmatrix} \qquad \textbf{(10.8)}$$

where

$$\begin{aligned} c_0 &= \frac{1}{\mu^4}(1 - \cos\mu \cosh\mu) \\ 4c_1 &= \frac{1}{\mu^3 c_0}(\sin\mu \cosh\mu - \cos\mu \sinh\mu) & 2c_2 &= \frac{1}{\mu^3 c_0}(\sinh\mu - \sin\mu) \\ 6c_3 &= \frac{1}{\mu^2 c_0}(\sin\mu \sinh\mu) & 6c_4 &= \frac{1}{\mu^2 c_0}(\cosh\mu - \cos\mu) \\ 12c_5 &= \frac{1}{\mu c_0}(\sin\mu \cosh\mu + \cos\mu \sinh\mu) & 12c_6 &= \frac{1}{\mu c_0}(\sin\mu + \sinh\mu) \end{aligned} \qquad \textbf{(10.9)}$$

Just as the stability functions ϕ_0 to ϕ_4 tend to unity as the axial force tend to zero, so the vibration functions c_0 to c_6 tend to unity as the angular frequency ω tends to zero. Equation (10.8) then gives the standard slope-deflexion equations. For low frequencies, they are given by

$$\begin{aligned} c_1 &\approx 1 - \frac{\lambda}{420} \,, & c_2 &\approx 1 + \frac{\lambda}{280} \,, & c_3 &\approx 1 - \frac{11\lambda}{1260} \\ c_4 &\approx 1 + \frac{13\lambda}{2520} \,, & c_5 &\approx 1 - \frac{13\lambda}{420} \,, & c_6 &\approx 1 + \frac{3\lambda}{280} \,. \end{aligned} \qquad \textbf{(10.10)}$$

where

$$\lambda = \mu^4 = \frac{\rho A \omega^2 l^4}{EI} \qquad \textbf{(10.10a)}$$

Values of these coefficients are tabulated in §A6.2 for values of λ in the range 0 to 4000.

For large values of μ (and λ), the following approximations apply.

$$\begin{aligned} 4c_1 &\approx \mu(1 - \tan\mu) \,, & 2c_2 &\approx -\mu\sec\mu \,, & 6c_3 &\approx -\mu^2\tan\mu \,, \\ 6c_4 &\approx -\mu^2\sec\mu \,, & 12c_5 &\approx -\mu^3(1 + \tan\mu) \,, & 12c_6 &\approx -\mu^3\sec\mu \,. \end{aligned} \qquad \textbf{(10.11)}$$

Example 10.1 Symmetric in-plane vibration of a portal frame

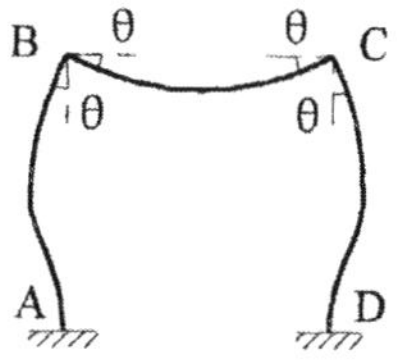

Figure 10.4 Symmetric vibration of a portal frame.

As a simple example of the application of these functions, the lowest natural frequency of the frame shown in Figure 10.4 will be sought. Finding the natural frequency of a frame with heavy members is similar to that of finding the buckling load. The equations of joint equilibrium are set up in terms of the joint deflexions. As there is zero external loading exciting the vibration, these equations are homogeneous. It is then a matter of equating the determinant of the coefficients of the joint deflexions to zero to find the criterion giving the natural frequencies of the frame. For the case shown, it will be assumed that the members are all of length l, have a bending stiffness EI and are axially rigid. The feet of the portal are fixed at A and D, so that for the mode shown, the only joint deflexions are those of the joints B and C through equal and opposite rotations θ. The bending moments at joint C are then given by (10.8) as

$$M_{CB} = \frac{EI}{l^3}(-2c_2 l^2\theta + 4c_1 l^2\theta) \,, \qquad M_{CD} = \frac{EI}{l^3}(4c_1 l^2\theta) \,. \qquad \textbf{(10.12)}$$

The sum of these two moments is equal to the external moment at C, which is zero. This then gives the natural frequency criterion

$$4c_1 - c_2 = 0 \tag{10.13}$$

This gives the value of λ corresponding to the lowest natural frequency as approximately equal to 160, or μ is approximately 3.556. Using the approximation given by (10.10) gives λ as 229, indicating that λ is well above the range of applicability of these expressions. However, using the expressions given by (10.11) gives μ as 3.566. The accuracy of this result is probably fortuitous.

10.3 The Natural Frequencies of Simple Beams

The natural frequencies of beams with simple end conditions (fixed, free or pinned) can readily be determined from (10.6) and (10.7). A fixed left-hand end is given by setting θ_A and v_A to zero in (10.6). Likewise, a fixed right-hand end is given by setting θ_B and v_B to zero. A free left-hand is given by setting M_A and S_A to zero in (10.7) and a free right-hand is given by setting M_B and S_B to zero. A pinned left-hand end is given by setting v_A to zero in (10.6) and M_A to zero in (10.7). Finally, a pinned right-hand end is given by setting v_B to zero in (10.6) and M_B to zero in (10.7). In all cases, the two pairs of end conditions give four homogeneous equations in A, B, C and D. If these terms are non-zero, the determinant of their coefficients in these equations must be zero. The natural frequencies can be found from the zeros of this determinant. The lowest frequency modes are shown in Figure 10.5. In case (d), there is an intermediate node 0.26l in from the left-hand end of the beam. In case (f), there are intermediate nodes 0.22l from the each end.

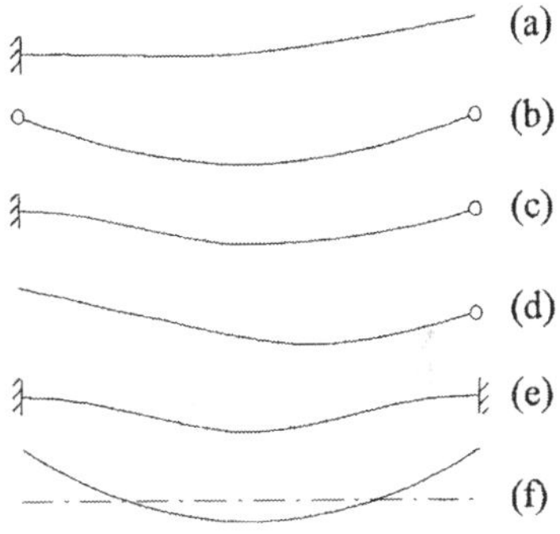

Figure 10.5 Fundamental vibration modes for simple beams.

(a) Vibration of a cantilever Using the process described above, the case shown in Figure 10.5a gives the following four equations

$$\begin{bmatrix} 0 \\ 0 \\ 0 \\ 0 \end{bmatrix} = \begin{bmatrix} 0 & 1 & 0 & 1 \\ 1 & 0 & 1 & 0 \\ -\cos\mu & -\sin\mu & \cosh\mu & \sinh\mu \\ -\sin\mu & \cos\mu & -\sinh\mu & -\cosh\mu \end{bmatrix} \begin{bmatrix} P \\ Q \\ R \\ T \end{bmatrix} \tag{10.14}$$

Setting the determinant of the matrix to zero yields the criterion for the natural frequencies

$$1 + \cos\mu \cosh\mu = 0 \tag{10.15}$$

(b) Vibration of a pin-ended beam By the same process, the natural frequencies are found from

$$\sin\mu \sinh\mu = 0 \tag{10.16}$$

which of course can be abbreviated to the condition that $\sin\mu$ is zero.

(c) Vibration of a fixed-pinned beam The natural frequencies are found from the equation

$$\sin\mu \cosh\mu - \sinh\mu \cos\mu = 0 \tag{10.17}$$

(d) Vibration of a free-pinned beam The criterion is again given by (10.17).

(e) Vibration of a fixed-ended beam In this case, the determinant of A in (10.6a) is zero. This has already been found in the process of inverting A to obtain (10.8). It yields the term c_0 in the denominators in (10.9). The natural frequencies can then be found from

$$\mu^4 c_0 = 1 - \cos\mu \cosh\mu = 0 \tag{10.18}$$

(f) Vibration of a free beam Such a case might arise in examining the oscillation of a satellite platform in space. In this case, the determinant of B in (10.7a) must be zero. However, by interchanging rows and changing the signs of rows and columns, this determinant can be turned into that of A. The criterion is again that given by (10.18). The natural frequencies can be found from the following table, μ_i and ω_i being related by (10.10a).

Table 10.1 Solutions to the Natural Frequency Equations

Cases	μ_1	μ_2	μ_3	μ_4	μ_5	μ_N
(a)	1.8751	4.6941	7.8548	10.9955	14.1372	π(N-0.5)
(b)	3.1416	6.2832	9.4248	12.5664	15.7080	Nπ
(c)&(d)	3.9266	7.0686	10.2102	13.3518	16.4934	π(N+0.25)
(e)&(f)	4.7300	7.8532	10.9956	14.1372	17.2788	π(N+0.5)

The last column gives μ_N for large values of N. In such cases, the criteria can be expressed sufficiently accurately without involving any hyperbolic functions. Note that μ_2 for case (e) is twice μ_1 for case (c). Likewise, μ_4 for case (e) is twice μ_2 for case (c). This is because the former correspond to antisymmetric modes of vibration of the fixed-fixed beam. It then behaves like two fixed-pinned beams end-to-end. The equivalents of these results and others will be found in Young (1989) Table 36. Perhaps a rather surprising conclusion that might be drawn is that although the natural frequencies increase with the stiffness of the beams, they do not necessarily increase with the degree of end fixity. However, it could be argued that the lowest natural frequencies in cases (d) and (f) are both zero. The corresponding modes are rigid-body motions of constant angular velocity in case (d) and constant linear or angular velocity in case (f).

10.4 The Axial Vibration of Beams

The purely axial vibration of a bar AB shown in Figure 10.6a can be analysed by examining a short length δx of it shown in Figure 10.6b. The tensile force P_x acting on it is equal to its area A times the axial stress. This stress is equal to Young's modulus times the local axial strain, so that

$$P_x = EA\frac{\partial u}{\partial x} \tag{10.19}$$

where u is the axial displacement of the centroid. Because of the inertia force resisting acceleration, this axial force varies along the bar, so that for the dynamic equilibrium of the element,

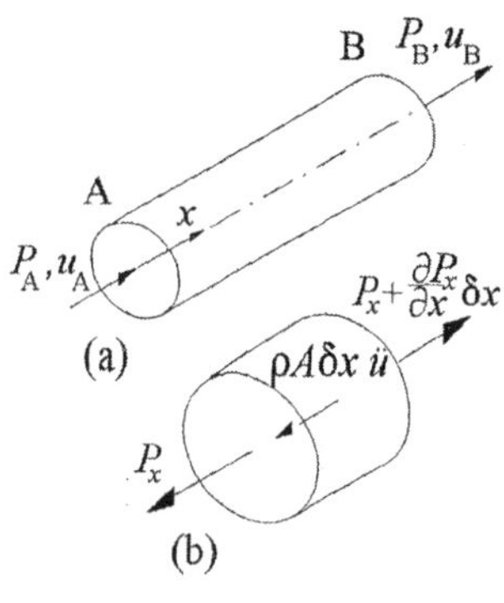

Figure 10.6 Axial vibration of a bar.

$$\frac{\partial P_x}{\partial x} = \rho A \frac{\partial^2 u}{\partial t^2} \qquad \textbf{(10.20)}$$

where ρ is the density of the material. Suppose that the bar vibrates axially, so that

$$u = U(x)\cos(\omega t + \phi) \qquad \textbf{(10.21)}$$

Then from (10.19) to (10.21) it follows that

$$EA\frac{d^2U}{dx^2} + \rho A\omega^2 U = 0 \qquad \textbf{(10.22)}$$

which has the general solution

$$U(x) = C\cos\mu_P\frac{x}{l} + D\sin\mu_P\frac{x}{l} \qquad \text{where} \quad \mu_P^2 = \frac{\rho\omega^2 l^2}{E} \quad (= \lambda \text{ in tables}) \qquad \textbf{(10.23)}$$

Setting P as $-P_A$ and u as u_A at $x=0$ and P as P_B and u as u_B at $x=l$ gives the equations

$$\begin{bmatrix} P_A \\ P_B \end{bmatrix} = \frac{EA}{l}\begin{bmatrix} d_1 & -d_2 \\ -d_2 & d_1 \end{bmatrix}\begin{bmatrix} u_A \\ u_B \end{bmatrix} \qquad \textbf{(10.24)}$$

where

$$d_1 = \mu_P\cot\mu_P \approx 1 - \lambda/3\,, \qquad d_2 = \mu_P\operatorname{cosec}\mu_P \approx 1 + \lambda/6\,. \qquad \textbf{(10.25)}$$

and the approximations apply for small λ. These coefficients are tabulated in terms of λ (cf. (10.23)) in §A6.2, where the appropriate expansions in terms of λ are also given.

Example 10.2 Forced axial vibration of a column

Figure 10.7 shows a column of length l which is free at its top end, A and fixed at its lower end, B. a sinusoidal axial exciting force of angular frequency ω and amplitude F_0 is applied at a as shown. The steady-state response is found from the first line (10.24) by setting u_B to zero, giving

$$u_A = \frac{F_0 l}{EA\,d_1}\sin\omega t \qquad \textbf{(10.26)}$$

Resonance occurs when d_1 is zero, that is, when $\omega_N = \frac{\pi}{l}(N + 0.5)\sqrt{\frac{E}{\rho}}$ where N is an integer.

Comparing this with the flexural vibration of a cantilever, the lowest natural frequencies are the same if I/Al^2 is about 0.2. This is too large for realistic beam problems, so that the lowest natural frequency will normally be associated with flexural, rather than axial, vibration.

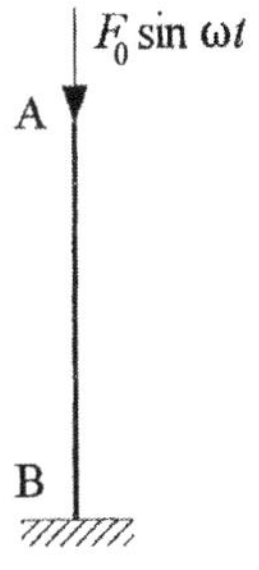

Figure 10.7 Vibration of a column.

10.5 The Torsional Vibration of Beams

The torsional vibration of bars can be analysed in a similar way to that in the previous section. Again, the dynamic equilibrium of a short length δx of the bar, shown in Figure 10.8b, is examined. The torque T_x acting on a section varies with the distance x along the beam. Assuming simple Saint-Venant torsion, this torque is related to the rate of twist by

$$T_x = GJ\frac{\partial\theta}{\partial x} \qquad \textbf{(10.27)}$$

The rotational inertia torque on the element δx is balanced by the increment in this torque, giving

$$\frac{\partial T_x}{\partial x} = \rho I_p \frac{\partial^2\theta}{\partial t^2} \qquad \textbf{(10.28)}$$

where I_p is the second moment of area of the section about its shear centre. Suppose that the bar vibrates torsionally, so that

$$\theta = \Theta(x)\cos(\omega t + \phi) \qquad \textbf{(10.29)}$$

Then from (10.27) to (10.29) it follows that

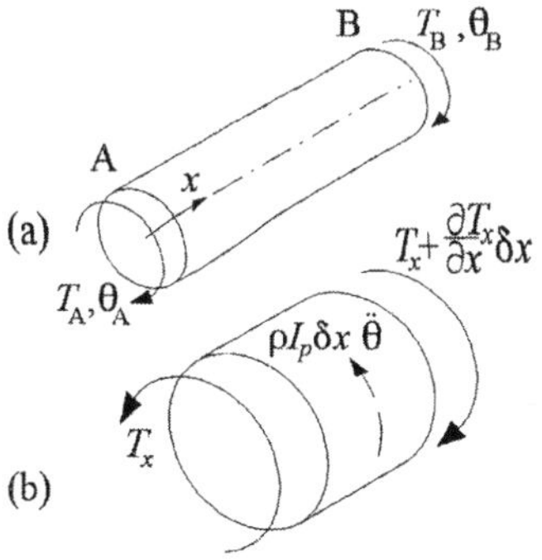

Figure 10.8 Torsional vibration of a bar.

$$GJ\frac{d^2\Theta}{dx^2} + \rho I_p \omega^2\Theta = 0 \qquad \textbf{(10.30)}$$

which has the general solution

$$\Theta(x) = C\cos\mu_T\frac{x}{l} + D\sin\mu_T\frac{x}{l} \qquad \text{where} \quad \mu_T^2 = \frac{\rho I_p \omega^2 l^2}{GJ} \quad (= \lambda \text{ in tables}) \qquad \textbf{(10.31)}$$

Setting T as $-T_A$ and θ as θ_A at $x=0$ and T as T_B and θ as θ_B at $x=l$ gives the equations

$$\begin{bmatrix} T_A \\ T_B \end{bmatrix} = \frac{GJ}{l}\begin{bmatrix} d_1 & -d_2 \\ -d_2 & d_1 \end{bmatrix}\begin{bmatrix} \theta_A \\ \theta_B \end{bmatrix} \qquad \textbf{(10.32)}$$

where

$$d_1 = \mu_T \cot\mu_T \quad , \quad d_2 = \mu_T \operatorname{cosec}\mu_T \qquad \textbf{(10.33)}$$

These functions are the same as those for axial vibration given in (10.25). They are tabulated in terms of λ, this time as defined by (10.31), in §A6.2.

Example 10.3 Natural vibration of rotor arm

The bar AB of length l shown in Figure 10.9 is rigidly held at B and rigidly attached to a rotor of inertia $k\rho I_p l$ at a (where k is a constant). If the rotation of the bar is given by (10.29), the angular acceleration of A is $-\omega^2$ times the rotation of A, so that from the first line of (10.32),

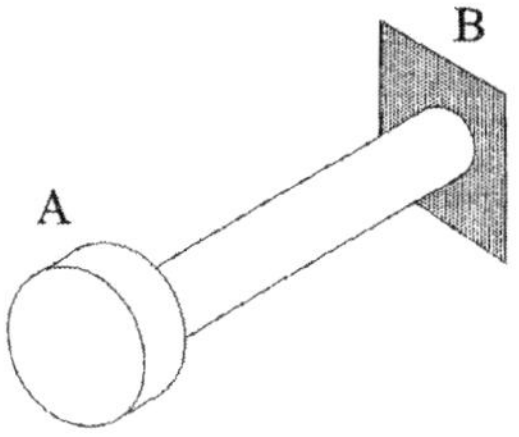

$$k\rho I_p l\omega^2\theta_A = T_A = \frac{GJ}{l}d_1\theta_A \qquad \textbf{(10.34)}$$

Figure 10.9 A rotor arm.

so that from (10.31) and (10.33),

$$\mu_T \tan\mu_T = \frac{1}{k} \qquad \textbf{(10.35)}$$

Clearly, the larger the inertia of the rotor, and hence the larger the value of k, the lower the lowest natural frequency of the system will become. As the inertia of the rotor falls to zero, the natural frequencies are given by

$$\omega_N = \frac{\pi}{l}(N + 0.5)\sqrt{\frac{GJ}{\rho I_p}}$$

which may be compared with the result found in Example 10.2. The lowest natural frequencies of axial vibration were found to be higher than those for flexural vibration, except for unrealistically short beams. However, making the same comparison between the flexural vibration and torsional vibration of beams, the same is not true, although the longer a beam becomes the more probable it is that the lowest frequency of vibration is in flexure.

10.6 Flexural Vibration with Large Axial Loading

Figure 10.10 shows the loading on a short length δx of a beam, allowing for large axial loading (cf. Figure 10.2). The equations of dynamic equilibrium are then

$$M = EI\frac{\partial^2 v}{\partial x^2}$$
$$\delta M + P\delta v + S\delta x = 0 \qquad \textbf{(10.36)}$$
$$\delta S = \rho A\frac{\partial^2 v}{\partial t^2}\delta x$$

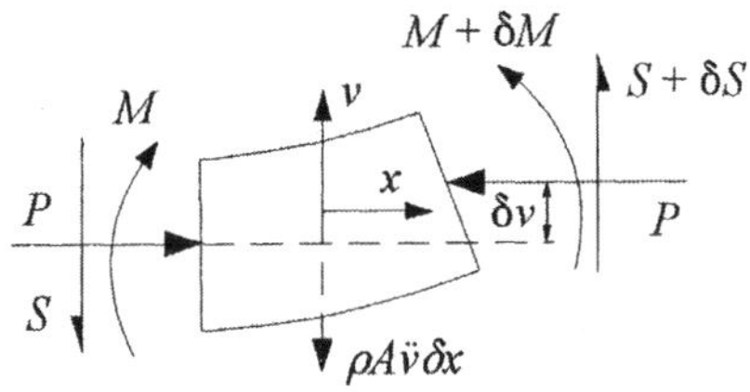

Figure 10.10 An axially-loaded vibrating beam element.

From these, the governing differential equation is given by

$$\frac{\partial^2}{\partial x^2}\left(EI\frac{\partial^2 v}{\partial x^2}\right) + P\frac{\partial^2 v}{\partial x^2} + \rho A\frac{\partial^2 v}{\partial t^2} = 0 \qquad \textbf{(10.37)}$$

Taking the form of v given by (10.2), the general solution for constant EI is given by

$$V(x) = B\cos\alpha x + C\sin\alpha x + D\cosh\beta x + F\sinh\beta x \qquad \textbf{(10.38)}$$

where

$$\alpha = \left\{\left[\left(\frac{P}{2EI}\right)^2 + \left(\frac{\rho A\omega^2}{EI}\right)\right]^{\frac{1}{2}} + \frac{P}{2EI}\right\}^{\frac{1}{2}}, \quad \beta = \left\{\left[\left(\frac{P}{2EI}\right)^2 + \left(\frac{\rho A\omega^2}{EI}\right)\right]^{\frac{1}{2}} - \frac{P}{2EI}\right\}^{\frac{1}{2}} \qquad \textbf{(10.39)}$$

A particularly simple solution can be found for the natural frequency of a simply-supported beam. This is given by taking B,D and F as zero and α as $N\pi/l$. This gives the condition

$$\frac{P}{P_{cN}} + \left(\frac{\omega}{\omega_{cN}}\right)^2 = 1 \qquad \textbf{(10.40)}$$

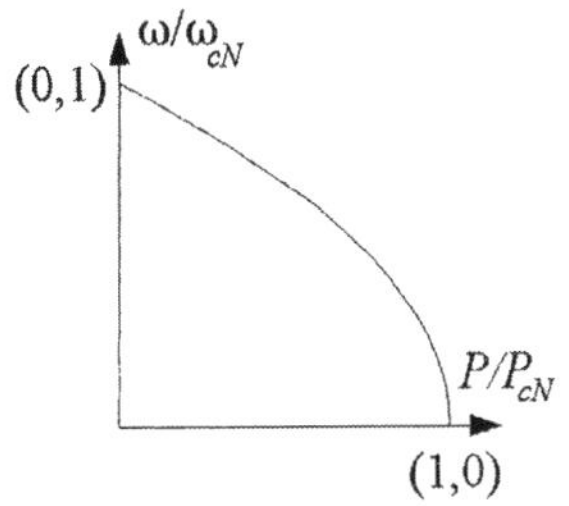

Figure 10.11 Critical surface for combined axial loading and vibration.

where P_{cN} is the Nth critical axial load in static conditions and ω_{cN} is the Nth natural frequency in the absence of a large axial force (cf. §9.2 and Table 10.1).

Figure 10.11 shows a plot of (10.40), illustrating how the natural frequency varies with axial loading, becoming zero when this loading is the critical loading. This is another example of the convex critical surfaces discussed in §9.9. If the vertical ordinate had been $(\omega/\omega_{cN})^2$ then this curve would have been a straight line.

Massonnet[1] shows that this is always the case where the modes of vibration and buckling are the same. He goes on to show that where this is not the case, this 'characteristic surface' is convex, so that equations of the kind given by (10.40) represent lower approximations to the true surface. This is shown to be true in general for the vibration of plates and shells and that similar relationships can be deduced for the effect of axial loading on the whirling speed of shafts.

{The general equations for the flexural-torsional vibration of thin-walled sections can be deduced by applying Hamilton's principle, as in Example 7.8. The kinetic energy terms per unit length are given by

$$T_l = \frac{1}{2}[\rho A(\dot{u}_0^2 + \dot{v}_0^2) + \rho I_p\dot{\theta}^2] \tag{10.41}$$

The Hamiltonian is then given by

$$L = \int_0^\tau\int_0^l\{\frac{1}{2}[\rho A(\dot{u}_0^2 + \dot{v}_0^2) + \rho I_p\dot{\theta}^2 - EI_{yy}u_0''^2 - EI_{xx}v_0''^2 - GJ\theta'^2 - EI_\omega\theta''^2 + P(u_0'^2 + v_0'^2 + 2y_0u_0'\theta' - 2x_0v_0'\theta' + r_0^2\theta'^2)]\}dz\,dt \tag{10.42}$$

where the expressions for the potential energy can be deduced from (9.24). Note that the work done by the shear forces in the x and y directions, S_x and S_y, and by the torque T have been not been included. This is because L must be minimised with respect to u_0, v_0 and θ, as the dynamic terms are not expressed in terms of the derivatives of these functions with respect to z. Minimising L in this way gives the three equations

$$\begin{aligned}\rho A\ddot{u}_0 + EI_{yy}u_0'''' + Pu_0'' + Py_0\theta'' &= 0\\ \rho A\ddot{v}_0 + EI_{xx}v_0'''' + Pv_0'' - Px_0\theta'' &= 0\\ \rho I_p\ddot{\theta} - GJ\theta'' + EI_\omega\theta'''' + Py_0u_0'' - Px_0v_0'' + Pr_0^2\theta'' &= 0\end{aligned} \tag{10.43}$$

These equations may be solved by taking u_0, v_0 and θ as in-phase sinusoidal functions of time and exponential functions of z. These exponential functions may be determined by methods similar to those given in §9.5. The simplest solution is given by taking these functions of z to be in-phase sinusoidal functions too.}

Further work on dynamic buckling will be found in Flügge (1962) Chapter 62 and Thompson and Hunt (1984).

{10.7 Response to Arbitrary Time-Dependent Loading}

{Figure 10.12a shows a mass m, attached to a spring of stiffness k, subject to a time-dependent force $P(t)$. The governing differential equation for the displacement of the mass, u, is

$$m\ddot{u} + ku = P(t) \qquad \text{or} \qquad \ddot{u} + \omega^2 u = Q(t) \tag{10.44}$$

where ω ($= \sqrt{k/m}$) is the natural frequency of the system and $Q(t)$ is $P(t)/m$. A force of constant intensity P_0 is applied at time $t = \tau$, so that $P(t)$ is given by the step function shown in Figure 10.12b. The initial conditions (at time $t-\tau=0$) are that the mass has zero displacement and velocity. The solution is then

[1] Massonnet, C. (1940) *Bulletin des Cours et des Laboratoires d'Essais des Constructions du Génie Civil et d'Hydraulique Fluviale* V.1 Nos. 1 & 2. Goemaere, Brussels.

$$u = \frac{P_0}{k}[1 - \cos\omega(t-\tau)] \qquad \textbf{(10.45)}$$

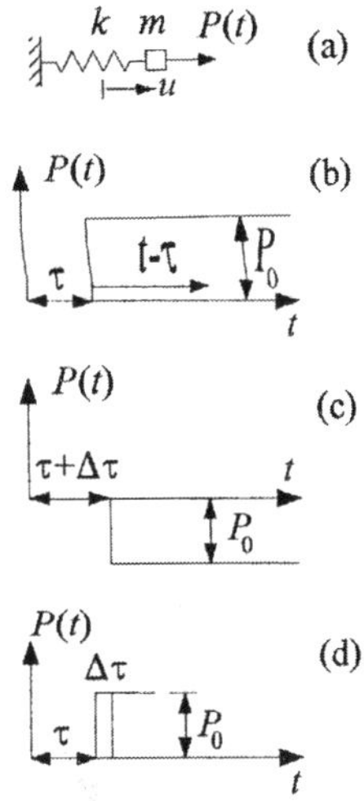

($t > \tau$). Similarly, if the loading had been of magnitude $-P_0$ and applied at a time $\tau+\Delta\tau$ as in Figure 10.12c, the displacement would be

$$u = -\frac{P_0}{k}[1 - \cos\omega(t-\tau-\Delta\tau)] \qquad \textbf{(10.46)}$$

The case shown in Figure 10.12d is that of a force of magnitude P_0 applied for a short time interval $\Delta\tau$. Its effect on the later response is given by the sum of the two previous cases, that is,

$$\begin{aligned} u &= \frac{P_0}{k}[1 - \cos\omega(t-\tau)] - \frac{P_0}{k}[1 - \cos\omega(t-\tau-\Delta\tau)] \\ &\approx \frac{P_0}{k}[\omega\Delta t \sin\omega(t-\tau)] \end{aligned} \qquad \textbf{(10.47)}$$

Figure 10.12 Response to a stepped excitation.

The response to a series of such short pulses can be found by summing the responses to each of them as deduced from (10.47). At the limit, the response to a variable excitation $P(t)$ is given by integrating the responses to the elementary strips of the kind shown in Figure 10.13, giving

$$u(t) = \frac{\omega}{k}\int_0^t P(\tau)\sin\omega(t-\tau)\,d\tau \qquad \textbf{(10.48)}$$

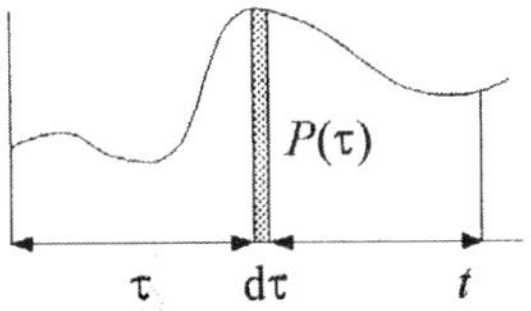

This is known as the *Duhamel* or *convolution integral.* In terms of the second form of equation (10.44), this becomes

$$u(t) = \frac{1}{\omega}\int_0^t Q(\tau)\sin\omega(t-\tau)\,d\tau \qquad \textbf{(10.49)}$$

Figure 10.13 Response to a variable excitation.

Example 10.4 Resonant harmonic excitation

Suppose that $P(t)$ is a harmonic force with a frequency equal to the natural frequency of the system, given by $P_0 \sin\omega t$, and applied for a time t_1. From (10.48), the response is

$$\begin{aligned} u(t) &= P_0\frac{\omega}{k}\int_0^{t_1}\sin\omega\tau\,\sin\omega(t-\tau)\,d\tau = P_0\frac{\omega}{2k}\int_0^{t_1}\cos\omega(t-2\tau) - \cos\omega t\,d\tau \\ &= \frac{P_0}{2k}[\sin\omega t_1\cos\omega(t-t_1) - \omega t_1\cos\omega t] \quad (t \ge t_1) \end{aligned} \qquad \textbf{(10.50)}$$

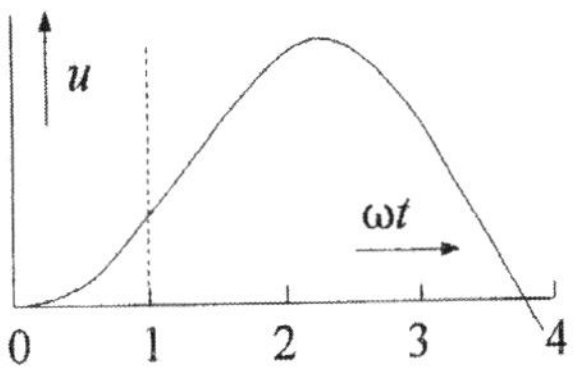

This response is plotted in Figure 10.14, taking ωt_1 as unity. Note that the maximum deflexion occurs a considerable time after the force has ceased to be applied.

{Although the Duhamel integral has been derived for a single degree of freedom system, it can be applied to the response of beams to random loading. Suppose that a non-uniform beam with workless end reactions has a distributed

Figure 10.14 Response to a short-term sinusoidal force.

load $p(x)$ which varies in amplitude with time. The differential equation for the deflexion v is then given by

$$\frac{\partial^2}{\partial x^2}\left(EI\frac{\partial^2 v}{\partial x^2}\right) + \rho A\frac{\partial^2 v}{\partial t^2} = p(x)f(t) \tag{10.51}$$

The solution will be expressed in the form

$$v = \sum_{i=1}^{\infty} V_i(x)\, q_i(t) \tag{10.52}$$

where $V_i(x)$ is the ith normal vibration mode associated with the beam and its end conditions (cf. §A9.1). Substituting this expression in (10.51), multiplying both sides by $V_j(x)$ and integrating over the length of the beam,

$$\int_0^l V_j \sum_{i=1}^{\infty} \frac{d^2}{dx^2}\left(EI\frac{d^2 V_i}{dx^2}\right) q_i\, dx + \int_0^l \rho A V_j \sum_{i=1}^{\infty} V_i \ddot{q}_i\, dx = \int_0^l V_j p(x) f(t)\, dx \tag{10.53}$$

From (A9.1.12) and (A9.1.15) the orthogonality conditions

$$\int_0^l V_j \frac{d^2}{dx^2}\left(EI\frac{d^2 V_i}{dx^2}\right) dx = 0 = \int_0^l \rho A V_j V_i\, dx \qquad (i \neq j) \tag{10.54}$$

can be deduced. Also, from (A9.1.11), for the particular normal mode $V_j(x)$,

$$\int_0^l V_j \frac{d^2}{dx^2}\left(EI\frac{d^2 V_j}{dx^2}\right) dx = \omega_j^2 \int_0^l \rho A V_j^2\, dx \tag{10.55}$$

where ω_j is the natural frequency associated with this mode. Using (10.54) and (10.55), (10.53) becomes

$$\ddot{q}_j + \omega_j^2 q_j = \frac{\int_0^l p(x) V_j\, dx}{\int_0^l \rho A V_j^2\, dx} f(t) = K_j f(t) \quad \text{(say)} \tag{10.56}$$

where K_j can be evaluated from the integrals, if the normal mode $V_j(x)$ is known. This corresponds to the second form of (10.44). Taking the beam to be at rest in its undeflected state at t=0, and applying (10.49),

$$q_j = \frac{K_j}{\omega_j}\int_0^t f(\tau) \sin \omega_j (t-\tau)\, d\tau \tag{10.57}$$

Evaluating $q_j(t)$ in this way for a sufficient number of normal modes $V_j(x)$, the solution can be found from (10.52). In the particular case where the loading is a point force of amplitude P applied a distance a from the left-hand end of the beam, (10.57) becomes

$$q_j = \frac{P V_j(a)}{\omega_j \int_0^l \rho A V_j^2(x)\, dx}\int_0^t f(\tau) \sin \omega_j (t-\tau)\, d\tau \tag{10.58}$$

However, in the case of a *uniform* beam, this case is best dealt with by treating it as two beams joined end-to end at the point of application of P.

The above method applies to time-dependent loading applied along the beam. a similar process can be applied to solve problems related to time-dependent end conditions. a function $g(x)$ is sought which satisfies the equation

$$\frac{d^2}{dx^2}\left(EI\frac{d^2g}{dx^2}\right) = 0 \tag{10.59}$$

For a uniform beam, this has the solution

$$g(x) = Ax^3 + Bx^2 + Cx + D \tag{10.60}$$

and in general there will be four constants of integration. The form of solution sought in this problem is

$$v(x,t) = f(t)g(x) + v_0(x,t) \tag{10.61}$$

where $v_0(x,t)$ satisfies a suitable set of workless end conditions and $f(t)g(x)$ satisfies the difference between these end conditions and the time-dependent end conditions. Substituting this form into the governing differential equation for the flexural vibration of a non-uniform beam gives

$$\frac{\partial^2}{\partial x^2}\left(EI\frac{\partial^2 v_0}{\partial x^2}\right) + \rho A\frac{\partial^2 v_0}{\partial t^2} = -\rho A g(x)\ddot{f}(t) \tag{10.62}$$

This is the same in form as (10.51) and so can be solved in the same way, where the normal modes $V_i(x)$ satisfy the same set of workless end conditions as v_0.

The response of a beam to a moving load can be dealt with approximately in two ways. If the mass of the moving load is large compared to the mass of the beam, it can be treated as a single degree of freedom problem with the beam behaving like a spring of varying stiffness. Timoshenko, Young and Weaver (1974) give this approximate solution. If the mass of the moving load is small in comparison to that of the beam, then it can be treated as a problem of the response of a beam to a moving force. This is covered by the solution given by (10.52) and (10.58). Warburton (1964) shows the effect of a force moving at a constant velocity U across a simply-supported uniform beam of length l. This varies with the non-dimensional parameter C which is equal to $\pi U/\omega_1 l$, where ω_1 is the lowest natural frequency of the beam. Up to a value of C equal to 0.2, the maximum central deflexion does not exceed 1.15 times the static central deflexion $Pl^3/48EI$. The worst case for $C \le 1.0$ is when C=0.617, when the maximum central deflexion is 1.743 times the static deflexion. Further amplification of these methods will be found in Warburton (ibid.) §3.4-§3.7 and §4.1-§4.2.}

The approximate analysis of the response to impact is based on equating the loss of energy of the impacting body to the gain in strain energy of the elastic body being struck. Thus suppose that a mass M is dropped from a height h above a rod which then shortens by an amount δ before the mass comes to rest. The potential energy lost by the mass is $Mg(h+\delta)$. Taking the rod to be uniform and equating this to the gain in its strain energy,

$$Mg(h + \delta) = \frac{EA\delta^2}{2l} \tag{10.63}$$

where E,A and l have the same meanings as earlier in this section (cf. (7.5)). This can be solved as a quadratic in δ. If h is zero, then the deflexion δ found is twice the static deflexion produced by gradually placing the mass on the end of the rod. In fact δ is the maximum resulting deflexion and the end of the rod oscillates about a mean position given by the static deflexion. If h is large in comparison with δ, then the left-hand side of (10.63) can be taken as Mgh giving δ more directly. The same principles may be used when the mass has a velocity v on striking the bar, in which case the kinetic energy $\frac{1}{2}Mv^2$ is used instead of the potential energy Mgh. Essentially the same method may be used to find the deflexion produced by a transverse impact on a beam. In this case, the bending strain energy induced by a displacement δ produced by a force at the point of impact is equated to the loss of energy of the impacting body. For example, the strain energy induced in a fixed-ended beam by a central perturbation is readily found to be $96EI\delta^2/l^3$. Such analyses take no account of the loss of energy on impact and the shock wave produced by the impact[1]. The travelling waves generated by impact are examined in the next section.}

10.8 Travelling Waves in Rods and Beams

The differential equations for travelling waves have already been formulated in the earlier part of this chapter. Here, a different form of solution will be sought. Taking the x axis to lie along the rod, deflexions of the form $f(x-ct)$ and $g(x+ct)$ will be examined where c is the speed at which the wave travels along the bar. This can be seen by incrementing x by a small amount δx and t by a small amount δt. Provided that δx is $c\delta t$, $f(x+\delta x-ct-c\delta t)$ is the same as $f(x-ct)$, so that the wave does not changed form but shifts in the positive x direction at a velocity c. Likewise, it can be seen that $g(x+ct)$ represents a wave form moving at a velocity c in the $-x$ direction. This velocity may or may not depend on the form of the wave. If c depends on the characteristics of the wave, the medium is said to be *dispersive.* This is because packets of waves of different wavelengths do not travel together through the medium, but become dispersed. As will be seen, axial and torsional waves travelling along a prismatic bar are not dispersed, c being a function of the properties of the bar only, but this is not true of flexural waves.

Two waves of similar frequency travelling in the same direction can produce a phenomenon called beats. This is shown in Figure 10.15. Suppose that waves (a) and (b) are given by

$$f_a(x,t) = \sin k(x - ct)\ ,$$
$$f_b(x,t) = \sin k'(x - c't). \tag{10.64}$$

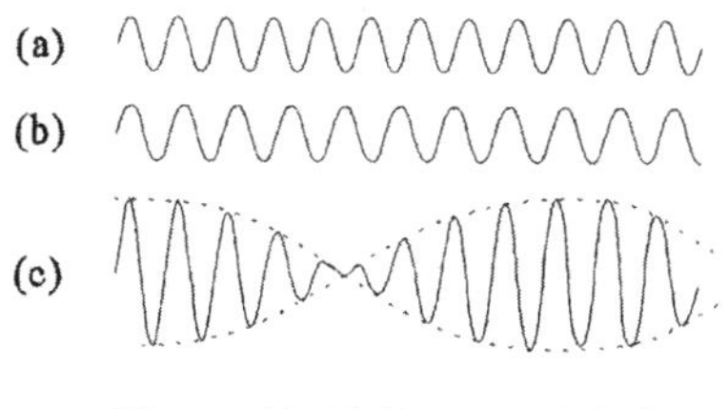

Figure 10.15 Beat modulation.

where k and k' differ by a small amount δk, and kc and $k'c'$ differ by a small amount δkc. (That is to say, the medium may be dispersive.) The combination of the two waves, shown in Figure 10.15c, is given by

$$\begin{aligned} f_a(x,t) + f_b(x,t) &= \sin k(x - ct) + \sin k'(x - c't) \\ &= 2\sin\tfrac{1}{2}[(k + k')x - (kc + k'c')t]\cos\tfrac{1}{2}[(k - k')x - (kc - k'c')t] \\ &\approx 2\sin k(x - ct)\cos\tfrac{1}{2}[\delta kx - \delta(kc)t] \end{aligned} \tag{10.65}$$

The first term is the original wave form, whose amplitude is now modified by the second wave form (shown as the dotted envelope in Figure 10.15c). This amplitude modulation produces beats which have their own velocity given by

$$c_g = \frac{\delta(kc)}{\delta k} \tag{10.66}$$

---

[1] More accurate analyses will be found in Timoshenko, Young and Weaver (1974). Longitudinal impact is discussed on pp. 373-387 and lateral impact on pp. 435-441.

The value of c_g at the limit, as δk tends to zero, is known as the *group velocity* of the waves. The rate at which the total energy (kinetic energy plus strain energy) is transported along a beam has been shown by Biot[1] to be generally proportional to the group velocity for linear, conservative media.

For bending waves in a beam, (10.1) applies. Taking EI as constant and v as given by $f(x\text{-}ct)$, this equation becomes

$$0 = EI\frac{\partial^4 f}{\partial x^4} + \rho A\frac{\partial^2 f}{\partial t^2} = EIf^{////} + \rho A c^2 f^{//} \tag{10.67}$$

where the primes indicate the differentiation of f with respect to $x\text{-}ct$, treated as a single parameter. The general solution of this is

$$v(x,t) = f(x - ct) = P\cos k(x - ct) + Q\sin k(x - ct) + R(x - ct) + S \tag{10.68}$$

where

$$k = c\sqrt{\frac{\rho A}{EI}} \tag{10.69}$$

The group velocity associated with sinusoidal waves then follows from (10.66). As k is a constant multiple of c, the group velocity is $2c$ in this case. The wavelength λ is given by $2\pi/k$. Thus for very short wavelengths, the velocity of propagation tends to become infinite. This apparent anomaly does not occur when the method is applied to Timoshenko beams (see Flügge (1962) §64.6). Exactly the same results follow when $v(x,t)$ is given by $g(x+ct)$, with the argument $(x\text{-}ct)$ being replaced by $(x+ct)$ in (10.68).

For travelling waves produced by axial loading, the governing equations are given by (10.19) and (10.20). These give the governing equation for u as

$$EA\frac{\partial^2 u}{\partial x^2} = \rho A\frac{\partial^2 u}{\partial t^2} \tag{10.70}$$

so that if $u(x,t)$ is given either by $f(x\text{-}ct)$ or $g(x+ct)$, and the wave speed is

$$c = \sqrt{\frac{E}{\rho}} \tag{10.71}$$

Note that in this case, the wave speed is independent of the wave form, which is transmitted without distortion, as there is no dispersion. Similarly, equations (10.27) and (10.28) for torsional vibration give the governing equation for θ as

$$GJ\frac{\partial^2\theta}{\partial x^2} = \rho I_p\frac{\partial^2\theta}{\partial t^2} \tag{10.72}$$

and $u(x,t)$ is given either by $f(x\text{-}ct)$ or $g(x+ct)$, and the wave speed is

$$c = \sqrt{\frac{GJ}{\rho I_p}} \tag{10.73}$$

When a travelling wave reaches the end of a beam or bar, an additional wave is usually generated at that end. This is such that the incident wave and the generated wave combine to give the correct end conditions. The generated wave travels in the opposite direction along the rod,

---

[1] Biot, M.A. (1957) General theorems of equivalence of group velocity and energy transport. *Phys. Rev.* V.105 p.1129.

and in the case of fixed-end conditions, may simply be a reflection of the incident wave. Such a reflection would have equal and opposite material displacements (and velocities) to the incident wave, so that the combination of the two gives zero displacement and velocity at the fixed end. When a compressive wave reaches a free end, it is matched by a tensile wave travelling in the opposite direction, so that the axial load at the end remains zero.

Example 10.5 Axial impact on a rod

Suppose that a mass m travelling at a velocity V strikes the free end of a cantilever of length l as shown in Figure 10.16a. It will be assumed that initially, the mass and the end of the rod remain in contact. The contact force $P_0(t)$ between the end of the rod and the mass is equal to the inertia force exerted by the mass and the elastic force produced by the local elastic strain in the rod. Thus at the free end ($x = 0$),

$$P_0 = -EA\frac{\partial u}{\partial x} = -m\frac{\partial^2 u}{\partial t^2} \qquad \textbf{(10.74)}$$

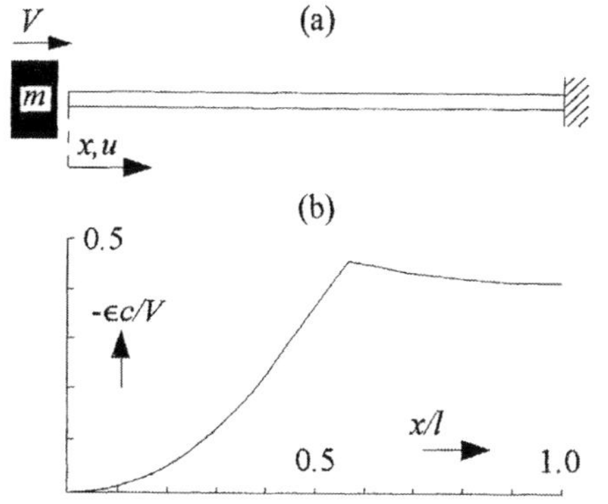

Figure 10.16 Axial impact on a cantilever.

Taking u to be expressed by $f(x\text{-}ct)$, and the differential $f'(x\text{-}ct)$ with respect to $(x\text{-}ct)$ to be $F(x\text{-}ct)$, this gives

$$P_0 = -EAf'(x-ct) = -mc^2f''(x-ct) \quad \left(c^2 = \frac{E}{\rho}\right) \quad \text{so that} \quad EAF = \frac{mE}{\rho}F' \qquad \textbf{(10.75)}$$

Taking the free end of the rod also to move with the velocity V immediately after impact, this means that $-cF(0)$ is V and the solution of (10.75) is

$$F(x-ct) = -\frac{V}{c}e^{\frac{A\rho}{m}(x-ct)} \qquad \textbf{(10.76)}$$

This wave reaches the fixed end at time t equal to l/c. a wave $g(x+ct)$ is then generated at the fixed end to nullify the end displacement. Likewise, its differential $G(x+ct)$ nullifies the fixed-end velocity. This resultant end velocity is given by

$$0 = -cF(l-ct) + cG(l+ct) \qquad \textbf{(10.77)}$$

so that

$$G(l+ct) = F(l-ct) = -\frac{V}{c}e^{\frac{A\rho}{m}(l-ct)} \qquad \textbf{(10.78)}$$

In general, G is a function of the parameter $(x+ct)$ and the above equation is for the specific case when x equals l. The general expression is then

$$G(x+ct) = -\frac{V}{c}e^{\frac{A\rho}{m}[2l-(x+ct)]} \qquad \textbf{(10.79)}$$

This wave moves back along the bar, reaching the free end at time t equal to $2l/c$. The condition given by (10.74) still has to be met, and this can only be done if a new wave $h(x\text{-}ct)$ is generated at the free end. Taking its differential to be $H(x\text{-}ct)$, the additional terms in (10.74) must satisfy the equation

$$EA[G(0+ct) + H(0-ct)] = mc^2[G'(0+ct) + H'(0-ct)] \tag{10.80}$$

(The function $F(x\text{-}ct)$ has already been chosen to satisfy (10.74)). Using the expression for $G(x+ct)$ given by (10.79), this leads to the differential equation

$$EAH(0-ct) - mc^2H'(0-ct) = \frac{V}{c}e^{\frac{A\rho}{m}(2l-ct)}\left(EA + mc^2\frac{A\rho}{m}\right) \tag{10.81}$$

As at the original contact between the mass and the end of the bar, we take there to be no impulsive force between the two when this new wave is generated. This means that there is no jump in the mass velocity (and end-of-bar velocity) at time $t=2l/c$ when this new wave begins to propagate. The solution of (10.81) is then

$$H(x-ct) = -\frac{2AV\rho}{mc}[2l + (x-ct)]e^{\frac{A\rho}{m}[2l+(x-ct)]} \tag{10.82}$$

As long as the end of the bar remains in compression, the mass will remain in contact with it. Unless the mass has somehow become glued to the bar, contact is lost as soon as the strain ϵ at the end of the bar becomes tensile. For time $t>2l/c$, this strain is given by

$$\epsilon = F(0-ct) + G(0+ct) + H(0-ct) = -\frac{V}{c}e^{-\frac{A\rho ct}{m}}\left[1 + e^{\frac{2A\rho l}{m}}\left(1 + \frac{2A\rho}{m}(2l - ct)\right)\right] \tag{10.83}$$

(where ϵ is tensile positive). Denoting the ratio of the mass of the bar to the end mass, $A\rho l/m$, by μ and the ratio of the actual time to the time taken for a wave to traverse the bar, ct/l, by τ, the condition for loss of contact (ϵ=0) can then be written as

$$\tau = 2 + \frac{1}{2\mu}\left(1 + e^{-2\mu}\right) \tag{10.84}$$

This result is only valid for the period when the strain ϵ is given by the combination of the three components in (10.83). When the third wave reaches the fixed end, it too is reflected and so a fourth wave reaches the free end at time t equal to $4l/c$. Thus (10.84) gives loss of contact in the time range $2<\tau<4$. At the lower limit, this corresponds to an infinite mass ratio, and at the upper limit, μ is 0.3694. The time of loss of contact when the two masses are the same is $2.567l/c$. The state of strain at this instant is shown in Figure 10.16b. For lower mass ratios, loss of contact involving further reflected waves must be examined.

10.9 The Whirling of Shafts

If a flexible shaft rotates at sufficient speed, it can deflect in such a way that the centrifugal forces induced balance the elastic constraints to its deformation. A beam whirling at an angular velocity ω under its own inertia loading can be analysed as if it had a distributed load of $-\rho A v\omega^2$ per unit length acting on it (if gyroscopic effects are ignored), where ρ is its density, A its cross-sectional area and v is the displacement of its centroid from the axis of rotation. Then from the last of equations (4.15),

$$\rho A\omega^2 v = EI\frac{d^4v}{dx^4} \tag{10.85}$$

which has the general solution

$$v = P\cos\frac{\mu x}{l} + Q\sin\frac{\mu x}{l} + R\cosh\frac{\mu x}{l} + T\sinh\frac{\mu x}{l} \quad \text{where} \quad \mu = \left(\frac{\rho A\omega^2}{EI}\right)^{1/4} l \tag{10.86}$$

However, this solution has already been given by (10.4). This means that the solutions for the natural frequencies of simple beams given in Table 10.1 also give the whirling speeds of the equivalent rotating shafts.

When relatively large masses are attached to the shaft, the inertia of the shaft is usually ignored. Figure 10.17a shows a disc of mass m attached to a light shaft. For the moment, it will be assumed that the centre of mass is on the axis of the shaft. If it is displaced by a distance δ from the axis, then it will exert a centrifugal force F on the shaft. At the whirling speed ω_n, this is $m\omega_n^2\delta$. The deflexion of the shaft at this point produced by a force F can be determined from the usual flexural analysis given in Chapter 4. Thus if the left-hand end of the shaft is in long rigid bearings, it can be treated as a fixed end. If the right-hand end of the shaft is in short flexible bearings, it can be treated as a spring-supported end. The analysis then gives the relationship $F=k\delta$. For example, if the force is applied at the middle of the shaft and both ends are simply supported, k is $48EI/l^3$ where EI is the bending stiffness of the shaft and l is its length. Then equating the force found in this way to the centrifugal force,

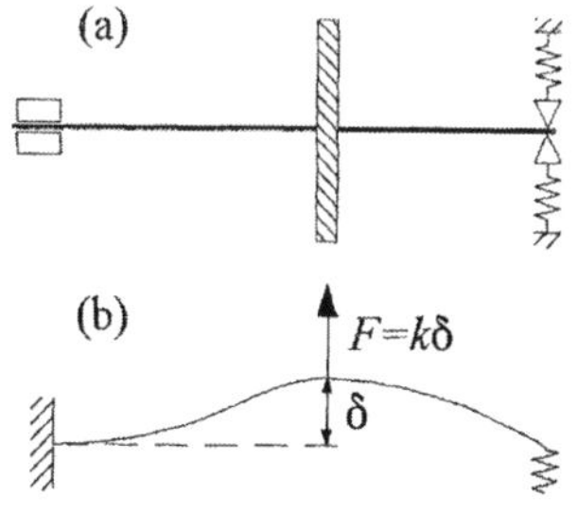

Figure 10.17 Whirling of a disc.

$$m\omega_n^2\delta = k\delta \qquad \text{so that} \qquad \omega_n^2 = \frac{k}{m} \tag{10.87}$$

Thus the whirling speed is the same in form as the natural frequency of a simple spring-mass system. In fact, the structural vibration analysis program described in Appendix 2 can be used to find the whirling speeds of compound shafts with lumped masses attached to them.

Suppose that the centre of mass had not been on the axis of the shaft but offset initially by an eccentricity e. Then the total distance of the centre of mass from the axis when it has a further elastic deflexion δ would be $\delta+e$. Then if the shaft is rotating at a speed ω, the first of equations (10.87) becomes

$$m\omega^2(\delta + e) = k\delta \tag{10.88}$$

Using ω_n as defined in (10.87), this gives

$$\frac{\delta + e}{e} = \frac{1}{1 - (\omega/\omega_n)^2} \tag{10.89}$$

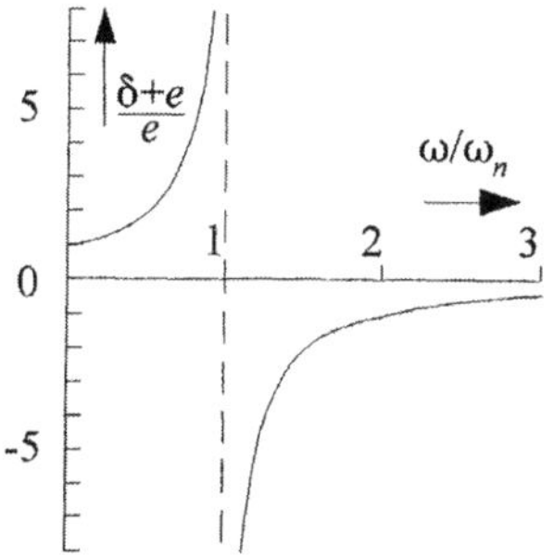

Figure 10.18 Total deflexion of an eccentric disc.

This gives the graph plotted in Figure 10.18. Note that once the whirling speed has been exceeded, the centre of mass becomes ever more closely aligned with the axis of rotation. Thus it is possible to drive a shaft at speeds in excess of its whirling speed provided that this speed is passed through rapidly, before large deflexions build up. Analyses allowing for viscous damping and gyroscopic effects are also possible (see Flügge (1962) §58.5 for example).

10.10 Approximate Methods

Rayleigh's method[1], first discussed in §9.8, was originally developed as an approximate method for determining natural frequencies. The structural system is assumed to be undamped and to vibrate freely with simple harmonic motion. No energy is then lost and the mode of deflexion can be represented by

$$v(x,t) = V(x)\cos(\omega t + \phi) \qquad \text{so that} \qquad \dot{v}(x,t) = -\omega V(x)\sin(\omega t + \phi) \tag{10.90}$$

The total energy of the system is given by the sum of the strain energy and the kinetic energy. The deflexion is zero when $(\omega t+\phi)$ is an odd multiple of $\pi/2$. As the strains, and hence the strain energy, are then also zero, the total energy is given by the kinetic energy at such instants. When $(\omega t+\phi)$ is a multiple of π, the system is instantaneously motionless. Then its kinetic energy is zero and the total energy is given by the strain energy. The natural frequency of the system can then be found by comparing the kinetic energy of the system when it is undeformed with the strain energy of the system when it is instantaneously at rest. For a simple beam of length l with workless end supports, the equation is

$$\frac{1}{2}\int_0^l \rho A \dot{v}_{max}^2 dx = \frac{1}{2}\omega^2\int_0^l \rho A (V(x))^2 dx = \frac{1}{2}\int_0^l EI\left(\frac{d^2 v_{max}}{dx^2}\right)^2 dx = \frac{1}{2}\int_0^l EI\left(\frac{d^2 V(x)}{dx^2}\right)^2 dx \tag{10.91}$$

giving

$$\omega^2 = \frac{\int_0^l EI\left(\frac{d^2V}{dx^2}\right)^2 dx}{\int_0^l \rho A V^2 dx} \tag{10.92}$$

where EI is the local bending stiffness and ρA is the local mass per unit length. The equality only holds true if $V(x)$ gives the form of the true vibration mode. If a reasonable approximation to the mode is used instead, the value of ω thus found will be an upper bound to the true value, just as in (9.89), the expression on the right-hand side is an upper bound to the true critical load. In both cases, the approximate mode can only be induced physically by imposing constraints on the system. These constraints make it appear stiffer than it actually is, this raising the critical load and the natural frequency. As discussed in Chapter 7, the approximate mode chosen must satisfy the geometrical end constraints relating to the end displacement and end slope.

Example 10.6 Flexural vibration of a conical beam

Figure 10.19 shows a cantilever in the form of a right circular cone, free at its tip, x=0, and fixed against rotation and displacement at the other end, x=l. The base radius of the cone is r. A vibration mode of the form

$$V(x) = c\left(\frac{x}{l} - 1\right)^2 \equiv c(X-1)^2 \tag{10.93}$$

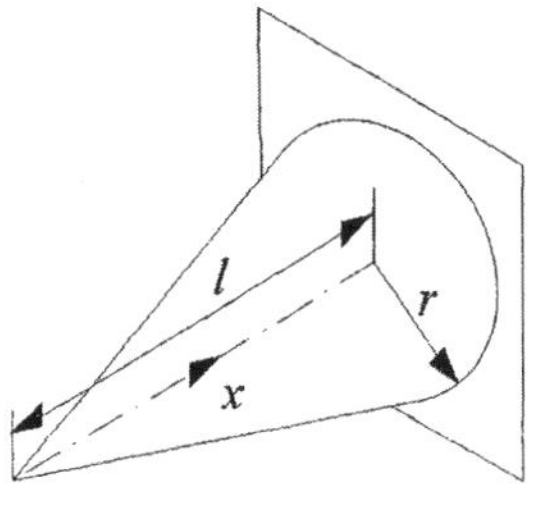

Figure 10.19 A vibrating cone.

---

[1] Lord Rayleigh gives the method in his book *Theory of Sound* Vol.1, pp.111 & 287 (1894).

will be assumed, where X is x/l. The bending stiffness and cross-sectional area are given by

$$EI = E\frac{\pi}{4}X^4r^4 \quad , \quad A = \pi X^2 r^2 \tag{10.94}$$

so that (10.92) becomes

$$\omega^2 = \frac{Er^2}{4\rho l^4}\frac{\int_0^1 X^4\left(\frac{d^2V}{dX^2}\right)^2 dX}{\int_0^1 X^2V^2\,dX} \tag{10.95}$$

Using the expression given by (10.93), this gives the lowest natural frequency as

$$\omega^2 = 21\frac{Er^2}{\rho l^4} \tag{10.96}$$

The exact solution found by Kirchhoff[1] is

$$\omega^2 = 19\frac{Er^2}{\rho l^4} \tag{10.97}$$

{The constraints on the function $V(x)$ chosen is that it must satisfy the geometric constraints. These are that the slope and displacement of the beam at the fixed end are both zero. It will be seen that the form given by (10.93) satisfies these conditions. It also satisfies the conditions of zero moment and shear force at the free end by virtue of the expression for the bending stiffness. Had this not been zero at the free end, modes of the form

$$V(X) = c_1(3 - 4X + X^4) \qquad \text{or} \qquad V(X) = c_2(11 - 15X + 5X^4 - X^5) \tag{10.98}$$

could have been used to satisfy all the end conditions. On substituting in (10.95), these would have given results which are 42% and 36% too high respectively.

The Ritz modification to Rayleigh's method can again be used to obtain more accurate results. As before, we take the mode to be given by a linear combination of modes,

$$V(x) = \sum_i c_i V_i(x) \tag{10.99}$$

where the coefficients c_i are constants and the functions $V_i(x)$ at least satisfy the geometric boundary conditions.

$$\omega^2 = \frac{\int_0^l EI\left(\sum_i c_i\frac{d^2V_i}{dx^2}\right)^2 dx}{\int_0^l \rho A\left(\sum_i c_iV_i\right)^2 dx} \tag{10.100}$$

As this expression gives an upper bound to the natural frequency, the coefficients c_i can be varied so as to minimise the value given. Partial differentiation then gives a set of equations equal to the number of unknown coefficients c_i of the form

[1] Kirchhoff, G.R. (1879) *Monatsberichte* p.815, Berlin.

$$\frac{\partial}{\partial c_j}\left\{\int_0^l EI\left(\sum_i c_i \frac{d^2V_i}{dx^2}\right)^2 - \omega^2\rho A\left(\sum_i c_i V_i\right)^2 dx\right\} = 0 \qquad \textbf{(10.101)}$$

or

$$\int_0^l EI\frac{d^2V}{dx^2}\frac{d^2V_j}{dx^2} - \omega^2\rho A V V_j\, dx = 0 \qquad \textbf{(10.102)}$$

where $V(x)$ is given by (10.99). This gives a set of homogeneous equations relating the terms c_i. The determinant of the coefficients of these terms gives a polynomial in ω^2, the lowest root of which gives an estimate of the lowest natural frequency.

Suppose that the end conditions are those for workless reactions and that each function $V_i(x)$ satisfies all the end conditions. Then

$$\int_0^l EI\frac{d^2V}{dx^2}\frac{d^2V_j}{dx^2}\, dx = \left|EI\frac{d^2V}{dx^2}\frac{dV_j}{dx}\right|_0^l - \int_0^l \frac{d}{dx}\left(EI\frac{d^2V}{dx^2}\right)\frac{dV_j}{dx}\, dx \qquad \textbf{(10.103)}$$

The first expression on the right-hand side is zero because either the bending moment or the slope is zero at the ends. The second term can also be integrated by parts, giving

$$\int_0^l \frac{d}{dx}\left(EI\frac{d^2V}{dx^2}\right)\frac{dV_j}{dx}\, dx = \left|\frac{d}{dx}\left(EI\frac{d^2V}{dx^2}\right)V_j\right|_0^l - \int_0^l \frac{d^2}{dx^2}\left(EI\frac{d^2V}{dx^2}\right)V_j\, dx \qquad \textbf{(10.104)}$$

Again, the first expression on the right-hand side is zero because either the shear force or the displacement is zero at the ends. Using these two results, (10.102) becomes

$$\int_0^l \left\{\frac{d^2}{dx^2}\left(EI\frac{d^2V}{dx^2}\right) - \omega^2\rho A V\right\}V_j\, dx = 0 \qquad \textbf{(10.105)}$$

For constant EI, the expression in curly brackets can be compared to the left-hand side of (10.3). These equations can be used in the same way as (10.102) to obtain the lowest natural frequency. They may also be obtained by the Galerkin method, which is expressed in terms of minimising error functions (see Flügge (1962) §§16.8, 44.2 & 65.5 and Sokolnikoff (1956) §115).

Example 10.7 a more exact analysis of the flexural vibration of a conical beam

Returning to the previous example, the Ritz-Galerkin method will be applied, taking

$$V(X) = c_1V_1(X) + c_2V_2(X) = c_1(X-1)^2 + c_2X(X-1)^2 \qquad \textbf{(10.106)}$$

Using this expression, (10.105) yields two equations which can be written in the form

$$\begin{aligned}(0.2 - \frac{\lambda}{105})c_1 + (0.1 - \frac{\lambda}{280})c_2 &= 0\\ (0.1 - \frac{\lambda}{280})c_1 + (\frac{6}{70} - \frac{\lambda}{630})c_2 &= 0\end{aligned} \qquad \textbf{(10.107)}$$

where λ is $\omega^2\rho l^4/Er^2$. Setting the determinant of the coefficients of c_1 and c_2 to zero gives

$$\omega^2 = 19.08Er^2/\rho l^4 \qquad \textbf{(10.108)}\}$$

Chapter 11 Matrix Analysis of Frames

11.1 Introduction

Prior to the development of computers, only methods suitable for manual analysis were considered. Most standard problems related to statically-determinate plane structures which were often taken to be pin-jointed. In analysing such frames, the first task was to determine the axial loads in the members and then the axial strains. Graphical techniques could be used both for this and then to determine the joint deflexions. A similar initial analysis was used for rigid-jointed frames capable of carrying the applied loading axially. In such cases, the deformations caused by the axial strains induced flexure of the members, so that a second analysis had to be carried out to determine the 'secondary stresses' induced by the corresponding bending moments. This process was only a second approximation to the exact solution of the implicit governing equations. Other techniques, such as moment distribution[1], were used when the applied loading was mainly carried by the flexure of the members. In such cases, simplifications could be made by ignoring the axial compressibility of the members. The complexity of the analyses rose with the degree of redundancy of the structure. Such statically-indeterminate problems were regarded as so difficult, that special books and courses were devoted to 'hyperstatic structures'.

With the availability of computers, systematic methods of structural analysis were sought which could be used to generate programs to analyse a wide variety of structures. For skeletal structures, two methods of matrix analysis were proposed. One is known as the *compatibility, force* or *flexibility matrix* method. Following the manual approach, the primary task is seen as the determination of the redundant internal loads and reactions. If the structure is statically indeterminate, imaginary cuts or releases are inserted to reduce it to a set of statically-determinate substructures. Corresponding to each degree of freedom introduced in this way, there will be a redundant internal interaction or reaction. Each substructure will have a set of statically-independent loads acting on it, which may include the above redundancies. (The rest will be reactions or interactions which can be determined from these by statics.) The state of stress, and hence the state of strain, can be determined from this set of loads. Let the 'stress resultant vector' $\boldsymbol{r}$ be a column vector of all such sets and the 'element deformation vector' $\boldsymbol{e}$ a column vector of all the corresponding substructure deformations through which only the loads $\boldsymbol{r}$ do work. From the elastic responses of each substructure, flexibility equations can now be written relating the set of strains $\boldsymbol{e}$ to the vector $\boldsymbol{r}$. As the substructure system is statically determinate, equilibrium equations can be written relating $\boldsymbol{r}$ to the set of external loads $\boldsymbol{p}$ on the whole structure and the set of redundancies $\boldsymbol{q}$. Finally, compatibility equations can be written which ensure that the substructure distortions given by $\boldsymbol{e}$ are related in such a way that the structure as a whole remains joined together. Using the previous equations, these compatibility equations yield relationships between the redundancies $\boldsymbol{q}$ and the external loads $\boldsymbol{p}$. The redundancies are then determined and the substructure loads $\boldsymbol{r}$ found from the earlier equilibrium equations.

A full description of this method will be found in Chapter 8 of Livesley (1975). It has some advantages over the stiffness method described below. The set of simultaneous equations to be solved is equal to the number of redundancies. The number of equations to be solved in the stiffness method is equal to the number of unknown joint deflexions, which may well be far larger. The flexibility method also permits the use of zero flexibilities. For example, the manual method

---

[1] A numerical method in which all the joints of the frame are initially considered to be rigidly held. These joint constraints are removed sequentially (a process known as 'relaxation') and the changes in bending moment determined.

of taking the beams to be axially rigid can be used, which considerably simplifies the equations. Zero flexibilities imply infinite stiffnesses, which cannot readily be incorporated into the stiffness method. Also, once the redundancies have been found, the internal state of stress can be determined directly. With the stiffness method, the joint deflexions are found first, and then the end loads on the members and finally their states of stress. Despite this, the stiffness method is almost universally preferred for automated structural analysis. The formation of the equations is more straightforward and does not require any matrix inversion. Also, as will be seen in the chapter on computer methods, use can be made of the symmetry and sparseness[1] of the stiffness matrix to increase the efficiency of the analysis.

The principles of the *equilibrium, displacement* or *stiffness matrix* method can be illustrated using the spring system shown in Figure 11.1. The linear spring is shown in Figure 11.1a. This has a spring stiffness k which relates the force F to the displacement δ by $F = k\delta$. (The spring flexibility f is such that $\delta = fF$.) The overall stiffness of a set of springs joined to the same point, and acting in the same direction, is the sum of their individual stiffnesses.

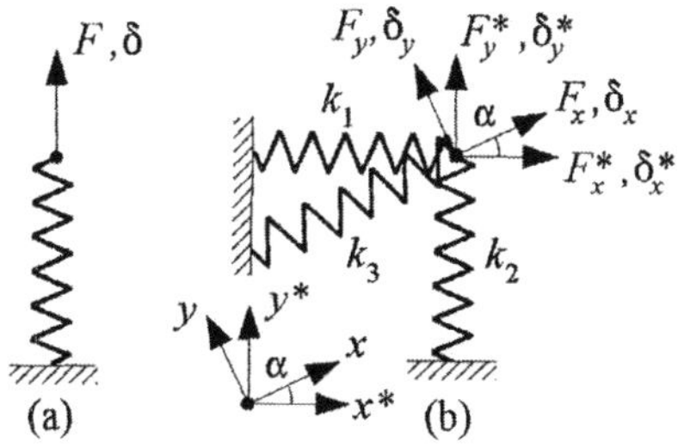

Figure 11.1 Spring systems.

In Figure 11.1b, three springs are joined at a common point. The horizontal one has a stiffness k_1 and the vertical one has a stiffness k_2. The combined stiffness of the two is given by

$$\begin{bmatrix} F_x^* \\ F_y^* \end{bmatrix} = \begin{bmatrix} k_1 & 0 \\ 0 & k_2 \end{bmatrix} \begin{bmatrix} \delta_x^* \\ \delta_y^* \end{bmatrix} \tag{11.1}$$

The third spring is at an angle α to the x^* axis. With respect to its local coordinate system, x,y its displacements δ_x and δ_y are related to its end forces F_x and F_y by

$$\begin{bmatrix} F_x \\ F_y \end{bmatrix}_3 = \begin{bmatrix} k_3 & 0 \\ 0 & 0 \end{bmatrix} \begin{bmatrix} \delta_x \\ \delta_y \end{bmatrix}_3 \quad \text{or} \quad \boldsymbol{P}_3 = \boldsymbol{K}_3 \boldsymbol{d}_3 \tag{11.2}$$

in abbreviated matrix notation, where $\boldsymbol{K}_3$ is known as a stiffness matrix. The load vector $\boldsymbol{P}_3$ and the deflexion vector $\boldsymbol{d}_3$ can be related to their equivalents $\boldsymbol{P}^*_3$ and $\boldsymbol{d}^*_3$ in terms of the global coordinate system x^*, y^* by

$$\begin{aligned} \begin{bmatrix} F_x^* \\ F_y^* \end{bmatrix}_3 &= \begin{bmatrix} \cos\alpha & -\sin\alpha \\ \sin\alpha & \cos\alpha \end{bmatrix} \begin{bmatrix} F_x \\ F_y \end{bmatrix}_3 \quad \text{or } \boldsymbol{P}_3^* = \boldsymbol{T}\boldsymbol{P}_3 \,, \\ \begin{bmatrix} \delta_x \\ \delta_y \end{bmatrix}_3 &= \begin{bmatrix} \cos\alpha & \sin\alpha \\ -\sin\alpha & \cos\alpha \end{bmatrix} \begin{bmatrix} \delta_x^* \\ \delta_y^* \end{bmatrix}_3 \quad \text{or } \boldsymbol{d}_3 = \boldsymbol{T}^t \boldsymbol{d}_3^* \,. \end{aligned} \tag{11.3}$$

where $\boldsymbol{T}$ is known as a *transformation matrix* and $\boldsymbol{T}^t$ is its transpose. In terms of the global coordinate system, (11.2) then becomes

---

[1] That is, many of the elements of the matrix will be zero and use can be made of this to reduce the number of numerical operations.

$$P_3^* = TP_3 = TK_3d_3 = TK_3T^td_3^* = K_3^*d_3^* \tag{11.4}$$

where

$$K_3^* = TK_3T^t = \begin{bmatrix} \cos\alpha & -\sin\alpha \\ \sin\alpha & \cos\alpha \end{bmatrix} \begin{bmatrix} k_3 & 0 \\ 0 & 0 \end{bmatrix} \begin{bmatrix} \cos\alpha & \sin\alpha \\ -\sin\alpha & \cos\alpha \end{bmatrix} = \begin{bmatrix} k_3\cos^2\alpha & k_3\sin\alpha\cos\alpha \\ k_3\sin\alpha\cos\alpha & k_3\sin^2\alpha \end{bmatrix} \tag{11.5}$$

As the joint is common to the ends of all three springs, d_3* is the same as the joint deflexion vector in (11.1). The total external load on the joint is the sum of the end loads on the first two springs, given by the joint load vector in (11.1), and that on the third spring, given by P_3* in (11.4). Using this compatibility condition and equilibrium condition, the stiffness matrix for all three springs is given by the sum of those in (11.1) and (11.5), giving

$$\begin{bmatrix} F_x^* \\ F_y^* \end{bmatrix} = \begin{bmatrix} k_1 + k_3\cos^2\alpha & k_3\sin\alpha\cos\alpha \\ k_3\sin\alpha\cos\alpha & k_2 + k_3\sin^2\alpha \end{bmatrix} \begin{bmatrix} \delta_x^* \\ \delta_y^* \end{bmatrix} \tag{11.6}$$

as the overall stiffness matrix equation for the complete system. Note that the matrix is symmetrical. This will always be the case provided that the joint deflexion vector is that of the corresponding deflexions through which the components of the joint load vector do work. This follows from Betti's theorem discussed in §3.4. Knowing the joint loading, the joint deflexions can be found from the overall stiffness matrix equation. The loads on individual components of the system can be found from member stiffness equations such as (11.4). The state of stress may then be determined, completing the solution.

Skeletal frameworks consist of a network of bars connected to joints at each of their two ends. Then instead of the elastic response of a component being expressed in terms of single relationship between the loads and the corresponding deflexions at one end, it is more convenient to write it as a pair of relationships for the two ends. Let these be ends A and B. If the load vectors at these ends are P_A and P_B respectively, and the corresponding deflexion vectors are d_A and d_B, these relationships can be written as

$$\begin{aligned} P_A &= K_{11}d_A + K_{12}d_B \\ P_B &= K_{21}d_A + K_{22}d_B \end{aligned} \tag{11.7}$$

It follows from Betti's theorem that

$$K_{11} = K_{11}^t \ , \quad K_{12} = K_{21}^t \ , \quad K_{22} = K_{22}^t \tag{11.8}$$

Most of this chapter deals with applications of the stiffness matrix method to the analysis of skeletal structures. The simplest example is the analysis of plane, pin-jointed frames dealt with in the next section. The example used covers in detail the method for assembling the stiffness matrices for the individual bars into the stiffness matrix for the whole frame. The plane, rigid-jointed frames discussed in §11.3 cover both frames which are loaded and deflect only in their own planes and grids, which are loaded and deflect only out of their own planes. These restrictions reduce the number of degrees of freedom at each joint to a maximum of three, thus simplifying the analysis. In §11.4, three-dimensional pin-jointed and rigid-jointed structures are examined. In the former case, there are still only three degrees of freedom at each joint (the three components of displacement), but in the latter case there are six, as the three components of rotation have to be included as well. In all cases, the general principles outlined above apply. In particular, the general form of the member stiffness matrix equations given by (11.7) and (11.8) holds. The stability and vibration of frames discussed in §11.5 and §11.6 can be analysed using modified stiffness matrices of the above kind. These employ the stability and vibration functions derived in Chapters 9 and 10. Applications of all these analyses are given in the computer programs provided.

The remainder of the chapter deals with special applications of the method. For example, in §11.8 it is shown that the elastic response of members with varying stiffness and curved members can be simulated exactly by assemblies of prismatic members. This is useful in using computer programs which will only accept data for prismatic members.

11.2 Plane Pin-Jointed Frames

Pin-ended bars behave like the above springs, with the spring stiffness k replaced by EA/l where E is Young's modulus, A is the cross-sectional area of the bar and l is its length. The bar shown in Figure 11.2 is oriented at an anticlockwise angle α to a *global* coordinate system (x^*, y^*). The *local* coordinate system (x,y) is related to the bar axes, the x axis being directed along the bar and the y axis at right-angles to it. In terms of the local coordinates,

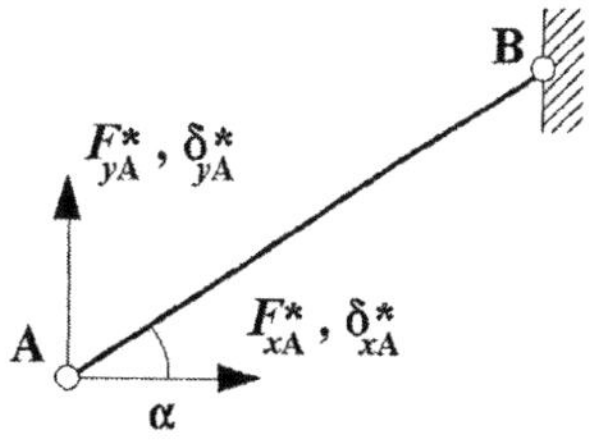

Figure 11.2 Global coordinates.

$$\begin{bmatrix} F_{xA} \\ F_{yA} \end{bmatrix} = \begin{bmatrix} EA/l & 0 \\ 0 & 0 \end{bmatrix} \begin{bmatrix} \delta_{xA} \\ \delta_{yA} \end{bmatrix} \quad \text{or} \quad \boldsymbol{P}_A = \boldsymbol{K}_{11} \boldsymbol{d}_A \,. \tag{11.9}$$

In terms of the global coordinate system, this becomes

$$\begin{bmatrix} F^*_{xA} \\ F^*_{yA} \end{bmatrix} = \boldsymbol{P}^*_A = \boldsymbol{K}^*_{11} \boldsymbol{d}^*_A = \frac{EA}{l} \begin{bmatrix} \cos^2\alpha & \sin\alpha\cos\alpha \\ \sin\alpha\cos\alpha & \sin^2\alpha \end{bmatrix} \begin{bmatrix} \delta^*_{xA} \\ \delta^*_{yA} \end{bmatrix} \tag{11.10}$$

where

$$\boldsymbol{K}^*_{11} = \boldsymbol{T}\boldsymbol{K}_{11}\boldsymbol{T}^t \quad , \quad \boldsymbol{T} = \begin{bmatrix} \cos\alpha & -\sin\alpha \\ \sin\alpha & \cos\alpha \end{bmatrix} \tag{11.11}$$

The new matrix $\boldsymbol{K}_{11}$* is necessarily symmetric because the old matrix $\boldsymbol{K}_{11}$ is symmetric. This also follows directly from Betti's theorem.

If there are no loads applied to a bar between its ends A and B, the end loads are equal and opposite, or $\boldsymbol{P}_B$ (= $\{F_{xB}, F_{yB}\}$) is equal to -$\boldsymbol{P}_A$. If the end B is free to move, the axial force in the bar is no longer given by the deflexion of A, $\boldsymbol{d}_A$, but the difference of deflexions at A and B, $\boldsymbol{d}_A - \boldsymbol{d}_B$. Then the equations for the bar become

$$\begin{aligned} \begin{bmatrix} F_{xA} \\ F_{yA} \end{bmatrix} &= \begin{bmatrix} EA/l & 0 \\ 0 & 0 \end{bmatrix} \begin{bmatrix} \delta_{xA} \\ \delta_{yA} \end{bmatrix} + \begin{bmatrix} -EA/l & 0 \\ 0 & 0 \end{bmatrix} \begin{bmatrix} \delta_{xB} \\ \delta_{yB} \end{bmatrix} \\ \begin{bmatrix} F_{xB} \\ F_{yB} \end{bmatrix} &= \begin{bmatrix} -EA/l & 0 \\ 0 & 0 \end{bmatrix} \begin{bmatrix} \delta_{xA} \\ \delta_{yA} \end{bmatrix} + \begin{bmatrix} EA/l & 0 \\ 0 & 0 \end{bmatrix} \begin{bmatrix} \delta_{xB} \\ \delta_{yB} \end{bmatrix} \end{aligned} \tag{11.12}$$

This conforms to the general form given by (11.7) and it can be seen that the four stiffness matrices satisfy (11.8). The equations for a particular bar can be transformed to relate the end loads and deflexions relative to the global coordinates of a framework, using the transformation previously used to derive $\boldsymbol{K}_{11}$* from $\boldsymbol{K}_{11}$. The end deflexion vectors are equal to the deflexion vectors of the joints to which they are attached. The sum of the load vectors acting on the ends of bars meeting at a joint is equal to the vector of external loads for that joint. These conditions of joint compatibility and joint equilibrium permit the equations for component bars to be assembled into the equations for the whole frame.

Example 11.1 A Cross-Braced Frame

All the members of the frame shown in Figure 11.3 have the same axial stiffness EA. For the bar AB, the local and global coordinates are the same, so that the submatrices in (11.12) can be used directly in assembling the frame stiffness matrix equation. As joints A and D are fixed, their displacements (and loads) do not appear in this equation. Thus for the bar AB, only the matrix K_{22}^* ($=K_{22}$) is required, which relates its end load at B to the end deflexion at B:

$$\begin{bmatrix} F_{xB}^* \\ F_{yB}^* \end{bmatrix} = \begin{bmatrix} EA/L & 0 \\ 0 & 0 \end{bmatrix} \begin{bmatrix} \delta_{xB}^* \\ \delta_{yB}^* \end{bmatrix} \qquad \textbf{(11.13)}$$

or

$$P_{B(AB)}^* = K_{22(AB)}^* \, d_{B(AB)}^*$$

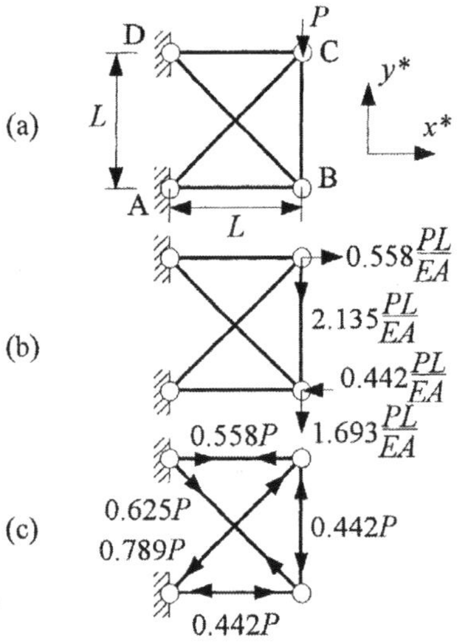

Figure 11.3 A cross-braced frame.

where (AB) has been added to the subscripts to indicate the bar under discussion. The same arguments hold for the bar DC, giving a similar equation to (11.13) with the subscript B being replaced by C and (AB) being replaced by (DC). For the same reason, only the K_{22}^* matrix is needed for the bars AC and DB. These are both of length $L\sqrt{2}$. In the former case, the angle α is 45° and in the latter case it is -45°. Transformations of the type given by (11.10) can be applied to all the stiffness matrices, so that

$$\begin{bmatrix} F_{xC}^* \\ F_{yC}^* \end{bmatrix} = \frac{EA}{L\sqrt{2}} \begin{bmatrix} \frac{1}{2} & \frac{1}{2} \\ \frac{1}{2} & \frac{1}{2} \end{bmatrix} \begin{bmatrix} \delta_{xC}^* \\ \delta_{yC}^* \end{bmatrix} \qquad \textbf{(11.14)}$$

or

$$P_{C(AC)}^* = K_{22(AC)}^* \, d_{C(AC)}^*$$

for the bar AC, and for the bar DB

$$\begin{bmatrix} F_{xB}^* \\ F_{yB}^* \end{bmatrix} = \frac{EA}{L\sqrt{2}} \begin{bmatrix} \frac{1}{2} & -\frac{1}{2} \\ -\frac{1}{2} & \frac{1}{2} \end{bmatrix} \begin{bmatrix} \delta_{xB}^* \\ \delta_{yB}^* \end{bmatrix} \qquad \textbf{(11.15)}$$

or

$$P_{B(DB)}^* = K_{22(DB)}^* \, d_{B(DB)}^* \, .$$

The bar BC is of length L and at an angle α of 90°. As the deflexions at both ends are sought, the full set of terms in (11.12) is required. Using similar transformations to (11.10), they become

$$\begin{bmatrix} F^*_{xB} \\ F^*_{yB} \end{bmatrix} = \begin{bmatrix} 0 & 0 \\ 0 & EA/L \end{bmatrix} \begin{bmatrix} \delta^*_{xB} \\ \delta^*_{yB} \end{bmatrix} + \begin{bmatrix} 0 & 0 \\ 0 & -EA/L \end{bmatrix} \begin{bmatrix} \delta^*_{xC} \\ \delta^*_{yC} \end{bmatrix}$$

$$\begin{bmatrix} F^*_{xC} \\ F^*_{yC} \end{bmatrix} = \begin{bmatrix} 0 & 0 \\ 0 & -EA/L \end{bmatrix} \begin{bmatrix} \delta^*_{xB} \\ \delta^*_{yB} \end{bmatrix} + \begin{bmatrix} 0 & 0 \\ 0 & EA/L \end{bmatrix} \begin{bmatrix} \delta^*_{xC} \\ \delta^*_{yC} \end{bmatrix} \qquad \textbf{(11.16)}$$

or

$$\boldsymbol{P}^*_{\mathrm{B(BC)}} = \boldsymbol{K}^*_{11} \boldsymbol{d}^*_{\mathrm{B(BC)}} + \boldsymbol{K}^*_{12} \boldsymbol{d}^*_{\mathrm{C(BC)}}$$

$$\boldsymbol{P}^*_{\mathrm{C(BC)}} = \boldsymbol{K}^*_{21} \boldsymbol{d}^*_{\mathrm{B(BC)}} + \boldsymbol{K}^*_{22} \boldsymbol{d}^*_{\mathrm{C(BC)}}$$

(cf. (11.7)). In this case, the compatibility of end deflexions gives

$$\begin{aligned} \boldsymbol{d}^*_{\mathrm{B}} &= \boldsymbol{d}^*_{\mathrm{B(AB)}} = \boldsymbol{d}^*_{\mathrm{B(BC)}} = \boldsymbol{d}^*_{\mathrm{B(DB)}} \\ \boldsymbol{d}^*_{\mathrm{C}} &= \boldsymbol{d}^*_{\mathrm{C(AC)}} = \boldsymbol{d}^*_{\mathrm{C(BC)}} = \boldsymbol{d}^*_{\mathrm{C(DC)}} \end{aligned} \qquad \textbf{(11.17)}$$

and the equilibrium of joint loads gives

$$\begin{aligned} \boldsymbol{P}^*_{\mathrm{B}} &= \boldsymbol{P}^*_{\mathrm{B(AB)}} + \boldsymbol{P}^*_{\mathrm{B(BC)}} + \boldsymbol{P}^*_{\mathrm{B(DB)}} \\ \boldsymbol{P}^*_{\mathrm{C}} &= \boldsymbol{P}^*_{\mathrm{C(AC)}} + \boldsymbol{P}^*_{\mathrm{C(BC)}} + \boldsymbol{P}^*_{\mathrm{C(DC)}} \end{aligned} \qquad \textbf{(11.18)}$$

It will be seen that there is a similarity between (11.17) and (11.18). This becomes clearer when they are rewritten in the form

$$\begin{bmatrix} \boldsymbol{d}^*_{\mathrm{B(AB)}} \\ \boldsymbol{d}^*_{\mathrm{C(AC)}} \\ \boldsymbol{d}^*_{\mathrm{B(BC)}} \\ \boldsymbol{d}^*_{\mathrm{C(BC)}} \\ \boldsymbol{d}^*_{\mathrm{B(DB)}} \\ \boldsymbol{d}^*_{\mathrm{C(DC)}} \end{bmatrix} = \begin{bmatrix} \boldsymbol{I} & \boldsymbol{0} \\ \boldsymbol{0} & \boldsymbol{I} \\ \boldsymbol{I} & \boldsymbol{0} \\ \boldsymbol{0} & \boldsymbol{I} \\ \boldsymbol{I} & \boldsymbol{0} \\ \boldsymbol{0} & \boldsymbol{I} \end{bmatrix} \begin{bmatrix} \boldsymbol{d}^*_{\mathrm{B}} \\ \boldsymbol{d}^*_{\mathrm{C}} \end{bmatrix}, \qquad \begin{bmatrix} \boldsymbol{P}^*_{\mathrm{B}} \\ \boldsymbol{P}^*_{\mathrm{C}} \end{bmatrix} = \begin{bmatrix} \boldsymbol{I} & \boldsymbol{0} & \boldsymbol{I} & \boldsymbol{0} & \boldsymbol{I} & \boldsymbol{0} \\ \boldsymbol{0} & \boldsymbol{I} & \boldsymbol{0} & \boldsymbol{I} & \boldsymbol{0} & \boldsymbol{I} \end{bmatrix} \begin{bmatrix} \boldsymbol{P}^*_{\mathrm{B(AB)}} \\ \boldsymbol{P}^*_{\mathrm{C(AC)}} \\ \boldsymbol{P}^*_{\mathrm{B(BC)}} \\ \boldsymbol{P}^*_{\mathrm{C(BC)}} \\ \boldsymbol{P}^*_{\mathrm{B(DB)}} \\ \boldsymbol{P}^*_{\mathrm{C(DC)}} \end{bmatrix} \qquad \textbf{(11.19)}$$

where $\boldsymbol{I}$ is a unit 2×2 submatrix and $\boldsymbol{0}$ is a zero 2×2 submatrix. It will be seen that the second matrix of these submatrices is the transpose of the first. It is known as a *connection matrix* and pairs of matrix equations of this general form are said to exhibit *contragredience.*

The right-hand sides of equations (11.18) can be expressed in terms of the bar end deflexions using (11.13) to (11.16). These bar end deflexions are given in terms of the joint deflexions using (11.17). This gives relationships between the joint loads and the joint deflexions of the form

$$\begin{bmatrix} \boldsymbol{P}^*_{B} \\ \boldsymbol{P}^*_{C} \end{bmatrix} = \begin{bmatrix} \boldsymbol{K}^*_{22\mathrm{(AB)}} + \boldsymbol{K}^*_{11\mathrm{(BC)}} + \boldsymbol{K}^*_{22\mathrm{(DB)}} & \boldsymbol{K}^*_{12\mathrm{(BC)}} \\ \boldsymbol{K}^*_{21\mathrm{(BC)}} & \boldsymbol{K}^*_{22\mathrm{(AC)}} + \boldsymbol{K}^*_{22\mathrm{(BC)}} + \boldsymbol{K}^*_{22\mathrm{(DC)}} \end{bmatrix} \begin{bmatrix} \boldsymbol{d}^*_{B} \\ \boldsymbol{d}^*_{C} \end{bmatrix} \qquad \textbf{(11.20)}$$

These equations could be derived formally using the connection matrix equations (11.19), but such a process is unnecessarily cumbersome, either by hand or with a computer. Expressing the submatrices and column vectors in full, (11.20) becomes

$$\begin{bmatrix} 0 \\ 0 \\ 0 \\ -P \end{bmatrix} = \begin{bmatrix} F_{xB}^{*} \\ F_{yB}^{*} \\ F_{xC}^{*} \\ F_{yC}^{*} \end{bmatrix} = \frac{EA}{L} \begin{bmatrix} 1+\frac{1}{2\sqrt{2}} & -\frac{1}{2\sqrt{2}} & 0 & 0 \\ -\frac{1}{2\sqrt{2}} & 1+\frac{1}{2\sqrt{2}} & 0 & -1 \\ 0 & 0 & 1+\frac{1}{2\sqrt{2}} & \frac{1}{2\sqrt{2}} \\ 0 & -1 & \frac{1}{2\sqrt{2}} & 1+\frac{1}{2\sqrt{2}} \end{bmatrix} \begin{bmatrix} \delta_{xB}^{*} \\ \delta_{yB}^{*} \\ \delta_{xC}^{*} \\ \delta_{yC}^{*} \end{bmatrix} \qquad (11.21)$$

Carrying out the Gaussian process of diagonalisation, this equation becomes

$$\begin{bmatrix} -0.598P \\ -2.135P \\ 0.755P \\ -P \end{bmatrix} = \frac{EA}{L} \begin{bmatrix} 1.3535 & 0 & 0 & 0 \\ 0 & 1.2612 & 0 & 0 \\ 0 & 0 & 1.3535 & \\ 0 & 0 & 0 & 0.4683 \end{bmatrix} \begin{bmatrix} \delta_{xB}^{*} \\ \delta_{yB}^{*} \\ \delta_{xC}^{*} \\ \delta_{yC}^{*} \end{bmatrix} \qquad (11.22)$$

giving the joint displacements shown in Figure 11.3b. The bar forces can then be determined from these deflexions by substituting them back into the member matrix equations, giving the results shown in Figure 11.3c.

11.3 Plane Rigid-Jointed Frames

If the joints of the frame are rigid, they are capable of transmitting moments as well as forces to the ends of the beams. For equilibrium, the vector sum of the moments so transmitted by a joint is equal to the total external moment acting on it. Corresponding to this equilibrium condition, is the compatibility condition that the rotations of all the beams meeting at a joint are common, and equal to the joint rotation. Equations (11.12) must be modified to take account of these end moments and rotations. The problem can still be considered as two-dimensional if the plane of the frame is also a plane of symmetry for all its members and the frame is only loaded in its own plane. Making use of (4.55) and (4.56), and using the notation for end loads and deflexions shown in Figure 11.4, these equations take the form

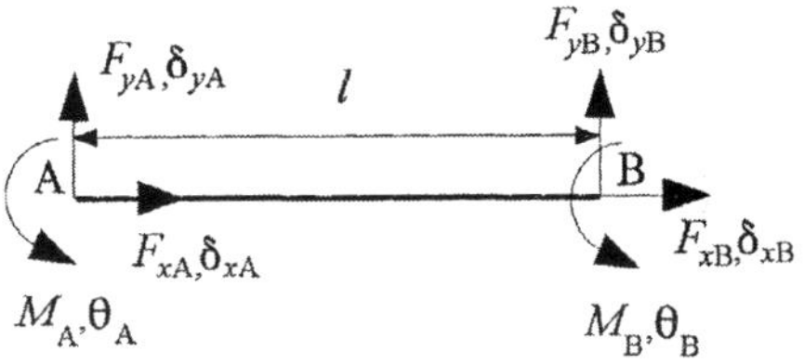

Figure 11.4 In-plane flexure of a beam.

$$\begin{bmatrix} F_{xA} \\ F_{yA} \\ M_A \end{bmatrix} = \begin{bmatrix} EA/l & 0 & 0 \\ 0 & 12EI/l^3 & 6EI/l^2 \\ 0 & 6EI/l^2 & 4EI/l \end{bmatrix} \begin{bmatrix} \delta_{xA} \\ \delta_{yA} \\ \theta_A \end{bmatrix} + \begin{bmatrix} -EA/l & 0 & 0 \\ 0 & -12EI/l^3 & 6EI/l^2 \\ 0 & -6EI/l^2 & 2EI/l \end{bmatrix} \begin{bmatrix} \delta_{xB} \\ \delta_{yB} \\ \theta_B \end{bmatrix}$$

$$\begin{bmatrix} F_{xB} \\ F_{yB} \\ M_B \end{bmatrix} = \begin{bmatrix} -EA/l & 0 & 0 \\ 0 & -12EI/l^3 & -6EI/l^2 \\ 0 & 6EI/l^2 & 2EI/l \end{bmatrix} \begin{bmatrix} \delta_{xA} \\ \delta_{yA} \\ \theta_A \end{bmatrix} + \begin{bmatrix} EA/l & 0 & 0 \\ 0 & 12EI/l^3 & -6EI/l^2 \\ 0 & -6EI/l^2 & 4EI/l \end{bmatrix} \begin{bmatrix} \delta_{xB} \\ \delta_{yB} \\ \theta_B \end{bmatrix} \qquad (11.23)$$

These two matrix equations again correspond to the general form given by (11.7).

Transforming from local to global coordinates is done in a similar fashion to that for a pin-ended bar, except that there is an extra term in the load and deflexion vectors, relating to the end moment and end rotation respectively. However, these are the same in both local and global coordinates. The transformation matrix is then

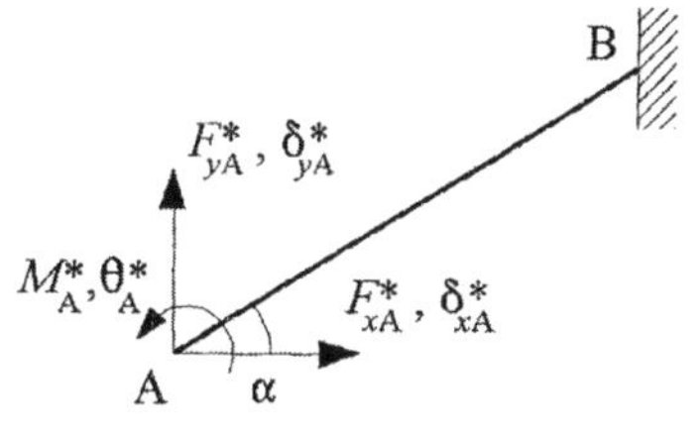

Figure 11.5 Global coordinates.

$$T = \begin{bmatrix} \cos\alpha & -\sin\alpha & 0 \\ \sin\alpha & \cos\alpha & 0 \\ 0 & 0 & 1 \end{bmatrix} \qquad \textbf{(11.24)}$$

As before, the stiffness matrix K_{11} is transformed to refer to global coordinates, it becomes K^*_{11} and is given by the matrix product $TK_{11}T^t$. The other stiffness matrices transform in the same way.

The slope-deflexion equations used to set up the stiffness matrices did not contain the additional terms associated with loading between the ends of the beams. These terms were discussed in §4.5. Using the principle of superposition, the total end reactions can be taken as those induced by the end rotations and displacements (as above) plus the reactions induced by the loading along the beam when end rotations and displacements are prevented (the fixed-end reactions). Conversely, in analysing the frame as a whole, the loading between the joints is replaced by loading on the joints equal to *minus* the sum of the fixed-end reactions. (To every reaction there is an equal and opposite action.)

Example 11.2 A Rigid-Jointed Frame with Beam Loading

The two beams shown in Figure 11.6 are of the same length l and have the same bending stiffness EI and axial stiffness EA. They are rigidly joined together at right-angles at A and encastré at their other ends, B and C. A downwards force F midway along AB induces a fixed-end moment $Fl/8$ anticlockwise and a fixed-end reaction of ½F upwards at A. These reactions of joint A are replaced by the equal and opposite actions on A, as shown. (As B is fixed, the actions on B do not affect this part of the analysis.) Taking the global coordinates to be aligned with the local coordinates for AB, The K_{11} matrix for this member is the same in terms of the global coordinates. For the member AC, the form of this matrix in global coordinates is

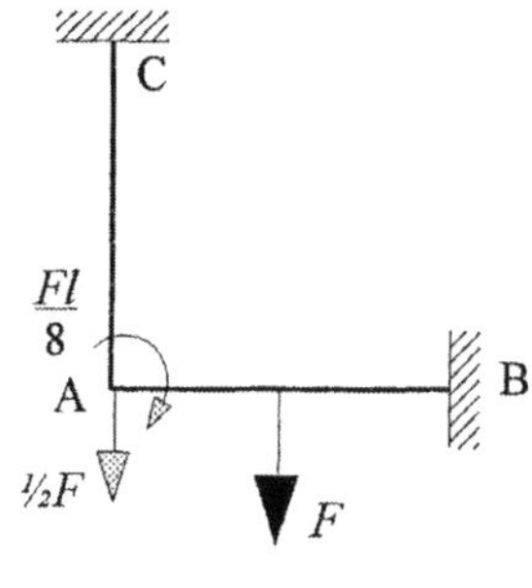

Figure 11.6 A beam-loaded frame.

$$K^*_{11(AC)} = \begin{bmatrix} 12EI/l^3 & 0 & -6EI/l^2 \\ 0 & EA/l & 0 \\ -6EI/l^2 & 0 & 4EI/l \end{bmatrix} \qquad \textbf{(11.25)}$$

Combining the two matrices and taking the value of A to be $100I/l^2$, the stiffness matrix equation for the complete structure becomes

$$\begin{bmatrix} 112EI/l^3 & 0 & -6EI/l^2 \\ 0 & 112EI/l^3 & 6EI/l^2 \\ -6EI/l^2 & 6EI/l^2 & 8EI/l \end{bmatrix} \begin{bmatrix} \delta^*_{xA} \\ \delta^*_{yA} \\ \theta^*_A \end{bmatrix} = \begin{bmatrix} F^*_{xA} \\ F^*_{yA} \\ M^*_A \end{bmatrix} = \begin{bmatrix} 0 \\ -F/2 \\ -Fl/8 \end{bmatrix} \tag{11.26}$$

Solving these equations gives the value of the deflexion vector $\{\delta^*_{xA}, \delta^*_{yA}, \theta^*_A\}$ as $\{-0.00071Fl^3/EI, -0.00375Fl^3/EI, -0.01335Fl^2/EI\}$. These values can be back-substituted into the original beam equations to find the end loads on the beams. For example, this gives the axial tension in AC as $0.375F$.

The general three-dimensional analysis of rigid-jointed frames involves six degrees of freedom at each joint. If the frame is plane and is loaded only in its plane, the above analysis is (usually) sufficient. If a plane frame is loaded by the other three possible components of joint load, they will induce out-of-plane deflexions. This is typical of floor- or roof-bearing grids.

For the beam AB shown in Figure 11.7 of length l, bending stiffness EI and torsional stiffness GJ, equations (11.7) take the form

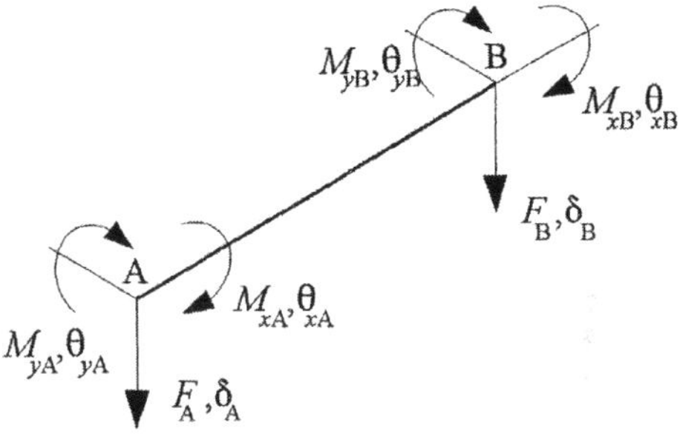

Figure 11.7 Typical member of a grid.

$$\begin{bmatrix} M_{xA} \\ M_{yA} \\ F_A \end{bmatrix} = \begin{bmatrix} GJ/l & 0 & 0 \\ 0 & 4EI/l & 6EI/l^2 \\ 0 & 6EI/l^2 & 12EI/l^3 \end{bmatrix} \begin{bmatrix} \theta_{xA} \\ \theta_{yA} \\ \delta_A \end{bmatrix} + \begin{bmatrix} -GJ/l & 0 & 0 \\ 0 & 2EI/l & -6EI/l^2 \\ 0 & 6EI/l^2 & -12EI/l^3 \end{bmatrix} \begin{bmatrix} \theta_{xB} \\ \theta_{yB} \\ \delta_B \end{bmatrix}$$

$$\begin{bmatrix} M_{xB} \\ M_{yB} \\ F_B \end{bmatrix} = \begin{bmatrix} -GJ/l & 0 & 0 \\ 0 & 2EI/l & 6EI/l^2 \\ 0 & -6EI/l^2 & -12EI/l^3 \end{bmatrix} \begin{bmatrix} \theta_{xA} \\ \theta_{yA} \\ \delta_A \end{bmatrix} + \begin{bmatrix} GJ/l & 0 & 0 \\ 0 & 4EI/l & -6EI/l^2 \\ 0 & -6EI/l^2 & 12EI/l^3 \end{bmatrix} \begin{bmatrix} \theta_{xB} \\ \theta_{yB} \\ \delta_B \end{bmatrix} \tag{11.27}$$

and if the beam AB is at an anticlockwise angle α to the global coordinate system, the transformation is given by

$$\begin{bmatrix} M^*_x \\ M^*_y \\ F^* \end{bmatrix} = \begin{bmatrix} \cos\alpha & -\sin\alpha & 0 \\ \sin\alpha & \cos\alpha & 0 \\ 0 & 0 & 1 \end{bmatrix} \begin{bmatrix} M_x \\ M_y \\ F \end{bmatrix} \quad \text{or} \quad \boldsymbol{P^* = TP} \tag{11.28}$$

so that the transformation matrix is the same as that given by (11.24). The matrix analysis of grids is based on the same principles as the previous analyses.

Example 11.3 A Laterally-Loaded Angle Frame

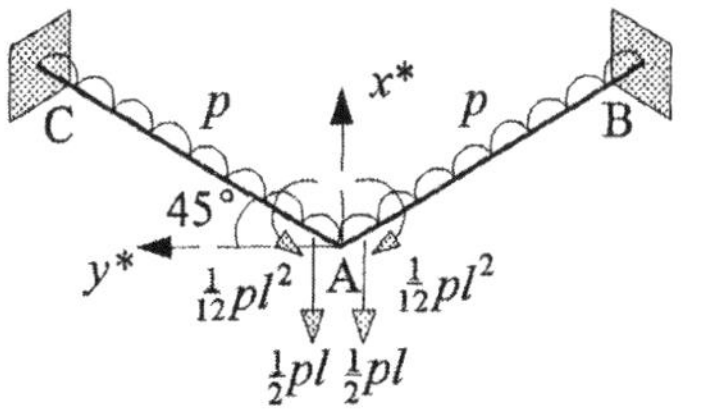

Figure 11.8 A laterally-loaded frame.

In the frame shown, AB and AC are both of length l, and have stiffnesses EI and GJ. They are rigidly joined at right-angles at A and encastré at their ends B and C. Both are loaded normal to the plane ABC by a uniform distributed load of intensity p. This induces the fixed-end actions shown at A. It is convenient to take the global coordinates x^* and y^* at 45° to the alignment of the beams, as shown. As AC is at an angle α of +45° to the global coordinate system, its $\boldsymbol{K}^*_{11}$ matrix is given by

$$\boldsymbol{K}^*_{11(AC)} = \boldsymbol{T}\boldsymbol{K}_{11(AC)}\boldsymbol{T}^t = \begin{bmatrix} GJ/2l + 2EI/l & GJ/2l - 2EI/l & -3\sqrt{2}EI/l^2 \\ GJ/2l - 2EI/l & GJ/2l + 2EI/l & 3\sqrt{2}EI/l^2 \\ -3\sqrt{2}EI/l^2 & 3\sqrt{2}EI/l^2 & 12EI/l^3 \end{bmatrix} \quad \textbf{(11.29)}$$

The $\boldsymbol{K}^*_{11}$ matrix for AB is found similarly, taking α as -45°. The matrix equation for the frame is then

$$\begin{bmatrix} 0 \\ \dfrac{pl^2}{6\sqrt{2}} \\ pl \end{bmatrix} = \begin{bmatrix} M^*_{xA} \\ M^*_{yA} \\ F^*_A \end{bmatrix} = \begin{bmatrix} GJ/l + 4EI/l & 0 & 0 \\ 0 & GJ/l + 4EI/l & 6\sqrt{2}EI/l^2 \\ 0 & 6\sqrt{2}EI/l^2 & 24EI/l^3 \end{bmatrix} \begin{bmatrix} \theta^*_{xA} \\ \theta^*_{yA} \\ \delta^*_A \end{bmatrix} \quad \textbf{(11.30)}$$

where the stiffness matrix is the sum of the two $\boldsymbol{K}^*_{11}$ matrices. Solving this equation gives

$$\theta^*_{xA} = 0 \ , \ \theta^*_{yA} = \frac{-pl^3}{3\sqrt{2}(GJ + EI)} \ , \ \delta^*_A = \frac{pl^4(GJ + 3EI)}{24EI(GJ + EI)} \quad \textbf{(11.31)}$$

{11.4 Three-Dimensional Frames}

{The stiffness matrices for pin-jointed bars in three dimensions can readily be written in terms of the global coordinates. The bar AB in Figure 11.9 is of length l and has projected lengths X, Y and Z in the directions x^*, y^* and z^* of a global coordinate system for the whole frame. If an axial force F_A is applied at the end A of AB, it will have components F_{xA}^*, F_{yA}^* and F_{zA}^* in the directions of the global coordinates. These components are related to F_A by

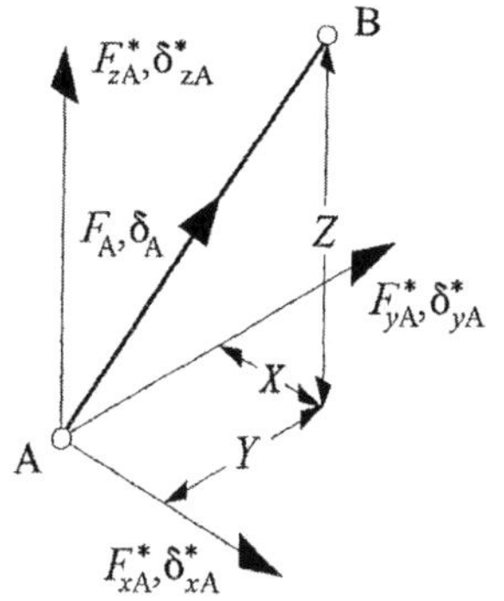

Figure 11.9 Spatial coordinates.

$$\begin{bmatrix} F^*_{xA} \\ F^*_{yA} \\ F^*_{zA} \end{bmatrix} = F_A \begin{bmatrix} X/l \\ Y/l \\ Z/l \end{bmatrix} \quad \textbf{(11.32)}$$

The axial displacement δ_A of the end A can be expressed in terms of its components δ_{xA}^*, δ_{yA}^* and δ_{zA}^* by

$$\delta_A = \begin{bmatrix} X/l & Y/l & Z/l \end{bmatrix} \begin{bmatrix} \delta^*_{xA} \\ \delta^*_{yA} \\ \delta^*_{zA} \end{bmatrix} \quad \textbf{(11.33)}$$

A similar equation can be written relating the axial displacement of end B, δ_B, to its components. Now taking the cross-sectional area of the bar to be A and its Young's modulus to be E,

$$F_A = \frac{EA}{l}(\delta_A - \delta_B) \tag{11.34}$$

From (11.32) to (11.34),

$$\begin{bmatrix} F^*_{xA} \\ F^*_{yA} \\ F^*_{zA} \end{bmatrix} = \begin{bmatrix} X/l \\ Y/l \\ Z/l \end{bmatrix} \frac{EA}{l} \begin{bmatrix} X/l & Y/l & Z/l \end{bmatrix} \begin{bmatrix} \delta^*_{xA} - \delta^*_{xB} \\ \delta^*_{yA} - \delta^*_{yB} \\ \delta^*_{zA} - \delta^*_{zB} \end{bmatrix}$$

$$= \frac{EA}{l^3} \begin{bmatrix} X^2 & XY & XZ \\ XY & Y^2 & YZ \\ XZ & YZ & Z^2 \end{bmatrix} \begin{bmatrix} \delta^*_{xA} \\ \delta^*_{yA} \\ \delta^*_{zA} \end{bmatrix} - \frac{EA}{l^3} \begin{bmatrix} X^2 & XY & XZ \\ XY & Y^2 & YZ \\ XZ & YZ & Z^2 \end{bmatrix} \begin{bmatrix} \delta^*_{xB} \\ \delta^*_{yB} \\ \delta^*_{zB} \end{bmatrix} \tag{11.35}$$

For equilibrium, the force on end B of the bar is equal and opposite to that at end A, so that

$$\begin{bmatrix} F^*_{xB} \\ F^*_{yB} \\ F^*_{zB} \end{bmatrix} = -\frac{EA}{l^3} \begin{bmatrix} X^2 & XY & XZ \\ XY & Y^2 & YZ \\ XZ & YZ & Z^2 \end{bmatrix} \begin{bmatrix} \delta^*_{xA} \\ \delta^*_{yA} \\ \delta^*_{zA} \end{bmatrix} + \frac{EA}{l^3} \begin{bmatrix} X^2 & XY & XZ \\ XY & Y^2 & YZ \\ XZ & YZ & Z^2 \end{bmatrix} \begin{bmatrix} \delta^*_{xB} \\ \delta^*_{yB} \\ \delta^*_{zB} \end{bmatrix} \tag{11.36}$$

Again, (11.35) and (11.36) correspond to the general form given by (11.7). The process of solution is as before.

Example 11.4 A Simple Space-Frame

The pin-jointed space frame shown in Figure 11.10 was previously examined in §2.7. The projected lengths X, Y, Z of the bars can be found from the coordinates of the joints shown next to the joint letters. Taking Young's modulus E to be common to all the bars, their cross-sectional areas A will be taken as roughly proportional to their axial loads. The following table gives the bar properties from which the stiffness matrices can be formed:

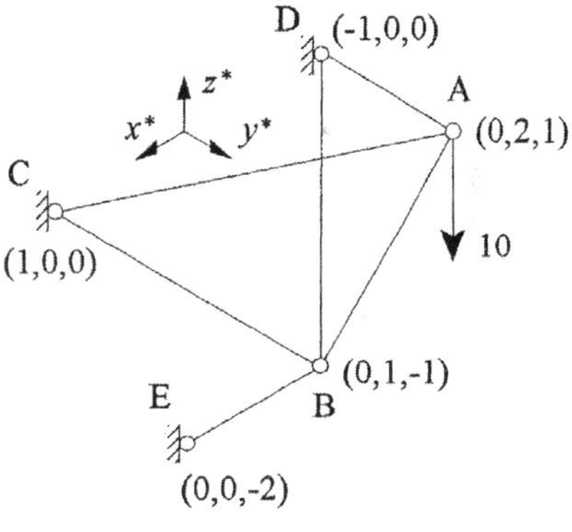

Figure 11.10 A simple space frame.

Bar	X	Y	Z	l	EA	EA/l^3
AB	0	-1	-2	$\sqrt{5}$	15	$3/\sqrt{5}$
AC	1	-2	-1	$\sqrt{6}$	4	$2/(3\sqrt{6})$
AD	-1	-2	-1	$\sqrt{6}$	4	$2/(3\sqrt{6})$
BC	1	-1	1	$\sqrt{3}$	3	$1/\sqrt{3}$
BD	-1	-1	1	$\sqrt{3}$	3	$1/\sqrt{3}$
BE	0	-1	-1	$\sqrt{2}$	14	$7/\sqrt{2}$

The K^*_{11} matrices found from this table are

$$K^*_{11(AB)} = \frac{3}{\sqrt{5}}\begin{bmatrix} 0 & 0 & 0 \\ 0 & 1 & 2 \\ 0 & 2 & 4 \end{bmatrix} , \; K^*_{11(AC)} = \frac{2}{(3\sqrt{6})}\begin{bmatrix} 1 & -2 & -1 \\ -2 & 4 & 2 \\ -1 & 2 & 1 \end{bmatrix} , \; K^*_{11(AD)} = \frac{2}{(3\sqrt{6})}\begin{bmatrix} 1 & 2 & 1 \\ 2 & 4 & 2 \\ 1 & 2 & 1 \end{bmatrix}$$

$$K^*_{11(BC)} = \frac{1}{\sqrt{3}}\begin{bmatrix} 1 & -1 & 1 \\ -1 & 1 & -1 \\ 1 & -1 & 1 \end{bmatrix} , \; K^*_{11(BD)} = \frac{1}{\sqrt{3}}\begin{bmatrix} 1 & 1 & -1 \\ 1 & 1 & -1 \\ -1 & -1 & 1 \end{bmatrix} , \; K^*_{11(BE)} = \frac{7}{\sqrt{2}}\begin{bmatrix} 0 & 0 & 0 \\ 0 & 1 & 1 \\ 0 & 1 & 1 \end{bmatrix} . \quad \textbf{(11.37)}$$

From (11.35) and (11.36) it will be seen that the K^*_{22} matrices are the same and the K^*_{12} and K^*_{21} matrices are equal to $-K^*_{11}$. Assembling these matrices into the complete stiffness matrix for the structure gives

$$\begin{bmatrix} P^*_A \\ P^*_B \end{bmatrix} = \begin{bmatrix} K^*_{11(AB)} + K^*_{11(AC)} + K^*_{11(AD)} & K^*_{12(AB)} \\ K^*_{21(AB)} & K^*_{22(AB)} + K^*_{11(BC)} + K^*_{11(BD)} + K^*_{11(BE)} \end{bmatrix} \begin{bmatrix} d^*_A \\ d^*_B \end{bmatrix} \quad \textbf{(11.38)}$$

or in terms of the above matrices and the load applied at joint A,

$$\begin{bmatrix} 0 \\ 0 \\ -10 \\ 0 \\ 0 \\ 0 \end{bmatrix} = \begin{bmatrix} \frac{4}{3\sqrt{6}} & 0 & 0 & 0 & 0 & 0 \\ 0 & \frac{3}{\sqrt{5}} + \frac{16}{3\sqrt{6}} & \frac{6}{\sqrt{5}} + \frac{8}{3\sqrt{6}} & 0 & -\frac{3}{\sqrt{5}} & -\frac{6}{\sqrt{5}} \\ 0 & \frac{6}{\sqrt{5}} + \frac{8}{3\sqrt{6}} & \frac{12}{\sqrt{5}} + \frac{4}{3\sqrt{6}} & 0 & -\frac{6}{\sqrt{5}} & -\frac{12}{\sqrt{5}} \\ 0 & 0 & 0 & \frac{2}{\sqrt{3}} & 0 & 0 \\ 0 & -\frac{3}{\sqrt{5}} & -\frac{6}{\sqrt{5}} & 0 & \frac{3}{\sqrt{5}} + \frac{2}{\sqrt{3}} + \frac{7}{\sqrt{2}} & \frac{6}{\sqrt{5}} - \frac{2}{\sqrt{3}} + \frac{7}{\sqrt{2}} \\ 0 & -\frac{6}{\sqrt{5}} & -\frac{12}{\sqrt{5}} & 0 & \frac{6}{\sqrt{5}} - \frac{2}{\sqrt{3}} + \frac{7}{\sqrt{2}} & \frac{12}{\sqrt{5}} + \frac{2}{\sqrt{3}} + \frac{7}{\sqrt{2}} \end{bmatrix} \begin{bmatrix} \delta^*_{xA} \\ \delta^*_{yA} \\ \delta^*_{zA} \\ \delta^*_{xB} \\ \delta^*_{yB} \\ \delta^*_{zB} \end{bmatrix} \quad \textbf{(11.39)}$$

This immediately gives δ^*_{xA} and δ^*_{xB} as zero and the remaining four equations give

$$\delta^*_{yA} = 7.2301 \;,\; \delta^*_{zA} = -8.3365 \;,\; \delta^*_{yB} = 0.43322 \;,\; \delta^*_{zB} = -2.4535 \,. \quad \textbf{(11.40)}$$

The member end loads in global coordinates can then be found from the member stiffness matrices. The axial loads are then found from these end loads using transformations of the type given by (11.33). These are the same as those found by statics in §2.7.

If the three-dimensional frame is rigid-jointed, the flexure and torsion of the members must also be taken into account. In this case, the orientation of a member is not given completely by the position of its ends. However, in many engineering structures, the major axes of the members are chosen to lie in a horizontal plane. Provided that the member itself is not vertical, its orientation is then defined by the position of its ends. Let X, Y and Z be the relative position of end B with respect to end A of a beam AB (as in the case of the above pin-jointed space frames).

Suppose for the moment that the major x axis of the beam lies in the x^*y^* plane of the global coordinate system. The projection of the length l of the beam on this plane will be $\sqrt{(X^2+Y^2)}$ which will be denoted by s as shown in Figure 11.11. The relationship between the local and global coordinate systems will be found in two stages. First, the components of a vector $\boldsymbol{F}$ relative to an intermediate coordinate system x', y', z' will be found in terms of the components relative to x,y,z. As shown in Figure 11.11c, the axes x and x' are coincident, but z' lies along AC. Then from Figure 11.11b, the components are

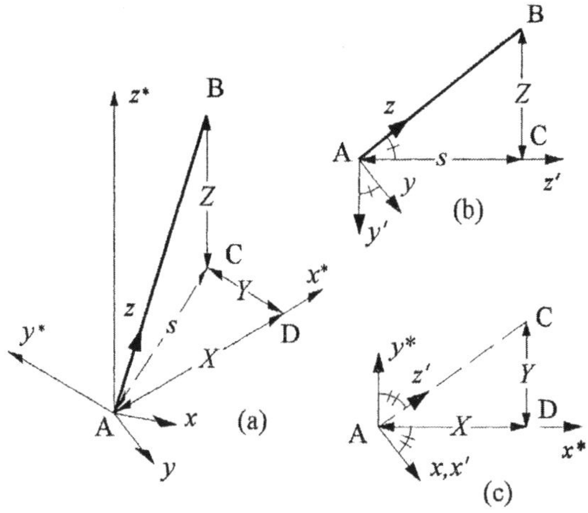

Figure 11.11 Orientation of beam coordinates relative to frame coordinates.

$$\begin{bmatrix} F_x' \\ F_y' \\ F_z' \end{bmatrix} = \begin{bmatrix} 1 & 0 & 0 \\ 0 & \dfrac{s}{l} & -\dfrac{Z}{l} \\ 0 & \dfrac{Z}{l} & \dfrac{s}{l} \end{bmatrix} \begin{bmatrix} F_x \\ F_y \\ F_z \end{bmatrix} \tag{11.41}$$

The relationship between the intermediate coordinates and the global coordinates is shown in Figure 11.11c, where the y' axis is in the $-z^*$ direction. (Note that right-hand coordinate systems are being used consistently.) The relationship between the components of $\boldsymbol{F}$ are then

$$\begin{bmatrix} F_x^* \\ F_y^* \\ F_z^* \end{bmatrix} = \begin{bmatrix} \dfrac{Y}{s} & 0 & \dfrac{X}{s} \\ -\dfrac{X}{s} & 0 & \dfrac{Y}{s} \\ 0 & -1 & 0 \end{bmatrix} \begin{bmatrix} F_x' \\ F_y' \\ F_z' \end{bmatrix} \tag{11.42}$$

The product of these two transformation matrices then gives the matrix for transformation from local to global coordinates. If the local coordinates are not such that the x axis lies in the x^*y^* plane, but the section has been rotated through a clockwise angle β[1] about the beam's axis from this position, then multiplication by a further transformation matrix is required, giving the overall transformation matrix as

[1] See Figure A2.2 of Appendix 2.

$$T = \begin{bmatrix} \frac{Y}{s} & \frac{XZ}{ls} & \frac{X}{l} \\ -\frac{X}{s} & \frac{YZ}{ls} & \frac{Y}{l} \\ 0 & -\frac{s}{l} & \frac{Z}{l} \end{bmatrix} \begin{bmatrix} \cos\beta & -\sin\beta & 0 \\ \sin\beta & \cos\beta & 0 \\ 0 & 0 & 1 \end{bmatrix} = \begin{bmatrix} xz\sin\beta + y\cos\beta & xz\cos\beta - y\sin\beta & xS \\ yz\sin\beta - x\cos\beta & yz\cos\beta + x\sin\beta & yS \\ -S\sin\beta & -S\cos\beta & z \end{bmatrix} \quad \textbf{(11.43)}$$

where

$$s^2 = X^2 + Y^2 \ , \quad x = \frac{X}{s} \ , \quad y = \frac{Y}{s} \ , \quad z = \frac{Z}{l} \ , \quad S = \frac{s}{l} \ . \quad \textbf{(11.44)}$$

Figure 11.12 shows the convention for the loads and deflexions of a typical member AB of a rigid-jointed three-dimensional framework. The major axis of the cross-section is x, y is the minor axis and z is the centroidal axis of the beam. The end forces and displacements are positive in the directions of these axes and the end moments and rotations are clockwise-positive about them. As before, the subscripts A or B will be added depending on whether they refer to end A or B of the beam. The load vectors in (11.7) will be taken to be of the form

$$P_A = \begin{bmatrix} M_{xA} \\ M_{yA} \\ M_{zA} \\ F_{xA} \\ F_{yA} \\ F_{zA} \end{bmatrix} \ , \quad P_B = \begin{bmatrix} M_{xB} \\ M_{yB} \\ M_{zB} \\ F_{xB} \\ F_{yB} \\ F_{zB} \end{bmatrix} \quad \textbf{(11.45)}$$

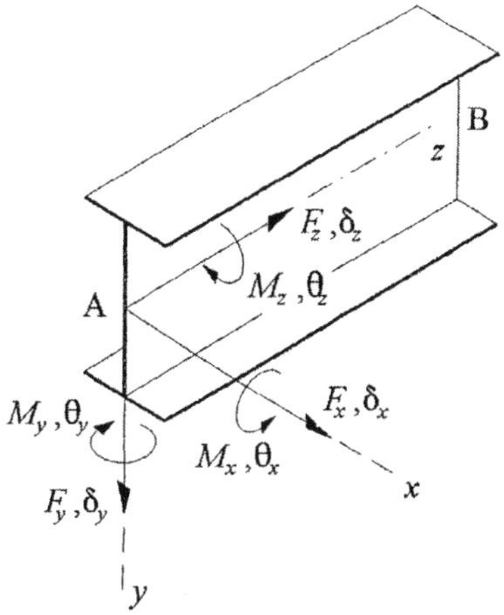

Figure 11.12 Loads and corresponding deflexions in three dimensions.

and the corresponding deflexion vectors to be of the form

$$d_A = \begin{bmatrix} \theta_{xA} \\ \theta_{yA} \\ \theta_{zA} \\ \delta_{xA} \\ \delta_{yA} \\ \delta_{zA} \end{bmatrix} \ , \quad d_B = \begin{bmatrix} \theta_{xB} \\ \theta_{yB} \\ \theta_{zB} \\ \delta_{xB} \\ \delta_{yB} \\ \delta_{zB} \end{bmatrix} \quad \textbf{(11.46)}$$

The coefficients of the stiffness matrices are found in a similar manner to those of the plane rigid-jointed frames. They are best expressed as arrays of 3×3 submatrices of the form

$$K_{11} = \begin{bmatrix} A_{11} & B_1 \\ B_1^t & C \end{bmatrix} , \quad K_{12} = K_{21}^t = \begin{bmatrix} A_{12} & -B_1 \\ -B_2^t & -C \end{bmatrix} , \quad K_{22} = \begin{bmatrix} A_{22} & B_2 \\ B_2^t & C \end{bmatrix} . \qquad (11.47)$$

where

$$A_{11} = A_{22} = \begin{bmatrix} 4EI_x/l & 0 & 0 \\ 0 & 4EI_y/l & 0 \\ 0 & 0 & GJ/l \end{bmatrix} , \quad A_{12} = \begin{bmatrix} 2EI_x/l & 0 & 0 \\ 0 & 2EI_y/l & 0 \\ 0 & 0 & -GJ/l \end{bmatrix}$$

$$B_1 = -B_2 = \begin{bmatrix} 0 & -6EI_x/l^2 & 0 \\ 6EI_y/l^2 & 0 & 0 \\ 0 & 0 & 0 \end{bmatrix} , \quad C = \begin{bmatrix} 12EI_y/l^3 & 0 & 0 \\ 0 & 12EI_x/l^3 & 0 \\ 0 & 0 & EA/l \end{bmatrix} . \qquad (11.48)$$

In this case, it will be seen that it is not strictly necessary to define A_{22} and B_2 as separate entities. However, there are more general problems[1] where all six submatrices are unrelated. The member stiffness matrices can be transformed to refer to the global coordinate system by transforming these submatrices. Thus

$$K_{11}^* = \begin{bmatrix} A_{11}^* & B_1^* \\ (B_1^*)^t & C^* \end{bmatrix} , \quad \text{where} \quad A_{11}^* = TA_{11}T^t \, , \; B_1^* = TB_1T^t \, , \; C^* = TCT^t \qquad (11.49)$$

and T is given by (11.43). The other member stiffness matrices transform in a similar fashion.}

{11.5 The Stability of Frames}

{The stability of frames can be analysed using modifications of the above stiffness matrix method, provided that the deflexions of the joints prior to buckling do not change the geometry of the frame significantly. A bifurcation is sought such that at a critical multiple λ_c of the vector of the given joint loads on the frame, the deflexions can change by an infinitesimal amount without a change in this loading. Suppose that the column vector of given joint loads is $\boldsymbol{P}$ and that induces a set of joint deflexions given by $\boldsymbol{d}$. These are related by the frame stiffness matrix K_P where the stiffness coefficients are taken to be dependent on the axial loading in the members, and hence on $\boldsymbol{P}$. At buckling, there is an adjacent state of equilibrium for which the deflexion vector is $\boldsymbol{d}+\delta\boldsymbol{d}$ but the loading $\boldsymbol{P}$ remains the same. Then

$$\lambda_c P = K_P d = K_P(d + \delta d) \qquad \text{so that} \qquad K_P \delta d = 0 \qquad (11.50)$$

As $\delta\boldsymbol{d}$ is a non-zero vector, the determinant of K_P must be zero. This is the buckling criterion for such problems. As discussed in Chapter 9, flexural and torsional stiffness may be modified by large axial loads. They also modify the stiffness matrices for pin-ended bars. In setting up (11.9), it was assumed that a pin-ended bar has no lateral stiffness. However, if the bar carries a large compressive axial force P, a lateral displacement δ_{yA} requires a force of $P\delta_{yA}/l$ in the opposite direction to maintain equilibrium. (This can be seen by taking moments about B.) In effect, this means that the bar has a negative stiffness $-P/l$ in the lateral direction. Equations (11.12) then

[1] Renton, J.D. (1962) Stability of Space Frames by Computer Analysis. *Proc. A.S.C.E.* V.88, No.ST4 p.81.

become

$$\begin{bmatrix} F_{xA} \\ F_{yA} \end{bmatrix} = \begin{bmatrix} EA/l & 0 \\ 0 & -P/l \end{bmatrix} \begin{bmatrix} \delta_{xA} \\ \delta_{yA} \end{bmatrix} + \begin{bmatrix} -EA/l & 0 \\ 0 & P/l \end{bmatrix} \begin{bmatrix} \delta_{xB} \\ \delta_{yB} \end{bmatrix}$$
$$\begin{bmatrix} F_{xB} \\ F_{yB} \end{bmatrix} = \begin{bmatrix} -EA/l & 0 \\ 0 & P/l \end{bmatrix} \begin{bmatrix} \delta_{xA} \\ \delta_{yA} \end{bmatrix} + \begin{bmatrix} EA/l & 0 \\ 0 & -P/l \end{bmatrix} \begin{bmatrix} \delta_{xB} \\ \delta_{yB} \end{bmatrix} \quad \textbf{(11.51)}$$

For a pin-jointed space frame, negative lateral stiffnesses of $-P/l$ have to introduced in both the bar's x and y directions (the z direction being along the axis of the bar). Taking end B of a bar AB to be prevented from displacing, the relationship between the forces and displacements at end A are given by

$$\begin{bmatrix} F_{xA} \\ F_{yA} \\ F_{zA} \end{bmatrix} = \begin{bmatrix} -P/l & 0 & 0 \\ 0 & -P/l & 0 \\ 0 & 0 & EA/l \end{bmatrix} \begin{bmatrix} \delta_{xA} \\ \delta_{yA} \\ \delta_{zA} \end{bmatrix} = \frac{(EA+P)}{l} \begin{bmatrix} 0 & 0 & 0 \\ 0 & 0 & 0 \\ 0 & 0 & 1 \end{bmatrix} \begin{bmatrix} \delta_{xA} \\ \delta_{yA} \\ \delta_{zA} \end{bmatrix} - \frac{P}{l} \begin{bmatrix} 1 & 0 & 0 \\ 0 & 1 & 0 \\ 0 & 0 & 1 \end{bmatrix} \begin{bmatrix} \delta_{xA} \\ \delta_{yA} \\ \delta_{zA} \end{bmatrix} \quad \textbf{(11.52)}$$

The transformation to global coordinates of the first expression on the final right-hand side of (11.52) is exactly that used to obtain (11.35). The second matrix is a unit matrix, $\boldsymbol{I}$. Taking the transformation matrix to be $\boldsymbol{T}$, this transforms to $\boldsymbol{TIT}^t$ in global coordinates. However, the transpose of any such transformation matrix is its inverse. Thus $\boldsymbol{TIT}^t$ is just $\boldsymbol{I}$. Then from (11.35) and (11.52) the $\boldsymbol{K}_{11}$ matrix for an axially-loaded pin-ended bar becomes

$$\boldsymbol{K}^*_{11} = \frac{(EA+P)}{l^3} \begin{bmatrix} X^2 & XY & XZ \\ XY & Y^2 & YZ \\ XZ & YZ & Z^2 \end{bmatrix} - \frac{P}{l} \begin{bmatrix} 1 & 0 & 0 \\ 0 & 1 & 0 \\ 0 & 0 & 1 \end{bmatrix} \quad \textbf{(11.53)}$$

in global coordinates. As before, the $\boldsymbol{K}^*_{22}$ matrix is equal to this and the $\boldsymbol{K}^*_{12}$ and $\boldsymbol{K}^*_{21}$ matrices are equal to $-\boldsymbol{K}^*_{11}$. The stiffness matrix for a pin-jointed plane or space frame can now be set up in the usual way and the buckling criterion found from its determinant. It must also be remembered that in addition to overall buckling of the frame, individual bars can buckle as pin-ended columns. Thus the axial compression in each bar must be checked to see that it does not exceed P_e as given by (9.6b).

Example 11.5 Buckling of a Built-Up Column

The pin-jointed frame shown in Figure 11.13 is designed to take an axial force F applied at joints A and E. These joints are prevented from moving (apart from a relative vertical motion to allow the frame to take the load F). The coordinates of each joint relative to the global system (x^*, y^*, z^*) are shown next the joint. Thus the joints B, C and D form an equilateral triangle of side $2\sqrt{3}$. The cross-sectional area of each bar is proportional to the amplitude of the axial load it carries. The bars meeting at the top and bottom joints A and E all have an axial compression P equal to $F\sqrt{5}/6$. The axial compression P in the horizontal bars BC, BD and CD is $-2F/3\sqrt{3}$. The axial stiffnesses of these two sets of bars are then $EA_0\sqrt{5}/6$ and

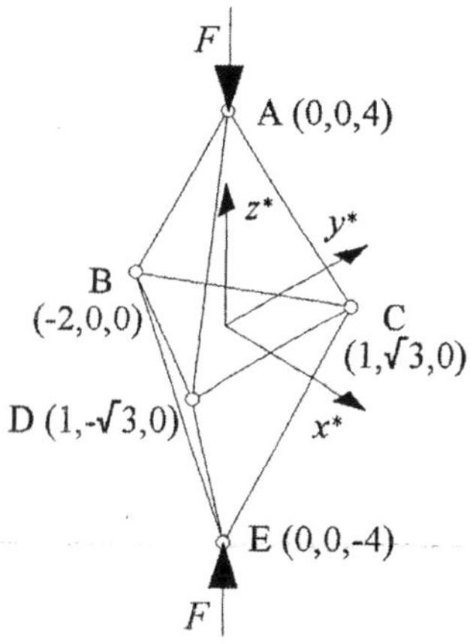

Figure 11.13 A built-up column.

$2EA_0/3\sqrt{3}$ respectively, where EA_0 is a reference stiffness. The ratio F/EA_0 will be taken as λ.

Assembling the bar stiffness matrices to give the stiffness matrix for the whole frame gives

$$\begin{bmatrix} P^*_B \\ P^*_C \\ P^*_D \end{bmatrix} = \begin{bmatrix} K^*_{22(AB)} + K^*_{11(BC)} + K^*_{11(BD)} + K^*_{11(BE)} & K^*_{12(BC)} & K^*_{12(BD)} \\ K^*_{21(BC)} & K^*_{22(AC)} + K^*_{22(BC)} + K^*_{11(CD)} + K^*_{11(CE)} & K^*_{12(CD)} \\ K^*_{21(BD)} & K^*_{21(CD)} & K^*_{22(AD)} + K^*_{22(BD)} + K^*_{22(CD)} + K^*_{11(DE)} \end{bmatrix} \begin{bmatrix} d^*_B \\ d^*_C \\ d^*_D \end{bmatrix} \quad (11.54)$$

The relationships between the forces and the displacements in the z direction are uncoupled from the other six equations, so that the above can be written as two matrix equations,

$$\begin{bmatrix} F^*_{xB} \\ F^*_{yB} \\ F^*_{xC} \\ F^*_{yC} \\ F^*_{xD} \\ F^*_{yD} \end{bmatrix} = EA_0 \begin{bmatrix} 72-28\lambda & 0 & -30-10\lambda & -\sqrt{3}(10-10\lambda) & -30-10\lambda & \sqrt{3}(10-10\lambda) \\ 0 & 20 & -\sqrt{3}(10-10\lambda) & -10-30\lambda & \sqrt{3}(10-10\lambda) & -10-30\lambda \\ -30-10\lambda & -\sqrt{3}(10-10\lambda) & 33-7\lambda & \sqrt{3}(13-7\lambda) & -40\lambda & 0 \\ -\sqrt{3}(10-10\lambda) & -10-30\lambda & \sqrt{3}(13-7\lambda) & 59-21\lambda & 0 & -40 \\ -30-10\lambda & \sqrt{3}(10-10\lambda) & -40\lambda & 0 & 33-7\lambda & -\sqrt{3}(13-7\lambda) \\ \sqrt{3}(10-10\lambda) & -10-30\lambda & 0 & -40 & -\sqrt{3}(13-7\lambda) & 59-21\lambda \end{bmatrix} \begin{bmatrix} \delta^*_{xB} \\ \delta^*_{yB} \\ \delta^*_{xC} \\ \delta^*_{yC} \\ \delta^*_{xD} \\ \delta^*_{yD} \end{bmatrix} \quad (11.55)$$

and

$$\begin{bmatrix} F^*_{zB} \\ F^*_{zC} \\ F^*_{zD} \end{bmatrix} = EA_0 \begin{bmatrix} 12+17\lambda & -10\lambda & -10\lambda \\ -10\lambda & 12+17\lambda & -10\lambda \\ -10\lambda & -10\lambda & 12+17\lambda \end{bmatrix} \begin{bmatrix} \delta^*_{zB} \\ \delta^*_{zC} \\ \delta^*_{zD} \end{bmatrix} \quad (11.56)$$

If the unit of length had been L, so that the distance from A to E became $8L$, then it would be only necessary to change the factor EA_0 to EA_0/L in these two equations. Equation (11.55) can be reduced to two sets of three simultaneous equations which relate symmetric joint loads to symmetric joint displacements and antisymmetric joint loads to antisymmetric joint displacements respectively, taking the x^*z^* plane as the plane of symmetry. Instability can be found by equating the determinants of the stiffness matrices in each case to zero. This gives the critical values of the load parameter, λ_c, as

$$\lambda_c = -\sqrt{26}-5, \quad -\tfrac{12}{7}, \quad -\tfrac{4}{9}, \quad 0, \quad \sqrt{26}-5 \quad \text{and} \quad 2.75. \quad (11.57)$$

The first, third and fifth values are repeated roots of the determinants, giving nine roots in all. The value zero arises because no constraint was introduced to prevent the column rotating about the z^* axis. Note that buckling is possible under tensile end loads F. The non-zero values of λ_c are so large that yielding or Euler buckling are likely to occur first. This is not always the case. For axially-loaded pin-jointed trusses with many bays, instability of this kind may well be the primary cause of failure.

The stability of rigid-jointed frames can be analysed by modifying the stiffness equations, using the stability functions introduced in §9.3. Thus for the fully three-dimensional case, (11.48) becomes

$$\boldsymbol{A}_{11} = \boldsymbol{A}_{22} = \begin{bmatrix} 4EI_x\phi_{3x}/l & 0 & 0 \\ 0 & 4EI_y\phi_{3y}/l & 0 \\ 0 & 0 & (GJ-Pr_0^2)/l \end{bmatrix}, \quad \boldsymbol{A}_{12} = \begin{bmatrix} 2EI_x\phi_{4x}/l & 0 & 0 \\ 0 & 2EI_y\phi_{4y}/l & 0 \\ 0 & 0 & -(GJ-Pr_0^2)/l \end{bmatrix}$$

$$\boldsymbol{B}_1 = -\boldsymbol{B}_2 = \begin{bmatrix} 0 & -6EI_x\phi_{2x}/l^2 & 0 \\ 6EI_y\phi_{2y}/l^2 & 0 & 0 \\ 0 & 0 & 0 \end{bmatrix}, \quad \boldsymbol{C} = \begin{bmatrix} 12EI_y\phi_{1y}/l^3 & 0 & 0 \\ 0 & 12EI_x\phi_{1x}/l^3 & 0 \\ 0 & 0 & EA/l \end{bmatrix}. \quad (11.58)$$

where the extra subscript x or y used for the ϕ functions indicates that they should be determined using the bending stiffnesses EI_x or EI_y respectively. The modified torsional stiffness $GJ\text{-}Pr_0^2$ is appropriate for the case when the beam may be treated as if the end bimoments were negligible. This can be deduced from (9.26) by setting θ'' to zero at z equal to 0 and l. Here, as previously, the beams are assumed to be bisymmetrical. More complex cases where this is not so and allowance may be needed for rigid gusset plates at the joints have also been examined by the author (ibid.).}

{11.6 The Vibration of Frames}

{The matrix analysis of frames subject to harmonic loading follows similar lines to the above stability analysis. The functions derived in Chapter 10 can be used to modify the stiffness matrices used for the ordinary linear analysis, such as those given by (11.47). In the case of a bisymmetrical prismatic beam they become

$$\boldsymbol{K}_{11} = \begin{bmatrix} \boldsymbol{A}_{11} & \boldsymbol{B}_1 \\ \boldsymbol{B}_1^t & \boldsymbol{C}_1 \end{bmatrix}, \quad \boldsymbol{K}_{12} = \boldsymbol{K}_{21}^t = \begin{bmatrix} \boldsymbol{A}_{12} & -\boldsymbol{B}_2 \\ \boldsymbol{B}_2^t & -\boldsymbol{C}_2 \end{bmatrix}, \quad \boldsymbol{K}_{22} = \begin{bmatrix} \boldsymbol{A}_{11} & -\boldsymbol{B}_1 \\ -\boldsymbol{B}_1^t & \boldsymbol{C}_1 \end{bmatrix}. \qquad \textbf{(11.59)}$$

where

$$\boldsymbol{A}_{11} = \begin{bmatrix} 4EI_x c_{1x}/l & 0 & 0 \\ 0 & 4EI_y c_{1y}/l & 0 \\ 0 & 0 & GJd_{1T}/l \end{bmatrix}, \quad \boldsymbol{A}_{12} = \begin{bmatrix} 2EI_x c_{2x}/l & 0 & 0 \\ 0 & 2EI_y c_{2y}/l & 0 \\ 0 & 0 & -GJd_{2T}/l \end{bmatrix},$$

$$\boldsymbol{B}_1 = \begin{bmatrix} 0 & -6EI_x c_{3x}/l^2 & 0 \\ 6EI_y c_{3y}/l^2 & 0 & 0 \\ 0 & 0 & 0 \end{bmatrix}, \quad \boldsymbol{B}_2 = \begin{bmatrix} 0 & -6EI_x c_{4x}/l^2 & 0 \\ 6EI_y c_{4y}/l^2 & 0 & 0 \\ 0 & 0 & 0 \end{bmatrix}, \qquad \textbf{(11.60)}$$

$$\boldsymbol{C}_1 = \begin{bmatrix} 12EI_y c_{5y}/l^3 & 0 & 0 \\ 0 & 12EI_x c_{5x}/l^3 & 0 \\ 0 & 0 & EAd_{1P}/l \end{bmatrix}, \quad \boldsymbol{C}_2 = \begin{bmatrix} 12EI_y c_{6y}/l^3 & 0 & 0 \\ 0 & 12EI_x c_{6x}/l^3 & 0 \\ 0 & 0 & EAd_{2P}/l \end{bmatrix}.$$

Note that it is necessary to have separate submatrices $\boldsymbol{B}_1$ and $\boldsymbol{B}_2$, $\boldsymbol{C}_1$ and $\boldsymbol{C}_2$. This is because there is dynamic loading along the beam and not just at its ends. The c and d functions are those given by (10.9), (10.25) and (10.33). The additional subscripts x and y refer to flexure about the principal axes x and y respectively. The subscripts P and T refer to axial and torsional vibration, discussed in §10.4 and §10.5 respectively.

In this case, the stiffness matrix $\boldsymbol{K}_\omega$ for the whole structure assembled from such submatrices relates the joint loads to the corresponding joint deflexions, where all the loads and deflexions are a common sinusoidal function of time, $\cos(\omega t+\phi)$. The response of a structure to synchronous harmonic loading of any given frequency[1] can then be found. The natural frequencies of a structure are given by analysing the cases when vibration of frame can take place without any forcing loads of the above kind. From the relationship between the loads and deflexions, we then have

[1] Note that in the software provided, this frequency is expressed in terms of ω. The value in hertz is given by $\omega/2\pi$.

$$0 = P = K_\omega d \tag{11.61}$$

so that for non-zero joint deflexions, the determinant of K_ω must be zero(cf. (11.50)). This is a transcendental function of ω with an infinite number of roots. The problem is similar to that of the stability of frames analysed in §11.5, although now a set of natural frequencies is likely to be of interest, rather than the lowest buckling load. In this case, it becomes particularly important to determine whether all the roots of the determinant in a particular range of values have been found. A method for doing so devised by Wittrick and Williams[1] will be discussed in the next chapter.

When a frame carries relatively large inertia loading, the mass of its beams may be neglected. Alternatively, the distributed mass of a beam might be approximated by inertia loads at its ends. In mechanics, the inertia of a rod is sometimes represented by taking half its mass to be at each end. This gives an error in the rotational inertia of the rod, which can be corrected by taking two thirds of its mass to be concentrated at the centre of the rod and one sixth at either end. Another way of correcting this error is to introduce negative rotational inertias at the ends of the bar, but this can lead to unreal solutions. Such approximations are introduced in order to give equations which are linear in ω^2, so that then methods associated with classical eigenvalue problems can be used. As will be seen, this can be achieved using the linearisation given by (10.10).

The linearised equations take the form

$$P = Kd - \omega^2 Md \tag{11.62}$$

where P is a column vector of loads applied to the frame, d is the column vector of corresponding deflexions, K is the normal static stiffness matrix for the frame and M is known as a mass matrix. Here, it is assumed that the loading and deflexions are harmonic with a frequency ω and in phase. As before, K is symmetrical and M is diagonal if only lumped inertias at the joints are considered, or at least symmetrical if the linear terms found from (10.10) are used as well. If no loads P are applied, then for each natural frequency ω_i there will be a corresponding non-zero deflexion $d_i \cos(\omega_i t+\phi)$. If K and M are $n\times n$ matrices, there will in general be n such eigenvectors and eigenvalues. It will be assumed that these are arranged in order, so that the lowest eigenvalue is ω_1 and the highest is ω_n. Let ω_i and ω_j be distinct eigenvalues. Then from (11.62),

$$Kd_i = \omega_i^2 Md_i \quad , \quad Kd_j = \omega_j^2 Md_j \tag{11.63}$$

Each equation can be premultiplied by the transpose of the other eigenvector, giving

$$d_j^t Kd_i = \omega_i^2 d_j^t Md_i \quad , \quad d_i^t Kd_j = \omega_j^2 d_i^t Md_j \tag{11.64}$$

As K and M are symmetrical, the transpose of the first of these equations gives

$$d_i^t Kd_j = \omega_i^2 d_i^t Md_j \tag{11.65}$$

which is the same as the second, apart from the scalar ω_i^2 in the one case and the scalar ω_j^2 in the other. As these two scalars are different, it follows that

$$d_i^t Kd_j = d_i^t Md_j = 0 \tag{11.66}$$

---

[1] Wittrick, W.H. and Williams, F.W. (1971) A general algorithm for computing natural frequencies of elastic structures. *Quart. J. Mech. Appl. Math.* Vol. 24, p.263.

The joint deflexion eigenvectors are then said to be orthogonal with respect to the mass and stiffness matrices. This orthogonality can be used to uncouple the original equations. Consider the column vector[1] $\boldsymbol{D}$ which is related to the joint deflexion vector $\boldsymbol{d}$ by

$$\boldsymbol{d} = \left[\boldsymbol{d}_1\ \boldsymbol{d}_2\ \ldots\ \boldsymbol{d}_n\right]\boldsymbol{D} = \boldsymbol{U}\boldsymbol{D} \tag{11.67}$$

where $\boldsymbol{U}$ is the $n \times n$ modal matrix made from the eigenvectors $\boldsymbol{d}_i$ (i=1 to n). Note that it is not a function of time. The more general form of (11.62), where no simple harmonic motion is assumed, is given by

$$\boldsymbol{P} = \boldsymbol{K}\boldsymbol{d} + \boldsymbol{M}\ddot{\boldsymbol{d}} \tag{11.68}$$

From (11.67), this becomes

$$\boldsymbol{P} = \boldsymbol{K}\boldsymbol{U}\boldsymbol{D} + \boldsymbol{M}\boldsymbol{U}\ddot{\boldsymbol{D}} \tag{11.69}$$

which on premultiplying by $\boldsymbol{U}^t$ gives

$$\boldsymbol{U}^t\boldsymbol{P} = \boldsymbol{U}^t\boldsymbol{K}\boldsymbol{U}\boldsymbol{D} + \boldsymbol{U}^t\boldsymbol{M}\boldsymbol{U}\ddot{\boldsymbol{D}} \tag{11.70}$$

Both $\boldsymbol{U}^t\boldsymbol{K}\boldsymbol{U}$ and $\boldsymbol{U}^t\boldsymbol{M}\boldsymbol{U}$ are diagonal matrices. This is because the off-diagonal terms are given by $\boldsymbol{d}_i^t\boldsymbol{K}\boldsymbol{d}_j$ and $\boldsymbol{d}_i^t\boldsymbol{M}\boldsymbol{d}_j$ ($i \neq j$) respectively. From (11.66), these terms are zero. Then (11.70) is a set of uncoupled equations from which each component of $\boldsymbol{D}$ can be determined individually from $\boldsymbol{U}^t\boldsymbol{P}$.

The lowest natural frequency, ω_1, may be determined from an arbitrary trial eigenvector $\boldsymbol{d}_T$, provided that it is a linear combination of the eigenvectors,

$$\boldsymbol{d}_T = \sum_{i=1}^{n} a_i \boldsymbol{d}_i \tag{11.71}$$

where at least a_1 is non-zero. From (11.63),

$$\boldsymbol{K}^{-1}\boldsymbol{M}\boldsymbol{d}_i = \frac{1}{\omega_i^2}\boldsymbol{d}_i \tag{11.72}$$

so that premultiplying $\boldsymbol{d}_T$ m times by $\boldsymbol{K}^{-1}\boldsymbol{M}$ gives

$$(\boldsymbol{K}^{-1}\boldsymbol{M})^m \boldsymbol{d}_T = \sum_{i=1}^{n} \frac{a_i}{\omega_i^{2m}} \boldsymbol{d}_i \tag{11.73}$$

Thus the higher the value of m the more dominant the first term in the series becomes, as ω_1 is the less than the other frequencies. When m is sufficiently large, $(\boldsymbol{K}^{-1}\boldsymbol{M})^m\boldsymbol{d}_T$ is virtually a scalar multiple of $(\boldsymbol{K}^{-1}\boldsymbol{M})^{m-1}\boldsymbol{d}_T$. This multiple is then $1/\omega_1^2$ and $(\boldsymbol{K}^{-1}\boldsymbol{M})^m\boldsymbol{d}_T$ is virtually a scalar multiple of $\boldsymbol{d}_1$. Similarly, when the distributed mass of the beams is entirely negligible, the highest natural frequency of the structural system can be found by premultiplying $\boldsymbol{d}_T$ by $\boldsymbol{M}^{-1}\boldsymbol{K}$ m times. However, when the distributed mass of the beams is taken fully into account, the number of natural frequencies of the structural system is infinite.

Example 11.6 Vibration of a Portal Frame

All the members of the portal frame ABCD shown in Figure 11.14 have the same length l and flexural stiffness EI. For simplicity, they will be taken as axially rigid. The feet A and D are encastré and the possible deflexions of B and C are the rotations θ_B and θ_C and a common horizontal displacement u as shown. The masses M attached to joints B and C are each taken to

[1] The components of the column vector $\boldsymbol{D}$ are known as the *principal coordinates* of the system.

include half the mass of the lintel, ρAl, where ρ is the density of the beam material and A is its cross-sectional area. Each of these two joints also has a rotational inertia of magnitude J. In terms of the vibration functions given by (10.9), the equations governing the vibration of the frame are given by

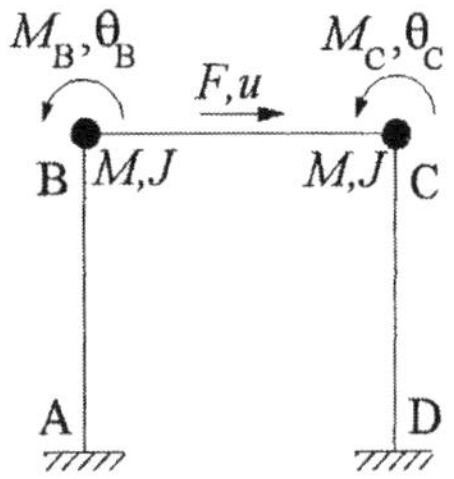

Figure 11.14 Vibration of a portal frame.

$$\begin{bmatrix} M_B \\ M_C \\ F \end{bmatrix} = \frac{EI}{l^3}\begin{bmatrix} 8l^2c_1 & 2l^2c_2 & 6lc_3 \\ 2l^2c_2 & 8l^2c_1 & 6lc_3 \\ 6lc_3 & 6lc_3 & 24c_5 \end{bmatrix}\begin{bmatrix} \theta_B \\ \theta_C \\ u \end{bmatrix} + \begin{bmatrix} J & 0 & 0 \\ 0 & J & 0 \\ 0 & 0 & 2M \end{bmatrix}\begin{bmatrix} \ddot{\theta}_B \\ \ddot{\theta}_C \\ \ddot{u} \end{bmatrix} \quad \textbf{(11.74)}$$

where M_B and M_C are the external moments applied to joints B and C and F is the total horizontal external force applied to BC. Suppose that ω is the frequency of harmonic vibration of the portal and that the non-dimensional constants m, j and λ are given by

$$\rho Alm = M \quad , \quad \rho Al^3 j = J \quad , \quad \lambda = \frac{\rho A\omega^2 l^4}{EI} \, . \qquad \textbf{(11.75)}$$

(Note that λ is as given by (10.10a).) It is then possible to express (11.74) in the non-dimensional form

$$\begin{bmatrix} \frac{M_B l}{EI} \\ \frac{M_C l}{EI} \\ \frac{Fl^2}{EI} \end{bmatrix} = \begin{bmatrix} 8c_1 & 2c_2 & 6c_3 \\ 2c_2 & 8c_1 & 6c_3 \\ 6c_3 & 6c_3 & 24c_5 \end{bmatrix}\begin{bmatrix} \theta_B \\ \theta_C \\ \frac{u}{l} \end{bmatrix} - \lambda\begin{bmatrix} j & 0 & 0 \\ 0 & j & 0 \\ 0 & 0 & 2m \end{bmatrix}\begin{bmatrix} \theta_B \\ \theta_C \\ \frac{u}{l} \end{bmatrix} \qquad \textbf{(11.76)}$$

Consider the particular case where no external loads are applied, j is zero and M is equal to half the mass of the beam BC. Suppose that λ is sufficiently small for the approximations given by (10.10) to apply. Then (11.76) becomes

$$\begin{bmatrix} 0 \\ 0 \\ 0 \end{bmatrix} = \begin{bmatrix} 8 & 2 & 6 \\ 2 & 8 & 6 \\ 6 & 6 & 24 \end{bmatrix}\begin{bmatrix} \theta_B \\ \theta_C \\ \frac{u}{l} \end{bmatrix} - \lambda\begin{bmatrix} \frac{2}{105} & -\frac{1}{140} & \frac{11}{210} \\ -\frac{1}{140} & \frac{2}{105} & \frac{11}{210} \\ \frac{11}{210} & \frac{11}{210} & \frac{61}{35} \end{bmatrix}\begin{bmatrix} \theta_B \\ \theta_C \\ \frac{u}{l} \end{bmatrix} \qquad \textbf{(11.77)}$$

The first 3×3 matrix on the right-hand side is the non-dimensional form of the stiffness matrix $\boldsymbol{K}$ and the second the non-dimensional form of a mass matrix $\boldsymbol{M}$. The non-dimensional form of $\boldsymbol{d}$ is $\{\theta_B, \theta_C, u/l\}$ and that of ω^2 is λ. The trial eigenvector $\boldsymbol{d}_T$ will be taken as {1, 1, 1}. Premultiplying it by $\boldsymbol{K}^{-1}\boldsymbol{M}$ six times yields an estimate of the lowest eigenvalue λ_1 of 10.31. The form of the eigenvector $\boldsymbol{d}_1$ is given as {1, 1, -1.809}. Using the exact expressions for the vibration functions gives the lowest eigenvalue as 10.27. (Iterating similarly with $\boldsymbol{M}^{-1}\boldsymbol{K}$ on the above trial eigenvector yields an eigenvalue of 1068 and an eigenvector of the form {1, 1, -0.0544}. However, as explained above, this cannot be thought of as the highest natural frequency of the system. Also, the approximation given by (10.10) loses validity for values of λ in excess of 100.)

Stiffness matrices for more complex problems, allowing for the vibration of Timoshenko beams under axial loading, have been formulated by Howson and Williams[1].}

---

[1] Howson, W.P. and Williams, F.W. (1973) *Journal of Sound and Vibration* Vol. 26 No. 4, p.503.

11.7 Equilibrium Matrices

Figure 11.15 shows a generalisation of a beam element. It is not necessarily uniform or straight, but is connected to a larger structural system at nodes (or joints) A and B. The coordinates of joint B relative to joint A are (X,Y,Z). The column vectors representing the loads and deflexions at these nodes will be taken to be of the same form as those given by (11.45) and (11.46). If the only loads are applied at A and B, then the equations of equilibrium can be written as

$$\begin{aligned} M_{xA} + ZF_{yA} - YF_{zA} + M_{xB} &= 0 \\ M_{yA} - ZF_{xA} + XF_{zA} + M_{xB} &= 0 \\ M_{zA} + YF_{xA} - XF_{yA} + M_{zB} &= 0 \\ F_{xA} + F_{xB} &= 0 \\ F_{yA} + F_{yB} &= 0 \\ F_{zA} + F_{zB} &= 0 \end{aligned} \tag{11.78}$$

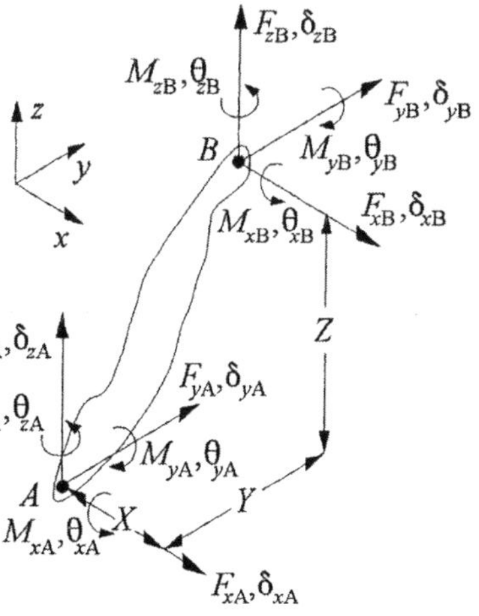

Figure 11.15 Equilibrium of a two-node element.

In matrix form, these become

$$\begin{bmatrix} 1 & 0 & 0 & 0 & Z & -Y \\ 0 & 1 & 0 & -Z & 0 & X \\ 0 & 0 & 1 & Y & -X & 0 \\ 0 & 0 & 0 & 1 & 0 & 0 \\ 0 & 0 & 0 & 0 & 1 & 0 \\ 0 & 0 & 0 & 0 & 0 & 1 \end{bmatrix} \begin{bmatrix} M_{xA} \\ M_{yA} \\ M_{zA} \\ F_{xA} \\ F_{yA} \\ F_{zA} \end{bmatrix} + \begin{bmatrix} M_{xB} \\ M_{yB} \\ M_{zB} \\ F_{xB} \\ F_{yB} \\ F_{zB} \end{bmatrix} = \begin{bmatrix} 0 \\ 0 \\ 0 \\ 0 \\ 0 \\ 0 \end{bmatrix} \tag{11.79}$$

or in terms of the notation of (11.45),

$$\boldsymbol{HP}_{\mathbf{A}} + \boldsymbol{P}_{\mathbf{B}} = 0 \tag{11.79a}$$

where $\boldsymbol{H}$ is known as an *equilibrium matrix*. The general form of the relationship between the end loads and the end deflexions of a linearly-elastic body is given by (11.7). Using the first of these equations and (11.79a), it follows that

$$\boldsymbol{P}_{\mathbf{B}} = -\boldsymbol{HP}_{\mathbf{A}} = -\boldsymbol{HK}_{\mathbf{11}}\boldsymbol{d}_{\mathbf{A}} - \boldsymbol{HK}_{\mathbf{12}}\boldsymbol{d}_{\mathbf{B}} \tag{11.80}$$

This must be identical to the second of equations (11.7). Also, the relationships given by (11.8) apply in general, so that

$$\boldsymbol{K}_{\mathbf{21}} = -\boldsymbol{HK}_{\mathbf{11}} \quad , \quad \boldsymbol{K}_{\mathbf{12}} = \boldsymbol{K}_{\mathbf{21}}^{t} = -\boldsymbol{K}_{\mathbf{11}}\boldsymbol{H}^{t} \quad , \quad \boldsymbol{K}_{\mathbf{22}} = -\boldsymbol{HK}_{\mathbf{12}} = \boldsymbol{HK}_{\mathbf{11}}\boldsymbol{H}^{t} . \tag{11.81}$$

Thus having found $\boldsymbol{H}$, which depends only on the relative positions of the two ends, all the stiffness matrices can be found in terms of the stiffness matrix $\boldsymbol{K}_{11}$. This can be found from the relationships between the loads and deflexions at end A when end B is encastré. This means that one elastic body connecting two joints of a frame can be replaced by another which also fits between these two joints, provided that they have the same elastic response at their free ends when treated as cantilevers. The *overall* response of the frame to joint loading will be the same in both cases. A useful application of this result is that a non-prismatic component of a frame can readily be replaced by an assembly of prismatic components. This allows standard structural analysis programs designed only to cope with straight, uniform beams to be used in such cases. Examples of this will be given in §11.8.

Example 11.7 Stiffness Matrices for a Beam with a Pinned End

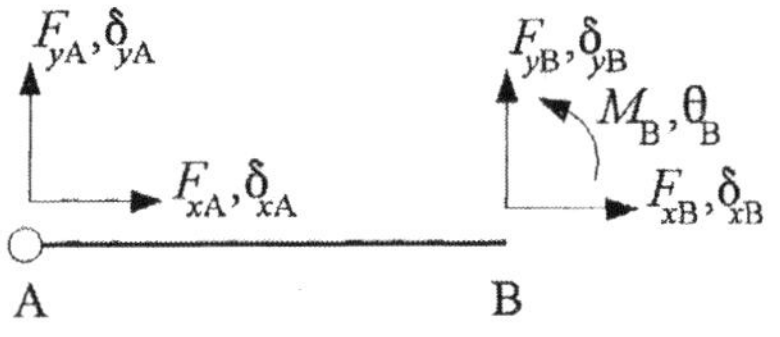

Figure 11.16 A beam attached to a pin and a rigid joint.

The beam shown in Figure 11.16 is to be used in a frame where its end A is attached to a pin joint and the end B to a rigid joint. This means that although end A may carry only the horizontal and vertical forces shown, the end B may carry an end moment M_B as well. If end B is held rigidly, the relationship between the loads and deflexions at end A are given by

$$\begin{bmatrix} F_{xA} \\ F_{yA} \end{bmatrix} = \begin{bmatrix} \frac{EA}{l} & 0 \\ 0 & \frac{3EI}{l^3} \end{bmatrix} \begin{bmatrix} \delta_{xA} \\ \delta_{yA} \end{bmatrix} \tag{11.82}$$

where the 2×2 matrix is the stiffness matrix $\boldsymbol{K}_{11}$. The equilibrium matrix equation is

$$\begin{bmatrix} 0 & -l \\ 1 & 0 \\ 0 & 1 \end{bmatrix} \begin{bmatrix} F_{xA} \\ F_{yA} \end{bmatrix} + \begin{bmatrix} M_B \\ F_{xB} \\ F_{yB} \end{bmatrix} = \begin{bmatrix} 0 \\ 0 \\ 0 \end{bmatrix} \tag{11.83}$$

This is the same in form as (11.79a). From (11.81), the other stiffness matrices are then

$$\boldsymbol{K}_{21} = \boldsymbol{K}_{12}^t = -\begin{bmatrix} 0 & -l \\ 1 & 0 \\ 0 & 1 \end{bmatrix} \begin{bmatrix} \frac{EA}{l} & 0 \\ 0 & \frac{3EI}{l^3} \end{bmatrix} = \begin{bmatrix} 0 & \frac{3EI}{l^2} \\ -\frac{EA}{l} & 0 \\ 0 & -\frac{3EI}{l^3} \end{bmatrix},$$

$$\boldsymbol{K}_{22} = \begin{bmatrix} 0 & -l \\ 1 & 0 \\ 0 & 1 \end{bmatrix} \begin{bmatrix} \frac{EA}{l} & 0 \\ 0 & \frac{3EI}{l^3} \end{bmatrix} \begin{bmatrix} 0 & 1 & 0 \\ -l & 0 & 1 \end{bmatrix} = \begin{bmatrix} \frac{3EI}{l} & 0 & -\frac{3EI}{l^2} \\ 0 & \frac{EA}{l} & 0 \\ -\frac{3EI}{l^2} & 0 & \frac{3EI}{l^3} \end{bmatrix} \tag{11.84}$$

Note that (11.81) still holds when the load and deflexion vectors are not of the same type at the two ends, resulting here in stiffness matrices being of different dimensions.

Suppose that the two-node element shown in Figure 11.15 is completely rigid. It then follows that the relationship between the end deflexions is given by

$$\begin{bmatrix} 1 & 0 & 0 & 0 & 0 & 0 \\ 0 & 1 & 0 & 0 & 0 & 0 \\ 0 & 0 & 1 & 0 & 0 & 0 \\ 0 & -Z & Y & 1 & 0 & 0 \\ Z & 0 & -X & 0 & 1 & 0 \\ -Y & X & 0 & 0 & 0 & 1 \end{bmatrix} \begin{bmatrix} \theta_{xB} \\ \theta_{yB} \\ \theta_{zB} \\ \delta_{xB} \\ \delta_{yB} \\ \delta_{zB} \end{bmatrix} = \begin{bmatrix} \theta_{xA} \\ \theta_{yA} \\ \theta_{zA} \\ \delta_{xA} \\ \delta_{yA} \\ \delta_{zA} \end{bmatrix} \quad \text{or} \quad \boldsymbol{H}^t\boldsymbol{d}_{\mathrm{B}} = \boldsymbol{d}_{\mathrm{A}} \,. \tag{11.85}$$

This geometrical relationship could have been found from virtual work considerations. If $\boldsymbol{d}_{\mathrm{A}}$ and $\boldsymbol{d}_{\mathrm{B}}$ had been vectors associated with small virtual rigid-body deflexions of AB, then the work done by $\boldsymbol{P}_{\mathrm{A}}$ and $\boldsymbol{P}_{\mathrm{B}}$ would have been zero. This means that

$$0 = \boldsymbol{P}_{\mathrm{A}}^t\boldsymbol{d}_{\mathrm{A}} + \boldsymbol{P}_{\mathrm{B}}^t\boldsymbol{d}_{\mathrm{B}} = \boldsymbol{P}_{\mathrm{A}}^t\boldsymbol{d}_{\mathrm{A}} - \boldsymbol{P}_{\mathrm{A}}^t\boldsymbol{H}^t\boldsymbol{d}_{\mathrm{B}} = \boldsymbol{P}_{\mathrm{A}}^t(\boldsymbol{d}_{\mathrm{A}} - \boldsymbol{H}^t\boldsymbol{d}_{\mathrm{B}}) \tag{11.86}$$

using (11.79a). As this is true for all $\boldsymbol{P}_{\mathrm{A}}{}^t$, the relationship given by (11.85) follows.

An application of these results is seen in Figure 11.17. The equilibrium matrix for the bar CD of length d shown in Figure 11.17a is given by

$$\begin{bmatrix} 1 & 0 & -d \\ 0 & 1 & 0 \\ 0 & 0 & 1 \end{bmatrix} \begin{bmatrix} M_C \\ F_{xC} \\ F_{yC} \end{bmatrix} + \begin{bmatrix} M_D \\ F_{xD} \\ F_{yD} \end{bmatrix} = \begin{bmatrix} 0 \\ 0 \\ 0 \end{bmatrix} \tag{11.87}$$

or

$$\boldsymbol{H}_d\boldsymbol{P}_{\mathrm{C}} + \boldsymbol{P}_{\mathrm{D}} = \boldsymbol{0} \tag{11.87a}$$

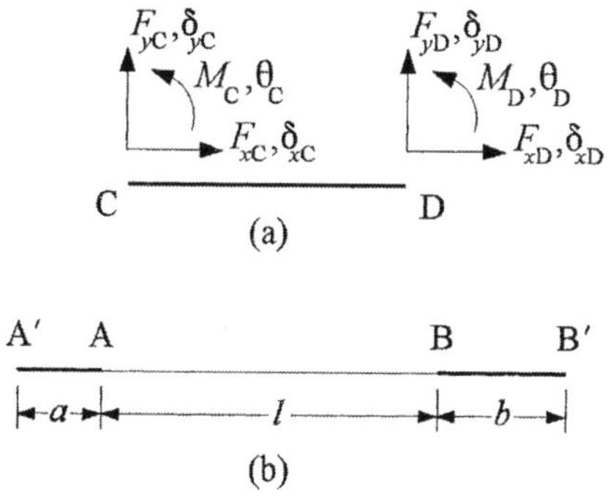

Figure 11.17 A beam with rigid end fittings.

The beam AB in Figure 11.17b is connected to a frame through rigid end fittings A′A and B′B of lengths a and b respectively. (These end fittings may be rigid gusset plates over several beams meeting at the same joints.) The centres of the frame joints are A′ and B′ and the frame deflexions will be expressed in terms of those at the joint centres. It will be taken that the stiffness matrices of the form given by (11.7) are known for AB. Including the extensions, those for A′B′ can be found from equations of the type given by (11.79a) and (11.85). Then relating the loads and deflexions at the ends of the fittings:

$$\begin{aligned} \boldsymbol{P}_{\mathrm{A'}} &= \boldsymbol{H}_{-a}\boldsymbol{P}_{\mathrm{A}} = \boldsymbol{H}_{-a}(\boldsymbol{K}_{11}\boldsymbol{d}_{\mathrm{A}} + \boldsymbol{K}_{12}\boldsymbol{d}_{\mathrm{B}}) = \boldsymbol{H}_{-a}\boldsymbol{K}_{11}\boldsymbol{H}_{-a}^t\boldsymbol{d}_{\mathrm{A'}} + \boldsymbol{H}_{-a}\boldsymbol{K}_{12}\boldsymbol{H}_b^t\boldsymbol{d}_{\mathrm{B'}} \\ \boldsymbol{P}_{\mathrm{B'}} &= \boldsymbol{H}_b\boldsymbol{P}_{\mathrm{B}} = \boldsymbol{H}_b(\boldsymbol{K}_{21}\boldsymbol{d}_{\mathrm{A}} + \boldsymbol{K}_{22}\boldsymbol{d}_{\mathrm{B}}) = \boldsymbol{H}_b\boldsymbol{K}_{21}\boldsymbol{H}_{-a}^t\boldsymbol{d}_{\mathrm{A'}} + \boldsymbol{H}_b\boldsymbol{K}_{22}\boldsymbol{H}_b^t\boldsymbol{d}_{\mathrm{B'}} \end{aligned} \tag{11.88}$$

where $\boldsymbol{H}_{-a}$ and $\boldsymbol{H}_b$ are of the same form as $\boldsymbol{H}_d$ except that $-a$ and b respectively replace d in (11.87). The stiffness matrices thus formed with respect to the central joint deflexions still have the symmetry properties given by (11.8). Thus $\boldsymbol{K}_{11}{}'$, given by $\boldsymbol{H}_{-a}\boldsymbol{K}_{11}\boldsymbol{H}_{-a}^t$, is still symmetrical.

{11.8 Non-Prismatic Members}

{Computer programs for the analysis of skeletal structures are usually designed to cope only with prismatic bars. If a system of straight uniform bars can be found with the same elastic properties as a non-uniform or curved bar, then the overall behaviour of a structure, with such systems replacing non-prismatic components, will not be changed. Standard computer programs can then be used for such problems. In the previous section, it was seen that all the necessary elastic properties of a structural component with two nodes could be deduced from its behaviour as a cantilever. Two examples will be considered here.

The first case is that of a straight bar of varying cross-section. Its overall axial and torsional stiffness can be matched by a uniform bar of the same length l and equivalent stiffnesses EA_e and GJ_e respectively, where

$$\frac{l}{EA_e} = \int_0^l \frac{dx}{EA} \quad , \quad \frac{l}{GJ_e} = \int_0^l \frac{dx}{GJ} \ . \qquad \textbf{(11.89)}$$

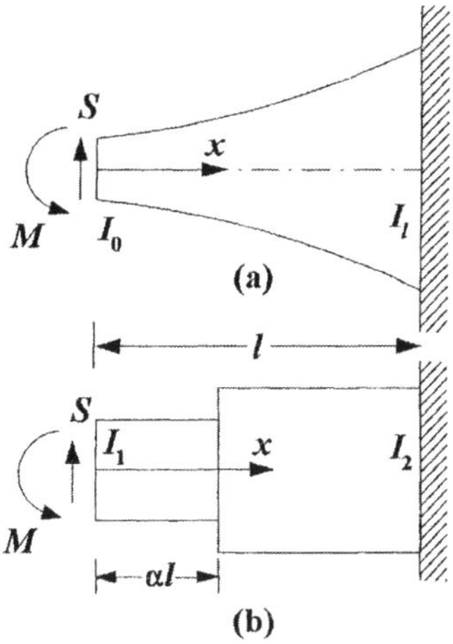

Figure 11.18 Equivalent substructure for a non-uniform bar.

This is because the overall flexibility of the beam is equal to the sum of the flexibilities of the short elements dx. The flexural response depends on the interaction between the end moment M and end shear force S. It cannot be represented by the behaviour of a single uniform beam.

Consider the non-uniform beam shown in Figure 11.18a. This has a flexural stiffness which varies from EI_0 at its left-hand end to EI_l at its right-hand end. It will be taken that the stiffness varies in accordance with a simple power law, so that EI is given by kX^n. A deep I-beam with a linearly-varying flange thickness would approximately correspond to $n=1$ and one with a thin web of linearly-varying depth to $n=2$. A rectangular section of constant width and linearly-varying depth corresponds to $n=3$ and a circular section of linearly-varying radius corresponds to $n=4$. The origin for the (dimensionless) X coordinate will be taken to be a distance el to the left of the left-hand end of the beam. The values of X at the two ends are e and $(1+e)$. Then

$$EI_0 = Eke^n \quad , \quad EI_l = Ek(1+e)^n \qquad \text{so that} \quad e = \frac{1}{\sqrt[n]{\dfrac{I_l}{I_0}} - 1} \ . \qquad \textbf{(11.90)}$$

The complementary strain energy of the beam under the action of the end moment M and shear force S is given by

$$\bar{U} = \int_0^l \frac{(M-Sx)^2}{2EI}\,dx = \int_e^{1+e} \frac{[M - Sl(X-e)]^2}{2EkX^n}\,l\,dX$$
$$= \frac{l}{2Ek}\left(M^2 f_{MM} - 2MSl f_{MS} + S^2 l^2 f_{SS}\right) \qquad \textbf{(11.91)}$$

where the flexibility coefficients are given in the following table.

n	f_{MM}	f_{MS}	f_{SS}
1	$\ln(1+\frac{1}{e})$	$1 - e\ln(1+\frac{1}{e})$	$e^2\ln(1+\frac{1}{e}) + \frac{1}{2} - e$
2	$\frac{1}{e(1+e)}$	$\ln(1+\frac{1}{e}) - \frac{1}{1+e}$	$\frac{1+2e}{1+e} - 2e\ln(1+\frac{1}{e})$
3	$\frac{1+2e}{2e^2(1+e)^2}$	$\frac{1}{2e(1+e)^2}$	$\ln(1+\frac{1}{e}) - \frac{3+2e}{2(1+e)^2}$
>3	$\frac{1}{n-1}\left[\frac{1}{e^{n-1}} - \frac{1}{(1+e)^{n-1}}\right]$	$\frac{1}{n^2-3n+2}\left[\frac{1}{e^{n-2}} - \frac{n+e-1}{(1+e)^{n-1}}\right]$	$\frac{1}{n^3-6n^2+11n-6}\left[\frac{2}{e^{n-3}} - \frac{2e^2+(n-1)(2e+n-2)}{(1+e)^{n-1}}\right]$

From §7.6, it then follows that the end response of the cantilever to a moment M and a shear force S is characterised completely by $l/2Ek$ and the coefficients f_{MM}, f_{MS} and f_{SS}. Matching coefficients are now sought using the compound cantilever shown in Figure 11.18b. This is also of length l, but has a constant flexural stiffness EI_1 over a length αl and a constant flexural stiffness EI_2 over the remaining length. The complementary strain energy induced by the same free end loading as before is given by

$$\bar{U} = \int_0^{\alpha l} \frac{(M - Sx)^2}{2EI_1} dx + \int_{\alpha l}^{l} \frac{(M - Sx)^2}{2EI_2} dx$$
$$= M^2 l \left[\frac{\alpha}{2EI_1} + \frac{1-\alpha}{2EI_2} \right] - MSl^2 \left[\frac{\alpha^2}{2EI_1} + \frac{1-\alpha^2}{2EI_2} \right] + S^2 l^3 \left[\frac{\alpha^3}{6EI_1} + \frac{1-\alpha^3}{6EI_2} \right] \quad \textbf{(11.92)}$$

By comparing the terms in (11.92) with those in (11.91), the two systems are found to be elastically equivalent if

$$I_1 = \frac{k(3f_{SS} - 2f_{MS})}{3f_{MM}f_{SS} + 2f_{MM}f_{MS} - f_{MM}^2 - 4f_{MS}^2},$$
$$I_2 = \frac{k(3f_{SS} + f_{MM} - 4f_{MS})}{3f_{MM}f_{SS} - 4f_{MS}^2}, \quad \textbf{(11.93)}$$
$$\alpha = \frac{3f_{SS} - 2f_{MS}}{2f_{MS} - f_{MM}}.$$

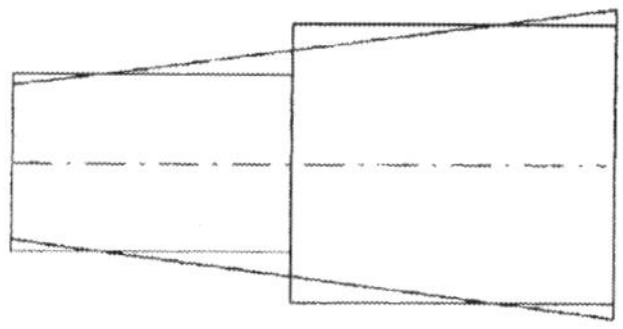

Figure 11.19 Elastically equivalent structural components.

Figure 11.19 shows the equivalent system for a beam with a flexural stiffness which varies linearly from EI_0 at the left-hand end to $2EI_0$ at the right-hand end. For the equivalent system,

$$I_1 = 1.140 I_0 \quad , \quad I_2 = 1.807 I_0 \quad , \quad \alpha l = 0.4313 l \,. \quad \textbf{(11.94)}$$

The stiffness distribution in the equivalent system is shown superimposed on that for the non-uniform beam.

A similar process can be used to replace a curved component by elastically equivalent prismatic components. The circular arch shown in Figure 11.20 will be taken as axially rigid and only in-plane behaviour will be considered. The complementary energy induced by a moment M and forces P and S shown in the figure is then given by

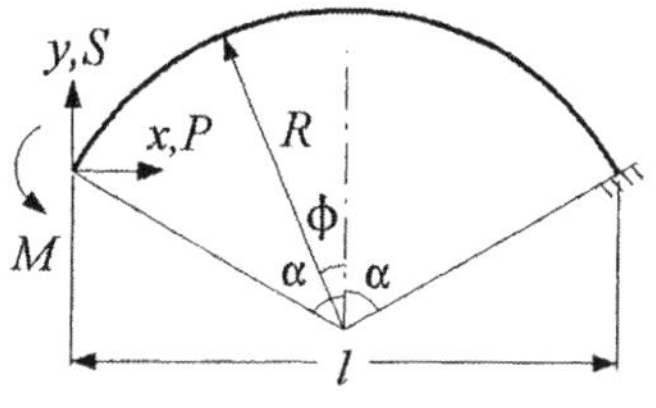

Figure 11.20 A circular arch.

$$\bar{U} = \int_s \frac{(M + Py - Sx)^2}{2EI} ds = \int_{-\alpha}^{\alpha} \frac{[M + PR(\cos\phi - \cos\alpha) + SR(\sin\phi - \sin\alpha)]^2}{2EI} R d\phi \quad \textbf{(11.95)}$$

Expressing this in terms of the span l of the arch (equal to $2R\sin\alpha$),

$$\bar{U} = \frac{M^2 l f_{MM}}{2EI} - \frac{MSl^2 f_{MS}}{EI} + \frac{S^2 l^3 f_{SS}}{2EI} + \frac{P^2 l^3 f_{PP}}{2EI} + \frac{MPl^2 f_{MP}}{EI} - \frac{SPl^3 f_{SP}}{EI} \quad \textbf{(11.96)}$$

where

$$f_{MM} = 2f_{MS} = \frac{\alpha}{\sin\alpha} \quad , \quad f_{PP} = \frac{\alpha + 2\alpha\cos^2\alpha - 3\sin\alpha\cos\alpha}{8\sin^3\alpha} \, ,$$
$$f_{MP} = 2f_{SP} = \frac{\sin\alpha - \alpha\cos\alpha}{2\sin^2\alpha} \quad , \quad f_{SS} = \frac{2\alpha\sin^2\alpha + \alpha - \sin\alpha\cos\alpha}{8\sin^3\alpha} \, . \qquad \textbf{(11.97)}$$

In general, the overall elastic properties of such an arch cannot be simulated by a rectangular portal frame made from axially-rigid beams. However, if the legs are at an equal and opposite slope, as shown in Figure 11.21, then an elastically-equivalent arch can be found. These legs are both of length jl ($j^2 = h^2 + k^2$) and of bending stiffness pEI. The lintel is of length $(1\text{-}2k)l$ and bending stiffness qEI. The complementary energy induced by the moment M and forces P and S are then

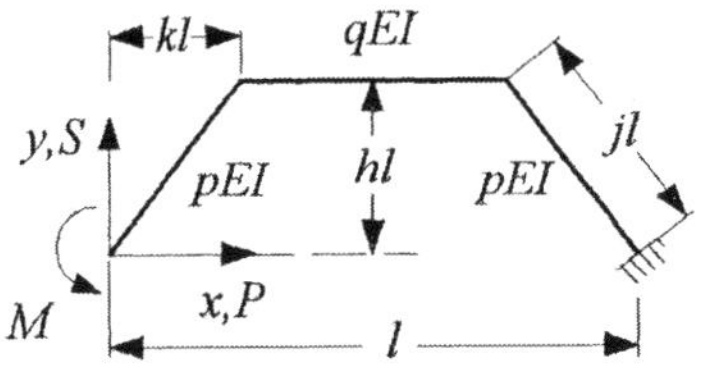

Figure 11.21 An equivalent arch.

$$\bar{U} = \int_0^{jl} \frac{1}{2pEI}\left[\left(M - \frac{ks}{j}S + \frac{hs}{j}P\right)^2 + \left(M - Sl + \frac{ks}{j}S + \frac{hs}{j}P\right)^2\right] ds + \int_{kl}^{l-kl} \frac{1}{2qEI}(M - Ss + Phl)^2 ds \qquad \textbf{(11.98)}$$

This gives an expression of the same form as (11.96) where

$$f_{MM} = 2f_{MS} = 2J + K \quad , \quad 3f_{PP} = h^2(2J + 3K) \, ,$$
$$f_{MP} = 2f_{SP} = h(J + K) \quad , \quad 3f_{SS} = (2k^2 - 3k + 3)J + (1 - k + k^2)K \qquad \textbf{(11.99)}$$
$$\text{and} \qquad J = \frac{j}{p} \quad , \quad K = \frac{(1 - 2k)}{q} \, .$$

Comparing this with (11.97) gives the conditions from which the four parameters h, k, p and q, describing the elastically-equivalent portal frame, can be found in terms of α. Some results are given in the following table.

α	h	k	p	q
2.903°	0.01589	0.4683	0.9998	1.0051
30°	0.1223	0.2600	1.0055	1.0090
45°	0.1893	0.2481	1.0103	1.0219
60°	0.2644	0.2298	1.0131	1.0440
90°	0.4626	0.1626	1.0008	1.1419
120°	0.8257	0.0058	0.9451	1.4726

The first line of the table is for the smallest angle α for which the values of the four parameters are real. As before, the forms of the substitute portals are similar to those of the original arches. Values for other angles α can be found using TRAPARCH.EXE.

The expression for $\bar{U}$ can be expanded to include the axial strain energy, for example. Then the axial compressibility of the arch can be allowed for in its equivalent. Note that the overall flexibilities are additive so that any extra flexibilities of the original can be expressed in terms of extra flexibilities in the equivalent.}

{11.9 Transfer Matrices}

{Structures consisting of a chain of modules linked end-to-end can be analysed in ways which minimise the amount of computation required. The deflexions $\boldsymbol{d}_B$ and loads $\boldsymbol{P}_B$ at end B of a module AB form the *state vector* $\boldsymbol{z}_B$ given by $\{\boldsymbol{d}_B, -\boldsymbol{P}_B\}$ at that end. (The convenience of the minus sign will become apparent later.) This can be related to the state vector $\boldsymbol{z}_A$ at end A of AB given by $\{\boldsymbol{d}_A, \boldsymbol{P}_A\}$ using a *transfer matrix* $\boldsymbol{U}$. From the general form of the stiffness matrix equations (11.7), (11.81) and the equilibrium equation (11.79a),

$$\boldsymbol{P}_A = \boldsymbol{K}_{11}\boldsymbol{d}_A - \boldsymbol{K}_{11}\boldsymbol{H}'\boldsymbol{d}_B \quad , \quad -\boldsymbol{P}_B = \boldsymbol{H}\boldsymbol{P}_A \tag{11.100}$$

This is given by

$$\begin{bmatrix} \boldsymbol{d}_B \\ -\boldsymbol{P}_B \end{bmatrix} = \begin{bmatrix} (\boldsymbol{H}')^{-1} & -(\boldsymbol{H}')^{-1}(\boldsymbol{K}_{11})^{-1} \\ \boldsymbol{0} & \boldsymbol{H} \end{bmatrix} \begin{bmatrix} \boldsymbol{d}_A \\ \boldsymbol{P}_A \end{bmatrix} \quad \text{or} \quad \boldsymbol{z}_B = \boldsymbol{U}\boldsymbol{z}_A \tag{11.101}$$

The form of $\boldsymbol{H}$ is given by $\boldsymbol{I}+\boldsymbol{J}$ where $\boldsymbol{I}$ is a unit matrix and $\boldsymbol{J}$ consists of elements which are the relative ordinates of B with respect to A. An alternative form of (11.79a) is

$$\boldsymbol{P}_A + \boldsymbol{H}^{-1}\boldsymbol{P}_B = \boldsymbol{0} \tag{11.102}$$

where $\boldsymbol{H}^{-1}$ consists of a unit matrix and a matrix of elements which are the relative ordinates of A with respect to B. That is,

$$\boldsymbol{H}^{-1} = \boldsymbol{I} - \boldsymbol{J} \qquad \text{so that} \qquad (\boldsymbol{H}')^{-1} = (\boldsymbol{H}^{-1})' = \boldsymbol{I} - \boldsymbol{J}' \tag{11.103}$$

The inverse of $\boldsymbol{K}_{11}$ is a flexibility matrix, $\boldsymbol{F}_{11}$ say. As an alternative to inverting the stiffness matrix, it may be found directly from the expression for the complementary strain energy of the form used in §11.8.

Example 11.8 Transfer Matrix for a Plane Rigid-Ended Beam

For the beam shown in Figure 11.4, it follows from (11.23) that

$$\boldsymbol{K}_{11} = \begin{bmatrix} EA/l & 0 & 0 \\ 0 & 12EI/l^3 & 6EI/l^2 \\ 0 & 6EI/l^2 & 4EI/l \end{bmatrix} \quad , \quad \boldsymbol{F}_{11} = \boldsymbol{K}_{11}^{-1} = \begin{bmatrix} l/EA & 0 & 0 \\ 0 & l^3/3EI & -l^2/2EI \\ 0 & -l^2/2EI & l/EI \end{bmatrix} \tag{11.104}$$

and the equilibrium equation is

$$\begin{bmatrix} 1 & 0 & 0 \\ 0 & 1 & 0 \\ 0 & -l & 1 \end{bmatrix} \begin{bmatrix} F_{xA} \\ F_{yA} \\ M_A \end{bmatrix} + \begin{bmatrix} F_{xB} \\ F_{yB} \\ M_B \end{bmatrix} = \begin{bmatrix} 0 \\ 0 \\ 0 \end{bmatrix} \quad \text{or} \quad \boldsymbol{H}\boldsymbol{P}_A + \boldsymbol{P}_B = \boldsymbol{0} \tag{11.105}$$

so that from (11.101),

$$\begin{bmatrix} \delta_{xB} \\ \delta_{yB} \\ \theta_B \\ -F_{xB} \\ -F_{yB} \\ -M_B \end{bmatrix} = \begin{bmatrix} 1 & 0 & 0 & -l/EA & 0 & 0 \\ 0 & 1 & l & 0 & l^3/6EI & -l^2/2EI \\ 0 & 0 & 1 & 0 & l^2/2EI & -l/EI \\ 0 & 0 & 0 & 1 & 0 & 0 \\ 0 & 0 & 0 & 0 & 1 & 0 \\ 0 & 0 & 0 & 0 & -l & 1 \end{bmatrix} \begin{bmatrix} \delta_{xA} \\ \delta_{yA} \\ \theta_A \\ F_{xA} \\ F_{yA} \\ M_A \end{bmatrix} \quad \text{or} \quad \boldsymbol{z}_{\mathrm{B}} = \boldsymbol{U}\boldsymbol{z}_{\mathrm{A}} . \tag{11.106}$$

The state vectors can be referred to a global coordinate system, using transformation matrices $\boldsymbol{T}$ as before. Thus

$$\boldsymbol{z}_{\mathrm{B}}^* = \begin{bmatrix} \boldsymbol{d}_{\mathrm{B}}^* \\ -\boldsymbol{P}_{\mathrm{B}}^* \end{bmatrix} = \begin{bmatrix} \boldsymbol{T} & \boldsymbol{0} \\ \boldsymbol{0} & \boldsymbol{T} \end{bmatrix} \begin{bmatrix} \boldsymbol{d}_{\mathrm{B}} \\ -\boldsymbol{P}_{\mathrm{B}} \end{bmatrix} , \quad \boldsymbol{z}_{\mathrm{A}} = \begin{bmatrix} \boldsymbol{d}_{\mathrm{A}} \\ \boldsymbol{P}_{\mathrm{A}} \end{bmatrix} = \begin{bmatrix} \boldsymbol{T}^t & \boldsymbol{0} \\ \boldsymbol{0} & \boldsymbol{T}^t \end{bmatrix} \begin{bmatrix} \boldsymbol{d}_{\mathrm{A}}^* \\ \boldsymbol{P}_{\mathrm{A}}^* \end{bmatrix} \tag{11.107}$$

where, in the above example, $\boldsymbol{T}$ is given by (11.24). With respect to global coordinates, (11.101) becomes

$$\begin{bmatrix} \boldsymbol{d}_{\mathrm{B}}^* \\ -\boldsymbol{P}_{\mathrm{B}}^* \end{bmatrix} = \begin{bmatrix} \boldsymbol{T}(\boldsymbol{H}^t)^{-1}\boldsymbol{T}^t & -\boldsymbol{T}(\boldsymbol{H}^t)^{-1}(\boldsymbol{K}_{11})^{-1}\boldsymbol{T}^t \\ \boldsymbol{0} & \boldsymbol{T}\boldsymbol{H}\boldsymbol{T}^t \end{bmatrix} \begin{bmatrix} \boldsymbol{d}_{\mathrm{A}}^* \\ \boldsymbol{P}_{\mathrm{A}}^* \end{bmatrix} \tag{11.108}$$

or, from (11.103),

$$\boldsymbol{z}_{\mathrm{B}}^* = \boldsymbol{U}^*\boldsymbol{z}_{\mathrm{A}}^* \quad \text{where} \quad \boldsymbol{U}^* = \begin{bmatrix} \boldsymbol{I} - \boldsymbol{T}\boldsymbol{J}^t\boldsymbol{T}^t & \boldsymbol{T}(\boldsymbol{F}_{11} - \boldsymbol{J}^t\boldsymbol{F}_{11})\boldsymbol{T}^t \\ \boldsymbol{0} & \boldsymbol{I} + \boldsymbol{T}\boldsymbol{J}\boldsymbol{T}^t \end{bmatrix} \tag{11.109}$$

Figure 11.22a shows a structure consisting of a chain of members (or modules) linked only at their common joints (or nodes) B to M. Consider a typical joint I shown in Figure 11.22b. This connects the member HI to the member IJ. The external loading applied at this joint is given by the column vector $\boldsymbol{P}_{\mathrm{I}}^*$, in terms of a global coordinate system. This must be equal to the sum of the load vectors $\boldsymbol{P}_{\mathrm{I(HI)}}^*$ and $\boldsymbol{P}_{\mathrm{I(IJ)}}^*$ acting at the ends I of HI and IJ. Then

$$\boldsymbol{P}_{\mathrm{I(IJ)}}^* = -\boldsymbol{P}_{\mathrm{I(HI)}}^* + \boldsymbol{P}_{\mathrm{I}}^* \tag{11.110}$$

At a normal joint, the deflexions at the ends I of these two members will be common, or

$$\boldsymbol{d}_{\mathrm{I(IJ)}}^* = \boldsymbol{d}_{\mathrm{I(HI)}}^* \tag{11.111}$$

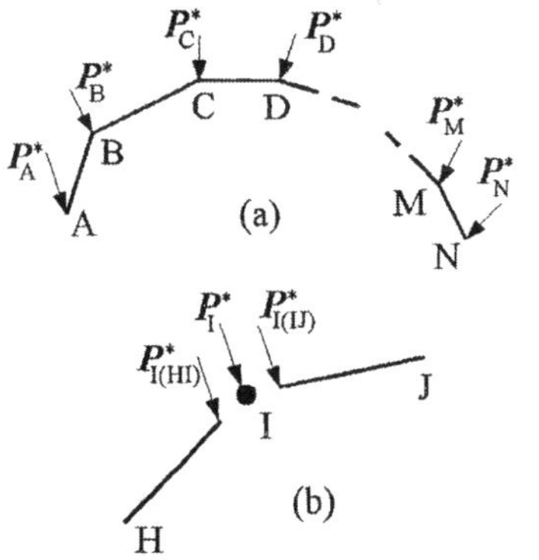

Figure 11.22 A structural chain.

This means that

$$\begin{bmatrix} \boldsymbol{d}_{\mathrm{I}}^* \\ \boldsymbol{P}_{\mathrm{I}}^* \end{bmatrix}_{\mathrm{(IJ)}} = \begin{bmatrix} \boldsymbol{d}_{\mathrm{I}}^* \\ -\boldsymbol{P}_{\mathrm{I}}^* \end{bmatrix}_{\mathrm{(HI)}} + \begin{bmatrix} \boldsymbol{0} \\ \boldsymbol{P}_{\mathrm{I}}^* \end{bmatrix} \quad \text{or} \quad \boldsymbol{z}_{\mathrm{I(IJ)}}^* = \boldsymbol{z}_{\mathrm{I(HI)}}^* + \boldsymbol{Z}_{\mathrm{I}}^* \tag{11.112}$$

(Note that a prescribed discontinuity in the end deflexions could be incorporated in $\boldsymbol{Z}_{\mathrm{I}}^*$, making its top component non-zero.) Starting at the left-hand end of the chain,

$$z^*_{A(AB)} = \begin{bmatrix} d^*_A \\ P^*_A \end{bmatrix} \tag{11.113}$$

where either a component of the load P_A* on joint A or its corresponding component in d_A*will be known. Using (1.109) and (1.112) sequentially,

$$\begin{aligned} z^*_{B(BC)} &= z^*_{B(AB)} + Z^*_B = U^*_{(AB)}z^*_{A(AB)} + Z^*_B \\ z^*_{C(CD)} &= z^*_{C(BC)} + Z^*_C = U^*_{(BC)}(U^*_{(AB)}z^*_{A(AB)} + Z^*_B) + Z^*_C \\ &\cdots\cdots\cdots \\ z^*_{N(MN)} &= U^*_{(MN)}\cdots U^*_{(BC)}U^*_{(AB)}z^*_{A(AB)} + Z^*_{BM} \end{aligned} \tag{11.114}$$

where, like $z_{A(AB)}$*, either a component of the load in $z_{N(MN)}$* or its corresponding component of deflexion will be known. The term Z_{BM}* is a sum of products of the terms Z_B* to Z_M* with the transfer matrices, both of which are known. As half the terms in $z_{A(AB)}$* and in $z_{N(MN)}$* are known, the last of matrix equations (11.114) represents a set of $2n$ equations with $2n$ unknowns, where n is the number of degrees of freedom at each (intermediate) joint. This remains the same, regardless of the number of joints in the structural chain. In a normal structural analysis, the number of simultaneous equations to be solved is equal to the *total* number of degrees of freedom of the structure, which increases with the number of free joints. The transfer matrix method would then seem to offer a considerable saving in computation. However, the saving is not as dramatic as it would appear at first sight, because of the optimisation methods which can be employed in standard structural analysis by computer, discussed in the next chapter. Also, there can be problems with cumulative rounding-off errors to which the transfer matrix method is prone.

More generally, the joints of the chain may be attached to other beams or springs which resist joint motion, as illustrated in Figure 11.23a. Suppose that the elastic properties of such an attachment at joint I is given by a stiffness matrix K_I*. That is, a load vector $K^*_I\, d^*_I$ is required to produce a deflexion vector d_I*. This means that when the joint I deflects by d_I* the attachment has a reaction on it of -K_I*d_I*. Equation (11.112) has to be modified to take account of this extra load. It now can be written in the form

$$\begin{bmatrix} d^*_I \\ P^*_I \end{bmatrix}_{(IJ)} = \begin{bmatrix} I & 0 \\ -K^*_I & I \end{bmatrix} \begin{bmatrix} d^*_I \\ -P^*_I \end{bmatrix}_{(HI)} + \begin{bmatrix} 0 \\ P^*_I \end{bmatrix} \tag{11.115}$$

$$\text{or} \quad z^*_{I(IJ)} = S^*_I z^*_{I(HI)} + Z^*_I\,.$$

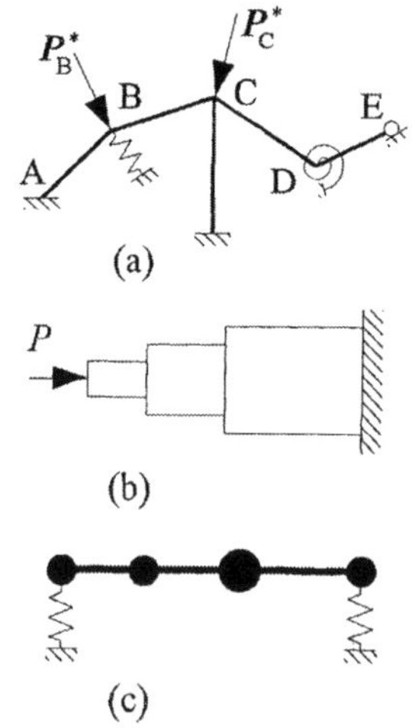

Figure 11.23 Transfer matrix problems.

The process followed in equations (11.114) can be repeated, leading to the same number of simultaneous equations to be solved as before. If loads are applied directly to the beams between the joints, they can be replaced by equivalent joint loads, as explained in §11.3. The stability of stepped columns such as that shown in Figure 11.23b can be analysed by the transfer matrix method. Pilkey (1994) Appendix III.1 examines this approach, together with the vibration of shafts with lumped masses, such as that shown in Figure 11.23c, or with continuously-distributed mass.}

11.10 Special Applications

The expressions for the end loads on a beam in (11.7) give only those components resulting from the end deflexions. If there is intermediate loading between the ends of a beam, fixed-end loads for this loading are added to these end loads, using the principle of superposition. Lack of fit and expansion due to temperature change can be dealt with in the same way. If a beam of length l and cross-sectional area A is made from a material with Young's modulus E is too long by an amount e, an additional compressive term for the axial end forces, EAe/l, must be added on at either end. These end loads induce equal and opposite end reactions from the beam, which must be added on to the end joint loads, as in Examples 11.2 and 11.3. If the lack of fit was due to a temperature rise T in the beam, and its material has a linear coefficient of expansion α, the lack of fit will be αTl and the end loads $EA\alpha T$. This can be illustrated with a simple example.

Example 11.9 Thermal Expansion of a Portal Frame

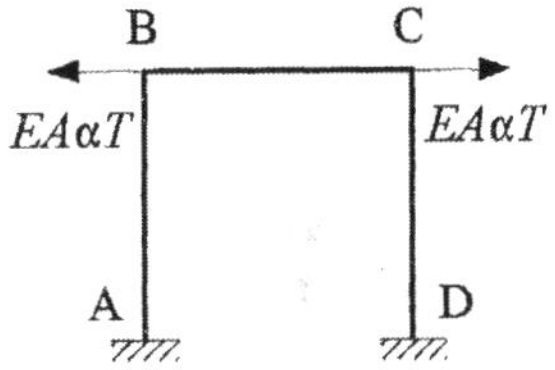

Figure 11.24 Temperature expansion of a portal lintel.

All the members of the portal frame shown in Figure 11.24 will be taken to have the same length l, flexural stiffness EI and axial stiffness EA. The lintel BC has linear coefficient of expansion α and undergoes a temperature increase T. This induces reaction loads $EA\alpha T$ on the joints B and C as shown. In terms of the notation shown in Figure 11.5, the stiffness matrix equation for the frame is then

$$\begin{bmatrix} \frac{EA}{l}+12\frac{EI}{l^3} & 0 & 6\frac{EI}{l^2} & -\frac{EA}{l} & 0 & 0 \\ 0 & \frac{EA}{l}+12\frac{EI}{l^3} & 6\frac{EI}{l^2} & 0 & -12\frac{EI}{l^3} & 6\frac{EI}{l^2} \\ 6\frac{EI}{l^2} & 6\frac{EI}{l^2} & 8\frac{EI}{l} & 0 & -6\frac{EI}{l^2} & 2\frac{EI}{l} \\ -\frac{EA}{l} & 0 & 0 & \frac{EA}{l}+12\frac{EI}{l^3} & 0 & 6\frac{EI}{l^2} \\ 0 & -12\frac{EI}{l^3} & -6\frac{EI}{l^2} & 0 & \frac{EA}{l}+12\frac{EI}{l^3} & -6\frac{EI}{l^2} \\ 0 & 6\frac{EI}{l^2} & 2\frac{EI}{l} & 6\frac{EI}{l^2} & -6\frac{EI}{l^2} & 8\frac{EI}{l} \end{bmatrix} \begin{bmatrix} \delta^*_{xB} \\ \delta^*_{yB} \\ \theta^*_B \\ \delta^*_{xC} \\ \delta^*_{yC} \\ \theta^*_C \end{bmatrix} = \begin{bmatrix} -EA\alpha T \\ 0 \\ 0 \\ EA\alpha T \\ 0 \\ 0 \end{bmatrix} \quad \textbf{(11.116)}$$

Solving this gives the joint deflexions

$$-\delta^*_{xB} = \delta^*_{xC} = \frac{Al^3\alpha T}{2Al^2+6I}, \quad \delta^*_{yB} = \delta^*_{yC} = 0, \quad \theta^*_B = -\theta^*_C = \frac{Al^2\alpha T}{2Al^2+6I} \quad \textbf{(11.117)}$$

The tensile axial force in BC, $P_{(BC)}$, is then that calculated from the above deflexions plus the fixed-end force. That is,

$$P_{(BC)} = \frac{EA}{l}(\delta^*_{xC} - \delta^*_{xB}) - EA\alpha T = -\frac{3EAI\alpha T}{Al^2+3I} \quad \textbf{(11.118)}$$

so that BC is in compression.

If pin joints are used in a frame which is mostly rigid-jointed, it is possible to modify the rigid-jointed analysis to allow for them. Figure 11.25a shows such a pin joint, I, of a plane frame linking two groups of bars rigidly jointed together. The first step in the analysis is to separate the two groups, giving them separate joints I′ and I″, as shown in Figure 11.25b. The loads and deflexions on these joints are related to those on the original joint I by the equilibrium and compatibility equations

$$F_{xI} = F_{xI}' + F_{xI}'' \;,\quad F_{yI} = F_{yI}' + F_{yI}'' \;,$$
$$\delta_{xI} = \delta_{xI}' = \delta_{xI}'' \;,\quad \delta_{yI} = \delta_{yI}' = \delta_{yI}'' \;. \qquad \textbf{(11.119)}$$

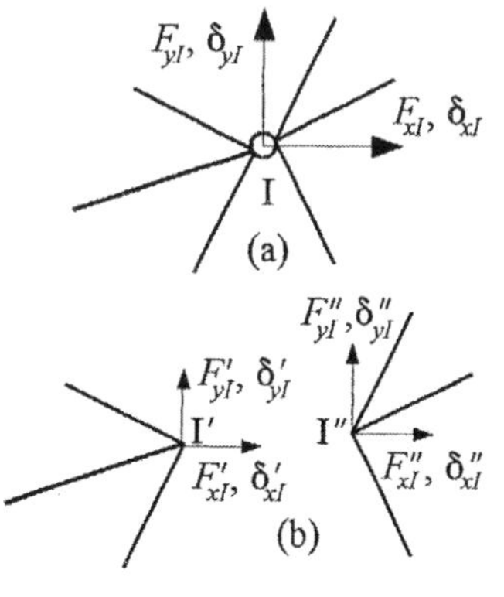

Figure 11.25 An inserted pin joint.

Suppose that the matrix equation has been formulated for the rigid-jointed frame with the coincident but separate joints I′and I″. Setting this up in the usual manner, relating the column vector of joint loads to the column vector of corresponding deflexions, results in a symmetrical stiffness matrix. Let rows *r, s, t* and *u* of the stiffness matrix relate to the joint loads F_{xI}', F_{yI}', F_{xI}'' and F_{yI}'' respectively. Then columns *r, s, t* and *u* of the stiffness matrix relate to the joint deflexions δ_{xI}', δ_{yI}', δ_{xI}'' and δ_{yI}'' respectively. The set of equations then take the form

$$\begin{aligned}
P_1 &= k_{11}d_1 + \ldots + k_{1r}\delta_{xI}' + k_{1s}\delta_{yI}' + k_{1t}\delta_{xI}'' + k_{1u}\delta_{yI}'' + \ldots k_{1n}d_n \\
&\cdots \\
F_{xI}' &= k_{r1}d_1 + \ldots + k_{rr}\delta_{xI}' + k_{rs}\delta_{yI}' + k_{rt}\delta_{xI}'' + k_{ru}\delta_{yI}'' + \ldots k_{rn}d_n \\
F_{yI}' &= k_{s1}d_1 + \ldots + k_{sr}\delta_{xI}' + k_{ss}\delta_{yI}' + k_{st}\delta_{xI}'' + k_{su}\delta_{yI}'' + \ldots k_{sn}d_n \\
F_{xI}'' &= k_{t1}d_1 + \ldots + k_{tr}\delta_{xI}' + k_{ts}\delta_{yI}' + k_{tt}\delta_{xI}'' + k_{tu}\delta_{yI}'' + \ldots k_{tn}d_n \\
F_{yI}'' &= k_{u1}d_1 + \ldots + k_{ur}\delta_{xI}' + k_{us}\delta_{yI}' + k_{ut}\delta_{xI}'' + k_{uu}\delta_{yI}'' + \ldots k_{un}d_n \\
&\cdots
\end{aligned} \qquad \textbf{(11.120)}$$

where P_i and d_i are the *i*th components of the load and deflexion column vectors respectively, and k_{ij} (= k_{ji}) is an element of the stiffness matrix (i,j = 1 to n). From (11.119), a new *r*th row can be formed from the sum of the *r*th and *t*th rows to give F_{xI} and a new *s*th row can be formed from the sum of the *s*th and *u*th rows to give F_{yI}. Also, the coefficients in the *r*th and *t*th columns can be combined, as they are both coefficients of δ_{xI}. Likewise, the coefficients in the *s*th and *u*th columns can be combined, as they are both coefficients of δ_{yI}. These combinations form new *r*th and *s*th columns respectively. The *t*th and *u*th columns are dropped, as are the *t*th and *u*th rows, which are redundant. Equations (11.120) now take the form

$$\begin{aligned}
P_1 &= k_{11}d_1 + \ldots + k_{1r}^*\delta_{xI} + k_{1s}^*\delta_{yI} + \ldots k_{1n}d_n \\
&\cdots \\
F_{xI} = F_{xI}' + F_{xI}'' &= k_{r1}^*d_1 + \ldots + k_{rr}^*\delta_{xI} + k_{rs}^*\delta_{yI} + \ldots k_{rn}^*d_n \\
F_{yI} = F_{yI}' + F_{yI}'' &= k_{s1}^*d_1 + \ldots + k_{sr}^*\delta_{xI} + k_{ss}^*\delta_{yI} + \ldots k_{sn}^*d_n \\
&\cdots
\end{aligned} \qquad \textbf{(11.121)}$$

where

$$k_{rr}^* = k_{rr} + 2k_{rt} + k_{tt} \;,\quad k_{ss}^* = k_{ss} + 2k_{su} + k_{uu} \;,\quad k_{rs}^* = k_{rs} + k_{ru} + k_{ts} + k_{tu} = k_{sr}^* \;,$$
$$\text{and}\quad k_{ir}^* = k_{ir} + k_{it} = k_{ri} + k_{ti} = k_{ri}^* \;,\quad k_{is}^* = k_{is} + k_{iu} = k_{si} + k_{ui} = k_{si}^* \quad (i \neq r,s) \qquad \textbf{(11.122)}$$

so that the symmetry of the modified stiffness matrix is retained.

A problem can arise in dealing with workless support conditions which are not oriented in the directions of the global coordinate system. For example, if at some support the joint can move freely in the x^* direction but is prevented from doing so in the y^* direction, the support conditions are that F_x^* and δ_y^* are zero. However, if these conditions apply relative to some axes x' and y'at an angle α to the global system, as shown in Figure 11.26, then the support conditions cannot be expressed so directly in terms of the components of load and deflexion. The relationships between the loads and deflexions at the joint I shown, relative to the two coordinate systems, are

$$F_{xI}' = F_{xI}^*\cos\alpha + F_{yI}^*\sin\alpha \;,\quad \delta_{xI}^* = \delta_{xI}'\cos\alpha - \delta_{yI}'\sin\alpha \,,$$
$$F_{yI}' = -F_{xI}^*\sin\alpha + F_{yI}^*\cos\alpha \;,\quad \delta_{yI}^* = \delta_{xI}'\sin\alpha + \delta_{yI}'\cos\alpha \,. \qquad \textbf{(11.123)}$$

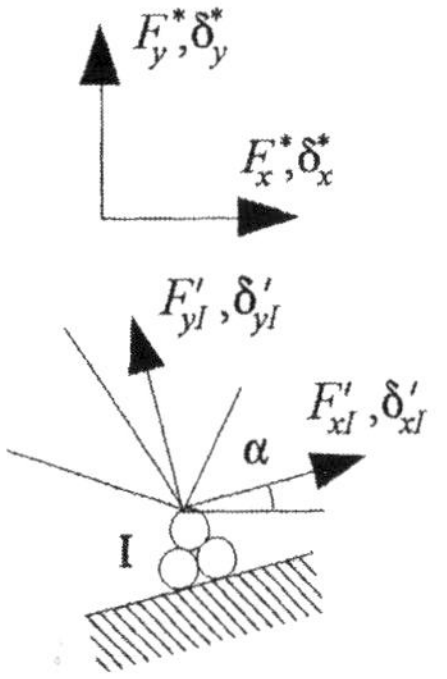

Figure 11.26 Local support conditions.

Initially, the stiffness matrix is formed so that all joint loads and deflexions are referred to the global coordinate system. Let the rth row refer to the force F_{xI}^* and the sth row to F_{yI}^*. Then the rth column of the stiffness matrix will contain the coefficients of the deflexion δ_{xI}^* and the sth column the coefficients of δ^*_{yI}. The equations are then initially of the form

$$\begin{aligned}
P_1 &= k_{11}d_1 + \ldots + k_{1r}\delta_{xI}^* + k_{1s}\delta_{yI}^* + \ldots\; k_{1n}d_n \\
&\ldots \\
F_{xI}^* &= k_{r1}d_1 + \ldots + k_{rr}\delta_{xI}^* + k_{rs}\delta_{yI}^* + \ldots\; k_{rn}d_n \\
F_{yI}^* &= k_{s1}d_1 + \ldots + k_{sr}\delta_{xI}^* + k_{ss}\delta_{yI}^* + \ldots\; k_{sn}d_n \\
&\ldots
\end{aligned} \qquad \textbf{(11.124)}$$

Expressing the displacements δ_{xI}^* and δ_{yI}^* in terms of δ_{xI}' and δ_{yI}',

$$\begin{aligned}
P_1 &= k_{11}d_1 + \ldots + (k_{1r}\cos\alpha + k_{1s}\sin\alpha)\delta_{xI}' + (-k_{1r}\sin\alpha + k_{1s}\cos\alpha)\delta_{yI}' + \ldots\; k_{1n}d_n \\
&\ldots \\
F_{xI}^* &= k_{r1}d_1 + \ldots + (k_{rr}\cos\alpha + k_{rs}\sin\alpha)\delta_{xI}' + (-k_{rr}\sin\alpha + k_{rs}\cos\alpha)\delta_{yI}' + \ldots\; k_{rn}d_n \\
F_{yI}^* &= k_{s1}d_1 + \ldots + (k_{sr}\cos\alpha + k_{ss}\sin\alpha)\delta_{xI}' + (-k_{sr}\sin\alpha + k_{ss}\cos\alpha)\delta_{yI}' + \ldots\; k_{sn}d_n \\
&\ldots
\end{aligned} \qquad \textbf{(11.125)}$$

In this new arrangement, the elements in the rth and sth columns, shown by the bracketed terms, are the coefficients of δ_{xI}' and δ_{yI}' respectively. Likewise, from (11.123), the rth and sth rows can be turned into expressions for F_{xI}' and F_{yI}'respectively, giving

$$\begin{aligned}
P_1 &= k_{11}d_1 + \ldots + k_{1r}'\delta_{xI}' + k_{1s}'\delta_{yI}' + \ldots\; k_{1n}d_n \\
&\ldots \\
F_{xI}' &= k_{r1}'d_1 + \ldots + k_{rr}'\delta_{xI}' + k_{rs}'\delta_{yI}' + \ldots\; k_{rn}'d_n \\
F_{yI}' &= k_{s1}'d_1 + \ldots + k_{sr}'\delta_{xI}' + k_{ss}'\delta_{yI}' + \ldots\; k_{sn}'d_n \\
&\ldots
\end{aligned} \qquad \textbf{(11.126)}$$

where

$$\begin{aligned}
k_{rr}' &= k_{rr}\cos^2\alpha + 2k_{rs}\sin\alpha\cos\alpha + k_{ss}\sin^2\alpha \,, \\
k_{rs}' &= (k_{ss} - k_{rr})\sin\alpha\cos\alpha + k_{rs}(\cos^2\alpha - \sin^2\alpha) = k_{sr}' \,, \\
k_{ss}' &= k_{rr}\sin^2\alpha - 2k_{rs}\sin\alpha\cos\alpha + k_{ss}\cos^2\alpha \,, \\
k_{ir}' &= k_{ir}\cos\alpha + k_{is}\sin\alpha = k_{ri}' \;,\quad k_{is}' = -k_{ir}\sin\alpha + k_{is}\cos\alpha = k_{si}' \quad (i \neq r,s).
\end{aligned} \qquad \textbf{(11.127)}$$

Thus the modified stiffness matrix remains symmetrical and the conditions at joint I can be expressed in terms of the local coordinate system. For the case shown, these conditions are that F_{xI}' and δ_{yI}' are zero. (The fact that the moment on this joint is zero can be expressed without any modification of the matrix.) The importance of retaining the symmetry of the stiffness matrix will be seen in the next chapter.

Chapter 12 Computer Analysis of Frames

12.1 Introduction

Computing machines have been around for a long time. The abacus, used by the Chinese, Greeks and Romans, has been in use for over two thousand years. The automatic counting machine described by Hero of Alexandria (2nd century AD) was a geared system attached to a carriage which could measure distance in stades. (One stade or stadium is about 200 metres.) Blaise Pascal adapted the gear system to produce the first digital counting machine in 1642. This was developed into the first commercial desktop calculator by C.X. Tomas in 1820. It could be used for all four arithmetic calculations on numbers. Charles Babbage took the idea further and designed a machine to carry out the repetitive calculations needed in producing mathematical tables. He produced a small demonstration model of his "difference engine" in 1822, but the full-scale machine was only completed 170 years later, for the Science Museum in London. In 1833, he gave up work on it in favour of a much more ambitious project for an "analytic engine". This was to have the essential properties of a modern programmable computer. It would have been capable of carrying out routines in an order which was determined from the conditions arising from the calculations, rather than in a preordained sequence. His assistant, Lady Ada Lovelace, developed the first concepts of computer programming. The project turned out to be enormously complex and expensive and was never completed.

An electromechanical computer was built by IBM in the late 1930s. The needs of World War II stimulated the development of two programmable electronic computers. The first to run was Colossus, in December 1943. It was used in Bletchley Park, England to decipher the most secure German codes. ENIAC (**E**lectronic **N**umerical **I**ntegrator **A**nd **C**omputer) was developed at the University of Pennsylvania for the analysis of new weapons systems. Programming was carried out by wiring up the components in the desired order. The mathematician John von Neumann took an interest in ENIAC and contributed fundamental ideas to the way in which computers should be organised and built. Alan Turing was another mathematician who worked at Bletchley Park. In 1936, he had described mathematically a theoretical 'universal computing machine', now known as a Turing machine. His work lead to the first computer which could be programmed electronically. The first electronically stored program ran on the Manchester University Mark I computer in June 1948. The same year saw the invention of the transistor by William Shockley. This lead to the integrated circuit which increased enormously the speed, power, compactness and efficiency of computers. Machines which had occupied whole floors of office blocks and required elaborate air conditioning for their thermionic valves were outclassed by pocket calculators. Previous assumptions about the division of purpose between mainframe and personal computers have become invalid. New kinds of computer, which no longer fit the theoretical models of Turing and von Neumann, have been developed. A bottleneck in the standard model is the central processing unit. All arithmetical operations are carried out sequentially by this unit. With a parallel processor[1], such operations are carried out simultaneously by a number of such units. The speed of operation then increases roughly in proportion to the number of units, but it creates problems for the programmer in synchronising their activity. Neural networks simulate some important features of the human brain. In a rather limited fashion, they are capable of self-programming or 'learning' from experience[2].

---

[1] Modi (1988) gives algorithms suitable for structural analysis using such processors.

[2] For applications, see Topping and Khan (1993) for example.

The earliest electronic programming was in binary machine code. This specified exactly how operations were to be carried out within a particular machine. Autocode and assembly language marked steps towards greater comprehensibility, but were also machine-specific. Higher-level languages allowed programs to be expressed in terms of the mathematical operations involved rather than the computer operations. To translate the former into the latter requires either an interpreter, which converts from one to the other during the running of the program, or a compiler, which produces a file containing the program in low-level code. The advantages of a compiler is that it can optimise the program before it is run and the code generated will work with the operating system on its own without the need for additional software. Specifying the program in terms of a higher-level language not only made it easier to write, being closer to natural language and mathematics, but made it possible to transfer such programs from one type of computer to another with little difficulty.

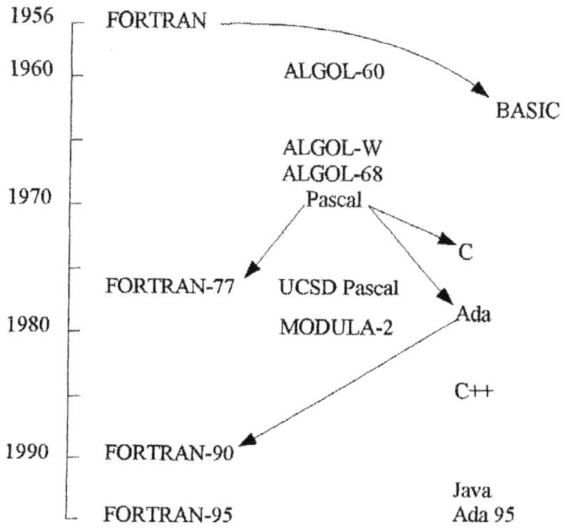

Figure 12.1 The evolution of computer languages.

The first such higher-level language was FORTRAN (**FOR**mula **TRAN**slator) which was introduced in 1956, as shown in Figure 12.1. Since then, hundreds of computer languages have been written. The figure shows some of those more relevant to engineers and the material in this book. Languages in the same columns tend to follow a particular line of development, but there are also cross-influences indicated by the arrows. FORTRAN was designed with the needs of scientists and engineers in mind, but computer scientists and mathematicians devised a more elegant language called ALGOL (**ALGO**rithmic **L**anguage). As can be seen, this went through several versions, eventually leading to Pascal (named after Blaise Pascal). This was the brain-child of Niklaus K. Wirth, who had earlier developed ALGOL-W. It was originally intended to be a teaching language, encouraging or enforcing good programming habits. However, it was so popular that it became a regular programming language in its own right and influenced all subsequent languages. BASIC (**B**eginners **A**ll-purpose **S**ymbolic **I**nstruction **C**ode) was also intended as an introduction to computing and became a popular language for personal computers. However, UCSD Pascal, was developed by the University of California at San Diego with the microchip in mind, and so became a competitor in this area. The 'C' language (so named for no better reason than it was the successor to a language called 'B') was designed for experienced programmers. It lacked some of the constraints of Pascal, regarded by some as pedantic. Wirth continued to devise new languages based on Pascal. The programs associated with this book are written in MODULA-2 (**MODU**lar **LA**nguage 2). This allows programs to be broken up into self-contained modules, including library modules. In this way, a large software package can be developed more easily by a team of programmers, each working on an individual module. It also includes low-level features not found in Pascal.

From the eighties onwards, languages began to incorporate object-oriented programming. Programs are written at a higher level of abstraction, operating on objects. What an operation on a particular object means is included in the definition of the object. For example, it would be possible to write a general-purpose structural analysis program without specifying whether it was two-dimensional or three-dimensional or whether the joints were pinned or rigid-jointed. Such details would be included in the object definitions. FORTRAN had been specified by the American Department of Defence and NASA as the programming language for its software. However, these organisations became increasingly aware of the shortcomings of FORTRAN and the Department of Defence commissioned a new language called Ada (after Ada Lovelace). This incorporated many of the current ideas about programming languages, including object-oriented programming, and became the new standard. A later version, Ada 95, was more heavily committed to object-oriented programming and it was incorporated in a new version of C, called C++. Elements of it also appeared in FORTRAN. With the coming of the Internet and the World-Wide Web, the need for code which could be transmitted through the network and work on any receiving platform became apparent. Sun Microsystems developed Java[1] in the early nineties for this purpose. It too was based on object-oriented programming and the first public release was in 1995. This is certainly not the end of the story. Developments in this area are so rapid that any book on the subject becomes obsolescent almost as soon as it is published. Some shortcomings of Java have already been noticed.

12.2 The Nature of the Frame Stiffness Matrix

The stiffness matrix $\boldsymbol{K}$ for the whole structure relates the column vector of joint loads to the corresponding column vector of joint deflexions. From Betti's theorem it follows that this matrix is symmetrical ($k_{ij} = k_{ji}$) and for a stable state it must also be positive-definite. Positive-definiteness implies that all the principal minors[2] of its determinant are positive. In particular, the terms on the leading diagonal and the determinant itself are positive. As the matrix is symmetric, it is sufficient to store only those elements k_{ij} on and above the leading diagonal (i.e. $j \geq i$). Also, such matrices tend to be sparse. That is, there are relatively few non-zero elements. Careful ordering of the equations can ensure that these elements are clustered around the leading diagonal. As will be seen, this minimises the storage and number of arithmetical operations required. The tightness of this banding is measured in terms of the half-bandwidth b. (The total bandwidth is $2b$-1.) In principle, all the possible joint numbering schemes could be examined, and one giving the lowest value of b chosen. However, if the order of the stiffness matrix is n, the number of such schemes is $n!$ so that this approach rapidly becomes impractical. Algorithms have been devised[3] which reduce the bandwidth by a suitable numbering of the joints. However, these do not guarantee that the optimum numbering will be found. Figure 12.2 shows three possible ways of numbering the joints of the same frame. In Figure 12.2a, the joints are numbered vertically downwards in columns, starting at the left of the frame. In Figure 12.2b, the joints are numbered horizontally to the right in rows, starting at the top of the frame. Lastly, in Figure 12.2c, the joints are numbered using a few simple rules. The numbers lie sequentially on the crests of waves radiating across the frame. These crests are shown by the heavier black lines used for the bars. The maximum number of joints to be found on any of the crests is kept to a minimum. Figure 12.3

---

[1] Java was named after a marketing person's favourite coffee to emphasise its stimulating properties.

[2] The principal minors are the determinants formed from the elements found in a given set of rows and the same set of columns of the original matrix. A sufficient proof of positive-definiteness is that every minor formed from the first m rows and columns of a real symmetric matrix is positive (for any m less than or equal to the order of the matrix). See Korn & Korn (1961) §13.5-6 for example.

[3] See for example Livesley, R.K. and Sabin, M.A. (1991) Algorithms for Numbering the Nodes of Finite-Element Meshes. *Computing Systems in Engineering* Vol.2 p.103.

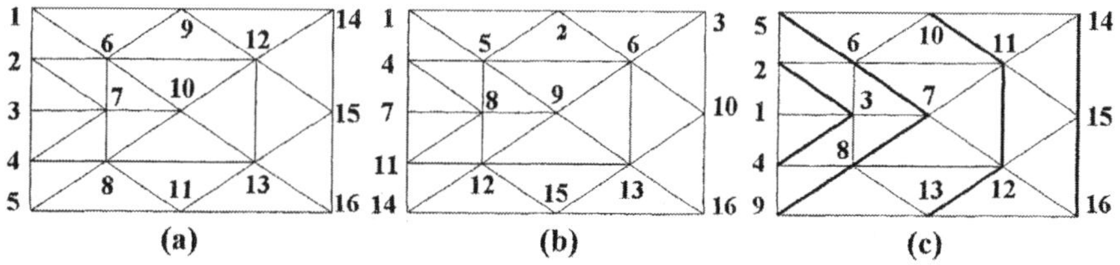

Figure 12.2 Joint numbering schemes.

shows the resulting distribution of non-zero beam submatrices (shown by black dots) in the frame stiffness matrix (shown by the grid). The half-bandwidth b is measured in terms of these

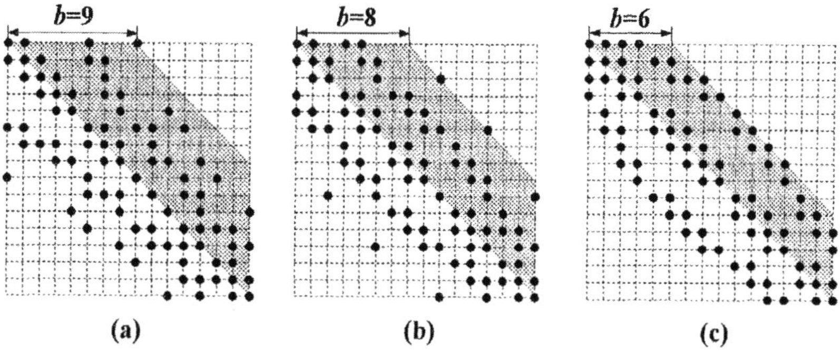

Figure 12.3 Banding for different joint numbering schemes.

submatrices. The upper half-bands are shown by the shaded areas. The total bandwidth is $2b$-1. The distributions (a), (b) and (c) correspond to the numbering schemes (a), (b) and (c) respectively in Figure 12.2. It will be seen that the third numbering scheme which follows the ordering rules minimises the bandwidth.

Some methods of analysis store and operate on only the submatrices (including the zero submatrices) within this half-bandwidth. A rectangular array is generated for this purpose. This array will usually contain zero submatrices which are not needed. The 'skyline' form of storage, shown in Figure 12.4, is more efficient. The array of submatrices is that previously shown in Figure 12.3b. The banding stores one hundred submatrices. (Even with the optimum banding in Figure 12.3c eighty one submatrices are stored.) The skyline method stores only the sixty one submatrices shown in the hatched area. This consists of the columns of submatrices, starting at the top non-zero matrix and ending with the matrix on the leading diagonal. The zero submatrices beneath the skyline are usually needed during the solution process.

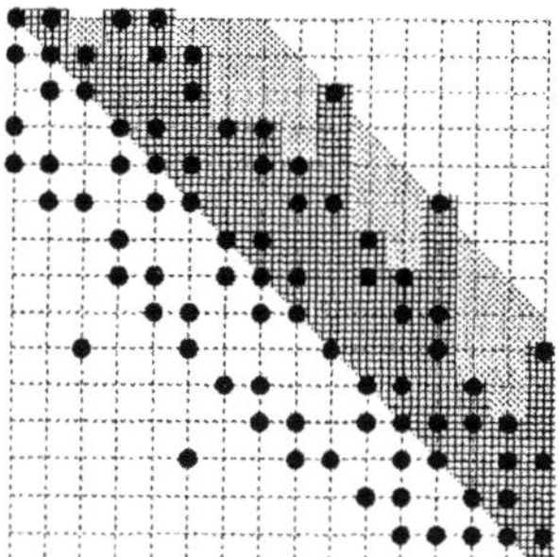

Figure 12.4 Skyline storage.

There are rare cases when some of the zero submatrices below the skyline are never used. Another method, using the dynamic storage of linked records, avoids creating space for even these unnecessary submatrices. Figure 12.5a shows a simple structure devised to compare the skyline and linked-record methods of storage. The skyline method stores three zero submatrices which are not used. These are shown by the empty circles in Figure 12.5b. The linked-record

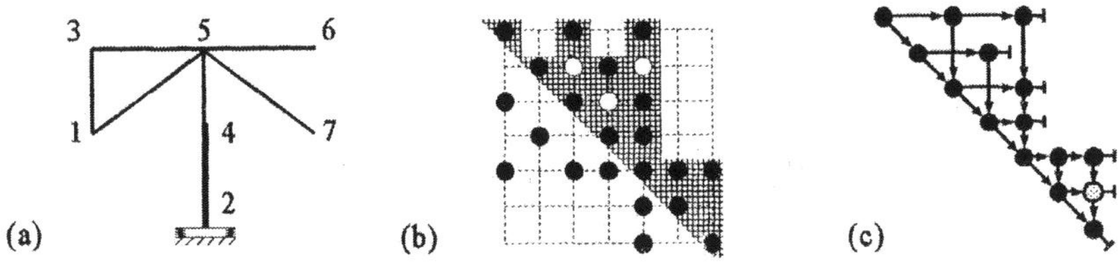

Figure 12.5 Skyline and linked-record storage.

method relates each leading diagonal submatrix to the corresponding joint and each submatrix above the leading diagonal to the corresponding bar. The corresponding joint is that related to the row (and column) in which the leading diagonal submatrix lies. The corresponding bar links the joints related to the row and the column in which the submatrix lies. (This relies on the fact that no two bars connect the same pair of joints.) Each record contains details of the joint or bar, the appropriate submatrix and two pointers to other records. These records are shown by the dots in Figure 12.5c and the pointers by arrows. The black dots contain non-zero submatrices and the shaded dot contains an empty submatrix which does not correspond to any bar. This has to be inserted because it is needed in the solution process. Each row and the leading diagonal are terminated by 'Nil' pointers. These are shown by a Tee form instead of an arrow. Comparing Figures 12.5b and 12.5c, it will be seen that the unnecessary zero submatrices are not generated. Also, both the banding and skyline methods store the submatrices in arrays. These require counters to establish the relationships between the submatrices. In the linked-record method, these relationships are established through the chain of pointers.

12.3 Data Input

The way in which input data is formatted is also affected by the method of storage. This is illustrated in the following table, which compares typical data input for the two methods.

Table 12.1 Input Data Files

Data	Array storage	Linked-record storage
Preliminary data:	Number of joints Number of section types Number of bars Number of load cases	*Not required*
Joint list:	Joint number Joint coordinates, ...	Joint identifier Joint coordinates [Extra joint details], ... |
Section list:	Material properties. Section number Section properties, ...	*Provided on a separate file:* *Scale factors, Material properties* | *Section number Section properties,* .... |
Bar list:	End joint Numbers Section Number, ...	End joint identifiers Section identifier [Orientation] .... |
Support list:	Joint number Support-type code	*Support conditions included in joint list.*
Load list:	[Load case number] Joint number Joint loads, ... *Repeat all above for each load case.*	[Joint identifier Joint loads] .... [Bar end joint identifiers Bar loading], ... | *Repeat all above for each load case.*

In the table, the dotted lines indicate repetition of the previous line for all instances (joints, sections, bars or loads). Square brackets indicate data which may or may not be present. Other information, such as a bar number or identifier, might also be included. For array storage, the preliminary data is needed to set the limits of the associated row and column counters. These are not required for linked-record storage, as new records and links are generated as new data about joints and bars is read. In this case, the end of a list is indicated by a terminator (indicated here by a vertical bar). The end of all the load cases can be detected by sensing the end of the file.

The joint list specifies the positions of all the joints, including the supports. From this, the length and direction of a bar connecting two joints can be found. This direction is usually enough to establish the relationship between the bar and frame coordinate systems. The exception is in the case of rigid-jointed three-dimensional frames, where the angle of rotation of the bar about its own axis, β, must also be specified (see §11.4). This orientation is specific to each bar and so is included in the bar list. If the joint is supported, these details may be included in the joint list or provided on a separate support list. This is relatively easy when a joint identifier is used instead of a joint number. The identifier may contain almost any keyboard symbol as well as numbers. Thus, for example, the use of the letters F, P and E in the identifier could indicate a fixed, pinned or elastic support. In the last case, the stiffnesses of the elastic support would have to be included as extra joint details. Other support conditions could include rollers (permitting rotation of the support about a given axis) or sliders (permitting the joint to displace in a given direction). Most analyses limit such conditions to rotation or displacement in relation to one of the overall frame coordinate axes, but this is not necessary (see §11.10). Again, if such support conditions are used, their orientation must be included in the extra joint details.

{Symmetry and antisymmetry have already been discussed in §1.5, where the theory is explained more fully. If a mirror-symmetrical frame is loaded in a similarly symmetrical fashion, the response will also be mirror-symmetrical. Figure 12.6a shows the plane of symmetry, marked M, and the mirror images of a set of displacements and rotations of some point in the frame. Because of the symmetry, the image point will also be a point on the frame. If the point lies on the plane of symmetry, the point will be coincident with its image point. However, the motions normal to the plane are equal and opposite, as are the rotations about axes parallel to the plane. If the frame is continuous at this point, this apparent contradiction is only satisfied if these deflexions are zero. The frame can be sectioned at this point and a workless constraint of the kind shown at the right of Figure 12.6a inserted at the section. This constraint prevents all motion out of the plane of symmetry and simulates the interaction between the two halves of the frame at this point. By the use of such constraints at the plane of symmetry, the frame can be sectioned at this plane and only one half of it need then be analysed. Likewise, if a mirror-symmetric frame is loaded in an antisymmetric fashion, the response is also antisymmetric. This is shown in Figure 12.6b where an antisymmetric mirror marked A is inserted at the plane of symmetry. The image of the motions of a typical point on the frame is then exactly the reverse of the mirror image. By the same arguments as before, the interaction between the two halves of the frame at the plane of symmetry can now be simulated by the workless constraint shown at the

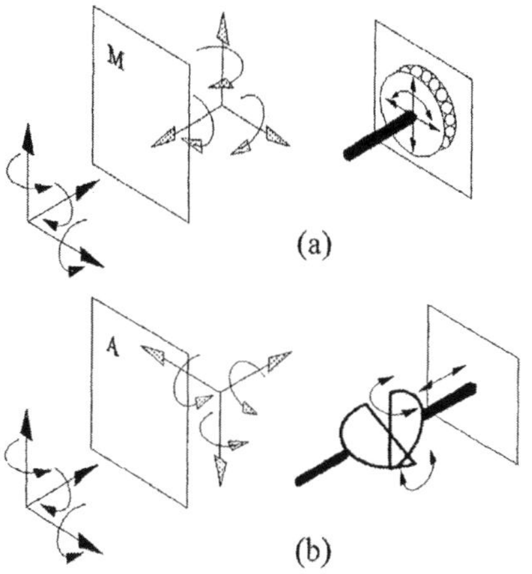

Figure 12.6 Symmetry and antisymmetry.

right of Figure 12.6b. This is a Hooke joint which is free to slide normal to the plane of symmetry. All out-of-plane motion is now permitted but in-plane motion is prevented.

Where a high degree of symmetry exists in the frame and its loading, the use of such sectioning and constraints very substantially reduces the data input and computation. Figure 12.7a shows a dome with sixteen-fold symmetry. Provided that the loading is similarly symmetrical, as might be the case with self-weight or snow loading, only the 22½° segment shown in Figure 12.7b need be analysed. The necessary mirror-symmetrical constraints are shown by round dots in both parts of Figure 12.7[1]. The apex is common to all sixteen segments and is only free to move vertically. The constraints at the base will depend on the fixity. For example, the base constraints at the segment edges of a fixed base remain fixed. Loads applied at joints on the common edges of two segments are equally divided between them. At the apex, one sixteenth of the vertical load is carried by each segment. Some bars may run along a common boundary between two segments. These are shown by the broken lines in Figure 12.7b. All the section properties of such bars should be divided by two when analysing the frame in terms of the behaviour of a segment. Mirror-symmetric and antisymmetric constraints are included in the software provided, so that such frames can be analysed in this way.}

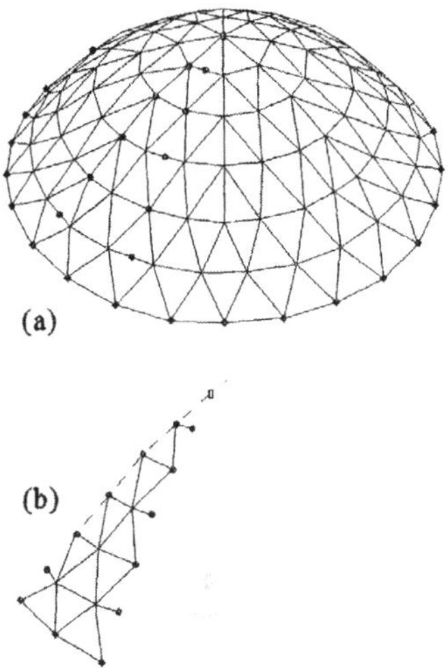

Figure 12.7 A symmetric dome.

There are advantages in providing the section data on a separate file. Most section data can be found in standard section tables which can be stored permanently on a section file. Reference can be made to this file rather than being written into the data file for each new structural problem. In this case, a facility must be included for input of data from more than one section file. Ideally, such a file would be a general-purpose one, including all the information needed for any projected use. Alternatively, its format and the method used to read it should be such that additional types of information can be added at a later date without confusing earlier programs. The units used to describe the main structure may differ from those for the sections. For example, the joint coordinates might be given in metres (or feet) and the section properties given in centimetres (or inches). A scale factor of 100 (or 12) should then be included at the beginning of the section file so that this difference can be taken in to account[2]. If a particular section file is restricted to sections made from one material, its properties can be listed at the beginning. These might include the shear and Young's moduli, the yield stress and density (for calculating frame mass and dynamic analysis). The use of a section identifier rather than a section number allows some indication to be given of the type of section used. Thus if the section is identified as a solid round bar, the only section data required is its radius. The program would then include a procedure to determine all the section properties from this.

The bar list usually consists of the numbers (or identifiers) of the two joints to which each bar is attached, and the section number (or identifier) associated with its bar properties. Its end fixity might also be described, Then bars with one or both ends pinned could be accommodated within a primarily rigid-jointed frame analysis or allowance made for the size of rigid gusset plates

---

[1] This problem is given by the data file R3DSIDOM.STR with the software provided.

[2] Thus using a factor of 39.37, section files written in inches could be used with structure files in metres.

at the joints (cf. §11.7). Normally, the members are assumed to be prismatic (straight and of constant section). Non-prismatic members can be replaced by elastically-equivalent assemblies of prismatic members, as described in §11.8. Any loading along the bar could be included at this stage, but visual checking of the file is easier if this information is given in the load list.

Two types of loading may be incorporated in the load list. Loading at the joints is given by a list of joint numbers (or identifiers) followed by a set of joint loads (including zero loads) corresponding to each degree of joint freedom[1]. If all the external loads at a particular joint are zero, then it should not be necessary to include the joint in the load list. Loading along a bar might also be taken into account. This can be replaced by equivalent joint loading as explained in §11.3 for example. However, if the program includes a stress analysis of the bars, more details of the loading is required. Consideration might be restricted to loads producing flexure about the bar's major axis, and some convention used for the positive sense of a force. Then a sufficient description of a point force along a beam would consist of the two end joint numbers (or identifiers), a character indicating that the load is a point force, its magnitude and its distance from the first joint. Similarly, a uniform distributed load would be described by a different character after the end joint numbers (or identifiers) followed by its intensity and the distance from the first joint at which it starts. This assumes that it continues until the second joint. If it does not, a second equal and opposite distributed load can be started at the point at which the first one terminates (cf. Figure 4.7b).

Further details of how data files are prepared for the software supplied are given in §A2.2. Sample structure files (*.STR files) and section files (*.SEC files) are also provided. Data may also be entered directly during the running of a programme, but this is not recommended for any but the simplest of problems.

12.4 Organisation of the Data

If the $n \times n$ stiffness matrix with elements k_{ij} is stored as a band in an $n \times b$ matrix, with elements K_{pq}, then the positions of the corresponding elements are as shown in Figure 12.8. The general rule is that the element corresponding to k_{ij} is $K_{i\,j-i+1}$. The band matrix is only filled within the shaded area as shown, leaving $\frac{1}{2}(b^2-b)$ elements empty. However, it still represents a considerable saving over storing the full matrix. Skyline storage is even more economical. Only the elements shown within the hatched areas in Figures 12.4 and 12.5 are stored. As this is an irregular set of columns, they could be stored in a one-dimensional array, together with an array of n-1 integers giving the number of elements in the second to nth columns. (The first column necessarily only contains one element.)

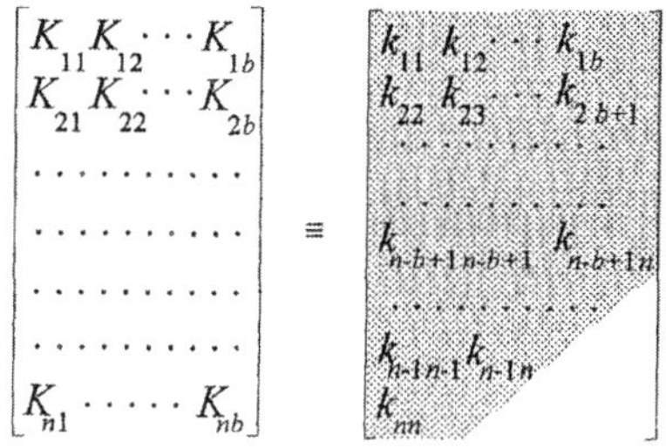

Figure 12.8 Band matrix storage.

The records shown in Figure 12.5c are generated dynamically as the data is read into the computer. The joint details are read first, and the spine of joint records created. These are linked by the pointers shown by the diagonal arrows in the figure. A separate pointer permanently points at the top left-hand record, but those within the records can be reversed to enable motion in either direction along the chain. When the bar details are read in, the off-diagonal records are created. A search along the spine is made until a match between a joint identifier and one of the bar end-joint identifiers is found. A new bar record is then created at the end of the row containing that

[1] The number of degrees of joint freedom is equal to the number of independent deflexions (rotations and displacements) an unconstrained joint in the frame would have.

joint. When all the bar records have been created, each row is sorted into the same order as a row of stiffness submatrices. Each bar record is now linked to one of its joints. It is now linked to its other joint by a chain of vertical pointers through other bar records as shown in Figure 12.5c. Starting at the first row, each subsequent row is examined to see whether it will be modified by it during the solution process. If so, the appropriate vertical links are established and new records with zero submatrices in are inserted where necessary. This process is repeated with subsequent rows, now linking any zero-submatrix records downwards too.

As will be seen in the solution process, it is convenient to store the load vectors for all m load cases in an $n \times m$ array. In the linked-record method, the load cases for each joint are stored in the appropriate joint record.

12.5 Methods of Solution

R.K. Livesley[1] published the earliest work on a general elastic frame analysis program. It was based on experiments with a program by J.M. Bennett in 1949 which contained the fundamentals of the method used. Many of the principles remain valid today. The general method of solution of the frame stiffness matrix equations using Gaussian elimination will be examined first. It will be assumed that these have been written so that a column vector of n joint deflexions, $\boldsymbol{d}$, is related to a column vector of n corresponding joint loads, $\boldsymbol{p}$, by an $n \times n$ frame stiffness matrix $\boldsymbol{K}$. If there are m different load cases $\boldsymbol{p}_1$ to $\boldsymbol{p}_m$, it is possible to avoid repeating the full analysis for a single case m times to obtain the resulting deflexions, $\boldsymbol{d}_1$ to $\boldsymbol{d}_m$. In matrix terms, these can be written as

$$\begin{gathered} \boldsymbol{K d}_1 = \boldsymbol{p}_1 \,,\; \boldsymbol{K d}_2 = \boldsymbol{p}_2 \,,\; \ldots \; \boldsymbol{K d}_m = \boldsymbol{p}_m \,, \\ \text{or} \quad \boldsymbol{K D} = \boldsymbol{P} \,, \quad \text{where} \quad \boldsymbol{D} = [\boldsymbol{d}_1 \ldots \boldsymbol{d}_m] \,,\; \boldsymbol{P} = [\boldsymbol{p}_1 \ldots \boldsymbol{p}_m] \,. \end{gathered} \tag{12.1}$$

The array $\boldsymbol{D}$ is found in terms of the array $\boldsymbol{P}$ in exactly the same way that $\boldsymbol{d}_k$ is found in terms of $\boldsymbol{p}_k$ ($k = 1$ to m), the difference being that the process for all m sets of equations is carried out simultaneously. In terms of its elements, the general form becomes

$$\begin{bmatrix} k_{11} & k_{12} & \cdots & k_{1j} & \cdots & k_{1n} \\ \cdots & \cdots & \cdots & \cdots & \cdots & \cdots \\ k_{i1} & k_{i2} & \cdots & k_{ij} & \cdots & k_{in} \\ \cdots & \cdots & \cdots & \cdots & \cdots & \cdots \\ k_{n1} & k_{n2} & \cdots & k_{nj} & \cdots & k_{nn} \end{bmatrix} \begin{bmatrix} d_{11} & \cdots & d_{1m} \\ \cdots & \cdots & \cdots \\ d_{j1} & \cdots & d_{jm} \\ \cdots & \cdots & \cdots \\ d_{n1} & \cdots & d_{nm} \end{bmatrix} = \begin{bmatrix} p_{11} & \cdots & p_{1m} \\ \cdots & \cdots & \cdots \\ p_{i1} & \cdots & p_{im} \\ \cdots & \cdots & \cdots \\ p_{n1} & \cdots & p_{nm} \end{bmatrix} \tag{12.2}$$

$$\text{or} \quad \sum_{j=1}^{n} k_{ij} d_{jk} = p_{ik} \quad (i = 1 \text{ to } n \,,\; k = 1 \text{ to } m)$$

where the terms k_{ij} are the elements of the symmetric matrix $\boldsymbol{K}$ ($k_{ij} = k_{ji}$), the terms d_{jk} are the elements of the matrix $\boldsymbol{D}$ and the terms p_{ik} are the elements of the matrix $\boldsymbol{P}$. It follows from this equation that

$$\sum_{j=1}^{n} (C k_{lj} + k_{ij}) d_{jk} = C p_{lk} + p_{ik} \quad (i, l = 1 \text{ to } n \,,\; k = 1 \text{ to } m) \tag{12.3}$$

where C is a constant. This means that a multiple of any row of $\boldsymbol{K}$ can be added to (or subtracted

---

[1] Livesley, R.K. (1953, 1954) Analysis of Rigid Frames by Electronic Digital Computer. *Engineering* Vol. 176, pp. 230 & 277 and Vol. 177, p.239.

from) any other row, and the equations will remain valid provided that the same operation is carried out with the rows of $\boldsymbol{P}$. The matrix $\boldsymbol{D}$ can be found by carrying out such operations on $\boldsymbol{K}$ and $\boldsymbol{P}$. The first element in row j of $\boldsymbol{K}$ can be made zero by subtracting k_{j1}/k_{11} times the first row from this row. The same process must be carried out with $\boldsymbol{P}$. If this is done for all j in the range 2 to n, the resulting arrays take the form

$$\boldsymbol{K}' = \left[\begin{array}{c|cccccc} k_{11} & \cdots & k_{1i} & \cdots & k_{1j} & \cdots & k_{1n} \\ \hline 0 & \cdots & \cdots & \cdots & \cdots & \cdots & \cdots \\ 0 & \cdots & k_{ii}' & \cdots & k_{ij} - \dfrac{k_{i1}}{k_{11}}k_{1j} & \cdots & k_{in}' \\ 0 & \cdots & \cdots & \cdots & \cdots & \cdots & \cdots \\ 0 & \cdots & k_{ji} - \dfrac{k_{j1}}{k_{11}}k_{1i} & \cdots & k_{jj}' & \cdots & k_{jn}' \\ 0 & \cdots & \cdots & \cdots & \cdots & \cdots & \cdots \\ 0 & \cdots & k_{ni}' & \cdots & k_{nj}' & \cdots & k_{nn}' \end{array}\right], \quad \boldsymbol{P}' = \begin{bmatrix} p_{11} & \cdots & p_{1m} \\ \cdots & \cdots & \cdots \\ p_{i1}' & \cdots & p_{im}' \\ \cdots & \cdots & \cdots \\ p_{j1}' & \cdots & p_{jm}' \\ \cdots & \cdots & \cdots \\ p_{n1}' & \cdots & p_{nm}' \end{bmatrix} \qquad \textbf{(12.4)}$$

where the modified elements are indicated by a prime, except for k_{ij}' which is given explicitly as $k_{ij} - k_{1j}k_{i1}/k_{11}$ and k_{ji}' which is given explicitly as $k_{ji} - k_{1i}k_{j1}/k_{11}$. The other modified elements can be found from these general expressions. Thus for example the form of k_{ni}' can be found by replacing j by n in the expression for k_{ji}'. The original array was symmetrical, so that k_{ij}, k_{i1} and k_{1j} are equal to k_{ji}, k_{1i} and k_{j1} respectively. From the above explicit expressions, it then follows that k_{ij}' and k_{ji}' are equal. As this true for all i and j in the range 2 to n it follows that the bottom right array bounded by the dashed lines is symmetrical. This means that the terms k_{ji}' ($j>i$) beneath the leading diagonal need not be formed, as the terms k_{ij}'have already been found. Also, if the terms k_{1i} first row of the original matrix are stored, the terms k_{i1} below k_{11} can be found from them. The process can now be repeated, eliminating the terms k_{i2}' below k_{22}'. In each case, the modified terms can be stored in place of the original ones without losing any information needed later in the solution. The modified elements $p_{ij}'$of $\boldsymbol{P}'$ take the form $p_{ij} - p_{1j}k_{i1}/k_{11}$ after the elimination of the terms k_{i1} ($i>1$) in the first column. These modified terms can also be stored in their original positions without losing useful information.

The process can be repeated to eliminate all the elements below the leading diagonal of $\boldsymbol{K}$. This process can be written in a form of MODULA-2 as

```
FOR r:= 1 TO n-1 DO
    FOR i:= r+1 TO n DO
        C:= k[r,i]/k[r,r];
        FOR j:= i TO n  DO  k[i,j]:= k[i,j] - C×k[r,j]  END;
        FOR j:= 1 TO m DO  p[i,j]:= p[i,j] - C×p[r,j]  END;
    END;
END;
```

where $k[i,j]$ is the current value of the term k_{ij}, $p[i,j]$ is the current value of the term p_{ij} and so on. (The form will be similar to this in most modern programming languages[1]). This process is known as *forward elimination* (or *reduction*) or *triangulation*. Row r is known as the *pivotal row* and k_{rr} is called the *pivotal element*. Note that only the elements on and above the leading diagonal of $\boldsymbol{K}$ are used, although (notionally) all the terms below the leading diagonal have been eliminated by this process. Subtracting multiples of one row from another does not change the value of the determinant. As all the elements below the leading diagonal are now zero, the determinant is now given simply as the product of the terms on the leading diagonal. In stability and vibration problems, the zeros of the determinant give the buckling loads and natural frequencies of the frame. {The number of arithmetical operations on the elements during triangulation is $(n^2\text{-}n)[(2n+5)/6 + m]$ of which $\frac{1}{2}(n^2\text{-}n)$ are divisions and the rest are equally divided between multiplications and subtractions. If the elements of the stiffness matrix are stored in the band matrix shown in Figure 12.6, then terms of the form $k[p,q]$ in the above process must be replaced by terms of the form $K[p,q\text{-}p+1]$ and on the fourth line 'FOR j:= i TO n DO' must be replaced by 'FOR j:= 1 TO b DO '. The number of operations involved in triangulation then becomes $(n^2\text{-}n)[b+\frac{1}{2}+m]$. In a more sophisticated coding, the upper limit b would be replaced by the lesser of b and $(n\text{-}i+1)$. This avoids redundant operations in the unused triangular area of the band }

It can now be further reduced to a unit matrix. If (12.2) is true then

$$\sum_{j=1}^{n} Ck_{ij}d_{jk} = Cp_{ik} \quad (i = 1 \text{ to } n\ , \quad k = 1 \text{ to } m) \tag{12.5}$$

is also true. This means that the equations remain true if any row of $\boldsymbol{K}$ is multiplied by a constant, C, and the same row of $\boldsymbol{P}$ is also multiplied by this constant. After the above process, the form of $\boldsymbol{K}$ is such that the last row of elements are all zero except for the last one, which has been modified to some value, k_{nn}'' say. Then this last term becomes unity if it, and the last row of elements in the modified form of $\boldsymbol{P}$, p_{nk}'' $(k = 1$ to $m)$, are divided by k_{nn}''. The other elements in the last column of $\boldsymbol{K}$ can now be made zero by working back upwards from the last row in the same way that elements in the first column were eliminated above. The last but one row of $\boldsymbol{K}$ now consists of zero elements apart from $k_{n\text{-}1\,n\text{-}1}''$. This is now made equal to unity as before, and the elements in the column above it made zero by the same process used for the last column. On repeating this process for the other columns, $\boldsymbol{K}$ becomes a unit matrix. This multiplied by $\boldsymbol{D}$ is now equal to the final form of $\boldsymbol{P}$. As multiplying $\boldsymbol{D}$ by a unit matrix has no effect on it, $\boldsymbol{D}$ is equal to the final form of $\boldsymbol{P}$. This form is given by the process

```
FOR r:= n TO 1 BY -1 DO
    C:= k[r,r];
    FOR j:= 1 TO m DO
        p[r,j]:= p[r,j]/C;
        S:= p[r,j]; (*Using S instead of p[r,j] can speed the process slightly.*)
        FOR i:= 1 TO r-1 DO  p[i,j]:= p[i,j] - S×k[i,r]  END;
    END;
END;
```

This process is known as *back substitution*. There is no need to carry out the operations on $\boldsymbol{K}$ explicitly because only the final form of $\boldsymbol{P}$ is of interest, as this is now equal to $\boldsymbol{D}$. {The number

[1] In MODULA-2, the FOR loop counter is incremented by one after each cycle, unless "BY x" is added, when the increment is x. If the upper limit of the loop is *less* than the lower limit, the loop is ignored.

of arithmetical operations carried out during back substitution is mn^2 of which mn are divisions and the rest are equally divided between multiplications and subtractions. If the band matrix is used instead, then again terms of the form $k[p,q]$ in the above process must be replaced by terms of the form $K[p,q\text{-}p+1]$. Back substitution only affects the load matrix and so the operations count remains the same.}

{Alternatively, the solution can be found by *Choleski's* (or the *Banachiewicz-Cholesky-Crout*) method. The Choleski decomposition expresses any square matrix $\boldsymbol{A}$ as the product $\boldsymbol{LU}$ of a lower triangular matrix $\boldsymbol{L}$ and an upper triangular matrix $\boldsymbol{U}$. All the elements of $\boldsymbol{L}$ above the leading diagonal are zero, as are all the elements of $\boldsymbol{U}$ beneath the leading diagonal. If $\boldsymbol{A}$ is symmetrical, then $\boldsymbol{L}$ is the same as $\boldsymbol{U}^{\mathrm{T}}$. Then

$$\boldsymbol{A} = \boldsymbol{U}^{\mathrm{T}}\boldsymbol{U} \quad \text{or} \quad \begin{bmatrix} a_{11} & a_{12} & \cdots & a_{1n} \\ a_{21} & a_{22} & \cdots & a_{2n} \\ \cdots & \cdots & \cdots & \cdots \\ \cdots & \cdots & \cdots & \cdots \\ a_{n1} & a_{n2} & \cdots & a_{nn} \end{bmatrix} = \begin{bmatrix} u_{11} & 0 & \cdots & 0 \\ u_{12} & u_{22} & 0 \; \cdots & 0 \\ \cdots & \cdots & \cdots & \cdots \\ \cdots & \cdots & \cdots & 0 \\ u_{1n} & u_{2n} & \cdots & u_{nn} \end{bmatrix} \begin{bmatrix} u_{11} & u_{12} & \cdots & u_{1n} \\ 0 & u_{22} & \cdots & u_{2n} \\ \cdots & \cdots & \cdots & \cdots \\ \cdots & \cdots & \cdots & \cdots \\ 0 & \cdots & 0 & u_{nn} \end{bmatrix} \tag{12.6}$$

From the product of these triangular matrices, we find that

$$\begin{aligned} a_{11} &= u_{11}^2 \,,\; a_{12} = u_{11}u_{12} \,,\; \ldots\; a_{1n} = u_{11}u_{1n} \,, \\ a_{22} &= u_{12}^2 + u_{22}^2 \,,\; \ldots\; a_{2j} = u_{12}u_{1j} + u_{22}u_{2j} \quad (j \geq 2), \\ a_{ij} &= \sum_{k=1}^{i} u_{ki}u_{kj} \quad (j \geq i). \end{aligned} \tag{12.7}$$

The elements u_{ij} can be found from this in the same order, giving

$$\begin{aligned} u_{11} &= \sqrt{a_{11}} \,,\; u_{12} = a_{12}/u_{11} \,,\; \ldots\; u_{1n} = a_{1n}/u_{11} \,, \\ u_{22} &= \sqrt{a_{22} - u_{12}^2} \,,\; \ldots\; u_{2j} = (a_{2j} - u_{12}u_{1j})/u_{22} \quad (j>2), \\ u_{ii} &= \sqrt{\Big(a_{ii} - \sum_{k=1}^{i-1} u_{ki}^2\Big)} \,,\; u_{ij} = \Big(a_{ij} - \sum_{k=1}^{i-1} u_{ki}u_{kj}\Big)/u_{ii} \quad (j>i). \end{aligned} \tag{12.8}$$

On examining these equations, it can be seen that u_{ij} does not exist unless a_{ij} exists, and that once u_{ij} has been formed no further reference is made to a_{ij}. This means that in decomposing the frame stiffness matrix $\boldsymbol{K}$ in this way, the elements of $\boldsymbol{U}$ can be stored where the corresponding elements of $\boldsymbol{K}$ had been stored. Then $\boldsymbol{U}$ can be generated and stored in $\boldsymbol{K}$ by the process

```
FOR i:= 1 TO n DO
    S:= k[i,i];
    FOR l:= 1 TO i-1 DO  S:= S - k[l,i]×k[l,i] END;
    k[i,i]:= sqrt(S);
    FOR j:= i+1 TO n DO
        S:= k[i,j];
        FOR l:= 1 TO i-1 DO  S:= S - k[l,i]×k[l,j] END;
        k[i,j]:= S/k[i,i]
    END;
END;
```

The function 'sqrt' takes the square root of its operand. In the above case, this is always positive for stable structures, because the stiffness matrix is positive definite. The number of arithmetical operations on the elements needed to form $\boldsymbol{U}$ in this way is $n(n+1)(2n+1)/6$, of which n are square roots, $\frac{1}{2}(n^2-n)$ are divisions and the other operations are equally divided between multiplications and subtractions. In terms of $\boldsymbol{U}$, (12.1) becomes

$$\boldsymbol{P} = \boldsymbol{K}\boldsymbol{D} = \boldsymbol{U}^{\mathrm{T}}\boldsymbol{U}\boldsymbol{D} \qquad \textbf{(12.9)}$$

so that

$$\boldsymbol{D} = (\boldsymbol{U}^{\mathrm{T}}\boldsymbol{U})^{-1}\boldsymbol{P} = \boldsymbol{U}^{-1}(\boldsymbol{U}^{\mathrm{T}})^{-1}\boldsymbol{P} = \boldsymbol{U}^{-1}\boldsymbol{F} \quad \text{where} \quad \boldsymbol{F} = (\boldsymbol{U}^{\mathrm{T}})^{-1}\boldsymbol{P} \qquad \textbf{(12.10)}$$

This is solved in two stages. The following routine generates $\boldsymbol{F}$ and stores the elements in $\boldsymbol{P}$.

```
FOR l:= 1 TO m DO
        FOR i:= 1 TO n DO
                S:= P[i,l];
                FOR j:= 1 TO i-1 DO  S:= S - P[j,l]×k[j,i] END;
                P[i,l]:=  S/k[i,i]
        END;
END;
```

The number of arithmetical operations carried out by the above routine is mn^2 of which mn are divisions and the rest are equally divided between multiplications and subtractions.

The following process generates $\boldsymbol{D} = \boldsymbol{U}^{-1}\boldsymbol{F}$ and stores it in $\boldsymbol{P}$.

```
FOR l:= 1 TO m DO
        FOR i:= n TO 1 BY -1 DO
                S:= P[i,l];
                FOR j:=i+1 TO n DO  S:= S - P[j,l]×k[i,j] END;
                P[i,l]:=  S/k[i,i]
        END;
END;
```

The number of arithmetical operations carried out by the above routine is mn^2 of which mn are divisions and the rest are equally divided between multiplications and subtractions. It can be seen that this is essentially the same process as back substitution. Comparing the number of arithmetical operations on the elements required by Gaussian elimination and Choleski's method, it will be found that the numbers of subtractions and multiplications are exactly the same in both cases. However nm more divisions are required by Choleski's method and it also takes n square roots which are not required in Gaussian elimination.

As can be seen from the routines, the process of producing the upper triangular matrix, $\boldsymbol{K}^{\nabla}$ say, in Gaussian elimination and of deriving the upper triangular matrix $\boldsymbol{U}$ in Choleski's method are very similar. Apart from an insignificant difference in the order of the operations, the resulting elements in any row of $\boldsymbol{K}^{\nabla}$ are just a multiple of those in the same row of $\boldsymbol{U}$, or

$$K^{\nabla} = \begin{bmatrix} k_{11}^{\nabla} & k_{12}^{\nabla} & \cdots & k_{1n}^{\nabla} \\ 0 & k_{22}^{\nabla} & \cdots & k_{2n}^{\nabla} \\ \cdots & \cdots & \cdots & \cdots \\ 0 & \cdots & 0 & k_{nn}^{\nabla} \end{bmatrix} = \begin{bmatrix} \sqrt{k_{11}^{\nabla}} & 0 & 0 \cdots & 0 \\ 0 & \sqrt{k_{22}^{\nabla}} & 0 \cdots & 0 \\ \cdots & \cdots & \cdots & \cdots \\ 0 & \cdots & 0 & \sqrt{k_{nn}^{\nabla}} \end{bmatrix} \begin{bmatrix} u_{11} & u_{12} & \cdots & u_{1n} \\ 0 & u_{22} & \cdots & u_{2n} \\ \cdots & \cdots & \cdots & \cdots \\ 0 & \cdots & 0 & u_{nn} \end{bmatrix} = RU \qquad (12.11)$$

Then (12.10) becomes

$$D = (U^{T}U)^{-1}P = ((R^{-1}K^{\nabla})^{T}R^{-1}K^{\nabla})^{-1}P = (K^{\nabla T}R^{-2}K^{\nabla})^{-1}P = (K^{\nabla})^{-1}R^{2}(K^{\nabla T})^{-1}P \qquad (12.12)$$

where R^2 is the diagonal matrix formed of the elements $k_{11}^{\nabla} \ldots k_{nn}^{\nabla}$ on the leading diagonal of K^{∇}.

There seems to be no advantage in using Choleski's method when all the loadings have been predetermined. However, if new ones are generated during the running of the program, and U has been retained, each new solution would only require a further $2n^2$ operations. From (12.12) it follows that there would be advantages in retaining K^{∇} under such circumstances, but each new solution would require a further $2n^2+n$ operations.}

12.6 Data Output

Standard data output consists of the joint deflexions and the loads at the ends of the bars. It is convenient to store this on a file which can be printed or displayed or used as a data file for post-processing by another program. The joint deflexions result immediately from the solution of the stiffness equations discussed in the previous section. These are in global (frame) coordinates and can be converted to local bar coordinates using transformations of the type given by the second of equations (11.3). The end loads, relative to local coordinates, then follow from equations of the type given by (11.7). (The fixed-end loads must be added onto these if a bar is loaded along its length.) As a check on the accuracy of the analysis, these end loads can be expressed in terms of global coordinates using transformations of the type given by the first of equations (11.3) to see that joint equilibrium is satisfied.

Graphic output is more readily absorbed than lists of numeric data. Some programs will plot bar bending-moment and shear-force diagrams of the type shown in Figure 2.18. However, for three-dimensional frames where there may be bending about two axes, these can become very complex and confusing. Their function is to indicate the state of stress in the members and this can be done more directly. If the section data includes the cross-sectional area and section moduli[1], the greatest axial stress, σ, and the greatest shear stress, τ, can be calculated. These will not necessarily occur at the same point on the cross section, but a conservative estimate of the worst state of stress can be made by assuming that they do. From the Mohr's circle diagram shown in Figure 12.9, the principal stresses are given by

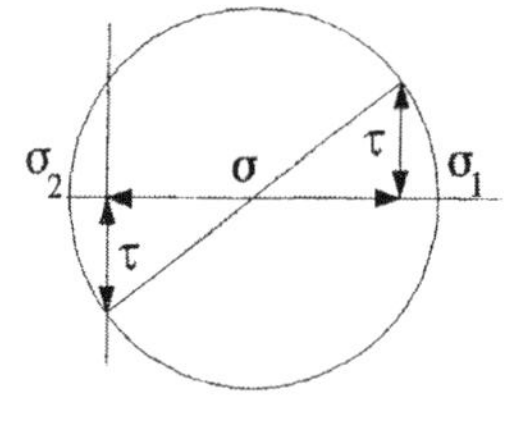

$$\sigma_1, \sigma_2 = \frac{\sigma}{2} \pm \sqrt{\frac{\sigma^2}{4} + \tau^2} \qquad (12.13)$$

Figure 12.9 Mohr's circle for axial and shear stress.

and the third principal stress is zero. If material failure is governed by von Mises' yield criterion, then from (3.55)

[1] See §4.2 and §A2.2.4.

$$\left(2\sqrt{\frac{\sigma^2}{4}+\tau^2}\right)^2 + \left(\frac{\sigma}{2}+\sqrt{\frac{\sigma^2}{4}+\tau^2}\right)^2 + \left(\frac{\sigma}{2}-\sqrt{\frac{\sigma^2}{4}+\tau^2}\right)^2 = 2\sigma_f^2 \qquad \textbf{(12.14)}$$

$$\text{or} \quad \sigma^2 + 3\tau^2 = \sigma_f^2$$

An equivalent axial stress, σ_e, can be defined as

$$\sigma_e = \sqrt{\sigma^2 + 3\tau^2} \qquad \textbf{(12.15)}$$

so that failure occurs when σ_e is equal to σ_f and the fraction σ_e/σ_f gives the ratio of the current loading to that at failure. For a graphical representation of the state of stress of the frame, it can be displayed with the members coloured according to the value of this ratio. A possible scheme is suggested in §A2.2.5.

If there is no loading along the members, the worst state of stress will be found at one or the other of their ends. If there are point loads along the beam, the bending moments at their points of application must be examined too. If distributed loads are also applied, a maximum (or minimum) may occur other than at a discontinuity in the loading. The beam then has to be broken up into strips in which the loading is continuous, the moment evaluated at the ends of each strip and the possibility of a maximum occurring within the strip also examined. Figure 12.10 shows a strip CD of length d with a uniform load of intensity p acting on it. Having found the moment and shear force at the left-hand end of the beam, the cumulative moment M and shear force S at C can be found from statics. Within the strip, the moment at a distance x from C is given by

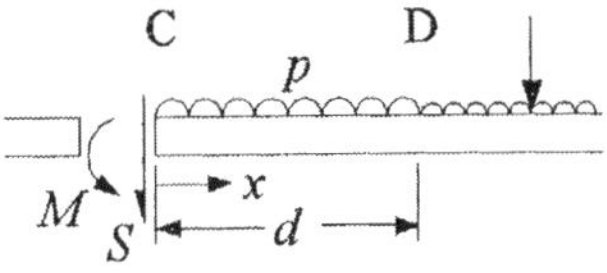

Figure 12.10 Maximum moment.

$$M_x = M + Sx + \tfrac{1}{2}px^2 \qquad \textbf{(12.16)}$$

This is a maximum (or minimum) of $(M - S^2/2p)$ when x is $-S/p$. The program must then check first whether this value of x lies within the limits $0<x<d$ and then whether this maximum (or minimum) is more significant than any found thus far. Note that if axial loading is taken into account, both the greatest positive and the greatest negative bending moments will have to found in determining the worst state of stress.

{The deformed state may also be illustrated graphically. This involves a degree of compromise. The deflexions found are based on small-deflexion theory and will probably be too small to be noticeable if displayed to scale. If amplified too greatly, the approximations in small-deflexion theory become apparent. It will be shown that the deflected form of a member[1] can be found in terms of the deflections of its ends and the vector joining these ends.

Figure 12.11 shows a beam AB of length l loaded by end moments and shear forces. The end rotations ϕ_{yA} and ϕ_{yB} are measured from the dashed line joining the displaced positions of these joints so that these rotations are related to the absolute rotations of the joints by

$$\phi_{yA} = \theta_{yA} - \frac{1}{l}(u_B - u_A)\,,$$
$$\phi_{yB} = \theta_{yB} - \frac{1}{l}(u_B - u_A)\,. \qquad \textbf{(12.17)}$$

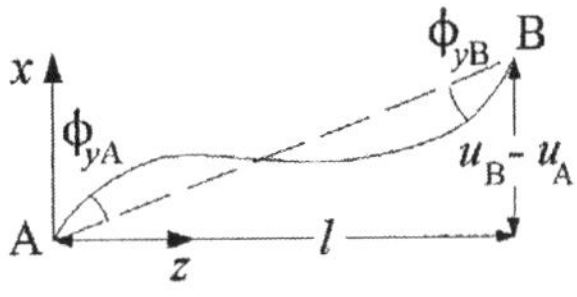

Figure 12.11 Flexure of a beam in the x-z plane.

where u_A and u_B are the end displacements in the x direction and the end rotations θ_{yA} and θ_{yB} are the clockwise rotations of the end joints about the y axis (out

[1] Assumed to be prismatic and loaded at its ends only.

of the paper). Then from (4.55) and (4.56), the moment and shear force at end A are given by

$$M_{yA} = \frac{2EI}{l}\left(2\phi_{yA} + \phi_{yB}\right) \ , \ S_{xA} = \frac{6EI}{l^2}\left(\phi_{yA} + \phi_{yB}\right). \tag{12.18}$$

so that from (4.15) and (12.18) the lateral displacement u from the dashed line is given by

$$EI\frac{d^2u}{dz^2} = S_{xA}z - M_{yA} \quad \text{or} \quad l\frac{d^2u}{dz^2} = 6\frac{z}{l}(\phi_{yA} + \phi_{yB}) - 4\phi_{yA} - 2\phi_{yB} \tag{12.19}$$

at a distance z from end A. This equation can be integrated twice, taking the gradient of u as ϕ_{yA} and ϕ_{yB} at the two ends and the end values of u as zero, giving

$$u = le(1 - e)[(1 - e)\phi_{yA} + e\phi_{yB}] \tag{12.20}$$

where e is z/l. A similar result can be obtained for flexure in the y-z plane. Allowing for uniform axial extension, the total displacement of the member in three dimensions is given by

$$\begin{bmatrix} u \\ v \\ w \end{bmatrix} = le(1 - e)^2\begin{bmatrix} \phi_{yA} \\ -\phi_{xA} \\ 0 \end{bmatrix} + le^2(1 - e)\begin{bmatrix} \phi_{yB} \\ -\phi_{xB} \\ 0 \end{bmatrix} + (1 - e)\begin{bmatrix} u_A \\ v_A \\ w_A \end{bmatrix} + e\begin{bmatrix} u_B \\ v_B \\ w_B \end{bmatrix} \tag{12.21}$$

Taking $\boldsymbol{k}$ as a unit vector in the z direction, this can be written in vector terms as

$$\boldsymbol{v} = le(1 - e)^2\boldsymbol{\phi}_A\times\boldsymbol{k} + le^2(1 - e)\boldsymbol{\phi}_B\times\boldsymbol{k} + (1 - e)\boldsymbol{v}_A + e\boldsymbol{v}_B$$

$$\text{where} \quad \boldsymbol{\phi}_A = \boldsymbol{\theta}_A - \frac{1}{l}(\boldsymbol{v}_A - \boldsymbol{v}_B)\times\boldsymbol{k} \ , \ \boldsymbol{\phi}_B = \boldsymbol{\theta}_B - \frac{1}{l}(\boldsymbol{v}_A - \boldsymbol{v}_B)\times\boldsymbol{k} \tag{12.22}$$

using (12.17) and its equivalent in the y-z plane. This vector equation can be expressed as readily in terms of the frame coordinate system as in terms of the member coordinate system. Using the properties of the vector triple product, it becomes in terms of joint rotations and displacements

$$\begin{aligned} \boldsymbol{v} = {} & e(1 - e)\{l[(1 - e)\boldsymbol{\theta}_A + e\boldsymbol{\theta}_B]\times\boldsymbol{k} - [\boldsymbol{k}\cdot(\boldsymbol{v}_A - \boldsymbol{v}_B)]\boldsymbol{k}\} \\ & + (1 - e^2)\boldsymbol{v}_A + e^2\boldsymbol{v}_B \end{aligned} \tag{12.23}$$

This gives the displacement of a point on the beam relative to its original position on the centerline. In plotting the deformed member, this original position, $\boldsymbol{r}_A + el\boldsymbol{k}$, must be added to $\boldsymbol{v}$ (where $\boldsymbol{r}_A$ gives the original position of joint A).

Plotting the exact deflexion curves for all the members of the frame is unnecessarily time-consuming. Provided that the deflected form is not over-amplified, it can be drawn as a set of polygonal lines with $\boldsymbol{v}$ evaluated to give the position of each vertex. As the bending moment varies linearly along the beam, the highest curvature will occur at the ends. More vertices are then needed at the ends to simulate the deflected form. A polygonal line with its ends at the two joints and vertices at values of e equal to 0.1, 0.25, 0.5, 0.75 and 0.9 has proved satisfactory.}

12.7 Stability and Vibration Problems

In §11.5 and §11.6, the overall buckling load and natural frequencies of a frame were seen to correspond to the zeros of Det($\boldsymbol{K}$), the determinant of frame stiffness matrix $\boldsymbol{K}$. Figure 12.12 shows how this determinant may vary with loading or frequency. The zeros are marked by circles at A,B,C,D,F and I. Asymptotes are also shown at E,G and H where the determinant becomes infinite. For this to happen, one or more of the elements of $\boldsymbol{K}$ must become infinite, which happens when a beam reaches its fixed-end buckling load or fixed-end natural frequency. This makes

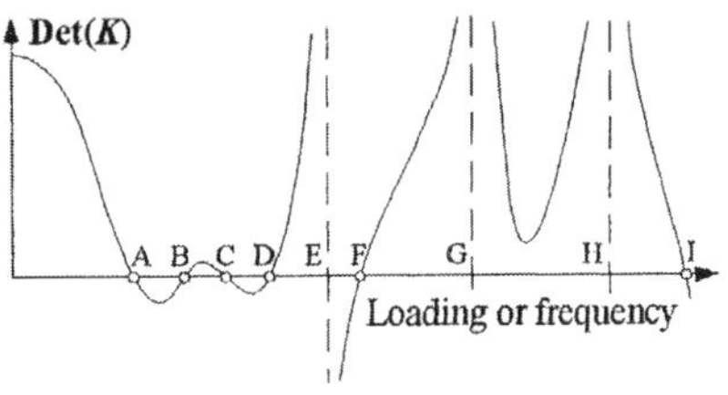

Figure 12.12 The variation of the stiffness determinant with loading or frequency.

deflexions of the beam possible without joint deflexions taking place, so that these singularities indicate local buckling or natural frequencies. If the frame is initially stable, then $\boldsymbol{K}$ is positive-definite so that Det($\boldsymbol{K}$) begins by being positive. How Det($\boldsymbol{K}$) varies with loading or frequency parameter will not be known a priori, so that the zeros must be sought by trial and error. If the parameter is increased by too large an increment, the zero at A may be missed and one of the higher zeros found instead. In vibration problems, it may be useful to know not just the lowest natural frequency of the frame, but all the natural frequencies within a range of frequencies. Some criteria are then required to avoid missing any zeros of Det($\boldsymbol{K}$). Wittrick and Williams have determined these for vibration[1] and stability[2]. The derivation of the criteria follows similar lines in both papers, and the earlier one will be summarised here.

The basic algorithm to be proved is that if $J(\Omega)$ is the number of natural frequencies of the structure which are less than some specified frequency, Ω, then

$$J(\Omega) = J_0(\Omega) + s(\boldsymbol{K}_\Omega) \tag{12.24}$$

where $J_0(\Omega)$ is the number of fixed-end frequencies, less than Ω, of the members of the structure, and $s(\boldsymbol{K}_\Omega)$ is the number of negative elements on the leading diagonal of the Gaussian upper-triangular form, $\boldsymbol{K}_\Omega^\nabla$, of the structure stiffness matrix $\boldsymbol{K}_\Omega$ at frequency Ω.

When applied to buckling problems, Ω is some specified multiple of a given loading on the structure. $J(\Omega)$ and $J_0(\Omega)$ are respectively the total number of buckling loads and the number of fixed-end instabilities of members occurring at load factors less than Ω, and $s(\boldsymbol{K}_\Omega)$ is the number of negative elements on the leading diagonal of the Gaussian upper-triangular form, $\boldsymbol{K}_\Omega^\nabla$, of the structure stiffness matrix $\boldsymbol{K}_\Omega$ for the load factor Ω.

{The proof relies on a theorem deduced by Rayleigh (1894, p.119) from energy considerations. It can be stated as follows.

Suppose that the natural frequencies of a structure are $\omega_1 \le \omega_2 \le \ldots \le \omega_r \le \ldots \le \omega_n$. *If one constraint is imposed on the structure, then the new natural frequencies of the structure, also arranged in order, will be such that the* r*th*, ω_r', *will lie between the* r*th and the* (r+1)*th original frequencies, or*

$$\omega_r \le \omega_r' \le \omega_{r+1}$$

(i) If none of the original natural frequencies are coincident, and no node of the original modes of vibration coincides with the constraint, then $\omega_r < \omega_r' < \omega_{r+1}$.

(ii) The term 'constraint' must refer to an inhibition of one of the degrees of freedom of the structure[3]. Here, it will be taken to remove that degree of freedom altogether.

(iii) Although Rayleigh's proof was for a finite system, it is also valid for an infinite system,

---

[1] Wittrick, W.H. and Williams, F.W. (1971) A General Algorithm for Computing Natural Frequencies of Elastic Structures. *Quart. Journ. Mech. and Applied Math.* Vol.24, Pt.3, p.263.

[2] Wittrick, W.H. and Williams, F.W. (1973) An Algorithm for Computing Critical Buckling Loads of Elastic Structures. *J. Struct. Mech.* Vol.1, Pt.4, p.497.

[3] Tarnai (op. cit. §9.9) has shown that some 'constraints', such as a pin-ended rod under axial compression, can have a destabilising effect. These are not constraints in the above sense.

provided that there is not a continuous spectrum of natural frequencies (i.e. the frequencies must be countable).

(iv) If the original structure has m coincident natural vibration modes at a frequency ω_c and a single constraint is applied, then the constrained structure will have m-1, m or m+1 modes at this frequency. This can be seen by applying the above theorem to m natural frequencies over a range which shrinks at the limit to the frequency ω_c. The converse is that if one constraint is removed from such a structure, there will be m-1, m or m+1 modes with frequency ω_c. This follows from taking the initial structure as the one with a constraint added.

(v) A corollary of the above theorem is that if a single constraint is removed from the structure, the number of natural frequencies, $J(\Omega)$, which lie below any given frequency Ω either remains the same or increases by one. More generally, if c constraints are removed, $J(\Omega)$ increases by b $(0 \le b \le c)$.

Let K_r be the rth principal minor of $\boldsymbol{K}$ (that is, the determinant formed from the first r rows and columns of $\boldsymbol{K}$) and K_0 will be defined as +1. The Gaussian upper-triangular form, $\boldsymbol{K}^\nabla$ has the same principal minors as $\boldsymbol{K}$ and these are given by the products of the elements k_{ii}'' on its leading diagonal. This means that K_r is given by the product of K_{r-1} and k_{rr}''. It follows that $s(\boldsymbol{K})$ as defined above is equal to the number of changes of sign in the sequence $K_0, K_1, \ldots K_n$.

The matrix equations are now augmented by adding loads at points along the members and relating them to their corresponding deflexions and the joint deflexions. The full set of N equations is now

$$\boldsymbol{p}^A = \boldsymbol{K}^A \boldsymbol{d}^A \quad \text{or} \quad \begin{bmatrix} \boldsymbol{p}^a \\ \boldsymbol{p} \end{bmatrix} = \begin{bmatrix} \boldsymbol{K}_{aa} & \boldsymbol{K}_{ab} \\ \boldsymbol{K}_{ba} & \boldsymbol{K}_{bb} \end{bmatrix} \begin{bmatrix} \boldsymbol{d}^a \\ \boldsymbol{d} \end{bmatrix} \tag{12.25}$$

where $\boldsymbol{p}^A$ consists of the column vector of additional loads, $\boldsymbol{p}^a$, and the joint loads, $\boldsymbol{p}$, the augmented displacement vector, $\boldsymbol{d}^A$, consists of the additional deflexions, $\boldsymbol{d}^a$, and the joint deflexions, $\boldsymbol{d}$. The stiffness matrix $\boldsymbol{K}^A$ consists of the array of submatrices shown which relate these load and deflexion vectors.

Suppose that the set of N-n additional degrees of freedom measured has been chosen so that if all N deflexions are prevented by rigid constraints, there are no natural frequencies of the constrained structure below some specified frequency Ω. This ensures that the elements of $\boldsymbol{K}_\omega^A$ $(\omega<\Omega)$ are finite, as by definition there are no fixed-end natural frequencies in this range. Moreover, all the principal leading minors $K_{\omega r}^A$ of $\boldsymbol{K}_\omega^A$ (r = 1 to N) must be finite. The condition for any natural frequency ω, $(\omega<\Omega)$, to arise in the *unconstrained* structure is now that Det($\boldsymbol{K}_\omega^A$) $(= K_{\omega N}^A)$ is zero.

All N constraints are now applied and then removed one by one, releasing the degrees of freedom in $\boldsymbol{d}^A$ in order, starting with the first. The condition for a natural frequency ω to arise in the structure after r degrees of freedom have been restored is that $K_{\omega r}^A$ is zero, and after r+1 degrees of freedom have been restored it is that $K_{\omega r+1}^A$ is zero. The variation of these two minors with ω is shown in Figure 12.13. Initially, as the structure is stable, $\boldsymbol{K}^A$ is positive definite so that all the principal minors are positive. The natural frequencies after r constraints have been removed are shown by the zeros on the upper graph, and after one more constraint has been removed by the zeros on the lower graph. From (v) above, removing one constraint will increase the number of natural frequencies below a given frequency by one or leave the number unchanged. This can be seen from the relative positions of the zeros on the two graphs.

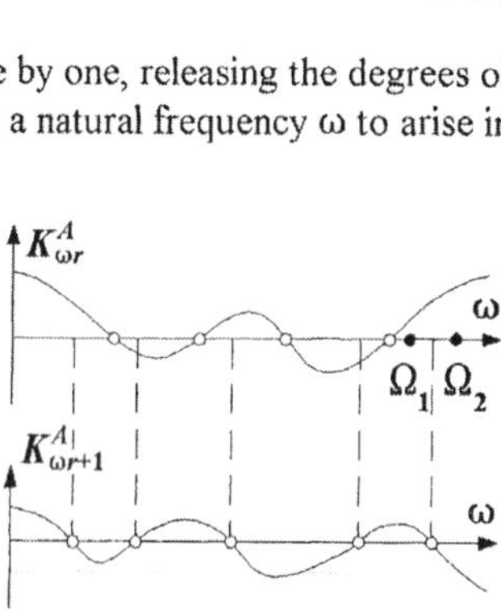

Figure 12.13 Variation of the leading principal minors of $\boldsymbol{K}_\omega^A$.

Two values for the specified natural frequency are chosen, as shown by the black dots on the upper axis. The lower, Ω_1, is such that both graphs have passed through the same number of zeros. The higher, at $\omega=\Omega_2$, is such that $K_{\omega r+1}{}^A$ has passed through one more zero than $K_{\omega r}{}^A$. Note that inevitably in the first case, both minors are of the same sign at frequency $\omega=\Omega_1$ and in the second case, they are of opposite sign at Ω_2. Thus the existence of one more natural frequency less than Ω associated with r+1 releases than there are associated with r releases (and from corollary (v) there can at most be one more) is detected by a change of sign between $K_{\Omega r}{}^A$ and $K_{\Omega r+1}{}^A$. Again from corollary (v) there can only be one natural frequency at most associated with the first release. This is detected by $K_{\Omega 1}{}^A$ being negative, i.e. there is a change of sign between $K_{\Omega 0}{}^A$ and $K_{\Omega 1}{}^A$. It follows that the number of natural frequencies of the augmented structural system below the specified frequency Ω is given by the number of changes in sign in the sequence of leading principal minors $K_{\Omega 0}{}^A$, $K_{\Omega 1}{}^A$, .. $K_{\Omega N}{}^A$ of $\boldsymbol{K}_\Omega{}^A$, i.e. $s(\boldsymbol{K}_\Omega{}^A)$.

The relationship between this and the original stiffness matrix will now be established. The set of augmented equations is given by (12.25), where its is understood that the stiffnesses are functions of frequency without using an additional subscript. In the original problem, no constraints $\boldsymbol{p}^a$ are applied. If partial Gaussian elimination is carried out to turn $\boldsymbol{K}_{aa}$ into its upper triangular form and eliminate $\boldsymbol{K}_{ba}$, the equations then become

$$\begin{bmatrix} \boldsymbol{0} \\ \boldsymbol{p} \end{bmatrix} = \begin{bmatrix} \boldsymbol{K}_{aa}^{\nabla} & \boldsymbol{K}_{ab}^{/} \\ \boldsymbol{0} & \boldsymbol{K} \end{bmatrix} \begin{bmatrix} \boldsymbol{d}^{a} \\ \boldsymbol{d} \end{bmatrix} \tag{12.26}$$

where $\boldsymbol{K}$ is the original stiffness matrix. Upper triangulation can be completed by operations on $\boldsymbol{K}$ only, giving

$$\boldsymbol{K}^{A\nabla} = \begin{bmatrix} \boldsymbol{K}_{aa}^{\nabla} & \boldsymbol{K}_{ab}^{/} \\ \boldsymbol{0} & \boldsymbol{K}^{\nabla} \end{bmatrix} \tag{12.27}$$

As the sign count is given by the number of negative terms on the leading diagonal of the upper triangular form, $s(\boldsymbol{K}^A)$ is the sum of $s(\boldsymbol{K}_{aa})$ and $s(\boldsymbol{K})$, or at the specified frequency Ω,

$$s(\boldsymbol{K}_\Omega^A) = s(\boldsymbol{K}_{aa\,\Omega}) + s(\boldsymbol{K}_\Omega) \tag{12.28}$$

Consider the case when constraints are applied to the joints so that $\boldsymbol{p}$ is such that $\boldsymbol{d}$ is zero and no other constraints are applied. From (12.25), we now have

$$\boldsymbol{p}^a = \boldsymbol{0} = \boldsymbol{K}_{aa\,\Omega}\boldsymbol{d}^a \tag{12.29}$$

and so the number of natural frequencies less than Ω associated with fixed-end vibration of the members of the structure, $J_0(\Omega)$, is given by $s(\boldsymbol{K}_{aa\Omega})$. The total number of natural frequencies of the structure, including the fixed-end natural frequencies of the members, which lie below Ω is given by $s(\boldsymbol{K}_\Omega{}^A)$ so that it can now be seen that (12.28) implies the algorithm given by (12.24).}

The software provided gives the 'zeros count', $J(\Omega)$ in (12.24), for the response to each frequency Ω examined. Up to point E in Figure 12.12, this will be the same as $J_0(\Omega)$, but just beyond E, $J_0(\Omega)$ decreases by one and the fixed-end natural frequency count increases from zero to one. Likewise, at the other asymptotes, the $J_0(\Omega)$ count decreases and the fixed-end natural frequency count increases but the total count remains the same. The proof distinguishes clearly between vibration involving joint motion and individual bar vibration where joint motion is prevented. In a pin-jointed frame, the degrees of freedom in the deflexion vector may be joint displacements only. Then what are referred to above as 'fixed-end natural frequencies' are not for encastré bars, but bars whose ends are free to rotate but not displace.

The arguments used to prove the equivalent algorithm for stability problems is similar, except that allowance has to be made for the possibility that buckling might occur at negative load factors. Also, the non-linearities of the stability problem mean that buckling can depend on the history of the loading. Even when the change in overall geometry of the frame is not taken into account, it becomes necessary to apply the loads in moderate increments and repeat the analysis at a given applied loading until the stiffness matrix for the assumed internal axial loads matches that for the actual internal loads. This is indicated by no change in the value of matrix determinant on successive cycles of analysis at the same applied loading.

Chapter 13 Influence Lines

13.1 Introduction

Influence lines illustrate the effects of loads travelling slowly[1] across beam-like structures. Such structures are usually ordinary beams or pin-jointed trusses. Some of the graphical techniques associated with influence lines may be replaced by more appropriate computer methods. However, the underlying concepts still remain relevant. Usually, an influence line diagram is first drawn for the effect at a fixed point on a structure produced by a unit moving load. From such a diagram, the effect produced by a train of loads or a distributed load can be found.

13.2 Elementary Beam Problems

Figure 13.1 shows the shear and bending influence lines for a point P on a simply-supported beam AB under the action of a unit load. The conventions used are those shown in Figure 4.5. The unit load is at a variable distance x from A. The influence lines in Figures 13.1b and 13.1c show how the induced shear force S_P and bending moment M_P at P vary with x. The shear force is given by

$$\begin{aligned} S_P &= R_B \quad (x<a), \\ &= R_B - 1 \quad (x>a). \end{aligned} \tag{13.1}$$

where the reaction at B is given by

$$R_B = \frac{x}{a+b} \tag{13.2}$$

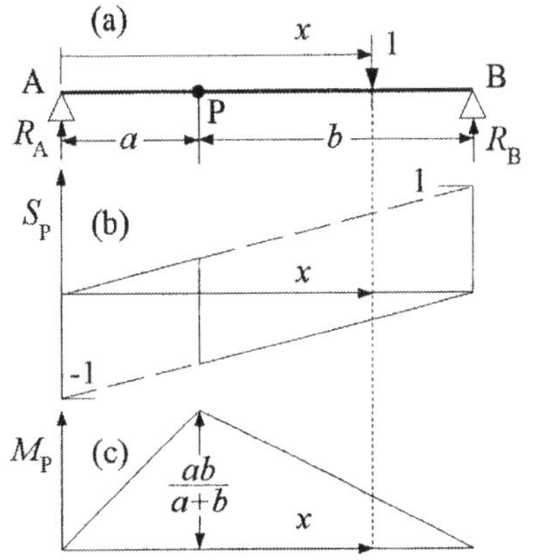

Figure 13.1 A simple beam.

The plot of the variation of R_B with x is shown by the upper sloping line in Figure 13.1b and the plot of R_B-1 (or $-R_A$) is given by the lower sloping line. The plot of S_P switches from the upper to the lower sloping line at $x=a$ where the travelling load crosses the point P. The plot of the variation of M_P with x is shown in Figure 13.1c. This is given by

$$\begin{aligned} M_P &= R_B b = \frac{xb}{a+b} \quad (x<a), \\ &= R_A a = a\left(1 - \frac{x}{a+b}\right) \quad (x>a). \end{aligned} \tag{13.3}$$

In this case, the influence line diagram happens to be the same as the bending moment diagram for a unit load at P, but this is a coincidence and not a general rule.

Suppose that at a distance x along the influence line diagram for a unit load, the ordinate is y, as shown in Figure 13.2a. If a load F is applied at this point, the effect produced at the point P is no longer y but is magnified to Fy. In the case of a distributed load of intensity q, an elementary strip δx of it produces a resultant force $q\delta x$ as shown in Figure 13.2b. If the ordinate

[1] Dynamic effects are not considered here.

of this strip on the influence line diagram is y, then the effect produced at P by this elementary strip is $qy\delta x$. The product $y\delta x$ corresponds to the area δA on the influence line diagram as shown in the figure. The total effect of the distributed load q is given by integrating $q\delta A$ over the length on which q acts. If q is constant, then the total effect is qA, where A is the area of the influence line diagram between the intercepts at the ends of the distribution as shown in Figure 13.2b.

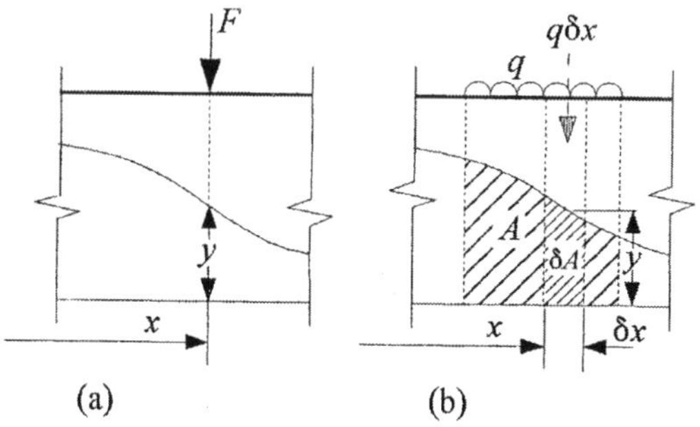

Figure 13.2 Influence of a distributed load.

Consider a train of loads, f_1 to f_5, at fixed distances apart, crossing a simply-supported beam as shown in Figure 13.3. The bending moment M_P induced at P can be taken as the sum of M_{PL}, the moment induced by the loads to the left of P, and M_{PR}, the moment induced by the loads to the right of P. The loads to the left induce a moment

$$M_{PL} = \sum_i (f_i y_i) = \sum_i (f_i x_i \tan\alpha) \qquad \textbf{(13.4)}$$

where α is the slope of the influence line as shown and the summation is from i=1 to 3 in the present case. The resultant of these loads, F_L, acts at a distance x_L from the left-hand end. The moment can be expressed in terms of this resultant, for the moment it produces about the left-hand end is the same as that produced by the loads f_i or

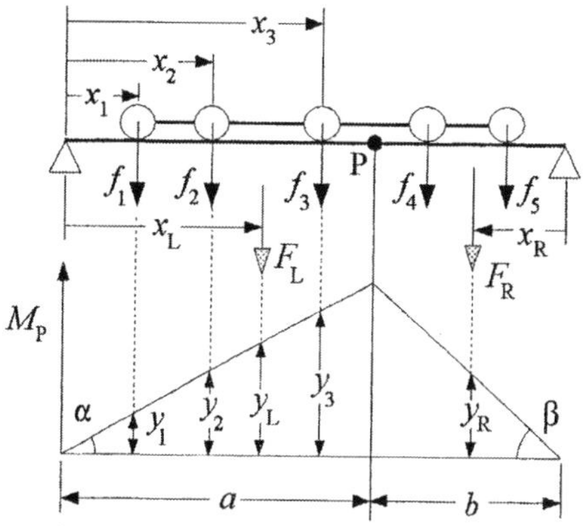

Figure 13.3 Moments induced by a load train.

$$M_{PL} = \sum_i (f_i x_i) \tan\alpha = F_L x_L \tan\alpha \qquad \textbf{(13.5)}$$

Similarly, the moment produced by the loads to the right of P is given by $F_R x_R \tan\beta$ where x_R is measured from the *right-hand* end of the beam and β is the slope of the influence line as shown. The total moment is then

$$M_P = F_L x_L \tan\alpha + F_R x_R \tan\beta \qquad \textbf{(13.6)}$$

If the train of loads shifts a small distance δx to the right, x_L is increased by δx and x_R is decreased by δx. The change in the total moment produced at P is then

$$\delta M_P = F_L \delta x \tan\alpha - F_R \delta x \tan\beta \qquad \textbf{(13.7)}$$

so that as long as

$$\frac{F_L}{F_R} > \frac{\tan\beta}{\tan\alpha} \left(= \frac{a}{b} \right) \qquad \textbf{(13.8)}$$

there is an increase in the moment produced. If the train of loads extends beyond the ends of the beam, F_L increases when another load enters the beam from the left and F_R decreases when a load leaves the beam at the right. Thus the ratio F_L/F_R does not decrease when loads enter or leave the beam as the train moves to the right. However, this ratio will decrease when a load immediately to the left of P crosses this point. It then contributes to the resultant F_R instead of the resultant F_L. A maximum value of M_P is then reached just as a load in the train crosses P which causes the inequality (13.8) no longer to hold. If the train is a long one, this may only be a local maximum and not an absolute maximum. Maxima found in this way may be used to design structures intended to carry moving trains of loads. Normally, such trains might cross in either direction. This means that the worst case must be found for the train of loads in the initial order and in the reverse order.

Most criteria of the above kind only establish local maxima, which must be compared to find the absolute maximum load or deflexion induced at some reference point P. There are some exceptions. Suppose that the train consist of only two loads, f_1 and f_2. It can be assumed without loss of generality that $f_1 > f_2$ and $a > b$. If the beam is simply supported, the maximum bending moment induced at P will be when f_1 is at P and f_2 is to the left of it. If a uniform distributed load is longer than such a beam, the worst bending moment occurs when the load covers the whole span. However, the maximum shear stress occurs when this load stretches from the left support to the point P, again assuming $a > b$.

Figure 13.4 shows a moving uniform distributed load of finite length displayed above an influence line diagram for some point P on a structure. As noted earlier, if the intensity of the loading is p, then the effect produced at P is given by pA where A is the hatched area shown. If the load moves a small distance δx to the right, then A increases by a small amount $y_2\delta x$ and diminishes by a small amount $y_1\delta x$. Thus as long as $y_2 > y_1$ the effect increases. A maximum is reached when $y_2 = y_1$. In the simply-supported beam problem, it follows that the worst bending moment at P induced by such a load occurs when the load is divided by P in the same proportion that P divides the beam. This can be seen from the properties of similar triangles applied to Figure 13.1c.

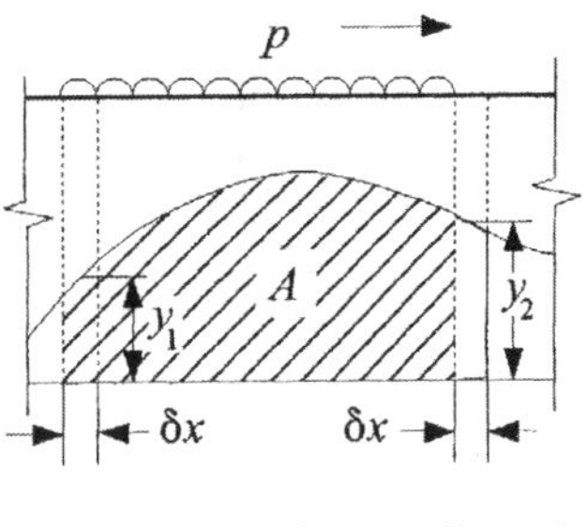

Figure 13.4 Maximum effect of a moving distributed load.

Consider the shear produced at P by the train load shown in Figure 13.5. The slopes of the two parts of the influence line are both $1/l$. This means that if the train moves a small distance δx to the right, a force f_i in the train contributes a positive increment to the shear at P of $f_i\delta x/l$, regardless of its position on the beam. Thus provided that no load crosses the point P, or enters or leaves the beam, the increment in shear produced by the train of loads is

$$\delta S_P = \sum_i f_i \,\delta x/l = F\,\delta x/l \qquad \textbf{(13.9)}$$

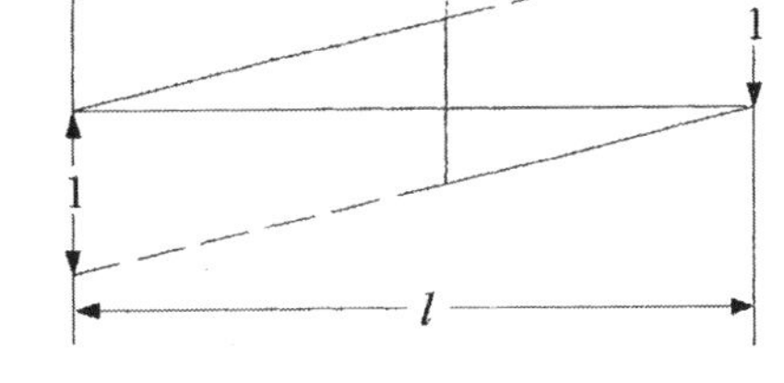

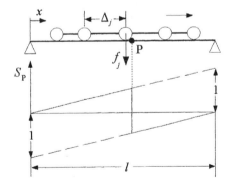

Figure 13.5 Maximum shear load of a train.

where F is the resultant load of the train. This is necessarily a positive increment if all the loads are acting downwards. A load just entering at the left of the beam will also make a positive contribution and a load just leaving the beam will make no contribution at all to the increment. Thus the positive shear will increase except when one of the loads in the train crosses the point P. If a load f_j is on the point of crossing P, a local maximum value in S_p is reached, as this shear force is reduced by the finite amount f_j during the crossing. The next local maximum is reached when the next load, Δ_j behind as shown, reaches P. If no load has entered or left the beam during this shift, this new maximum is greater if $F\Delta_j/l - f_j$ is positive. This can be used as a criterion in looking for the absolute maximum shear force produced. Of course, the maximum *amplitude* of the shear force is normally sought, regardless of its sign. This means that the most negative shear force must also be examined.

13.3 Envelope Diagrams

In designing a span, it is useful to know the maximum bending moment or shear force sustained at every section for any position of the loads crossing the span. Figures 13.6a and 13.6b show respectively the plots of the maximum bending moments and shear forces at each point of a simply supported beam when a unit load crosses it. The plots of the maxima (shown by solid lines) form envelopes of the diagrams[1] (shown dotted) for the load at each point along the span. In the case of the shear force diagram, the maximum positive and the maximum negative shear forces are shown by the two sloping lines. Thus the maximum *absolute* shear force is given by two lines forming a Vee shape, with end values of unity and a central value of a half.

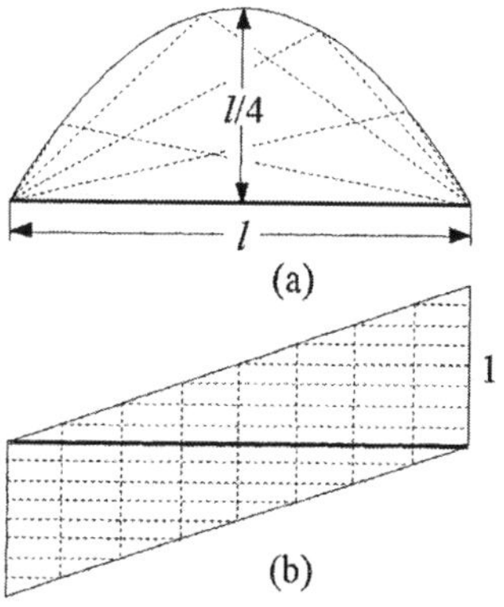

Figure 13.6 Bending moment and shear force envelopes.

Some methods of determining the maxima for more complex loading were given in the last section. A further method is particularly useful in finding bending moment envelopes for load trains. It has already been seen that the maximum bending moment at a point P will occur when one of the loads in that train is over that point. The unit-load bending moment influence line diagram for P shown in Figure 13.7 can be thought of as a straight line from zero at the left-hand end to b at the right-hand end, from which a straight line of unit slope, starting at P, must be subtracted. Suppose that the resultant load F of the complete train acts at a distance d to the left of P and the resultant of the loads in the train to the right of P, F_R, acts a distance c to the right of P. The bending moment at P is then

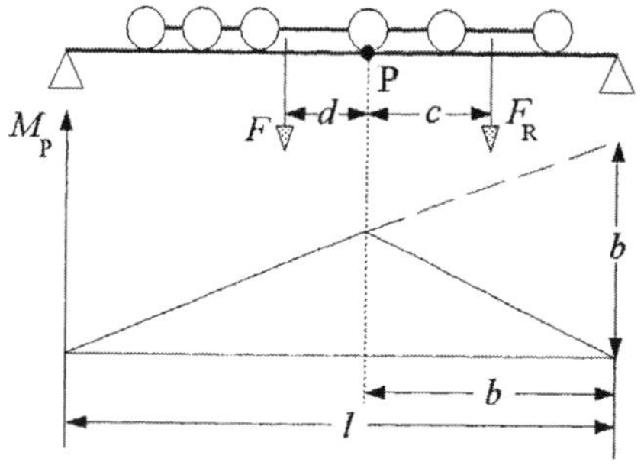

Figure 13.7 Absolute maximum bending moment.

$$M_P = \frac{Fb}{l}(l - b - d) - F_R c \qquad (13.10)$$

The two expressions on the right-hand side of this equation result from the two straight lines mentioned above. It follows that the bending moment M_P at a point beneath a given (moving) load varies parabolically with b. By differentiating this expression with respect to b, the maximum value of M_P occurs when

[1] That is, the ordinary bending moment and shear force diagrams discussed in §2.8.

$$\frac{l}{2} = b + \frac{d}{2} \tag{13.11}$$

Physically, this means that the midpoint of the beam lies midway between P and the resultant F. This gives a maximum for a given set of train loads on the beam. The bending moment envelope for the load train can be formed from parabolae of the type given by (13.10).

13.4 Trusses

The influence lines for the bar forces induced by loads crossing a truss are usually found from the method of sections. The loads are not normally applied directly to the truss but act on a roadway or railway which is supported at the truss joints. The joints are taken to act as simple supports and the loads transmitted to them are equal and opposite to the support reactions. Figure 13.8 shows how a unit load at a distance d from a joint would be transmitted to a single truss. The end reactions on a length l of the supporting way are $1\text{-}d/l$ and d/l so that the loads on the joints are equal and opposite to this. (Normally, a bridge would consist of two similar trusses and any travelling loads would be taken as being shared equally between them.)

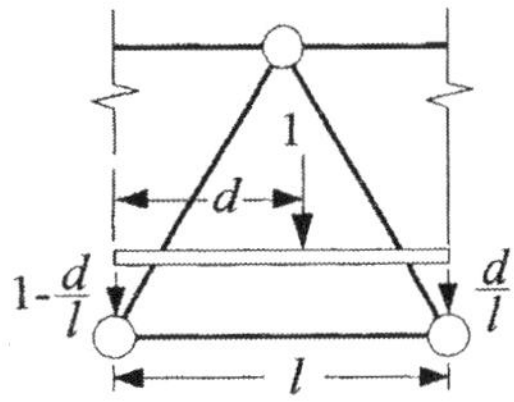

Figure 13.8 Transfer of a unit load to a pin-jointed truss.

The influence lines for axial tension in the bars CD, DE and EF of the equilateral triangular Warren truss shown in Figure 13.9a can be found in this way. Sectioning through these three bars, the bar forces are found to be linear functions of the position, x, of the unit load, either when it is to the left of E or to the right of F. Also, the bar forces are zero when the load is over the supports A and B. When the load is between E and F, it is transmitted to these joints in the manner discussed in the previous paragraph. If the bar force n takes the values n_E when the load is at E and n_F when it is at F, it takes the value

$$n = n_E\left(1 - \frac{d}{l}\right) + n_F\frac{d}{l} \tag{13.12}$$

when the load is d to the right of E ($d<l$).

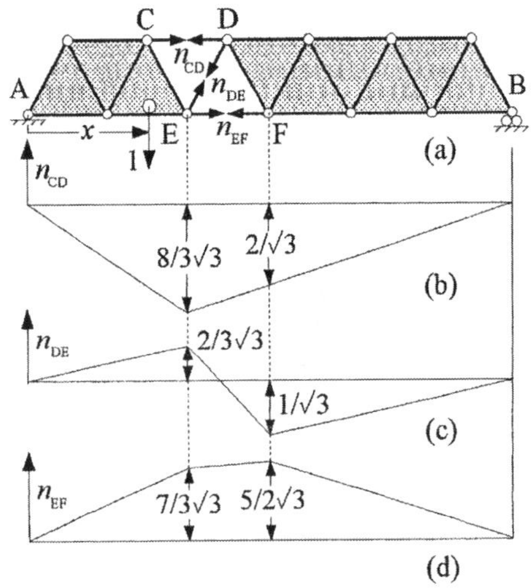

Figure 13.9 Bar force influence lines.

This follows from the proportions in which the unit load is transmitted to joints E and F, as shown in Figure 13.8. From (13.12), it follows that the bar force influence line varies linearly between n_E and n_F as the load moves from one joint to the other. In general then, it is only necessary to determine the ordinates of the influence lines at either end of the sectioned lower chord (n_E and n_F in this case) and the full influence line can be drawn. It consists of three straight lines. Here, these are one from zero at A to n_E at E, one from zero at B to n_F at F and a third linking the first two at E and F. The influence lines for CD, DE and EF are shown in Figures 13.9b to 13.9d. Note that in the first case, the third line is simply a continuation of the second line.

13.5 Müller-Breslau's Theorem

Müller-Breslau's theorem is derived from Betti's reciprocal theorem (see §3.4). It can be used to find the influence line for a workless reaction or internal load induced by a unit force crossing the structure. A release is introduced which permits the reaction or internal load to move by a unit amount[1] through its corresponding deflexion. The deflexion curve produced by this is the same as the required influence line.

This statement is Müller-Breslau's theorem, and the proof consists of comparing two problems. In both cases, the structure is given the above release. In the real case, a unit force (L_1=1) which induces the reaction or internal load L_2 which prevents any deflexion of the release (D_2=0). In the second problem, the release is moved through a unit deflexion (D_2'=1) by the application of a suitable reaction or internal load L_2' and no other load is applied (L_1'=0). Applying (3.34),

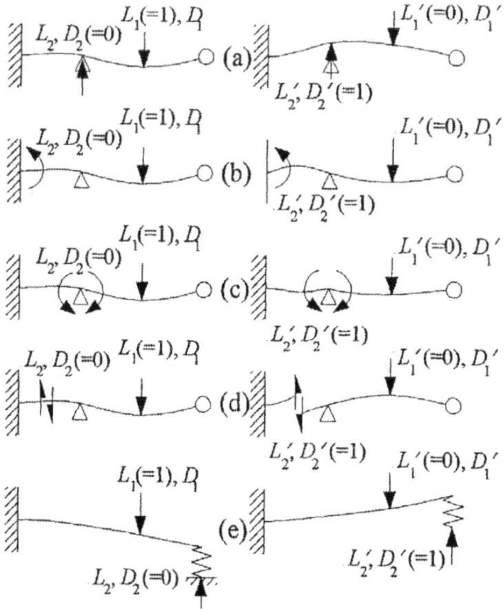

Figure 13.10 Applications of Müller-Breslau's theorem.

$$L_1 D_1' + L_2 D_2' = L_1' D_1 + L_2' D_2 \ ,$$

$$\text{or} \quad 1.D_1' + L_2 .1 = 0.D_1 + L_2'.0 \ , \qquad \textbf{(13.13)}$$

$$\text{hence} \quad L_2 = -D_1' \ .$$

The deflexion D_1' occurs at the same position as the unit force L_1. Thus the plot of D_1' for all positions of L_1 is the deflexion curve of the structure in the second problem. The plot of L_2 for all positions of L_1 is the influence line diagram in the first problem. Then from (13.13), the influence line for the first problem is the deflexion curve for the second problem. (Note that a positive L_2 ordinate is taken upwards, whereas positive D_1' is downwards. Thus the two diagrams are the same way up rather than one being the inverse of the other, as might be inferred from (13.13)).

The left-hand column of Figure 13.10 shows five real problems where an influence line diagram is sought for the action of a moving unit force L_1. The matching displacement curves for a unit deflexion are paired with them in the right-hand column. In Figure 13.10a, the influence line for the reaction at the knife edge is found from the displacement curve induced by a unit upwards displacement at this support. In Figure 13.10b, the influence line for the fixed-end moment at the left-hand end is found from the curve given by a unit rotation of this end. In Figure 13.10c, the influence line for the internal moment at the knife edge is found from the curve produced by giving the beam a unit relative rotation at this point by applying a pair of equal and opposite moments. Similarly, in Figure 13.10d a pair of equal and opposite shear forces are applied at some section inducing a unit relative displacement, D_2', either side of the release. (Note that this is a single release rather than a real cut; the slope of the beam on either side of the release is the

[1] The unit amount (displacement or rotation) is considered to be a small deflexion.

same.) The resulting displacement curve is the influence line for the internal shear force, L_2, shown in the left-hand diagram. In Figure 13.10e, the reaction at an elastic support is required. The reaction at the top of the spring is not a workless reaction, but that supporting its base is. This means that the required displacement curve, giving the influence line for this reaction, is found by giving the *base* of the spring a unit upwards displacement D_2'.

Example 13.1 Influence Line for the Reaction at a Spring Support

Suppose that the beam shown in Figure 13.10e is of length l and bending stiffness EI and that the spring has a stiffness equal to EI/l^3. Measuring x from the fixed left-hand end, the bending moment induced by an upwards spring force L_2' is given by

$$M = EI\frac{d^2v}{dx^2} = L_2'(l - x) \tag{13.14}$$

where the displacement of the beam, v, is upwards-positive as usual. At the fixed end, v and its gradient are both zero, so that on integrating (13.14) twice,

$$v = \frac{L_2'}{EI}\left(\frac{lx^2}{2} - \frac{x^3}{6}\right) \tag{13.15}$$

The upwards displacement of the base of the spring is then given by

$$D_2' = \frac{L_2'}{EI}\left(\frac{l^3}{2} - \frac{l^3}{6}\right) + \frac{L_2'l^3}{EI} = \frac{4L_2'l^3}{3EI} \tag{13.16}$$

However, D_2' is a unit displacement, so that from (13.16) and (13.15),

$$L_2' = \frac{3EI}{4l^3} \quad , \quad v = \frac{x^2}{8l^3}(3l - x) \tag{13.17}$$

This gives the displacement curve for the second problem and hence the influence line diagram for the initial problem.

Exactly the same method can be applied to statically-determinate structures. In such cases, a single release turns the structure into a mechanism and the unit deflexion produces a determinate rigid-body motion. Thus to find the bending moment influence line diagram for the point P in Figure 13.11a, a virtual pin is inserted and AP and PB given a relative unit rotation. This produces the deflected form shown in Figure 13.11b, which is also the required influence line. The shear force influence line diagram is found by giving the two points on either side of P a unit relative displacement, but maintaining a common slope on the two sides, as in Figure 13.11c. This method can also be applied to finding the influence lines for bar tensions in statically-determinate pin-jointed trusses, giving a unit bar shortening. However, it is generally simpler to use the method in §13.4.

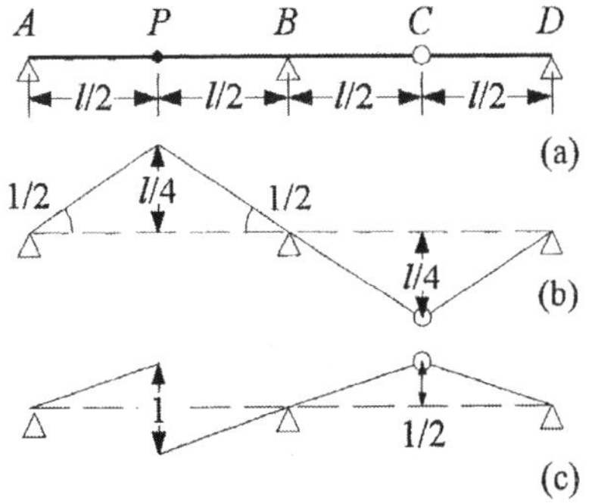

Figure 13.11 Statically-determinate influence lines.

Chapter 14 Optimum Structures

14.1 Introduction

The first work in this area was concerned with finding structures using the least volume of material to support loads in known positions with given support conditions, given limiting stresses in tension and compression. An introduction to the subject will be found in Parkes (1965) Chapter 5. More advanced texts on this classical theory of optimum structures include Cox (1965) and Hemp (1973). The subject is now taken to include optimisation on the basis of other criteria such as stability and dynamic effects. Haug and Arora (1979) and Morris (1982) consider some of these. Most structural optimisation is carried out by a series of approximations to the ideal solution. Such iterative processes are well suited to computer analysis. A trial structure is defined and modified to meet certain objectives such as minimising mass or cost, maximising stiffness and carrying capacity or controlling the range of natural frequencies. The modification is carried out by altering a set of variables (member properties, joint positions or frame topology) within the limits imposed by certain constraints, such as stress or displacement limits or the zone in space which the structure is allowed to occupy.

14.2 Maxwell's Theorem

In structural mechanics, Maxwell's *reciprocal* theorem[1] is well known. This is a particular case of Betti's theorem, discussed in §3.4. Only two loads, P_1, P_2 and their corresponding deflexions, D_1, D_2 are considered. The second set of loads and deflexions are marked by primes. We take

$$P_1 = 0\ ,\ P_2 = 1\ ,\ P_1' = 1\ ,\ P_2' = 0 \qquad \textbf{(14.1)}$$

as shown in Figures 14.1a and 14.1b respectively. Substituting this in (3.34) gives

$$0\times D_1' + 1\times D_2' = 1\times D_1 + 0\times D_2 \quad \text{or} \quad D_2' = D_1 \qquad \textbf{(14.2)}$$

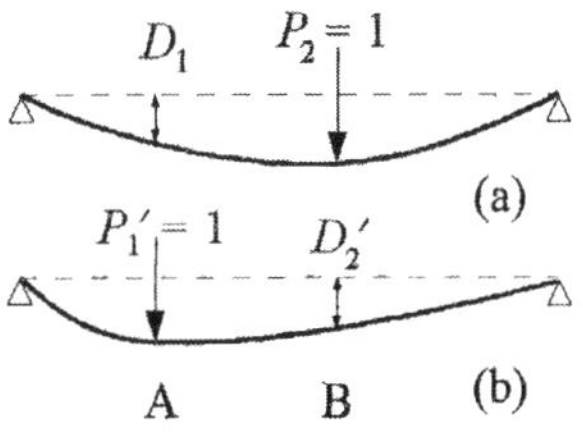

Figure 14.1 Maxwell's reciprocal theorem.

This can be expressed as "*The deflexion produced at a point A by a unit load at point B is the same as the deflexion at point B produced by a unit load at point A*" it being understood that at each point the loads and deflexions correspond to one another.

Maxwell's theorem of framework design[2] can be used in seeking optimum structures. It is an application of tension coefficients to pin-jointed frames (cf. §2.7). Suppose that the external load applied to joint i of such a frame is given by the force vector $\boldsymbol{F}_i$ and its position by $\boldsymbol{r}_i$. It is linked by a bar ij to a joint j with a position vector $\boldsymbol{r}_j$. The force exerted by the bar on joint i is then given by $t_{ij}(\boldsymbol{r}_j - \boldsymbol{r}_i)$ where t_{ij} is the tension coefficient for ij. The equilibrium of joint i is given by

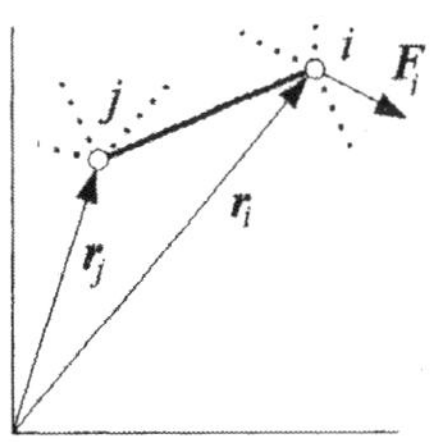

Figure 14.2 Typical bar ij of a pin-jointed frame.

$$\sum_{adj\, j} t_{ij}(\boldsymbol{r}_j - \boldsymbol{r}_i) + \boldsymbol{F}_i = \boldsymbol{0} \qquad \textbf{(14.3)}$$

[1] See W.D. Niven (1890) Vol. 1 p. 598 for example.

[2] Ibid. Vol. 2 p. 175.

where the summation is over all joints j linked to joint i.

Rearranging (14.3), taking a scalar product and summing over all the joints

$$\sum_i \left(\sum_{adj\,j} t_{ij}(\boldsymbol{r}_i - \boldsymbol{r}_j)\bullet\boldsymbol{r}_i \right) = \sum_i \boldsymbol{F}_i\bullet\boldsymbol{r}_i \tag{14.4}$$

In summing this expression over all the joints (including the supports) each bar force appears twice; once for each joint. Summing the left-hand side of (14.4) over the **bars** ij instead of the joints i, (14.4) becomes

$$\sum_{bars\ ij} t_{ij}[(\boldsymbol{r}_i - \boldsymbol{r}_j)\bullet\boldsymbol{r}_i + (\boldsymbol{r}_j - \boldsymbol{r}_i)\bullet\boldsymbol{r}_j] = \sum_i \boldsymbol{F}_i\bullet\boldsymbol{r}_i \tag{14.5}$$

Now

$$t_{ij}[(\boldsymbol{r}_i - \boldsymbol{r}_j)\bullet\boldsymbol{r}_i + (\boldsymbol{r}_j - \boldsymbol{r}_i)\bullet\boldsymbol{r}_j] = t_{ij}(\boldsymbol{r}_i - \boldsymbol{r}_j)\bullet(\boldsymbol{r}_i - \boldsymbol{r}_j) = t_{ij}l_{ij}^2 = P_{ij}l_{ij} \tag{14.6}$$

where l_{ij} is the length of bar ij and P_{ij} is the tensile axial force in it.

Suppose that each bar is either at its limiting tensile stress σ_t or at its limiting compressive stress σ_c and that the volume (length times cross-sectional area) of bar ij is V_{ij}. Then

$$P_{ij}l_{ij} = \sigma_t V_{ij}\ (P_{ij}\ \text{tensile})\quad \text{or}\quad = -\sigma_c V_{ij}\ (P_{ij}\ \text{compressive}) \tag{14.7}$$

Then (14.5) becomes

$$\sum_{bars\ ij} \left(\sigma_t V_{ij}\ \text{ or } -\sigma_c V_{ij}\right) = \sigma_t V_t - \sigma_c V_c = \sum_i \boldsymbol{F}_i\bullet\boldsymbol{r}_i \tag{14.8}$$

where V_t is the total volume of the tensile members and V_c is the total volume of the compressive members. Equation (14.8) is Maxwell's theorem. It is possible to draw the following conclusions from it.

(i) *If a known set of forces ($\boldsymbol{F}_i$) is to be carried by a framework at a known set of points (given by the position vectors $\boldsymbol{r}_i$) then whatever frame is designed, $\sigma_t V_t - \sigma_c V_c$ will always be the same.*

(ii) *The minimum total volume may be obtained by minimising either V_t or V_c.*

(iii) *A framework whose members are entirely in tension (or compression) is always an optimum frame.*

Note that the external forces include the (known) reactions, as would be the case if they were statically determinate. The frame itself may or may not be statically determinate.

The choice of origin for the vectors $\boldsymbol{r}_i$ in (14.8) is arbitrary. The fact that the sum of the scalar products does not change when a different origin is chosen can be deduced from the equilibrium condition that $\sum \boldsymbol{F}_i$ is zero. If $\sum \boldsymbol{F}_i\cdot\boldsymbol{r}_i$ is zero then $\sigma_t V_t$ is equal to $\sigma_c V_c$, and if the two limiting stresses are the same, the volumes of material in tension and compression are the same. Examples of such *equivolume frames* are shown in Figure 14.3. This is most readily seen by taking the black dots shown as the origin of the position vectors in each case. In Figures 14.3a and 14.3b, these vectors are normal to the applied loads. In Figure 14.3c, the distance separating the pair of unit loads is twice that separating the other pair of loads. Note that in this case, the frame is redundant, which implies that there is no unique choice of cross-sectional area of the bars in the limiting-stress design.

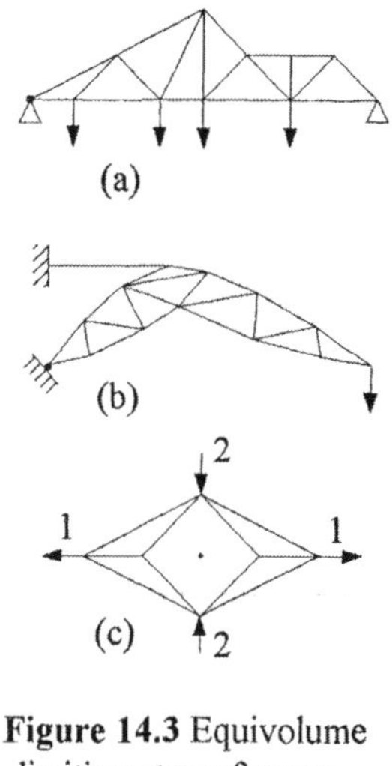

Figure 14.3 Equivolume limiting-stress frames.

Conclusion (ii) above can be drawn, because from (14.8) it follows that any change of design resulting in a reduction of the tensile volume must be accompanied by a reduction in the compressive volume (and vice-versa). The limit to which this can be taken is when either the tensile volume or the compressive volume becomes zero. Any such frame is then an optimum structure, leading to conclusion (iii).

Example 14.1 Optimum Frames for a System of Three Forces

Figure 14.4 shows three tensile frames designed to carry three forces F of equal magnitude at equal angles to one another at a distance L from a central origin. In the third case, the angle θ shown must be less than 30° for all the members to be in tension. From the right-hand side of (14.8), the volume of each tensile structure should be $3FL/\sigma_t$. This is immediately obvious in case (a). In case (b), the length of each bar is $L\sqrt{3}$ and the force in it is $F/\sqrt{3}$, so that the volume is the same. In case (c), suppose that the six bars like AB are of length l. The length of the three bars like BC is then $2l\sin\theta$ and L is $l\cos\theta + (l\sin\theta)/\sqrt{3}$. The force in the three bars like BC is $F\sec\theta\sin(30° - \theta)/\sqrt{3}$. The volume of this frame is then given by

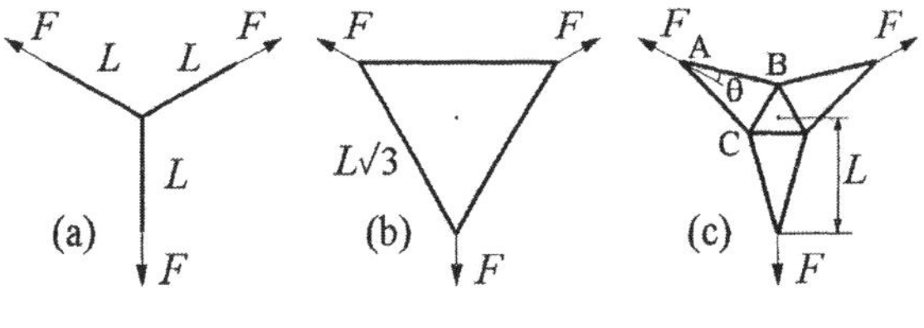

Figure 14.4 Various optimum frames for a given loading.

$$V_t = \frac{3Fl}{\sigma_t}\sec\theta + \frac{6Fl}{\sigma_t\sqrt{3}}\sec\theta\sin(30° - \theta)\sin\theta = \frac{3Fl}{\sigma_t}\left(\cos\theta + \frac{1}{\sqrt{3}}\sin\theta\right) \qquad (14.9)$$

so that, given the above relationship between L and l, the total volume is again $3FL/\sigma_t$.

Another frame could be made by combining those in Figures 14.4a and 14.4b. This is redundant, but provided that the tension T in the inner bars is less than F, the outer bars will have a tensile force $(F-T)/\sqrt{3}$ in them, yielding an optimum tensile structure. It will be assumed that the material has a Young's modulus E and that there is no lack of fit. The inner bars will be taken to have a cross-sectional area A and the outer ones to have a cross-sectional area a. Then applying Castigliano's theorem and assuming that all the bars are fully stressed,

$$0 = \frac{\partial \bar{U}}{\partial T} = \frac{1}{2}\frac{\partial}{\partial T}\left(3\frac{T^2L}{EA} + \frac{(F-T)^2L\sqrt{3}}{Ea}\right) \quad \text{where} \quad \sigma_t A = T \;,\quad \sigma_t a = \frac{F-T}{\sqrt{3}} \qquad (14.10)$$

However, the first condition is immediately satisfied by the remaining two. This means that this combined frame represents a range of optimum structures ranging from A=0 (T=0) to a=0 (T=F).

14.3 Michell Structures

Michell[1] developed Maxwell's theorem to provide a basis for the generation of certain types of optimum frame. Consider all possible frames, with the same limiting stresses σ_t and σ_c, which will support a given set of forces $\boldsymbol{F}_i$ at positions $\boldsymbol{r}_i$. The total volume V of any frame will be the sum of V_t and V_c. We seek to minimise V, which is the same as minimising a constant (positive) multiple of V plus another constant. From (14.8), such an expression is given by

$$2\sigma_t\sigma_c V + (\sigma_t - \sigma_c)(\sigma_t V_t - \sigma_c V_c) = (\sigma_t + \sigma_c)(\sigma_t V_t + \sigma_c V_c) = (\sigma_t + \sigma_c)\sum_{\text{bars } ij} |P_{ij}|\, l_{ij} \qquad (14.11)$$

[1] Michell, A.G.M. (1904) The Limits of Economy of Material in Frame-structures. *Phil. Mag.* V.8 p.589.

Thus minimising V is the same as minimising $\sum|P_{ij}|l_{ij}$, even when σ_t and σ_c are not the same.

Consider the joints of all possible frames to be embedded in a deformable space. Suppose that this strained in such a way that the magnitude of any direct strain does not exceed some infinitesimal amount ϵ. Then the change in length of any bar ij, $e_{ij}\, l_{ij}$, will be such that the magnitude of e_{ij} is no greater than ϵ. As the frames are in equilibrium, the virtual work done by the external loads, δW, during this deformation will be equal to the virtual work done by the internal loads (see §2.2 and §2.4). Then for any possible frame,

$$\delta W = \sum_{bars\ ij} P_{ij} e_{ij} l_{ij} \le \sum_{bars\ ij} |P_{ij}|\,|e_{ij}|\,l_{ij} \le \epsilon \sum_{bars\ ij} |P_{ij}|\,l_{ij} \tag{14.12}$$

Suppose that a frame can be found in which all the bars are strained by exactly ϵ, and in the same sense as their axial loads. Then

$$\delta W = \sum_{bars\ ij} P_{ij}^* e_{ij}^* l_{ij}^* = \epsilon \sum_{bars\ ij} |P_{ij}^*|\,l_{ij}^* \tag{14.13}$$

where the asterisk denotes this particular frame. Comparing (14.12) and (14.13), we see that this particular frame has the lowest possible value of $\sum|P_{ij}|l_{ij}$ and hence from (14.11), the lowest possible volume for the given loading.

For example, suppose that the strain field is one of uniform tensile strain ϵ in all directions. Then any frame carrying the loading with all its bars in tension will be a minimum-volume frame, as in Figure 14.4. This volume will be given by

$$V^* = \sum_{bars\ ij} \frac{P_{ij}^*}{\sigma_t} l_{ij}^* = \frac{\delta W}{\sigma_t \epsilon} = \frac{1}{\sigma_t \epsilon} \sum_i \boldsymbol{F}_i \cdot \epsilon \boldsymbol{r}_i = \frac{1}{\sigma_t} \sum_i \boldsymbol{F}_i \cdot \boldsymbol{r}_i \tag{14.14}$$

as in Example 14.1.

However, as has already been seen, some loading conditions dictate that there must be both tensile and compressive bars in the frame. In the case of optimum frames for loadings such that $\sum \boldsymbol{F}_i \bullet \boldsymbol{r}_i$ is zero (as shown in Figure 14.3 for example) it follows from (14.8), (14.11) and (14.13) that

$$V^* = \frac{1}{2}\left(\frac{1}{\sigma_t} + \frac{1}{\sigma_c}\right) \sum_{bars\ ij} |P_{ij}^*|\,l_{ij}^* = \frac{1}{2}\left(\frac{1}{\sigma_t} + \frac{1}{\sigma_c}\right)\frac{\delta W}{\epsilon} \tag{14.15}$$

Frames are sought in the which the bars have overall strains of $\pm\epsilon$ induced by such strain fields. As these are the principal strains of the field, the tensile and compressive bars of such frames will be orthogonal to one another. Such frames are known as Michell structures.

{The general compatibility equations governing strain fields of this kind are most readily expressed in tensor notation (see Renton (1987) for example). Orthogonal curvilinear coordinates (y^1, y^2, y^3) must satisfy the metric equation

$$ds^2 = (h_1 dy^1)^2 + (h_2 dy^2)^2 + (h_3 dy^3)^2 \tag{14.16}$$

(ibid. (3.78)) where ds is the distance between two adjacent points with an infinitesimal difference in coordinates (dy^1, dy^2, dy^3). For example, in cylindrical polar coordinates,

$$y^1 = r\ ,\ y^2 = \theta\ ,\ y^3 = z\ ,\ h_1 = 1\ ,\ h_2 = r\ ,\ h_3 = 1\ . \tag{14.17}$$

If the (physical) components of displacement in the directions of the coordinate lines are (u_1, u_2, u_3) the expressions for the physical components of strain are given by

$$e_{aa} = \frac{1}{h_a}\left(\frac{\partial u_a}{\partial y^a} + \frac{u_b}{h_b}\frac{\partial h_a}{\partial y^b} + \frac{u_c}{h_c}\frac{\partial h_a}{\partial y^c}\right)\ ,\quad \gamma_{ab} = \frac{h_a}{h_b}\frac{\partial}{\partial y^b}\left(\frac{u_a}{h_a}\right) + \frac{h_b}{h_a}\frac{\partial}{\partial y^a}\left(\frac{u_b}{h_b}\right) \tag{14.18}$$

(ibid. (A1.12))[1] where e_{aa} and γ_{ab} are the normal and shear strains respectively (cf. (3.9) to (3.13)) and a, b and c are any permutation of 1, 2 and 3. These then represent six equations. If the normal strains are principal strains equal to $\pm\epsilon$ so that the shear strains are zero, (14.18) imposes conditions on the possible strain fields. The plane-strain equivalents are given by dropping the terms in c, giving only three equations.}

Plane-strain problems where the principal strains are everywhere $+\epsilon$ and $-\epsilon$ correspond to incompressible flow. These are associated with slip-line fields in plastic flow described by Hencky[2] nets. Consider the case where there is a radial principal strain of $+\epsilon$ and a circumferential principal strain of $-\epsilon$. Expressed in terms of plane polar coordinates, (14.18) becomes

$$e_{11} = \frac{\partial u_1}{\partial r} = \epsilon \quad , \quad e_{22} = \frac{1}{r}\left(\frac{\partial u_2}{\partial \theta} + u_1\right) = -\epsilon \quad , \quad \gamma_{12} = \frac{1}{r}\frac{\partial u_1}{\partial \theta} + r\frac{\partial}{\partial r}\left(\frac{u_2}{r}\right) = 0 . \qquad \mathbf{(14.19)}$$

This yields the following general solution for the radial and circumferential displacements of the field:

$$u_1 = \epsilon r + A\sin\theta + B\cos\theta \quad , \quad u_2 = -2\epsilon r\theta + A\cos\theta - B\sin\theta + Cr \qquad \mathbf{(14.20)}$$

where A,B and C are arbitrary constants associated with rigid-body motion. With respect to the coordinates shown in Figure 14.5, the vertical displacement of a point in this field is given by

$$\begin{aligned} u_v &= u_1\cos\theta - u_2\sin\theta \\ &= \epsilon r(\cos\theta + 2\theta\sin\theta) + B - Cr\sin\theta \end{aligned} \qquad \mathbf{(14.21)}$$

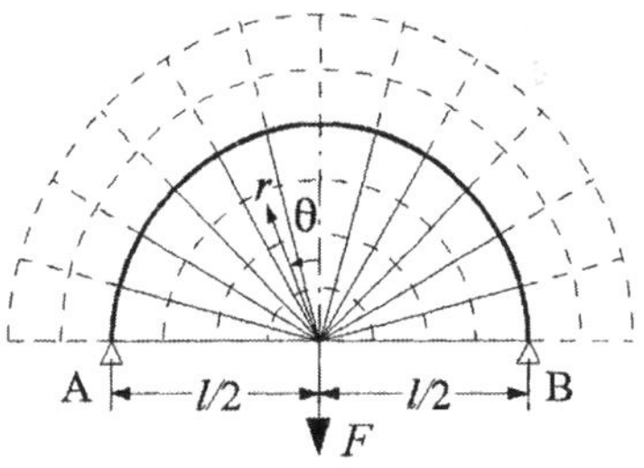

Figure 14.5 Half-plane Michell frame.

This field is applied to finding a structure to support a force F midway between two knife-edge supports A and B, l apart as shown. Note that u_2 is not the same when θ=0 and θ=2π, so that this field cannot be used to describe the whole plane. A suitable frame will then be sought in the upper half-plane. At the supports A and B, u_v must be zero. This implies that C is zero and that B is $-\pi\epsilon l/2$. This gives the displacement of F at its point of application (r=0) and hence from (14.15) the volume of the corresponding Michell frame is

$$V^* = \frac{1}{2}\left(\frac{1}{\sigma_t} + \frac{1}{\sigma_c}\right)\frac{\delta W}{\epsilon} = \frac{\pi}{4}\left(\frac{1}{\sigma_t} + \frac{1}{\sigma_c}\right)Fl \qquad \mathbf{(14.22)}$$

The members of the frame must lie in the directions of the principal strains and carry the load to the supports, so they form the semicircular fan shown by solid lines in Figure 14.5. The radial members are in tension and the semicircular arch (shown bold) is in compression.

As this was a strain-field solution for an incomplete plane, this was not necessarily the optimal frame that could be found for a complete, continuous field. This field can be piecewise-continuous, matching

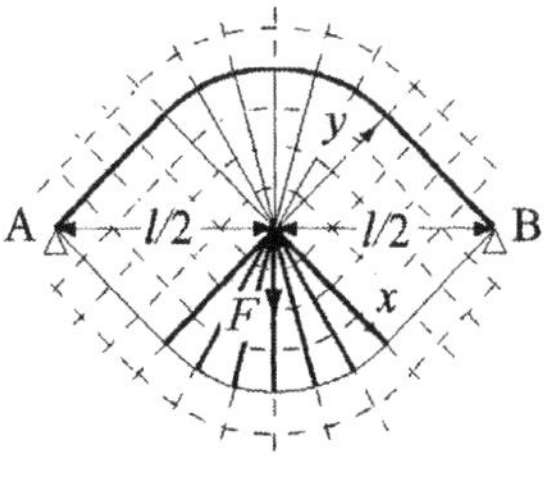

Figure 14.6 Full-plane Michell frame.

---

[1] Love (1952) uses the *inverse* definition of the coefficients h_i (cf. Chapter 1 Equations (32) and (36)), but the above are the currently accepted definitions.

[2] Hencky, H. (1923) Über einige statisch bestimmte Fälle des Gleichgewichts in plastischen Körpern. *Z. angew. Math. Mech.* V.3 p.241.

the displacements of different fields on their common boundaries. Figure 14.6 shows such a field, made from matching polar and cartesian strain fields. The upper and lower polar fans subtend angles of 90° and the rest of the field is filled with two cartesian fields with axes at 45° to the horizontal as shown. General expressions for the strain fields are obtained as before and the conditions of compatibility on their common boundaries and zero displacement at A and B are then satisfied. Taking the polar coordinates in the upper field to be as before, and those in the right-hand field to be x and y as shown, the displacements in these two fields are given by

$$u_r = \epsilon(r - \sqrt{2}D\cos\theta) \quad , \quad u_\theta = -\epsilon(2r\theta - \sqrt{2}D\sin\theta)$$
$$u_x = -\epsilon(x - D + \tfrac{1}{2}\pi y) \quad , \quad u_y = \epsilon(y - D + \tfrac{1}{2}\pi x) \qquad \textbf{(14.23)}$$
$$\text{where} \quad D = \frac{l}{2\sqrt{2}}\left(1 + \frac{\pi}{2}\right)$$

This gives a vertical displacement at the point of application of the force F of $\epsilon\sqrt{2}D$ downwards, so that from (14.15) the volume of the corresponding Michell frame is

$$V^* = \frac{1}{2}\left(\frac{1}{\sigma_t} + \frac{1}{\sigma_c}\right)\frac{\delta W}{\epsilon} = \frac{1}{4}\left(\frac{1}{\sigma_t} + \frac{1}{\sigma_c}\right)Fl\left(1 + \frac{\pi}{2}\right) \qquad \textbf{(14.24)}$$

It will be seen that this volume is 18% less than that given by (14.22).

At first sight, these two solutions are impractical. They both require fans with an infinite number of members, and that shown in Figure 14.6 is unstable. However, they could provide guidance for reinforcement in fibre-reinforced materials, and frames with a finite number of members can be found which approximate to them. Some of these are shown in the following table.

Table 14.1 Frames to Support a Central Point Force

Case	Structure	$2V\sigma_t\sigma_c/Fl(\sigma_t + \sigma_c)$	V_1^*/V (%)	V_2^*/V (%)
1	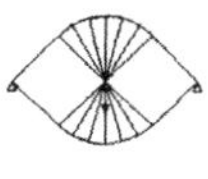	$\frac{1}{2}\left(1 + \frac{\pi}{2}\right) = 1.2854$	100	122.2
1a	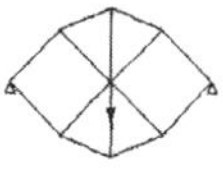	$2\sqrt{2} - 1.5 = 1.3284$	96.8	118.2
2		$\frac{\pi}{2} = 1.5708$	81.8	100
2a		$\sqrt{3} = 1.7321$	74.2	90.7
2b	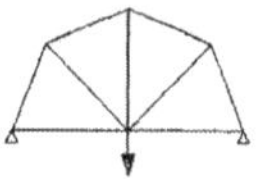	$4(\sqrt{2} - 1) = 1.6569$	77.6	94.8

In all cases, the fans have been divided into equal segments. In columns four and five of the table, the volumes of the two Michell structures, V_1* and V_2*, are expressed as ratios of the volumes V of the other structures to give a measure of their efficiencies.

If the principal strains are $\pm\epsilon$, then Mohr's circle is as shown in Figure 14.7. Then it is possible to define this state in terms of coordinates at 45° to the directions of these strains. In terms of this orientation, the shear strains are $\pm 2\epsilon$ and the normal strains are zero. Taking these conditions relative to polar coordinates, (14.19) becomes

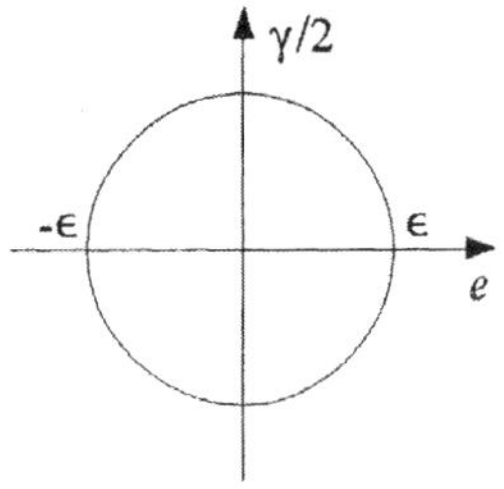

Figure 14.7 Mohr's strain circle for pure shear.

$$e_{11} = \frac{\partial u_1}{\partial r} = 0 \quad , \quad e_{22} = \frac{1}{r}\left(\frac{\partial u_2}{\partial \theta} + u_1\right) = 0,$$
$$\gamma_{12} = \frac{1}{r}\frac{\partial u_1}{\partial \theta} + r\frac{\partial}{\partial r}\left(\frac{u_2}{r}\right) = \pm 2\epsilon. \qquad \textbf{(14.25)}$$

The general solution for the displacements of the strain field is given by

$$u_1 = A\sin\theta + B\cos\theta \quad , \quad u_2 = A\cos\theta - B\sin\theta + Cr \pm 2\epsilon r \ln r \qquad \textbf{(14.26)}$$

where A,B and C correspond to the same rigid-body motions as before. Note that this displacement field will fill the whole of two-dimensional space without the problems associated with the solution given by (14.20).

Consider the problem of a torque T applied to a ring of radius r_o which transmits it to an inner ring of radius r_i, as shown in Figure 14.8a. It will be assumed that the torque forces are applied tangentially, so that the structure transmitting this torque is an equivolume one. For no displacements at the inner ring, (14.26) becomes

$$u_1 = 0 \quad , \quad u_2 = 2\epsilon r \ln\left(\frac{r}{r_i}\right) \qquad \textbf{(14.27)}$$

The volume of the Michell structure for this field is then found from (14.15) to be

$$V^* = \frac{1}{2}\left(\frac{1}{\sigma_t} + \frac{1}{\sigma_c}\right)\frac{\delta W}{\epsilon} = \left(\frac{1}{\sigma_t} + \frac{1}{\sigma_c}\right)T\ln\left(\frac{r_o}{r_i}\right) \qquad \textbf{(14.28)}$$

The members of the Michell structure lie along the trajectories of the principal strains of this field. These form the spirals shown, which are at ±45° to the radius vectors from the centre of the disc.

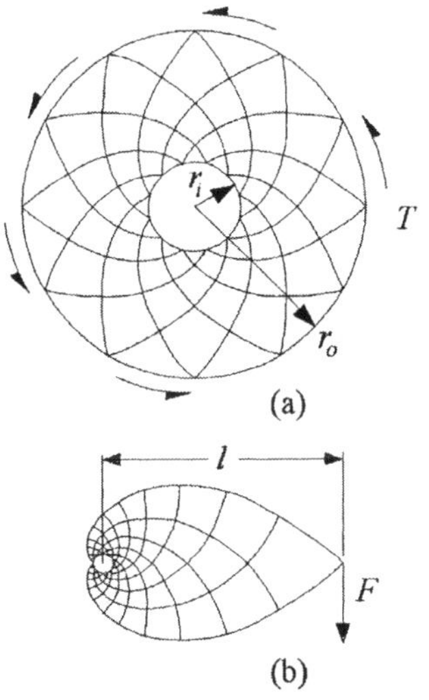

Figure 14.8 Orthogonal spiral Michell structures.

A similar problem is shown in Figure 14.8b. Here, the inner ring is to carry a point force F at a distance l from its centre. The required Michell frame is the spiral structure shown and its volume is given by

$$V^* = \left(\frac{1}{\sigma_t} + \frac{1}{\sigma_c}\right)Fl\ln\left(\frac{r_o}{r_i}\right) \qquad \textbf{(14.29)}$$

($r_o = l$ cf. (14.28)). Variations on these structures for other loading and support conditions will be found in Hemp (1973) for example.

Some three-dimensional Michell structures have also been found. Again, it is necessary to find a strain field with principal strains of $\pm\epsilon$ so that the Michell structure lying in it has its members strained to these limits. Alternatively, it is again possible to seek fields in which the normal strains are zero and the shear strains are $\pm 2\epsilon$.

{One such field can be described in terms of spherical polar coordinates

$$y^1 = r \ , \ y^2 = \theta \ , \ y^3 = \phi \ , \ h_1 = 1 \ , \ h_2 = r \ , \ h_3 = r\sin\theta . \qquad \textbf{(14.30)}$$

with the displacements

$$u_1 = u_2 = 0 \ , \ u_3 = 2\epsilon r \sin\theta \ln\cot(\theta/2) \qquad \textbf{(14.31)}$$

Then appplying (14.18), the only non-zero strain is γ_{23}, which is -2ϵ (cf. Renton (1987) §A1.2). This implies principal strains of $\pm\epsilon$ at 45° to the y^2 and y^3 coordinate lines on spherical surfaces of constant y^1 (i.e. constant r). The trajectories of these principal strains are known as rhumb lines.}

Figure 14.9 shows a problem to which the above solution can be applied. Equal and opposite torques T are applied around the peripheries of two circles of radius R, a distance l apart. The origin of the spherical polar coordinates is taken midway between these two circles, on their line of centres. These two circles subtend an angle θ_o to this line at the origin, as shown. The work done by the torques, δW, during the displacements given by (14.31) is given by T times the relative angle of twist, Φ, of the two circles. This angle is given by

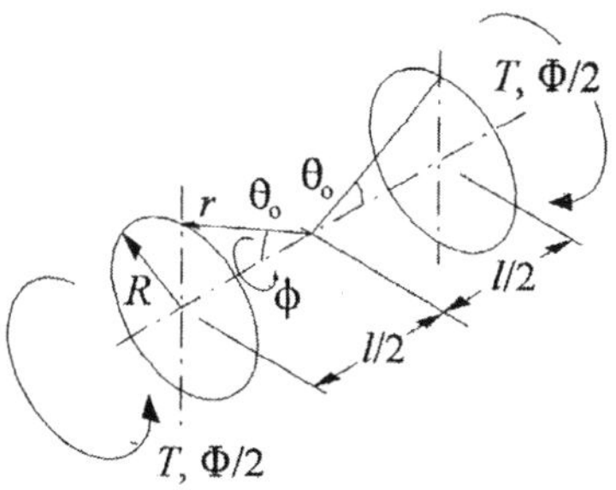

Figure 14.9 Longitudinal torsion problem.

$$\begin{aligned}\Phi &= u_3(r,\theta_o)/R - u_3(r,\pi-\theta_o)/R \\ &= 4\epsilon\ln\cot(\theta_o/2) \qquad \{R = r\sin\theta_o\}.\end{aligned} \qquad \textbf{(14.32)}$$

The shear forces exerted by the torques on the peripheries of the circles are normal to the radius vectors from the origin, so that (14.15) applies, giving the optimum volume

$$\begin{aligned}V^* &= \frac{1}{2}\left(\frac{1}{\sigma_t}+\frac{1}{\sigma_c}\right)\frac{\delta W}{\epsilon} \\ &= 2T\left(\frac{1}{\sigma_t}+\frac{1}{\sigma_c}\right)\ln\cot(\theta_o/2) \qquad \textbf{(14.33)} \\ &\{\ \theta_o = \tan^{-1}(2R/l)\ \}\end{aligned}$$

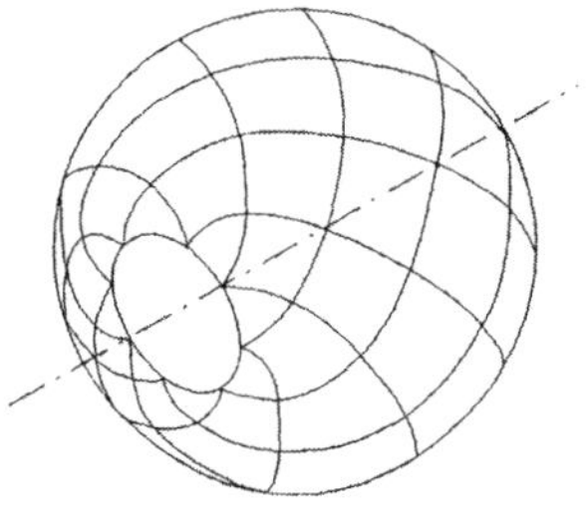

Figure 14.10 Spherical Michell structure.

The optimum structure giving this volume lies on the surface of a sphere which passes through the two circles. The pattern of its members is shown in Figure 14.10.

It is perhaps surprising that the structure does not lie on a cylindrical surface of radius R, as shown in Figure 14.11. A strain field with principal strains of $\pm\epsilon$ on this surface can be found relatively easily. It has trajectories at $\pm 45°$ to the axis of the cylinder, as shown in the figure. However, the principal strains are either more or less than these values in the rest of

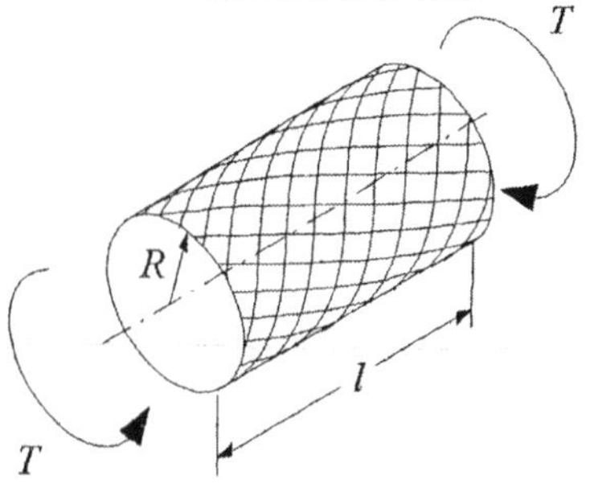

Figure 14.11 Cylindrical structure.

three-dimensional space, and so applying the theory only yields an optimum for this surface. This has members lying along the trajectories shown and has a volume given by

$$V_c^* = \left(\frac{1}{\sigma_t} + \frac{1}{\sigma_c}\right)\frac{Tl}{R} \tag{14.34}$$

This value is always greater than that given by (14.33). The ratio V_c^*/V^* varies from 2.162 when l/R is 10 to 1.039 when l/R is 1.0.

Other strain fields have been used to develop Michell structures for particular problems. Thus Hemp (ibid. §4.8) uses principal strain trajectories formed by orthogonal cycloids to generate cantilever frames confined to a rectangular zone and Ghista and Resnikoff[1] use bipolar coordinates for a proposed reentry body configuration. Other suitable strain fields may be developed from the orthogonal coordinate systems described in Korn and Korn (1961) §6.5-1 for example.

Another problem of practical interest is that of finding the stiffest structure, using a given volume of material, which will carry a specified loading. Here, maximising the stiffness is defined as minimising the work done by the loading and hence minimising the strain energy stored. The problem is then that of minimising the strain energy of the structure subject to the conditions that its volume is to remain constant and that it will carry the given loading. These conditions can be expressed by the use of Lagrange multipliers. It was first shown by Cox (1965) that the stiffest structure in this sense is a Michell structure. Constrained optimisation will be discussed further in the next section.

{14.4 Linear Programming}

{Linear programming is one of the fundamental mathematical techniques of optimisation. It relies on the governing equations being linear, although it can be modified to minimise quadratic functions (see Haug and Arora (1979) §2.5). The standard form of the problem is to minimise the *objective function* V where

$$V = \mathbf{c}^t\mathbf{x} \qquad \text{or} \qquad V = \sum_{j=1}^{n} c_j x_j \tag{14.35}$$

$\mathbf{c}$ is a column vector of constant, non-zero elements c_j and $\mathbf{x}$ a column vector of variable elements x_j which are greater than or equal to zero. These variables are subject to the constraints

$$\mathbf{A}\mathbf{x} = \mathbf{b} \qquad \text{or} \qquad \sum_{j=1}^{n} a_{ij}x_j = b_i \qquad (i = 1 \text{ to } m, \quad m<n.) \tag{14.36}$$

Other problems can be reduced to this standard form by noting the following.

(i) Maximising W is the same as minimising $-W$.

(ii) Any variable which may be both positive and negative can be expressed as the difference of two positive variables.

(iii) Any condition that a function of $\mathbf{x}$ is less than or equal to a given value can be expressed as an equality to that value, by the addition of an extra variable to the column vector $\mathbf{x}$, This is known as a *slack variable* and is added to the left-hand side of the inequality.

(iv) Likewise, if an expression in $\mathbf{x}$ is greater than or equal to a fixed value, it can be turned into an equality by subtracting a slack variable.

---

[1] Ghista, D.N. and Resnikoff, M.M. (1968) Development of Michell Minimum Weight Structures. *N.A.S.A. Technical Note D-4345*. However, this is used for a purely compressive field where the principal strains would be the same in all directions.

Example 14.2a Reduction to the Standard Form

Suppose that the problem is to maximise W where

$$\begin{aligned} W &= -7y_1 + 8y_2 + 3y_3 \quad \text{subject to} \\ 6y_1 &+ 5y_2 \geq 14 \\ 2y_1 &+ 9y_2 - y_3 \leq 8 \end{aligned} \tag{14.37}$$

where y_1 and y_2 are positive variables and the variable y_3 can be positive or negative. This can be changed to the standard form of minimising V where

$$\begin{aligned} V &= 7x_1 - 8x_2 - 3x_3 + 3x_4 \quad \text{subject to} \\ 6x_1 &+ 5x_2 - x_5 = 14 \\ 2x_1 &+ 9x_2 - x_3 + x_4 + x_6 = 8 \quad \text{where} \\ x_1 &= y_1 \;, \quad x_2 = y_2 \;, \quad x_3 - x_4 = y_3 \;. \end{aligned} \tag{14.38}$$

and the variables x_j are all positive.

In (14.36), there must be fewer independent equations than variables x_j $(m<n)$.If this were not so, these variables would be determined, and so would V, and so no minimisation could take place. There is then no unique solution to these equations. First, the possible solutions are found, using the Gauss-Jordan process. This is very similar to normal Gaussian elimination.

First, the coefficients a_{ij} and the terms b_i are arranged into a single array or tableau:

$$\begin{bmatrix} a_{11} & a_{12} & \cdots & a_{1p} & a_{1p+1} & \cdots & a_{1n} & b_1 \\ \cdots & \cdots & \cdots & \cdots & \cdots & \cdots & \cdots & \cdots \\ \cdots & \cdots & \cdots & \cdots & \cdots & \cdots & \cdots & \cdots \\ \cdots & \cdots & \cdots & \cdots & \cdots & \cdots & \cdots & \cdots \\ a_{m1} & a_{m2} & \cdots & a_{mp} & a_{mp+1} & \cdots & a_{mn} & b_m \end{bmatrix} \tag{14.39}$$

Then for each row i in turn (i=1 to m) the following process is carried out.

1) Find the first non-zero term in row i, a_{ik}' say. (The prime indicates that this term may already have been changed during this process.)
2) If *all* the elements in row i are zero, discard it. It is simply a linear combination of the previous rows. (If all the elements except b_i' are zero, then the equation is inconsistent with the rest and there is an error in the way the problem has been posed.)
3) If $k \neq i$ then interchange the ith and kth columns, renaming x_i and x_k as x_k' and x_i'.
4) Divide *all* the elements in row i by the current a_{ii}', so that a_{ii}' itself now becomes unity.
5) Eliminate all elements a_{li}' both above it ($l<i$) and below it ($l>i$) by subtracting a_{li}' times row i from row l.

There will now be p rows remaining ($p \leq m$), as m-p rows were discarded at step 2. These have the form:

$$\left[\begin{array}{cccc|cccc} 1 & 0 & \cdots & 0 & a'_{1p+1} & \cdots & a'_{1n} & b'_1 \\ \cdots & \cdots & \cdots & \cdots & \cdots & \cdots & \cdots & \cdots \\ \cdots & \cdots & \cdots & \cdots & \cdots & \cdots & \cdots & \cdots \\ 0 & 0 & \cdots & 1 & a'_{pp+1} & \cdots & a'_{pn} & b'_p \end{array}\right] \tag{14.40}$$

The original set of equations is satisfied by taking x_1' to x_p' to be b_1' to b_p' and the remaining variables x_{p+1}' to x_n' to be zero. The column vector $\{b_1' ... b_p'\}$ is known as a *basic solution* x_B. If b_1' to b_p' are all positive, then it is known as a *basic feasible solution,* because the variables x_j are defined as all positive. The variables x_1' to x_p' are known as *key variables* and the variables x_{p+1}' to x_n' are known as *free variables.* The remaining solutions can be found by setting each of the free variables to unity in turn, and the remainder to zero, and solving the (modified) equations for the case when the coefficients b_1' to b_p' are all zero. Any multiple of these *complementary solutions,* when added onto the basic solution, is also a solution of the original equations.

Example 14.2b Finding the Full Set of Solutions

Taking the problem given in Example 14.2a, the array corresponding to (14.39) is

$$\begin{bmatrix} 6 & 5 & 0 & 0 & -1 & 0 & 14 \\ 2 & 9 & -1 & 1 & 0 & 1 & 8 \end{bmatrix} \quad \textbf{(14.41)}$$

Using the Gauss-Jordan process, this is reduced to the form

$$\begin{bmatrix} 1 & 0 & \frac{5}{44} & -\frac{5}{44} & -\frac{9}{44} & -\frac{5}{44} & \frac{43}{22} \\ 0 & 1 & -\frac{3}{22} & \frac{3}{22} & \frac{1}{22} & \frac{3}{22} & \frac{5}{11} \end{bmatrix} \quad \textbf{(14.42)}$$

A basic feasible solution is then

$$x_B = \{\tfrac{43}{22}\ \tfrac{5}{11}\ 0\ 0\ 0\ 0\} \quad \textbf{(14.43)}$$

and the complementary solutions, which can also be read directly from (14.42), are

$$\begin{aligned} x_{C3} &= \{-\tfrac{5}{44}\ \tfrac{3}{22}\ 1\ 0\ 0\ 0\} \\ x_{C4} &= \{\tfrac{5}{44}\ -\tfrac{3}{22}\ 0\ 1\ 0\ 0\} \\ x_{C5} &= \{\tfrac{9}{44}\ -\tfrac{1}{22}\ 0\ 0\ 1\ 0\} \\ x_{C6} &= \{\tfrac{5}{44}\ -\tfrac{3}{22}\ 0\ 0\ 0\ 1\} \end{aligned} \quad \textbf{(14.44)}$$

Further basic feasible solutions can be found by adding positive multiples of these complementary solutions to the original basic solution in such a way as to make one or more of its coefficients zero. In the present example, the only other basic solutions which can be formed in this way are found by adding a multiple of a single complementary solution to the basic solution, giving

$$\begin{aligned} x_{B3} &= x_B + \tfrac{86}{5} x_{C3} = \{0\ \tfrac{14}{5}\ \tfrac{86}{5}\ 0\ 0\ 0\} \\ x_{B4} &= x_B + \tfrac{10}{3} x_{C4} = \{\tfrac{7}{3}\ 0\ 0\ \tfrac{10}{3}\ 0\ 0\} \\ x_{B5} &= x_B + 10 x_{C5} = \{4\ 0\ 0\ 0\ 10\ 0\} \\ x_{B6} &= x_B + \tfrac{10}{3} x_{C6} = \{\tfrac{7}{3}\ 0\ 0\ 0\ 0\ \tfrac{10}{3}\} \end{aligned} \quad \textbf{(14.45)}$$

If there is an optimal feasible solution, then there is an optimal basic feasible solution. The proof of this relies on showing that it must be possible to change the optimal feasible solution into an optimal basic feasible solution without changing the value of V. The proof will be found in Haug and Arora (ibid.) for example. This restricts the search for the optimum solution to the set of basic feasible solutions.

In general, there will be $n!/(n-p)!p!$ basic solutions of which the basic feasible solutions

($x_i' \geq 0$, all i) will be a subset. Each of these solutions can be examined, or it is possible to proceed from one solution to the next one giving a lower value of V, using the *simplex method*. Returning to the array given by (14.40), the set of equations that this represents means that

$$x_i' = b_i' - \sum_{j=p+1}^{n} a_{ij}' x_j' \quad (i = 1 \text{ to } p) \tag{14.46}$$

The basic solution is given by taking all x_j' ($j = p+1$ to n) as zero, giving the value of V found from (14.35) as

$$V = \sum_{i=1}^{p} c_i' b_i' = V_\text{B} \tag{14.47}$$

where the primes on the coefficients c_i'indicate that they will have been renumbered in the same way as the variables x_i', if that had been necessary. The general expression for V is found from (14.35), (14.46) and (14.47) to be

$$\begin{aligned} V &= \sum_{i=1}^{p} c_i' x_i' + \sum_{j=p+1}^{n} c_j' x_j' = \sum_{i=1}^{p} c_i' \left(b_i' - \sum_{j=p+1}^{n} a_{ij}' x_j' \right) + \sum_{j=p+1}^{n} c_j' x_j' \\ &= V_\text{B} + \sum_{j=p+1}^{n} (c_j' - v_j') x_j' \quad \text{where} \quad v_j' = \sum_{i=1}^{p} c_i' a_{ij}' . \end{aligned} \tag{14.48}$$

Thus a new value of V can be found which is less than V_B if for some j, (c_j'- v_j') is negative and a feasible (i.e. positive) x_j' can be included in the basic variables. The form given by (14.48) then indicates whether the optimum solution has been found, and if not, what new basis to search for. The usual process is to include the x_j' with the most negative (c_j'- v_j') in the new basis, forming it in the way used to obtain (14.45).

Example 14.2c Minimising the Objective Function by the Simplex Method

Continuing with the problem given in Example 14.2a and using the basic feasible solution given by (14.43), the expression for V given by (14.48) can be determined from the coefficients c_i' in (14.38) and the coefficients a_{ij}' in (14.42).

$$\begin{aligned} V &= V_\text{B} + \sum_{j=p+1}^{n} (c_j' - v_j') x_j' \\ &= \left(7 \times \frac{43}{22} - 8 \times \frac{5}{11}\right) + \left[-3 - \left(7 \times \frac{5}{44} - 8 \times \frac{-3}{22}\right)\right] x_3' \\ &+ \left[3 - \left(7 \times \frac{-5}{44} - 8 \times \frac{3}{22}\right)\right] x_4' + \left[0 - \left(7 \times \frac{-9}{44} - 8 \times \frac{1}{22}\right)\right] x_5' + \left[0 - \left(7 \times \frac{-5}{44} - 8 \times \frac{3}{22}\right)\right] x_6' \\ &= \frac{221}{22} - \frac{215}{44} x_3' + \frac{215}{44} x_4' + \frac{79}{44} x_5' + \frac{83}{44} x_6' \end{aligned} \tag{14.49}$$

This indicates that x_3' should be included in the basic solution to reduce V below V_B (= 221/22). The new basic solution is then $\boldsymbol{x}_{\text{B}3}$ as given in (14.45). The value of V_B associated with this basic solution is (-8×14-3×86)/5 which is -74. The process can be repeated, starting from this new basis. The expression for V given by (14.48) starting from this new basic solution is

$$V = -74 + 43 x_1' + 0 x_4' - 7 x_5' - 3 x_6' \tag{14.50}$$

indicating that V could be reduced if x_5' or x_6' could form part of a new basic feasible solution. However, this would require adding positive multiples of the new complementary solutions

$$x_{C5} = \{0 \ \tfrac{1}{5} \ \tfrac{9}{5} \ 0 \ 1 \ 0\}$$
$$x_{C6} = \{0 \ 0 \ 1 \ 0 \ 0 \ 1\} \qquad (14.51)$$

to x_{B3} to produce a new basic feasible solution. As none of the elements in these vectors is negative, this is not possible and the minimum V has been found.

This process can readily be automated within a computer program.A problem which can arise occurs when a *degenerate* basic solution is found. This happens when one of the basic variables is zero. Unless this solution is trapped, this iterative process will get stuck in a closed loop. Fortunately, such cases are rare.

Example 14.3 Application of Linear Programming to Structure Optimisation

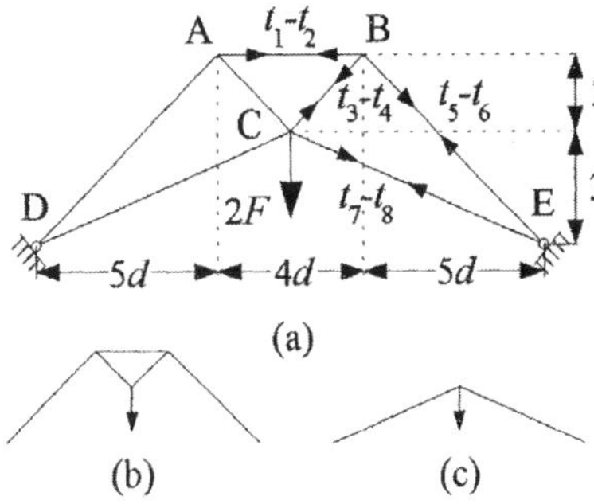

Figure 14.12 A redundant pin-jointed frame.

The method can be used to optimise pin-jointed frames with joints in predetermined positions. These must be initially redundant, and the optimisation process removes the non-optimal members. Lack-of-fit will be considered admissible, if this contributes towards optimality of the frame. Then only equilibrium conditions and not compatibility conditions need to be applied to the analysis. Figure 14.12a shows a simple problem of this kind. The bar forces will be expressed in terms of tension coefficents. As these may be positive or negative, they will be expressed as the difference of two positive parameters. Thus, for example, the tension coefficient for the bar AB is t_1-t_2. For simplicity, the limiting stress will be taken as σ in both tension and compression. As each member will be stressed to its limit under the applied loading, its cross-sectional area is given by its axial load divided by σ. The volume of the member is then given by the modulus of its tension coefficient multiplied by the square of its length divided by σ. In the case of the bar AB, this modulus will be either t_1 or t_2, depending on whether it is in tension or compression, and the other parameter will be zero. Assuming symmetrical behaviour, the volume of the frame can then be written as

$$V = \frac{d^2}{\sigma}[16(t_1+t_2) + 16(t_3+t_4) + 100(t_5+t_6) + 116(t_7+t_8)] \qquad (14.52)$$

This is subject to two equilbrium conditions at B and a vertical equilibrium condition at C:

$$\begin{aligned} 4(t_1-t_2) + 2(t_3-t_4) - 5(t_5-t_6) &= 0 \\ 2(t_3-t_4) + 5(t_5-t_6) &= 0 \\ 2(t_3-t_4) - 3(t_7-t_8) &= \frac{F}{d} \end{aligned} \qquad (14.53)$$

These conditions only apply to the differences of pairs of parameters. The function V to be minimised relates to their sums. These sums can be reduced, without changing their differences, until one or other of the pair is zero. Thus the minimisation process ensures that one or other of each pair is zero. On removing a factor of F/d from the last column, the array corresponding to (14.39) is then

$$\begin{bmatrix} 4 & -4 & 2 & -2 & -5 & 5 & 0 & 0 & 0 \\ 0 & 0 & 2 & -2 & 5 & -5 & 0 & 0 & 0 \\ 0 & 0 & 2 & -2 & 0 & 0 & -3 & 3 & 1 \end{bmatrix} \quad \textbf{(14.54)}$$

Using the Gauss-Jordan process, this reduces to

$$\begin{bmatrix} 1 & 0 & 0 & 0 & -1 & 0 & \frac{3}{2} & -\frac{3}{2} & -\frac{1}{2} \\ 0 & 1 & 0 & -1 & 0 & 0 & -\frac{3}{2} & \frac{3}{2} & \frac{1}{2} \\ 0 & 0 & 1 & 0 & 0 & -1 & \frac{3}{5} & -\frac{3}{5} & -\frac{1}{5} \end{bmatrix} \quad \textbf{(14.55)}$$

where interchanges of the second and third columns and then of the third and fifth columns took place during the process. This gives the basic solution as

$$\{t_1\ t_2\ t_3\ t_4\ t_5\ t_6\ t_7\ t_8\} = \{t_1'\ t_5'\ t_2'\ t_4'\ t_3'\ t_6'\ t_7'\ t_8'\} = \frac{F}{d}\{-\tfrac{1}{2}\ 0\ \tfrac{1}{2}\ 0\ -\tfrac{1}{5}\ 0\ 0\ 0\} \quad \textbf{(14.56)}$$

which is not a feasible solution because it contains negative terms. This can be rectified by expressing the negative terms by the other parameter in each pair. The array given by (14.55) is modified accordingly by multiplying the first and third rows by -1 and swapping the first and third columns with the fifth and sixth columns respectively, giving

$$\begin{bmatrix} 1 & 0 & 0 & 0 & -1 & 0 & -\frac{3}{2} & \frac{3}{2} & \frac{1}{2} \\ 0 & 1 & 0 & -1 & 0 & 0 & -\frac{3}{2} & \frac{3}{2} & \frac{1}{2} \\ 0 & 0 & 1 & 0 & 0 & -1 & -\frac{3}{5} & \frac{3}{5} & \frac{1}{5} \end{bmatrix} \quad \textbf{(14.57)}$$

giving the basic feasible solution

$$\{t_1\ t_2\ t_3\ t_4\ t_5\ t_6\ t_7\ t_8\} = \{t_5'\ t_1'\ t_2'\ t_4'\ t_6'\ t_3'\ t_7'\ t_8'\} = \frac{F}{d}\{0\ \tfrac{1}{2}\ \tfrac{1}{2}\ 0\ 0\ \tfrac{1}{5}\ 0\ 0\} \quad \textbf{(14.58)}$$

From (14.52), the corresponding volume is

$$V_B = \frac{Fd}{\sigma}\left(16\times\frac{1}{2} + 16\times\frac{1}{2} + 100\times\frac{1}{5}\right) = 36\frac{Fd}{\sigma} \quad \textbf{(14.59)}$$

The general expression for V given by (14.48) is

$$\begin{aligned} V &= V_B + \frac{Fd}{\sigma}[(16+16)t_4 + (16+16)t_1 \\ &\quad + (100+100)t_5 + (116+24+24+60)t_7 + (116-24-24-60)t_8] \\ &= V_B + \frac{Fd}{\sigma}[32t_4 + 32t_1 + 200t_5 + 224t_7 + 8t_8] \end{aligned} \quad \textbf{(14.60)}$$

so that the use of any alternative parameter only increases V and the optimum has been found. Note that if there is an optimum solution there is always an optimum basic solution. This means that the number of bar forces is equal to the number of equilibrium conditions. Thus, if there is an optimum frame there must be an optimum statically-determinate frame. The basic solution given by (14.58) indicates that AB and BE are in compression, BC is in tension and CE has no force in it (and so does not exist). This frame is shown in Figure 14.12b[1]. From conclusion (iii)

[1] Note that although this frame satisfies the criteria used, it forms a four-bar chain mechanism, so that some extra stabilisation is necessary. The same is true of some Michell structures, such as that shown in Figure 14.6.

to Maxwell's theorem (14.8), it might be assumed that the frame composed of DC and CE shown in Figure 14.12c would be the optimum one. However, the note following this conclusion indicates that this is only true if all the loads on the frame, *including the reactions*, are known. In the present case, the reactions are statically indeterminate.If the point C is moved down towards the line DE the forces in DC and CE would become infinite, as would the volume of the frame composed of these two members.}

14.5 General Methods of Optimisation

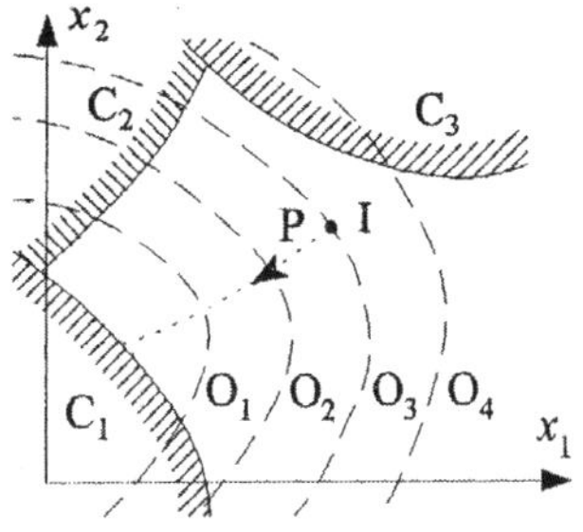

Figure 14.13 Optimisation space.

Figure 14.13 illustrates the general optimisation problem. An *objective function* or *cost function* is to be minimised by finding the optimum values of certain *design variables* or *optimisation variables.* In the figure, these variables are x_1 and x_2 . In the previous section, these variables were the tension coefficients and the objective function was the volume of the structure. The dashed lines O_1 to O_4 are contours where the objective function is constant. The analysis starts with an initial design marked I. This is modified by altering the design variables. The *path of steepest descent*, marked P, is sought along which the modification proceeds most rapidly towards the optimum design. In the previous section, this path was found by use of the simplex method. There may be certain *constraints* on the values of the design variables. These are shown by the hatched lines C_1 to C_3 in the figure. The minimisation process may then find a true minimum of the objective function or be limited by the constraints, as shown. If the problem is defined by continuous functions, the constraints may be incorporated into a Lagrangean function whose stationary values are sought. The process is then similar to that in Example 7.8.

The objectives may be to minimise weight, maximise stiffness (which is equivalent to minimising a strain energy function) controlling the natural frequencies of the structure or maximising its stability. In all cases, a suitable function to minimise is chosen. Where there may be more than one objective, a compound objective function may be used, where the values of the different objectives are weighted. The constraints on a design may be similar, for example that the structure should not exceed a certain weight, its stresses and displacements under load should be limited, its natural frequencies and stability should be controlled. In addition, the space which the structure is to occupy may specified. The path of steepest descent is determined by a *sensitivity analysis.* This may be carried out by differentiating the objective function with respect to the design variables and hence determining the steepest downwards gradient locally.

The objective function may sometimes be a function of the displacements $\boldsymbol{d}$ of the structure. These will be a function of the stiffness of the structure which in turn will be a function of the design variables. The stiffness matrix $\boldsymbol{K}$ of the structure can be recalculated for each change of a design variable and a new set of displacements $\boldsymbol{d}$ calculated. This will involve about bn^2 operations, where n is the order of $\boldsymbol{K}$ and b is its bandwidth (cf. §12.5). However, a change in a particular design variable may cause a change in the properties of relatively few components of the structure. Instead of recalculating $\boldsymbol{K}$ completely, it is then possible to find the effects of this change more directly, involving only the order of bn operations[1].

---

[1] See for example Livesley, R.K. and Modi, J.J. (1991) The Re-Analysis of Linear Structures on Serial and Parallel Computers. *Computer Systems in Engineering* V.2 p.379.

Ramm, Bletzinger and Maute[1] classify structural optimisation using four categories. These are:

(a) Topology Optimisation (organisation of the layout of the components),
(b) Shape Optimisation (position of the joints of a skeletal structure or form of a continuum),
(c) Section Sizing (choice of section used in the components) and
(d) Material Optimisation (which may not just involve the choice of a material but how it is reinforced).

Not all these problems involve the minimisation of a smooth function. From the practical point of view, the sections may be limited to those listed in section tables. Likewise, there may only be a finite number options in choosing the material. In such cases, evolutionary methods, or more specifically genetic algorithms, may have to be used.

As in Michell structures, (a) and (b) may merge into a single optimisation process. It was seen in Example 14.3 that it is possible to start from a redundant structure and optimise it by removing the right members. This is a form of topological optimisation; no geometrical change in form is involved. Likewise, beginning with a continuum of a given form, material can be removed from it during the optimisation process. This can be done by dividing up the continuum into *design patches*. In the simplest approach, the design variables are zero or unity depending on whether a design patch is either absent or present. A more effective method is to assign a density of material to each patch. The structure is then optimised by keeping its total mass constant and and varying the densities of the patches. The material is then thought of as a 'structural foam' with a stiffness which varies from zero at zero density to the full material stiffness when there are no voids in the material. This variation is made non-linear, so that the stiffness increases more rapidly with higher density. This corresponds to the actual behaviour of structural foams, but also biases the optimisation in favour of solid material. Figure 14.14 is an illustration of this process given by Ramm *et al.* (ibid.). It shows part of a uniformly-loaded bridge deck supported by columns. Mirror images of this part exist about its left- and right-hand sides, so that the full structure forms a series of arches. In the initial design, Figure 14.14a, there is a uniform distribution of material throughout the arch. Successive cycles of optimisation cause the material to migrate to where it is most needed, leading to the design shown in Figure 14.14b.

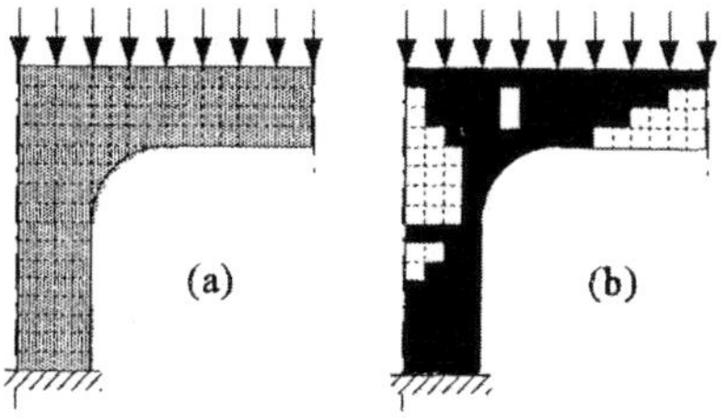

Figure 14.14 Optimisation of a substructure.

The above process leads to irregular, discontinuous forms because of the nature of the design patches. Nevertheless, the result can be seen to imply a smooth structural form. Instead of taking design patches as the design variables, weighted shape functions can be used instead. Such shape functions are used in finite-element analyses and computer-aided design. The best known are those devised by P. Bézier in 1972. A third-order curve, given by the vector function $\boldsymbol{P}(t)$, is expressed in terms of four position vector variables, $\boldsymbol{p}_1$ to $\boldsymbol{p}_4$, and takes the form

$$P(t) = (1-t)^3 p_1 + 3t(1-t)^2 p_2 + 3t^2(1-t)p_3 + t^3 p_4 \qquad \textbf{(14.61)}$$

where the parameter t varies from zero to unity (cf. 12.23)). A curve generated by these variables is shown in Figure 14.15. Higher-order curves are similar in form; again the numerical coefficients

[1] Ramm, E., Bletzinger, K.U. and Maute, K. (1998) Structural Optimisation *Proc. I.A.S.S. Colloquium, Current and Emerging Technologies of Shell and Spatial Structures. Madrid, April 1997* p.201.

are given by the binomial theorem. Surfaces in three dimensions are generated by two-parameter functions and can be used to optimise shell forms.

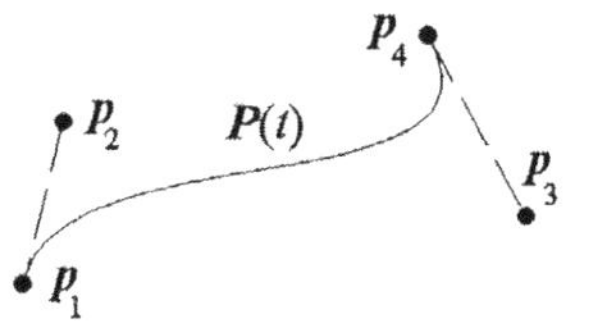

Figure 14.15 A Bézier curve.

Optimal shell forms have been generated experimentally using the funicular technique first proposed by the architect A. Gaudi. This is based on the observation that a cable or membrane with little flexural stiffness will adopt a distorted form which is in pure tension when allowed to hang under its own weight. If this form is inverted, so that in effect the direction of gravity has been reversed, the form will be one which is in pure compression. This form is well suited to making concrete structures. H. Isler[1] hung elastic membranes soaked in epoxy resin which retained their form once the resin had set. Accurate measurements of these forms were used to design concrete shell rooves for a wide variety of purposes. Such shapes can also be generated using the methods described in the previous paragraph. The form found for a square membrane supported at its corners only is shown in Figure 14.16. The upwards-curving lip at the free edge of the membrane is typical of such structures and has a stabilising effect.

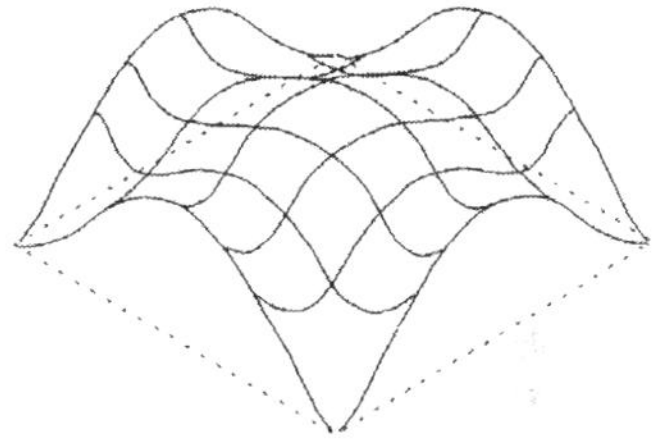

Figure 14.16 An optimum membrane.

Tensile structures are now widely used for roofing large areas. These may be cable nets or membranes. The tension may be induced by the supports or internal pressure. For example, the Tokyo dome (1988)[2] with a span of 200m is used as a permanent cover for a sports field and grandstand. It consists of a PTFE-coated glass fibre double membrane. (The outer layer is 0.8mm thick and the inner one is 0.35mm thick.) This is retained by a square grid of 80mm diameter steel cables. An internal excess pressure of only 300 Pa is required to maintain its shape. Optimum tensile structures will be in a state of uniform tensile stress and usually their self-weight is relatively unimportant. Such forms occur naturally as soap films which are minimal surface-area membranes. The surface area can then be used as the objective function in optimising the form.

Genetic algorithms are used to generate and select the 'fittest' designs in a Darwinian sense (see Goldberg (1989) for example). A population of possible designs is created in a random fashion, subject to whatever criteria are imposed on acceptable solutions. The best designs are then allowed to 'breed' at random, producing 'offspring' with characteristics of both 'parents'. Even here, a degree of randomness is introduced, so that further offspring from the same parents are not identical, but result from the selection of different parental characteristics. In order to prevent these characteristics being limited to those found in the first generation, small random variations ('mutations') are permitted between generations. Succeeding generations develop 'clusters' with similar characteristics. These may be thought of as 'species' which are sufficiently differentiated to have become incapable of breeding with other species. A cull is made of the least successful designs in each generation, paying attention also to preventing any cluster from becoming overcrowded. Such an approach to optimisation may be used even when the objective function does vary smoothly with the design variables. When these are too numerous, it may prove preferable to finding the path of steepest descent. An example is given in the next section.

---

[1] Isler, H. (1993) Generating Shell Shapes by Physical Experiments. *Bulletin of the I.A.S.S.* V.34 p.53.

[2] Ishii, K. (1995) p.128, (1999) p.236.

14.6 Vibration Optimisation

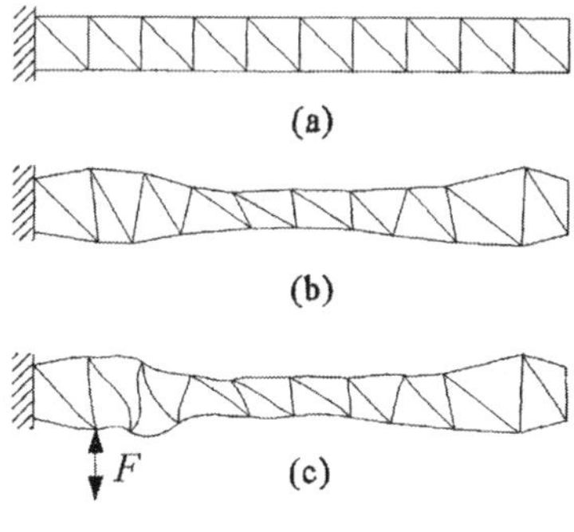

Figure 14.17 Vibration isolation.

Figure 14.17 illustrates the optimisation of a truss to minimise the transmission of vibration over a specified frequency range. The initial design, a regular cantilevered Pratt truss, is shown in Figure 14.17a. This problem has been examined by Keane[1] in a series of papers. The design variables are the joint positions, which are allowed to vary within acceptable limits. The analysis assumes viscous damping in the members, so that the governing equations are similar to those for A.C. circuits. This means that some of the concepts in transmission line theory can be used. One joint of the truss is excited with an oscillatory force F as shown in Figure 14.17c. The energy flow along the truss is found using receptance methods, and the objective function to be minimised is the vibrational energy level in vertical beam at the free end of the truss, averaged over the range 150-250 Hz.

The number of design variables and the complexity of the analysis were considered to be too great for sensitivity analysis, so that genetic algorithms of the kind described at the end of the last section were used. A constant population of 300 designs was allowed to evolve over 15 generations, requiring 4,500 evaluations, giving the best design[2] as that shown in Figure 14.17b. One of the effects of the optimisation process was to reduce the number of natural frequencies in the above range from 31 to 20. Figure 14.17c shows a computer-generated image of the response to the oscillating force F shown near the twelfth natural frequency, 207.01 Hz. It can be seen that most of the kinetic energy is related to the flexure of members in the immediate vicinity of the force and that the free end remains relatively undisturbed.

The kinetic energy of a beam oscillating in flexure can be determined from the expressions in §10.2. The mean kinetic energy, T_{mf} say, is found from (10.2) to be

$$T_{mf} = \frac{\omega}{2\pi}\int_{x=0}^{l}\int_{t=0}^{2\pi/\omega}\tfrac{1}{2}\rho A\dot{v}^2\,dt\,dx = \frac{M\omega^2}{4l}\int_{x=0}^{l}V(x)^2\,dx \qquad \textbf{(14.62)}$$

where $V(x)$ is given by (10.4). The constants P, Q, R and T are related to the end deflexions by (10.6). After some considerable manipulation, the mean kinetic energy may be expressed in terms of the amplitudes of these deflexions by

$$\begin{aligned} T_{mf} = \frac{EI}{4l^3}[& l^2(c_2^2 - c_1)(\Theta_A{}^2 + \Theta_B{}^2) + l^2(4c_1c_2 - 3c_4 - c_2)\Theta_A\Theta_B \\ & + 9(c_4^2 - c_5)(V_A{}^2 + V_B{}^2) + 18(c_6 - c_3c_4)V_AV_B \\ & + 6l(c_3 - c_2c_4)(\Theta_BV_B - \Theta_AV_A) + 6l(c_2c_3 - c_4)(\Theta_BV_A - \Theta_AV_B)] \end{aligned} \qquad \textbf{(14.63)}$$

---

[1] See Keane,A.J. (1995) Passive Vibration Control via Unusual Geometries: The Application of Genetic Algorithm Optimisation to Structural Design. *Journal of Sound and Vibration* V.185 p.441.

[2] The descriptions of the initial and final designs are listed on the files R3DSIAJK.VIB and R3DSIBJK.VIB respectively. These were made out of aluminium and tested experimentally. The section used is listed on the section file ALUMINSI.SEC. Their dynamic responses can be examined using the software provided (see §A2.1.3).

where the functions c_1 to c_6 are given by (10.9) and are listed in §A6.2. The end deflexions are related to their amplitudes by

$$\begin{aligned} \theta_A &= \Theta_A\cos(\omega t+\phi) \ , \quad \theta_B = \Theta_B\cos(\omega t+\phi) \\ v_A &= V_A\cos(\omega t+\phi) \ , \quad v_B = V_B\cos(\omega t+\phi). \end{aligned} \tag{14.64}$$

Similarly, the expression for the mean kinetic energy associated with axial vibration, T_{mp} say, are found from (10.21) to (10.25) to be

$$T_{mp} = \frac{EA}{8l}[(d_2^2 - d_1)(U_A^2 + U_B^2) + 2(d_2 - d_1 d_2)U_A U_B] \tag{14.65}$$

where

$$u_A = U_A\cos(\omega t+\phi) \ , \quad u_B = U_B\cos(\omega t+\phi) \ . \tag{14.66}$$

The mean kinetic energy of the beams of the trusses shown in Figure 14.17 were found from these expressions, which make no allowance for damping. The vibration isolation of the end beam was measured by comparing its mean kinetic energy to that of the whole truss. This ratio has been calculated at 10Hz intervals over the range of interest and is shown in Figure 14.18. The white dots connected by the broken line show the response of the original truss in Figure 14.17a. The black dots connected by the continuous line show the response of the optimised truss in Figure 14.17b. It can be seen that the optimised truss is several orders of magnitude more effective in isolating vibration from the end beam over most of the range.

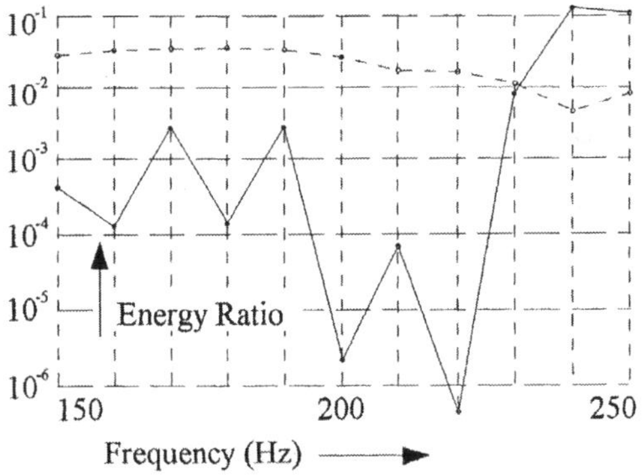

Figure 14.18 Vibration isolation of the end beam.

Methods of optimal design for constrained dynamic response are given in Haug and Arora (1979) §5.1. The problem can be defined as follows. The structural response may be expressed in terms of a vector of *state variables* $\boldsymbol{s}$ ($\boldsymbol{s} = \{s_1, s_2, \ldots s_n\}$) such as a set of displacements $\boldsymbol{u}$ or a set of stresses σ. These values of these state variables will depend on the design variables $\boldsymbol{x}$ ($\boldsymbol{x} = \{x_1, x_2, \ldots x_k\}$). The objective (or cost) function $O(\boldsymbol{x})$ may more readily be expressed in terms of both the design variables and the state variables. Typically, if it used to limit the maximum response of the system in a time interval ($0 \le t \le \tau$) it will take the form

$$O(\boldsymbol{x}) = Max\{f_o(t,\boldsymbol{s}(t),\boldsymbol{x})\} \ , \quad (0 \le t \le \tau) \tag{14.67}$$

where the state variables are governed by some differential equations of time. For example, they might be n first-order differential equations of the form

$$\boldsymbol{F}(\boldsymbol{x})\dot{\boldsymbol{s}}(t) = \boldsymbol{f}(t,\boldsymbol{s}(t),\boldsymbol{x}) \quad (0 \le t \le \tau) \tag{14.68}$$

where $\boldsymbol{F}(\boldsymbol{x})$ is an $n \times n$ matrix and $\boldsymbol{f}(t,\boldsymbol{s}(t),\boldsymbol{x})$ is a column vector of elements. There will also be initial conditions on the state variables at the beginning of the interval, such as

$$\boldsymbol{s}(0) = \boldsymbol{s}^0 \tag{14.69}$$

There may also be m constraints within the time interval of the form

$$C_j(t,\boldsymbol{s}(t),\boldsymbol{x}) \le 0 \ , \quad (j = 1 \text{ to } m) \tag{14.70}$$

and possibly functional constraints relating to the total time interval such as

$$\int_0^\tau g(t,s(t),x) \leq 0 \qquad \textbf{(14.71)}$$

The design variables may have upper and lower acceptable limits given by

$$x_i^L \leq x_i \leq x_i^U \quad , \quad (i = 1 \text{ to } k) \qquad \textbf{(14.72)}$$

The normal non-linear programming method usually divides the time interval ($0 \leq t \leq \tau$) into a set of r equal subintervals Δt ($\tau = r\Delta t$) and the above conditions expressed in terms of the usual finite-difference approximations. The objective function $O(x)$ is replaced by an *artificial objective function* $\bar{O}(x)$ so that the condition given by (14.67) is replaced by the conditions at the ends of of the r time intervals

$$f_o(t_i,s(t_i),x) - \bar{O}(x) \leq 0 \quad (t_i = i\Delta t, \quad i = 0 \text{ to } r) \qquad \textbf{(14.73)}$$

where for at least one value of i the inequality becomes and equality. The governing differential equation (14.68) is evaluated at the end of each interval, using the usual finite-difference approximations for the derivatives. Thus (14.68) becomes

$$F(x)[s(t_i) - s(t_{i-1})] = \Delta t \times f(t_i,s(t_i),x) \quad (i = 1 \text{ to } r) \qquad \textbf{(14.74)}$$

The functional constraint given by (14.71) may also be integrated numerically, using the same time intervals for compatible accuracy. The minimisation of $\bar{O}(x)$ then involves a total of $(nr+k+1)$ variables with $[(n+m+1)r+m+2k+1]$ constraints.

An improvement on this approach in which all the conditions over the time interval ($0 \leq t \leq \tau$) are replaced by functional constraints of the type give by (14.71). This is done by noting that the condition that $\phi(t)$ is less than or equal to zero over the time interval ($0 \leq t \leq \tau$) is equivalent to

$$\int_{t=0}^{\tau} \langle\phi(t)\rangle dt = 0 \quad \text{where} \quad \left\{ \begin{array}{ll} \langle\phi(t)\rangle = 0, & \phi(t) < 0 \\ \langle\phi(t)\rangle = \phi(t), & \phi(t) \geq 0 \end{array} \right\} \qquad \textbf{(14.75)}$$

As the argument of the integral is then either positive or zero, the integral itself can only be zero if $\phi(t)$ is less than or equal to zero at every instant during the interval. The authors also suggest a state-space optimisation method which uses integrals of this kind and an adjoint column-vector function $\lambda(t)$ such that

$$\int_{t=0}^{\tau} \lambda(t)[F(x)\dot{s}(t) - f(t,s(t),x)]dt = 0 \qquad \textbf{(14.76)}$$

(cf. (14.68)). Using variational methods, it is then possible to express the governing equations of the problem in terms which no longer require explicit reference to the state variables $s(t)$. The above authors find that this approach gives an order of magnitude reduction in the computing time required for optimisation.

14.7 Optimisation with Buckling Problems

Optimising a structural system, allowing for instability effects, can pose considerable difficulties. For example, it has been assumed that optimal conditions are reached when overall and local buckling are equally possible. Thompson and Lewis[1] point out that where there are coincident buckling loads, the structure becomes exceptionally sensitive to imperfections, which reduce the maximum loading possible. Equation (14.12) of Morris (1982) implies that a suitable constraint against buckling might be to ensure that none of the members of a structure exceeded its pin-ended buckling load (9.6b). A more satisfactory criterion might be to use the empirical limits on the slenderness ratios[2] laid out in various standards and codes of practice. This still does not address the issue of the overall buckling of the structure. Consider the case where the stiffness matrix is a linear function of the axial loading so that the general form (11.50) may be written as

$$\lambda \boldsymbol{P} = \boldsymbol{K}_P \boldsymbol{d} = (\boldsymbol{K} + \lambda \boldsymbol{K}_G)\boldsymbol{d} \tag{14.77}$$

where $\boldsymbol{K}$ is the normal linear stiffness matrix for the problem and $\boldsymbol{K}_G$ is called the *geometric stiffness matrix*. This is because it depends on the the geometric properties of the beam and not its elastic properties. This is directly applicable to pin-ended members. For example, in (11.51),

$$\boldsymbol{K}_{11P} = \begin{bmatrix} EA/l & 0 \\ 0 & -P/l \end{bmatrix} = \begin{bmatrix} EA/l & 0 \\ 0 & 0 \end{bmatrix} + P\begin{bmatrix} 0 & 0 \\ 0 & -1/l \end{bmatrix} = \boldsymbol{K}_{11} + P\boldsymbol{K}_{11G} \tag{14.78}$$

The same separation process can be used for beams in flexure, provided that Pl^2/EI is small enough for the approximations given by (9.10) can be used. Allowing for large axial loading, the first stiffness matrix in (11.23) becomes

$$\begin{aligned} \boldsymbol{K}_{11P} &= \begin{bmatrix} EA/l & 0 & 0 \\ 0 & 12EI\phi_1/l^3 & 6EI\phi_2/l^2 \\ 0 & 6EI\phi_2/l^2 & 4EI\phi_3/l \end{bmatrix} \\ &\approx \begin{bmatrix} EA/l & 0 & 0 \\ 0 & 12EI/l^3 & 6EI/l^2 \\ 0 & 6EI/l^2 & 4EI/l \end{bmatrix} + P\begin{bmatrix} 0 & 0 & 0 \\ 0 & -6/5l & -1/10 \\ 0 & -1/10 & -2l/15 \end{bmatrix} = \boldsymbol{K}_{11} + P\boldsymbol{K}_{11G} \end{aligned} \tag{14.79}$$

The approximations to the other submatrices are similar in form, where

$$\boldsymbol{K}_{12G} = \boldsymbol{K}_{21G}^T = \begin{bmatrix} 0 & 0 & 0 \\ 0 & 6/5l & -1/10 \\ 0 & 1/10 & l/30 \end{bmatrix}, \quad \boldsymbol{K}_{22G} = \begin{bmatrix} 0 & 0 & 0 \\ 0 & -6/5l & 1/10 \\ 0 & 1/10 & -2l/15 \end{bmatrix}. \tag{14.80}$$

The advantage of this form is that if optimisation of a structure is restricted to changing the elastic properties of its components, then only the variation of the $\boldsymbol{K}_{11}$ matrix need be considered. Khot[3] *et al.* examine minimum weight design under such circumstances and show that the optimality criterion can be taken as implying that the ratio of a strain-energy density function to a mass-density function is the same for all the components.

Expressions such as the above can be used to find the optimal form of columns. Figure

---

[1]Thompson, J.M.T. and Lewis, G.M. (1972) On the Optimum Design of Thin-Walled Compression Members. *J. Mech. Phys. Solids* V.20 p.101.

[2] The slenderness ratio of a beam is given by $\sqrt{(I/Al^2)}$ where I is the least second moment of area of its section, A its area and l is the length of the beam.

[3] Khot, N.S., Venkayya, V.B. and Berke, L. (1976) Optimum Structural Design with Stability Constraints. *Int. J. Num. Meth. Engg.* V.10 p.1097.

14.19 shows a column of variable flexural stiffness broken up into a set of very short uniform columns, so that Pl^2/EI for each of them is very small. The simplifying approximation to their stiffness matrices given by (14.79) and (14.80) can then be used for these short columns and the buckling load for the set found. The properties of each elementary column may be described in terms of parameters which act as the design variables of the problem. Often, the cross-sectional area A is used as a design variable. If the section is circular, the value of I is then given by $A^2/4\pi$. The problem may then be described as finding the minimum-volume column which will support a given axial load P, given that the yield stress σ_y is not exceeded and that P is less than the buckling load P_c . If the overall height of the column is L and it is divided into n elementary columns, these conditions are given by minimising V where

$$V = \sum_{i=1}^{n} A_i \frac{L}{n} \quad \text{and} \quad P - P_c \leq 0 \ , \quad P - A_i \sigma_y \leq 0 \quad (i = 1 \text{ to } n). \qquad \textbf{(14.81)}$$

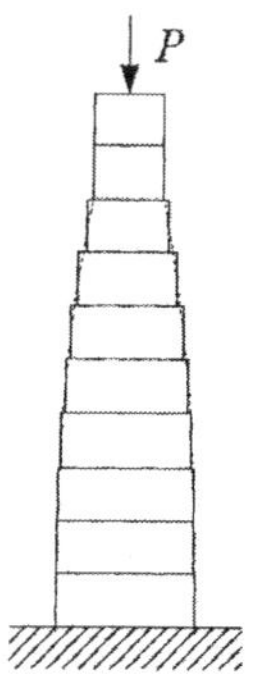

Figure 14.19 Finite elements for column optimisation.

Using the above approximations, the stiffness matrix equation for the assemblage of elementary columns can be written in the form

$$\boldsymbol{P} = \boldsymbol{K}_P \boldsymbol{d} = (\boldsymbol{K} + P\boldsymbol{K}_G)\boldsymbol{d} \qquad \textbf{(14.82)}$$

The buckling load P_c is found from the condition that $\boldsymbol{P}$ is zero for non-zero $\boldsymbol{d}$ (cf. §11.5). This can be found from the zeros of the determinant of $\boldsymbol{K}_P$ as usual or (14.82) can be turned into a standard eigenvalue problem of the form

$$(\boldsymbol{K}_G^{-1}\boldsymbol{K} + P\boldsymbol{I})\boldsymbol{d} = \boldsymbol{0} \qquad \textbf{(14.83)}$$

Figure 14.19 shows a solution obtained Haug and Arora (ibid. §4.2). The true profile of the column is shown by the dotted outline. It is optimised for a load P for which the cross-sectional area of the top of the column is constrained by the need to avoid exceeding the limiting stress σ_y . Theoretical solutions have been found[1] where there is no limiting stress constraint. They indicate a saving of about 13% in volume over the optimum uniform column. This is confirmed by the above numerical solutions for lower values of P.

The problem of designing the tallest possible column subject to buckling under its self-weight has been examined by Keller and Niordson[2]. They also take I to be a multiple of A^2 where the optimum shape is defined as a function of A. The shape of the column is found by iteration and is not given explicitly, although near the top it forms a cusp, with A varying as the cube of the distance from the top. The height of the optimum column is found to be 2.034 time that of the tallest possible uniform column. The problem of a uniform column buckling under its self-weight was analysed by Greenhill[3]. The critical self-weight per unit length is given by

$$q_c = \frac{7.837EI}{l^3} \qquad \textbf{(14.84)}$$

This can be increased by restraining the column with guy cables for example. If these could be

---

[1] See Timoshenko and Gere (1961) p.132 for example.

[2] Keller, J.B. and Niordson, F. I. (1966) The Tallest Column. *Journal of Mathematics and Mechanics* V.16 No.5 p.433.

[3] Greenhill, A.G. (1881) *Proc. Cambridge Phil. Soc.* V.4.

used to prevent the lateral displacement of one point on the column, it can be shown[1] that the optimum position of this point is the node of the second buckling mode, which is $0.4806l$ above the base. The critical self-weight is then raised to $55.98\ EI/l^3$.

---

[1] See Chilver (1967) Chapter 1 §8 p.49.

Chapter 15 Regular Structures

15.1 Introduction

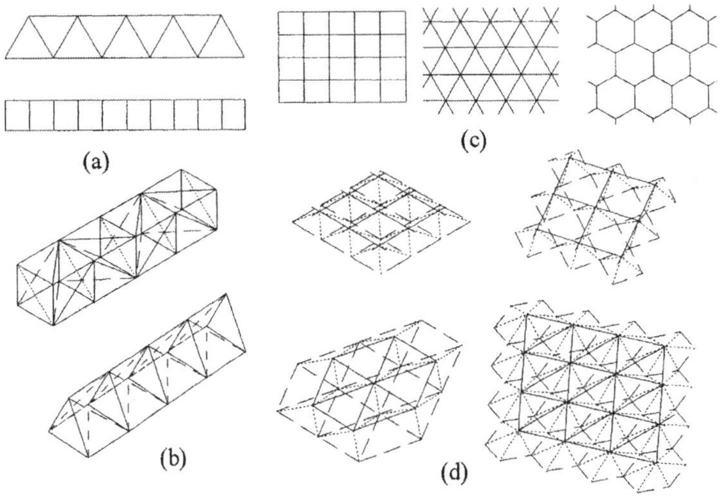

Figure 15.1 Regular trusses and grids.

A regular structure consists of a set of modules, each of which is the same, with respect to its local coordinate system, as every other module. Regularity is similar to symmetry, as discussed in §1.5, although often a regular structure is a bounded portion of an infinite symmetrical form. Figure 15.1 shows the main groups of regular structures; others will be considered later in this chapter. Plane trusses are shown in Figure 15.1a. Other forms are shown in Figure 2.10. They may have pinned joints, like the Warren truss shown at the top or rigid joints, like the Vierendeel truss shown below it. Space trusses are so called because they are essentially three-dimensional, although they are used for space probes and space stations. Examples are shown in Figure 15.1b. They are generally rectangular or triangular in cross-section, but they may also have hexagonal cross-sections. Plane grids are usually rectangular, but may also be triangular or hexagonal in form, as shown in Figure 15.1c. The four most common forms of space grid are shown in Figure 15.1d. These consist of two layers of plane grids connected by diagonals, shown dotted in the figure. Triple-layer grids have also been used. These have the advantage of separating the top and bottom layers yet further, and so increasing flexural stiffness of the grid. The additional layer in the middle helps to stabilise the grid.

Such regular structures are often used because they employ large numbers of identical parts, thus simplifying fabrication. Empirically, it was recognised that there must be some simplification possible in their analysis. Thus trusses were seen to behave rather like beams and grids to behave like plates. Methods of finding the appropriate 'equivalent continuum' to a regular structure were then proposed, some of which had a degree of validity. For the reasons discussed in §8.5, such analogies should not be taken too far. Using finite difference calculus, the exact equations governing such structures can be written and solved. The earliest work on this was by Bleich and Melan (1927). More recent solutions will be found in Wah and Calcote (1970) and Dean (1976) and others will be given here and in Appendix 5.

15.2 Mathematical Preliminaries

It has long been realised that the joint deflexions of a regular (periodic) structure might be described by analytic functions. However, these functions only become meaningful at discrete intervals (i.e. at the joints) and are of no significance otherwise. Consider the simple pin-jointed mechanism shown in Figure 15.2 which is supported on knife edges at equal intervals of l. The displacement of the X<u>th</u> pin could be expressed equally well by the function $\delta \sin \pi x/l$, where x is the distance from the left-hand end, or by $-(-1)^X\delta$. Differential calculus is used for functions of continuously variable parameters such as x, but for functions of discrete parameters such as X, finite difference calculus is normally used.

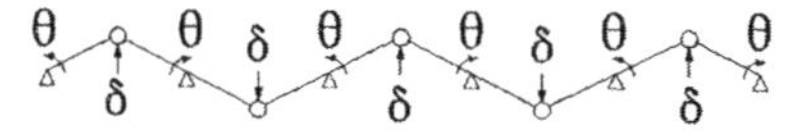

Figure 15.2 Deflexions of a mechanism.

The theory of finite-difference calculus can be traced back to 1717, but its application to structural analysis is comparatively recent. The theory is given by Jordan (1965) and is summarised in Chapter 13 of Brown (1965). Finite-difference operators have already been discussed in §8.5. The *shift operator* E_X, will be used here. It is defined by

$$E_X f(X) = f(X+1) \qquad \text{or more generally} \qquad E_X^m f(X) = f(X+m) \tag{15.1}$$

where m is usually an integer, but may be any rational number. The effects of this operator on various functions are given in the following table.

Table 15.1 Shift Operations on Common Functions

$f(X)$	$(E_X+E_X^{-1})f(X)$	$(E_X-E_X^{-1})f(X)$	$f(X)$	$(E_X+E_X^{-1})f(X)$	$(E_X-E_X^{-1})f(X)$
X	$2X$	2	$\sin kX$	$2\cos k \sin kX$	$2\sin k \cos kX$
X^2	$2X^2+2$	$4X$	$\cos kX$	$2\cos k \cos kX$	$-2\sin k \sin kX$
X^3	$2X^3+6X$	$6X^2+2$	$\sinh kX$	$2\cosh k \sinh kX$	$2\sinh k \cosh kX$
X^4	$2X^4+12X^2+2$	$8X^3+8X$	$\cosh kX$	$2\cosh k \cosh kX$	$2\sinh k \sinh kX$
X^5	$2X^5+20X^3+10X$	$10X^4+20X^2+2$	$a^{\pm X}$	$(a+1/a)a^{\pm X}$	$\pm(a-1/a)a^{\pm X}$

Some inverse operations on polynomials were given by (8.36). This inverse process is akin to integration, as is summation (see Jordan (1965) Chapter III). A correspondence between integration and summation suggests a possible form of Fourier series for finite-difference functions. It rests on the following summations[1]

$$\sum_{X=0}^{N-1} \sin 2kX = \frac{\sin kN \sin k(N-1)}{\sin k} \quad , \quad \sum_{X=0}^{N-1} \cos 2kX = \frac{\sin kN \cos k(N-1)}{\sin k} \; . \tag{15.2}$$

where X and N are integers and k is a real number which is not a multiple of π. The full series is listed in §A5.4. The most useful application for grids involves using only the sine series. If k is $\pi R/2N$, then the second of the above sums is zero if R is an even integer, and unity if R is odd. Then if P and Q are two different integers,

[1] Jordan (1965) §43 incorrectly gives the upper limit of these summations as N.

$$\sum_{X=0}^{N-1} \sin\frac{P\pi X}{N}\sin\frac{Q\pi X}{N} = \frac{1}{2}\sum_{X=0}^{N-1}\left(\cos(P-Q)\frac{\pi X}{N} - \cos(P+Q)\frac{\pi X}{N}\right) = 0 \qquad \textbf{(15.3)}$$

because either the sum of each term on the right-hand side is zero or they cancel each other out. If Q is equal to P, the first term on the right-hand side is unity and the sum of the second terms is zero, so that the value of the expression is ½N. Note that these summations remain the same if the lower limit is raised to X=1. If the finite-difference function $f(X)$ can be represented by

$$f(X) = \sum_{P=1}^{N-1} a_P \sin\frac{P\pi X}{N} \qquad \textbf{(15.4)}$$

in the range 0<X<N, then from the above

$$\sum_{X=1}^{N-1} f(X)\sin\frac{Q\pi X}{N} = \sum_{P=1}^{N-1} a_P \sum_{X=1}^{N-1} \sin\frac{P\pi X}{N}\sin\frac{Q\pi X}{N} = \tfrac{1}{2}a_Q N \qquad \textbf{(15.5)}$$

For example, if $f(X)$ is unity, from the first of equations (15.2)

$$a_Q = \frac{2}{N}\sum_{X=0}^{N-1}\sin\frac{Q\pi}{N}X = \frac{2}{N}\,\frac{\sin\frac{Q\pi}{2}\sin\left(\frac{Q\pi}{2}-\frac{Q\pi}{2N}\right)}{\sin\frac{Q\pi}{2N}} = \begin{cases} 0 & (\text{Q even}) \\ \frac{2}{N}\cot\frac{Q\pi}{2N} & (\text{Q odd}) \end{cases} \qquad \textbf{(15.6)}$$

In (8.26), the *forward difference operator* Δ and the *backward difference operator* ∇ were also defined, but their more common usage of these symbols as the Laplace operator and del (or nabla) respectively will be employed here. In two-dimensional cartesian coordinates, they are

$$\Delta f = \nabla^2 f = \left(\frac{\partial^2}{\partial x^2} + \frac{\partial^2}{\partial y^2}\right) f \qquad \textbf{(15.7)}$$

Example 15.1 A Continuous Beam

The continuous beam shown in Figure 15.3 has n equal spans of length l and is subject to a distributed load of linearly-varying intensity from zero at the left-hand end to nq at the right-hand end.

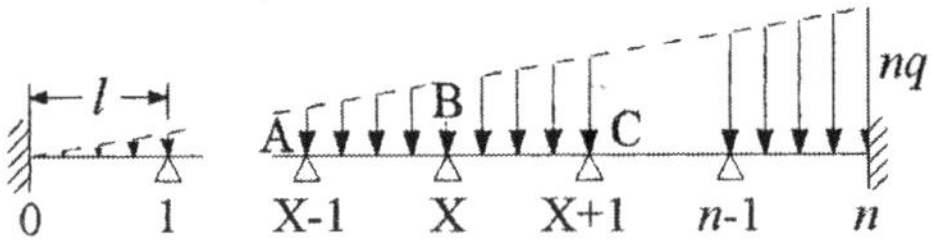

Figure 15.3 An equal-span continuous beam.

The two ends are encastré, otherwise each span has knife-edge supports. The equilibrium conditions at a typical intermediate support, B, can be expressed in terms of the rotation of the beam at B and the rotations at the adjacent supports, A and C, using the slope-deflexion equations developed in §4.5. The moments at end B of AB and BC are

$$\begin{aligned} M_{BA} &= \frac{2EI}{l}(2\theta_B + \theta_A) + M^f_{BA} \qquad \left(M^f_{BA} = \frac{ql^2}{30} - \frac{qXl^2}{12}\right) \\ M_{BC} &= \frac{2EI}{l}(2\theta_B + \theta_C) + M^f_{BC} \qquad \left(M^f_{BC} = \frac{ql^2}{30} + \frac{qXl^2}{12}\right) \end{aligned} \qquad \textbf{(15.8)}$$

where B is the X<u>th</u> joint, as shown. As there is no external moment at joint B,

$$0 = M_B = \frac{2EI}{l}(\theta_A + 4\theta_B + \theta_C) + M^f_{BA} + M^f_{BC} \qquad \textbf{(15.9)}$$

Timoshenko and Young (1965) refer to this as a *three-angle equation*[1]. The fixed-end moments are given by (15.8). Using these and rearranging (15.9), the final form is the finite-difference equation

$$\frac{2EI}{l}(E_X^{-1} + 4 + E_X)\theta_X = -\frac{ql^2}{15} \qquad \textbf{(15.10)}$$

where θ_{X-1}, θ_X and θ_{X+1} are θ_A, θ_B and θ_C. Finite-difference equations can be solved in a similar manner to differential equations. A 'particular integral' is sought which satisfies the right-hand side of (15.10). In addition, a 'complementary function' is needed which satisfies the homogeneous form of this equation (right-hand side zero) so that the boundary conditions can be met. From Table 15.1, the general form of the solution is then found to be

$$\theta_X = -\frac{ql^4}{180EI} + A(-2+\sqrt{3})^X + B(-2-\sqrt{3})^X \qquad \textbf{(15.11)}$$

where the constants A and B in the 'complementary function' are determined from the boundary conditions that the rotation is zero at X=0 and X=n, so that

$$\begin{aligned} 0 = \theta_0 &= -\frac{ql^4}{180EI} + A + B \\ 0 = \theta_n &= -\frac{ql^4}{180EI} + A(-2+\sqrt{3})^n + B(-2-\sqrt{3})^n \end{aligned} \qquad \textbf{(15.12)}$$

For grids, the joint numbering will be in terms of two parameters, X and Y, and the state of the joints may be expressed as a function of both of them. It is then necessary to use two shift operators, E_X and E_Y, where in general

$$E_x^m E_y^n f(X,Y) = f(X+m, Y+n) \qquad \textbf{(15.13)}$$

where m and n are rational numbers. If the number X corresponds to a position x equal to aX, it is possible to relate $f(x)$ to a function f(X) which takes the same values at the joints. Then

$$E^m f(X) = f(X+m) = f(x+ma) = \left(1 + ma\frac{d}{dx} + \frac{(ma)^2}{2}\frac{d^2}{dx^2} + \ldots \frac{(ma)^r}{r!}\frac{d^r}{dx^r}\right) f(x) \qquad \textbf{(15.14)}$$

using Taylor's series. The two-dimensional equivalent may readily be inferred from this. It will be seen later that such relationships can be used to compare grids and plates.

{Some grids are rotationally symmetrical about an axis normal to their planes. Thus a square grid may be invariant with respect to a rotation of 90°, a triangular grid invariant under a rotation of 60° and a hexagonal grid invariant under a rotation of 120° (see Figure 15.1c). The equations governing their behaviour must have similar invariant properties. Suppose that the cartesian coordinate system xy is rotated through the angle θ shown in Figure 15.4 to become the coordinates $x'y'$. Then the change δf in a function $f(x,y)$ between O and P is given by

$$\delta f = \frac{\partial f}{\partial x}\delta x + \frac{\partial f}{\partial y}\delta y = \frac{\partial f}{\partial x'}\delta x' + \frac{\partial f}{\partial y'}\delta y' \qquad \textbf{(15.15)}$$

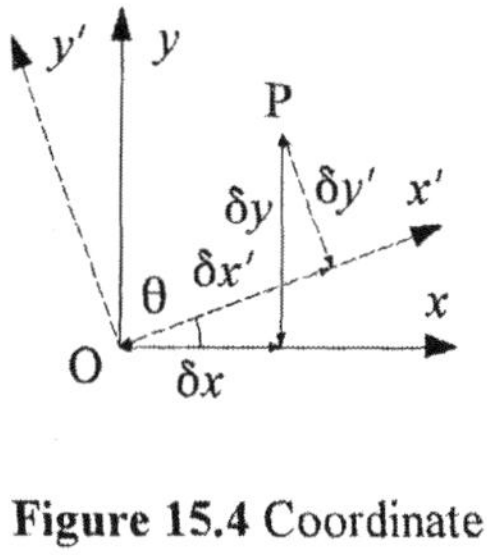

Figure 15.4 Coordinate transformation.

[1] Ibid. §9.3. The better-known *three-moment equation* is given in the same section.

$$\text{where} \quad \delta x' = \delta x \cos\theta + \delta y \sin\theta \quad , \quad \delta y' = -\delta x \sin\theta + \delta y \cos\theta . \tag{15.16}$$

As these equations hold for all δx and δy, it follows that

$$\frac{\partial}{\partial x} = \cos\theta \frac{\partial}{\partial x'} - \sin\theta \frac{\partial}{\partial y'} \quad , \quad \frac{\partial}{\partial y} = \sin\theta \frac{\partial}{\partial x'} + \cos\theta \frac{\partial}{\partial y'} . \tag{15.17}$$

Using these relationships between the differential operators, the invariant forms of linear differential expressions can be found. The most general second-order form which is invariant under a rotation θ of 60° (or one of either 90° or 120°) is given by the harmonic expression Δ in (15.7). The biharmonic Δ^2 is the most general fourth-order form which is invariant under rotations of 60° or 120°. Both are invariant under rotation through any angle, as is any power of the operator Δ. The most general fourth-order form which is invariant under rotations of 90° is given by

$$I_4(90°) = A\left(\frac{\partial^4}{\partial x^4} + \frac{\partial^4}{\partial y^4}\right) + B\frac{\partial^4}{\partial x^2 \partial y^2} + C\left(\frac{\partial^4}{\partial x^3 \partial y} - \frac{\partial^4}{\partial x \partial y^3}\right) \tag{15.18}$$

The biharmonic is the particular case when B is equal to $2A$ and C is zero. The above forms cover all the invariant differential expressions which will used later to describe grids, but higher-order differential invariants can be found by the same process. Thus

$$I_6(60°) = I_6(120°) = A\left(\frac{\partial^2}{\partial x^2} + \frac{\partial^2}{\partial y^2}\right)^3 + B\left(\frac{\partial^6}{\partial x^6} - 15\frac{\partial^6}{\partial x^4 \partial y^2} + 15\frac{\partial^6}{\partial x^2 \partial y^4} - \frac{\partial^6}{\partial y^6}\right) \tag{15.19}$$

and

$$I_6(90°) = A\left(\frac{\partial^6}{\partial x^6} + \frac{\partial^6}{\partial y^6}\right) + B\left(\frac{\partial^6}{\partial x^4 \partial y^2} + \frac{\partial^6}{\partial x^2 \partial y^4}\right) + C\left(\frac{\partial^6}{\partial x^5 \partial y} - \frac{\partial^6}{\partial x \partial y^5}\right) \tag{15.20)\}}$$

15.3 Regular Plane Trusses

Two main categories of plane truss will be examined. These are those which can be treated as pin-jointed, such as the Warren truss in the upper part of Figure 15.1a and those which are essentially rigid-jointed, such as the Vierendeel truss shown in the lower part. On the whole, the former are preferable, as applied loads can be carried more efficiently by structural members which are in pure tension or compression rather than in flexure. However, multi-storey building frames are often similar in their behaviour to Vierendeel trusses.

Dean and Tauber[1] have analysed such trusses for special loading cases and other solutions have been given by the author[2]. This chapter will be primarily concerned with the methods of obtaining solutions. For the most part, the solutions themselves are given in §A5.2. The process of forming the governing equations for a truss is similar to that described in Chapter 11. The elastic response of typical members is expressed in terms of stiffness matrix equations related to a global coordinate system for the truss. The joint equilibrium equations for typical joints of the truss are then found in terms of these stiffness matrices. These equations will contain the

---

[1] Dean, D.L. and Tauber, S. (1959) Solutions for One-dimensional Structural Lattices. *Proc. ASCE, Engineering Mechanics Division* V.85, No. EM4, p.31.

[2] Renton, J.D. (1969) Behavior of Howe, Pratt and Warren Trusses. *Proc. ASCE, Structural Division* V.95, No. ST2, p.183.

deflexions of joints adjacent to the typical joints. These deflexions are expressed in terms of the deflexions of the typical joints, using the shift operator. Solutions of these finite-difference equations are then sought for various loading and support conditions.

Example 15.2 A Howe (or Pratt) Truss

Figure 15.5 shows a truss[1] which begins at joints 0 and ends at joints N. The governing equations will be formed for the X<u>th</u> joints. The lower one has external horizontal and vertical forces H and V applied to it and displaces u horizontally and v vertically. The loads and displacements of the X<u>th</u> upper joint, immediately above the X<u>th</u> lower joint, are denoted similarly, using primes. Each bay is of length m and of height h. It is convenient to define the elastic properties of the component bars in terms of their axial coefficients. The *axial coefficient* of a bar is given by (Young's modulus)×(cross-sectional area)/(length)³. These coefficients will be denoted by a, c and d for the vertical, horizontal and diagonal bars respectively. If the length of the diagonal members is D, the form of (11.10) becomes

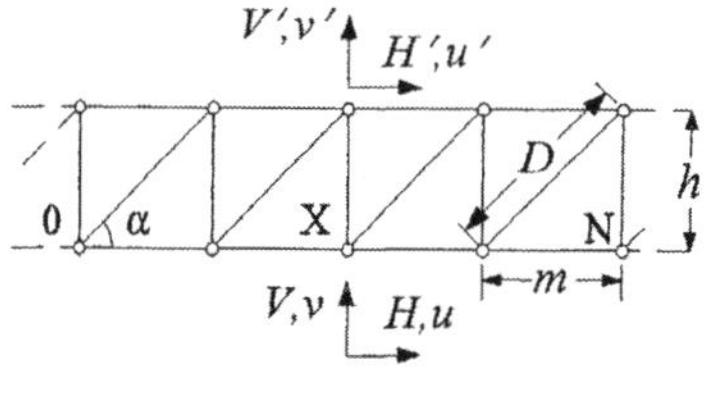

Figure 15.5 Notation for a Howe truss.

$$\begin{bmatrix} F^*_{xA} \\ F^*_{yA} \end{bmatrix} = d \begin{bmatrix} D^2\cos^2\alpha & D^2\sin\alpha\cos\alpha \\ D^2\sin\alpha\cos\alpha & D^2\sin^2\alpha \end{bmatrix} \begin{bmatrix} \delta^*_{xA} \\ \delta^*_{yA} \end{bmatrix} = d \begin{bmatrix} m^2 & hm \\ hm & h^2 \end{bmatrix} \begin{bmatrix} \delta^*_{xA} \\ \delta^*_{yA} \end{bmatrix} \tag{15.21}$$

where x and y are in the horizontal and vertical directions as before. The other stiffness matrices for the bars can be formed in a similar fashion. Using these relationships to express the joint equilibrium conditions at the X<u>th</u> lower joint,

$$\begin{aligned} \begin{bmatrix} H(\mathrm{X}) \\ V(\mathrm{X}) \end{bmatrix} = d \begin{bmatrix} m^2 & hm \\ hm & h^2 \end{bmatrix} \begin{bmatrix} u(\mathrm{X}) - u'(\mathrm{X}+1) \\ v(\mathrm{X}) - v'(\mathrm{X}+1) \end{bmatrix} &+ c \begin{bmatrix} m^2 & 0 \\ 0 & 0 \end{bmatrix} \begin{bmatrix} 2u(\mathrm{X}) - u(\mathrm{X}-1) - u(\mathrm{X}+1) \\ 2v(\mathrm{X}) - v(\mathrm{X}-1) - v(\mathrm{X}+1) \end{bmatrix} \\ &+ a \begin{bmatrix} 0 & 0 \\ 0 & h^2 \end{bmatrix} \begin{bmatrix} u(\mathrm{X}) - u'(\mathrm{X}) \\ v(\mathrm{X}) - v'(\mathrm{X}) \end{bmatrix} \end{aligned} \tag{15.22}$$

using the notation shown in Figure 15.5 for the terms in the force and displacement column vectors. Using the shift operator, this can be written as the pair of equations

$$\begin{aligned} H &= m^2[d(u - \mathrm{E_X} u') + c(2 - \mathrm{E_X^{-1}} - \mathrm{E_X})u] + dhm(v - \mathrm{E_X} v') \\ V &= dhm(u - \mathrm{E_X} u') + h^2[(d+a)v - d\mathrm{E_X} v' - av'] \end{aligned} \tag{15.23}$$

it being understood that H, V, u, v, u' and v' are functions of X. Likewise

$$\begin{aligned} H' &= m^2[d(u' - \mathrm{E_X^{-1}} u) + c(2 - \mathrm{E_X^{-1}} - \mathrm{E_X})u'] + dhm(v' - \mathrm{E_X^{-1}} v) \\ V' &= dhm(u' - \mathrm{E_X^{-1}} u) + h^2[(d+a)v' - d\mathrm{E_X^{-1}} v - av] \end{aligned} \tag{15.24}$$

These equations can be written in matrix form as

---

[1] The Howe truss (patented 1840) was designed to take tension in the vertical bars. The Pratt truss (patented 1842) was designed to take tension in the diagonal bars.

$$\begin{bmatrix} d+c(2-\mathrm{E_X}-\mathrm{E_X^{-1}}) & d & -d\mathrm{E_X} & -d\mathrm{E_X} \\ d & d+a & -d\mathrm{E_X} & -d\mathrm{E_X}-a \\ -d\mathrm{E_X^{-1}} & -d\mathrm{E_X^{-1}} & d+c(2-\mathrm{E_X}-\mathrm{E_X^{-1}}) & d \\ -d\mathrm{E_X^{-1}} & -d\mathrm{E_X^{-1}}-a & d & d+a \end{bmatrix} \begin{bmatrix} um \\ vh \\ u'm \\ v'h \end{bmatrix} = \begin{bmatrix} H/m \\ V/h \\ H'/m \\ V'/h \end{bmatrix} \qquad \textbf{(15.25)}$$

It can be seen that the matrix is symmetric in the stiffness coefficients and antisymmetric in the powers of the shift operator. This is true in general and is a useful check on the correct formulation of the equations.

A number of solutions exist for which the load vector on the right-hand side of (15.25) is zero. These are the three rigid-body motions

$$um = u'm = U_0 \;\; ; \;\; vh = v'h = V_0 \;\; ; \;\; um = -u'm = \Theta_0 \;\text{ and }\; vh = v'h = 2\mathrm{X}\Theta_0 \, . \qquad \textbf{(15.26)}$$

There are also three modes corresponding to uniform extension, bending and shear. Uniform extension of the upper and lower chords is given by the deflexions

$$um = u'm = \mathrm{X}U_1 \;\; , \;\; vh = v'h = -\mathrm{X}U_1 \, . \qquad \textbf{(15.27)}$$

A pure flexural mode is given by

$$um = -u'm = \mathrm{X}\Theta_1 \;\; , \;\; vh = v'h = \mathrm{X}^2\Theta_1 \, , \qquad \textbf{(15.28)}$$

and a mode which includes shear by

$$um = -u'm = \tfrac{1}{2}(\mathrm{X}^2 + \tfrac{c}{a} + \tfrac{c}{d} - \tfrac{1}{6})\Theta_2 \;,\; vh = (\tfrac{1}{3}\mathrm{X}^3 - \tfrac{c}{2a})\Theta_2 \;,\; v'h = (\tfrac{1}{3}\mathrm{X}^3 + \tfrac{c}{2a})\Theta_2 \, . \qquad \textbf{(15.29)}$$

Note that the flexural mode may be deduced from this mode by differentiating with respect to X. Likewise, the rigid-body rotation is the differential of the flexural mode. This can be shown to be a general property of the solutions which results from the interchangeability of the order of differential and shift operations. The above solutions (15.26) to (15.29) represent the 'complementary functions' which can be used to satisfy the boundary conditions. In addition, a 'particular integral' solution will be required to satisfy the joint equilibrium conditions for the actual loading. If this loading is a polynomial function of X, then a solution in terms of polynomials in X should be sought. Likewise, loading which is a sinusoidal function of X will be matched by a sinusoidal response. For example, suppose that the only loading applied is a constant downwards force P applied to the lower joints. In this case, it is convenient to rearrange (15.25) by multiplying the last two rows by $-\mathrm{E_X}$ and redefining the deflexion vector to give the form

$$\begin{bmatrix} c(2-\mathrm{E_X}-\mathrm{E_X^{-1}}) & 2d+c(2-\mathrm{E_X}-\mathrm{E_X^{-1}}) & d(1-\mathrm{E_X}) & d(1+\mathrm{E_X}) \\ 0 & 2d & d(1-\mathrm{E_X}) & d(1+\mathrm{E_X})+2a \\ -c(2-\mathrm{E_X}-\mathrm{E_X^{-1}}) & 2d+c(2-\mathrm{E_X}-\mathrm{E_X^{-1}}) & d(1-\mathrm{E_X}) & d(1+\mathrm{E_X}) \\ 0 & 2d & d(1-\mathrm{E_X}) & d(1+\mathrm{E_X})+2a\mathrm{E_X} \end{bmatrix} \begin{bmatrix} m(u+\mathrm{E_X}u')/2 \\ m(u-\mathrm{E_X}u')/2 \\ h(v+v')/2 \\ h(v-v')/2 \end{bmatrix} = \begin{bmatrix} 0 \\ -P/h \\ 0 \\ 0 \end{bmatrix} \qquad \textbf{(15.30)}$$

The first and third rows and the second and fourth rows then yield the pair of equations

$$c(2-\mathrm{E_X}-\mathrm{E_X^{-1}})(u+\mathrm{E_X}u')m = 0 \;\; , \;\; a(1-\mathrm{E_X})(v-v')h = -P/h \, . \qquad \textbf{(15.31)}$$

As the 'complementary function' solutions have already been found, no new arbitrary terms are sought in the solution of (15.31). The solution of these equations then takes the form

$$(u+\mathrm{E_X}u')m = 0 \;\; , \;\; (v-v')h = P\mathrm{X}/ah \, . \qquad \textbf{(15.32)}$$

The remainder of the solution then follows readily, giving

$$um = -\frac{P}{6ch}(X^3 + 3X^2) \quad , \quad u'm = \frac{P}{6ch}(X^3 - 3X + 2) \,,$$
$$vh = v'h + \frac{P}{ah}X = \frac{P}{2h}\left[\left(\frac{1}{a} + \frac{1}{d}\right)(X^2 + X) - \frac{1}{6c}(X^4 + 2X^3 - 5X^2 + 2X)\right] . \tag{15.33}$$

This, combined with the earlier complementary functions gives the solution for a constant loading P at intermediate lower joints for given end conditions. For the cantilever truss with N bays shown in Figure 15.6[1], there are four geometric conditions at the left-hand end. These are

$$u(0) = v(0) = 0 \,,$$
$$u'(0) \ \{= v'(0)\ \} = 0 \,. \tag{15.34}$$

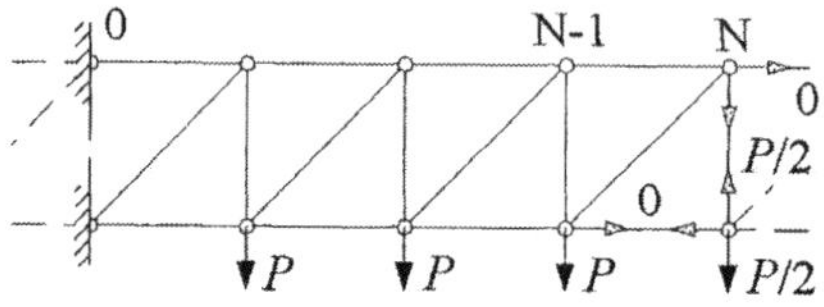

Figure 15.6 A cantilevered Howe truss.

At the right-hand end, the conditions relate to the loading. These end-load conditions may be expressed in more than one way. For convenience, they are expressed by taking the forces in the Nth bottom chord bar and the (N+1)th top chord bar as zero and that in the Nth vertical bar as $P/2$, giving

$$u(N) - u(N-1) = 0 \quad , \quad u'(N+1) - u'(N) = 0 \quad , \quad ah[v'(N) - v(N)] = P/2h \,. \tag{15.35}$$

It can be seen that seven conditions are imposed by (15.34) and (15.35), but there are only six arbitrary constants associated with the complementary functions which are to be determined from these conditions. However, it must be remembered that the governing equations relate to an infinite truss but the solution sought is for a finite truss. In the infinite truss, the 0th vertical bar must carry the resultant vertical load, so that the 0th upper and lower joints must move apart due to its extension. In the finite truss, there is no such bar and both joints remain fixed. In fact, a small vertical motion of the real 0th upper joint has no effect on the rest of the truss. Thus a local correction can be added to the solution for v' (by adding a Kronecker delta function to it for example) to satisfy the condition in curly brackets in (15.34). The remaining six conditions give the following values for the constants in the complementary functions.

$$U_0 = -\frac{P}{6ch} \quad , \quad \Theta_0 = \frac{P}{24h}\left[\frac{1}{c}(2N + 5) - 6(2N + 1)\left(\frac{1}{a} + \frac{1}{d}\right)\right] ,$$
$$\Theta_1 = -\frac{P}{6ch}(3N^2 + 2) \quad , \quad 2cU_1 = 2aV_0 = c\Theta_2 = \frac{P}{h}(N + \tfrac{1}{2}) \,. \tag{15.36}$$

The joint displacements can be deduced from this, (15.26) to (15.29) and (15.33). For example, the vertical deflexion of the Nth lower joint is given by the expression

$$v(N) = -\frac{P}{2h^2}\left[\frac{1}{2c}(N^4 + N^2) + N^2\left(\frac{1}{a} + \frac{1}{d}\right)\right] \tag{15.37}$$

This may be compared with the (upwards) displacement of a cantilevered beam of length L, bending stiffness EI and shear stiffness GkA subject to a (downwards) uniform load per unit length p. This is readily found from energy methods to be

---

[1] Note that the load at the free end has been taken as $P/2$. This corresponds to a uniform distributed load of P/m applied to each bay and transferred to the joints.

$$v(L) = -\frac{pL^4}{8EI} - \frac{pL^2}{2GkA} \tag{15.38}$$

The slope-deflexion equations for modular beams were derived in §8.5. These require the determination of the module flexibilities[1] f_{mm} and f_{ss} which correspond to $1/EI$ and $1/GkA$ for continuous beams. From (8.28), these can be found from the strain energy U_m in the Xth module. The shear mode given by (15.29) induces bar forces which imply a resultant axial force on the truss, but taken in combination with that given by (15.27), where

$$2U_1 = \Theta_2 = S/ch \tag{15.39}$$

gives the bar forces in the Xth module shown in Figure 15.7, where the Xth joints are at the right of the module. The forces exerted by the bars to the right of the module induce a resultant shear force S and a resultant moment SmX. However, using this moment as the effective moment on the module results in coupling between flexural and shear effects. Taking the module moment to be that exerted at the midpoint of the module (the *flexural centre* referred to in §8.5) uncouples the shear and bending effects. This moment, M, is then $Sm(X-½)$. The expression for the strain energy in the bars forming the module is then

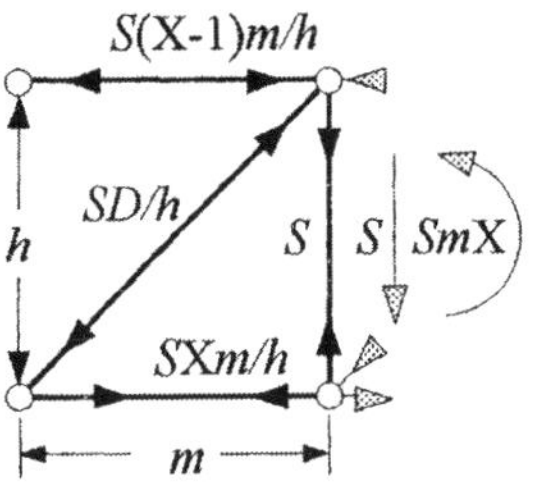

Figure 15.7 A Howe module.

$$\begin{aligned} U_m &= \frac{1}{2}\left\{\left[S\frac{m}{h}(X-1)\right]^2\frac{1}{cm^2} + \left[S\frac{D}{h}\right]^2\frac{1}{dD^2} + \left[S\frac{m}{h}X\right]^2\frac{1}{cm^2} + \frac{S^2}{ah^2}\right\} \\ &= \frac{1}{2}\left\{[S(X-\tfrac{1}{2})]^2\frac{2}{ch^2} + \frac{S^2}{h^2}\left[\frac{1}{a} + \frac{1}{2c} + \frac{1}{d}\right]\right\} = \frac{m}{2}(M^2 f_{mm} + S^2 f_{ss}) \end{aligned} \tag{15.40}$$

where the final expression is given by (8.28), the module length being m in this example. Comparing terms in the two expressions gives

$$f_{mm} = \frac{2}{ch^2m^3}\ , \quad f_{ss} = \frac{1}{h^2m}\left(\frac{1}{a} + \frac{1}{2c} + \frac{1}{d}\right)\ . \tag{15.41}$$

The point loads P are equivalent to a distributed load p of intensity P/m and the overall length of the truss, L, is Nm. Using (15.41) and the equivalent flexibilities $1/EI$ and $1/GkA$ respectively, the deflexion of the equivalent beam given by (15.38) is exactly the same as that given by (15.37) for the real truss.

This is not always the case. The general solution, where the flexural centre is not midway along the module, is given by (A5.1.2) of Appendix 5. Even then, this is not always the exact solution. It assumes that the deformation of the truss is given by the characteristic response to flexure and shear. If the truss is statically-determinate, as in the above case, these are only possible responses to such loading. Also, it is clear from Figure 15.7 that the applied shear load acts at the bottom right-hand joint of the module, so that the response, which is found from energy methods, relates to the lower joints. Also, the other solutions in §A5.1 show that the flexural response of trusses may give rise to shear-like terms not accounted for in the simple beam analogy. For example, from (8.38) it follows that the end displacement of a cantilevered modular beam in the direction of an end force F is given by

---

[1] Note that here and in Appendix 5 it will be assumed that the uncoupled flexibilities (with reference to the flexural centre) are used. In Chapter 8, a prime was used to distinguish these specific forms from the general forms.

$$v = \frac{Fm^3 f_{mm}}{12}[4N^3 - N] + FmNf_{ss} \tag{15.42}$$

in the case where e is zero. The equivalent expression for a continuous beam of length L (=mN) is

$$v = \frac{FL^3}{3EI} + \frac{FL}{GkA} \tag{15.43}$$

so that there is no equivalent of the second term in (15.42) in the continuous beam case.

The flexibilities of other forms of plane truss are given by equations (A5.2.1) to (A5.2.12) of Appendix 5. It can be seen from these equations that flexural stiffness of a truss is often (but not always) the same as that of an ordinary beam consisting of the upper and lower chords of the truss only[1]. Trusses tend to be relatively very weak in shear, so that the inclusion of the shear terms in the expressions for deflexion becomes more significant than for ordinary beams. When the truss is statically-determinate, the characteristic bar forces under combined flexure and shear can be determined directly without going through the full matrix analysis described in Example 15.2. The characteristic modes for the statically-indeterminate trusses are not the immediate responses to resultant loads, but the responses towards which the trusses tend at a number of bays from the application of these loads. This is of course true for ordinary beams, although the rate of convergence on the characteristic response is usually more rapid.

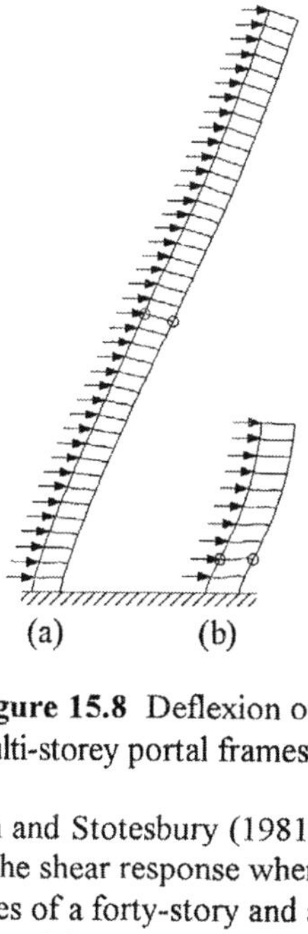

(a) (b)

Figure 15.8 Deflexion of multi-storey portal frames.

A Vierendeel truss with its axis vertical instead of horizontal becomes a multi-storey portal frame. As can be seen from (A5.2.7), its shear stiffness depends on the flexural stiffness of its component members. Usually, this means that the shear response (sometimes called 'racking') is even more significant than for ordinary trusses. Lin and Stotesbury (1981) state that the flexural response of such structures is less significant than the shear response when they have fewer than forty stories. Figure 15.8 shows the deflexion modes of a forty-story and a ten-story portal frame subject to a uniform wind load[2]. Each storey is 10m wide and 5m high and made from a beam of with a cross-sectional area of 100cm^2 and a major second moment of area of 40,000cm^4. Taking Young's modulus as 200 GPa, the stiffnesses EI and GkA found from (A5.2.7) are 100.2 GNm^2 and 15.40 MN respectively. Using these values in (15.38), the contributions of the flexural and shear deformations to the lateral deflexion at the top of each structure can be found. For the forty-storey portal, the flexural component of deflexion is 1.54 times that of the shear component, but in the case of the ten-storey portal, the shear component is 10.4 times the flexural component. The shear effect dominates in any such frame with less than 33 stories. These effects can be seen in the modes shown in Figure 15.8. Opposite curvatures are produced by each mode. In the lower part of each frame, below the ringed joints, flexural curvature dominates and in the upper part, shear curvature dominates.

[1] Saka, T. and Heki, K. (Dec.1981) *Memoirs of the Faculty of Engineering, Osaka City University* V.22 pp.167-173 obtain the correct flexural stiffnesses by treating the trusses as continua, but this process does not yield the correct shear stiffnesses owing to the essentially discrete nature of frameworks.

[2] These modes were generated from the data files R2DSISKY.STR and R2DSIS10.STR.

The axial stiffness (or flexibility) of a truss relates the total axial force on a truss to the overall stretch of each module in a uniform extension mode. To avoid coupling between extension and flexure, the line of action of the axial force is defined so that no relative rotation of the modules is induced. A suitable uniform extension mode for a Pratt (or Howe) truss is given by (15.27). This mode indicates that axial extension is coupled with a shear deflexion. This coupling will be discussed further in the next section. For most trusses, the axial stiffness is simply the sum of the stiffnesses of the upper and lower chords. The cross-braced truss shown in Figure A5.2c is an exception. In this case the axial stiffness is given by

$$EA = m^3\left(b + t + \frac{2ad}{a + 2d}\right) \tag{15.44}$$

so that the stiffnesses of all the component bars contribute to the axial stiffness in this case. Out-of-plane flexural and torsional flexibilities of plane frames can be deduced by treating the joints as rigid. Simplified expressions for these flexibilities are given in §A5.2. These properties are usually more significant in the case of space trusses.

15.4 Regular Space Trusses

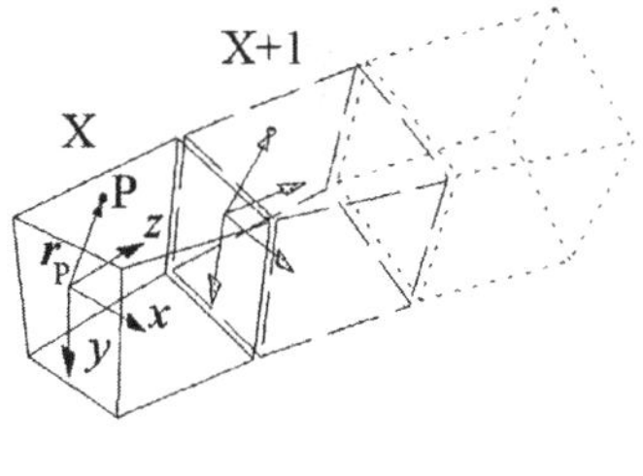

Figure 15.9 Uniform strain of a modular structure.

Examples of space trusses are shown in Figure 15.1b. In three dimensions, the torsion and extension of a truss are commonly considered, in addition to its flexure. The appropriate elastic properties can be found by considering a modular structure in a state of uniform strain. {That is, the strains in the members of module X+1 are the same as those in the corresponding members of module X. This state is the characteristic response to pure bending, torsion and extension discussed in Chapter 8. Suppose that each module has a cartesian coordinate system[1] x,y,z embedded within it which rotates and displaces with the module. Then under uniform strain, the change in position, $\boldsymbol{e}_P$, of a point P within the module relative to the module coordinates is the same for all modules. Likewise, the relative change in position, $\boldsymbol{e}_0$, of the origin and relative rotation, $\boldsymbol{\theta}$, of the coordinates in the next module appears to be the same.

Suppose that the absolute displacement of the origin of the coordinate system in the X<u>th</u> module is $\boldsymbol{u}_X$ and its rotation is $\boldsymbol{\theta}_X$. Then from small deflexion theory, the absolute displacement and rotation of the coordinates in the X+1<u>th</u> module are given by

$$\boldsymbol{u}_{X+1} = \boldsymbol{u}_X + \boldsymbol{e}_0 + \boldsymbol{\theta}_X \times m\boldsymbol{k} \quad , \quad \boldsymbol{\theta}_{X+1} = \boldsymbol{\theta}_X + \boldsymbol{\theta} \tag{15.45}$$

where m is the length of a module. It follows that $\boldsymbol{\theta}_X$ is a linear function of X, $\boldsymbol{\theta}_0$+X$\boldsymbol{\theta}$ say, but $\boldsymbol{u}_X$ is quadratic (its *increment* being linear). The same will be true of the displacement $\boldsymbol{v}_X$ of P, which is given by

$$\boldsymbol{v}_X = \boldsymbol{u}_X + \boldsymbol{e}_P + \boldsymbol{\theta}_X \times \boldsymbol{r}_P \tag{15.46}$$

where $\boldsymbol{r}_P$ is the initial position of P with respect to the embedded coordinates. From this and (15.45), the change in the absolute displacement of P between successive modules is then

$$\boldsymbol{v}_{X+1} - \boldsymbol{v}_X = \boldsymbol{e}_0 + (\boldsymbol{\theta}_0 + X\boldsymbol{\theta}) \times m\boldsymbol{k} + \boldsymbol{\theta} \times \boldsymbol{r}_P \tag{15.47}$$

---

[1] Note that as shown in Figure 15.9, the z axis is along the truss, although the module numbering along the truss is still in terms of X.

Suppose that the displacement of P is a known quadratic function of X,

$$\boldsymbol{v}_X = \boldsymbol{v}_0 + X\boldsymbol{v}_1 + X^2\boldsymbol{v}_2 \tag{15.48}$$

Let

$$\begin{aligned} &\boldsymbol{e}_0 + \boldsymbol{\theta}_0 \times m\boldsymbol{k} = m(\varepsilon_x \boldsymbol{i} + \varepsilon_y \boldsymbol{j} + \varepsilon_z \boldsymbol{k}) \ , \ \boldsymbol{\theta} = m(\vartheta_x \boldsymbol{i} + \vartheta_y \boldsymbol{j} + \vartheta_z \boldsymbol{k}) \ , \\ &\boldsymbol{r}_P = m(x_P \boldsymbol{i} + y_P \boldsymbol{j} + z_P \boldsymbol{k}) \ , \ \boldsymbol{v}_1 = u_1 \boldsymbol{i} + v_1 \boldsymbol{j} + w_1 \boldsymbol{k} \ , \ \boldsymbol{v}_2 = u_2 \boldsymbol{i} + v_2 \boldsymbol{j} + w_2 \boldsymbol{k} \ . \end{aligned} \tag{15.49}$$

Then on substituting (15.48) into (15.47) and comparing like coefficients on both sides of the equation,

$$\begin{aligned} 2u_2 &= m^2\vartheta_y \ , \quad 2v_2 = -m^2\vartheta_x \ , \quad w_2 = 0 \ , \\ u_1 &= m(\varepsilon_x + \vartheta_y z_P - \vartheta_z y_P) - \tfrac{1}{2}m^2\vartheta_y \ , \\ v_1 &= m(\varepsilon_y - \vartheta_x z_P + \vartheta_z x_P) + \tfrac{1}{2}m^2\vartheta_x \ , \\ w_1 &= m(\varepsilon_z + \vartheta_x y_P - \vartheta_y x_P) \ . \end{aligned} \tag{15.50}$$

Now $m\varepsilon_z$ is the longitudinal component of $\boldsymbol{e}_0$ and so ε_z is the longitudinal strain[1]. This and the rates of rotation ϑ_x, ϑ_y and ϑ_z are the distortion terms in (8.12).

The related flexibility coefficients can then be found as follows. The stiffness matrix equations for the joints of a typical module of the truss are formed[2] for the case where no joint loads are applied. The most general solution of these equations is sought in terms of the joint displacements as quadratic functions of X. From the stiffness matrix relationships, the bar forces can be found and hence the resultant loads on the truss, M_x, M_y, T and P in (8.12),determined in terms of coefficients of these quadratics. Using (15.50), the relationships between the resultant loads and the distortion terms are then found.} This process has been fully automated[3] and the results so found for trusses with a regular polygonal cross-section, as well as some earlier results obtained manually, are given in Appendix 5.

The shear flexibility coefficients are found by the strain-energy methods described in §8.4. The characteristic modes for shear (with flexure) have strains which vary linearly with X which result from the varying bending moment. Suppose that the characteristic response of the X<u>th</u> module to a shear force S applied at the end is $\mathcal{S}_X$. The characteristic response to an equal and opposite shear force applied one module further along will be $-\mathcal{S}_{X-1}$. The characteristic response to both applied simultaneously is then $(\mathcal{S}_X - \mathcal{S}_{X-1})$ or $(1-E_X^{-1})\mathcal{S}_X$. This must be the characteristic response to a moment of magnitude Sm. The response to a unit moment, $\mathcal{M}_X$, is then $(1-E_X^{-1})\times \mathcal{S}_X/Sm$. Conversely, if $\mathcal{M}_X$ is known, $\mathcal{S}_X$ can be found by operating on $\mathcal{M}_X$ with $(1-E_X^{-1})^{-1}$. The required inverse operations are given by (8.36). The additional unknown constant terms which arise are evaluated by satisfying the joint equilibrium equations. In three dimensions, there may coupling between flexure about an axis and shear in the direction of that axis. In this case, use of the flexural centre does not uncouple these effects.

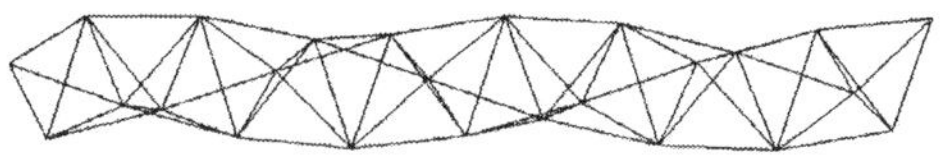

Figure 15.10 A tetrahedral truss.

---

[1] As ε_x and ε_y are related to the rigid-body rotation $\boldsymbol{\theta}_0$, they do not necessarily correspond to γ_x and γ_y.

[2] For pin-jointed frames, these are most conveniently expressed in terms of the axial coefficients, EA/l^3, using the matrix equations given by (11.35) and (11.36).

[3] Renton, J.D. (1995) Automated Derivation of Structural Formulae. *Computers & Structures* V.56, p.959. (This paper also lists some results for trusses where the height/width ratio of the cross-section is variable.)

There are regular trusses for which the beam-like elastic response can only be partially specified. Figure 15.10 shows a truss made from bars of equal length l forming a series of regular tetrahedra, each sharing a common face with the next. The pattern is repeated along the axis of the truss but always at a different orientation. In this case, the only overall elastic response which can be determined relates to torsion and extension. Suppose that all the bars have the same cross-sectional area A and Young's modulus E. Then this relationship is given by

$$\begin{bmatrix}\vartheta_z\\ \varepsilon_z\end{bmatrix} = \frac{1}{450EAl^2\sqrt{10}}\begin{bmatrix}9800 & 690l\sqrt{2}\\ 690l\sqrt{2} & 639l^2\end{bmatrix}\begin{bmatrix}T\\ P\end{bmatrix} \qquad \textbf{(15.51)}$$

Figure 15.11 shows a similar truss known as an inverted batten beam. In this case, the pattern is repeated at the same orientation every two bays, so that it is possible to define a two-bay module and determine all the overall elastic characteristics of the truss in terms of it. The truss consists of a series of alternating equilateral triangles in planes normal to its axis. The bars forming these triangles each have an axial coefficient b. The bars linking these triangles, shown dotted in Figure 15.11, each have an axial coefficient d. The non-zero flexibility coefficients in (8.12) are then

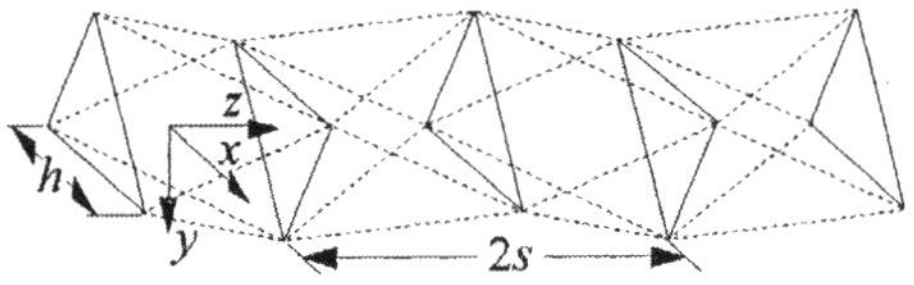

Figure 15.11 An inverted batten beam.

$$f_{11}=f_{22}=\frac{4}{9h^2s^3}\left(\frac{2}{b}+\frac{3}{d}\right),\quad f_{33}=\frac{2}{h^4sd},\quad f_{44}=f_{55}=\frac{1}{h^2sd},\quad f_{66}=\frac{1}{54s^3}\left(\frac{2}{b}+\frac{9}{d}\right). \qquad \textbf{(15.52)}$$

Hoff[1] examined the transmission of load systems without a resultant through space trusses. The first example[2] he takes is shown in Figure 15.12. The diagonal bars are shown dashed (or dotted). The only two diagonal bars across the section brace the end faces. Without these, the truss would be a mechanism, the cross-section being free to form a rhombus. The truss is loaded by the four axial forces shown, which have no resultant. The frame is statically-determinate and the tensions in the right-hand chord members are as shown. The linear decay can be expressed by the form $\pm(A+BX)$ where A is 100, B is -25 and the count of X starts from zero or one. Hoff then cross-braces all the face panels, including the end ones, adding eighteen redundancies to the structure. The tensions in the chords again decay linearly where A is now 87.5 and B is again -25.

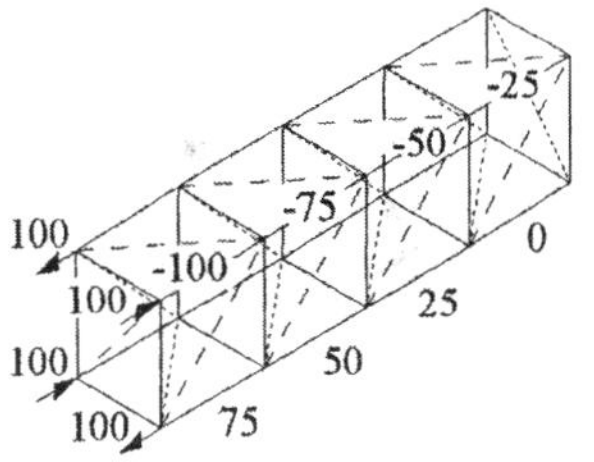

Figure 15.12 Warping of Hoff's truss.

For Saint-Venant's principle to apply, so that the characteristic responses of the truss to resultants typify its behaviour, a stronger rate of decay is necessary. The problem with this truss is that the cross-section is only braced at the ends. The intermediate modules have no resistance to distortion of their cross-sections. Vlasov (1961) illustrates the equivalent problem for a continuum in his Figure 151. A bimoment of exactly the kind applied to the truss is applied to a

---

[1] Hoff, N.J. (1945) The Applicability of Saint-Venant's Principle to Airplane Structures. *Journal of Aeronautical Sciences*, V.12 p.455.

[2] This is given by the data file P3DSIHOF.STR. In all configurations, the bars are made from the same material but the diagonal bars have half the cross-sectional area of the rest.

very thin circular tube. This causes it to become oval, narrowing along the lines of action of the compressive pair of loads and widening where the tensile pair are applied.

This lack of resistance to distortion of the cross-section is why Saint-Venant's principle does not apply to this truss. The analysis of §5.9 would indicate that normally, the response of a beam to an end bimoment should be one of *exponential* decay of the stresses and strains. In the present case, this can be achieved by bracing the transverse panels. Suppose that there are diagonals in all these panels, in the same direction as the first panel, adding three redundancies to the initial configuration. Decay of the axial loads is now far more rapid, in the compression chords being -100, -48, -20.9 and -7.4. An exponential decay could be inferred from this. Exponential decay solutions of the finite-difference equations of the truss can be found from the last line of Table 15.1. However, the value of a is usually complex, implying damped harmonic decay. For example, if all the bays (both transverse and on the faces) are cross-braced, a is $(3.074 \pm 2.149i)$.

This indicates that in order to use the characteristic response of trusses to resultant loads, as in this and the last section, each module must have structural integrity. In the first two examples, the internal modules formed mechanisms which meant that they could not individually sustain load systems without resultants. For structural integrity, each module must be capable of resisting any such loading, so that its effects decay rapidly along the truss. All the modules listed in §A5.3 have such structural integrity.

15.5 Membranes and Cable Nets

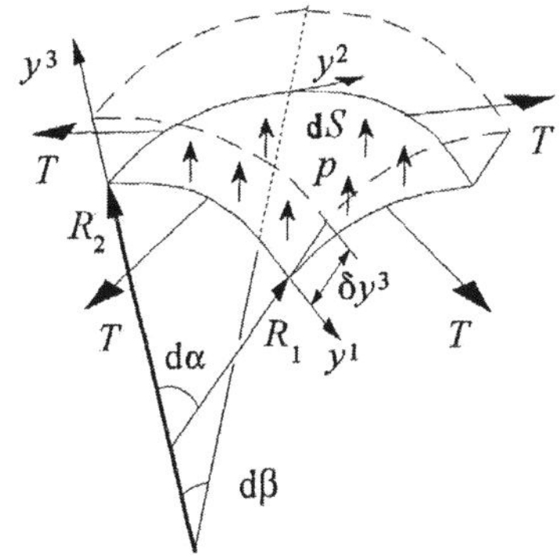

Figure 15.13 A membrane element under pressure.

Membranes and cable nets are often used to provide lightweight roofing structures. They derive their stiffness from being in tension. This is usually provided by pretensioning at the supports or by air-pressure loading. The general equations of equilibrium of membranes are most readily expressed in terms of tensor notation (see Green & Zerna (1968) Chapter 12 or Renton (1987) §3.9 for example).

The simplest form of tension membrane is a soap film. It may be taken as weightless and to have a uniform surface tension. This is given by the membrane force per unit length T in Figure 15.13. This shows a small element of surface of area dS bounded by the coordinate lines y^1, y^2 of an orthogonal coordinate system. The lines of the third ordinate y^3 are then normal to this surface. Consider the equilibrium of this element under the action of a normal pressure, p, in the direction of y^3. Suppose that the element is given a virtual displacement so that its coordinates change by δy^3 in the y^3 direction only. The work done by the pressure is then $p\, dS\, \delta y^3$. This may be equated to the work done against the surface tension in stretching the element to its new area shown by the broken lines in Figure 15.13[1]. This is given by the change of area of dS multiplied by T, leading to the equation

$$p\, dS\, \delta y^3 = T \frac{\partial (dS)}{\partial y^3} \delta y^3 \tag{15.53}$$

For example, in the case of a sphere of radius R, where (y^1, y^2, y^3) are the spherical coordinates

[1] This is similar to the process of equating the external work done to the internal work done on plastic hinges in plastic collapse analyses.

(θ,ϕ,r),

$$dS = R^2 \sin\theta \, d\theta \, d\phi \;, \quad \frac{\partial(dS)}{\partial y^3} = 2R \sin\theta \, d\theta \, d\phi \;, \qquad \textbf{(15.54)}$$

so that (15.53) gives T as $pR/2$.

Integrating (15.53) over the whole surface gives

$$p\delta V = T\delta S \qquad \textbf{(15.55)}$$

where δV is the change in the pressurised volume and δS is the change in surface area, as measured from the equilibrium state. Suppose that p is zero. Then for *any* small perturbation from the equilibrium state, δS is zero. The equilibrium state is then one of minimum surface area for the given boundary conditions[1]. Moreover, from (15.53), the change in area of the element dS in moving out by δy^3 is zero. If the radii of curvature of the y^1 and y^2 coordinate lines are R_1 and R_2 respectively, and the element sides subtend angles $d\alpha$ and $d\beta$ as shown in Figure 15.13, then

$$dS = R_1 d\alpha \, R_2 d\beta \;,$$
$$\frac{\partial(dS)}{\partial y^3} = \lim_{\delta y^3 \to 0} \{[(R_1 + \delta y^3)(R_2 + \delta y^3) - R_1 R_2] d\alpha \, d\beta / \delta y^3\} = (R_1 + R_2) d\alpha \, d\beta \,. \qquad \textbf{(15.56)}$$

Then if p is zero, the variation of dS is zero so that R_1 and R_2 are equal and opposite. This means that the principal radii of curvature, which are the greatest and least (or most negative) local radii of curvature, are opposite in sign. Then the *Gaussian curvature,* which is the product of the inverse principal radii, is negative. Such membranes are typically saddle-shaped. These forms have been generated numerically[2] and give a close estimate of the optimum shape of membrane and cable-net structures. If there is a pressure p it follows from (15.53) and (15.56) that the sum of the radii of curvature is positive. Also, (15.55) implies that the surface area is a minimum for a given pressurised volume (δV=0) and given boundary conditions. Again, a spherical soap bubble is a simple example of this.

The principal uses of cable nets are either as the main structural component of roofing or as stiffening for a membrane[3]. In the former case, the tension in the cable mainly arises from pre-tensioning. In the latter case, it arises mainly due to its interaction with the membrane, so that its elastic response is more significant in determining the state of stress.

When the cable net is the main structural component, it behaves in an analogous manner to a membrane. Consider a two-way net where the distance d between the nodes and the tension T in the cables is the same throughout. The tension coefficient t (=T/d) is then the same for all components. Figure 15.14 shows the loading at a typical node of such a cable net. Three nodes in any one plane form the base vertices of identical adjoining equilateral triangles. The common sides of these two pairs of triangles lie on

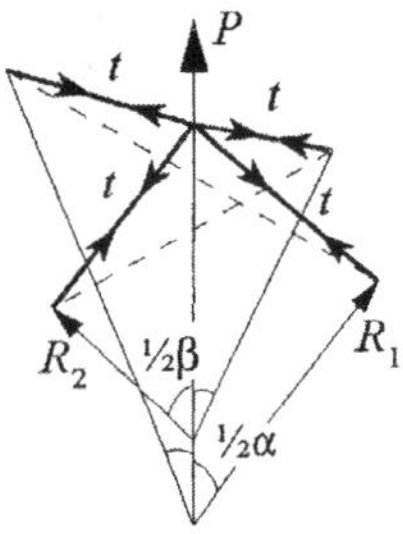

Figure 15.14 Equilibrium of elements of a cable net.

[1] This was first noted by Plateau in 1873.

[2] See for example Lewis, W.J. and Lewis, T.S. (1996) Application of Formian and Dynamic Relaxation to the Form-Finding of Minimal Surfaces. *Journal of the I.A.S.S.* V.37 No.3 p.165.

[3] See Saitoh, M. (1998) Recent developments in Tension Structures. *Proc. I.A.S.S. Colloquium, Current and Emerging Technologies of Shell and Spatial Structures. Madrid, April 1997* p.105

a line through the common node. A force P acts at the node radially outwards along this line. The equilibrium of the node is then given by

$$P = 2t[R_1(1 - \cos\alpha/2) + R_2(1 - \cos\beta/2)] \tag{15.57}$$

where the lengths of the equal sides of the triangles are R_1 and R_2 and subtend the angles α and β respectively (cf. (15.53) and (15.56)). Again, in the absence of P, R_1 and R_2 must then be equal and opposite in sign.

A closer analogy between membranes and cable nets can be drawn when the normal loading induces only small displacements w from an initially flat form. Figure 15.15 shows a curve of radius of curvature R whose slope and vertical position is given by θ and w respectively. Then

$$\frac{dw}{dx} = \tan\theta\ , \quad \frac{d^2w}{dx^2} = \frac{d\theta}{dx}\sec^2\theta = \frac{d\theta}{dx}\left[1 + \left(\frac{dw}{dx}\right)^2\right],$$

$$R\delta\theta = \sqrt{1 + \left(\frac{dw}{dx}\right)^2}\,\delta x, \quad \text{so that} \quad \frac{1}{R} = \frac{\dfrac{d^2w}{dx^2}}{\left[1 + \left(\dfrac{dw}{dx}\right)^2\right]^{3/2}} \tag{15.58}$$

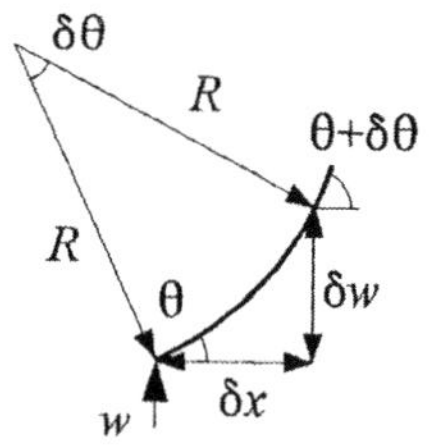

Figure 15.15 Gradients and curvature.

If the slope dw/dx is small, then the denominator in the last expression may be omitted. Taking the y^1 and y^2 directions to be x and y in the initial plane, it follows from (15.53) and (15.56) that

$$p = T\left(\frac{1}{R_1} + \frac{1}{R_2}\right) \approx -T\left(\frac{\partial^2 w}{\partial x^2} + \frac{\partial^2 w}{\partial y^2}\right) \tag{15.59}$$

noting that the senses of the radii of curvature are different in Figures 15.13 and 15.15.

The exact equilibrium equation for the X<u>th</u> node of a cable in terms of its deflexion w_X and that of the adjacent nodes can be found from Figure 15.16. If the cable tension coefficient is t, then

$$P = t(w_X - w_{X-1}) + t(w_X - w_{X+1}) = t(2 - E_X - E_X^{-1})w_X \tag{15.60}$$

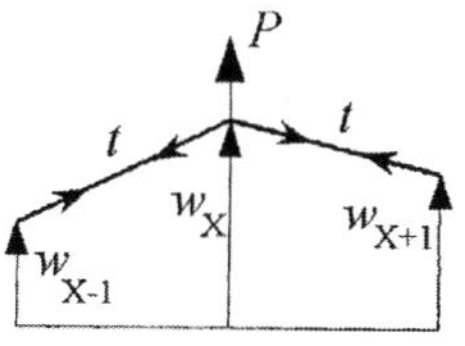

Figure 15.16 Equilibrium of a cable node.

The shaded area in Figure 15.17a shows a module of a square cable net. A second counter, Y, is introduced so that an increment of one in X or Y corresponds to a shift of l in the directions of cartesian axes x or y. Then using (15.60),

$$P = [t(2 - E_X - E_X^{-1}) + t(2 - E_Y - E_Y^{-1})]w \tag{15.61}$$

where E_Y is the shift operator in the Y direction and w is understood to be a function of X and Y. The shaded module is of area l^2 and so the point load P corresponds to the resultant of a pressure p of intensity P/l^2. Then on using expansions of the kind given by (15.14) and taking l to be small,

$$p = -t\left(\frac{\partial^2 w}{\partial x^2} + \frac{\partial^2 w}{\partial y^2}\right) + O(l^2) \tag{15.62}$$

which may be compared with the membrane equation (15.59).

A triangular form of cable net can be analysed in terms of the unit shown in Figure 15.17b. Again, the lengths of cable between nodes will be taken as l. The hexagonal shaded zone defines the module associated with a typical node and has an area of $\sqrt{3}l^2/2$. Unit steps in the x and y directions are $l/2$ and $\sqrt{3}l/2$ respectively. The finite difference equation of equilibrium is now

$$P = t[6 - E_X^2 - E_X^{-2} - E_XE_Y - E_X^{-1}E_Y^{-1} - E_XE_Y^{-1} - E_X^{-1}E_Y]w \qquad \textbf{(15.63)}$$

Taking l as small, the shift operators can again be expanded to give

$$p = -\sqrt{3}t\left(\frac{\partial^2 w}{\partial x^2} + \frac{\partial^2 w}{\partial y^2}\right) + O(l^2) \qquad \textbf{(15.64)}$$

Hexagonal nets have two types of node which differ in orientation. The first kind, loaded by forces P, are shown by the black dots in Figure 15.17c. The second kind, loaded by forces Q, are shown by white dots. In general, P and Q can be different functions of X and Y, as can their displacements w and v respectively. The net consists of repetitions of the basic module shown by the shaded rhombus. Taking all the cable lengths between nodes to be of length l, the area of this rhombus is $3\sqrt{3}l^2/2$ and the unit steps in the x and y directions are $\sqrt{3}l/2$ and $l/2$ respectively. There are now two equilibrium equations for the module, one for each typical joint:

$$\begin{aligned} P(X,Y) &= t[3w - (E_XE_Y + E_X^{-1}E_Y + E_Y^{-2})v] \\ Q(X+1,Y+1) &= t[3E_XE_Yv - (1 + E_X^2 + E_XE_Y^3)w] \end{aligned} \qquad \textbf{(15.65)}$$

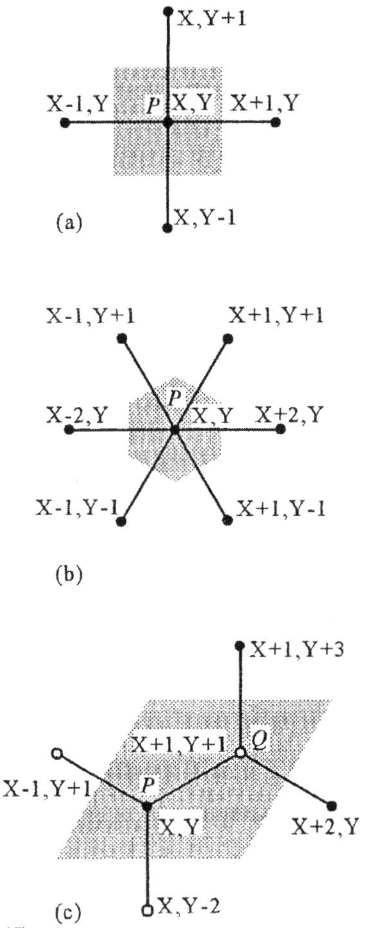

Figure 15.17 Cable net modules.

If Q is the same function as P, the first equation can be operated on by E_XE_Y and compared to the second equation, giving

$$3E_XE_Yv + (E_X^2E_Y^2 + E_Y^2 + E_XE_Y^{-1})v = 3E_XE_Yw + (1 + E_X^2 + E_XE_Y^3)w \qquad \textbf{(15.66)}$$

Expanding the shift operators as before gives the same second-order differential expressions for v and w in (15.66). Then v and w can be taken as the same functions to the same degree of accuracy as before, leading to the differential equation

$$p = -\frac{t}{\sqrt{3}}\left(\frac{\partial^2 w}{\partial x^2} + \frac{\partial^2 w}{\partial y^2}\right) + O(l) \qquad \textbf{(15.67)}$$

It will be seen that (15.62), (15.64) and (15.67) are all harmonic equations. As noted in §15.2, it follows from symmetry considerations that this is the only form these second-order equations could take. Although the coefficients of the differentials differ, all three networks are equally structurally efficient. That is, assuming that the working stress is the same in all three cases, the volume of cable per unit net area required to produce the same resistance to deflexion is the same for each. This might be inferred from (iii) of §14.2 and the work of Cox (1965) mentioned at the end of §14.3. However, it is not true of grids, which are not pure tension structures.

The existence of an approximate differential equations governing the behaviour of discrete structures suggests that solutions of these equations may indicate the sort of solutions to look for in solving the original exact finite difference equations. For example, the solution of (15.62) for a constant pressure p_o with a fixed circular boundary of radius r is given by

$$w = -\frac{p_o}{4t}\left(x^2 + y^2 - r^2\right) \qquad \textbf{(15.68)}$$

Similar solutions exist for the three types of cable net. Taking R to be r/l and P to have a constant value P_o at all nodes, the finite difference equation (15.61) for the square net is satisfied by

$$w = -\frac{P_o}{4t}(X^2 + Y^2 - R^2) \qquad \textbf{(15.69)}$$

Likewise, for the triangular net, (15.63) is satisfied by

$$w = -\frac{P_o}{18t}(X^2 + 3Y^2 - 4R^2) \qquad \textbf{(15.70)}$$

and when P and Q both take the constant value P_o, (15.65) for the hexagonal net is satisfied by

$$v = w = -\frac{P_o}{12t}(3X^2 + Y^2 - 4R^2) \qquad \textbf{(15.71)}$$

Of course, not all the nodes will lie on a circle of radius r, but the above three solutions will still give accurate estimates of the deflexions of these nets supported on such a circular boundary. There are other exact solutions for cable nets which are the equivalents of membrane solutions, such as deflexion under constant pressure with support on an elliptic boundary and the response to sinusoidal loading with a rectangular boundary.

15.6 Plane Grids

Elllington and McCallion[1] were the first to use finite-difference calculus to analyse grillages. These were rectangular grillages in which the torsional stiffness of the members was ignored. Thein Wah[2] and the author[3] extended the analysis to take account of the torsional response of the beams. When they have no torsional stiffness, the two sets of beams only have a shear interaction at the joints. This means that the angle at which they cross has no effect on the solution. Here, the torsional stiffness of the beams will be taken into account and it will be assumed that the two sets of beams forming the grid cross orthogonally, as shown in Figure

---

[1] Ellington, J.P. and McCallion, H. (1957) Moments and Deflections of a Simply-Supported Beam Grillage. *Aeronautical Quarterly* V.8, p.360.

[2] Wah, T. (1964) Analysis of Laterally Loaded Gridworks. *Proc. A.S.C.E.* V.90, No. EM2, p.83.

[3] Renton, J.D. (1964) A Finite Difference Analysis of the Flexural-Torsional Behaviour of Grillages. *Int. J. Mech. Sci.* V.6, p.209.

15.18. Each joint has three degrees of freedom, the rotations ϕ and ψ about the X and Y axes and the downwards displacement w. Gridworks are usually loaded by forces normal to their planes and these can be represented by joint loads P. These degrees of joint freedom and the joint loads will be taken as functions of X and Y. The stiffness matrix equations for grids are given by (11.27). For beams in the X direction, ϕ,ψ and w correspond to θ_x, $-\theta_y$ and δ. For beams in the Y direction, ϕ,ψ and w correspond to θ_y, θ_x and δ. Then in terms of the step functions E_X and E_Y, the equilibrium of a typical joint is given by

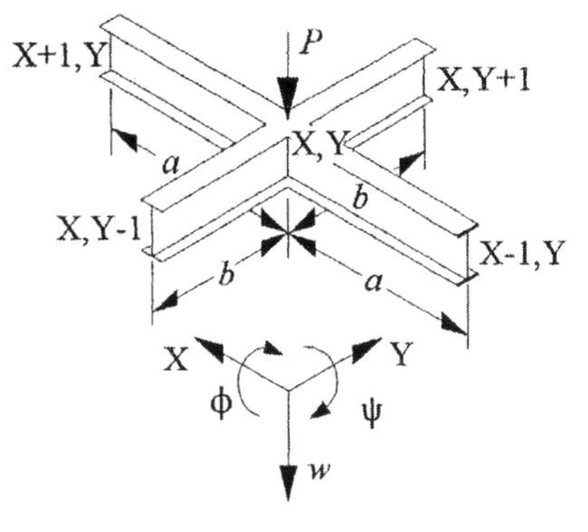

Figure 15.18 Detail of a rectangular grid.

$$\begin{bmatrix} -\dfrac{GJ_1}{a}\Delta_X + \dfrac{2EI_2}{b}(6+\Delta_Y) & 0 & \dfrac{6EI_2}{b^2}(E_Y^{-1}-E_Y) \\ 0 & -\dfrac{GJ_2}{b}\Delta_Y + \dfrac{2EI_1}{a}(6+\Delta_X) & \dfrac{6EI_1}{a^2}(E_X-E_X^{-1}) \\ \dfrac{6EI_2}{b^2}(E_Y-E_Y^{-1}) & \dfrac{6EI_1}{a^2}(E_X^{-1}-E_X) & -\dfrac{12EI_1}{a^3}\Delta_X - \dfrac{12EI_2}{b^3}\Delta_Y \end{bmatrix} \begin{bmatrix} \phi \\ \psi \\ w \end{bmatrix} = \begin{bmatrix} 0 \\ 0 \\ P \end{bmatrix} \qquad \textbf{(15.72)}$$

where $\Delta_X = E_X + E_X^{-1} - 2$, $\Delta_Y = E_Y + E_Y^{-1} - 2$ (cf. (8.26)).

and the subscripts 1 and 2 refer to the stiffnesses of the beams in the X and Y directions respectively. As in the last section, the finite-difference operators can be expanded as a series of differential operators to give an approximate differential equation relating w to the equivalent normal pressure p on the grillage[1] (equal to P/ab). This takes the form

$$\frac{EI_1}{b}\frac{\partial^4 w}{\partial x^4} + \left(\frac{GJ_1}{b} + \frac{GJ_2}{a}\right)\frac{\partial^4 w}{\partial x^2 \partial y^2} + \frac{EI_2}{a}\frac{\partial^4 w}{\partial y^4} + O(\lambda^2) = p \qquad \textbf{(15.73)}$$

where λ may be a or b. This is the form of the differential equation for a laterally-loaded orthotropic plate and if

$$\frac{EI_1}{b} = \frac{1}{2}\left(\frac{GJ_1}{b} + \frac{GJ_2}{a}\right) = \frac{EI_2}{a} \qquad \textbf{(15.74)}$$

it becomes the equation for an isotropic plate (see (A8.1.8)). This is biharmonic in form, which as noted in §15.2 is the most general fourth order equation to be rotationally symmetric for all angles of rotation. This similarity to the plate equation suggests that solutions to plate problems may indicate solutions to grid problems and this is indeed the case.

The use of Fourier series can be illustrated by examining the problem of a simply-supported rectangular grid. From (15.73), it would seem probable that if the normal loading can be expressed by sinusoidal functions, the normal displacement of the grid, w, may also be expressed by sinusoidal functions of the same wavelength. Suppose that w is given by the product $\sin r\mathrm{X} \sin s\mathrm{Y}$. From Table 15.1, operating on this function with $(E_X - E_X^{-1})$ produces a cosine

[1] Renton, J.D. (1965) On the gridwork analogy for plates. *J. Mech. Phys. Solids*, V.13 p.413. This analogy has been noted by other authors, for example Bareš and Massonnet (1966) §5.2.

function of X and operating on it with $(E_Y - E_Y^{-1})$ produces a cosine function of Y. However, operations with $\boldsymbol{\Delta}_X$ or $\boldsymbol{\Delta}_Y$ only change the coefficients of these sine functions. Then looking at the first two lines of (15.72) it follows that ϕ will be given by some multiple of $\sin rX \cos sY$ and ψ by some multiple of $\cos rX \sin sY$. The general form of such a solution is then

$$\phi(X,Y) = \phi_{rs} \sin rX \cos sY \quad , \quad \psi(X,Y) = \psi_{rs} \cos rX \sin sY$$
$$w(X,Y) = w_{rs} \sin rX \sin sY \quad , \quad P(X,Y) = P_{rs} \sin rX \sin sY \tag{15.75}$$

Substituting these forms into (15.72) yields the relationships

$$\left(\frac{GJ_1}{a}(1 - \cos r) + \frac{EI_2}{b}(2 + \cos s) \right) \phi_{rs} = \left(\frac{6EI_2}{b^2} \sin s \right) w_{rs}$$
$$\left(\frac{GJ_2}{b}(1 - \cos s) + \frac{EI_1}{a}(2 + \cos r) \right) \psi_{rs} = -\left(\frac{6EI_1}{a^2} \sin r \right) w_{rs} \tag{15.76}$$
$$-\left(\frac{EI_2}{b^2} \sin s \right) \phi_{rs} + \left(\frac{EI_1}{a^2} \sin r \right) \psi_{rs} + 2\left(\frac{EI_1}{a^3}(1 - \cos r) + \frac{EI_2}{b^3}(1 - \cos s) \right) w_{rs} = \frac{P_{rs}}{12}$$

The first two equations can be used to find ϕ_{rs} and ψ_{rs} in terms of w_{rs}. The third equation can then be used to find w_{rs} in terms of P_{rs}. Suppose that the solution is to be used to analyse a rectangular grid with M spans in the X direction and N spans in the Y direction. Simple supports can be simulated by antisymmetrical boundary conditions on X=0 or M and Y=0 or N. These are given by taking r as $\pi R/M$ and s as $\pi S/N$ in (15.75), where R and S are integers. The use of finite Fourier series was explained in §15.2. Extending this to two dimensions, any function P(X,Y) can be expressed as a finite sum of functions like that in (15.75) by

$$P(X,Y) = \sum_{R=1}^{M} \sum_{S=1}^{N} P_{rs} \sin rX \sin sY \quad \text{where} \quad P_{rs} = \frac{4}{MN} \sum_{X=1}^{M} \sum_{Y=1}^{N} P(X,Y) \sin rX \sin sY \tag{15.77}$$

and r and s are as defined above. This gives P as zero on the boundary (where the value of P is irrelevant). In particular, if P takes the constant value P_0, from (15.6),

$$P_{rs} = \frac{4}{MN} \cot\frac{r}{2} \cot\frac{s}{2} \quad \text{(R and S both odd)} \tag{15.78}$$

If either R or S is even, P_{rs} is zero. Having found P_{rs} , the deflected form can be found as described above. This is equivalent to the Navier solution for plates (see Timoshenko and Woinowsky-Krieger (1959) §28 for example).

In many cases, the torsional stiffness of the beams forming the grid is negligible in comparison with their flexural stiffness. Also, the beams and their spacing are often the same in both directions, so that EI_1/b and EI_2/a are both equal to D^* say. Then (15.73) becomes

$$D^* \left(\frac{\partial^4 w}{\partial x^4} + \frac{\partial^4 w}{\partial y^4} \right) \approx p \tag{15.79}$$

If the x and y axes are not aligned with the beams but at 45° to them, from (15.17) this becomes

$$\tfrac{1}{2} D^* \left(\frac{\partial^4 w}{\partial x^4} + 6 \frac{\partial^4 w}{\partial x^2 \partial y^2} + \frac{\partial^4 w}{\partial y^4} \right) \approx p \tag{15.80}$$

A Navier-type solution of these equations can be found for rectangular boundaries of side lengths A by B. It uses a double Fourier sine series for w which automatically gives simply-supported boundary conditions on x=0, x=A, y=0 and y=B. This takes the form

$$w = \sum_{m=1}^{\infty}\sum_{n=1}^{\infty} w_{mn} \sin\frac{m\pi x}{A} \sin\frac{n\pi y}{B} \tag{15.81}$$

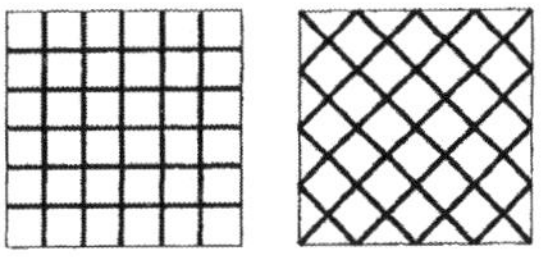

(a) (b)

Figure 15.19 Grid layouts.

If the normal pressure p is expressed in terms of a similar series with coefficients p_{mn}, then on substituting the two series into (15.79) and comparing them term by term gives

$$w_{mn} = \frac{p_{mn}}{\pi^4 D^*\left(\dfrac{m^4}{A^4} + \dfrac{n^4}{B^4}\right)} \tag{15.82}$$

and doing the same with (15.80) gives

$$w_{mn} = \frac{2p_{mn}}{\pi^4 D^*\left(\dfrac{m^4}{A^4} + 6\dfrac{m^2 n^2}{A^2 B^2} + \dfrac{n^4}{B^4}\right)} \tag{15.83}$$

Consider the particular case when the sides of the grid are both of length A and a uniform load per unit area p_0 is applied. The coefficient p_{mn} is then given by

$$p_{mn} = \frac{16p_0}{\pi^2 mn} \tag{15.84}$$

The two layouts described by (15.79) and (15.80) are shown by (a) and (b) respectively in Figure 15.19. From this analysis, the central deflexions are given by

$$\text{(a)}: \; w = 0.008204\frac{p_0 A^4}{D^*} \quad \text{and} \quad \text{(b)}: \; w = 0.004016\frac{p_0 A^4}{D^*} \tag{15.85}$$

respectively, so that a diagonal grid is more than twice as stiff as square grid using the same beams and spacing. The higher efficiency of such grids is noted by Morgan (1968) p.217.

Other solutions of the finite-difference equations for grids are listed in §A5.4. These include square grids with uniform loading and circular boundaries, triangular grids with rectangular and triangular boundaries, and the continuum approximations to these and other grids. Note that the ratio of the equivalent plate stiffness of an equilateral triangular grid to that of a regular hexagonal grid made from the same beams is given by

$$\frac{D_{\text{tri}}}{D_{\text{hex}}} = \frac{3(3EI + GJ)(EI + GJ)}{2EI(EI + 3GJ)} \tag{15.86}$$

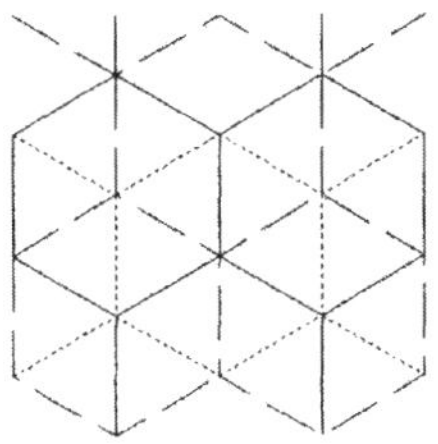

A
s
c

Figure 15.20 Triangles and hexagons.

an be seen from Figure 15.20, a pattern of equilateral triangles can be produced from overlaying three patterns of regular hexagons of the same side length. Then if this ratio is to give a structural efficiency in terms of stiffness for a given amount of material per unit area, it must be divided by three. The ratio of GJ to EI varies between zero and one. At the lower limit, a triangular grid is then 1.5 times as efficient as a hexagonal grid and at the upper limit they are equally efficient.

15.7 Space Grids

Space grids are three-dimensional versions of plane grids. Makowski (1965) and Borrego (1968) list many types of space grid. Sometimes, the definition is confined to grids which are essentially three-dimensional, rather than being an assemblage of plane structures. Four basic forms of these are shown in Figure 15.1d. These consist of a top deck of bars forming regular polygons (shown by solid lines), a bottom deck of bars also forming regular polygons (shown by broken lines) and diagonal bars forming pyramids linking the two decks (shown dotted). Stressed-skin construction is sometimes used, in which some (or all) of the polygonal faces are covered with plates which form an integral part of the structure. The parallel square mesh (PSM) truss shown at the top left of Figure 15.1d is also shown in Figure 15.22. Proprietary versions of this are the Unistrut and Space Deck systems. In its skeletal form, it has one degree of freedom which allows it to form a mechanism. The top and bottom square meshes can deform into rhombuses of opposite orientation, causing warping of the upper and lower planes. The addition of a single diagonal bar to one of these squares, or the imposition of simply-supported boundary conditions are enough to prevent this. An inclined square mesh (ISM) grid is shown to the right of the PSM grid. Here, the upper square mesh has a side length $\sqrt{2}$ times as long as the lower mesh and oriented at 45° to it. In its free state, there are modes corresponding to the warping mode of the PSM grid. An additional mode is shown in Figure 15.21. Each pyramid twists about its apex by an equal and opposite amount to that of its neighbours. This rotation can be represented by a function $(-1)^{X+Y}\theta$, which alternates at the same pitch as the modules of the grid (cf. Figure 15.2). Because of this, it cannot be represented by the behaviour of an equivalent continuum, because the scale of this behaviour becomes smaller as the fineness of the mesh increases. Both the remaining two grids shown in Figure 15.1d are formed by linking tetrahedra and are not mechanisms in their unsupported states. At bottom left is shown a parallel triangular mesh (PTM) grid. The Octetruss and Triodetic system are proprietary versions of this. Lastly, the T-H grid at the bottom right of the figure consists of a triangular upper grid forming the bases of downwards-pointing tetrahedra, linked at their apexes by a hexagonal lower grid. This grid is often used in constructing geodesic domes.

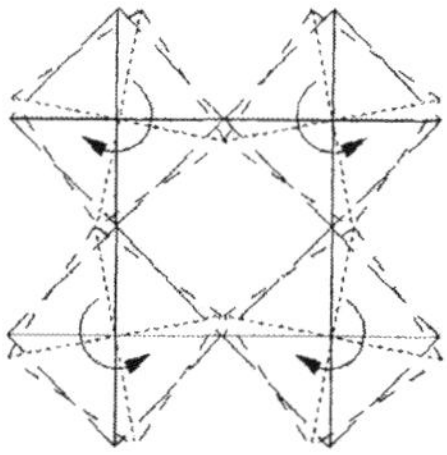

Figure 15.21 Twisting of the ISM grid pyramids.

The equations governing the elastic response of a PSM grid can be expressed in terms of the equilibrium of two typical joints, A in the lower layer and B in the upper layer. These and the adjacent joints are shown in the lower diagram of Figure 15.22. Also shown are the finite difference operators associated with these adjacent joints. An alternative approach is to imagine a phantom joint B directly above A, so that the operator associated with the real joint B is $E_X^{1/2}\,E_Y^{1/2}$. This simplifies the expressions for the deflexions of joint B. Using the notation shown, the governing equations can be written in the form

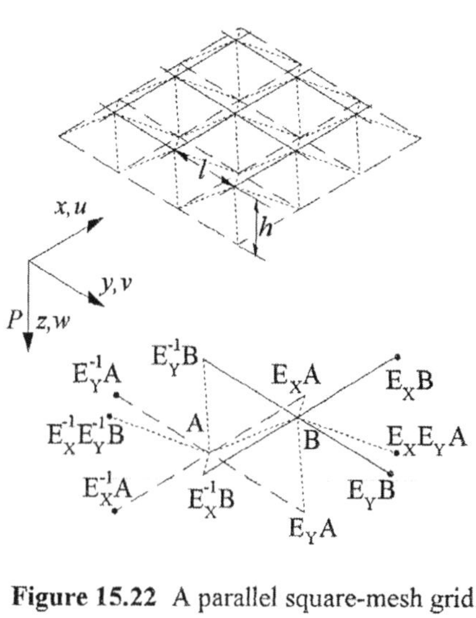

Figure 15.22 A parallel square-mesh grid.

$$\begin{bmatrix} 4bD_X\bar{D}_X+4d & 0 & 0 & -d\bar{S}_X\bar{S}_Y & -d\bar{D}_X\bar{D}_Y & d\bar{D}_X\bar{S}_Y \\ 0 & 4bD_Y\bar{D}_Y+4d & 0 & -d\bar{D}_X\bar{D}_Y & -d\bar{S}_X\bar{S}_Y & d\bar{S}_X\bar{D}_Y \\ 0 & 0 & 4d & d\bar{D}_X\bar{S}_Y & d\bar{S}_X\bar{D}_Y & -d\bar{S}_X\bar{S}_Y \\ -dS_XS_Y & -dD_XD_Y & dD_XS_Y & 4tD_X\bar{D}_X+4d & 0 & 0 \\ -dD_XD_Y & -dS_XS_Y & dS_XD_Y & 0 & 4tD_Y\bar{D}_Y+4d & 0 \\ dD_XS_Y & dS_XD_Y & -dS_XS_Y & 0 & 0 & 4d \end{bmatrix} \begin{bmatrix} u_A\frac{l}{2} \\ v_A\frac{l}{2} \\ w_Ah \\ u_B\frac{l}{2} \\ v_B\frac{l}{2} \\ w_Bh \end{bmatrix} = \begin{bmatrix} 0 \\ 0 \\ P_A/h \\ 0 \\ 0 \\ P_B/h \end{bmatrix} \qquad \textbf{(15.87)}$$

where b, d and t are the axial coefficients of the bottom, diagonal and top bars respectively. The subscripts A and B refer to the deflexions u, v and w and the vertical loads P at joints A and B respectively. The lengths of the top and bottom bars are l and the distance between the top and bottom layers is h. For compactness, the following abbreviations have been used:

$$\begin{aligned} D_{\mathrm{X\ or\ Y}} &= 1 - \mathrm{E}_{\mathrm{X\ or\ Y}} \;, \quad \bar{D}_{\mathrm{X\ or\ Y}} = 1 - \mathrm{E}^{-1}_{\mathrm{X\ or\ Y}} \\ S_{\mathrm{X\ or\ Y}} &= 1 + \mathrm{E}_{\mathrm{X\ or\ Y}} \;, \quad \bar{S}_{\mathrm{X\ or\ Y}} = 1 + \mathrm{E}^{-1}_{\mathrm{X\ or\ Y}} \end{aligned} \qquad \textbf{(15.88)}$$

The warping mode of the PSM grid mentioned earlier can be expressed by the deflexions

$$hw_A = C\mathrm{XY} \;,\; lu_B = C(\mathrm{Y}+\tfrac{1}{2}) \;,\; lv_B = C(\mathrm{X}+\tfrac{1}{2}) \;,\; hw_B = C(\mathrm{X}+\tfrac{1}{2})(\mathrm{Y}+\tfrac{1}{2}) . \qquad \textbf{(15.89)}$$

This is a mechanism inducing no internal strains and requiring no external loads.

As with the rectangular plane grid analysed in §15.6, a Navier-type solution can be found for rectangular boundaries providing the equivalent of simple supports[1]. (In fact, antisymmetrical conditions are induced about the planes of these boundaries.) As in (15.77), the normal loadings P_A and P_B can be expressed as a double sine series and the deflexions resulting from each term found by substitution in (15.87). Taking

$$\begin{aligned} u_A &= \sum_{\mathrm{R}=1}^{\mathrm{M}}\sum_{\mathrm{S}=1}^{\mathrm{N}} u^A_{rs}\cos r\mathrm{X}\sin s\mathrm{Y} \;, & u_B &= \sum_{\mathrm{R}=1}^{\mathrm{M}}\sum_{\mathrm{S}=1}^{\mathrm{N}} u^B_{rs}\cos r(\mathrm{X}+\tfrac{1}{2})\sin s(\mathrm{Y}+\tfrac{1}{2}) \;, \\ v_A &= \sum_{\mathrm{R}=1}^{\mathrm{M}}\sum_{\mathrm{S}=1}^{\mathrm{N}} v^A_{rs}\sin r\mathrm{X}\cos s\mathrm{Y} \;, & v_B &= \sum_{\mathrm{R}=1}^{\mathrm{M}}\sum_{\mathrm{S}=1}^{\mathrm{N}} v^B_{rs}\sin r(\mathrm{X}+\tfrac{1}{2})\cos s(\mathrm{Y}+\tfrac{1}{2}) \;, \\ w_A &= \sum_{\mathrm{R}=1}^{\mathrm{M}}\sum_{\mathrm{S}=1}^{\mathrm{N}} w^A_{rs}\sin r\mathrm{X}\sin s\mathrm{Y} \;, & w_B &= \sum_{\mathrm{R}=1}^{\mathrm{M}}\sum_{\mathrm{S}=1}^{\mathrm{N}} w^B_{rs}\sin r(\mathrm{X}+\tfrac{1}{2})\sin s(\mathrm{Y}+\tfrac{1}{2}) \;, \\ P_A &= \sum_{\mathrm{R}=1}^{\mathrm{M}}\sum_{\mathrm{S}=1}^{\mathrm{N}} P^A_{rs}\sin r\mathrm{X}\sin s\mathrm{Y} \;, & P_B &= \sum_{\mathrm{R}=1}^{\mathrm{M}}\sum_{\mathrm{S}=1}^{\mathrm{N}} P^B_{rs}\sin r(\mathrm{X}+\tfrac{1}{2})\sin s(\mathrm{Y}+\tfrac{1}{2}) \;. \end{aligned} \qquad \textbf{(15.90)}$$

where as before r is πR/M and s is πS/N. Using the following notation for compactness,

$$s_r = \sin\frac{r}{2} \;,\quad s_s = \sin\frac{s}{2} \;,\quad c_r = \cos\frac{r}{2} \;,\quad c_s = \cos\frac{s}{2} \;, \qquad \textbf{(15.91)}$$

the coefficients of the expressions for the deflexions in (15.90) are then given by term-by-term comparison by substituting these expressions in (15.87):

---

[1] See Hussey, M.J.L. Tarzi, A.I and Theron, W.F.D. (1971) Simply Supported Rectangular Double-Layered Grids. *Proc. ASCE, Structural Division*, V.97, No.ST3, p.753.

$$\begin{bmatrix} 4bs_r^2+d & 0 & 0 & -dc_rc_s & ds_rs_s & ds_rc_s \\ 0 & 4bs_s^2+d & 0 & ds_rs_s & -dc_rc_s & dc_rs_s \\ 0 & 0 & d & -ds_rc_s & -dc_rs_s & -dc_rc_s \\ -dc_rc_s & ds_rs_s & -ds_rc_s & 4ts_r^2+d & 0 & 0 \\ ds_rs_s & -dc_rc_s & -dc_rs_s & 0 & 4ts_s^2+d & 0 \\ ds_rc_s & dc_rs_s & -dc_rc_s & 0 & 0 & d \end{bmatrix} \begin{bmatrix} u_{rs}^A \frac{l}{2} \\ v_{rs}^A \frac{l}{2} \\ w_{rs}^A h \\ u_{rs}^B \frac{l}{2} \\ v_{rs}^B \frac{l}{2} \\ w_{rs}^B h \end{bmatrix} = \begin{bmatrix} 0 \\ 0 \\ P_{rs}^A/4h \\ 0 \\ 0 \\ P_{rs}^B/4h \end{bmatrix} \qquad \textbf{(15.92)}$$

This then gives solutions for a rectangular grid with side lengths Ml and Nl by the same method used for the plane rectangular grid analysed in §15.6.

The equations governing other kinds of space grids are given in §A5.5. As with plane grids, equivalent differential equations can be found[1] which characterise the macroscopic behaviour of space grids. If the top and bottom bars of a PSM grid have the same axial coefficient b, then this equation takes the form

$$\frac{\partial^4 w}{\partial x^4} + \frac{\partial^4 w}{\partial y^4} = \frac{2p}{bh^2l^2} \qquad \textbf{(15.93)}$$

where w is the lateral displacement of the grid and p is the equivalent normal load per unit area on the surface of the grid. For PTM grids and T-H grids, the governing differential equation is like the biharmonic plate equation,

$$\frac{\partial^4 w}{\partial x^4} + 2\frac{\partial^4 w}{\partial x^2 \partial y^2} + \frac{\partial^4 w}{\partial y^4} = \frac{p}{D} \qquad \textbf{(15.94)}$$

where for a PTM grid with top and bottom bar axial coefficients b and t and lengths l, separated by a height h,

$$D = \frac{3\sqrt{3}\, bth^2l^2}{4(b+t)} \qquad \textbf{(15.95)}$$

and using the same notation for a T-H grid, where the top bars form a triangular grid and the bottom ones a hexagonal grid and the diagonal linking the two have the axial coefficient d,

$$D = \frac{bh^2l^2(81td + 54bt + bd)}{6\sqrt{3}(27td + 54bt + 11bd + 4b^2)} \qquad \textbf{(15.96)}$$

The difference equations for an ISM grid reduce to an eighth order equation rather than a fourth order equation, and this is given by (A5.5.4). Because they do not all behave in the same way, direct comparisons between these grids can only be made in special circumstances. This has been done[1] for grids 'simply-supported' on square boundaries of side of length a loaded by a uniform load of intensity p per unit area. In all cases, sinusoidal solutions can be found to both the finite-difference equations and the approximate differential equations, giving rise to Navier-type analyses. Taking a PSM grid, with members all of length l and axial coefficient EA/l^3, the central deflexion, as the mesh becomes infinitely fine, is $0.03282pa^4/EAl$ (when the top and bottom bars are parallel to the boundaries). With the same grid turned through 45°, so that the top and bottom

[1] Renton, J.D. (1970) General Properties of Space Grids. *Int. J. Mech. Sci.* V.12, p.801.

bars are oriented as in Figure 15.19b instead of as in Figure 15.19a, this becomes $0.01607pa^4/EAl$. The stiffness of the latter is then 2.042 times that of the former, exactly as in the case of the two plane grids compared in (15.85). For the purposes of comparison, it will be assumed that the other grids have the same density of material in both the upper and lower layers and in the layer of diagonal bars linking them. Matching also the separation of the layers, h, the central deflexions of the PTM and T-H grids are $0.02167pa^4/EAl$ and $0.02605pa^4/EAl$. With an infinitely fine ISM grid, the central deflexion is $0.02418pa^4/EAl$ regardless of whether the bars upper or lower layer are parallel to the boundaries. In these examples, the PSM grid skewed at 45° to the boundaries is significantly stiffer than the others. Advantage of this orientation is also taken in a square grid of Vierendeel trusses known as a Diagrid. Makowski (1965) notes that this diagonal arrangement results in a more uniform stress distribution and greater rigidity, because the beams are roughly aligned with the principal stress trajectories. Other work on optimising space trusses includes that of El-Sheikh[1]. For those trusses considered, the optimum span/depth and span/panel-width ratios for corner-supported space trusses were 8 and 12 respectively. For edge-supported trusses, both these ratios were either 12 or 16.

15.8 Reticulated Barrel Vaults

Barrel vaults are vaults which have cross-sections in the form of circular arches. The surface of the vault is developable, that is, it can be laid out as a flat sheet without inducing membrane strains. This means that a regular plane gridwork of bars can be used to brace such a surface. Double curvature increases the rigidity of such a structure, but then the pattern of bars becomes distorted and is no longer truly regular. The theoretical and practical aspects of barrel vault design have been considered by Makowski (1985). The length/span ratio of barrel vaults is usually less than 2.0 and the rise/span ratio greater than 0.2. For stiffness and stability, good rise/length ratios vary between 0.17 and 0.3 according to Shougun, Jianheng and Hengxi[2]. They state that reducing the length/span ratio or increasing the rise/span ratio improves the structure according to these criteria. Figure 15.23 shows some standard layouts of barrel-vault grids. These are shown in the developed state, the length being horizontal. A simple rectangular grid without diagonal bracing is relatively weak and unstable. In 1890, Föppl attempted unsuccessfully to patent the design shown in Figure 15.23a. Makowski (1965) says that the bracing in the 'lightweight' system shown in Figure 15.23c follows the principal stress trajectories. It has been used in conjunction with lightweight concrete, the grid acting as part of the formwork, or with panels as infilling which act

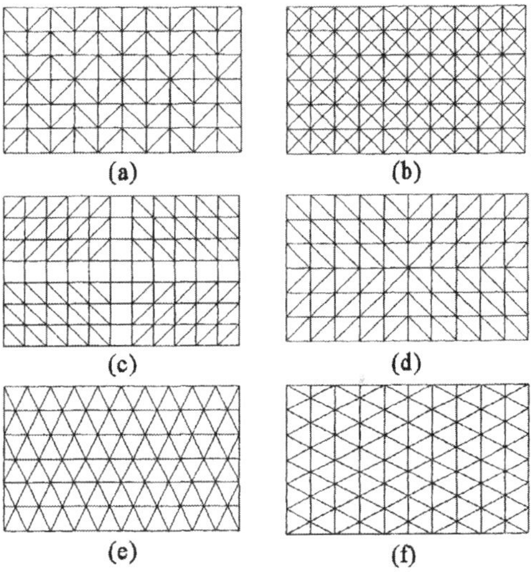

Figure 15.23 Bar layouts for barrel vaults.

[1] El-Sheikh, A. (1998) Optimum Design of Space Trusses. *Journal of the I.A.S.S.* V.39 No.3 p.159.

[2] Du Shougun, Sun Jianheng and Xia Hengxi (1993) Stability Behaviour Investigation of Braced Barrel Vaults. *Space Structures 4* V.1 p.62. Thomas Telford Services, London.

as an integral part of the structure. The triangular grid shown in Figure 15.23e is known as the TeZet system and has been used by E. Torroja and S. du Château and is also the basis of the Wuppermann system. For every triangular grid there exists a corresponding hexagonal grid, as shown in Figure 15.24. Thus the hexagonal grid corresponding to the triangular grid shown in Figure 15.23e has a third of its bars oriented in the hoop sense and that corresponding to the grid shown in Figure 15.23f has a third of its bars oriented in the longitudinal sense. Shougun, Jianheng and Hengxi (ibid.) have examined the layouts shown in Figures 15.23c, d and e and conclude that the triangular arrangement in Figure 15.23e is considerably more stable than the others.

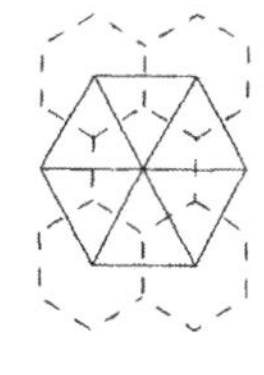

Figure 15.24

Finite difference calculus can be used to analyse barrel vaults. Figure 15.25 shows part of a TeZet vault (cf Figure 15.23e). The analysis used here[1] is the equivalent of Navier-type solution discussed in §15.6. Here, the operator E_X is used to give a step of length b along the truss and E_Θ produces a step of d in the hoop sense, which subtends an angle 2ϕ at the axis of curvature of the vault. Thus joint B in Figure 15.25 could also be labelled $E_X E_\Theta A$, using the earlier notation.

The governing equations are expressed in terms of the local coordinates of each joint. The local x,y and z axes are in the longitudinal, tangential and radially-inwards directions. Thus x_A and x_B are in the same direction but y_B and z_B are at an angle 2ϕ to y_A and z_A. In expressing the stiffness equations for the bar AB in terms of these coordinates, two transformations must be carried out. First, the equations in terms of the bar coordinates must be referred to the plane shown in Figure 15.25b by a rotation through α and then referred to the local coordinates at each end by rotations of $\pm\phi$.

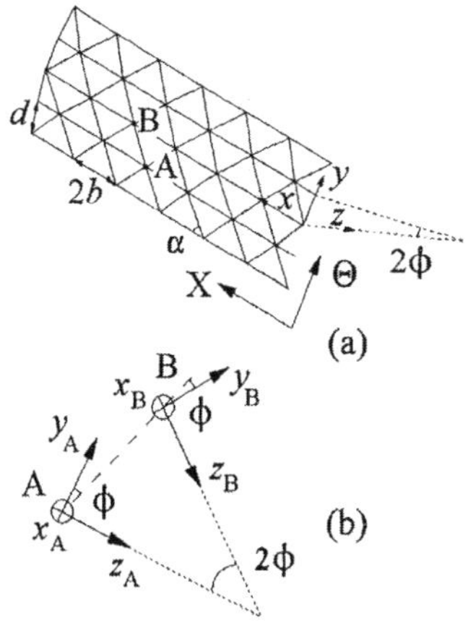

Figure 15.25 Notation used for the analysis of a TeZet vault.

The local coordinates x,y and z for the bar AB are as shown in Figure 11.12. The (major) x axis is taken to lie in the plane of triangles containing AB. In terms of these coordinates, the stiffness equations are given by (11.7) and (11.45) to (11.48). These must now be expressed in terms of the loads and deflexions at A and B relative to their local coordinates (x_A, y_A, z_A) and (x_B, y_B, z_B). This is done by first relating the vectors to a joint-like coordinate system (x_o, y_o, z_o) for which ϕ is zero and then to the joint coordinates at A and B, allowing for the angles $\pm\phi$. The required transformation matrices are

$$\boldsymbol{u}_o = \boldsymbol{T}_\alpha \boldsymbol{u} \quad \text{or} \quad \begin{bmatrix} u_o \\ v_o \\ w_o \end{bmatrix} = \begin{bmatrix} -\sin\alpha & 0 & \cos\alpha \\ \cos\alpha & 0 & \sin\alpha \\ 0 & 1 & 0 \end{bmatrix} \begin{bmatrix} u \\ v \\ w \end{bmatrix} = \begin{bmatrix} -d/l & 0 & b/l \\ b/l & 0 & d/l \\ 0 & 1 & 0 \end{bmatrix} \begin{bmatrix} u \\ v \\ w \end{bmatrix},$$

$$\boldsymbol{u}_A = \boldsymbol{T}_\phi \boldsymbol{u}_o \quad \text{or} \quad \begin{bmatrix} u_A \\ v_A \\ w_A \end{bmatrix} = \begin{bmatrix} 1 & 0 & 0 \\ 0 & \cos\phi & -\sin\phi \\ 0 & \sin\phi & \cos\phi \end{bmatrix} \begin{bmatrix} u_o \\ v_o \\ w_o \end{bmatrix}. \qquad \textbf{(15.97)}$$

[1] See Renton, J.D. (1991) Analysis of Reticulated Barrel Vaults by Finite Difference Calculus. *Spatial Structures at the Turn of the Millennium, Proc. I.A.S.S.* V.3 p.135.

so that the overall transformation matrix is

$$\boldsymbol{u}_A = \boldsymbol{T}_\phi \boldsymbol{T}_\alpha \boldsymbol{u}_o = \boldsymbol{T}(\phi,b,d)\boldsymbol{u} \quad \text{or} \quad \begin{bmatrix} u_A \\ v_A \\ w_A \end{bmatrix} = \begin{bmatrix} -d/l & 0 & b/l \\ b/l\cos\phi & -\sin\phi & d/l\cos\phi \\ b/l\sin\phi & \cos\phi & d/l\sin\phi \end{bmatrix} \begin{bmatrix} u \\ v \\ w \end{bmatrix} . \qquad \textbf{(15.98)}$$

The transformation matrix $\boldsymbol{T}(\phi,\mathrm{b},\mathrm{d})$ then gives the transformed $\boldsymbol{K}_{11}{}^*$ matrix as detailed in (11.49). However, for the vectors related to joint B, the transformation matrix $\boldsymbol{T}(-\phi,\mathrm{b},\mathrm{d})$ should be used, where ϕ is replaced by $-\phi$ (or $\sin\phi$ by $-\sin\phi$) in (15.98). The other stiffness matrices are then given by

$$\boldsymbol{K}_{12}^* = \mathrm{T}(\phi,b,d)\boldsymbol{K}_{12}\mathrm{T}^t(-\phi,b,d) \ , \quad \boldsymbol{K}_{21}^* = \mathrm{T}(-\phi,b,d)\boldsymbol{K}_{21}\mathrm{T}^t(\phi,b,d) \ ,$$

$$\boldsymbol{K}_{22}^* = \mathrm{T}(-\phi,b,d)\boldsymbol{K}_{22}\mathrm{T}^t(-\phi,b,d) \ , \quad \text{where} \quad \mathrm{T}(\phi,b,d) = \begin{bmatrix} \boldsymbol{T}(\phi,b,d) & \mathbf{0} \\ \mathbf{0} & \boldsymbol{T}(\phi,b,d) \end{bmatrix} . \qquad \textbf{(15.99)}$$

The necessary transformation matrices T for the other bars connecting joint A to the five other adjacent joints are similar, allowance being made for their particular orientation. For the other hoop bars, they are, in terms of the joints connected, $\mathrm{E_X E_\theta^{-1}A}$: $\mathrm{T}(-\phi,b,-d)$, $\mathrm{E_X^{-1}E_\theta^{-1}A}$: $\mathrm{T}(-\phi,-b,-d)$, $\mathrm{E_X^{-1}E_\theta A}$: $\mathrm{T}(\phi,-b,d)$, and for the longitudinal bars they are $\mathrm{E_X^{2}A}$: $\mathrm{T}(0,l,0)$, $\mathrm{E_X^{-2}A}$: $\mathrm{T}(0,-l,0)$.

It is now possible to set up the stiffness matrix equilibrium equation for joint A:

$$\begin{bmatrix} k_{11} & k_{12} & k_{13} & k_{14} & k_{15} & k_{16} \\ k_{21} & k_{22} & k_{23} & k_{24} & k_{25} & k_{26} \\ k_{31} & k_{32} & k_{33} & k_{34} & k_{35} & k_{36} \\ k_{41} & k_{42} & k_{43} & k_{44} & k_{45} & k_{46} \\ k_{51} & k_{52} & k_{53} & k_{54} & k_{55} & k_{56} \\ k_{61} & k_{62} & k_{63} & k_{64} & k_{65} & k_{66} \end{bmatrix} \begin{bmatrix} \theta_{xA} \\ \theta_{yA} \\ \theta_{zA} \\ \delta_{xA} \\ \delta_{yA} \\ \delta_{zA} \end{bmatrix} = \begin{bmatrix} M_{xA} \\ M_{yA} \\ M_{zA} \\ F_{xA} \\ F_{yA} \\ F_{zA} \end{bmatrix} \qquad \textbf{(15.100)}$$

Defining
$$\Sigma_{\mathrm{X\,or\,\Theta}} = \mathrm{E_{X\,or\,\Theta}} + \mathrm{E_{X\,or\,\Theta}^{-1}} \ , \quad \Delta_{\mathrm{X\,or\,\Theta}} = \mathrm{E_{X\,or\,\Theta}} - \mathrm{E_{X\,or\,\Theta}^{-1}} \ ,$$
$$s = \sin\phi \ , \quad c = \cos\phi \ , \quad S = \sin\alpha \ , \quad C = \cos\alpha, \qquad \textbf{(15.101)}$$

and using the following abbreviations for the hoop bar stiffnesses,

$$a = EA/l \ , \ j = GJ/l \ , \ m = 2EI_x/l \ , \ n = 6EI_x/l^2 \ ,$$
$$o = 12EI_x/l^3 \ , \ p = 2EI_y/l \ , \ q = 6EI_y/l^2 \ \ r = 12EI_y/l^3 \ , \qquad \textbf{(15.102)}$$

and using the corresponding capital letters for the longitudinal bars, the matrix coefficients are

$$k_{11} = 4C^2j + 8S^2m + (S^2m - C^2j)\Sigma_X\Sigma_\Theta - J\Sigma_X^2 \ ; \ k_{12} = -cSC(j+m)\Delta_X\Delta_\Theta \ ;$$
$$k_{13} = sSC(j+m)\Delta_X\Sigma_\Theta \ ; \ k_{14} = 0 \ ; \ k_{15} = -sSn(4 + \Sigma_X\Sigma_\Theta) \ ; \ k_{16} = -cSn\Sigma_X\Delta_\Theta \ ;$$
$$k_{22} = 4c^2(S^2j + 2C^2m) + 8s^2p + [c^2(C^2m - S^2j) - s^2p]\Sigma_X\Sigma_\Theta + M(2 + \Sigma_X^2) \ ;$$
$$k_{23} = cs(S^2j - C^2m - p)\Sigma_X\Delta_\Theta \ ; \ k_{24} = sSq(4 - \Sigma_X\Sigma_\Theta) \ ; \ k_{25} = scC(n+q)\Delta_X\Delta_\Theta \ ;$$
$$k_{26} = C(c^2n - s^2q)\Delta_X\Sigma_\Theta + N\Sigma_X\Delta_X \ ;$$
$$k_{33} = 4s^2(S^2j + 2C^2m) + 8c^2p + [s^2(S^2j - C^2m) + c^2p]\Sigma_X\Sigma_\Theta + P(2 + \Sigma_X^2) \ ; \qquad \textbf{(15.103)}$$
$$k_{34} = cSq\Sigma_X\Delta_\Theta \ ; \ k_{35} = C(s^2n - c^2q)\Delta_X\Sigma_\Theta - Q\Sigma_X\Delta_X \ ; \ k_{36} = scC(n+q)\Delta_X\Delta_\Theta \ ;$$
$$k_{44} = (C^2a + S^2r)(4 - \Sigma_X\Sigma_\Theta) - A\Sigma_X^2 \ ; \ k_{45} = cSC(r-a)\Delta_X\Delta_\Theta \ ;$$
$$k_{46} = -sSC(r-a)\Delta_X\Sigma_\Theta \ ; \ k_{55} = c^2(S^2a + C^2r)(4 - \Sigma_X\Sigma_\Theta) + s^2o(4 + \Sigma_X\Sigma_\Theta) - R\Sigma_X^2 \ ;$$
$$k_{56} = sc(S^2a + C^2r + o)\Sigma_X\Delta_\Theta \ ; \ k_{66} = s^2(S^2a + C^2r)(4 + \Sigma_X\Sigma_\Theta) + c^2o(4 - \Sigma_X\Sigma_\Theta) - O\Sigma_X^2 \ .$$

and k_{ji} is the same as k_{ij} except that $\Delta_{X \text{ or } \Theta}$ is replaced by $-\Delta_{X \text{ or } \Theta}$. The Δ operator turns even functions (X^{2n}, cos X, cosh X, etc.) into odd functions (X^{2n-1}, sin X, sinh X, etc.) and vice versa. The Σ operator leaves even functions as even and odd functions as odd. Examining (15.103) shows that if the operations on the deflexions are to produce terms of a similar type in a given row, θ_{xA}, δ_{yA} and δ_{zA} must be even functions of X and θ_{yA}, θ_{zA} and δ_{xA} odd functions (or vice versa). Likewise, θ_{xA}, θ_{zA} and δ_{yA} must be even functions of Θ and θ_{yA}, δ_{xA} and δ_{zA} odd functions (or vice versa). A Navier-type solution of the form

$$\begin{aligned}
\theta_{xA} &= \sum_{m=1}^{M-1}\sum_{n=1}^{N-1} a_{mn}\sin\mu X\cos\nu\Theta \;, & \theta_{yA} &= \sum_{m=1}^{M-1}\sum_{n=1}^{N-1} b_{mn}\cos\mu X\sin\nu\Theta \;, \\
\theta_{zA} &= \sum_{m=1}^{M-1}\sum_{n=1}^{N-1} c_{mn}\cos\mu X\cos\nu\Theta \;, & \delta_{xA} &= \sum_{m=1}^{M-1}\sum_{n=1}^{N-1} d_{mn}\cos\mu X\sin\nu\Theta \;, \\
\delta_{yA} &= \sum_{m=1}^{M-1}\sum_{n=1}^{N-1} e_{mn}\sin\mu X\cos\nu\Theta \;, & \delta_{zA} &= \sum_{m=1}^{M-1}\sum_{n=1}^{N-1} f_{mn}\sin\mu X\sin\nu\Theta \;.
\end{aligned} \qquad \textbf{(15.104)}$$

where μ is $m\pi$/M and ν is $n\pi$/N, gives antisymmetrical conditions on the ends X=0 and M and on the sides Θ=0 and N. Equal vertical loads w_o at each joint are given by

$$F_{yA} = w_o\sin(2\Theta - \mathrm{N})\phi \;, \quad F_{zA} = w_o\cos(2\Theta - \mathrm{N})\phi \;. \qquad \textbf{(15.105)}$$

If N is even, these can be expressed in terms of suitable double Fourier series as

$$F_{yA} = \sum_{m=1}^{M-1}\sum_{n=1}^{N-1} p_{mn}\sin\mu X\cos\nu\Theta \;, \quad F_{zA} = \sum_{m=1}^{M-1}\sum_{n=1}^{N-1} q_{mn}\sin\mu X\sin\nu\Theta \qquad \textbf{(15.106)}$$

where p_{mn} and q_{mn} are zero if either m or n is even, otherwise

$$p_{mn} = \frac{4w_o(\cot\frac{1}{2}\mu\,\cos\mathrm{N}\phi)}{\mathrm{MN}(\cos\nu - \cos 2\phi)}\sin 2\phi \;, \quad q_{mn} = -\frac{4w_o(\cot\frac{1}{2}\mu\,\cos\mathrm{N}\phi)}{\mathrm{MN}(\cos\nu - \cos 2\phi)}\sin\nu \;. \qquad \textbf{(15.107)}$$

(cf. (15.6)). If N is odd, a further term, $\frac{1}{2}p_{mN}\sin\mu X\cos\pi\Theta$, must be added to the summation over m in the expression for F_{yA}. Substituting the expressions for the deflexions and the loads given by (15.104) and (15.106) into (15.100) yields six equations relating the coefficients a_{mn} to f_{mn} to p_{mn} and q_{mn}, the functions of X and Θ on both sides of the equations cancelling out.

The resulting solution is generic for the barrel-vault type and has been programmed as TRIBARL.EXE (see Appendix 3). It is for any circular vault made from a triangular grid of the above type, loaded and supported in the above way. The properties of the mesh and its component bars can be changed instantaneously, facilitating its use as a design tool. If only approximate answers are required, the summations can be restricted to less than M-1 and N-1, thus speeding the analysis.

15.9 Braced Domes

The types of dome construction considered here will be confined to those which are essentially based on a structural skeleton. Stressed-skin and folded-plate domes are sometimes similar in form, but derive their strength from membrane or plate action, which is outside the intended scope of this book. Figure 15.26 illustrates some common structural forms. Figure 15.26a shows a ribbed dome which may include an edge ring to take the lateral thrust. Such domes are rigid-jointed and rely on the flexural stiffness of their component beams. The Schwedler dome shown in Figure 15.26b dates from the mid-19th century and has light ties cross-bracing a rigid-jointed frame. As shown in Engel (1968) p.98, bracing on one diagonal only may

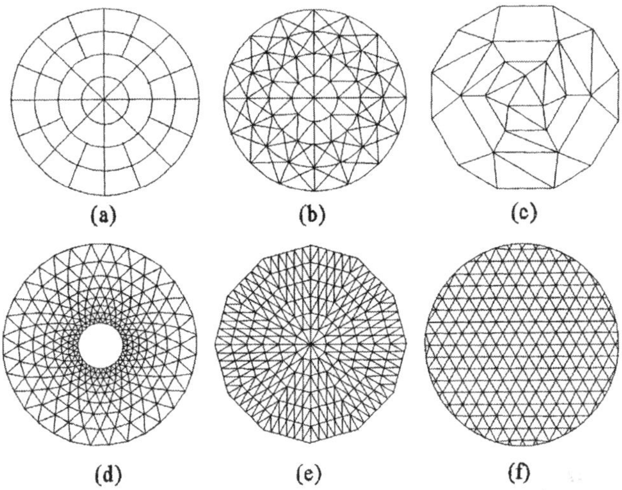

Figure 15.26 Some types of lattice dome.

also be used, provided that it can take both tension and compression. Rigid-jointed domes tend to be heavier than pin-jointed domes because the stress distribution in flexure does not make full use of the potential of the material. A Zimmermann dome is shown in Figure 15.26c. Starting from the central triangle, triangles are inserted in alternate tiers to double the number of sides on the polygon. Note that diagonal bracing is only used on alternate segments. Föppl analysed domes of the type shown in Figure 15.26d and showed them to be statically-determinate and stable only if the number of sides on the ring polygons was odd. Makowski (1965) refers to such forms as network domes. The Exhibition Hall dome in Brno (span 93.5m, rise 19.1 m) is a notable example of this type. A way of avoiding the need for a large number of bars of different length is to use continuous members spiralling out from the centre, harnessing them and the circumferential bars together at the joints. However, a dome similar to the Brno dome was built in Bucharest in 1961 and collapsed within two years of construction. This was partly due to the inability of the joint harnesses to prevent relative slip between the bars. Figure 15.26e shows the (patented) layout of a Kiewitt dome. One such dome at Houston has a 196m span. It consists of six identical segments, each of which is divided into two subsegments which are mirror images of each other[1]. The sides of the triangles in each subdivision are parallel to one or other of its sides. Three-way grid domes of the type shown in Figure 15.26f were pioneered by S. du Château and Fujio Matsushita. Rotational symmetry plays no part in the layout of such grids, but laying the bars along the paths of great circles appears to simplify the connections.

. A geodesic is the shortest path between any two points on a surface. Great circles are geodesics on spherical surfaces, and are usually unique[2]. Madrid and New York lie on almost the same latitude, but the shortest route between them is not along their mean line of latitude. Instead, the shortest route for an aircraft is along a great circle which deviates up to 680km north of this line. Geodesic domes were patented by R. Buckminster Fuller in 1947 and by 1970 some 3,000 had been built[3]. His object was to triangulate a spherical surface in as regular a manner as possible. There are only four ways in which this can be done with complete regularity. These are using tetrahedra (F=4, V=4, E=6), octahedra (F=8, V=6, E=12), dodecahedra (F=12, V=20,

---

[1] The symmetry can be used to reduce the analysis of such domes under gravitational load to that of a 15° segment, as in the data file R3DSIKWT.STR provided.

[2] The exception is when the two points are at opposite ends of the diameter of a sphere, when an infinite number of great circles pass through both of them.

[3] Geodesic structures had been used before, notably in the design of the fuselage of the Wellington bomber by Neville Barnes Wallis.

E=30) or icosahedra (F=20, V=12, E=30) where F is the number of faces (or triangles), V is the number of vertices (or joints) and E is the number of edges (or bars). Any polyhedron must satisfy Euler's formula ($F+V-E=2$). Should one try to triangulate a closed surface with n triangles with six bars meeting at each joint, the number of joints would be $3n/6$ and the number of bars would be $3n/2$, giving zero on the right-hand side of Euler's formula. Suppose that m of the joints had only 5 triangles meeting at them. Then $m/6$ must be added to the vertex count. To satisfy Euler's formula, the value of m must then be 12 (and n is arbitrary). Fuller takes these five-way joints to be the vertices of a regular icosahedron, and all the other joints to lie on the sphere circumscribing this figure. A geodesic dome may be developed in the manner shown in Figure 15.27. Great circles are drawn on the surface of the sphere, linking the vertices of the icosahedron. Corresponding to each of its faces, there is now a spherical triangle. The midpoints of each of the sides of this spherical triangle are now linked. If this is done by bars, each triangular face of the icosahedron, shown at the bottom of Figure 15.27, is replaced by the four triangles shown to the right of it. This results in a polyhedron with 80 triangular faces, 42 joints and 120 bars. Proceeding anticlockwise around Figure 15.27, three further stages of this recursive process[1] are shown. At each stage, new joints are inserted at the midpoints of the great-circle sides of each spherical triangle linking the existing joints. Linking these new joints with bars produces four triangles where there was one before. Thus the next three stages successively yield a total of 320, 1280, and 5120 triangles[2]. If $\boldsymbol{R}_A$ and $\boldsymbol{R}_B$ are the position vectors of two joints on the surface of the sphere of radius R, this process requires finding the length l_{AB} linking them, or the position $\boldsymbol{R}_C$ of a new joint C midway along the geodesic linking them. These are given by

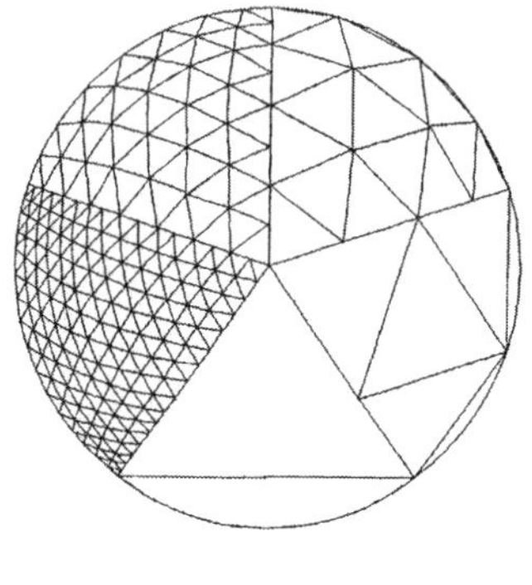

Figure 15.27 A geodesic triangulation.

$$l_{AB} = |\boldsymbol{R}_A - \boldsymbol{R}_B| \quad , \quad \boldsymbol{R}_C = \frac{R(\boldsymbol{R}_A + \boldsymbol{R}_B)}{|\boldsymbol{R}_A + \boldsymbol{R}_B|} \quad \textbf{(15.108)}$$

This does not produce triangles with equal sides. After the final (fourth-order) triangulation given above, the shortest bar is of length 0.06918R and the longest is 0.08264R, 19% longer.

An alternative triangulation, shown in Figure 15.28, is given in Chapter 3 of Margarit and Buxadé (1972). In this case, no bars lie along the sides of the original triangle, although some cross the sides. Initially, the triangle is divided by great circles from each apex to the midpoint of the opposite side (as shown by the solid lines). The

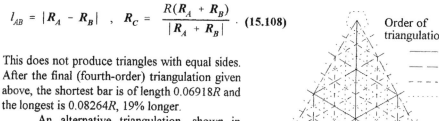

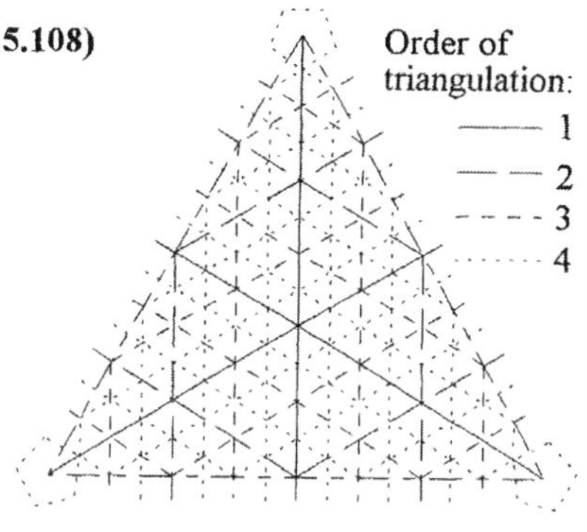

Figure 15.28 An alternative triangulation.

---

[1] Recursion is a powerful programming tool and has been used in the program GEOTRI.EXE to find the joint positions and bar lengths. The source code is given in GEOTRI.MOD.

[2] The United States pavilion for Montreal's Expo '67 was based on a fourth-order triangulation. The triangles formed the upper layer of a steel T-H grid with the perspex panels used as infilling in the hexagonal layer. It was 76m diameter and 61m high.

second, third and fourth order subdivision are shown by broken lines, dashed lines and dotted lines respectively. Again, allowing for the curvature of the surface on which the joints lie, the bars linking them vary in length between $0.07477R$ and $0.09505R$, some 27% longer. The authors state that such a dome can be built from four different bar lengths, allowing for the tolerance of the circular gusset plates used at the joints.

As shown in Figure 15.24, for every triangular pattern with six-way joints a complementary hexagonal pattern can be found. Thus a closed surface may be broken up into complementary patterns of hexagons apart from the twelve pentagons needed to satisfy Euler's formula. Kaiser domes, usually of 60m to 120m span, are based on this geometry, being formed mostly of double-layer hexagonal grids. They have five-fold symmetry, an apex of the icosahedron from which they are formed occurring at the apex of the dome. Despite the essential irregularities of such triangular and hexagonal grids, continuum analogies have been used in analysing these structures. This requires both their equivalent plate-like behaviour (as discussed earlier in this chapter) and their membrane-like behaviour to be known. Heki[1] gives equivalent membrane stiffnesses for a number of regular grids. A pin-jointed triangular mesh of equal bars of length l and axial stiffness EA is found to behave like an isotropic membrane. If Young's modulus for the equivalent membrane is taken as the same, the equivalent membrane thickness is $2A/\sqrt{3}l$ and the equivalent Poisson's ratio is a third. The equivalent shear modulus is consistent with these values, being $3E/8$. For a regular hexagonal grid made from the same bars, equivalent isotropic stiffnesses are proposed but it is noted that both the pin-jointed grid and its equivalent continuum are unstable.

15.10 Stability Considerations

The stability of a braced domes has been analysed in terms of an equivalent isotropic shell, although the consequences can be disastrous, as in the case of the 51m diameter C.W. Post dome, Long Island University, 1970. The analysis was based on an equivalent membrane thickness, calculated on the basis of giving it the same volume of material per unit area as that in the actual dome. For the triangular mesh grid discussed at the end of the last section, this would give an equivalent thickness of $2\sqrt{3}A/l$, *three times* the equivalent thickness given by Heki's analysis. Even this equivalent thickness is only a guide for membrane action; the flexural stiffness is given by (A5.4.12). The actual type of bracing used is illustrated in Figure 15.29. This has no isotropic continuum equivalent. The only loading conditions considered were symmetric dead and superimposed loading, and antisymmetric wind loading[2]. The dome collapsed as a result of snap-through under asymmetric loading conditions[3] (cf. Figure 9.1). The use of even the correct equivalent shell stiffnesses in buckling analyses is questionable. Experimental values of the buckling load for spherical shells under external pressure are often one tenth of that

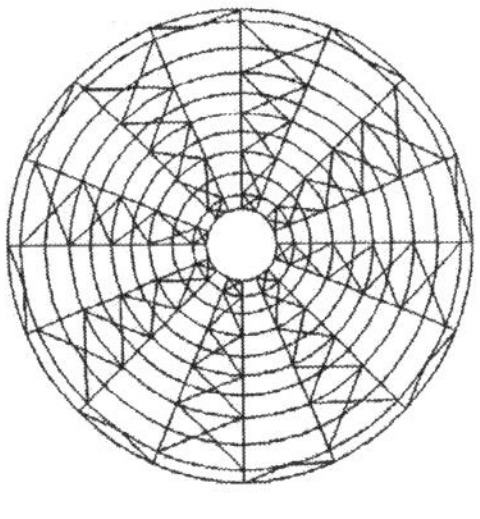

Figure 15.29 Type of bracing in C.W.Post Dome.

---

[1] I.A.S.S. Working Group on Spatial Steel Structures (1984) *Analysis, Design and Realization of Space Frames a State-of-the Art Report.* Bulletin of the I.A.S.S., V.25, No. 1/2, p.96.

[2] More realistic design loading is given by Ohmori, H. and Yamamoto, K. (1998) *Shape optimisation of shell and spatial structures for specified stress distribution.* Journal of the I.A.S.S., V.39, No. 1, p.3.

[3] See Levy (1992) for further details.

predicted for perfect shells, owing to small imperfections. As explained in the previous section, any reticulation of a dome is inherently non-uniform and so imperfect. Also, the buckling mode may be a wave pattern with a half-wave length equal to the length of the bars, so that no joint displacement is involved (see Wah and Calcote (1970) Figure 4.7 for example). This could not be predicted by any continuum analogy. It suggests that the applied load at buckling is unlikely to induce more than the pin-ended Euler load in any bar of the reticulated shell. Single-layer reticulated shells are particularly prone to local snap-through instability[1] (see Figures 9.1 & 9.2). The I.A.S.S. Working Group (ibid. §3.5c) propose various approximate formulae for such cases. More rigorous methods of dealing with such problems use incremental iterative procedures. These take full account of the non-linearity caused by the change in geometry of the structure under load, see for example Yang and Kuo (1994).

The use of an equivalent beam approach in examining the stability of trusses is more appropriate than the equivalent shell approach discussed above. Consider for example the lateral buckling of a rigid-jointed Warren truss. The out-of-plane bending stiffness and torsional stiffness are given by (A5.2.8). If the truss is used as a cantilever, an estimate of the end shear force required to cause buckling can then be obtained from (9.56). Suppose that all the members of the truss have the same length l and out-of-plane bending stiffness EI ($=GJ$).Then from these equations it follows that the critical shear force applied at the end of a cantilever of N bays is

$$S_c \approx 6.68\frac{EI}{\mathrm{N}^2 l^2} \qquad \textbf{(15.109)}$$

This gives a safe lower bound to the lateral buckling load for such trusses with two or more bays[2].

Timoshenko and Gere (1961) suggest in §2.18 treating the instability of trusses subject to large axial loads as equivalent to beams buckling under bending and shear. This was dealt with in §9.4. For comparison with their results, the case of a pin-ended strut will be considered. Then from (9.15) and (9.20b),

$$P_c = \frac{\pi^2 EI}{l^2}\,\frac{1}{1+\dfrac{\pi^2 EI}{l^2}\left(\dfrac{1}{GkA}\right)} \qquad \textbf{(15.110)}$$

The authors define I in this expression as the second moment of area of the cross section of the (compound) strut. This can be interpreted to mean that the values of EI are those given by $1/f_{22}$ for such trusses in Appendix 5. However, their values of GkA differ somewhat from the values of $1/f_{44}$ given there. For the pin-jointed trusses, the expressions given in the appendix have extra terms involving the axial coefficients, b and t, of the upper and lower chords. Their effective shear stiffness of a Vierendeel truss, GkA^* given by (2-63), and that given by (A5.2.7), GkA, are

$$\frac{1}{GkA^*} = \frac{1}{12mh^2}\left(\frac{1}{2i}+\frac{1}{j}\right), \quad \frac{1}{GkA} = f_{44} = \frac{1}{12mh^2}\left(\frac{1}{2i}+\frac{c^2}{j(4i+c)^2}\right) \qquad \textbf{(15.111)}$$

where the notation is defined in Appendix 5.

On their own, such trusses would tend to buckle out-of-plane. The above authors consider them to form opposite pairs of faces of a three-dimensional truss in which the bracing between the pairs plays no role in resisting buckling. Alternatively, the appropriate flexural and shear stiffnesses for space trusses given in §A5.3 may be used. It should be noted that in some cases, cross-coupling between stiffnesses may give rise to behaviour outside the scope of ordinary

---

[1] Undulations in such reticulated surfaces, like the corrugations in Nissen huts, improves their stability.

[2] This can be tested using and modifying the data file R3DMMLAT.STB with the software provided.

isotropic beam theory.

A more direct way[1] of determining the buckling loads of such trusses does not require drawing analogies with the behaviour of solid beams. The exact finite difference equations for such trusses can be set up, as in Example 15.2, but modifying the equations to allow for large axial forces as in (11.51). Suppose that in the above example, the top and bottom chord members each take an axial force of $P/2$. Then (15.25) becomes

$$\begin{bmatrix} d - c\nabla_X & d & -dE_X & -dE_X \\ d & d + a + \rho\nabla_X & -dE_X & -dE_X - a \\ -dE_X^{-1} & -dE_X^{-1} & d - c\nabla_X & d \\ -dE_X^{-1} & -dE_X^{-1} - a & d & d + a + \rho\nabla_X \end{bmatrix} \begin{bmatrix} um \\ vh \\ u'm \\ v'h \end{bmatrix} = \begin{bmatrix} H/m \\ V/h \\ H'/m \\ V'/h \end{bmatrix} \tag{15.112}$$

where ∇_X is defined in (15.72) and ρ is $P/2mh^2$. Buckling may occur when the joint loads on the right-hand side of the equation are zero but the joint deflexions on the left-hand side are not. In some sense then, the determinant of the matrix on the left-hand side of the equation is zero. This condition is given by

$$c\nabla_X^2\{4\rho ad + \nabla_X[cad - 2\rho c(a + d) + 2\rho^2 d] - \nabla_X^2\rho^2 c\}\phi(X) = 0 \tag{15.113}$$

where $\phi(X)$ is some function of X. Polynomial solutions of

$$\nabla_X^2\phi(X) = 0 \tag{15.114}$$

correspond to rigid-body motions of the truss and its characteristic responses to extension flexure and shear. The remaining solutions of (15.113) result from the operations of the terms in curly brackets on $\phi(X)$ giving zero. Suppose that

$$\phi(X) = A\sin(KX + \alpha) \tag{15.115}$$

Then

$$\begin{aligned} \nabla_X\phi(X) &= \nabla_X[A\sin(KX + \alpha)] = -2(1 - \cos K)A\sin(KX + \alpha) \\ &= -2k\phi(X) \qquad (k = 1 - \cos K) \end{aligned} \tag{15.116}$$

Substituting this result into (15.113) gives a quadratic in ρ from which the lowest buckling load, P_c, can be found. The deflexion mode corresponding to this can be written as

$$\begin{aligned} mu &= A\cos K(X + \tfrac{1}{2}) \ , \quad \tfrac{1}{2}h(v + v') = \frac{A}{\rho k}(2d + ck)\sin\tfrac{1}{2}K\sin KX, \\ mu' &= -A\cos K(X - \tfrac{1}{2}) \ , \quad \tfrac{1}{2}h(v - v') = -\frac{A}{d}(d + ck)\cos K(X + \tfrac{1}{2}). \end{aligned} \tag{15.117}$$

The total bending moment M at any section through the joints can be deduced from this to be

$$M = 2Am^2hc\sin\tfrac{1}{2}K\sin KX \tag{15.118}$$

By taking K as π/N, this moment is then zero, corresponding to a truss with zero mean lateral displacement and zero resultant moment at X equal to zero and N. The buckling load found from the above quadratic is then that for a pin-ended lattice column of length L (=mN). Using the approximation

---

[1] Renton, J.D. (1973) Buckling of Long, Regular Trusses. *Int. J. Solids Structures* V.9 p.1489.

$$k \approx \frac{\pi^2}{2N^2}\left(1 - \frac{\pi^2}{12N^2}\right) \qquad \textbf{(15.119)}$$

makes it possible to compare the results obtained with those given by the beam analogy. Taking EI as given by (A5.2.3), then for a Pratt or Howe truss,

$$P_c = \frac{\pi^2 EI}{l^2}\frac{1}{1 + \dfrac{\pi^2 EI}{l^2}\dfrac{1}{mh^2}\left(\dfrac{1}{a} + \dfrac{1}{d} + \dfrac{1}{6c}\right)} \qquad \textbf{(15.120)}$$

which is of the form of (15.110) but does not conform completely to the expression for GkA. Using similar methods for a Warren truss gives

$$P_c = \frac{\pi^2 EI}{l^2}\frac{1}{1 + \dfrac{\pi^2 EI}{l^2}\dfrac{1}{mh^2}\left(\dfrac{2}{d} - \dfrac{1}{12c}\right)} \qquad \textbf{(15.121)}$$

which again is not identical to that found using the expression for GkA given by (A5.2.2). In the case of a cross-braced truss, large axial forces are induced in all the bars and the form is not the same as that for a beam in bending and shear. It is given by

$$P_c = \frac{\pi^2 EI}{l^2}\frac{1}{1 + \dfrac{s}{2} + \dfrac{\pi^2 EI}{l^2}\dfrac{1}{mh^2}\left(\dfrac{1}{d(2+s)} - \dfrac{(4+5s)}{12c}\right)}, \quad s = \frac{adh^2}{m^2(ac+ad+cd)} \qquad \textbf{(15.122)}$$

(cf. (A5.2.5)).

The forms given by (15.120) to (15.122) are not identical to those given by an 'equivalent beam' approach for the reasons discussed in §8.5. Solutions for modular beams can generate extra shear-like terms associated with their flexural behaviour, as in (A5.1.3) to (A5.1.6) for example. It can be seen that here again the differences lie in the shear-like terms. The above three solutions for P_c should then be taken as the more accurate approximations, as they take into account the modular nature of the trusses.

15.11 Vibration Problems

Several methods can be used to find the natural frequencies of regular structures. They may be illustrated by analysing the rectangular grid shown in Figure 15.30. The notation used conforms to that shown in Figure 15.18. Here, a square grid of beams, each of length a with M spans in the x direction and N spans in the y direction, will be examined. Finite difference methods were first applied to the vibration of rectangular grids by Ellington and McCallion[1]. They considered only the flexural stiffness of the beams and took half the mass of each beam to be concentrated at the joints at either end. Wah and Calcote (1970) §4.3 use the vibration functions derived in §10.2 and §10.5 to analyse the flexural-torsional vibration of rectangular grids without lumping the distributed masses of the beams at the joints.

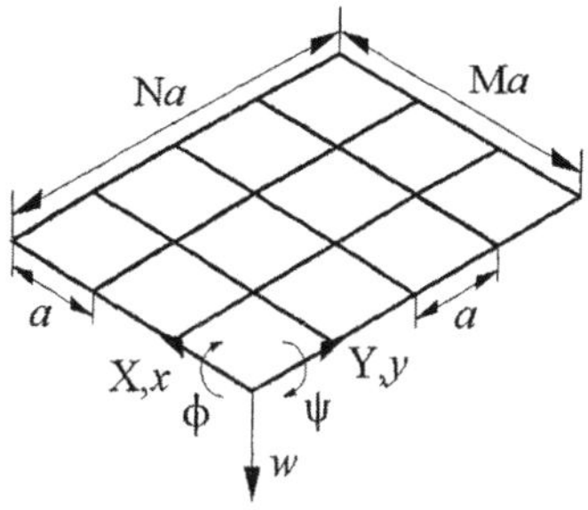

Figure 15.30 Notation for a rectangular grid.

[1] Ellington, J.P. and McCallion, H. (1959) The Free Vibration of Grillages. *ASME Journal of Applied Mechanics*, V.26, p.603.

Two approximate methods may also be used. Using the approximations to the vibration functions given by (10.10), the equations become a linear function of the parameter λ and can be analysed in the same way as the portal frame in Example 11.6. For comparison with the other results, all the beams of the grid will be taken to have the same length a and the same section with flexural stiffness EI and negligible torsional stiffness. The finite-difference equations for a typical joint vibrating freely are then

$$\frac{EI}{a^3}\begin{bmatrix} 8c_1+2c_2(\mathrm{E_Y}+\mathrm{E_Y^{-1}}) & 0 & 6c_4(\mathrm{E_Y^{-1}}-\mathrm{E_Y}) \\ 0 & 8c_1+2c_2(\mathrm{E_X}+\mathrm{E_X^{-1}}) & 6c_4(\mathrm{E_X}-\mathrm{E_X^{-1}}) \\ 6c_4(\mathrm{E_Y}-\mathrm{E_Y^{-1}}) & 6c_4(\mathrm{E_X^{-1}}-\mathrm{E_X}) & 48c_5-12c_6(\mathrm{E_X}+\mathrm{E_X^{-1}}+\mathrm{E_Y}+\mathrm{E_Y^{-1}}) \end{bmatrix}\begin{bmatrix} a\phi \\ a\psi \\ w \end{bmatrix} = \begin{bmatrix} 0 \\ 0 \\ 0 \end{bmatrix} \qquad \mathbf{(15.123)}$$

(cf. 15.72). Suppose that the joint deflexions are given by

$$a\phi = A\sin\frac{m\pi\mathrm{X}}{\mathrm{M}}\cos\frac{n\pi\mathrm{Y}}{\mathrm{N}}\ ,\quad a\psi = B\cos\frac{m\pi\mathrm{X}}{\mathrm{M}}\sin\frac{n\pi\mathrm{Y}}{\mathrm{N}}\ ,\quad w = C\sin\frac{m\pi\mathrm{X}}{\mathrm{M}}\sin\frac{n\pi\mathrm{Y}}{\mathrm{N}}\ , \qquad \mathbf{(15.124)}$$

multiplied by a sinusoidal function of time, sin $(\omega t+\alpha)$. Then making use of Table 15.1 and the approximations given by (10.10), (15.123) becomes

$$\begin{bmatrix} 8+4c_n & 0 & -12s_n \\ 0 & 8+4c_m & 12s_m \\ -12s_n & 12s_m & 24(2-c_m-c_n) \end{bmatrix} u = \frac{\lambda}{420}\begin{bmatrix} 8-6c_n & 0 & 26s_n \\ 0 & 8-6c_m & -26s_m \\ 26s_m & -26s_n & 12(52+9c_m+9c_n) \end{bmatrix} u \qquad \mathbf{(15.125)}$$

where

$$s_m = \sin\frac{m\pi}{\mathrm{M}}\ ,\quad c_m = \cos\frac{m\pi}{\mathrm{M}}\ ,\qquad s_n = \sin\frac{n\pi}{\mathrm{N}}\ ,\quad c_n = \cos\frac{n\pi}{\mathrm{N}}\ ,\qquad u = \begin{bmatrix} a\phi \\ a\psi \\ w \end{bmatrix}. \qquad \mathbf{(15.126)}$$

This can be solved as in Example 11.6 or by finding the zeros of the determinant of the difference of the first matrix and trial multiples of the second.

An even faster approximate solution is given by using the equivalent continuum equation (15.73). The mass of the grid within a typical modular area a^2 is $2\rho Aa$, where ρ is the density of the material and A is the area of a cross section. The inertia force p per unit area is then $-2\rho A\ddot{w}/a$ and taking the lateral displacement as

$$w = C\sin\frac{m\pi x}{\mathrm{M}a}\sin\frac{n\pi y}{\mathrm{N}a}\sin(\omega t+\alpha)\ , \qquad \mathbf{(15.127)}$$

(15.73) gives

$$\mu^4 = \lambda = \frac{\rho A\omega^2 a^4}{EI} \approx \frac{\pi^4}{2}\left(\frac{m^4}{\mathrm{M}^4}+\frac{n^4}{\mathrm{N}^4}\right). \qquad \mathbf{(15.128)}$$

These results can be compared with the natural frequencies in terms of μ given by Wah and Calcote in Table 4.1 (noting that their u is the same as μ above). Two sets of results, giving the greatest variation, are listed in the following table, where ‘E.&M.’ are Ellington and McCallion’s results and ‘W.&C.’ are those of Wah and Calcote.

M	N	m	n	E.&M.	W.&C.	(15.125)	(15.128)
4	4	3	3	2.281	2.356	2.378	2.356
4	5	3	4	2.341	2.436	2.461	2.439

In these examples, it is clear that the results given by the continuum analogy are more than adequate.

The continuum analogy has also been applied to the vibration of pin-jointed frames. The response of the bars to axial forces is given by (10.24). The response of pin-ended bars to shear forces can be deduced from (10.8). Setting M_A and M_B to zero gives the relationship

$$\begin{bmatrix} S_A \\ S_B \end{bmatrix} = -\frac{m\omega^2}{6c_3}\begin{bmatrix} 2c_1 & c_2 \\ c_2 & 2c_1 \end{bmatrix}\begin{bmatrix} v_A \\ v_B \end{bmatrix} \qquad \textbf{(15.129)}$$

where m is the mass of the bar, ρAl. For an infinitely stiff beam ($EI = \infty$) c_1, c_2 and c_3 become equal to unity and the response is that of a thin rigid rod. More commonly, half the mass of the bar is lumped at each end. This overestimates its rotational inertia, leading to underestimates in the natural frequency as in Ellington and McCallion's results above. (The square of the natural frequency is proportional to the stiffness and inversely proportional to inertia.)

Noor, Anderson and Greene[1] estimate the natural frequencies of both space trusses and space grids, using the equivalent continuum approach. Their plate-like behaviour of space grids is based on assumptions equivalent to those in Kirchhoff plate theory. Likewise, the beam-like properties of space grids are derived on an assumption equivalent to taking plane sections to remain plane. This can result in different flexibilities to those given here. For example the properties of the inverted batten beam given by (15.52) are given by the above authors as

$$f_{11}=f_{22}=\frac{4}{9h^2s^3}\left(\frac{3}{2b}+\frac{3}{d}\right),\quad f_{33}=\frac{2}{h^4sd},\quad f_{44}=f_{55}=\frac{1}{h^2sd},\quad f_{66}=\frac{1}{54s^3}\left(\frac{5}{2b}+\frac{9}{d}\right). \qquad \textbf{(15.130)}$$

Because trusses are relatively weak in shear, Timoshenko beam theory, as given in Example 7.8, is appropriate. Taking the mass of the truss to be represented by masses m at each joint, and treating each cross section through the joints as rigid, the linear and rotational inertias per unit length are $3m/s$ and $mh^2/2s$ respectively. Alternatively, the exact solution of the finite difference equations for the problem can be found[2]. A vibration mode with a half-wave length λs will be assumed and the particular case $b=d$, $h=s$ examined. Then the lowest values of $\omega^2 m/s^2 d$ found from the finite difference equation and the variations from it are given by

λ	Exact solution	(15.52)	(15.130)
10	1.427×10^{-3}	-3.44%	+6.80%
30	1.799×10^{-5}	-0.42%	+10.6%
50	2.336×10^{-6}	-0.15%	+10.9%

---

[1] Noor, A.K., Anderson, M.S. and Greene, W.H. (1978) Continuum Models for Beam- and Platelike Lattice Structures. *AIAA Journal*, V.16, p.1219.

[2] Renton, J.D. (1984) The Beam-Like Behavior of Space Trusses. *AIAA Journal*, V.22, p.273.

Appendix 1: Common Notation

A	Area of a cross-section.
B	A bimoment defined in §5.9.
B,C	Shear coefficients, as in (6.61) and §A4.1.
$\boldsymbol{C}$	A stiffness matrix, as in §3.11.
c	A wave velocity in §10.8.
$c_{1...6}$	Flexural vibration functions listed in Appendix 6.2.
c_{ij}	A stiffness coefficient referred to in §3.11.
D	Plate bending stiffness as defined by (A8.1.4).
$\boldsymbol{d}$	A column vector of deflexions, usually corresponding to $\boldsymbol{P}$.
d_1,d_2	Axial and torsional vibration functions listed in Appendix 6.2.
$\mathrm{E_X}$	A finite difference operator which increments X by one (see (8.26)).
E	Young's modulus, the stress required to produce a unit strain in the same direction.
F	Usually a force.
f_{ij}	A flexibility coefficient, as in (8.12).
G	Shear modulus, the shear stress required to produce a unit shear strain.
g	Acceleration due to gravity (approximately 9.81 m/s^2).
I	Second moment of area about an axis indicated by a subscript.
I_ω	The warping constant defined by (5.81).
J	Torsional constant defined in §5.2, but not necessarily the second polar moment of area.
$\boldsymbol{K}$	A stiffness matrix as in §11.1.
k	A shear constant given by (6.4).
k_{ij}	A stiffness coefficient, as in (11.120).
L,l	Usually lengths.
M	A moment.
m	A mass.
$\boldsymbol{P}$	A column vector of loads, usually corresponding to $\boldsymbol{d}$.
P	An axial force.
p	A uniform force per unit length or per unit area.
q	A distributed force per unit length or a shear flow defined in §5.3.
R	A radius of (flexural) curvature.
r	A radial coordinate.
$\boldsymbol{S}$	A compliance matrix, as in (3.63).
s_{ij}	A compliance coefficient, as in (3.63).
$\boldsymbol{T}$	A transformation matrix, as in §11.1.
t	Time, or thickness of a plate or shell.
t_{AB}	The tension coefficient for a bar AB (see §2.7).
$U_{()}$	Strain energy (per unit volume, V, per unit length, l) see §7.1.
u,v,w	Usually displacements in the x,y and z directions.
W	Work done, as in (2.13) and (7.10) or a weight.
x,y,z	Cartesian coordinates (right-handed convention used).
α,β	Angles, or the functions defined by (5.90).
γ	Usually a shear strain as in (3.11), or a rate of overall shear displacement, as in (8.12).
Δ	A forward difference operator defined by (8.26) or del as in (15.7).
δ	A displacement or an indication of an increment.
ϵ	A normal strain, as defined by (3.8).
ε	A rate of overall longitudinal displacement, as in (8.12).

θ	An angle or rotation.
ϑ	A rate of rotation (twist or curvature) along a (generalised) beam.
λ,μ	Lamé constants defined by (3.77), or
λ	some (dimensionless) parameter (cf. §A6.2),
μ	a parameter defined by (10.4), see also (10.23) and (10.31).
ν	Poisson's ratio as defined in §3.3.
Π	The total potential energy of a body and its loads, defined by (7.11).
π	3.14159265358979323846264338327950288419716939937510582097494459... etc.[1]
ρ	Density (mass per unit volume).
Σ	A summation.
σ	Usually a normal stress, as defined in §3.1.
τ	Usually a shear stress, as defined in §3.1.
$\boldsymbol{\Phi}$	A harmonic vector in §A7.1.
Φ	Potential, as in §2.3 or a function used in §6.5.
ϕ	An angle, or Prandtl's stress function in (5.23) or the stability functions in (9.8).
χ	A stress function defined by Love (1952) §229.
Ψ	An Airy function in §6.7 or a harmonic scalar in §A7.1.
Ω,ω	Rates of rotation with time.
ω	Warping function used in §5.9.
∇	A backward difference operator defined by (8.26), or nabla as in (15.7).
$\Delta\!\!\!\!\nabla$	A finite difference operator defined by (8.26).

[1]For the next 9,937 digits, see www.crosswinds.net/~bohabohu/

Appendix 2: How to Use the Structural Analysis Programs

A2.1 Getting Started

A2.1.1 Running a linear analysis

Details of the system are given later. This section shows it in operation. The following examples can be run from the disk. You can copy the disk to a new folder and run from there to alter files or create new ones. The best display is given by running in purely MS-DOS mode.

Running the programs is presented as a dialogue between the screen prompts, labelled 'PC' and your response at the keyboard labelled 'You:'. The exact keystrokes you need to use will be shown. If you need to press the 'Enter' or 'Carriage Return' key, this will be shown as [Enter]. The function keys to press are also shown in square brackets as [F1] etc.

The four initial prompts are shown blinking in display frames. The later prompts appear white on a black background or vice-versa.

You: RUN [Enter]

PC: `Press C to continue or H to halt`

You: C

PC: `CHOOSE PROGRAM BY PRESSING A FUNCTION KEY`

You: [F3]

This picks out the Plane Rigid-Jointed Framework Analysis program. The Plane Grid Frame and Plane Pin-Jointed Frame programs operate similarly by pressing [F1] or [F2].[1]

PC: `CHOOSE STRUCTURE BY PRESSING A FUNCTION KEY`

You: [F2] {Initially opposite `R2DSIPTL.STR` on the menu}

This picks up the data for a structure and its loading from the file listed by the [F2] key. If there are files listed by the other function keys, these keys can be pressed instead of [F2].[2]

PC: `CHOOSE SECTION DATA BY PRESSING A FUNCTION KEY`

You: [F2] {Initially opposite `NORMALSI.SEC` on the menu}

This picks up the data for the sections used by the structure. Normally, several section files would be listed. A file naming convention is used to show only those section files which are in the same units as the structures files. If more than one section file is needed, or some of the section data is going to be entered from the keyboard, this will be prompted for later.[2] *The structures program you chose now analyses the structures problem you chose.*

PC: `Show mass of structure? Key Y or N.`

You: Y

Here and later, "Y" is for Yes and "N" is for No. Thus pressing "Y" will cause the total mass of the structure to be shown a little later (in kilograms if SI units are used).

---

[1] The normal three-dimensional analysis programs, accessed by pressing [F4] or [F5], operate similarly apart from the displays of the frames. An additional prompt appears before the displays, requesting the projection to be used: "Choose mapping. Key I,X,Y or Z". Pressing key "I" then shows an isometric projection of the frame, with the X axis to the right, the Y axis to the left and the Z axis upwards. Pressing keys "X", "Y" or "Z" shows the views of the frame as seen looking along the X, Y or Z axes. The operation of the remaining two programs, initiated by pressing [F6] and [F7], is documented separately.

[2] Only files which follow a certain naming convention (see later) and are on the same (sub)directory as the rest will be shown. The option to pick a file which has not been listed can be chosen by pressing [F6]. If it is on another directory or drive, the full path should precede the filename. All data can also be entered from the keyboard. This option is chosen by pressing [F7].

PC: `Show the frame? Key Y or N.`

You: Y

If the machine does not have VGA graphics or better, press "N". The unloaded frame is now shown with each joint labelled with either one or two symbols, usually letters or numbers.[3]

PC: `Press C, H or F to continue, halt or file.`

You: C

The next prompt is displayed after the analysis has been completed.

PC: `Frame mass is .........`
`Show the affected frame? Key Y or N.`

You: Y

PC: `For Load Case:`

You: 1 [Enter]

The program analyses the structure for all the loadings (up to three) specified on the structures file. However, on the sample files provided, only one load case is listed.

PC: `Amplification factor:`

You: 1 [Enter]

The amplification factor scales the given loading up or down, depending on whether a number greater or less than 1 is used. The deflected form is shown to scale, so that it is usually necessary to enter a high amplification factor for the mode of deformation to be seen. The bars are coloured according to their highest state of stress. The colours go from black (low stress) to violet (high stress) in the order given by the stress spectrum at the bottom right of the display.

PC: `Press C, H or F to continue, halt or file.`

You: C

The machine now returns to the "Show the affected frame? Key Y or N." prompt above, to allow you to repeat the display with different parameters. If you press the "N" key, you now move on to the next stage.

PC: `Output the deflections? Key Y or N.`

You: Y

The deflections can now be shown on the screen or sent in the same format to a file called NORMAL.DFL for printing out or further analysis.

PC: `To NORMAL.DFL file? Key Y or N.`

You: N

The deflection of the joints (displacements and rotations) are now listed after the joint labels (which were used at the joints in the display of the unloaded structure).

You: {Press any key to continue}

When all the deflections have been listed, the bar end loads (moments and forces) can be shown on the screen or sent in the same format to a file called NORMAL.LOD.

PC: `Output the bar end loads in bar coordinates? Key Y or N.`

You: Y

PC: `To NORMAL.LOD file? Key Y or N.`

You: N

The end joint labels for each bar are now shown, followed by the critical stress ratio (see separate section on this). The bar end loads at each joint are then listed after its joint label.

PC: `Press any key to continue.`

A screenful of data is shown each time a key is pressed. The final keystroke ends the program.

---

[3] The three-dimensional structural analysis programs return to the above step to allow you to view the structure from another projection (see note 1).

A2.1.2 Running a stability analysis

The way in which a stability analysis is run is very similar to that just described for linear analyses. The main difference is that the initial loading is multiplied by various load factors to find a multiple of the loading at which buckling can occur. This analysis is non-linear, so that the loading history, given by the sequence of load factors chosen, affects the results. If only very small deflections are induced prior to buckling, this history will not be significant. Otherwise, an accurate estimate can only be obtained by increasing the loading in small increments.

The dialogue follows the style previously described.

You: RUN [Enter]

PC: `Press C to continue or H to halt`

You: C

PC: `CHOOSE PROGRAM BY PRESSING A FUNCTION KEY`

You: [F6]

This picks out the Stability Analysis (Rigid 3-D Frame) program.

PC: `CHOOSE STRUCTURE BY PRESSING A FUNCTION KEY`

You: [F4] {Initially opposite `R3DSIPYR.STB` on the menu}

This picks up the data for a structure and its loading from the file listed by the [F4] key. If there are files listed by the other function keys, these keys can be pressed instead of [F4].

PC: `CHOOSE SECTION DATA BY PRESSING A FUNCTION KEY`

You: [F2] {Initially opposite `NORMALSI.SEC` on the menu}

This picks up the data for the sections used by the structure. The program now analyses the structures problem you chose.

PC: `Show mass of structure? Key Y or N.`

You: N

PC: `Show the frame? Key Y or N.`

You: Y

It is assumed that the machine has VGA graphics or better. If not, press "N" instead of "Y", as suggested in §A2.1.1.

PC: `Choose mapping. Key I,X,Y or Z`

You: I

An isometric view of the unloaded frame is now shown. The alternatives are explained in note 1 of §A2.1.1.

PC: `Press C, H or F to continue, halt or file.`

You: C

The program now returns to the prompt "Show the frame? Key Y or N." to allow the frame to be viewed using a different projection. If you now key "N", the next prompt is displayed after two analyses have been completed, calculating the determinant of the stiffness matrix for the whole structure under zero loading and that for the modified stiffness matrix, allowing for the axial bar forces under the specified loading.

PC:
```
Zeros count:0 Scaled determinant:.. Load factor:0.000000E+0
Zeros count:0 Scaled determinant:.. Load factor:1.000000E+0
Key E:End; S:Show frame; R:to reset or T: try load factor ..
```

You: T

From the two analyses already carried out, the computer has extrapolated an estimate of the multiple of the specified loading at which buckling first occurs (at which point the determinant should be zero). Keying "T" repeats the analysis using this load factor. When this new analysis has been completed, the above prompt will show the new value of the (scaled) determinant found.

"T" can be keyed repeatedly until the next load factor to try is negligibly different from the one just tried[4].

PC: `Key E:End; S:Show frame; R:to reset or T: try load factor ..`

You: S

PC: `Amplification factor:`

You: 0.5 [Enter]

The amplification factor scales the deflection mode up or down, depending on whether a number greater or less than 1 is used. The above value may be too high, in which case the deformed frame will be shown as a jagged jumble of lines. If the factor is too high, it can be reset when the "S:Show frame" option is chosen again later. If the value is too low, the frame will show no obvious signs of distortion. The value can be reset in the same way as when it is too high. However, if the value of the amplification required to show the buckling mode is not significantly less than unity, the mode is unlikely to be a true representation of the buckling mode.

PC: `Choose mapping. Key I,X,Y or Z`

You: I

The screen now shows the frame in its first buckling mode. The bars are coloured according to what proportion of their fixed-end buckling load has been reached. This is indicated in the reference spectrum at the bottom right of the screen.

PC: `Press C, H or F to continue, halt or file.`

You: C

PC: `Key E:End; S:Show frame; R:to reset or T: try load factor ..`

If the view of the buckling mode was unsatisfactory, you can now try a different amplification factor or projection by pressing "S", or try a load factor closer to the buckling value. Alternatively, higher buckling loads can be examined by pressing "R" (see footnote 4). Finally,

You: E

leads to the option of outputting the unamplified deflections for the current loading (Load Case 2) and the previous loading (Load Case 1). This operates in the same way as with a linear analysis, after which the program terminates.

If the axial load in a beam exceeds the fixed-end flexural buckling load, the number of times by which it does so is added to a fixed-end buckling load count, which is displayed. The significance of this and the zeros count is explained in §12.7.

A2.1.3 Running a vibration analysis

The way in which a vibration analysis is run is very similar to that just described for a stability analysis, with the additional options of joint deflection and bar load output, in the same form as for linear analyses.

The dialogue follows the style previously described.

You: RUN [Enter]

PC: `Press C to continue or H to halt`

You: C

PC: `CHOOSE PROGRAM BY PRESSING A FUNCTION KEY`

You: [F7]

This picks out the Vibration Analysis (Rigid 3-D Frame) program.

PC: `CHOOSE STRUCTURE BY PRESSING A FUNCTION KEY`

---

[4] Alternatively, "R" can be pressed to reset the next load factor manually. This may be used to enter an estimate which you think will be better than the extrapolated (or interpolated) value, or to examine buckling loads and modes higher than the fundamental one.

You: [F3] {Initially opposite R3DSIPTL.VIB on the menu}

This picks up the data for a structure and its loading from the file listed by the [F3] key. If there are files listed by the other function keys, these keys can be pressed instead of [F3].

PC: CHOOSE SECTION DATA BY PRESSING A FUNCTION KEY

You: [F2] {Initially opposite NORMALSI.SEC on the menu}

This picks up the data for the sections used by the structure. The program now analyses the structures problem you chose.

PC: Show mass of structure? Key Y or N.

You: N

PC: Show the frame? Key Y or N.

You: Y

It is assumed that the machine has VGA graphics or better. If not, press "N" instead of "Y", as suggested in §A2.1.1.

PC: Choose mapping. Key I,X,Y or Z

You: I

An isometric view of the unloaded frame is now shown. The alternatives are explained in note 1 of §A2.1.1.

PC: Press C, H or F to continue, halt or file.

You: C

The program now returns to the prompt "Show the frame? Key Y or N." to allow the frame to be viewed using a different projection. If you now key "N", the next prompt is displayed after the static (zero frequency) analysis has been completed.

```
PC: Zeros count:0 Scaled determinant:.. Frequency:0.000000E+0
      Enter your value of the next frequency to try:
```

You: 4 [Enter][5]

```
PC: Zeros count:0 Scaled determinant:.. Frequency:4.000000E+0
    Key E:End, S:Show frame, R:to reset,
    N:New resonant frequency sought or T:try frequency..
```

You: T

Repeat keying "T" at the prompt until the next frequency to try does not differ significantly from the one just tried.

You: S

PC: Amplification factor:

You: 0.005 [Enter][5]

The amplification factor scales the deflection mode up or down, depending on whether a number greater or less than 1 is used. The above value may be too high, in which case the deformed frame will be shown as a jagged jumble of lines. If the factor is too high, it can be reset when the "S:Show frame" option is chosen again later. If the value is too low, the frame will show no obvious signs of distortion. The value can be reset in the same way as when it is too high. However, if the value of the amplification required to show the vibration mode is not significantly less than unity, the mode is unlikely to be a true representation of the natural frequency mode.

PC: Choose mapping. Key I,X,Y or Z

You: I

The screen now shows the first natural mode of frame vibration. The bars are coloured

---

[5] These are suggested values for the structure file R3DSIPTL.VIB. The frequency used is ω radians per second, as in Chapter 10. To obtain the frequency in Hertz, divide ω by 2π.

according to what proportion of their fixed-end natural frequency has been reached. This is indicated in the reference spectrum at the bottom right of the screen.

PC: `Press C, H or F to continue, halt or file.`

You: C

PC: `Key E:End, S:Show frame, R:to reset,`
`N:New resonant frequency sought or T:try frequency..`

If the view of the vibration mode was unsatisfactory, you can now try a different amplification factor or projection by pressing "S", or try a frequency closer to the natural frequency. Here, the next natural frequency will be sought.

You: N

PC: `Enter resonance sought (lowest=1):`

You: 2 [Enter]

PC: `Enter your value of the next frequency to try:`

You: 7 [Enter][5]

PC: `Key E:End, S:Show frame, R:to reset,`
`N:New resonant frequency sought or T:try frequency..`

You: T

As before, "T" is keyed until the next resonant frequency to try is not significantly different from the one just tried. The mode of vibration can be examined as before, this time an amplification factor of 0.01 is likely to be more appropriate[1]. Next, the response to a forced vibration will be examined.

PC: `Key E:End, S:Show frame, R:to reset,`
`N:New resonant frequency sought or T:try frequency..`

You: R

PC: `Enter your value of the next frequency to try:`

You: 10 [Enter]

PC: `Key E:End, S:Show frame, R:to reset,`
`N:New resonant frequency sought or T:try frequency..`

You: E

PC: `Output the deflections? Key Y or N.`

You: Y

PC: `To NORMAL.DFL file? Key Y or N.`

You: N

The computer now displays the name of the structures file used and the frequency of the applied loading. The joints are then listed, followed by the amplitudes of their rotations (Xrot, Yrot, Zrot) and displacements (Xdisp, Ydisp, Zdisp) in the directions of the global coordinates (X,Y,Z) used.

PC: `Press any key to end display`

You: {Any key}

PC: `Output bar end loads in bar coordinates? Key Y or N.`

You: Y

PC: `To NORMAL.LOD file? Key Y or N.`

You: N

The computer now displays the name of the structures file used and the frequency of the applied loading. The bars are then listed (in terms of their end joints) followed by their frequency ratios. (These ratios are of the frequency applied, to the bars' natural fixed-end frequencies, and are used to determine the colours chosen earlier from the frequency spectrum for displaying the bars.) After each bar joint, the amplitudes of the moments (Mx, My, Mz) and forces (Fx, Fy, Fz) at that end of the bar are given in terms of the bar coordinates.

PC: `Press any key to end display`

You: {Any key}
This terminates the program. If the frequency exceeds the fixed-end natural frequency of a beam in flexure, the number of times by which it does so is added to a fixed-end natural frequency count, which is displayed. The significance of this and the zeros count is explained in §12.7.

A2.2 Data Files

A2.2.1 Entering data

In principle, all the data for an analysis can be entered directly from the keyboard, although it is normally more convenient to store it in files which are then read by the programs. These files can be generated by any simple text editor, such as MS-DOS 'edit', Windows 'Notepad' or the Norton editor. NB: a normal word processor will generate information about the page layout and character style. This will *not* be understood by the program; only a simple ASCII text editor should be used. Two kinds of data file are needed, a structures problem data file (*.STR file for example) and the section data file (*.SEC file). Try printing the sample files provided and listed in §A2.4 to see the format required.

The structures problem data specifies the structure and its loadings. (In the linear analyses, up to three separate loadings can be analysed simultaneously for the same structure.) If this data is stored in an appropriately named file which is in the working directory, then it will appear under the screen prompt `CHOOSE STRUCTURE BY PRESSING A FUNCTION KEY` and can be chosen by pressing the appropriate key, [F1] to [F5], listed against the file name. If there are more than five such files in the working directory, or the file is not appropriately named, then [F6] should be pressed and the name entered, as indicated by the prompt. File naming is discussed in the next section. If the file is in another directory or drive, the full path should precede the name, in the usual manner. Finally most of the programs allow all the structures problem data can be entered from the keyboard by pressing [F7].

The structures problem data contains no details of the section properties of the individual bars. Instead, a section label, consisting of two letters and a number, is used. The section data associated with this label is entered separately, usually from a section file. Thus permanent section files of standard sections can be referred to for a wide variety of structural problems. If the naming convention for files is adhered to, a list of section files using the same units (imperial, metric etc.) as the structures problem file just chosen will appear under the screen prompt `CHOOSE SECTION DATA BY PRESSING A FUNCTION KEY`. As before, the appropriate file can be chosen by pressing the function key, [F1] to [F5], listed against it, or by pressing [F6] and entering the file name (and its path, if it is not in the current directory). Also, the section data can be entered entirely from the keyboard if [F7] is pressed. If the section data is stored in more than one file, the option to pick another file will be given by a prompt during the running of the program. All remaining section data not listed in any file may be entered from the keyboard.

A2.2.2 File naming conventions

The first three letters of a program name indicate the type of structure it is intended to analyse. These are

- `GRD` Plane grid frame
- `P2D` Pin-jointed plane truss
- `R2D` Rigid-jointed plane frame
- `P3D` Pin-jointed three-dimensional truss
- `R3D` Rigid-jointed three-dimensional frame.

The same first three letters should be used for the corresponding structures problem file. The next (4th) letter of a program name indicates the type of analysis used.

N	Normal (linear) analysis
S	Stability analysis
V	Vibration analysis.

The corresponding last three (extension) letters of the structures problem file name are

.STR	Normal structural problem
.STB	Structural stability problem
.VIB	Structural vibration problem.

Together with the matching first three letters, this enables the package to display a list of structures problem files appropriate to a particular program. The remaining letters of the program names are for reference purposes only and do not affect the running of the package.

The 4th and 5th letters of the structures problem file name (SI or AM for example) indicate the units in which the data is expressed (e.g. SI or American). These are matched with the 7th and 8th letters of the section data files (NORMAL**SI**.SEC and NORMAL**AM**.SEC for example). The package uses these letters, together with the extension .SEC to display a list of section files appropriate to the chosen structures problem file.

The 6th to 8th letters of a structures problem file name can be used to indicate the kind of structure described (WRN for a Warren truss for example).

A2.2.3 Structures problem data files

All the input and output data files are simple alphanumeric files which can be viewed and printed using the usual MS-DOS commands `type` and `print`. They may also be viewed and printed from Windows Notepad. A structures problem data file consists of optional comments in square brackets at the beginning, followed by joint data, bar data, load data and in the case of vibration problems, joint inertia data. Each set of data is terminated by a vertical bar "|".

A2.2.3a Joint data

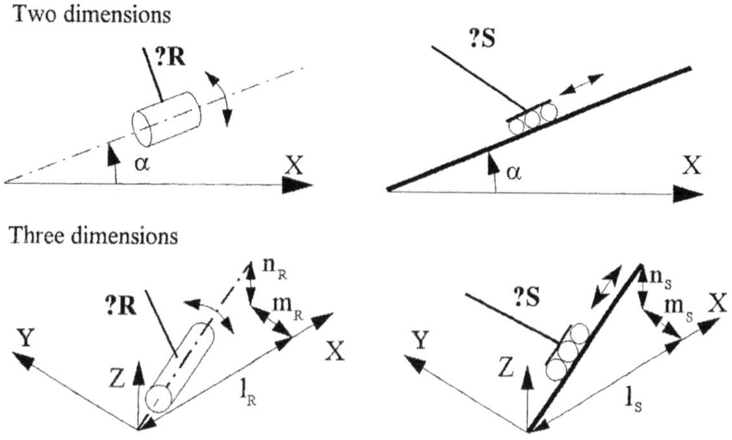

Figure A2.1 Rocker and Slider joints in two and three dimensions.

The joint data consists of a list of joints followed by their global coordinates, X, Y and in the case of the three-dimensional analyses, Z. Each joint is specified by two characters, which can be letters, numbers or any other printable character, excluding the terminator "|" already mentioned. The second of the two specifying characters can be just a space (giving the appearance that only one character has been used). Upper and lower case characters are treated as distinct. Thus the twelve joints specified by the character pairs

`Aa aA 32 q  ww A1 0  6k 7? ## &  £$`

are all recognised as distinct joints, although normally only letters and numerals are used. Certain capital letters used as a second character have special meanings in particular analyses. These are listed in the following table, where the column headings follow the naming conventions given in §A2.2.2.

Analysis	GRD	P2D	R2D	P3D	R3DN	R3DS	R3DV
Antisymmetry					A	A	A
Elastic support	E	E	E	E	E	E	E
Fixed support	F		F		F	F	F
Mirror symmetry					M	M	M
Normal reaction					N	N	N
Pinned support	P	P	P	P	P	P	P
Rocker + Slider			Q		Q	Q	Q
Rocker support	R				R	R	R
Slider support		S	S	S	S	S	S

Some apparent omissions from the table are intentional. Thus for example the second letter `F` is not appropriate for pin-jointed frame analysis, as the maximum fixity permitted for a joint is given by pinning it to a rigid support. Also, the second letter `M` is not needed in the plane, rigid-jointed frame analysis to indicate a joint on a plane of mirror symmetry. This can be achieved by using a slider joint instead.

Only the second letters `F` and `P` are in themselves complete descriptions of joint constraint. All the others require further data which is given after the joint coordinates. The required data is listed in the following table. Elastic supports (indicated by the second letter `E`) require all the possible rotational and displacement stiffnesses to be listed, even if some of them are zero. The use of large numbers to represent infinite stiffnesses is not recommended, as this may lead to computational errors. The use of other types of joint constraint avoids this problem.

The remaining types of joint constraint are referred to as 'vectored' constraints, as they act in specified directions. In the case of the plane frame analyses carried out by the first three programs, the direction is given by an angle α, measured anticlockwise from the X axis. In three dimensions, it is described by the components (l,m,n) of a vector, which need not be a unit vector, in the required direction. For a **R**ocker joint, the direction is that of the axis about which the joint is free to rotate, all other joint deflections being prevented. For a **S**lider joint, the direction is that along which the joint is free to displace, again all other joint deflections being prevented (see Figure A2.1). The second letter `Q` indicates a rotating and sliding support for rigid-jointed frames. For the rigid plane frame analysis, only the angle of sliding need be indicated, as the rotation must be about an axis normal to the plane of the frame. However, in the three-dimensional case, the components of the directions of rotation (l_R,m_R,n_R) and sliding (l_S,m_S,n_S) must both be given, and in that order.

The remaining support vectors are normal to the planes of **A**ntisymmetry, **M**irror symmetry, or to the smooth rigid plane providing only a Normal reaction. The directions of these vectors are expressed by the components (l,m,n) as these support cases are available only for the three-dimensional rigid-frame analysis. If a structure is both symmetrical and symmetrically loaded, a mirror-like plane can be introduced, cutting the structure in half. One half of the structure and its loading then appears as the mirror image of the other half, imagining it to be reflected in this plane. Mirror-symmetrical joints can be inserted where this plane intersects the structure, and only one half of the structure described and analysed. If more than one plane of symmetry exists, as in a dome under gravitational loading for example, these can be used to section the structure further, and only the smallest slice analysed. Antisymmetry exists if the structure is symmetrical, but its loading is the exact opposite of a symmetrical loading. Antisymmetrical joints can now be inserted on the intersection of the structure and the plane of symmetry, and just one half of the structure analysed as before. Mirror symmetry and antisymmetry must be used with care in stability and vibration analyses, as modes of buckling and free vibration may be overlooked. For example, antisymmetrical in-plane sidesway buckling of a portal frame will normally occur at lower buckling loads than symmetrical in-plane buckling, even though the loading is symmetrical axial loading of the stanchions.

These conditions are summarised in the following table, which lists the additional joint data required (after the joint coordinates) for supports described by various second characters. In the Elastic support row, r indicates a rotational stiffness about a particular axis and k a linear stiffness in the direction of a particular axis.

Analysis	GRD	P2D	R2D	P3D	R3D
Antisymmetry					l,m,n
Elastic support	r_X,r_Y,k_Z	k_X,k_Y	k_X,k_Y,r_Z	k_X,k_Y,k_Z	r_X,r_Y,r_Z,k_X,k_Y,k_Z
Mirror symmetry					l,m,n
Normal reaction					l,m,n
Rocker + Slider			α		l_R,m_R,n_R,l_S,m_S,n_S
Rocker support	α				l_R,m_R,n_R
Slider support		α	α	l_S,m_S,n_S	l_S,m_S,n_S

A2.2.3b Bar data

The bar data consists of a listing of bars, denoted by their 'near' joint character pair followed by their 'far' joint character pair. (The longitudinal axis of a bar is taken to go from the 'near' joint to the 'far' joint. In three-dimensional problems it is the bar's z axis.) These pairs are followed by the bar's section label (see §A2.2.4 'Section data files'). For the R3D programs only, this is followed by the bar's orientation about its z axis. In general, this is the clockwise angle, β degrees, through which it has been rotated from the orientation such that the bar's x axis is parallel to the plane containing the frame's X and Y axes and its y axis has a component in the -Z direction (see Figure A2.2). This method of analysis was introduced because most structures are designed so that β is zero. When the z axis of the bar coincides with the Z axis of the frame, the initial orientation ($\beta = 0$) is taken to be such that the x and y axes coincide with the X and Y axes respectively. If the z and Z axes are in opposite directions, the x and X axes are taken to be opposite directions and the y and Y axes in the same direction when β is zero.

As shown, the bar's x axis is parallel to the flanges of the section and the y axis parallel to its web. Normally, these are also the major and minor principal axes of the section. However, for angle sections this is not the case. Here, the major and minor principal axes will be denoted by u and v respectively. Further information on these local axes will be found in §A2.2.4.

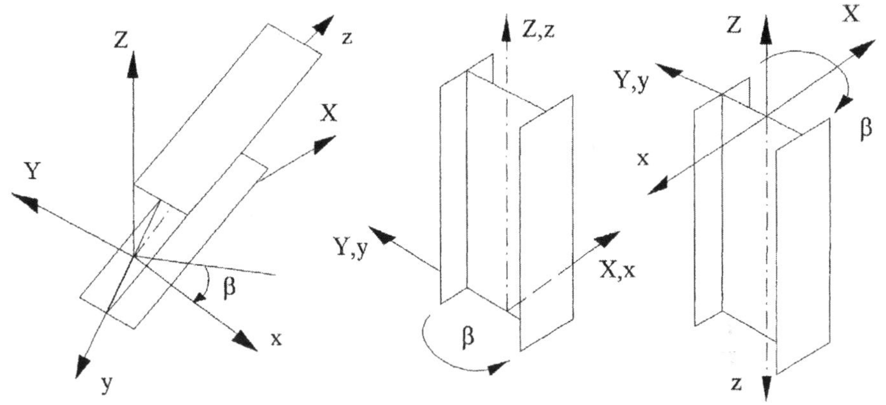

Figure A2.2 Bar rotation angle β

A2.2.3c Load data

The load data for normal analyses consists of up to three different load cases, each terminated with a vertical bar "|". All the load cases are analysed simultaneously. The data files for stability and vibration problems may contain only one load case each. Every load case consists of a list of the loaded joints and the loads acting on them. In addition, for normal rigid-jointed frame analysis (`GRD, R2D & R3DN`) descriptions of loading along the bars can be included. Each loaded joint is denoted by its joint character pair, followed by the full set of external loads acting on it, including zero values for those which are not applied. The following diagram shows the sets of loads required as data by the analyses listed on the first line. The second line shows the order in which these loads are entered. M indicates a moment about a particular axis and F a force in the direction of a particular axis.

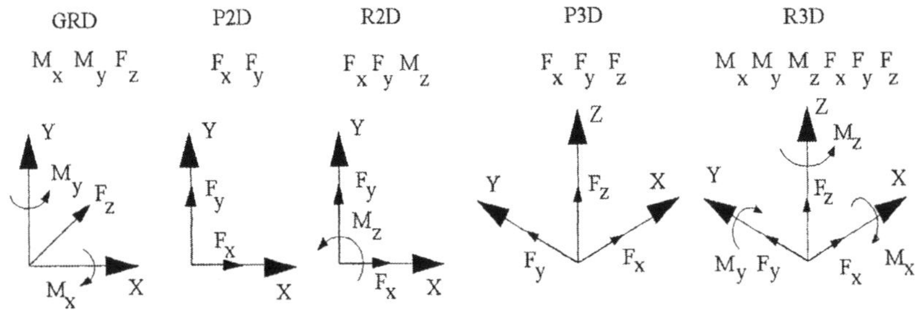

Figure A2.3 Loading conventions for various analyses.

Note that a right-handed coordinate system has been chosen in all cases except the first one, where it is more natural to view the grid from above and consider F_Z to be acting downwards. (In a right-handed system, F_Z would be upwards.)

As already mentioned for the plane grid, rigid-jointed plane frame and rigid-jointed three-dimensional frame analyses, in addition to the loads at the joints, point loads and distributed loading along the bars can be specified too. For example, the file `R2DSIPTL.STR` has a bar loading description:

```
B  *  C    Q  1E3  0     P  1E4  2     P  1E4  8    |
```

This indicates that the bar BC has a distributed load (`Q`) of 1kN/m starting at B and two point loads (`P`) of 10kN applied at 2m and 8m from B. As can be seen, each load is specified by three components: its type letter (`P` or `Q`), its magnitude and its position (or starting position) measured from the first joint. The loads are taken to act in the opposite direction to the beam's y axis, thus producing bending about the section's major axis. A sequence of up to five such loads can be specified for each bar. Each sequence, as above, is terminated with a "|". The two joints must be listed in the same order as the initial joint listing and the loadings are in the order of their distance from the first of these two joints. Note that the symbol "|" terminating a sequence of bar loads is *in addition* to the "|" terminating the list of loadings for a particular load case.

The vibration analysis automatically allows for the distributed inertia loads generated by the beams. In addition, the inertia of rigid bodies attached to the joints can also be taken into account. This information is listed after the load data. Each joint with such a body attached is denoted by its joint character pair, and this is followed by the inertias I_x, I_y & I_z and the mass M of the body. If the body is to be treated as a point mass, the three rotational inertias must be listed as zero. As with the lists of joints, bars and loads, this list must also be terminated with "|".

A2.2.4 Section data files

Examples of section data files will be found in §A2.4. Any number of section data files can be used in analysing a given problem, provided that a common set of global units is used and that the same section label, as described later, is not used for more than one section. The files are meant to correspond closely to section tables, where the units used (inches or cm) may differ from the global units used in describing the structure (feet or m). A section file starts with optional comments in square brackets, followed by properties which apply to all the sections in the file. These are

L_G/L_S E G σ_R ρ F/ma

L_G/L_S is the (global length unit)/(section length unit) ratio. A common set of global units (feet or metres for example) is used throughout the input and output of data, except in describing section properties in a section file. Tables will often list these in a different unit of length, L_S, from the global unit of length, L_G. The conversion factor L_G/L_S allows the section file data to be listed in the same units as the tables. Thus for example in a section file using american units, such as `NORMALAM.SEC`, L_G/L_S is the foot/inch ratio (12) and in a section file using SI units, such as `NORMALSI.SEC`, L_G/L_S is the metre/centimetre ratio (100).

E is Young's modulus in global units.

G is the shear modulus in global units.

σ_R is the reference stress (maximum working stress or yield stress) in global units. This is used in calculating the critical stress ratio (see §A2.2.5).

ρ is the density of the section material in global units.

F/ma is the global ratio of the force units to the mass-acceleration units. Thus in an american section file such as `NORMALAM.SEC`, force is measured in kips, mass in pounds and acceleration in feet per second^2, giving 32,200. In `NORMALSI.SEC`, SI units are used and so the ratio is unity.

The individual section data are listed on the lines following the above common section data. Each bisymmetric section (such as an I-beam) can be described as follows:

S.L. A I_x I_y Z_x Z_y J Z_t

S.L. is the **section label**. This consists of two letters and a number. Capital and lower case letters are treated as being distinct from one another. The letter pairs ?A, ?T, ?G, ?R, CH, Ch and SH are reserved for other types of section described later. (The query can be any letter.) The number can be any positive integer less than 65536.

A is the area of the section.

I_x is the second moment of area of the section about its x axis (here the major axis).

I_y is the second moment of area of the section about its y axis (here the minor axis).

Z_x is the elastic section modulus about the x axis. This is I_x/D_x, where D_x is the furthest distance of any point on the section from the x axis.

Z_y is the elastic section modulus about the y axis. This is I_y/D_y, where D_y is the furthest distance of any point on the section from the y axis.

J is the Saint-Venant torsional constant for the section. That is, GJ gives the section's torsional stiffness, when a consistent set of units is used.

Z_t is the torsional section modulus, which is the torsional equivalent of the elastic section moduli for bending. It is the ratio T/τ_m where T is the torque applied and τ_m is the maximum shear stress it induces in the section.

Other types of section require different data to describe their properties. These are illustrated in Figure A2.4.

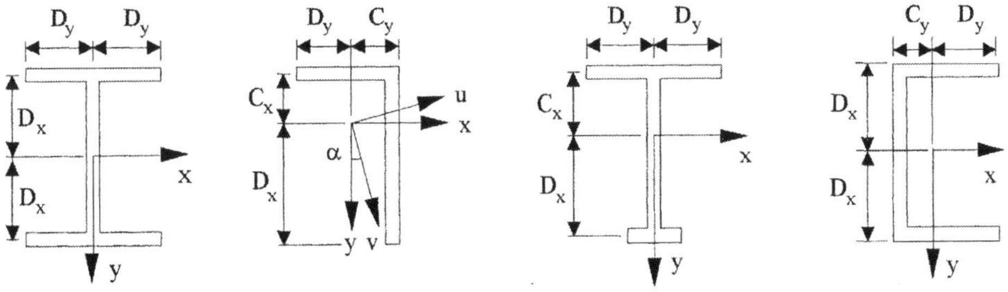

Figure A2.4 Section data.

If there is no symmetry about the section's x axis (as in Tee sections) then a correct stress analysis may need to use the distance C_x of the furthest point from the x axis measured in the opposite direction to D_x (given above). Likewise, if the section is not symmetric about its y axis (as in Channel sections) the distance C_y may be required. In **all** cases, the x axis is parallel to the flanges of the section and the y axis parallel to its web. Except for angle sections, these will also be the respective major and minor principal axes of the section. For angle sections however, the major and minor principal axes, u and v, are distinct from the x and y axes and at an angle α to them. For equal-angle sections, $\alpha = 45°$, but for unequal-angle sections the value of tan α, as listed in section tables, must be entered. For equal-angle sections, I_y is necessarily equal to I_x and instead of I_y the value of the second moment of area about the minor principal axis, I_v, must be entered.

The data required for these special types of section are listed below. Following the section type, the special letters in the section label are given and then the section properties in the order in which they must appear on the section file.

Equal-angle section	`EA..`	A I_x I_v Z_x J Z_t C_x
Unequal-angle section	`?A..`	A I_x I_y Z_x Z_y J Z_t C_x C_y tanα
Tee section	`?T..`	A I_x I_y Z_x Z_y J Z_t C_x
Gantry girder	`?G..`	A I_x I_y Z_x Z_y J Z_t C_x
Channel section	`Ch..`	A I_x I_y Z_x Z_y J Z_t C_y
Circular hollow bar	`CH..`	A I_x Z_x
Round bar	`?R..`	A I_x Z_x
Square hollow section	`SH..`	I_x Z_x J Z_t

Note that an unequal angle section cannot have E as the first letter of its section label and that the equal angle section is the only one requiring a section property, I_v, not related to the x and y axes. The choice of required data has been made to conform to that most readily available in British and American section tables. In American tables, S is used instead of Z to denote the section modulus.

NB All the above section properties are in section length units (L_S) when listed in a file. However, when the data are entered from the keyboard, global units (L_G) are used.

A2.2.4a Section data entered from the keyboard

Any section data not listed on a section file can be entered from the keyboard. Having searched any section files indicated by the user, the programs will then prompt for the data on sections not found. The screen will display the section label and then request the section data corresponding to it. This must be entered in global units. Depending on the program, all or some of the following prompts for data will appear:

```
EA   EIx   EIy   EIv   Dx   Dy   GJ   EZt   Cx   Cy   Tanα   Ref.stress/E   M
```

where all the terms are defined in the previous section apart from 'Ref.stress' which is the reference stress, σ_R, and M which is the mass per unit length of section.

A2.2.5 Output data

The programs are designed to give their most significant results in the form of graphical output. This is done by showing the deformed frame with the bars colour coded, usually according to their state of stress. Under normal loading conditions, the deformations are too small to be discernible on the screen. By increasing the amplification factor used in the display (see §A2.1.1 to §A2.1.3), the deflections can be increased proportionally to show the mode of deformation. It should be noted that small deflection theory is implicit in the analysis, so that over-amplification of the deformations will produce unrealistic displays. The bars will then be shown as broken straight lines rather than curves.

The state of stress in a bar is given by the **critical stress ratio**. In the absence of shear stresses, this is the ratio of the maximum axial stress in a bar, σ, to its reference stress σ_R (see §A2.2.4). However, if torsion produces a maximum shear stress τ, an equivalent axial stress, σ_e, is calculated where

$$\sigma_e = \sqrt{(\sigma^2 + 3\tau^2)}$$

and the critical stress ratio is then given by σ_e/σ_R. If σ_R is the yield stress, the critical stress ratio then gives the proportion of the state of stress reached to that which would cause yield according to von Mises' criterion. This ratio is shown on the displays of the normal analyses by the colour of the bar (see the colour coding below). Changing the amplification of such displays not only amplifies the deformations but also the critical stress ratios as displayed in colour.

The colour coding of the bars, when used for stability analysis displays, shows the **buckling load ratio**. This is the ratio of the axial force in a bar to its fixed-end Euler buckling load, $4\pi^2EI_v/l^2$, where l is the length of the bar. Note that I_v is used (which is usually the same as I_y) as this gives the lowest fixed-end buckling load. Changing the amplification of a stability analysis display affects only the buckling mode shown and not the buckling load ratios displayed.

The colour coding of the bars, when used for vibration analysis displays, shows the **natural frequency ratio**. This is the ratio of the frequency at which a bar is vibrating to the lowest natural frequency of flexural vibrations of the bar with its ends fixed. Again, flexure is taken to be about the minor principal axis of the section. Changing the amplification of the vibration analysis display affects only the vibration mode shown and not the natural frequency ratios displayed.

The **colour coding** of bars in the display of the affected frame represents one of the above ratios. The sequence of these ratios, from low to high, is shown by the colour spectrum at the bottom right-hand corner of the screen. The colours represent the different ranges of these ratios as follows:

Black < 0.1 ≤ Grey < 0.2 ≤ Dark Blue < 0.3 ≤ Blue < 0.4 ≤ Dark Green < 0.5
0.5 ≤ Light Green < 0.6 ≤ Yellow < 0.7 ≤ Orange < 0.8 ≤ Red < 0.9
0.9 ≤ Magenta < 1.0 ≤ Violet.

Numerical results can also be shown on the screen or sent to a file for printing or further analysis. The formats of the screen displays and the files are the same. Single precision arithmetic is used in the standard versions so that no more than five significant digits are shown in the output. (Higher precision versions are available but they impose penalties on the speed of operation and storage capacity of the computer.) Engineering notation is used, so that all the powers of ten are given in multiples of three. For example, `21.851E-3` is 0.021851. The capital letter `E` should be used to indicate an exponent in the input data in the same way. The compiler used, TopSpeed Modula-2, indicates numerical overflow by generating the number `99.999E+0`. If this appears in the output, you should suspect faulty input data or a corrupted program. Zero values in the output may appear as such or, owing to rounding-off errors, as values which are of the order of a millionth of comparable non-zero values.

A2.2.5a Deflexion output

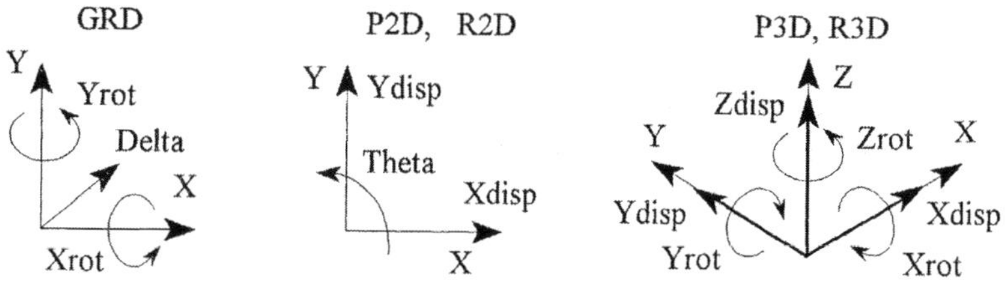

Figure A2.5 Joint deflexions for different analyses.

This can be shown on the screen or stored in a file called `NORMAL.DFL`. Note that if the file already exists, it will be overwritten. For each load case, the deflexions of each joint are listed after the joint characters. The diagrams in Figure A2.5 show the sense of the joint displacements (`Xdisp, Ydisp, Zdisp & Delta`) and the joint rotations (`Xrot, Yrot, Zrot & Theta`) which appear in the output of different analyses. The axes X, Y and Z are the global coordinate axes used to define the joint positions in the input data.

A2.2.5b Load output

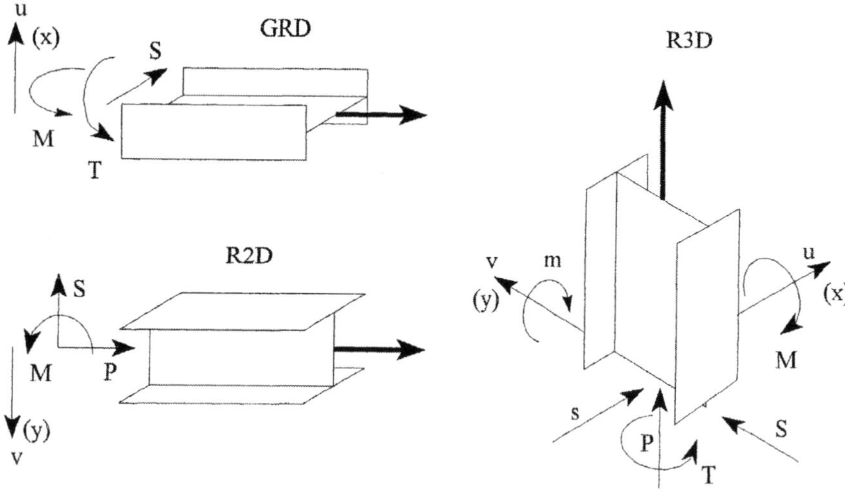

Figure A2.6 Bar loads

Load output can be shown on the screen or stored in a file called `NORMAL.LOD`. Note that if the file already exists, it will be overwritten. For each load case, the 'near' and 'far' joint characters of a bar are listed, followed by data for that bar. This consists of the critical stress ratio described above and the bar end loads, listed after the end joint characters. These loads are related to the bar's *local* coordinate system in a similar way as those shown in Figure A2.3 are related to the global system. Here the local coordinate system is defined in terms of the *principal* axes of the section and the longitudinal axes of the bar as shown by the bold arrow in Figure A2.6. `P` is a force acting in the direction of this axis and `T` is a clockwise torque about it. If `P` is positive at the first joint listed, then the bar is in compression; if it is positive at the second joint, the bar is in tension (the two values should be equal and opposite). `M` is the bending moment about the major axis of the section and `S` a shear force inducing bending about this axis. Finally `m` is a bending moment about the minor axis and `s` a shear force inducing bending about the minor axis.

A2.3 Common Errors in Producing Files

Errors in producing files will normally lead to error messages appearing which indicate the fault. However, some can cause the program to crash. Then even after correcting the error, the program may fail to operate properly. This can lead to a failure of the graphical display or, if this is bypassed, to the output of numbers which are incorrect by a very large order of magnitude. In this case, reboot the P.C. This can be done by switching off the machine and starting again, or pressing the 'Reset' button, if your machine has one.

A2.3.1 Errors common to all files

1) Comments can be put in square brackets "[" & "]" at the beginning of files, but these brackets can only appear at the beginning and end of the comments.
2) Vertical bars "|" must be used to separate or terminate groups of data. (This symbol is the upper case of "\".) On a structural problem file, it comes after the joint list, after the bar list and after the load list for each load case. **In addition**, if bar loading is described, a vertical bar must be put after the set of loadings **for each bar**. In a section file, a vertical bar must be put at the end of the general properties line and at the end of the section file.
3) **Do not** use the TAB key in writing a file.
4) **Do not** use the letters O & l instead of the numbers 0 & 1 (and vice-versa).

A2.3.2 Structural problem files

1) All joints are represented by two characters, the second of which may be a space. Certain capital letters (`A,E,F,M,N,P,Q,R,S`) have special meanings as second characters, and should only be used in the right context.
2) All structures must be continuous. The programs will not analyse two separate structures unlinked by any bar, if they are both on the same structural problem file.
3) No two joints can have more than one bar directly linking them. This is not only physically nonsensical, but it will also cause the programs to suspend operation.
4) In describing the bar loading, as in the latter part of §A2.2.3c, the end joints **must** be listed in the order in which they appear in the joint list.
5) The incorrect amount of data on a line will cause the computer to misread the file, leading to reports such as `No joint found` .... In the joint list, the special second letters (`A,E,M,N,Q,R,S`) imply further information which should appear after the joint coordinates. (Do not forget to enter two coordinates in two-dimensional analyses and three coordinates in three-dimensional analyses.) In rigid three-dimensional analyses, a number must be entered after the section label to describe the rotation of the bar about its own axis.
6) Zero bending stiffnesses can be used to simulate pin-ended bars in rigid-jointed analyses, provided that not all the bars meeting at a joint are of this kind. This would cause the joint to have

an indeterminate rotation, and failure of the analysis.

7) Sometimes the structure may appear to be correctly joined up but is not. Try amplifying the deflected form to see if it comes apart.

A2.3.3 Section files

1) Make sure that the correct number of data items appear on each line, even if some of them are irrelevant to the particular analysis you want to carry out. Entering zero as a value can be dangerous, as the program may divide by this zero at some stage (see §A2.3.2 (6) above). In particular, if the program is going to work out a bending stress or a shear stress, the corresponding Z-value must not be zero.

2) Section labels must be two **letters** and a number less than 65536. The second letters (`A,G,h,H,T`) have special meanings and should only be used in the right context.

A2.4 The Data Files Provided

The files associated with particular programs are listed below. The structure files all have the associated section files listed in the comments in square brackets at the beginning of the file. The linking letters used by the menu system to make these associations are highlighted.

Plane Grid Frame:

GRDSICRS.STR	NORMAL**SI**.SEC	Illustrates features of the data input.

Plane Pin-Jointed Frame:

P2DSIWRN.STR	NORMAL**SI**.SEC	Illustrates features of the data input.

Plane Rigid-Jointed Framework:

R2DSIPTL.STR	NORMAL**SI**.SEC	Illustrates features of the data input.
R2DMMCLR.STR	JDRDBT**MM**.SEC	Millimetre units used for a model cantilever.
R2DSIS10.STR	SKYSCR**SI**.SEC	A 10-storey portal frame. See Figure 15.8.
R2DSISKY.STR	SKYSCR**SI**.SEC	A 40-storey portal frame. See Figure 15.8.

Three-Dimensional Pin-Jointed Frame:

P3DSITPR.STR	NORMAL**SI**.SEC	Illustrates features of the data input.
P3DSIGTT.STR	NORMAL**SI**.SEC	A tetrahedral truss. See Figure 15.10.
P3DSIHOF.STR	VARIED**SI**.SEC	Hoff's truss. See Figure 15.12.

Three-Dimensional Rigid-Jointed Frame:

R3DSIDOM.STR	NORMAL**SI**.SEC	Illustrates features of the data input.
R3DMMTRS.STR	JDRDBT**MM**.SEC	Millimetre units used for a model truss.
R3DSITOR.STR	VARIED**SI**.SEC	A bridge structure.
R3DSIKWT.STR	NORMAL**SI**.SEC	A Kiewitt dome. See Figure 15.26e.

Stability Analysis (Rigid 3-D Frame):

R3DSIPT1.**STB**	NORMAL**SI**.SEC	In-plane sidesway buckling of a portal.
R3DSIPT2.**STB**	NORMAL**SI**.SEC	Out-of-plane buckling of a portal.
R3DSIPYR.**STB**	NORMAL**SI**.SEC	Torsional buckling of a tetrapod.
R3DMMLAT.**STB**	JDRDBT**MM**.SEC	Lateral buckling. See (15.109).

Vibration Analysis (Rigid 3-D Frame):

R3DSIPTL.**VIB**	NORMAL**SI**.SEC	Vibration modes of a portal (listed on file).
R3DSIAJK.**VIB**	ALUMIN**SI**.SEC	See Figure 14.17a.
R3DSIBJK.**VIB**	ALUMIN**SI**.SEC	See Figure 14.17b.

Section Files:	**Section material:**	**Force units:**	**Structure units:**	**Section units:**
NORMALSI.SEC	Steel	Newtons	metres	centimetres
VARIEDSI.SEC	Steel	Newtons	metres	centimetres
SKYSCRSI.SEC	Steel	Newtons	metres	centimetres
ALUMINSI.SEC	Aluminium	Newtons	metres	centimetres
WOODENSI.SEC	Wood	Newtons	metres	centimetres
JDRDBTMM.SEC	Steel	Newtons	millimetres	millimetres
NORMALAM.SEC	Steel	kips	feet	inches

Appendix 3: How to Use the Other Progams

AMISECS.EXE (Generates section files from the data in the AISC tables for I-sections)

This uses the data available in the american AISC tables. The notation used for the input is compatible with these tables. The format generated is like that of the I-sections in NORMALAM.SEC.

BARFLEX.EXE (Two-bar flexural equivalent of a non-uniform beam)

The non-uniform beam is taken to vary in flexural stiffness from EI_0 at one end to rEI_0 at the other. This variation is proportional to x^n where x is the distance from some origin and n is a positive integer. This is replaced by two uniform beams, joined end-to-end to give the same total length, L. The program gives the flexural stiffness of each and the position of their junction. (See §11.8).

GEOTRI.EXE (Variation of bar lengths in a geodesic dome)

This gives the least and greatest lengths of bars in a geodesic dome, triangulated as shown in Figure 15.27. The order of triangulation is requested as input. As shown in the figure, the order of triangulation is zero at the bottom and increases in successive segments anticlockwise up to four. The results are expressed as the ratio of bar length to dome radius. The source code is also provided by the file GEOTRI.MOD. This may be modified so that individual joint positions can be output, given the initial nodal positions ***P***,***Q*** and ***R*** for example.

HPPIC.EXE (Prints a *.PIC file directly using a Hewlett-Packard printer)

This takes a *.PIC file created when the 'Press ... F ... to .. file' option is chosen in the graphics displays of the Structural Analysis programs of Appendix 2. It then prints the display using any printer which understands HPGL2 commands. The height and width of the printed image can be controlled and the joint labels can be shown or hidden. This is not a screen dump, and so the image printed may be considerably enhanced.

MOHRSCIR.EXE (Mohr's circles for stress and strain for strain gauge rosettes)

This takes the strain readings given by an arbitrary rosette of three gauges, 1, 2 and 3 (as shown in the figure). It shows the Mohr's circles for strain and stress, the principal strains and stresses, stresses in the directions of the gauges and the orientation of the principal directions I and II relative to the rosette. Any consistent set of units can be used. However, if Young's modulus is expressed in TPa (e.g. 0.21 for steel) and the strains are entered as microstrains, then the stresses given by the program are in MPa (N/mm^2). The screen prompts are:

(i) Enter Young's modulus: (*e.g. 0.21*)
(ii) Enter Poisson's ratio: (*e.g. 0.3*)
(iii) Enter angle between gauges 1 & 2: (α *degrees*)
(iv) Enter angle between gauges 1 & 3: (β *degrees*)
(v) Enter strains in gauges 1, 2 & 3: (e_1, e_2 *and* e_3)

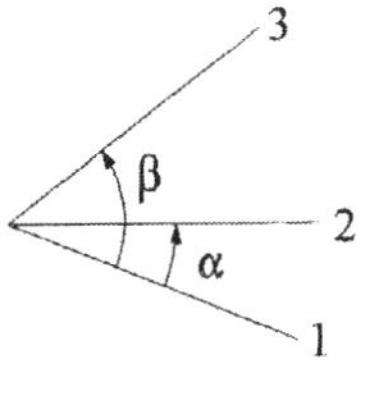

The Mohr's circle for strain is then shown, together with the values of the principal strains. At the top right, the orientation of the rosette relative to the principal strains is shown. On pressing 'C' to continue, Mohr's circle for stress is shown, together with the principal stresses and the tensile stresses in the directions of the gauges. Pressing 'R' then repeats stage (v) or pressing 'E' ends the program.

PICTOHPG.EXE (Creates a *.HPG graphics file from a *.PIC file)

The full name of a *.PIC file must be entered at the prompt (see HPPIC.EXE). It must be on the same directory as the program. It then creates a *.HPG graphics file on the same directory, suitable for any word-processing package accepting such files. In order to be compatible with all such packages, this reduces the image to a series of straight lines.

STABVIB.EXE (Lists the values of stability and vibration functions for a given λ)

See Appendix 6.2 for usage.

TORCONS.EXE (Gives the values of J and Z_t for certain standard sections)

This program calculates the above section constants for I-beams, Tee, Channel, Unequal Angle and Equal Angle sections. The formulae in Table 20 of Young (1989) are used. The root radius requested is that of the curve at the junction of the web and flange. For variable thicknesses, the values at this junction should be used.

TRAPARCH.EXE (Gives the three-bar frame flexurally equivalent to a circular bar)

The program asks for the angle α shown in Figure 11.20 and gives the constants h, k, p and q for the equivalent three-bar frame shown in Figure 11.21. Note that the axial flexibility of the original circular bar and the three prismatic bars which replace it is ignored.

TRIBARL.EXE (Triangulated barrel vault analysis)

This uses the finite-difference calculus techniques for barrel vaults given in §15.8. It implements the general solution for TeZet barrel vaults given in that section for the case of equal vertical loads at each joint.The layout of the grid of bars is shown in Figure 15.23e. As shown in Figure 15.25, equal bars run lengthwise along the vault. Two sets of transverse bars run around the vault, forming a grid of isosceles triangles. These transverse bars are identical, but may have different properties from the longitudinal bars. The support conditions along the sides and ends of the vault are taken to be antisymmetrical, approximating pinned edges. The program permits changes to the number of bars used and their properties during a single run. The overall shape of the vault and the total applied load are kept constant. These data are requested first. Any consistent set of units can be used. The initial prompts are "Total load:"(Enter the total vertical load on the vault), "Barrel radius:"(Enter the radius of curvature of the vault), "Angle subtended (degrees):"(Enter the angle subtended by the span of the vault at the centre of curvature) and "Barrel length:"(Enter the length of the barrel vault). The bar triangulation data for the grid used are then entered. The required accuracy, expressed as the number of terms in the finite double Fourier series used, is also requested. Exact solutions are given if this matches the number of triangles, but speedier solutions are obtained by using a smaller number of terms. The prompts related to this are "Hoop triangles (an even number <=36):"(for the grid shown in Figure 15.23e, this is 6), "Hoop terms (<= Hoop triangles):"(see above), "Longitudinal triangles (a number <100):"(for the grid shown in Figure 15.23e, this is 10), "Longitudinal terms (<= Longitudinal triangles):"(again, see above). It will be assumed that the flexural and torsional behaviour of the bars is governed by the equations

(a) (b) (c)

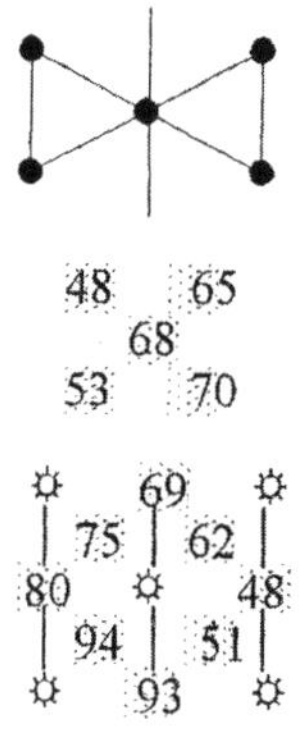

$$\frac{M}{I} = \frac{\sigma_{max}}{y_{max}} = E\frac{d^2v}{dx^2} \quad , \quad \frac{T}{J} = \frac{\tau_{max}}{r_{max}} = G\vartheta$$

(cf. (4.15) and (5.5)). In the latter case, appropriate values of J and r_{max} can be found which make the equation true for non-circular sections. Taking the major axes of the bars to lie in the plane vault, the program first prompts for the section properties of the hoop bars and then of the longitudinal bars in the following order: E, G, r_{max}, A (the area of the section), I (major), y_{max} (major), J, I (minor) and y_{max} (minor).

The program makes use of the extended ASCII character set and display to show the distribution of joint deflexions and maximum bar stresses in the developed form of the barrel vault. The largest deflexion or stress is shown at the top of the screen and the others as shown as a percentage of this in coloured boxes laid out below in a pattern reflecting the positions of the joints or bars in one quadrant of the vault. (As the structure and loading are symmetrical, the other quadrants are mirror images of this.) The top left-hand corner of the display is a corner of the grid. The half-span of the vault is shown across the screen and its half-length down it. The colour of a box reflects the percentage of the maximum value anywhere in the vault. These colours are: 0-19%: blue, 20-39%: cyan, 40-59%: green, 60-69%: orange, 70-79%: red, 80-89%: violet and 90-100%: white. The exact percentage is the number within the box. Figure (a) shows part of the actual structure, with five joints shown by dots and eight bars shown by lines. Figure (b) shows the layout of the percentages of the maximum deflexion for each of these joints. The joint deflexions shown are first the rotations θ_x, θ_y and θ_z about the local x,y and z axes (see Figure 15.25) and then the displacements δ_x, δ_y and δ_z in the directions of these axes. The equivalent maximum stress under combined loading, according to von Mises' criterion, is given by $\sqrt{(\sigma_{max}^2 + 3\tau_{max}^2)}$. These equivalents are then shown for each bar, according to the layout shown in Figure (c). The program then returns to the "Hoop triangles:" prompt and exits if 0 is entered.

Appendix 4: Section Properties and Related Formulae

In the following formulae,

A	is the area of a cross-section,
B,C	are the constants in $1/k = B + C\nu^2/(1+\nu)^2$ (cf. (6.61)),
e	defines the offset of the shear centre, as shown in the diagrams,
I_{xx}, I_{yy}	are the second moments of a section area about the x and y axes,
I_ω	is the warping constant defined by (5.81),
J	is the torsional constant defined in §5.2,
k	The shear constant in the shear stiffness GkA (cf. (6.4)),
M_x, M_y	are moments about the x and y axes respectively,
o	defines the offset of the centroid, as shown in the diagrams,
S	a shear force parallel to the y axis, acting through the shear centre,
T	a torque,
$t_{()}$	a thickness,
x,y,z	Cartesian coordinate axes (right-handed convention used),
Z_x, Z_y	M_x/σ_{max}, M_y/σ_{max},
Z_s	S/τ_{max},
Z_t	T/τ_{max},
ν	Poisson's ratio as defined in §3.3,
σ_{max}	the maximum axial stress induced by a particular moment,
τ_{max}	the maximum shear stress induced by a torque or shear force.

A4.1 Exact Formulae for Solid Sections

A4.1a A rectangular section

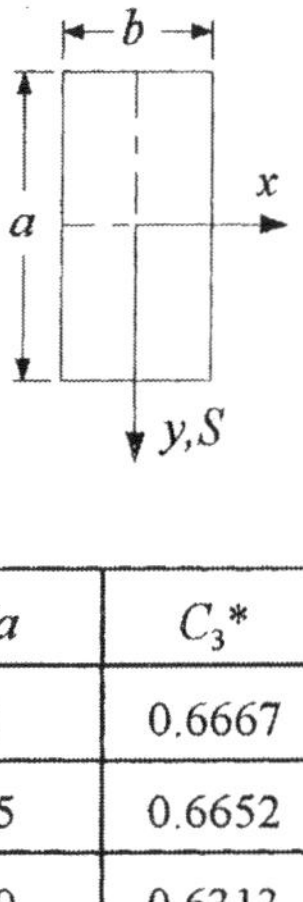

$A = ba$; $I_{xx} = ba^3/12$; $I_{yy} = b^3a/12$; $Z_x = ba^2/6$; $Z_y = b^2a/6$;
$J = \tfrac{1}{3}ab^3[1 - D_1 b/a]$; $Z_t = \tfrac{1}{3}ab^2[1 - D_2 b/a]$;

$$B = 1.2 \; ; \; C = C_1\left(\frac{b}{a}\right)^4 \text{ or } C_2\left(\frac{b}{a}\right)^2 \; ; \; 1/Z_s = \frac{1.5}{ab}\left[1 + \frac{\nu}{1+\nu}\left(\frac{b}{a}\right)^2 C_3\right].$$

b/a	D_1	D_2	b/a	C_1	C_2	b/a	C_3*
0	0.6302	0.6302	0	0.2	-	0	0.6667
0.2	0.6302	0.6275	0.1	0.1939	-	0.5	0.6652
0.4	0.6298	0.5681	0.2	0.1878	-	1.0	0.6313
0.5	0.6279	0.5247	0.5	0.1695	-	1.5	0.5651
0.6	0.6236	0.4845	1.0	0.1392	0.1392	2.0	0.4955
0.7	0.6163	0.4497	2.0	-	0.3511	5.0	0.2570
0.8	0.6060	0.4205	5.0	-	0.6699	10.0	0.1385
0.9	0.5931	0.3960	10.0	-	0.8229	15.0*	0.0945
1.0	0.5783	0.3755	∞	-	1.0	∞	0

*The formula is based on the maximum shear stress being τ_{yz} at $(\pm b/2,0)$. This is true for values of b/a up to 15. For higher values of b/a, τ_{xz} near $(\pm(b/2-a),\pm a/2)$ is somewhat higher, see Timoshenko and Goodier (1970) p.366.

A4.1b A solid circular section

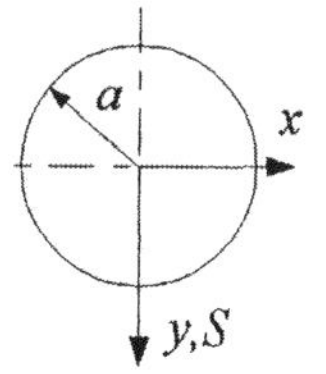

$$A = \pi a^2 \;;\; I_{xx} = I_{yy} = \pi a^4/4 \;;\; Z_x = Z_y = \pi a^3/4 \;;\; J = \pi a^4/2 \;;$$
$$Z_t = \pi a^3/2 \;;\; B = 7/6 \;;\; C = 1/6 \;;\; Z_s = 2\pi a^2(1+\nu)/(3+2\nu).$$

The maximum shear stress in shear is τ_{yz} at (0,0).

A4.1c A hollow circular section

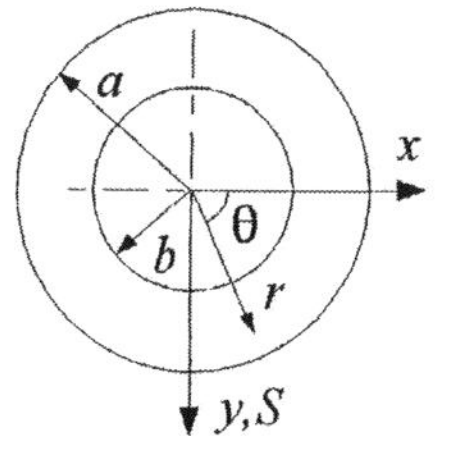

$$A = \pi(a^2-b^2) \;;\; I_{xx} = I_{yy} = \pi(a^4-b^4)/4 \;;\; Z_x = \pi(a^4-b^4)/4a \;;$$
$$Z_y = \pi(a^4-b^4)/4b \;;\; J = \pi(a^4-b^4)/2 \;;\; Z_t = \pi(a^4-b^4)/2a \;;$$
$$B = (7a^4+34a^2b^2+7b^4)/6(a^2+b^2)^2 \;;\; C = (a^2-b^2)^2/6(a^2+b^2)^2 \;;$$
$$Z_s = 4I_{xx}(1+\nu)/[3a^2+b^2+2\nu(a^2+b^2)] .$$

The maximum shear stress in shear is τ_{yz} at $(\pm b,0)$. Note that this means that as b tends to zero, the maximum shear becomes *twice* that predicted for a solid section. This implies the existence of an infinitesimal stress raiser with a stress concentration factor of two. This effect is highly localised. In terms of the polar coordinates (r,θ) shown, the shear stresses are

$$\tau_{rz} = \frac{S}{8(1+\nu)I_{xx}}(3+2\nu)\left(r^2 - a^2 - b^2 + \frac{a^2b^2}{r^2}\right)\sin\theta \;,$$

$$\tau_{\theta z} = \frac{S}{8(1+\nu)I_{xx}}\left[(3+2\nu)\left(a^2 + b^2 + \frac{a^2b^2}{r^2}\right) - (1-2\nu)r^2\right]\cos\theta \;.$$

These are derived from the χ function in Love (1952) p.335.

A4.1c A semicircular section

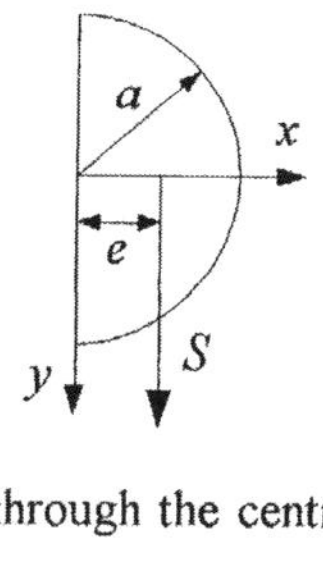

$$A = \pi a^2/2 \;;\; I_{xx} = \pi a^4/8 \;;\; I_{yy}{}^* = (9\pi^2-64)a^4/72\pi \;;\; Z_x = \pi a^3/8 \;;$$
$$Z_y = (9\pi^2-64)a^3/24(3\pi-4) \;;\; J = 0.2976a^4 \;;\; Z_t = 0.3506a^3 \;;$$
$$B = 7/6 \;;\; C = 0.01453 \;;\; Z_s = \pi a^2/[3 + 0.5071\nu/(1+\nu)] \;;$$
$$e = 8a/5\pi \;.$$

The second moment of area $I_{yy}{}^*$ is about the axis parallel to the y axis through the centroid, located at $(4a/3\pi,0)$. The maximum shear stress in torsion or shear is τ_{yz} at (0,0).

A4.1d An elliptic section

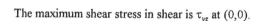

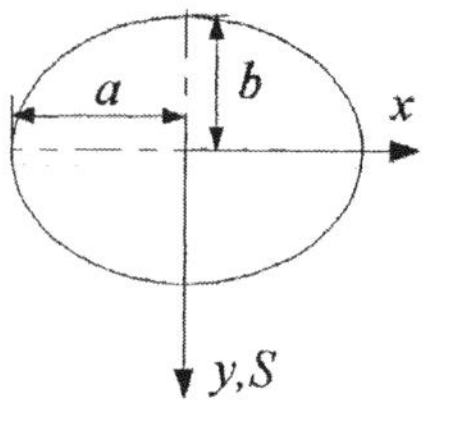

$$A = \pi ab \;;\; I_{xx} = \pi ab^3/4 \;;\; I_{yy} = \pi a^3b/4 \;;\; Z_x = \pi ab^2/4 \;;\; Z_y = \pi a^2b/4 \;;$$
$$J = \pi a^3b^3/(a^2+b^2) \;;\; Z_t = \pi ab^2/2 \;\; (a>b) \;;\; B = 1+(a^2+b^2)/3(a^2+3b^2) \;;$$
$$C = 2a^4/3b^2(a^2+3b^2) \;;\; Z_s = \pi ab(a^2+3b^2)(1+\nu)/[2a^2+4b^2(1+\nu)] .$$

The maximum shear stress in shear is τ_{yz} at (0,0).

A4.1e An equilateral triangular section

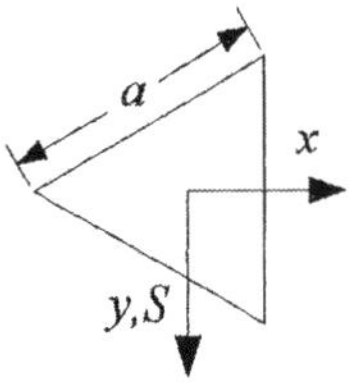

$A = \sqrt{3}a^2/4$; $I_{xx} = I_{yy} = a^4/(32\sqrt{3})$; $Z_x = a^3/(16\sqrt{3})$; $Z_y = a^3/32$;
$J = \sqrt{3}a^4/80$; $Z_t = a^3/20$;
{$\nu = 0.5$ only: $k = 3/4$; $Z_s = \sqrt{3}a^2/8$ } .

The worst shear stress in torsion (or shear) is at the midpoint of the (vertical) side. If S acts in the x direction, the shear stiffness is the same but Z_s becomes $a^2/2$, the worst shear stress being a the midpoints of the sloping sides.

A4.1f Torsion of a circular section with a cutout

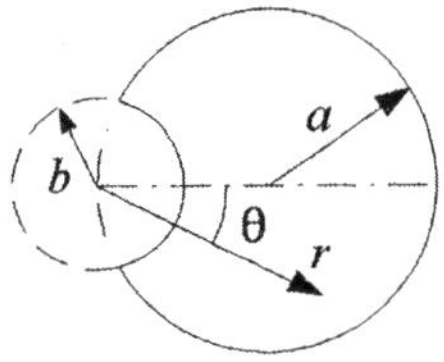

Prandtl's stress function for the section shown is

$$\phi = G\vartheta(r^2 - b^2)\left(\frac{a}{r}\cos\theta - \frac{1}{2}\right)$$

(see §5.4). This gives a torsional constant of

$$J = \frac{1}{2}\cos^{-1}\left(\frac{b}{2a}\right)(2a^4 - 4a^2b^2 - b^4) + \frac{b}{8}\sqrt{4a^2 - b^2}\,(2a^2 + 9b^2)$$

and a maximum stress $\tau_{r\theta}$ at $(b,0)$ of $G\vartheta(2a\text{-}b)$. Note that as b tends to zero, the maximum stress becomes *twice* that for a solid circular section. This is another example of an infinitesimal stress raiser discussed in §A4.1c.

A4.2 Approximate Formulae for Thin Sections

In this section, any thickness $t_{()}$ will be taken as small in comparison with the overall dimensions of the section. Accurate values of the torsional constants J and Z_t can be found using the program TORCONS.EXE (see Appendix 3). This makes use of the formulae in Table 20 of Young (1989). Some sections have an almost negligible warping constant, I_ω, based on the warping across the thickness of the section. This is listed in bold curly brackets.

A4.2a A thin strip

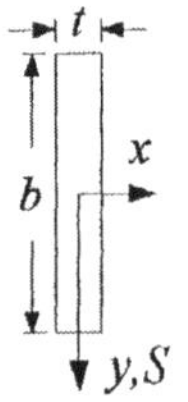

$A = bt$; $I_{xx} = tb^3/12$; $I_{yy} = t^3b/12$; $Z_x = tb^2/6$; $Z_y = t^2b/6$;
$J = bt^3/3$; $Z_t = bt^2/3$; $k = 5/6$; $Z_s = 2bt/3$; { $I_\omega = b^3t^3/36$ } .

A4.2b An I-beam

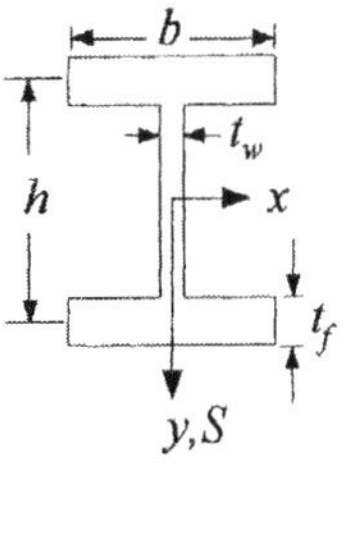

$A = 2bt_f + ht_w$; $I_{xx} = bt_fh^2/2 + t_wh^3/12$; $Z_x = 2I_{xx}/(h+t_f)$;
$I_{yy} = (2t_fb^3 + t_w^3h)/12$; $Z_y = 2I_{yy}/b$; $J = (2bt_f^3 + ht_w^3)/3$;
$k = 120t_wI_{xx}^2/Ah^2(t_w^2h^3 + 10t_ft_wbh^2 + 30t_f^2hb^2 + 5t_ft_wb^3)$;
$Z_s = 8I_{xx}t_w/(t_wh^2 + 4t_fbh)$; $I_\omega = h^2t_fb^3/24$.

A4.2c A channel section

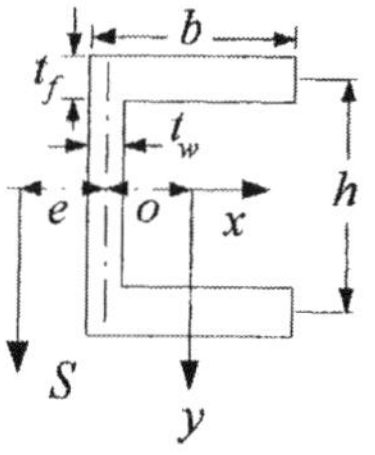

$$A = 2bt_f + ht_w \;;\; I_{xx} = bt_f h^2/2 + t_w h^3/12 \;;\; Z_x = 2I_{xx}/(h+t_f) \;;$$
$$I_{yy} = b^3 t_f(bt_f + 2ht_w)/3(2bt_f + ht_w) \;;\; Z_y = b^2 t_f(bt_f + 2ht_w)/3(bt_f + ht_w) \;;$$
$$J = (2bt_f^3 + ht_w^3)/3 \;;\; o = b^2 t_f/(2bt_f + ht_w) \;;\; e = 3b^2 t_f/(6bt_f + ht_w) \;;$$
$$k = 120 t_w I_{xx}^2/Ah^2(t_w^2 h^3 + 10 t_f t_w b h^2 + 30 t_f^2 h b^2 + 20 t_f t_w b^3) \;;$$
$$Z_s = 8 I_{xx} t_w/(t_w h^2 + 4 t_f b h) \;;\; I_\omega = h^2 t_f b^3(3bt_f + 2ht_w)/12(6bt_f + ht_w) \,.$$

A4.2d A tee section

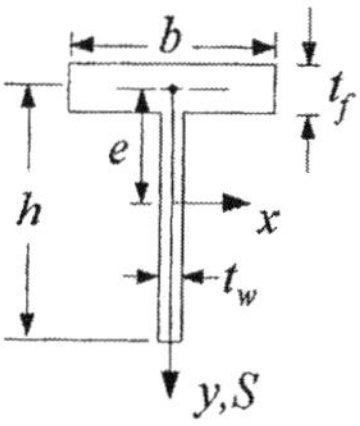

$$A = t_w h + t_f b \;;\; I_{xx} = t_w h^3(t_w h + 4t_f b)/12(t_w h + t_f b) \;;$$
$$Z_x = 2I_{xx}(t_w h + t_f b)/(t_w h + 2t_f b) \;;\; I_{yy} = (t_w^3 h + t_f b^3)/12 \;;$$
$$Z_y = 2I_{yy}/b \;;\; J = (t_w^3 h + t_f^3 b)/3 \;;\; e = t_w h^2/2(t_w h + t_f b) \;;$$
$$k = 240 A I_{xx}^2/t_w h^4(2t_w^2 h^3 + 14 t_f t_w b h^2 + 32 t_f^2 h b^2 + 5 t_f t_w b^3) \;;$$
$$Z_s = 2t_w h[1 + t_w t_f bh/(t_w h + 2t_f b)^2]/3 \;;\; \{ I_\omega = (4t_w^3 h^3 + t_f^3 b^3)/144 \} \,.$$

A4.2e An equal angle section

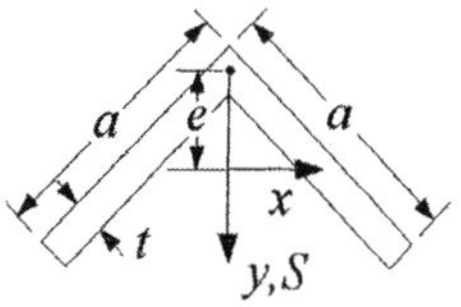

$$A = 2at \;;\; I_{xx} = ta^3/12 \;;\; Z_x = ta^2/3\sqrt{2} \;;\; I_{yy} = ta^3/3 \;;$$
$$Z_y = ta^2\sqrt{2}/3 \;;\; J = 2at^3/3 \;;\; e = a/2\sqrt{2} \;;$$
$$k = 5/12 \;;\; Z_s = 2\sqrt{2}\,ta/3 \;;\; \{ I_\omega = a^3 t^3/18 \} \,.$$

For a shear force in the x direction, k and Z_s remain the same.

A4.2f An unequal angle section ($b > a$)

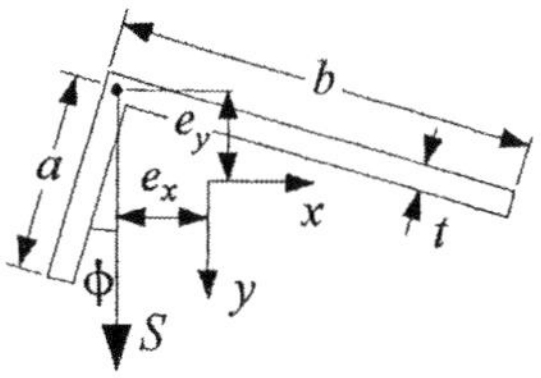

$$K = \sqrt{\{[(a+b)^4 - 6a^2b^2]^2 - 16a^3b^3(a+b)^2\}} \;;\; \phi = \tfrac{1}{2}\sin^{-1}(6a^2b^2/K);$$
$$e_x = (b^2\cos\phi - a^2\sin\phi)/2(a+b);\; e_y = (b^2\sin\phi + a^2\cos\phi)/2(a+b);$$
$$A = t(a+b) \;;\; I_{xx} = t[(a+b)^4 - 6a^2b^2 - K]/24(a+b) \;;$$
$$Z_x = I_{xx}/(a\cos\phi - e_y) \;;\; I_{yy} = t[(a+b)^4 - 6a^2b^2 + K]/24(a+b) \;;$$
$$Z_y = I_{yy}/(b\cos\phi - e_x) \;;\; J = (a+b)t^3/3 \;;$$
$$Z_s = 2I_{xx}\cos\phi/(a\cos\phi - e_y)^2 \;;\; \{ I_\omega = (a^3+b^3)t^3/36 \} \;;$$
$$k = 60 I_{xx}^2/At[8(a^5\cos^2\phi + b^5\sin^2\phi) - 25(a^4\cos\phi + b^4\sin\phi)e_y + 20(a^3+b^3)e_y^2] \,.$$

A4.2g A box section

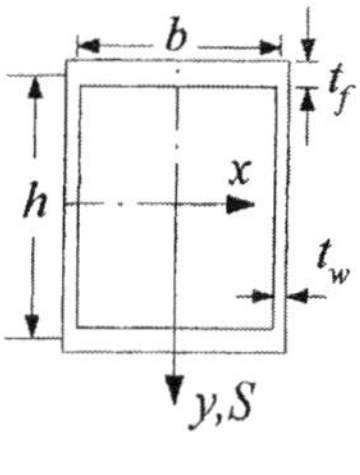

$$A = 2(bt_f + ht_w) \;;\; I_{xx} = bt_f h^2/2 + t_w h^3/6 \;;\; Z_x = 2I_{xx}/(h+t_f) \;;$$
$$I_{yy} = t_f b^3/6 + t_w h b^2/2 \;;\; Z_y = 2I_{yy}/(b+t_w) \;;\; J = 2b^2 t_f h^2 t_w/(bt_w + ht_f) \;;$$
$$Z_t = 2hbt_w{}^* \;;\; Z_s = 8I_{xx} t_w/h(t_w h + 2t_f b)^* \;;$$
$$k = 120 t_w I_{xx}^2/Ah^2(2t_w^2 h^3 + 10 t_f t_w b h^2 + 15 t_f^2 h b^2 + 5 t_f t_w b^3) \;;$$

*The worst shear stress may occur at a sharp internal corner (see §5.5).

Further properties to be found in Young (1989) are as follows. Table 1: $A, I_{xx}, I_{yy}, Z_x, Z_y$; Table 16: curved beams; Table 20: J, Z_t; Table 21: e, J, I_ω. Pilkey (1994) also lists the following constants. Table 2-1: A, I_{xx}, I_{yy}; Table 2-2: Z_x, Z_y; {Table 2-4: $1/k$}; Table 2-5: J; Table 2-6: e, I_ω.

Appendix 5: Properties of Regular Structures and Standard Solutions

A5.1 General Solutions for Modular Beams

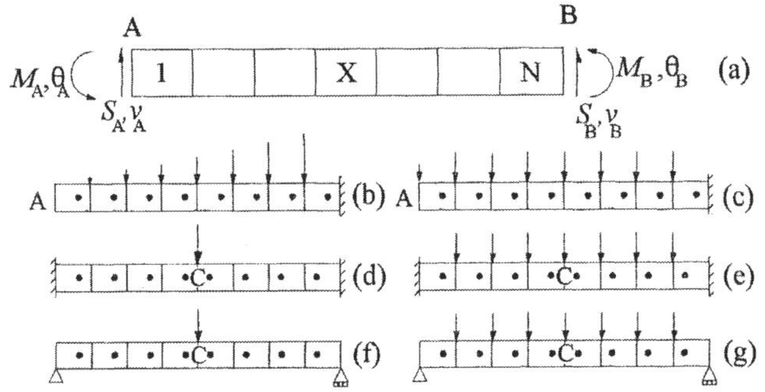

Figure A5.1 Standard cases of modular beams.

The governing equations for modular beams are given in §8.5. In particular, the slope-deflexion equations are given by (8.38). The end deflexions resulting from an end-loaded cantilever can be found directly from these. Further solutions[1] are listed here for the modular beam AB shown in Figure A5.1a. Suppose that the cantilever shown in Figure A5.1b is loaded by a linearly-varying load P_x equal to XP at the right of the Xth module. The upwards displacement at this junction is

$$v_x = \frac{-m^3 P f_{mm}}{120}[\mathrm{X}^5 - \mathrm{X}(5\mathrm{N}^4 - 5\mathrm{N}^2 + 1) + 4\mathrm{N}^5 - 5\mathrm{N}^3 + \mathrm{N} + 20e(\mathrm{N}^4 - \mathrm{N}^3\mathrm{X} + \mathrm{NX} - \mathrm{N}^2) + 20e^2(\mathrm{N}^3 - \mathrm{X}^3 + \mathrm{X} - \mathrm{N})] + \frac{mPf_{ss}}{6}(\mathrm{X}^3 - \mathrm{N}^3 + \mathrm{N} - \mathrm{X}) \quad \textbf{(A5.1.1)}$$

where there are N modules of width m, the flexural centre is me to the right of the centre of the module, and f_{mm} and f_{ss} correspond to the inverses $1/EI$ and $1/GkA$ of the flexural and shear stiffnesses in the simplest cases (see §8.5).

Suppose that the cantilever is loaded by a downwards force P per module, divided into ½P applied at each end of it. This will result in a load of P at each junction, with only ½P applied at the tip, as in Figure A5.1c. The tip deflexion is then given by

$$v_\mathrm{A} = -\frac{Pm^3 f_{mm}}{8}[\mathrm{N}^4 + 4e\mathrm{N}^2(e + \mathrm{N})] - \tfrac{1}{2}Pm\mathrm{N}^2 f_{ss} \quad \textbf{(A5.1.2)}$$

The remaining standard cases are for symmetrical modular beams with 2N modules. The flexural centre is me to the right of the centre in each of the N modules in the left-hand half of the beam and me to the left of centre in the remaining N modules (as indicated by the dots). The vertical deflexions of the central junction, v_C upwards, are as follows.

For a fixed-ended beam with a central force F acting downwards (Figure A5.1d),

$$v_\mathrm{C} = \frac{Fm^3 f_{mm}}{24}(\mathrm{N} - \mathrm{N}^3) - \tfrac{1}{2}Fm\mathrm{N}f_{ss} \quad \textbf{(A5.1.3)}$$

(Note that the central deflexion is not a function of e in this case.)

---

[1] See Renton, J.D. (1996) Generalized Beam Theory and Modular Structures. *International Journal of Solids and Structures* V.33, p.1425.

For a fixed-ended beam with a uniform load P at each junction, (Figure A5.1e),

$$v_C = \frac{Pm^3 f_{mm}}{24}[2eN(N^2-1) - N^4 + N^2] - \tfrac{1}{2}PmN^2 f_{ss} \qquad \textbf{(A5.1.4)}$$

For a pin-ended beam with a central force F acting downwards (Figure A5.1f),

$$v_C = \frac{Fm^3 f_{mm}}{24}[N - 4N^3 - 12eN(e+N)] - \tfrac{1}{2}FmNf_{ss} \qquad \textbf{(A5.1.5)}$$

For a pin-ended beam with a uniform load P at each junction, (Figure A5.1g),

$$v_C = \frac{Pm^3 f_{mm}}{24}[2N^2 - 5N^4 - 12eN^2(e+N)] - \tfrac{1}{2}PmN^2 f_{ss} \qquad \textbf{(A5.1.6)}$$

A5.2 Plane Trusses

Figure A5.2 shows the members (marked bold) of the modules of some standard plane frames. Where a member is marked by a broken line, half of it is assigned to each contiguous module. Each module is of breadth m and height h. With respect to the axes x,y z shown in Figure A5.2, the flexibilities f_{mm} and f_{ss} correspond to the flexibilities f_{22} and f_{44} in (8.12). Where there is no coupling with other modes of behaviour (such as torsion for example) these flexibilities can be equated to the inverse flexural and shear stiffnesses, $1/EI$ and $1/GkA$. These flexibilities and the eccentricity e used in the above equations can be expressed in terms of the axial coefficients (Young's modulus × cross-sectional area / length³) t,b,d and a for the top, bottom, diagonal and (vertically) across members of the pin-jointed frames (a) to (d). The line of action of the resultant axial force on the truss, P, is chosen so that it produces no flexure (f_{16} and f_{26} are zero). The axial flexibility of a truss is normally given by

$$f_{66} = 1/[m^3(b+t)] \qquad \textbf{(A5.2.1)}$$

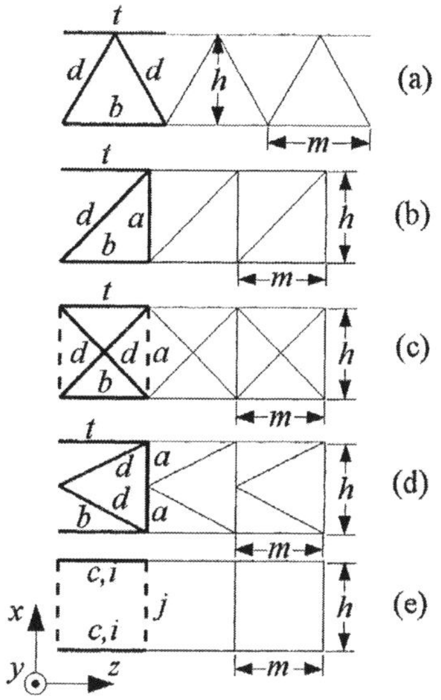

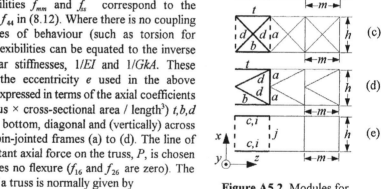

Figure A5.2 Modules for standard plane frames.

Where this is not so, the correct expression will be given.

For the Warren truss module shown in Figure A5.2a, (lower joint shear dominant),

$$(f_{mm} =)\ f_{22} = \frac{1}{m^3h^2}\left(\frac{1}{b} + \frac{1}{t}\right), \quad (f_{ss} =)\ f_{44} = \frac{1}{mh^2}\left(\frac{2}{d} + \frac{1}{4t}\right), \quad e = 0. \qquad \textbf{(A5.2.2)}$$

For the Howe (or Pratt) truss shown in Figure A5.2b,

$$f_{22} = \frac{1}{m^3h^2}\left(\frac{1}{b} + \frac{1}{t}\right), \quad f_{44} = \frac{1}{mh^2}\left(\frac{1}{a} + \frac{1}{d} + \frac{1}{b+t}\right), \quad e = \frac{t-b}{2(b+t)}. \qquad \textbf{(A5.2.3)}$$

In this case, there are coupling terms f_{46} and f_{64} between the shear and extension giving

$$\gamma_x = \frac{1}{mh^2}\left(\frac{1}{a}+\frac{1}{d}+\frac{1}{b+t}\right)S_x - \frac{1}{m^2h(b+t)}P$$
$$\varepsilon_z = -\frac{1}{m^2h(b+t)}S_x + \frac{1}{m^3(b+t)}P \qquad \textbf{(A5.2.4)}$$

For the cross-braced truss shown in Figure A5.2c,

$$f_{22} = \frac{1}{m^3h^2}\left\{\frac{2[(a+2d)(b+t)+2ad]}{ad(b+t)+2bt(a+2d)}\right\}, \qquad e=0,$$
$$f_{44} = \frac{1}{mh^2}\left\{\frac{1}{2d}+a\left[\frac{d(t-b)}{ad(b+t)+2bt(a+2d)}\right]^2\right\}, \; f_{66} = \frac{1}{m^3}\left\{\frac{a+2d}{(b+t)(a+2d)+2ad}\right\}. \qquad \textbf{(A5.2.5)}$$

(Note that if $t=b=c$ then f_{22} and f_{44} become $2/m^3h^2c$ and $1/2mh^2d$ respectively.)

For the K-truss shown in Figure A5.2d,

$$f_{22} = \frac{1}{m^3h^2}\left(\frac{1}{b}+\frac{1}{t}\right), \; f_{44} = \frac{2}{mh^2}\left(\frac{1}{a}+\frac{1}{d}\right), \qquad e = -\frac{1}{2}. \qquad \textbf{(A5.2.6)}$$

Note that here a is the axial coefficient for each vertical half-member.

A rigid-jointed Vierendeel truss is shown in Figure A5.2e. The above flexibilities are now expressed in terms of the axial coefficient c $(=b=t)$ for the equal upper and lower chords and the bending stiffnesses EI of the chords and vertical beams, given by im^3h^2 and jm^2h^3 respectively.

$$f_{22} = \frac{2}{m^3h^2}\frac{1}{(4i+c)}, \quad f_{44} = \frac{1}{12mh^2}\left(\frac{1}{2i}+\frac{c^2}{j(4i+c)^2}\right), \quad e=0. \qquad \textbf{(A5.2.7)}$$

The axial coefficient of the vertical bar is not required, as the characteristic responses do not involve the extension of the vertical bars.

The above flexibility coefficients for pin-jointed plane frames can be regarded as reasonable approximations to their values when the joints are rigid. The effect of allowing for the flexure of the component bars will be proportional to the square of their slenderness ratios. The out-of-plane flexural and torsional constants for such rigid-jointed plane frames have been found[1] for cases when the upper and lower chords are made from identical bars. These expressions tend to be rather complex if account is taken of the torsional stiffness GJ of the bars as well as their (out-of-plane) flexural stiffness EI_o. As a rule $GJ \ll EI_o$, and ignoring the torsional stiffness gives the following formulae. These are expressed in terms of the *flexural coefficients* EI_o/l^3 of the component bars. Here, the top and bottom bars will be taken as identical, with a flexural coefficient C. The vertical and diagonal bars will be taken to have flexural coefficients A and D respectively. For the Warren truss shown in Figure A5.2a,

$$\left(\frac{1}{EI_o}=\right) f_{11} = \frac{1}{2m^3C}, \qquad \left(\frac{1}{GJ}=\right) f_{33} = \frac{1}{24mh^2}\left(\frac{1}{C}+\frac{12}{D}\right). \qquad \textbf{(A5.2.8)}$$

For the Howe (or Pratt) truss shown in Figure A5.2b, there are coupling terms f_{13} and f_{31} between flexure and torsion giving

[1] Renton, J.D. (1982) On the Behaviour of Plane Trusses. *Oxford University Engineering Laboratory Report* No. 1409/82.

$$
\begin{aligned}
\vartheta_x &= \frac{1}{2m^3C}M_x - \frac{1}{4m^2hC}T \\
\vartheta_z &= -\frac{1}{4m^2hC}M_x + \frac{1}{mh^2}\left(\frac{1}{4A} + \frac{1}{6C} + \frac{1}{4D}\right)T
\end{aligned}
\tag{A5.2.9}
$$

For the cross-braced truss shown in Figure A5.2c,

$$
f_{11} = \frac{1}{2m^3}\left(\frac{A+2D}{AC+2D(A+C)}\right) \quad , \quad f_{33} = \frac{1}{24mh^2}\left(\frac{1}{C}+\frac{3}{D}\right) . \tag{A5.2.10}
$$

For the K-truss shown in Figure A5.2d,

$$
f_{11} = \frac{1}{8m^3C}\left(\frac{4C(A+D)+AD}{C(A+D)+AD}\right) \quad , \quad f_{33} = \frac{2(A+D)}{3mh^2AD}\left(\frac{3C(A+D)+AD}{4C(A+D)+AD}\right) . \tag{A5.2.11}
$$

where again A is evaluated for each vertical half-member. The properties of the Vierendeel truss shown in Figure A5.2e are in this case just a particular case of those given by (A5.2.10),

$$
f_{11} = \frac{1}{2m^3C} \quad , \quad f_{33} = \frac{1}{24mh^2C} . \tag{A5.2.12}
$$

A5.3 Space Trusses

The elastic properties of space trusses with a cross-section in the form of a regular polygon of side h will be listed here. The properties of some other space trusses are given in §15.4. Each module of a truss has either one or two bays. The results will be expressed in terms of the bay length s. The module length m is then either s or $2s$. Unless stated otherwise, the flexural centre is midway along the module, so that e is zero. Figures A5.3 to A5.5 show the basic modules consisting of transverse breadth bars, forming the outline of the cross-section and longitudinal chord bars, linking the corresponding apexes of these outlines. The axial coefficients of these bars will be denoted by b and c respectively. There will also be diagonal bars across the transverse faces of the trusses. Their axial coefficients will be denoted by d. In the case of two-bay modules, there will be a second set of breadth bars, such as A″B″,B″C″ and C″A″ and a second set of chord bars linking the triangles as shown in Figure A5.3. In this case, the diagonal bars in both bays will be specified. In the case of square and hexagonal trusses, there will also be across bars, with axial coefficients a, running from one apex to another on the cross-section. The z axis lies along the centroids of the cross-sections. The symmetry of the matrix in (8.12) will be assumed, so that f_{ij} will be

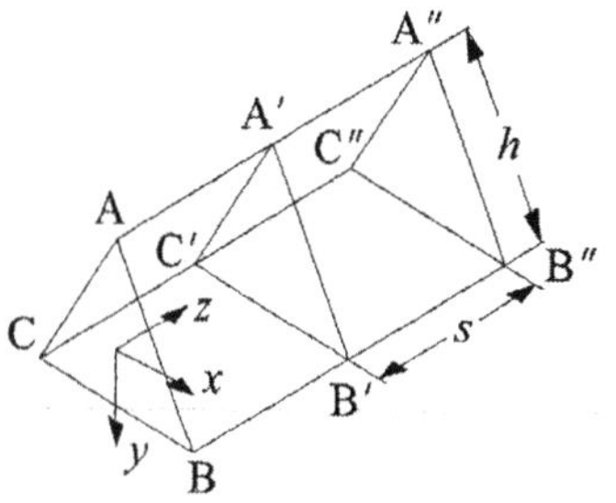

Figure A5.3 Triangular space trusses.

given, but not f_{ji}.

Single-bay triangular trusses

In such cases, $m=s$ and the layout in only one bay is specified. Three configurations are given by adding the listed diagonal bars.
(T1) AB′, BC′ and CA′ forming a clockwise spiral:

$$f_{11} = f_{22} = \frac{2}{h^2 s^3 c} , \quad f_{14} = f_{25} = \frac{2}{\sqrt{3} h^2 s^2 c} , \quad f_{33} = \frac{4}{h^4 s}\left(\frac{1}{b} + \frac{1}{c} + \frac{1}{d}\right) ,$$
$$f_{36} = -\frac{2}{\sqrt{3} h^2 s^2 c} , \quad f_{44} = f_{55} = \frac{2}{3h^2 s}\left(\frac{1}{b} + \frac{1}{4c} + \frac{1}{d}\right) , \quad f_{66} = \frac{1}{3s^3 c} . \tag{A5.3.1}$$

(T2) AC′, BA′ and BC′. (This was the configuration for a Boeing solar power satellite design.)

$$f_{11} = f_{22} = \frac{2}{h^2 s^3 c} , \quad f_{13} = -\frac{4}{3h^3 s^2 c} , \quad f_{14} = \frac{1}{3\sqrt{3} h^2 s^2 c} , \quad f_{25} = -\frac{1}{\sqrt{3} h^2 s^2 c} ,$$
$$f_{33} = \frac{4}{h^4 s}\left(\frac{1}{b} + \frac{1}{3c} + \frac{1}{d}\right) , \quad f_{34} = \frac{2}{3\sqrt{3} h^3 sc} , \quad f_{36} = \frac{2}{3\sqrt{3} h^2 s^2 c} , \tag{A5.3.2}$$
$$f_{44} = \frac{2}{3h^2 s}\left(\frac{1}{b} + \frac{11}{12c} + \frac{1}{d}\right) , \quad f_{46} = \frac{4}{9hs^2 c} , \quad f_{55} = \frac{2}{3h^2 s}\left(\frac{1}{b} + \frac{1}{4c} + \frac{1}{d}\right) , \quad f_{66} = \frac{1}{3s^3 c} .$$

The existence of f_{34} implies S_x produces torsion. From (8.22), this can be eliminated by offsetting its line of action by

$$y_0 = \frac{bdh}{2\sqrt{3}(bd + 3bc + 3cd)} \tag{A5.3.3}$$

(T3) AB′, BA′, BC′, CB′, CA′ and AC′ cross-bracing each face:

$$f_{11} = f_{22} = \frac{4}{h^2 s^3}\left(\frac{b + 2d}{2bc + bd + 4cd}\right) , \quad f_{33} = \frac{2}{h^4 sd} ,$$
$$f_{44} = f_{55} = \frac{1}{h^2 s}\left(\frac{1}{3d} + \frac{2bd^2}{[2c(b+2d)+bd]^2}\right) , \quad f_{66} = \frac{1}{3s^3}\left(\frac{b + 2d}{bc + 2bd + 2cd}\right) . \tag{A5.3.4}$$

Double-bay triangular trusses

(T4) Diagonal bars AB′, BC′, CA′, A′C″, B′A″ and C′B″ spiralling in opposite directions in each bay:

$$f_{11} = f_{22} = \frac{2}{h^2 s^3 c} , \quad f_{33} = \frac{4}{h^4 s}\left(\frac{1}{c} + \frac{1}{d}\right) ,$$
$$f_{44} = f_{55} = \frac{1}{6h^2 s}\left(\frac{1}{c} + \frac{4}{d}\right) , \quad f_{66} = \frac{1}{3s^3 c} . \tag{A5.3.5}$$

Single-bay square trusses

The module consists of only one bay, (m=s). Four examples will be considered.

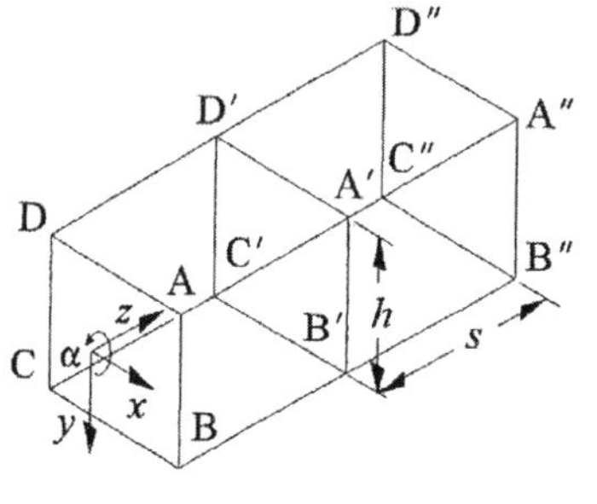

Figure A5.4 Square space trusses.

(S1a) Diagonal bars AD′,BA′,CB′ and DC′form an anti-clockwise spiral. There is an across bar BD:

$$f_{11}=f_{22}=\frac{1}{h^2s^3c}\ ,\ f_{14}=f_{25}=-\frac{1}{2h^2s^2c}\ ,$$
$$f_{33}=\frac{1}{h^4s}\left(\frac{1}{b}+\frac{1}{c}+\frac{1}{d}\right)\ ,\ f_{36}=\frac{1}{2h^2s^2c}\ ,$$
$$f_{44}=f_{55}=\frac{1}{2h^2s}\left(\frac{1}{b}+\frac{1}{2c}+\frac{1}{d}\right)\ ,\ f_{66}=\frac{1}{4s^3c}\ . \qquad \textbf{(A5.3.6)}$$

(S1b) This is as (S1a) but with the addition of a second across bar, AC. The results are as before except that

$$f_{33}=\frac{1}{h^4s}\left(\frac{1}{2a+b}+\frac{1}{c}+\frac{1}{d}\right)\ . \qquad \textbf{(A5.3.7)}$$

(S2) Diagonal bars AB′, CB′, DA′ and DC′give parallel pairs in opposite faces. The results are given relative to the principal axes given by rotating x and y anticlockwise about z through α=45°. One across bar, BD, is used.

$$f_{11}=f_{22}=\frac{1}{h^2s^3c}\ ,\ f_{23}=\frac{1}{\sqrt{2}h^3s^2c}\ ,\ f_{33}=\frac{1}{h^4s}\left(\frac{1}{b}+\frac{1}{2c}+\frac{1}{d}\right)\ ,$$
$$f_{44}=\frac{1}{2h^2s}\left(\frac{1}{b}+\frac{1}{d}\right)\ ,\ f_{55}=\frac{1}{2h^2s}\left(\frac{a+b}{(2a+b)^2}+\frac{5}{4c}+\frac{1}{d}\right)\ , \qquad \textbf{(A5.3.8)}$$
$$f_{56}=-\frac{1}{2\sqrt{2}hs^2c}\ ,\ f_{66}=\frac{1}{4s^3c}\ .$$

(S3) Diagonal bars AB′, BA′, BC′, CB′, CD′, DC′, DA′and AD′cross-brace all the side panels. Across bars AC and BD cross-brace the transverse panels too.

$$f_{11}=f_{22}=\frac{1}{h^2s^3}\left(\frac{b+2d}{bc+bd+2cd}\right)\ ,\ f_{33}=\frac{1}{2h^4sd}\ ,$$
$$f_{44}=f_{55}=\frac{1}{4h^2s}\left(\frac{1}{d}+\frac{2bd^2}{[c(b+2d)+bd]^2}\right)\ ,\ f_{66}=\frac{1}{4s^3}\left(\frac{2a+b+2d}{(2a+b)(c+2d)+2cd}\right)\ . \qquad \textbf{(A5.3.9)}$$

(S4) Diagonal bars AB′, AD′,CB′ and CD′ and an across bar AC, so that the structure is symmetrical about the y axis after x,y have been rotated through α=45°.

$$f_{11}=f_{22}=\frac{1}{h^2s^3c}\ ,\ f_{33}=\frac{1}{h^4s}\left(\frac{1}{b}+\frac{1}{d}\right)\ ,\ f_{44}=f_{55}=\frac{1}{2h^2s}\left(\frac{1}{b}+\frac{1}{d}\right)\ ,\ f_{66}=\frac{1}{4s^3c}\ \cdot \qquad \textbf{(A5.3.10)}$$

The flexural centre is offset by e= -½ for flexure about the x axis (along AC) and by e= +½ for flexure about the y axis (cf. the K-truss earlier).

Double-bay square trusses

(S5) Diagonal bars AD′, BA′, BC′, CD′, A′B″, C′B″, D′C″ and D′A″and across bars BD and B′D′, giving symmetry about the y axis after x,y have been rotated through α=45°. This is the West German SPAS system.

$$f_{11} = \frac{2}{h^2s^3}\left(\frac{ab+2ad+bd}{abd+2c(ab+2ad+bd)}\right),\ f_{22} = \frac{1}{h^2s^3c},\ f_{33} = \frac{1}{h^4s}\left(\frac{1}{2c}+\frac{1}{d}\right),$$
$$f_{44} = \frac{1}{2h^2sd},\ f_{55} = \frac{1}{2h^2s}\left[\frac{1}{c}+\frac{1}{d}+\frac{1}{c}\left(\frac{abd}{abd+2c(ab+2ad+bd)}\right)^2\right],\ f_{66} = \frac{1}{4s^3c}. \qquad \textbf{(A5.3.11)}$$

Most of the above results were obtained manually[1]. The process for finding the constant elastic response of space trusses has been automated[2] and used to find the following flexibility coefficients for hexagonal trusses. This process yields results for a variable height/width ratio of the cross section, but only the results for regular polygonal cross sections are given here. Note that the shear coefficients cannot be determined this way.

Single-bay hexagonal trusses

The module used is shown in Figure A5.5. To brace the section, radial across bars from A to the apexes of the section are used.

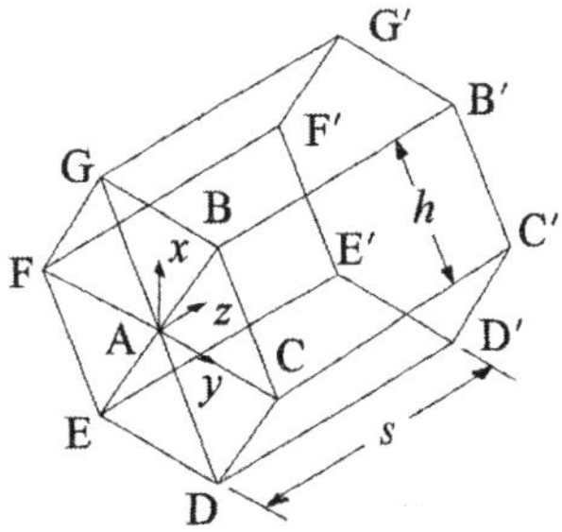

Figure A5.5 Hexagonal space trusses.

(H1) Diagonal bars BC′, CD′, DE′, EF′, FG′ and GB′ form a clockwise spiral.

$$f_{11} = f_{22} = \frac{1}{3h^2s^3c},\quad f_{36} = -\frac{1}{3\sqrt{3}h^2s^2c},$$
$$f_{33} = \frac{2}{9h^4s}\left(\frac{1}{a+b}+\frac{1}{c}+\frac{1}{d}\right),\ f_{66} = \frac{1}{6s^3c} \qquad \textbf{(A5.3.12)}$$

(H2) Diagonal bars BC′, DC′, DE′, FE′, FG′ and BG′ form a zigzag pattern.

$$f_{11} = f_{22} = \frac{1}{3h^2s^3c},\quad f_{33} = \frac{2}{9h^4s}\left(\frac{1}{b}+\frac{1}{d}\right),\quad f_{66} = \frac{1}{6s^3c}. \qquad \textbf{(A5.3.13)}$$

(H3) Diagonal bars GF′, BG′, CB′, CD′, DE′ and EF′give two sets of three diagonals sloping in opposite directions.

---

[1]Renton, J.D. (1984) The Beam-Like Behaviour of Space Trusses. *A.I.A.A. Journal* V.22, p.273.

[2]Renton, J.D. (1995) Automated Derivation of Structural Formulae. *Computers & Structures* V.56, p.959.

$$f_{11} = f_{22} = \frac{1}{3h^2s^3c} \,, \quad f_{23} = -\frac{2}{9h^3s^2c} \,,$$
$$f_{33} = \frac{2}{9h^4s}\left(\frac{1}{b} + \frac{2}{3c} + \frac{1}{d}\right) \,, \quad f_{66} = \frac{1}{6s^3c} \,. \qquad \textbf{(A5.3.14)}$$

(H4) Diagonal bars BC′, CB′, CD′, DC′, DE′, ED′, EF′, FE′, FG′, GF′, GB′, G′B cross-brace each surface panel.

$$f_{11} = f_{22} = \frac{2(b+2d)}{3h^2s^3[2c(b+2d)+3bd]} \,, \quad f_{33} = \frac{1}{9h^4sd} \,,$$
$$f_{66} = \frac{a+b+2d}{6s^3[(a+b)(c+2d)+2cd]} \,. \qquad \textbf{(A5.3.15)}$$

Instead of bracing the cross-section with six radial bars, three bars, BD, DF and FB can be used. In the above four cases, the flexibilities are the same as before except that a should be set zero, where it occurs in the expressions.

A5.4 Solutions for Plane Grids

Where solutions for loading expressed as sinusoidal functions of X are known, solutions for other loading functions can be found by expressing them in terms of finite Fourier series. Such series are finite because the series is only required to match the function at unit steps of X (Jordan (1965) §146, Wah & Calcote (1970) §2.5 and Dean (1976) Lecture III and Appendix III). The full finite Fourier series is given by

$$f(\mathrm{X}) = a_0 + \sum_{\mathrm{M}=1}^{n} a_{\mathrm{M}}\cos\frac{2\mathrm{M}\pi\mathrm{X}}{\mathrm{N}} + b_{\mathrm{M}}\sin\frac{2\mathrm{M}\pi\mathrm{X}}{\mathrm{N}} \,, \quad \text{where}$$
$$a_{\mathrm{M}} = \frac{\mathrm{K}}{\mathrm{N}}\sum_{\mathrm{X}=0}^{\mathrm{N}-1} f(\mathrm{X})\cos\frac{2\mathrm{M}\pi\mathrm{X}}{\mathrm{N}} \,, \quad (\mathrm{K}=1 \text{ for } \mathrm{M}=0 \text{ or } \mathrm{M}=\tfrac{1}{2}\mathrm{N}\,, \text{ else } \mathrm{K}=2), \qquad \textbf{(A5.4.1)}$$
$$b_{\mathrm{M}} = \frac{2}{\mathrm{N}}\sum_{\mathrm{X}=0}^{\mathrm{N}-1} f(\mathrm{X})\sin\frac{2\mathrm{M}\pi\mathrm{X}}{\mathrm{N}} \,.$$

and n is ½N (if N is even) or ½(N-1) (if N is odd). This holds true for the points X from 0 to N inclusive, provided that $f(0)$ and $f(\mathrm{N})$ are the same.

The following cosine and sine half-wave series are commonly used.

$$f(\mathrm{X}) = \tfrac{1}{2}(a_0 + a_{\mathrm{N}}\cos\pi\mathrm{X}) + \sum_{\mathrm{M}=1}^{\mathrm{N}-1} a_{\mathrm{M}}\cos\frac{\mathrm{M}\pi\mathrm{X}}{\mathrm{N}} \,, \quad \text{where}$$
$$a_{\mathrm{M}} = \frac{2}{\mathrm{N}}\left(\tfrac{1}{2}(f(0) + f(\mathrm{N})\cos\mathrm{M}\pi) + \sum_{\mathrm{X}=1}^{\mathrm{N}-1} f(\mathrm{X})\cos\frac{\mathrm{M}\pi\mathrm{X}}{\mathrm{N}}\right) \,, \qquad \textbf{(A5.4.2)}$$
$$f(\mathrm{X}) = \sum_{\mathrm{M}=1}^{\mathrm{N}-1} b_{\mathrm{M}}\sin\frac{\mathrm{M}\pi\mathrm{X}}{\mathrm{N}} \,, \quad \text{where} \quad b_{\mathrm{M}} = \frac{2}{\mathrm{N}}\sum_{\mathrm{X}=1}^{\mathrm{N}-1} f(\mathrm{X})\sin\frac{\mathrm{M}\pi\mathrm{X}}{\mathrm{N}} \,.$$

and the sine series function is zero at X=0 and X=N (see §15.2). The orthogonality relationships, such as (15.3), needed to determine the expressions for the coefficients a_{M} and b_{M} in (A5.4.1) and (A5.4.2) can be derived from (15.2). A particularly useful result is given by (15.5). For a constant function equal to unity, this gives the coefficients in the sine half-wave series in (A5.4.2) as

$$b_{\mathrm{M}} = 0 \quad \text{(if M is even) and} \quad b_{\mathrm{M}} = \frac{2}{\mathrm{N}} \cot \frac{\mathrm{M}\pi}{2\mathrm{N}} \quad \text{(if M is odd)} . \qquad \textbf{(A5.4.3)}$$

The use of double Fourier series for grids is a natural extension of this. A Navier-type solution for rectangular grids with rectangular boundaries is given in §15.6. This is also given by Wah & Calcote (1970) §4.2.1. In §4.2.2, they also give a Lévy-type solution for the case when one pair of rectangular boundaries is simply-supported and the other pair are clamped. The solution for a square grid with a clamped circular boundary with a radius of R bays subject to a constant joint load P_0 is given by the author[1]. Using the notation of §15.6, this is given by

$$\phi = \frac{P_0 a^2 \mathrm{Y}}{4(3EI + GJ)}(\mathrm{X}^2 + \mathrm{Y}^2 - \mathrm{R}^2) , \quad \psi = -\frac{P_0 a^2 \mathrm{X}}{4(3EI + GJ)}(\mathrm{X}^2 + \mathrm{Y}^2 - \mathrm{R}^2)$$
$$w = \frac{P_0 a^3}{16(3EI + GJ)}(\mathrm{X}^2 + \mathrm{Y}^2 - \mathrm{R}^2)\left(\mathrm{X}^2 + \mathrm{Y}^2 - \mathrm{R}^2 - \frac{GJ}{3EI}\right) \qquad \textbf{(A5.4.4)}$$

Of course, all the joints of the grid will not normally lie on the boundary so that this solution is necessarily an approximation. However, for R=7, the maximum error in the vertical displacement of the grid was found to be only 2.7% of the central displacement. For the case where $GJ=EI$, the solution for a pinned circular boundary of radius Ra is given by

$$\phi = \frac{P_0 a^2 \mathrm{Y}}{16EI}(\mathrm{X}^2 + \mathrm{Y}^2 - 3\mathrm{R}^2 + 1) , \quad \psi = -\frac{P_0 a^2 \mathrm{X}}{16EI}(\mathrm{X}^2 + \mathrm{Y}^2 - 3\mathrm{R}^2 + 1)$$
$$w = \frac{P_0 a^3}{64EI}(\mathrm{X}^2 + \mathrm{Y}^2 - \mathrm{R}^2)(\mathrm{X}^2 + \mathrm{Y}^2 - 5\mathrm{R}^2 + \tfrac{5}{3}) \qquad \textbf{(A5.4.5)}$$

For triangular grids, solutions have been found for rectangular, circular and triangular boundaries[2]. In these solutions, it is assumed that all the component beams are of the same length a and bending stiffness EI and that the torsional stiffness GJ is negligible. The grid is then formed from equilateral triangles, as shown in Figure A5.6. The joint numbering scheme is the same as that for the triangular mesh net shown in Figure 15.17b. This then corresponds to the Cartesian coordinate system x,y shown. A normal load P is applied in the direction into the paper to the joint (X,Y) which induces a deflexion w in the same direction and rotations θ_x and θ_y about the x and y axes as shown. For compactness, the following notation will be used.

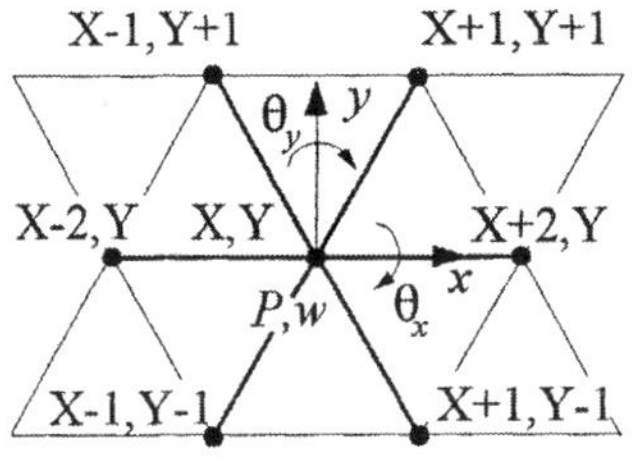

Figure A5.6 Conventions for a triangular grid.

$$\overset{+}{E}_{\mathrm{X\,or\,Y}} = E_{\mathrm{X\,or\,Y}} + E^{-1}_{\mathrm{X\,or\,Y}} , \quad \overset{-}{E}_{\mathrm{X\,or\,Y}} = E_{\mathrm{X\,or\,Y}} - E^{-1}_{\mathrm{X\,or\,Y}} . \qquad \textbf{(A5.4.6)}$$

The equations of equilibrium of a typical joint then take the form

---

[1] Renton, J.D. (1964) A Finite Difference Analysis of the Flexural-Torsional Behaviour of Grillages. *Int. J. Mech. Sci.* V.6. p.209.

[2] Renton, J.D. (1966) On the Analysis of Triangular Mesh Grillages. *Int. J. Solids Structures* V.2 p.307.

$$\begin{bmatrix} 8+E_X^+E_Y^+ & -E_X^-E_Y^- & 6E_X^+E_Y^- \\ -E_X^-E_Y^- & 16+E_X^+E_Y^+ +4(E_X^+)^2 & -6E_X^-E_Y^+ - 12E_X^+E_X^- \\ -E_X^+E_Y^- & E_X^-E_Y^+ +2E_X^+E_X^- & 4[8-E_X^+E_Y^+ -(E_X^+)^2] \end{bmatrix} \begin{bmatrix} \sqrt{3}\theta_X \\ \theta_Y \\ \dfrac{w}{a} \end{bmatrix} = \begin{bmatrix} 0 \\ 0 \\ \dfrac{a^2P}{3EI} \end{bmatrix} \qquad \textbf{(A5.4.7)}$$

Some solutions of these equations are given in the following table.

Table A5.4.1 Solutions for Triangular Grids

$\sqrt{3}\theta_X$	θ_Y	w/a	$Pa^2/2EI$
0	$10X^4 - 44/3$	X^5	1080X
$-10Y^4 + 4/3$	0	Y^5	120Y
$\frac{8}{9}(-9X^4Y^3+6Y^3+36X^2Y-14Y)$	$\frac{8}{27}(27X^3Y^4-6X^3-36XY^2+34X)$	X^4Y^4	$24(X^4+36X^2Y^2+9Y^4+\frac{5}{27})$
0	$K_1 \sin mX$	$\cos mX$	$K_2 \cos mX$
$K_3 \sin nY$	0	$\cos nY$	$K_4 \cos nY$
$K_5 \cos mX \sin nY$	$K_6 \sin mX \cos nY$	$\cos mX \cos nY$	$K_7 \cos mX \cos nY$

where

$$K_1 = \frac{-6(\sin m + \sin 2m)}{2\cos 2m + \cos m + 6}\ , \quad K_2 = -12\left(\cos 2m + 2\cos m - 3 + \frac{3(\sin m + \sin 2m)^2}{2\cos 2m + \cos m + 6}\right)$$

$$K_3 = \frac{6\sin n}{\cos n + 2}\ , \quad K_4 = \frac{12(1-\cos n)^2}{\cos n + 2}\ ,$$

$$K_5 = \frac{6\sin n(2\cos^3 m + 6\cos m + \cos n)}{4\cos^3 m\cos n + 9\cos^2 m + \cos^2 n + 6\cos m\cos n + 7}\ , \qquad \textbf{(A5.4.8)}$$

$$K_6 = \frac{-6\sin m(2\cos^2 m\cos n + 5\cos m + 2\cos n)}{4\cos^3 m\cos n + 9\cos^2 m + \cos^2 n + 6\cos m\cos n + 7}\ ,$$

$$K_7 = -6[2\cos 2m + 4\cos m\cos n - 6 + K_5\cos m\sin n - K_6\sin m(2\cos m + \cos n)]\ .$$

Further solutions can be obtained by differentiating those in the table. For example, differentiating the first solution twice with respect to X shows that a solution exists for no rotation θ_X or load P where θ_Y is $6X^2$ and w is aX^3. Combining such solutions leads to the terms required for the double Fourier type of analysis already described and the following polynomial solutions.

For a triangular mesh grid with a clamped circular boundary of radius Ra subject to a uniform load P_0 at each joint,

$$\theta_X = -\frac{P_0a^2Y}{48\sqrt{3}EI}(X^2+3Y^2-4R^2)\ , \quad \theta_Y = \frac{P_0a^2X}{144EI}(X^2+3Y^2-4R^2)\ ,$$

$$w = \frac{P_0a^3}{1152EI}(X^2+3Y^2-4R^2)^2\ . \qquad \textbf{(A5.4.9)}$$

For a triangular mesh grid with an equilateral triangular boundary of side Aa,

$$\theta_X = \frac{-P_0 a^2}{192\sqrt{3}AEI}[(X^2+3Y^2+2AY)^2-24Y^4-12Y^2-4(\tfrac{1}{3}A^2-1)(X^2+\tfrac{8}{3}AY)-\tfrac{8}{3}AY]$$

$$\theta_Y = \frac{P_0 a^2 X}{48AEI}(Y+\tfrac{1}{3}A)(X^2+Y^2+\tfrac{2}{3}AY-\tfrac{8}{9}A^2+2) \qquad \textbf{(A5.4.10)}$$

$$w = \frac{P_0 a^3}{384AEI}[Y^3-X^2Y-\tfrac{1}{3}A(X^2+3Y^2)+\tfrac{4}{27}A^3](\tfrac{4}{3}A^2-4-X^2-3Y^2)$$

The expressions for w in the above two solutions tend to those for the equivalent isotropic plate when the mesh size becomes infinitely fine in comparison with the overall dimensions of the grid. The equivalent plate stiffness, D, is given by $3\sqrt{3}\,EI/4a$.

The general form of a triangular grid is shown in Figure A5.7. It is made from three sets of beams, labelled 1 to 3, with flexural and torsional stiffnesses EI_i and GJ_i and lengths l_i, (i = 1 to 3). These form triangles with the internal angles α,β,γ and a circle through the apexes of diameter λ as shown. The differential equation giving a first approximation to this grid is

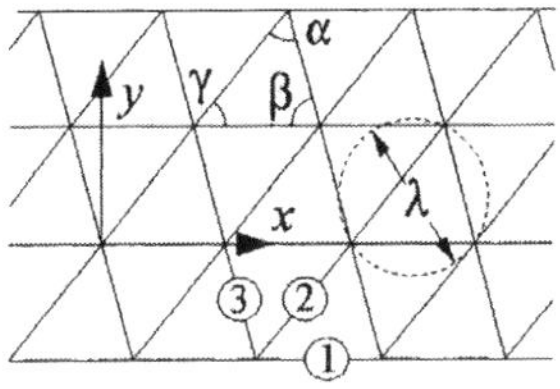

Figure A5.7 A general triangular grid.

$$p\lambda^2\sin\alpha\sin\beta\sin\gamma = \Bigg[l_1\frac{\partial^2}{\partial x^2}\left(EI_1\frac{\partial^2}{\partial x^2}+GJ_1\frac{\partial^2}{\partial y^2}\right) +$$

$$l_2\left(\cos\gamma\frac{\partial}{\partial x}+\sin\gamma\frac{\partial}{\partial y}\right)^2\left(EI_2\left\{\cos\gamma\frac{\partial}{\partial x}+\sin\gamma\frac{\partial}{\partial y}\right\}^2+GJ_2\left\{\sin\gamma\frac{\partial}{\partial x}-\cos\gamma\frac{\partial}{\partial y}\right\}^2\right) + \qquad \textbf{(A5.4.11)}$$

$$l_3\left(\cos\beta\frac{\partial}{\partial x}-\sin\beta\frac{\partial}{\partial y}\right)^2\left(EI_3\left\{\cos\beta\frac{\partial}{\partial x}-\sin\beta\frac{\partial}{\partial y}\right\}^2+GJ_3\left\{\sin\beta\frac{\partial}{\partial x}+\cos\beta\frac{\partial}{\partial y}\right\}^2\right)\Bigg]w + O(\lambda^4)$$

If all the beams are the same, this reduces to the biharmonic plate equation where the plate stiffness is given by

$$D = \frac{\sqrt{3}}{4l}(3EI + GJ) \qquad \textbf{(A5.4.12)}$$

(cf. (A8.1.4)). Yettram and Just[1] showed that another particular case of (A5.4.11) also represents an isotropic plate. This is when

$$EI_1 = GJ_1 = \lambda D\cos\alpha \;,\; EI_2 = GJ_2 = \lambda D\cos\beta \;,\; EI_3 = GJ_3 = \lambda D\cos\gamma\,. \qquad \textbf{(A5.4.13)}$$

The process of deriving the approximate continuum equation for grids has been automated[2] and produced equivalent isotropic plate stiffnesses for the combinations of triangular and hexagonal grids shown in Figure A5.8. Using the abbreviations e for EI/l and g for GJ/l and the subscripts t for the triangular grid members (shown bold) and h for the hexagonal grid members, these are

[1] Yettram, A.L. and Just, D.J. (1965) The representation of Edge-Supported Plates in Flexure by Triangulated Grid-Frameworks. *Int. J. Mech. Sci.* V.7 p.415.

[2] Renton, J.D. (1974) Algebra Manipulating Programs for Structural Analogies. *Computers & Structures* V.4 p. 993.

$$D = \frac{\sqrt{3}}{4}\left(\frac{e_h(e_h^2 + 6e_h g_h + 11e_h e_t + 9g_h e_t)}{3(e_h^2 + e_h g_h + 4e_h e_t + g_h e_t)} + 3e_t + g_t\right) \quad \textbf{(A5.4.14)}$$

for the case shown in Figure A5.8a and

$$D = \frac{\sqrt{3}}{4}\left(\frac{2e_h(11e_h + 9g_h)}{3(4e_h + g_h)} + 3e_t + g_t\right) \quad \textbf{(A5.4.15)}$$

for the case shown in Figure A5.8b. A particular case of the former gives the equivalent plate stiffness of a simple hexagonal grid as

$$D = \frac{EI(EI + 3GJ)}{2\sqrt{3}\,l(EI + GJ)} \quad \textbf{(A5.4.16)}$$

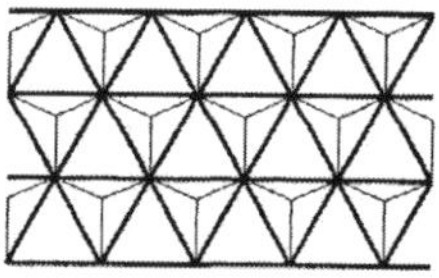
(a)

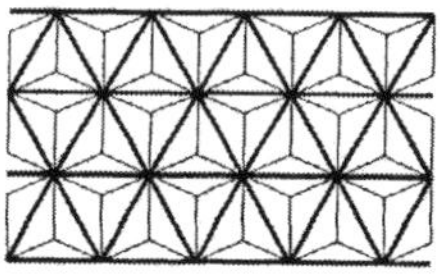
(b)

Figure A5.8 Combined triangular and hexagonal grids.

Gutkowski and Obrębski[1] formulate the general equations for hexagonal grids and give a number of solutions for members with no torsional stiffness. These are for specific rhomboidal, circular and rectangular boundary conditions.

A5.5. Solutions for Space Grids

The second-order continuum approximation to the PSM grids analysed in §15.7 is

$$\begin{aligned}
&\{d[4dbt(\partial_x^8 + \partial_y^8) + 16bt(b+t)(\partial_x^6\partial_y^2 + \partial_x^2\partial_y^6) + 4d(b^2+t^2)\partial_x^4\partial_y^4] + \tfrac{1}{3}l^2[4d^2bt(\partial_x^{10} + \partial_y^{10})\\
&12dbt(b+t)(\partial_x^8\partial_y^2 + \partial_x^2\partial_y^8) + (2b^2d^2 + 2dt(2b^2 + 2bt + dt) - 48b^2t^2)(\partial_x^6\partial_y^4 + \partial_x^4\partial_y^6)]\}h^2l^2w\\
&\quad = \{4d(b+t)[d(\partial_x^4 + \partial_y^4) + 4(b+t)\partial_x^2\partial_y^2] + \tfrac{1}{3}l^2[5bd^2 + 2td^2 - 12bdt)(\partial_x^6 + \partial_y^6)\\
&\qquad + (16b^2d + 4t^2d - 48bt(b+t) + 20bdt + 3d^2t)(\partial_x^4\partial_y^2 + \partial_x^2\partial_y^4)\}p
\end{aligned} \quad \textbf{(A5.5.1)}$$

where

$$\partial_x = \frac{\partial}{\partial x} \quad , \quad \partial_y = \frac{\partial}{\partial y} \, . \quad \textbf{(A5.5.2)}$$

The general first-order approximation is given by dropping the tenth-order differentials on the left-hand side of the equation and the sixth-order differentials on the right-hand side.

---

[1] Gutkowski, W. and Obrębski, J. (1971) The Hexagonal grid. *Bulletin de l'Académie Polonaise des Sciences*, V.19 No.5 p.29.

The equations for the other space grids in Figure 15.1d are given below. As before, these will be expressed in terms of the axial coefficients (EA/l^3 values) t, b, and d for the top and bottom bars, and the diagonal bars connecting the two meshes. The joint displacements u, v and w are in the directions of Cartesian axes x, y and z. The joints are taken to be loaded with forces P acting in the z direction. Finite-difference operators E_X and E_Y give steps of length l in the x and y directions. The vertical separation of the upper and lower meshes is taken to be h.

In the case of the ISM grid shown in Figure A5.9, it is necessary to define three typical joints, A,B and C. Joints B and C differ only in the orientation of the diagonal bars meeting at them. Using the notation of (15.88), the form of the finite-difference equations is

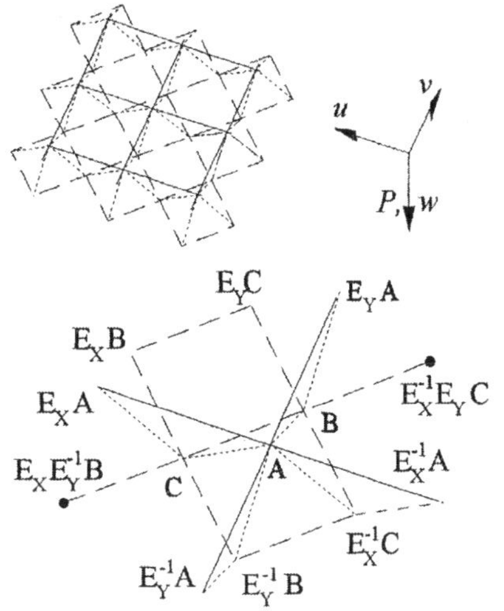

Figure A5.9 An Inclined Square Mesh grid.

$$\begin{bmatrix} 4b & 0 & 0 & -b\bar{S}_X S_Y & b\bar{D}_X D_Y & 0 & 0 & 0 & 0 \\ 0 & 4b+2d & 0 & b\bar{D}_X D_Y & -b\bar{S}_X S_Y & 0 & 0 & -dS_Y & -dD_Y \\ 0 & 0 & 2d & 0 & 0 & 0 & 0 & -dD_Y & -dS_Y \\ -bS_X\bar{S}_Y & bD_X\bar{D}_Y & 0 & 4b+2d & 0 & 0 & -dS_X & 0 & -dD_X \\ bD_X\bar{D}_Y & -bS_X\bar{S}_Y & 0 & 0 & 4b & 0 & 0 & 0 & 0 \\ 0 & 0 & 0 & 0 & 0 & 2d & -dD_X & 0 & -dS_X \\ 0 & 0 & 0 & -d\bar{S}_X & 0 & -d\bar{D}_X & 2d+4tD_X\bar{D}_X & 0 & 0 \\ 0 & -d\bar{S}_Y & -d\bar{D}_Y & 0 & 0 & 0 & 0 & 2d+4tD_Y\bar{D}_Y & 0 \\ 0 & -d\bar{D}_Y & -d\bar{S}_Y & -d\bar{D}_X & 0 & -d\bar{S}_X & 0 & 0 & 4d \end{bmatrix} \begin{bmatrix} l\,u_B \\ l\,v_B \\ h\,w_B \\ l\,u_C \\ l\,v_C \\ h\,w_C \\ l\,u_A \\ l\,v_A \\ h\,w_A \end{bmatrix} = \begin{bmatrix} 0 \\ 0 \\ P_B/h \\ 0 \\ 0 \\ P_C/h \\ 0 \\ 0 \\ P_A/h \end{bmatrix} \qquad \textbf{(A5.5.3)}$$

The solution of these equations given by $u_B = -v_C = C(-1)^{X+Y}$, where C is an arbitrary constant and all the other joint displacements are zero, corresponds to the mode shown in Figure 15.21 (cf. Figure 15.2). Taking P_B and P_C to be zero, and using Taylor's expansions of the type given by (15.14), a differential relationship between w_A and the load per unit area p $(=P_A\ /l^2)$ can be formed. To the first approximation, this is eighth-order and unlike the other grids there appear to be no special cases in which it reduces to a fourth order equation. Using the notation given by (A5.5.2) for compactness, the second approximation takes the form

$$\begin{aligned} &\{3bdt[4b(\partial_x^8+\partial_y^8)+8(2t-b)(\partial_x^6\partial_y^2+\partial_x^2\partial_y^6)+8(b+12t)\partial_x^4\partial_y^4]+4b^2dtl^2(\partial_x^{10}+\partial_y^{10}) \\ &+btl^2[3(4dt-8bt-bd)(\partial_x^8\partial_y^2+\partial_x^2\partial_y^8)+8(100dt+24bt-bd)(\partial_x^6\partial_y^4+\partial_x^4\partial_y^6)]\}\,w_A h^2 l^2 \\ &=\{3d[(b^2+4bt)(\partial_x^4+\partial_y^4)+2(4bt+8t^2-b^2)\partial_x^2\partial_y^2] \\ &+l^2[1.25b^2d-6b^2t+2bdt)(\partial_x^6-\partial_x^4\partial_y^2-\partial_x^2\partial_y^4+\partial_y^6)+4(4bdt-12bt^2+dt^2)(\partial_x^4\partial_y^2+\partial_x^2\partial_y^4)]\}\,p \end{aligned} \qquad \textbf{(A5.5.4)}$$

Again, the first approximation is given by dropping the tenth-order differentials on the left-hand side of the equation and the sixth-order differentials on the right-hand side. Because all these equations only contain even-order differentials, sinusoidal solutions are always possible.

A parallel triangular mesh (PTM) grid is shown in Figure A5.10. The finite difference equations are expressed in terms of the deflexions of typical joints A and B in the upper and lower layers and the adjacent joints shown. The operators E_X and E_Y give steps of $\sqrt{3}l/2$ and $l/2$ in the x and y directions respectively. Using the notation of (A5.4.6), the finite-difference equations for the grid can be written as

$$\begin{bmatrix} \boldsymbol{B} & \boldsymbol{D} \\ \bar{\boldsymbol{D}} & \boldsymbol{T} \end{bmatrix} \begin{bmatrix} \boldsymbol{d}_B \\ \boldsymbol{d}_A \end{bmatrix} = \begin{bmatrix} \boldsymbol{F}_B \\ \boldsymbol{F}_A \end{bmatrix} \qquad \textbf{(A5.5.5)}$$

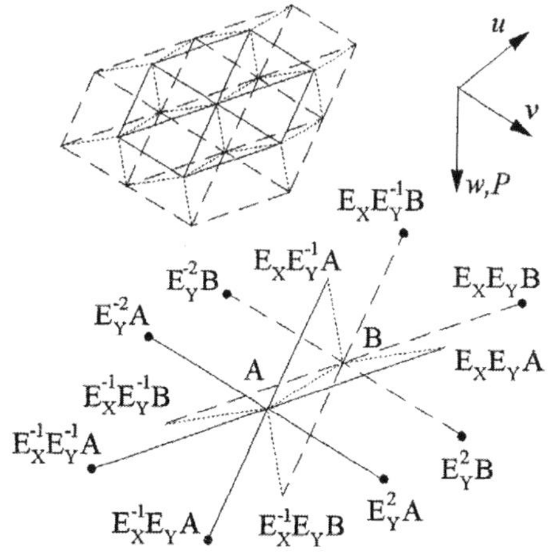

Figure A5.10 Notation for a PTM grid.

where the vectors and submatrices are

$$\boldsymbol{d}_{A\text{ or }B} = \begin{bmatrix} u_{A\text{ or }B}\dfrac{l}{2\sqrt{3}} \\ v_{A\text{ or }B}\dfrac{l}{2} \\ w_{A\text{ or }B} \end{bmatrix}, \quad \boldsymbol{F}_{A\text{ or }B} = \begin{bmatrix} 0 \\ 0 \\ \dfrac{P_{A\text{ or }B}}{h} \end{bmatrix}, \quad \boldsymbol{B} = \begin{bmatrix} 6d+9b(4-E_X^+E_Y^+) & -3bE_X^-E_Y^- & 0 \\ -3bE_X^-E_Y^- & 2d+b[4-E_X^+E_Y^+-4(E_Y^-)^2] & 0 \\ 0 & 0 & 3d \end{bmatrix}$$

$$\boldsymbol{D} = \begin{bmatrix} -d(4+E_XE_Y^+) & -dE_XE_Y^- & -d(2-E_XE_Y^+) \\ -dE_XE_Y^- & -dE_XE_Y^+ & dE_XE_Y^- \\ -d(2-E_XE_Y^+) & dE_XE_Y^- & -d(1+E_XE_Y^+) \end{bmatrix}, \quad \boldsymbol{T} = \begin{bmatrix} 6d+9t(4-E_X^+E_Y^+) & -3tE_X^-E_Y^- & 0 \\ -3tE_X^-E_Y^- & 2d+t[4-E_X^+E_Y^+-4(E_Y^-)^2] & 0 \\ 0 & 0 & 3d \end{bmatrix}. \qquad \textbf{(A5.5.6)}$$

and $\bar{\boldsymbol{D}}$ is given by changing the signs of the powers of E_X and E_Y in $\boldsymbol{D}$. From (A5.4.6), it will be seen that this has no effect on $E^+_{X\text{ or }Y}$ but reverses the sign of $E^-_{X\text{ or }Y}$. The second-order approximation to the relationship between w_A and a uniform load p on the top layer is then given by the differential equation

$$\begin{aligned} &9\sqrt{3}\,bt[4d(b+t)\nabla^8 + l^2(bd+dt-3bt)\nabla^{10}]h^2l^2w_A = \\ &2\{24d(b+t)^2\nabla^4 + l^2[7b^2d+10bdt+3dt^2-72bt(b+t)]\nabla^6 \\ &\quad -\tfrac{2}{3}dl^2(b^2+6bt+t^2)(9\partial_x^4\partial_y^2 - 6\partial_x^2\partial_y^4 + \partial_y^6)\}p \end{aligned} \qquad \textbf{(A5.5.7)}$$

where

$$\nabla^2 = \frac{\partial^2}{\partial x^2} + \frac{\partial^2}{\partial y^2} \qquad \textbf{(A5.5.8)}$$

The final group of sixth-order differentials on the right-hand side of (A5.5.7) reflects the fact that the structure is symmetric under rotations of 60° about the z axis but is not truly isotropic (cf (15.19)).

Figure A5.11 shows a triangular-hexagonal (T-H) grid. The upper mesh is of equilateral triangles with sides of length l and the lower mesh is of regular hexagons with sides of length $l/\sqrt{3}$. The two meshes are separated by a vertical distance h. As before, the axial coefficients of the top, diagonal and bottom bars will be denoted by t, d and b respectively. The finite-difference operators $\mathrm{E_X}$ and $\mathrm{E_Y}$ give steps of $l/2\sqrt{3}$ and $l/2$ in the x and y directions. A typical joint A characterises all the joints in the upper mesh. The lower mesh has two typical joints, B and C, which differ by 60° in the orientation of the bars meeting at them. In this example, B and C are imaginary joints immediately below A. The joints actually used in formulating the finite difference equations for joint A are shown in the figure. The complete set of equations takes the form

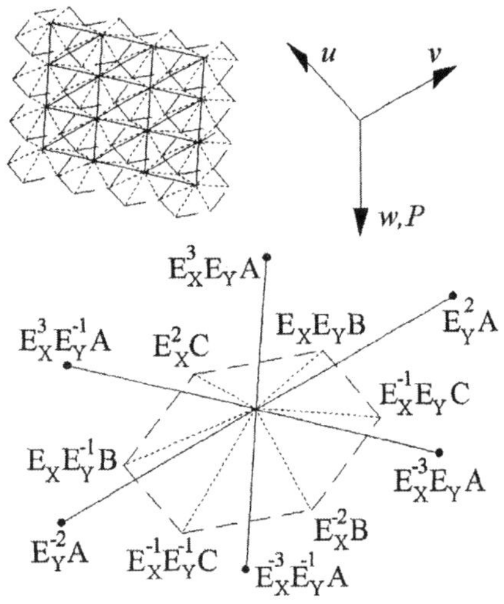

Figure A5.11 Notation for a T-H grid.

$$\begin{bmatrix} \boldsymbol{B} & \boldsymbol{C} & \boldsymbol{D}_1 \\ \bar{\boldsymbol{C}} & \boldsymbol{B} & \boldsymbol{D}_2 \\ \bar{\boldsymbol{D}}_1 & \bar{\boldsymbol{D}}_2 & \boldsymbol{T} \end{bmatrix} \begin{bmatrix} \boldsymbol{d}_B \\ \boldsymbol{d}_C \\ \boldsymbol{d}_A \end{bmatrix} = \begin{bmatrix} \boldsymbol{F}_B \\ \boldsymbol{F}_C \\ \boldsymbol{F}_A \end{bmatrix} \qquad \textbf{(A5.5.9)}$$

where the column vectors are and the stiffness submatrices are given by

$$\boldsymbol{d}_{A,B\,\mathrm{or}\,C} = \begin{bmatrix} u_{A,B\,\mathrm{or}\,C}\dfrac{l}{2\sqrt{3}} \\ v_{A,B\,\mathrm{or}\,C}\dfrac{l}{2} \\ w_{A,B\,\mathrm{or}\,C} \end{bmatrix}, \quad \boldsymbol{F}_{A,B\,\mathrm{or}\,C} = \begin{bmatrix} 0 \\ 0 \\ \dfrac{P_{A,B\,\mathrm{or}\,C}}{h} \end{bmatrix}, \quad \boldsymbol{B} = \begin{bmatrix} 6(b+d) & 0 & 0 \\ 0 & 2(b+d) & 0 \\ 0 & 0 & 3d \end{bmatrix},$$

$$\boldsymbol{C} = -b\begin{bmatrix} 4\mathrm{E_X^{-2}} + \mathrm{E_XE_Y^+} & \mathrm{E_XE_Y^-} & 0 \\ \mathrm{E_XE_Y^-} & \mathrm{E_XE_Y^+} & 0 \\ 0 & 0 & 0 \end{bmatrix}, \quad \boldsymbol{D}_1 = d\begin{bmatrix} -(4\mathrm{E_X^2} + \mathrm{E_X^{-1}E_Y^+}) & \mathrm{E_X^{-1}E_Y^-} & 2\mathrm{E_X^2} - \mathrm{E_X^{-1}E_Y^+} \\ \mathrm{E_X^{-1}E_Y^-} & -\mathrm{E_X^{-1}E_Y^+} & \mathrm{E_X^{-1}E_Y^-} \\ 2\mathrm{E_X^2} - \mathrm{E_X^{-1}E_Y^+} & \mathrm{E_X^{-1}E_Y^-} & -(\mathrm{E_X^2} + \mathrm{E_X^{-1}E_Y^+}) \end{bmatrix},$$

$$\boldsymbol{D}_2 = d\begin{bmatrix} -(4\mathrm{E_X^{-2}} + \mathrm{E_XE_Y^+}) & -\mathrm{E_XE_Y^-} & -(2\mathrm{E_X^{-2}} - \mathrm{E_XE_Y^+}) \\ -\mathrm{E_XE_Y^-} & -\mathrm{E_XE_Y^+} & \mathrm{E_XE_Y^-} \\ -(2\mathrm{E_X^{-2}} - \mathrm{E_XE_Y^+}) & \mathrm{E_XE_Y^-} & -(\mathrm{E_X^{-2}} + \mathrm{E_XE_Y^+}) \end{bmatrix}, \qquad \textbf{(A5.5.10)}$$

$$\boldsymbol{T} = \begin{bmatrix} 9t[4 - (\mathrm{E_X^3} + \mathrm{E_X^{-3}})\mathrm{E_Y^+}] + 12d & 3t(\mathrm{E_X^3} - \mathrm{E_X^{-3}})\mathrm{E_Y^-} & 0 \\ 3t(\mathrm{E_X^3} - \mathrm{E_X^{-3}})\mathrm{E_Y^-} & t[4 - (\mathrm{E_X^3} + \mathrm{E_X^{-3}})\mathrm{E_Y^+} - 4(\mathrm{E_Y^-})^2] + 4d & 0 \\ 0 & 0 & 6d \end{bmatrix}.$$

and the bars over the submatrices have the same significance as previously. The second-order approximation to the relationship between w_A and a uniform load p on the top layer is then given by the differential equation

$$
\begin{aligned}
& bdh^2l^2[36(bd+54bt+81dt)(bd+2bt+dt)\nabla^4 \\
& \quad +9(b^2d^2+82bd^2t+56b^2dt+81d^2t^2+216bdt^2+108b^2t^2)l^2\nabla^6 \\
& \quad +2(78bd^2t+36b^2dt-b^2d^2-81d^2t^2-108bdt^2-56b^2t^2)l^2\partial_y^2(\partial_y^2-3\partial_x^2)^2]\nabla^4 w_A \\
& = 3\sqrt{3}[72d(11bd+54bt+27dt+4b^2)(bd+2bt+dt)\nabla^4 \\
& \quad +3(81b^2d^3+28b^3d^2+162bd^3t+352b^2d^2t-52b^3dt \\
& \quad +81d^3t^2+324bd^2t^2+216b^2dt^2-216b^3t^2)l^2\nabla^6 \\
& \quad +2d(5b^2d^2+4b^3d+42bd^2t+24b^2dt-27d^2t^2-12b^3t-54bdt^2-54b^2t^2)l^2\partial_y^2(\partial_y^2-3\partial_x^2)^2]p
\end{aligned}
\tag{A5.5.11}
$$

The first approximations to equivalent continua for PTM and T-H grids are particular cases of (A5.5.7) and (A5.5.11) in which the tenth-order differentials are dropped from the left-hand sides and the sixth-order differentials are dropped from the right-hand sides of these equations. Any solution of the biharmonic equation given by (15.94), where D is given by (15.95) for PTM grids and by (15.96) for T-H grids, is then a first-approximation solution.

An alternative approach[1] to analysing the characteristic response of pin-jointed trusses and grids is to compare the strain energy stored in a module with the work done by the resultant load pairs acting on the edges of it. If α is the axial coefficient of a bar and τ its tension coefficient, its axial strain energy is $\tau^2/2\alpha$. The characteristic response is such that τ will either be the same in each module or vary linearly in the presence of a shear force. If the truss is statically-determinate, the tension coefficients are readily found in terms of the resultants acting on it. The expression for U_l so found can be compared with that given by (8.13) to yield the elements f_{ij}.

A generalisation of Kirchhoff's plate theory can be made by stating that it is based on the steady-state characteristic response to moments and torques per unit length. This concept can be applied to grids as well. For statically-determinate pin-jointed grids, the analysis follows exactly the same lines as that just described for trusses, yielding relationships between the moments per unit length and the rates of change of rotation through which they do work. If these rates are interpreted as curvatures of a representative vertical displacement of the grid, plate-like grid stiffnesses can be derived. For the T-H grid discussed above, this yields an equivalent isotropic plate stiffness

$$
D = \frac{bh^2l^2t(27td+18bt+11bd)}{6\sqrt{3}(b+9t)(td+2bt+bd)} \tag{A5.5.12}
$$

This is always slightly less than the expression given by (15.96), which is more explicitly related to the response of the top joints to vertical loads.

[1] Renton, J.D. (2001) The fundamentals of beam theory and how they can be applied to frames. *IASS Symposium on Theory, Design and Realization of Shell and Spatial Structures, Nagoya.*

Appendix 6: Tables

A6.1 Formulae for Simple Beam Problems

The following table shows the end **reaction** (shown bold) or the corresponding end deflexion (multiplied by EI) if there is no reaction to prevent it. The deflexion curve can be determined from the coefficients J, K and L. In cases *1a* to *6a* a downwards load P is applied at a distance a from the left-hand end, A. In cases *1b* to *6b* a uniform distributed load of intensity p acting downwards is applied, starting at a distance a from A and extending a further distance b to the right-hand end B. The total length of the beam, $(a+b)$ is denoted by l.
The deflexion curve for the point load P is given by $EIv = Jx^3 + Kx^2 + Lx + N - P\{X\}^3/6$ where v is the upwards displacement.
For the distributed load of intensity p it is given by $EIv = Jx^3 + Kx^2 + Lx + N - p\{X\}^4/24$. In both cases, the origin of x is at A and the origin of X is a along the beam where the loading is applied. The curly brackets are Macaulay brackets. The coefficient N is only non-zero in cases *1a* and *1b* when it is equal to the tabulated value of EIv_A.

Case	$M_A \sim EI\theta_A$	$M_B \sim EI\theta_B$	$S_A \sim EIv_A$	$S_B \sim EIv_B$	J	K	L
1a	$Pb^2/2$	$\mathbf{-Pb}$	$-Pb^2(2l+a)/6$	$\mathbf{P}$	0	0	$Pb^2/2$
1b	$pb^3/6$	$\mathbf{-pb^2/2}$	$-pb^3(3l+a)/24$	$\mathbf{pb}$	0	0	$pb^3/6$
2a	$-Pab(l+b)/6l$	$Pab(l+a)/6l$	$\mathbf{Pb/l}$	$\mathbf{Pa/l}$	$Pb/6l$	0	$-Pab(l+b)/6l$
2b	$-pb^2(2l^2-b^2)/24l$	$pb^2(l+a)^2/24l$	$\mathbf{pb^2/2l}$	$\mathbf{pb(l+a)/2l}$	$pb^2/12l$	0	$-pb^2(2l^2-b^2)/24l$
3a	$-Pab^2/4l$	$\mathbf{-Pab(l+a)/2l^2}$	$\mathbf{Pb^2(2l+a)/2l^3}$	$\mathbf{Pa(3l^2-a^2)/2l^3}$	$Pb^2(2l+a)/12l^3$	0	$-Pab^2/4l$
3b	$-pb^3(l+3a)/48l$	$\mathbf{-pb^2(l+a)^2/8l^2}$	$\mathbf{pb^3(3l+a)/8l^3}$	$\mathbf{pb(8l^3-4lb^2+b^3)/8l^3}$	$pb^3(3l+a)/48l^3$	0	$-pb^3(l+3a)/48l$
4a	$\mathbf{Pa}$	$-Pa^2/2$	$\mathbf{P}$	$-Pa^2(2l+b)/6$	$P/6$	$-Pa/2$	0
4b	$\mathbf{pb(l+a)/2}$	$-pb(3la+b^2)/6$	$\mathbf{pb}$	$-pb(2l^3+6al^2+b^3)/24$	$pb/6$	$-pb(l+a)/4$	0
5a	$\mathbf{Pab(l+b)/2l^2}$	$Pa^2b/4l$	$\mathbf{Pb(3l^2-b^2)/2l^3}$	$\mathbf{Pa^2(2l+b)/2l^3}$	$Pb(3l^2-b^2)/12l^3$	$-Pab(l+b)/4l^2$	0
5b	$\mathbf{pb^2(2l^2-b^2)/8l^2}$	$pb^2(6a^2+4ab+b^2)/48l$	$\mathbf{pb^2(6l^2-b^2)/8l^3}$	$\mathbf{pb(8l^3-6l^2b+b^3)/8l^3}$	$pb^2(6l^2-b^2)/48l^3$	$-pb^2(2l^2-b^2)/16l^2$	0
6a	$\mathbf{Pab^2/l^2}$	$\mathbf{-Pa^2b/l^2}$	$\mathbf{Pb^2(l+2a)/l^3}$	$\mathbf{Pa^2(l+2b)/l^3}$	$Pb^2(l+2a)/6l^3$	$-Pab^2/2l^2$	0
6b	$\mathbf{pb^3(l+3a)/12l^2}$	$\mathbf{-pb^2(6a^2+4ab+b^2)/12l^2}$	$\mathbf{pb^3(l+a)/2l^3}$	$\mathbf{pb(2al^2+2abl+b^3)/2l^3}$	$pb^3(l+a)/12l^3$	$-pb^3(l+3a)/24l^2$	0

Appendix A6.2 Stability and Vibration Functions

Livesley's stability functions discussed in Chapter 9 and Armstrong's vibration functions discussed in Chapter 10 are listed later in this appendix. They can also be generated by running the program STABVIB.EXE. This is done by entering the value of the non-dimensional parameter λ at the prompt 'Set Lambda:' where

$$\lambda_{buckling} = \frac{Pl^2}{EI}, \quad \lambda_{flexural\ vibration} = \frac{\rho A \omega^2 l^4}{EI},$$

$$\lambda_{axial\ vibration} = \frac{\rho \omega^2 l^2}{E}, \quad \lambda_{torsional\ vibration} = \frac{\rho I_p \omega^2 l^2}{GJ}$$

where P is the axial compressive[1] force in buckling problems, ω is the vibration frequency in radians per second, EI is the appropriate bending stiffness, l is the length of the beam, ρ is the density of the beam material, E is Young's modulus, A is the cross-sectional area of the beam, GJ is the torsional stiffness of the beam and I_p is the polar moment of area of the cross-section about its shear centre (S.I. units are assumed). The output gives the stability functions ϕ_1 to ϕ_4 after the prompts "Phi1:" to "Phi4:", the flexural vibration functions c_1 to c_6 after the prompts "C1:" to "C6:" and the axial or torsional vibration functions d_1 and d_2 after the prompts "D1:" and "D2:".

The stability function ϕ_0 is generated from the expansion

$$\phi_0 = 1 + 2m - \left[2\sum_{r=1}^{m} \frac{1}{1 - \dfrac{\lambda}{(2r\pi)^2}}\right] - \left(\frac{1}{12} - 2\sum_{r=1}^{m}\frac{1}{(2r\pi)^2}\right)\lambda$$
$$- \left(\frac{1}{720} - 2\sum_{r=1}^{m}\frac{1}{(2r\pi)^4}\right)\lambda^2 \ldots - \left(\frac{|B_{2n}|}{(2n)!} - 2\sum_{r=1}^{m}\frac{1}{(2r\pi)^{2n}}\right)\lambda^n \ldots$$

where B_n is the nth Bernoulli coefficient. The choice of m is arbitrary, but the larger the value of m, the swifter will be the convergence of the power series in λ. The other stability functions are then found from

$$6\phi_2 = \frac{\lambda}{2(1 - \phi_0)}, \quad \phi_1 = \phi_0\phi_2, \quad 4\phi_3 = 3\phi_2 + \phi_0, \quad 2\phi_4 = 3\phi_2 - \phi_0$$

The flexural vibration functions are generated from the following series.

$$c_0 = 4\left[\frac{1}{4!} - \frac{4\lambda}{8!} + \frac{(4\lambda)^2}{12!} \ldots + \frac{(-4\lambda)^n}{(4n+4)!}\right]$$

$$4c_1 = \frac{4}{c_0}\left[\frac{1}{3!} - \frac{4\lambda}{7!} + \frac{(4\lambda)^2}{11!} \ldots + \frac{(-4\lambda)^n}{(4n+3)!}\right], \quad 2c_2 = \frac{2}{c_0}\left[\frac{1}{3!} + \frac{\lambda}{7!} + \frac{\lambda^2}{11!} \ldots + \frac{\lambda^n}{(4n+3)!}\right],$$

$$6c_3 = \frac{2}{c_0}\left[\frac{1}{2!} - \frac{4\lambda}{6!} + \frac{(4\lambda)^2}{10!} \ldots + \frac{(-4\lambda)^n}{(4n+2)!}\right], \quad 6c_4 = \frac{2}{c_0}\left[\frac{1}{2!} + \frac{\lambda}{6!} + \frac{\lambda^2}{10!} \ldots + \frac{\lambda^n}{(4n+2)!}\right],$$

$$12c_5 = \frac{2}{c_0}\left[1 - \frac{4\lambda}{5!} + \frac{(4\lambda)^2}{9!} \ldots + \frac{(-4\lambda)^n}{(4n+1)!}\right], \quad 12c_6 = \frac{2}{c_0}\left[1 + \frac{\lambda}{5!} + \frac{\lambda^2}{9!} \ldots + \frac{\lambda^n}{(4n+1)!}\right].$$

The axial and torsional vibration functions are generated from the following series.

---

[1] If the axial force is tensile, negative values of λ can be entered.

$$d_1 = 1 - \frac{\lambda}{3} - \frac{\lambda^2}{45} - \frac{2\lambda^3}{945}\cdots - \frac{(4\lambda)^n}{(2n)!}|B_{2n}| = 1 + 2\lambda\sum_{n=1}^{\infty}\frac{1}{\lambda - n^2\pi^2}$$

$$d_2 = 1 + \frac{\lambda}{6} - \frac{7\lambda^2}{360} - \frac{31\lambda^3}{15120}\cdots - \frac{(4^n-2)\lambda^n}{(2n)!}|B_{2n}| = 1 + 2\lambda\sum_{n=1}^{\infty}\frac{(-1)^n}{\lambda - n^2\pi^2}$$

where B_{2n} is the 2nth Bernoulli number. If ϕ_0 is generated, it can also be used to evaluate d_1 and d_2 using the relationships

$$d_1(\lambda) = \phi_0(4\lambda)\ ,\quad d_2^2(\lambda) = \lambda + d_1^2(\lambda)\ .$$

It should be noted that there are different definitions of Bernoulli numbers. What is written here as the modulus of the 2nth Bernoulli number is defined elsewhere as the nth Bernoulli number (or coefficient). The form used here is tabulated on page 810 of Abramowitz and Stegun (1965) up to B_{60}. The only non-zero odd Bernoulli number is B_1, which has the value of -0.5. After that, all Bernoulli numbers with indices divisible by four are negative, and the rest are positive. It will be seen that only the moduli of the Bernoulli numbers are used in the above formulae, so that their sign is not significant. Here is a list of the first few.

$$B_0 = 1 \quad B_2 = \frac{1}{6} \quad B_4 = -\frac{1}{30} \quad B_6 = \frac{1}{42}$$

$$B_8 = -\frac{1}{30} \quad B_{10} = \frac{5}{66} \quad B_{12} = -\frac{691}{2730} \quad B_{14} = \frac{7}{6}$$

The magnitude of all the higher Bernoulli numbers is greater than unity, B_{60} having a value of approximately -2.14×10^{34}.

Stability Functions (Compression)

λ	$12\phi_1$	$6\phi_2$	$4\phi_3$	$2\phi_4$
0	12.0000	6.0000	4.0000	2.0000
1	10.7986	5.8993	3.8649	2.0344
2	9.5942	5.7971	3.7260	2.0710
3	8.3867	5.6933	3.5832	2.1101
4	7.1761	5.5880	3.4361	2.1519
5	5.9622	5.4811	3.2844	2.1967
6	4.7449	5.3724	3.1278	2.2446
7	3.5241	5.2620	2.9659	2.2962
8	2.2996	5.1498	2.7982	2.3516
9	1.0713	5.0357	2.6242	2.4115
10	-0.1609	4.9195	2.4434	2.4761
11	-1.3972	4.8014	2.2552	2.5462
12	-2.6378	4.6811	2.0588	2.6223
13	-3.8828	4.5586	1.8534	2.7052
14	-5.1324	4.4338	1.6381	2.7957
15	-6.3868	4.3066	1.4118	2.8948
16	-7.6463	4.1769	1.1731	3.0037
17	-8.9110	4.0445	0.9206	3.1239
18	-10.1811	3.9094	0.6526	3.2568
19	-11.4570	3.7715	0.3669	3.4046
20	-12.7389	3.6306	0.0609	3.5697
21	-14.0270	3.4865	-0.2684	3.7549
22	-15.3217	3.3392	-0.6246	3.9638
23	-16.6232	3.1884	-1.0126	4.2010
24	-17.9320	3.0340	-1.4382	4.4722
25	-19.2484	2.8758	-1.9087	4.7845
26	-20.5728	2.7136	-2.4339	5.1475
27	-21.9055	2.5472	-3.0263	5.5735
28	-23.2472	2.3764	-3.7030	6.0795
29	-24.5982	2.2009	-4.4877	6.6886
30	-25.9590	2.0205	-5.4138	7.4342
31	-27.3304	1.8348	-6.5303	8.3651
32	-28.7127	1.6436	-7.9127	9.5563
33	-30.1068	1.4466	-9.6829	11.1295
34	-31.5134	1.2433	-12.0515	13.2948
35	-32.9332	1.0334	-15.4174	16.4508
36	-34.3670	0.8165	-20.6375	21.4540
37	-35.8159	0.5920	-29.9519	30.5440
38	-37.2808	0.3596	-51.6606	52.0202
39	-38.7629	0.1185	-163.4559	163.5744
40	-40.2634	-0.1317	152.7909	-152.9226

Stability Functions (Tension)

λ	$12\phi_1$	$6\phi_2$	$4\phi_3$	$2\phi_4$
0	12.0000	6.0000	4.0000	2.0000
-1	13.1986	6.0993	4.1316	1.9677
-2	14.3944	6.1972	4.2600	1.9372
-3	15.5876	6.2938	4.3852	1.9086
-4	16.7781	6.3891	4.5076	1.8815
-5	17.9662	6.4831	4.6272	1.8559
-6	19.1518	6.5759	4.7442	1.8317
-7	20.3350	6.6675	4.8587	1.8088
-8	21.5160	6.7580	4.9709	1.7871
-9	22.6947	6.8474	5.0809	1.7665
-10	23.8713	6.9357	5.1887	1.7469
-11	25.0458	7.0229	5.2946	1.7283
-12	26.2183	7.1091	5.3986	1.7106
-13	27.3887	7.1944	5.5007	1.6937
-14	28.5573	7.2787	5.6010	1.6776
-15	29.7240	7.3620	5.6997	1.6623
-16	30.8889	7.4444	5.7968	1.6476
-17	32.0519	7.5260	5.8924	1.6336
-18	33.2133	7.6067	5.9865	1.6202
-19	34.3730	7.6865	6.0792	1.6073
-20	35.5310	7.7655	6.1705	1.5950
-21	36.6875	7.8438	6.2605	1.5832
-22	37.8424	7.9212	6.3493	1.5719
-23	38.9958	7.9979	6.4368	1.5611
-24	40.1477	8.0738	6.5232	1.5506
-25	41.2981	8.1491	6.6085	1.5406
-26	42.4472	8.2236	6.6926	1.5310
-27	43.5948	8.2974	6.7757	1.5217
-28	44.7412	8.3706	6.8578	1.5128
-29	45.8862	8.4431	6.9389	1.5042
-30	47.0299	8.5149	7.0191	1.4959
-31	48.1724	8.5862	7.0983	1.4879
-32	49.3136	8.6568	7.1767	1.4801
-33	50.4536	8.7268	7.2541	1.4727
-34	51.5925	8.7962	7.3308	1.4655
-35	52.7302	8.8651	7.4066	1.4585
-36	53.8668	8.9334	7.4816	1.4518
-37	55.0023	9.0012	7.5559	1.4453
-38	56.1367	9.0684	7.6294	1.4390
-39	57.2701	9.1350	7.7022	1.4329
-40	58.4024	9.2012	7.7742	1.4270

Flexural Vibration Functions

λ	$4c_1$	$2c_2$	$6c_3$	$6c_4$	$12c_5$	$12c_6$
0	4.00000	2.00000	6.00000	6.00000	12.0000	12.0000
10	3.90310	2.07303	5.46838	6.31689	8.24852	13.3194
20	3.80275	2.14940	4.92049	6.64916	4.41970	14.7090
30	3.69873	2.22933	4.35534	6.99779	0.50884	16.1734
40	3.59080	2.31305	3.77183	7.36388	-3.48918	17.7178
50	3.47871	2.40082	3.16875	7.74861	-7.57991	19.3476
60	3.36218	2.49292	2.54481	8.15329	-11.7694	21.0688
70	3.24090	2.58964	1.89861	8.57932	-16.0642	22.8880
80	3.11454	2.69133	1.22855	9.02825	-20.4716	24.8124
90	2.98273	2.79835	0.53295	9.50179	-24.9997	26.8499
100	2.84507	2.91111	-0.19007	10.0018	-29.6571	29.0091
110	2.70110	3.03004	-0.94260	10.5304	-34.4534	31.2997
120	2.55035	3.15564	-1.72690	11.0897	-39.3994	33.7321
130	2.39226	3.28845	-2.54553	11.6824	-44.5068	36.3182
140	2.22622	3.42909	-3.40129	12.3113	-49.7886	39.0711
150	2.05157	3.57821	-4.29731	12.9795	-55.2595	42.0051
160	1.86756	3.73658	-5.23712	13.6904	-60.9359	45.1365
170	1.67332	3.90503	-6.22462	14.4480	-66.8358	48.4836
180	1.46793	4.08452	-7.26425	15.2568	-72.9800	52.0668
190	1.25030	4.27612	-8.36097	16.1216	-79.3915	55.9093
200	1.01921	4.48104	-9.52045	17.0481	-86.0966	60.0372
210	0.77329	4.70066	-10.7491	18.0428	-93.1252	64.4803
220	0.51096	4.93656	-12.0543	19.1130	-100.511	69.2728
230	0.23040	5.19056	-13.4444	20.2670	-108.295	74.4536
240	-0.07047	5.46474	-14.9292	21.5147	-116.520	80.0680
250	-0.39409	5.76152	-16.5199	22.8672	-125.239	86.1683
260	-0.74327	6.08374	-18.2297	24.3377	-134.514	92.8154
270	-1.12135	6.43471	-20.0740	25.9416	-144.416	100.081
280	-1.53223	6.81834	-22.0710	27.6971	-155.030	108.050
290	-1.98058	7.23931	-24.2424	29.6259	-166.456	116.822
300	-2.47200	7.70320	-26.6143	31.7541	-178.817	126.519
350	-5.87491	10.9663	-42.8727	46.7768	-261.096	195.346
400	-12.5764	17.5243	-74.4676	77.1065	-414.703	335.268
450	-32.3564	37.1568	-166.848	168.191	-850.858	757.492
500	-3543.26	3547.91	-16483.4	16483.5	-76685.5	76577.9
600	27.0900	-22.7569	107.997	-110.740	405.541	-542.271
700	16.7893	-12.7907	59.1396	-64.7879	164.229	-331.309
800	13.2216	-9.57804	41.5365	-50.2521	67.8642	-266.577
900	11.3210	-8.05499	31.6409	-43.6020	6.99330	-238.751
1000	10.0777	-7.21435	24.7568	-40.1609	-40.2430	-226.126
1100	9.15497	-6.72198	19.3142	-38.3802	-81.1701	-221.540
1200	8.40751	-6.43585	14.6330	-37.6048	-118.993	-221.987
1300	7.76169	-6.28603	10.3648	-37.5157	-155.376	-226.032
1400	7.17552	-6.23480	6.30762	-37.9445	-191.316	-232.945
1500	6.62253	-6.26079	2.33020	-38.7999	-227.486	-242.364
1600	6.08454	-6.35177	-1.66170	-40.0345	-264.401	-254.146
1700	5.54783	-6.50115	-5.74329	-41.6291	-302.501	-268.289
1800	5.00111	-6.70617	-9.98084	-43.5846	-342.202	-284.904
1900	4.43420	-6.96697	-14.4384	-45.9180	-383.934	-304.201
2000	3.83709	-7.28618	-19.1827	-48.6618	-428.167	-326.488
2200	2.50876	-8.12249	-29.8406	-55.5918	-526.412	-381.846
2400	0.91094	-9.28950	-42.7325	-65.0082	-642.821	-456.183
2600	-1.11546	-10.9271	-59.0790	-78.0405	-786.860	-558.513
2800	-3.84538	-13.2977	-81.0215	-96.7702	-975.267	-705.284
3000	-7.81930	-16.9331	-112.795	-125.390	-1241.27	-929.505
3200	-14.2870	-23.0769	-164.212	-173.683	-1661.95	-1308.10
3400	-26.9694	-35.4453	-264.511	-270.862	-2467.07	-2070.70
3600	-64.2017	-72.3701	-557.789	-561.007	-4789.12	-4349.68
3800	-4289.93	-4297.80	-33769.8	-33769.9	-265829	-265347
4000	87.6193	80.0570	634.013	637.160	4542.69	5070.34

Axial and Torsional Vibration Functions

λ	d_1	d_2	λ	d_1	d_2
0.00	**1.0000**	**1.0000**	**0**	**1.0000**	**1.0000**
0.01	0.9967	1.0017	10	152.8615	-152.8879
0.02	0.9933	1.0033	20	1.1083	-4.6043
0.03	0.9900	1.0050	30	-5.2374	-7.5913
0.04	0.9867	1.0067	40	152.8152	152.9207
0.05	**0.9833**	**1.0084**	**50**	**7.0677**	**9.9752**
0.06	0.9800	1.0101	60	0.8780	7.7914
0.07	0.9766	1.0118	70	-4.6643	9.6006
0.08	0.9732	1.0135	80	-17.1082	19.3502
0.09	0.9699	1.0152	90	152.7376	-152.9749
0.10	**0.9665**	**1.0169**	**100**	**15.4867**	**-18.3813**
0.11	0.9631	1.0186	110	5.9017	-12.0003
0.12	0.9598	1.0203	120	0.5269	-10.9635
0.13	0.9564	1.0220	130	-4.8213	-12.4113
0.14	0.9530	1.0237	140	-13.0210	-17.6594
0.15	**0.9496**	**1.0254**	**150**	**-36.9959**	**-39.0605**
0.16	0.9462	1.0272	160	152.6295	153.0518
0.17	0.9428	1.0289	170	25.6465	28.6746
0.18	0.9394	1.0306	180	11.8992	17.8574
0.19	0.9360	1.0324	190	5.2008	14.6904
0.20	**0.9326**	**1.0341**	**200**	**0.0566**	**14.1423**
0.25	0.9154	1.0429	250	152.4891	-153.1490
0.30	0.8981	1.0518	300	-0.5332	-17.3351
0.35	0.8807	1.0608	350	-131.3239	-132.8691
0.40	0.8632	1.0699	400	9.1935	21.9073
0.45	0.8456	1.0791	450	-21.2460	30.2260
0.50	**0.8278**	**1.0885**	**500**	**58.0479**	**-61.9088**
0.60	0.7919	1.1075	600	-32.6667	-41.1344
0.70	0.7554	1.1270	700	7.0878	27.2792
0.80	0.7184	1.1469	800	2846.9114	-2846.5439
0.90	0.6809	1.1674	900	-4.1125	-30.3620
1.00	**0.6427**	**1.1884**	**1000**	**151.3239**	**153.9728**
1.10	0.6040	1.2099	1100	-5.3209	33.7087
1.20	0.5646	1.2320	1200	414.6743	-415.3586
1.30	0.5246	1.2547	1300	3.4544	-36.1496
1.40	0.4839	1.2780	1400	-127.9969	-134.2051
1.50	**0.4426**	**1.3019**	**1500**	**24.1712**	**45.1581**
1.60	0.4005	1.3265	1600	-34.7874	53.6837
1.70	0.3577	1.3517	1700	101.2833	-108.3529
1.80	0.3141	1.3777	1800	0.5099	-42.4288
1.90	0.2697	1.4043	1900	-103.8449	-113.7330
2.00	**0.2245**	**1.4317**	**2000**	**50.3392**	**66.3930**
2.20	0.1316	1.4889	2200	-208.6136	215.1861
2.40	0.0350	1.5496	2400	-13.3685	-51.2002
2.60	-0.0655	1.6138	2600	59.2373	76.9195
2.80	-0.1704	1.6821	2800	-96.9384	112.0062
3.00	**-0.2798**	**1.7548**	**3000**	**13.3280**	**-55.9470**
3.20	-0.3943	1.8322	3200	2848.1553	2846.6885
3.40	-0.5143	1.9149	3400	-9.0583	59.3811
3.60	-0.6403	2.0032	3600	189.7616	-196.8392
3.80	-0.7729	2.0979	3800	-22.4286	-66.4565
4.00	**-0.9128**	**2.1995**	**4000**	**146.5938**	**157.3297**

Appendix 7: The Characteristic Response of Cones

A7.1 Formulae for Spherical Polar Coordinates

Formulae for the elastic behaviour of continua relative to cylindrical polar coordinates were given in §3.10. Figure A7.1 shows the position of an infinitesimal element located at P in terms of the spherical polar coordinates r, θ and ϕ. The radius r is the distance of P from some pole O. This radius is at an angle θ to some axis z through O. The plane containing this axis and radius is at an angle ϕ to some axis x through O and perpendicular to z. This angle is clockwise-positive as viewed along the z axis. The material displacements of the body are u, v and w in the directions of increasing r, θ and ϕ respectively, as shown in the diagram. The faces of the infinitesimal element are all negative faces, in the sense described in §3.1. As shown in the detail, the top face of the element is a negative θ face, the lower left face is a negative r face and the lower right face is a negative ϕ face. Then all the stresses shown act in the negative directions. The same subscripts will be used for the corresponding strains. In terms of the material displacements, these are

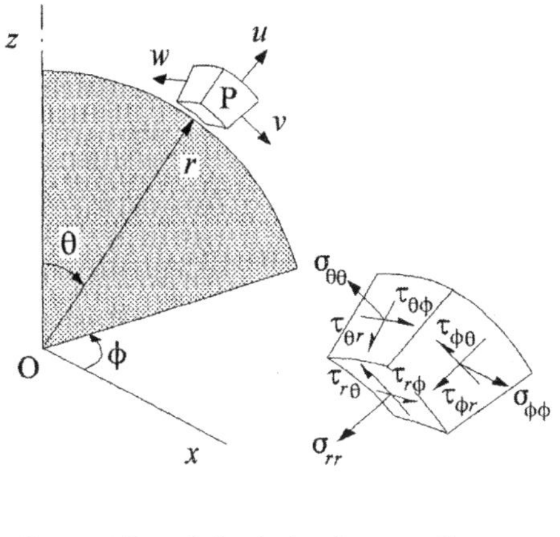

Figure A7.1 Spherical polar coordinates.

$$\epsilon_{rr} = \frac{\partial u}{\partial r} \,,\quad \gamma_{r\theta} = \frac{1}{r}\left(\frac{\partial u}{\partial \theta} + r\frac{\partial v}{\partial r} - v \right) ,\quad \gamma_{r\phi} = \frac{1}{r}\left(\frac{\partial u}{\partial \phi}\operatorname{cosec}\theta + r\frac{\partial w}{\partial r} - w \right)$$

$$\epsilon_{\theta\theta} = \frac{1}{r}\left(\frac{\partial v}{\partial \theta} + u \right) ,\quad \gamma_{\theta\phi} = \frac{1}{r}\left(\frac{\partial v}{\partial \phi}\operatorname{cosec}\theta + \frac{\partial w}{\partial \theta} - w\cot\theta \right) ,$$

$$\epsilon_{\phi\phi} = \frac{1}{r}\left(\frac{\partial w}{\partial \phi}\operatorname{cosec}\theta + v\cot\theta + u \right) . \qquad \textbf{(A7.1.1)}$$

The equations of static equilibrium are

$$\frac{\partial \sigma_{rr}}{\partial r} + \frac{1}{r}\left(2\sigma_{rr} + \frac{\partial \tau_{\theta r}}{\partial \theta} + \frac{\partial \tau_{\phi r}}{\partial \phi}\operatorname{cosec}\theta - \sigma_{\theta\theta} - \sigma_{\phi\phi} + \tau_{\theta r}\cot\theta \right) + R = 0$$

$$\frac{\partial \tau_{r\theta}}{\partial r} + \frac{1}{r}\left(3\tau_{r\theta} + \frac{\partial \sigma_{\theta\theta}}{\partial \theta} + \frac{\partial \tau_{\phi\theta}}{\partial \phi}\operatorname{cosec}\theta + (\sigma_{\theta\theta} - \sigma_{\phi\phi})\cot\theta \right) + \Theta = 0$$

$$\frac{\partial \tau_{r\phi}}{\partial r} + \frac{1}{r}\left(3\tau_{r\phi} + \frac{\partial \tau_{\theta\phi}}{\partial \theta} + \frac{\partial \sigma_{\phi\phi}}{\partial \phi}\operatorname{cosec}\theta + 2\tau_{\theta\phi}\cot\theta \right) + \Phi = 0 \qquad \textbf{(A7.1.2)}$$

where R, Θ and Φ are the body forces in the directions of increasing r, θ and ϕ respectively.

For isotropic materials, these equilibrium conditions can be satisfied by expressing the displacement field in terms of the Papkovich-Neuber functions (see Renton (1987) Appendix A3.1 for example). If $\boldsymbol{u}$ is the vector giving the material displacement of any point in the body, then this expression takes the form

$$G\boldsymbol{u} = \boldsymbol{\Phi} - \frac{1}{4(1-\nu)}\nabla[\Psi + \boldsymbol{r}.\boldsymbol{\Phi}] \qquad \textbf{(A7.1.3)}$$

where G is the shear modulus, ν is Poisson's ratio, $\boldsymbol{\Phi}$ is a harmonic vector $\{\Phi_r, \Phi_\theta, \Phi_\phi\}$, so that $\nabla^2\boldsymbol{\Phi}$ is zero, or

$$r^2 \sin\theta \nabla^2 \Phi_r - 2\left[\Phi_r \sin\theta + \frac{\partial}{\partial\theta}(\Phi_\theta \sin\theta) + \frac{\partial \Phi_\phi}{\partial\phi}\right] = 0$$

$$r^3 \sin^2\theta \nabla^2\left(\frac{1}{r}\Phi_\theta\right) + 2\left[\left(\frac{\partial\Phi_r}{\partial\theta} + r\frac{\partial\Phi_\theta}{\partial r}\right)\sin^2\theta - \frac{1}{2}\Phi_\theta - \frac{\partial\Phi_\phi}{\partial\phi}\cos\theta\right] = 0 \qquad \textbf{(A7.1.4)}$$

$$r^3 \sin^2\theta \nabla^2\left(\frac{\Phi_\phi}{r\sin\theta}\right) + 2\left[\frac{\partial\Phi_r}{\partial\phi} + \frac{\partial\Phi_\theta}{\partial\phi}\cot\theta + r\frac{\partial\Phi_\phi}{\partial r}\sin\theta - \Phi_\phi \operatorname{cosec}\theta + \frac{\partial\Phi_\phi}{\partial\theta}\cos\theta\right] = 0$$

Ψ is a harmonic scalar so that $\nabla^2\Psi$ is zero where

$$\nabla^2 f = \frac{1}{r^2}\left[\frac{\partial}{\partial r}\left(r^2\frac{\partial f}{\partial r}\right) + \frac{\partial^2 f}{\partial\theta^2} + \frac{\partial f}{\partial\theta}\cot\theta + \frac{\partial^2 f}{\partial\phi^2}\operatorname{cosec}^2\theta\right] \qquad \textbf{(A7.1.5)}$$

and $\boldsymbol{r}$ is the radius vector $\{r,0,0\}$ of the material point.
Solutions in terms of the harmonic scalar Ψ lead to no solutions which are independent of those found in terms of $\boldsymbol{\Phi}$ and so will be ignored in the rest of this appendix.

A7.2 The General Method

The general theory of beams was discussed in Chapter 8. It relies on there being a characteristic response of prismatic elastic systems to resultant loading, as inferred from Saint-Venant's principle. Some non-prismatic systems, such as cones or curved beams, also have such characteristic responses. Those for a solid cone under axial loading or shear[1], torsion[2] and flexure[3] were found separately but a general method[4] will be described here which includes the responses of hollow cones.

Spherical polar coordinates will be used, taking the origin at the apex of the cone. The solutions of the individual problems will now be considered. In all cases, the faces of the cone will be taken as generated by surfaces of constant θ, where $\theta=\alpha$ for the outer surface and $\theta=\beta$ for the inner one. Föppl's solution for the torsion of a cone applies equally well to hollow cones and solid cones, because no tractions are induced on any conical surface of constant θ. In principle, a total of six independent expressions for the material displacements (and hence the stresses) are required to satisfy the conditions of zero traction on the inner and outer surfaces of the cone. The form of these expressions as functions of r and ϕ can be determined from the vector nature of the applied load. The solution is facilitated by using t, the tangent of the half-angle $\theta/2$, instead of θ. The hollow cone problem can be dealt with by expressing the equations in terms of T (or t^{-1}), the tangent of the complement of the half-angle (i.e. the cotangent of $\theta/2$). If this is combined with redefining ν as $-\nu$, the equations governing the form of the deflections are invariant under such a change of parameter. The physical significance of this transformation can be seen by noting that the angle of the radius vector to the cone axis can be defined as θ or $(\pi-\theta)$. The sense of positive

---

[1] Michell, J.H. (1900) London Math. Soc. Proc. V.32 p.29.

[2] Föppl, A. (1905) Sitzungsberichte Bayer, Akademie Wissenschaften V.35 p.249.

[3] Renton, J.D. (1997) J. Mech. Phys. Solids V.45 p.753.

[4] Renton, J.D. (1998) Journal of Elasticity V.49 p.101.

ν changes between the two cases, otherwise, the governing equations remain the same. This invariance can be seen from the original form of the equations and noting the relationship between the differentials given by

$$t\frac{d}{dt} = -T\frac{d}{dT} \quad \left(= \sin\theta\frac{d}{d\theta} \right) \tag{A7.2.1}$$

It means that solutions in terms of t immediately yield solutions in terms of T. In some cases, these two solutions are identical and then the method of Frobenius is applied. In all cases, this yields a sufficient number of independent solutions to describe the response to the tip loading when all the surface tractions are zero.

A7.3 An Axially-loaded Hollow Cone

For this case, the harmonic vector $\mathbf{\Phi}$ takes the form

$$\mathbf{\Phi} = \frac{1}{r}\begin{bmatrix} f(t) \\ g(t) \\ 0 \end{bmatrix} \quad \text{where} \quad \nabla^2\mathbf{\Phi} = \mathbf{0} \tag{A7.3.1}$$

After some re-arrangement, the harmonic conditions to be satisfied by this vector are

$$\begin{aligned} (1+2t^2+t^4)t\frac{d}{dt}\left(t\frac{df}{dt}\right) - 8t^2 f - 4(1+t^2)t^2\frac{dg}{dt} - 4t(1-t^2)g = 0 \\ 4t^2\frac{df}{dt} + (1+t^2)t\frac{d}{dt}\left(t\frac{dg}{dt}\right) - (1+t^2)g = 0 \end{aligned} \tag{A7.3.2}$$

Taking $f(t)$ and $g(t)$ to be expressed by series in terms positive powers of t yields the solutions

$$\begin{aligned} &\text{A:} \quad f(t) = 1 \;, \quad g(t) = -t \\ &\text{B:} \quad f(t) = \frac{1-t^2}{1+t^2} = \cos\theta \;, \quad g(t) = \frac{-2t}{1+t^2} = -\sin\theta \end{aligned} \tag{A7.3.3}$$

These correspond to the solutions used by Michell for an axially-loaded solid cone. Rewriting (A7.3.2) in terms of T and redefining g as $-g$ yields the same form. Solution A now becomes

$$\text{C:} \quad f(T) = f(t) = 1 \;, \quad g(T) = +T \quad \text{or} \quad g(t) = 1/t \tag{A7.3.4}$$

This means that the solutions given by (A7.3.3) yield new solutions in terms of T. However, the new solution derived from B is the same as the old one. The method of Frobenius is now applied to find a fourth independent solution of the form

$$f(t) = \cos\theta\ln t + \sum_{i=0}^{\infty} f_i t^{2i} \;, \quad g(t) = -\sin\theta\ln t + \sum_{i=0}^{\infty} g_i t^{2i+1} \tag{A7.3.5}$$

In this particular case, both power series are zero. The general form of the material deformation found from (A7.1.3) is then

$$G\mathbf{u} = \frac{1}{r}\begin{bmatrix} 4(1-\nu)(A + B + C\cos\theta + D\cos\theta\ln t) \\ 4(1-\nu)(-At + \frac{B}{t}) - (3-4\nu)(C\sin\theta + D\sin\theta\ln t) - D\cot\theta \\ 0 \end{bmatrix} \tag{A7.3.6}$$

This gives the stresses

$$\sigma_{rr} = -\frac{2}{r^2}[\,4(1-\nu)(A+B) + 2(2-\nu)\cos\theta(C+D\ln t) + 2\nu D]$$
$$\sigma_{\theta\theta} = \frac{2}{r^2}[2(1-\nu)\{A(1-t^2)+B(1-t^{-2})-D\} + (1-2\nu)\cos\theta(C+D\ln t) + D\cot^2\theta]$$
$$\sigma_{\phi\phi} = \frac{2}{r^2}[2(1-\nu)\{A(1+t^2)+B(1+t^{-2})\} + (1-2\nu)\cos\theta(C+D\ln t) - D\cot^2\theta - 2\nu D]$$
$$\tau_{r\theta} = \frac{2}{r^2}[2(1-\nu)(2At-2Bt^{-1}) + (1-2\nu)\sin\theta(C+D\ln t) + (3-2\nu)D\cot\theta] \qquad \textbf{(A7.3.7)}$$
$$\tau_{r\phi} = \tau_{\theta\phi} = 0$$

Given that these stresses are inversely proportional to r^2 and are not functions of ϕ, the equilibrium equations (A7.1.2) yield

$$\frac{d}{d\theta}(\bar{\tau}_{r\theta}\sin\theta) = (\bar{\sigma}_{\theta\theta} + \bar{\sigma}_{\phi\phi})\sin\theta \quad , \quad \frac{d}{d\theta}(\bar{\sigma}_{\theta\theta}\sin\theta) = \bar{\sigma}_{\phi\phi}\cos\theta - \bar{\tau}_{r\theta}\sin\theta \qquad \textbf{(A7.3.8)}$$

where the bar indicates that part of the expression for the stress given by (A7.3.7) which is in terms of θ only (i.e. the term r^{-2} is omitted). From these equations it follows that

$$\bar{\sigma}_{\theta\theta}\sin\theta - \bar{\tau}_{r\theta}\cos\theta = K\operatorname{cosec}\theta \qquad \textbf{(A7.3.9)}$$

where K is a constant. This indicates that a suitable linear relationship between the constants A to D will make K zero. In fact, it is only necessary to make D zero. Then if $\sigma_{\theta\theta}$ is zero for a particular value of θ, $\tau_{r\theta}$ will be too. This reduces the conditions imposed on the constants A to C to two, namely that $\sigma_{\theta\theta}$ should be zero at θ equal to α and β. This gives two conditions relating these constants and the remaining condition is given by evaluating the axial tensile force P:

$$\begin{aligned} P &= \int_{\beta}^{\alpha}[\sigma_{rr}\cos\theta - \tau_{r\theta}\sin\theta]2\pi r\sin\theta\, r d\theta \\ &= 4\pi\,\Big|_{\beta}^{\alpha}\cos\theta[2(1-\nu)(2A-2B+C) - C\sin^2\theta]\Big| \end{aligned} \qquad \textbf{(A7.3.10)}$$

This outlines the general method applied to the other problems, although it could be expressed more simply in this particular case. However, the inclusion of terms in D permits the solution of all problems associated with the displacement field given by (A7.3.1). The other cases can be dealt with similarly, so that only the outline solutions will be given.

A7.4 A Hollow Cone with a Tip Shear Force

A displacement field of the same form as that used by Michell for the problem of shear on a solid cone will be sought. This means that the harmonic vector $\mathbf{\Phi}$ is expressed by

$$\mathbf{\Phi} = \frac{1}{r}\begin{bmatrix}\cos\phi\, f(t)\\ \cos\phi\, g(t)\\ \sin\phi\, h(t)\end{bmatrix} \qquad \textbf{(A7.4.1)}$$

In terms of t, the harmonic conditions give rise to the equations

$$(1+t^2)^2\left[t\frac{d}{dt}\left(t\frac{df}{dt}\right)-f\right]-8ft^2-4t\left[(1+t^2)\left(t\frac{dg}{dt}+h\right)+(1-t^2)g\right]=0$$
$$(1+t^2)\left[t\frac{d}{dt}\left(t\frac{dg}{dt}\right)-2g\right]+4t^2\frac{df}{dt}-2h(1-t^2)=0 \qquad \textbf{(A7.4.2)}$$
$$(1+t^2)\left[t\frac{d}{dt}\left(t\frac{dh}{dt}\right)-2h\right]-4ft-2g(1-t^2)=0$$

These yield the solutions

$$\begin{aligned}
&\text{A:}\quad f(t)=t\ ,\quad g(t)=1\ ,\quad h(t)=-1\\
&\text{B:}\quad f(t)=0\ ,\quad g(t)=(1+t^2)\ ,\quad h(t)=-(1+t^2)\\
&\text{C:}\quad f(t)=\sin\theta\ ,\quad g(t)=\cos\theta\ ,\quad h(t)=-1\\
&\text{D:}\quad f(t)=t^{-1}\ ,\quad g(t)=-1\ ,\quad h(t)=-1\\
&\text{E:}\quad f(t)=0\ ,\quad g(t)=-(1+t^{-2})\ ,\quad h(t)=-(1+t^{-2})\\
&\text{F:}\quad f(t)=\sin\theta\ln t\ ,\quad g(t)=\cos\theta\ln t\ ,\quad h(t)=-\ln t
\end{aligned} \qquad \textbf{(A7.4.3)}$$

The general form of the displacement field in this case is

$$G\boldsymbol{u}=\frac{1}{r}\begin{bmatrix}\cos\phi[4(1-\nu)\sin\theta(A+F\ln t)+Ct+Et^{-1}]\\ \cos\phi[(3-4\nu)\cos\theta(A+F\ln t)+B(1+t^2)-D(1+t^{-2})+C-E-F]\\ \sin\phi[-(3-4\nu)(A+F\ln t)-B(1+t^2)-D(1+t^{-2})-C-E]\end{bmatrix} \qquad \textbf{(A7.4.4)}$$

This gives the stresses

$$\sigma_{rr}=-\frac{2\cos\phi}{r^2}[(4-2\nu)\sin\theta(A+F\ln t)+Ct+Et^{-1}-2\nu F\cot\theta]$$
$$\sigma_{\theta\theta}=\frac{2\cos\phi}{r^2}[(1-2\nu)\sin\theta(A+F\ln t)+B(t+t^3)+Ct+D(t^{-1}+t^{-3})+Et^{-1}+(3-2\nu)F\cot\theta]$$
$$\sigma_{\phi\phi}=\frac{2\cos\phi}{r^2}[(1-2\nu)\sin\theta(A+F\ln t)-B(t+t^3)-D(t^{-1}+t^{-3})-(1-2\nu)F\cot\theta]$$
$$\tau_{r\theta}=\frac{\cos\phi}{r^2}[-(2-4\nu)\cos\theta(A+F\ln t)-2B(1+t^2)+\tfrac{1}{2}C(t^2-3)+2D(1+t^{-2}) \qquad \textbf{(A7.4.5)}$$
$$-\tfrac{1}{2}E(t^{-2}-3)+(6-4\nu)F]$$
$$\tau_{\theta\phi}=\frac{\sin\phi}{r^2}[-2B(t+t^3)-Ct+2D(t^{-1}+t^{-3})+Et^{-1}-2(1-2\nu)F\operatorname{cosec}\theta]$$
$$\tau_{r\phi}=\frac{\sin\phi}{r^2}[(2-4\nu)(A+F\ln t)+2B(1+t^2)+\tfrac{1}{2}C(3-t^2)+2D(1+t^{-2})+\tfrac{1}{2}E(3-t^{-2})]$$

The constants A to F are determined from the surface tractions. If these are zero, F is zero and the other surface conditions determine the ratios of the constants A to E. Their magnitudes are determined from the tip shear force S. Taking this to act in the sense of positive θ in the plane $\phi=0$:

$$S=-\int_\beta^\alpha\int_0^{2\pi}[(\tau_{r\theta}\cos\theta+\sigma_{rr}\sin\theta)\cos\phi-\tau_{r\phi}\sin\phi]r^2\sin\theta\,d\phi\,d\theta \qquad \textbf{(A7.4.6)}$$
$$=-\pi\Big|_\beta^\alpha\cos\theta\{A[8(1-\nu)+2\sin^2\theta]+4B+3C+4D+3E\}\Big|$$

A7.5 A Hollow Cone in Torsion

For this case, the appropriate form of the harmonic vector is given by

$$\Phi = \frac{1}{r^2}\begin{bmatrix} 0 \\ 0 \\ h(t) \end{bmatrix} \tag{A7.5.1}$$

The first two harmonic conditions given by (A7.1.4) are satisfied automatically and the third is

$$2h\sin^2\theta + \sin\theta\left[\frac{d}{d\theta}\left(\sin\theta\frac{dh}{d\theta}\right)\right] - h = 0 \tag{A7.5.2}$$

One solution[5] of this is

$$h = A\sin\theta \tag{A7.5.3}$$

which gives rise to only one stress,

$$\tau_{r\phi} = -\frac{3A}{r^3}\sin\theta \tag{A7.5.4}$$

This means that all conical surfaces of constant θ are stress-free. The constant A is found from the torque T applied:

$$T = \int_\beta^\alpha 2\pi\tau_{r\phi}r^3\sin^2\theta\,d\theta = 2A\pi[3(\cos\alpha - \cos\beta) + \cos^3\beta - \cos^3\alpha] \tag{A7.5.5}$$

A7.6 A Hollow Cone in Flexure

For this case, the appropriate form of the harmonic vector is given by

$$\Phi = \frac{1}{r^2}\begin{bmatrix} \sin\phi\, f(t) \\ \sin\phi\, g(t) \\ \cos\phi\, h(t) \end{bmatrix} \tag{A7.6.1}$$

In terms of t, the harmonic conditions are

$$\begin{aligned}
&(1+t^2)^2\left(t^2\frac{d^2f}{dt^2} + t\frac{df}{dt} - f\right) - 4\left((1+t^2)t^2\frac{dg}{dt} + t(1-t^2)g\right) + 4t(1+t^2)h = 0 \\
&4t^2(1+t^2)\frac{df}{dt} + (1+t^2)^2\left(t^2\frac{d^2g}{dt^2} + t\frac{dg}{dt}\right) - 2(1-t^2)^2g + 2(1-t^4)h = 0 \\
&2t(1+t^2)f + (1-t^4)g + \tfrac{1}{2}(1+t^2)^2\left(t^2\frac{d^2h}{dt^2} + t\frac{dh}{dt}\right) - (1-t^2)^2h = 0
\end{aligned} \tag{A7.6.2}$$

The solutions found are

---

[5] The other solution can be formulated as a power series in t or $\sin\theta$.

$$\begin{aligned}
&A\colon f(t) = 2t \ , \quad g(t) = h(t) = -(1+t^2)\ ;\\
&B\colon f(t) = 0 \ , \quad g(t) = 1 \ , \quad h(t) = \frac{2}{1+t^2} - 1 = \cos\theta\ ;\\
&C\colon f(t) = \frac{t(1-t^2)}{(1+t^2)^2} = \tfrac{1}{2}\sin\theta\cos\theta \ , \quad g(t) = \frac{-2t^2}{(1+t^2)^2} = -\tfrac{1}{2}\sin^2\theta \ , \quad h(t) = 0\ ;\\
&D\colon f(t) = 2t^{-1} \ , \quad g(t) = (1+t^{-2}) \ , \quad h(t) = -(1+t^{-2})\ ;\\
&E\colon f(t) = \operatorname{cosec}\theta \ , \quad g(t) = \ln t \ , \quad h(t) = 1 + \cos\theta\ln t\ ;\\
&F\colon f(t) = \tfrac{1}{2}(\sin\theta - \operatorname{cosec}\theta + \sin\theta\cos\theta\ln t) \ , \quad g(t) = \tfrac{1}{2}(\cos\theta - \sin^2\theta\ln t) \ , \quad h(t) = 0
\end{aligned} \tag{A7.6.3}$$

These yield the general form of the displacement field given by

$$\begin{aligned}
Gu &= \frac{\sin\phi}{r^2}[2At + 2Dt^{-1} + (5-4\nu)\{\sin\theta\cos\theta(C+F\ln t) + (E-F)\operatorname{cosec}\theta + F\sin\theta\}]\\
Gv &= \frac{\sin\phi}{r^2}[-A(1+t^2) + B - C + D(1+t^{-2}) - (2-4\nu)\{\sin^2\theta(C+F\ln t) - F\cos\theta\}\\
&\qquad + (E-F)\cot\theta\operatorname{cosec}\theta + \{E(4-4\nu) - F\}\ln t]\\
Gw &= \frac{\cos\phi}{r^2}[-A(1+t^2) + (B-C)\cos\theta - D(1+t^{-2}) + (1+\cos\theta\ln t)\{E(4-4\nu) - F\}\\
&\qquad + (F-E)\operatorname{cosec}^2\theta]
\end{aligned} \tag{A7.6.4}$$

The stresses resulting from this displacement field are

$$\begin{aligned}
\sigma_{rr} &= -\frac{4\sin\phi}{r^3}[2At + 2Dt^{-1} + (5-\nu)\sin\theta\{\cos\theta(C+F\ln t) + F\}\\
&\qquad + \{(E-F)(5-4\nu) - \nu F\}\operatorname{cosec}\theta]\\
\sigma_{\theta\theta} &= \frac{2\sin\phi}{r^3}[A(t-t^3) + (1-2\nu)\{\sin\theta\cos\theta(C+F\ln t) + F\sin\theta\}\\
&\qquad + D(t^{-1} - t^{-3}) + \operatorname{cosec}\theta\{(1-\nu)(8E-6F) + 2\cot^2\theta(F-E) + F\}]\\
\sigma_{\phi\phi} &= \frac{2\sin\phi}{r^3}[A(3t+t^3) + D(3t^{-1} + t^{-3}) + 2(E-F)\operatorname{cosec}^3\theta\\
&\qquad + (1-2\nu)\{3\sin\theta(C\cos\theta + F + F\cos\theta\ln t) - F\operatorname{cosec}\theta\}]\\
\tau_{r\theta} &= \frac{\sin\phi}{r^3}[4A(1+t^2) - 4D(1+t^{-2}) - 3B + (8-4\nu)\{C + (F-E)\cot\theta\operatorname{cosec}\theta + F\ln t\}\\
&\qquad - (4+4\nu)\{\sin^2\theta(C+F\ln t) - F\cos\theta\} - 12E(1-\nu)\ln t]\\
\tau_{r\phi} &= \frac{\cos\phi}{r^3}[4A(1+t^2) + 4D(1+t^{-2}) - 3B\cos\theta - 12E(1-\nu)(1+\cos\theta\ln t)\\
&\qquad + (8-4\nu)\{\cos\theta(C+F\ln t) - F\cot^2\theta + E\operatorname{cosec}^2\theta\}]\\
\tau_{\theta\phi} &= -\frac{2\cos\phi}{r^3}[A(t+t^3) - D(t^{-1} + t^{-3}) + (1-2\nu)\{\sin\theta(C+F\ln t) - F\cot\theta\}\\
&\qquad - 2(E-F)\cot\theta\operatorname{cosec}^2\theta]
\end{aligned} \tag{A7.6.5}$$

The conditions of zero surface traction require that F be taken as equal to $2E$. There then remain four independent conditions of zero surface traction to be satisfied, giving relationships between the constants A to E. The bending moment about an axis in the plane $\phi=0$ and normal to that of the cone is then given by

$$M = \int_0^{2\pi}\int_\beta^\alpha (r\cos\theta\,\tau_{r\phi}\cos\phi + r\tau_{r\theta}\sin\phi)r^2\sin\theta\, d\theta\, d\phi$$
$$= \pi \big|_\beta^\alpha \cos\theta\{8(D-A)+(B-4C)(3+\cos^2\theta)+(1+\nu)[8C+4E(\sin^2\theta\ln t-\cos\theta)]\} + 2E(1+\nu)\big| \tag{A7.6.6}$$

A7.7 The Unloaded Cone Paradox

Carothers[6] found a solution for the bending of a plane wedge by a couple applied about its vertex. If this solution is taken to apply to wedge angles in excess of π radians, the problem becomes that of a notch. In particular, for a wedge angle of about 4.492 radians, a state of stress can exist when no couple is applied to the tip. In the case of the cone loaded by a bending couple M at its tip, analysed in §A7.6, A, E and F must be taken as zero to avoid singularities at $\theta=\pi$. The condition that M is zero is found from (A7.6.6) to be

$$[3B - 4C(1-2\nu) + 8D](1+\cos\beta) + (B-4C)(1+\cos^3\beta) = 0 \tag{A7.7.1}$$

The conditions that $\sigma_{\theta\theta}$ and $\tau_{r\theta}$ are zero on $\theta=\beta$ are given by

$$(1-2\nu)C\sin\beta\cos\beta + D(t^{-1} - t^{-3}) = 0$$
$$3A + C[4(1+\nu)\sin^2\beta - 8 + 4\nu] + 4D(1+t^{-2}) = 0 \tag{A7.7.2}$$

The condition that $\tau_{\theta\phi}$ is zero on this surface is then automatically satisfied, because E and F are zero. For a non-zero state of stress, either t is unity or

$$9(1+\nu)t^6 + (9-3\nu)t^4 + (7-5\nu)t^2 - (1+\nu) = 0 \tag{A7.7.3}$$

This has one positive real root for t which depends on the value of Poisson's ratio. When ν is zero, the corresponding conical notch angle, 2β, is 1.342 radians and when ν is 0.5, 2β is 1.755 radians. The solution for t equal to unity implies that the ratios of the constants are

$$B = 4C \quad , \quad D = -(1+\nu)C \tag{A7.7.4}$$

Arguing along the lines of Sternberg and Koiter[7], the solution of §A7.6 should not be used beyond the first pathological case, i.e. when the cone becomes a half-space (t=1).

---

[6] Carothers, S.D. (1912) Plane strain in a wedge. *Proc. Roy. Soc. Edinburgh* V.23 p.292.

[7] Sternberg, E. and Koiter, W.T. (1958) The wedge under a concentrated couple: a paradox in the two-dimensional theory of elasticity. *J. Appl. Mech.* V.25 p.575.

Appendix 8: Elements of Plate Theory

A8.1 The Governing Equations for Isotropic Plates

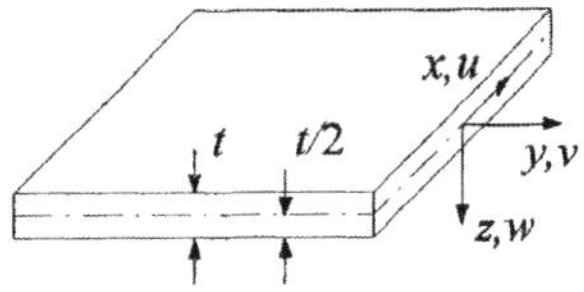

Figure A8.1 A plate element.

The usual assumptions of small-deflexion theory will be taken to apply. Figure A8.1 shows an element of a plate of thickness t. The directions of the x and y axes lie in the middle surface of the plate and the z axis is normal to it. The displacements u and v of material points within the plate are in the directions of the x and y axes. The deflexion of the middle surface normal to its plane is denoted by w. This surface is assumed not to deform within its own plane. Fibres perpendicular to the middle surface of the plate are assumed to remain perpendicular to this surface during deformation. This is known as *Kirchhoff's hypothesis*. Then due to the rotation of these fibres with the slope of the middle surface,

$$u = -z\frac{\partial w}{\partial x} \quad , \quad v = -z\frac{\partial w}{\partial y} \qquad \textbf{(A8.1.1)}$$

The strains within the plate are then given by

$$\epsilon_{xx} = \frac{\partial u}{\partial x} = -z\frac{\partial^2 w}{\partial x^2} \; , \quad \epsilon_{yy} = \frac{\partial v}{\partial y} = -z\frac{\partial^2 w}{\partial y^2} \; , \quad \gamma_{xy} = \frac{\partial u}{\partial y} + \frac{\partial v}{\partial x} = -2z\frac{\partial^2 w}{\partial x \partial y} \qquad \textbf{(A8.1.2)}$$

As in the elementary theory of bending, the stress σ_{zz} normal to the middle surface is taken to be negligible. The material is then in a state of plane stress, so that

$$\begin{aligned}
\sigma_{xx} &= \frac{E}{1-\nu^2}[\epsilon_{xx} + \nu\epsilon_{yy}] = -\frac{Ez}{1-\nu^2}\left(\frac{\partial^2 w}{\partial x^2} + \nu\frac{\partial^2 w}{\partial y^2}\right) \\
\sigma_{yy} &= \frac{E}{1-\nu^2}[\epsilon_{yy} + \nu\epsilon_{xx}] = -\frac{Ez}{1-\nu^2}\left(\frac{\partial^2 w}{\partial y^2} + \nu\frac{\partial^2 w}{\partial x^2}\right) \\
\tau_{xy} &= G\gamma_{xy} = -\frac{Ez}{1+\nu}\frac{\partial^2 w}{\partial x \partial y}
\end{aligned} \qquad \textbf{(A8.1.3)}$$

These stresses can be used to find the moments per unit length applied to an element as shown in Figure A8.2. These are given by

$$M_x = \int_{-t/2}^{t/2} z\sigma_{xx}\,dz = -D\left(\frac{\partial^2 w}{\partial x^2} + \nu\frac{\partial^2 w}{\partial y^2}\right)$$

$$M_y = \int_{-t/2}^{t/2} z\sigma_{yy}\,dz = -D\left(\frac{\partial^2 w}{\partial y^2} + \nu\frac{\partial^2 w}{\partial x^2}\right)$$

$$-M_{xy} = M_{yx} = \int_{-t/2}^{t/2} z\tau_{xy}\,dz = -D(1-\nu)\frac{\partial^2 w}{\partial x \partial y}$$

where $D = \dfrac{Et^3}{12(1-\nu^2)}$ **(A8.1.4)**

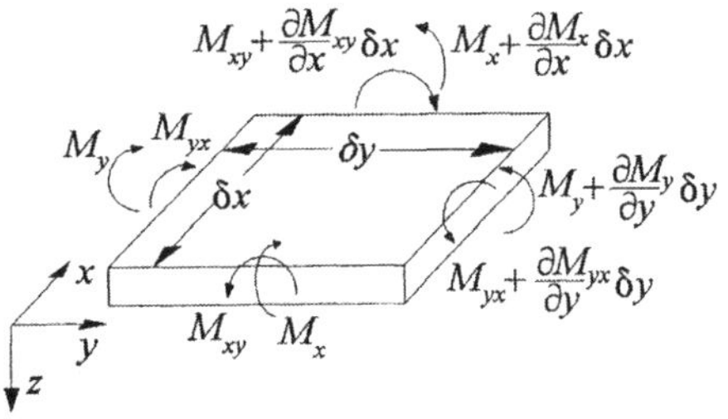

Figure A8.2 Moments acting on an element.

The moments per unit length may be related to the internal shear forces per unit length shown in Figure A8.2 by taking moments about the y and x axes respectively, giving

$$\begin{aligned} \frac{\partial M_x}{\partial x} + \frac{\partial M_{yx}}{\partial y} &= S_x \\ \frac{\partial M_y}{\partial y} - \frac{\partial M_{xy}}{\partial x} &= S_y \end{aligned} \qquad \textbf{(A8.1.5)}$$

Figure A8.3 shows the shear forces and a distributed load of intensity p per unit area acting on the surface of the plate. Resolving the forces in the z direction gives

$$\frac{\partial S_x}{\partial x} + \frac{\partial S_y}{\partial y} + p = 0 \qquad \textbf{(A8.1.6)}$$

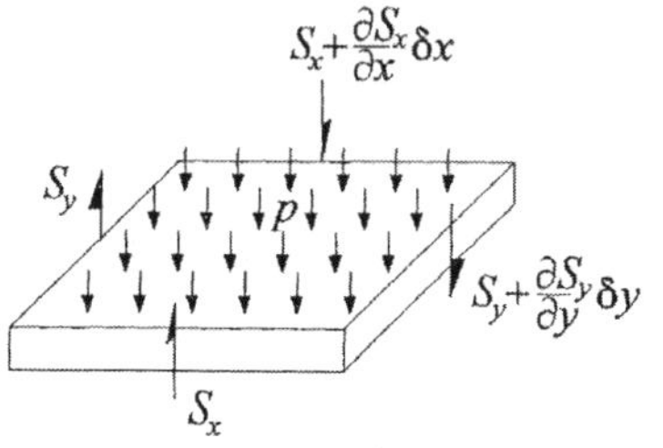

Figure A8.3 Shear forces on a plate element.

so that from (A8.1.5),

$$\frac{\partial^2 M_x}{\partial x^2} + \frac{\partial^2 M_{yx}}{\partial x \partial y} - \frac{\partial^2 M_{xy}}{\partial y \partial x} + \frac{\partial^2 M_y}{\partial y^2} + p = 0 \qquad \textbf{(A8.1.7)}$$

and on using (A8.1.4),

$$D\nabla^4 w = D\left(\frac{\partial^4 w}{\partial x^4} + 2\frac{\partial^4 w}{\partial x^2 \partial y^2} + \frac{\partial^4 w}{\partial y^4}\right) = p \qquad \textbf{(A8.1.8)}$$

A8.2 Boundary Conditions

Only two conditions can be imposed over any portion of the boundary. The moment about the local tangent to this portion can be specified, or its rotation about this tangent, but not both. Likewise, the normal displacement w of the portion of boundary can be specified, or the loading on it which does work during this displacement. (This loading is not simply an internal shear force of the kind shown in Figure A8.3.)

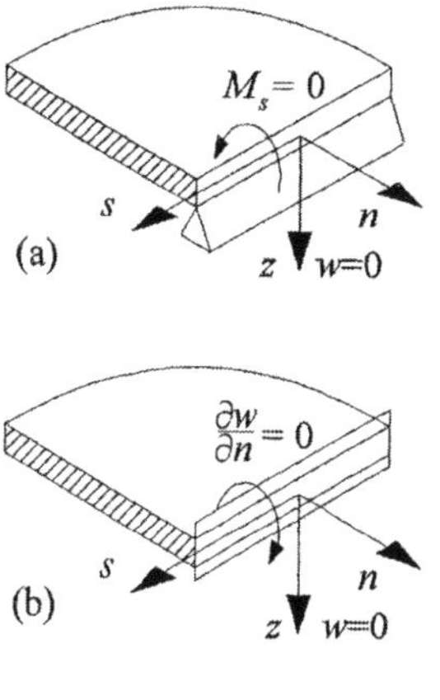

Figure A8.4 Support conditions.

The simplest support conditions are shown in Figure A8.4. The s axis runs along the boundary and the n axis is the outward normal to it. In Figure A8.4a, the boundary is supported on a knife edge, which exerts no moment about the boundary but prevents any displacement w along it. This means that the derivatives of w with s are zero locally. The moment M_s about the boundary is zero, and from the first two of equations (A8.1.4) it follows that this is given in terms of the second derivatives of w with respect to s and with n. As it has just been seen that the second derivative of w with respect to s is zero, it follows that the second derivative of w with respect to n is zero. The boundary conditions are then

$$w = \frac{\partial^2 w}{\partial n^2} = 0 \quad \text{at a knife edge.} \qquad \textbf{(A8.2.1)}$$

In the case of a clamped edge, both displacement in the z direction and rotation about the boundary are prevented, as shown in Figure A8.4b. The boundary conditions are then

$$w = \frac{\partial w}{\partial n} = 0 \qquad \text{at a clamped edge.} \qquad \textbf{(A8.2.2)}$$

Free edge conditions require that no physical constraints are placed on either w or $\partial w/\partial n$. It was seen in Figure 4.12 that a distributed moment m per unit length is similar to an end force of magnitude m. Similarly, a distributed moment which varies in intensity at a rate m' per unit length behaves like a distributed load of intensity m'. In the case of the edge of a plate, this distributed moment is M_{yx} along an edge parallel to the x axis and M_{xy} along an edge parallel to the y axis. The rates of variation of these moments produce an equivalent edge shear which has to be added to the edge shears given by (A8.1.5) and (A8.1.4). Free edge conditions are then given by setting the edge moment M_s to zero and this total edge shear, S^T to zero, or

$$\begin{aligned} M_y &= -D\left(\frac{\partial^2 w}{\partial y^2} + \nu\frac{\partial^2 w}{\partial x^2}\right) = 0\,, \\ S_y^T &= \frac{\partial M_y}{\partial y} - \frac{\partial M_{xy}}{\partial x} + \frac{\partial M_{yx}}{\partial x} = -D\left(\frac{\partial^3 w}{\partial y^3} + (2-\nu)\frac{\partial^3 w}{\partial x^2 \partial y}\right) = 0\,. \end{aligned} \qquad \textbf{(A8.2.3)}$$

for a free edge parallel to the x axis, and for a free edge parallel to the y axis,

$$\begin{aligned} M_x &= -D\left(\frac{\partial^2 w}{\partial x^2} + \nu\frac{\partial^2 w}{\partial y^2}\right) = 0\,, \\ S_x^T &= \frac{\partial M_x}{\partial x} + \frac{\partial M_{yx}}{\partial y} - \frac{\partial M_{xy}}{\partial y} = -D\left(\frac{\partial^3 w}{\partial x^3} + (2-\nu)\frac{\partial^3 w}{\partial x \partial y^2}\right) = 0\,. \end{aligned} \qquad \textbf{(A8.2.4)}$$

A further explanation of these conditions will be found in the next section.

A8.3 Strain Energy Analysis

The strain energy δU in the small volume of plate shown in Figure A8.2 can be determined from (3.37) and (A8.1.3).

$$\begin{aligned} \delta U &= \int_{-t/2}^{t/2} \frac{1}{2E}[\sigma_{xx}^2 + \sigma_{yy}^2 - 2\nu\sigma_{xx}\sigma_{yy} + 2(1+\nu)\tau_{xy}^2]dz\,\delta x\,\delta y \\ &= \tfrac{1}{2}D\left[\left(\frac{\partial^2 w}{\partial x^2}\right)^2 + \left(\frac{\partial^2 w}{\partial y^2}\right)^2 + 2\nu\frac{\partial^2 w}{\partial x^2}\frac{\partial^2 w}{\partial y^2} + 2(1-\nu)\left(\frac{\partial^2 w}{\partial x \partial y}\right)^2\right]\delta x\,\delta y \end{aligned} \qquad \textbf{(A8.3.1)}$$

The work done by the loads on the plate shown in Figure A8.5 is given by

$$W = \int_A pw\,dA + \oint\left(M_n\frac{\partial w}{\partial s} - M_s\frac{\partial w}{\partial n} + Sw\right)ds \qquad \textbf{(A8.3.2)}$$

where A is the area of the plate, M_n, M_s and S are the moments and shear force per unit length acting on the edge of the plate as shown and the line integral is around the total periphery. The first term in its integrand becomes

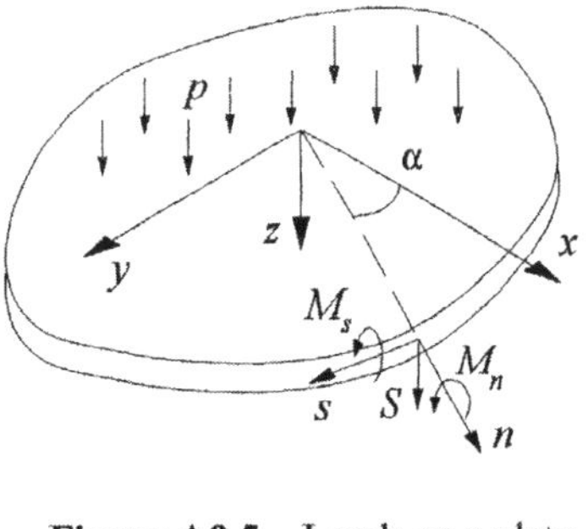

Figure A8.5 Loads on a plate.

$$\oint M_n \frac{\partial w}{\partial s}\, ds = \left| M_n w \right| - \oint \frac{\partial M_n}{\partial s} w\, ds = -\oint \frac{\partial M_n}{\partial s} w\, ds \qquad \textbf{(A8.3.3)}$$

where the first term on the right-hand side vanishes because integration begins and ends at the same point. Equation (A8.3.2) now becomes

$$W = \int_A p w\, dA + \oint \left[-M_s \frac{\partial w}{\partial n} + \left(S - \frac{\partial M_n}{\partial s} \right) w \right] ds \qquad \textbf{(A8.3.4)}$$

so that the term in round brackets represents an equivalent edge shear, S^T, as discussed in the previous section.

{The strain energy in an elementary area of the plate can be integrated over the whole plate to give the total strain energy. The effect of a small variation δw of w is now examined. The effect on the first term in the expression for the strain energy is

$$\delta \int_A \left(\frac{\partial^2 w}{\partial x^2} \right)^2 dA = 2 \int_A \frac{\partial^2 w}{\partial x^2} \frac{\partial^2 (\delta w)}{\partial x^2}\, dA = 2 \int_A \left[\frac{\partial}{\partial x} \left(\frac{\partial^2 w}{\partial x^2} \frac{\partial (\delta w)}{\partial x} - \frac{\partial^3 w}{\partial x^3} \delta w \right) + \frac{\partial^4 w}{\partial x^4} \delta w \right] dA \qquad \textbf{(A8.3.5)}$$

Green's theorem, applied to two functions $F(x,y)$ and $G(x,y)$ takes the form

$$\int_A \left(\frac{\partial F}{\partial x} - \frac{\partial G}{\partial y} \right) dA = \oint (G dx + F dy) = \oint (-G \sin\alpha + F \cos\alpha) ds \qquad \textbf{(A8.3.6)}$$

Also,

$$\frac{\partial}{\partial x} = \cos\alpha \frac{\partial}{\partial n} - \sin\alpha \frac{\partial}{\partial s} \quad , \quad \frac{\partial}{\partial y} = \sin\alpha \frac{\partial}{\partial n} + \cos\alpha \frac{\partial}{\partial s} \, . \qquad \textbf{(A8.3.7)}$$

Applying these equations to the term in round brackets on the right-hand side of (A8.3.5),

$$\begin{aligned} &\int_A \frac{\partial}{\partial x} \left(\frac{\partial^2 w}{\partial x^2} \frac{\partial (\delta w)}{\partial x} - \frac{\partial^3 w}{\partial x^3} \delta w \right) dA \\ &= \oint \left[\frac{\partial^2 w}{\partial x^2} \left(\cos\alpha \frac{\partial (\delta w)}{\partial n} - \sin\alpha \frac{\partial (\delta w)}{\partial s} \right) - \frac{\partial^3 w}{\partial x^3} \delta w \right] \cos\alpha\, ds \end{aligned} \qquad \textbf{(A8.3.8)}$$

Integrating by parts,

$$\oint \frac{\partial (\delta w)}{\partial s} \frac{\partial^2 w}{\partial x^2} \sin\alpha \cos\alpha\, ds = \left| \delta w \frac{\partial^2 w}{\partial x^2} \sin\alpha \cos\alpha \right| - \oint \delta w \frac{\partial}{\partial s} \left(\frac{\partial^2 w}{\partial x^2} \sin\alpha \cos\alpha \right) ds \qquad \textbf{(A8.3.9)}$$

where, as in (A8.3.3), the first term on the right-hand side is zero. Then from (A8.3.5) to (A8.3.9),

$$\begin{aligned} \delta \int_A \left(\frac{\partial^2 w}{\partial x^2} \right)^2 dA &= 2 \int_A \frac{\partial^4 w}{\partial x^4} \delta w\, dA + 2 \oint \frac{\partial^2 w}{\partial x^2} \cos^2\alpha \frac{\partial (\delta w)}{\partial n}\, ds \\ &+ 2 \oint \left[\frac{\partial}{\partial s} \left(\frac{\partial^2 w}{\partial x^2} \sin\alpha \cos\alpha \right) - \frac{\partial^3 w}{\partial x^3} \cos\alpha \right] \delta w\, ds \end{aligned} \qquad \textbf{(A8.3.10)}$$

Applying the same methods to the other terms in the expression for the variation in the strain energy, subtracting the variation in the work done given by (A8.3.2) and making use of the inverse of equations (A8.3.7) gives the variation of the total potential with a small variation in w

as

$$\delta\Pi = \int_A (D\nabla^4 w - p)\delta w \, dA + \oint \left[D\left((1-\nu)\frac{\partial^2 w}{\partial n^2} + \nu\nabla^2 w \right) + M_s \right] \frac{\partial(\delta w)}{\partial n} ds$$

$$+ \oint \left[-D\left((1-\nu)\frac{\partial^3 w}{\partial n \partial s^2} + \frac{\partial}{\partial n}\nabla^2 w \right) - \left(S - \frac{\partial M_n}{\partial s} \right) \right] \delta w \, ds = 0 \qquad \textbf{(A8.3.11)}$$

The total potential can be minimised independently with respect to arbitrary variations of w in the interior of the plate, the slope $\partial w/\partial n$ on the boundary and the displacement w on the boundary. Each of the three integrands in the above equation must then be zero, giving

$$D\nabla^4 w - p = 0$$

$$D\left((1-\nu)\frac{\partial^2 w}{\partial n^2} + \nu\nabla^2 w \right) + M_s = 0 \qquad \textbf{(A8.3.12)}$$

$$D\left((1-\nu)\frac{\partial^3 w}{\partial n \partial s^2} + \frac{\partial}{\partial n}\nabla^2 w \right) + \left(S - \frac{\partial M_n}{\partial s} \right) = 0$$

where the second bracketed term in the last equation is the total edge shear, S^T. These equations may be compared with (A8.1.8), (A8.2.3) and (A8.2.4).}

Appendix 9: Further Theoretical Considerations

A9.1 Orthogonality of Normal Modes

The normal modes of buckling and vibration of beams have orthogonal properties which can be used to advantage in some analyses. The will be derived for the flexure of non-uniform beams with workless end supports, but the method can be generalised to show orthogonality of the normal flexural-torsional modes of complete frameworks with workless supports and joints.

The governing differential equation for the flexure of a beam of variable stiffness EI with an axial load P is

$$\frac{d^2}{dx^2}\left(EI\frac{d^2v}{dx^2}\right) + P\frac{d^2v}{dx^2} = 0 \qquad \textbf{(A9.1.1)}$$

in the absence of any lateral loading. For given (workless) end conditions, this has a set of normal modes of deflexion (buckling) $v_i(x)$ corresponding to critical values of the axial load, P_{ci}. Each such mode will be taken to correspond to a unique buckling load, where the lowest load is P_{c1}. Then

$$\frac{d^2}{dx^2}\left(EI\frac{d^2v_i}{dx^2}\right) + P_{ci}\frac{d^2v_i}{dx^2} = 0 \qquad \textbf{(A9.1.2)}$$

Multiplying this by the jth mode and integrating over the length l of the beam gives

$$\int_0^l\left[\frac{d^2}{dx^2}\left(EI\frac{d^2v_i}{dx^2}\right) + P_{ci}\frac{d^2v_i}{dx^2}\right]v_j\,dx = 0 \qquad \textbf{(A9.1.3)}$$

Integrating this by parts,

$$\left|\left[\frac{d}{dx}\left(EI\frac{d^2v_i}{dx^2}\right) + P_{ci}\frac{dv_i}{dx}\right]v_j\right|_0^l - \int_0^l\left[\frac{d}{dx}\left(EI\frac{d^2v_i}{dx^2}\right) + P_{ci}\frac{dv_i}{dx}\right]\frac{dv_j}{dx}dx = 0 \qquad \textbf{(A9.1.4)}$$

The term in square brackets represents the end shear (cf. (9.2)) which is either zero or the end displacement v is zero for workless end reactions. Thus the first expression in (A9.1.4) is zero. Integrating the first term in the integrand of the second expression by parts,

$$\int_0^l\left[\frac{d}{dx}\left(EI\frac{d^2v_i}{dx^2}\right)\right]\frac{dv_j}{dx}dx = \left|EI\frac{d^2v_i}{dx^2}\frac{dv_j}{dx}\right|_0^l - \int_0^l EI\frac{d^2v_i}{dx^2}\frac{d^2v_j}{dx^2}dx \qquad \textbf{(A9.1.5)}$$

The first expression on the right-hand side gives the product of an end moment and an end rotation. For workless end reactions, one or the other of these is zero at each end. Then from (A9.1.4) and (A9.1.5), (A9.1.3) becomes

$$\int_0^l EI\frac{d^2v_i}{dx^2}\frac{d^2v_j}{dx^2}dx = P_{ci}\int_0^l\frac{dv_i}{dx}\frac{dv_j}{dx}dx \qquad \textbf{(A9.1.6)}$$

This process could be repeated, starting with the differential equation (A9.1.2) for v_j and multiplying it by v_i, giving

$$\int_0^l EI \frac{d^2 v_i}{dx^2} \frac{d^2 v_j}{dx^2} dx = P_{cj} \int_0^l \frac{dv_i}{dx} \frac{dv_j}{dx} dx \tag{A9.1.7}$$

As (A9.1.6) and (A9.1.7) differ only by the multipliers P_{ci} and P_{cj}, each side must be separately zero, giving the orthogonality conditions

$$\int_0^l EI \frac{d^2 v_i}{dx^2} \frac{d^2 v_j}{dx^2} dx = 0 = \int_0^l \frac{dv_i}{dx} \frac{dv_j}{dx} dx \quad (i \neq j) \tag{A9.1.8}$$

The governing equation for the vibration of a non-uniform beam is

$$\frac{\partial^2}{\partial x^2}\left(EI \frac{\partial^2 v}{\partial x^2} \right) + \rho A \frac{\partial^2 v}{\partial t^2} = 0 \tag{A9.1.9}$$

The mode associated with the ith natural frequency is of the form

$$v = V_i(x) \sin(\omega_i t + \alpha_i) \tag{A9.1.10}$$

Substituting this in (A9.1.9) gives

$$\frac{d^2}{dx^2}\left(EI \frac{d^2 V_i}{dx^2} \right) = \rho A \omega_i^2 V_i \tag{A9.1.11}$$

Multiplying this by the mode associated with the jth natural frequency and integrating over the length l of the beam gives

$$\int_0^l \left[\frac{d^2}{dx^2}\left(EI \frac{d^2 V_i}{dx^2} \right) \right] V_j \, dx = \int_0^l \rho A \omega_i^2 V_i V_j \, dx \tag{A9.1.12}$$

Integrating the left-hand side by parts twice, as before, and taking the end shear forces or the end displacements as zero, and the end moments or the end rotations as zero, gives

$$\int_0^l EI \frac{d^2 V_i}{dx^2} \frac{d^2 V_j}{dx^2} dx = \omega_i^2 \int_0^l \rho A \, V_i V_j \, dx \tag{A9.1.13}$$

Substituting the jth mode into (A9.1.9) and integrating the product with the ith mode gives

$$\int_0^l EI \frac{d^2 V_i}{dx^2} \frac{d^2 V_j}{dx^2} dx = \omega_j^2 \int_0^l \rho A \, V_i V_j \, dx \tag{A9.1.14}$$

As ω_i^2 and ω_j^2 are taken to be different, (A9.1.13) and (A9.1.14) can only both be true if

$$\int_0^l EI \frac{d^2 V_i}{dx^2} \frac{d^2 V_j}{dx^2} dx = 0 = \int_0^l \rho A \, V_i V_j \, dx \quad (i \neq j) \tag{A9.1.15}$$

These are the orthogonality conditions for flexural vibration modes. Both cases examined here are examples of the orthogonality of eigenfunctions in Sturm-Liouville problems.

References and Suggested Reading

Abramowitz, M. and Stegun, I.A. (*Eds.*) (1965) *Handbook of Mathematical Functions.* Dover Publications, New York.

Armstrong, I.D. (1969) *dynamic stability functions for continuous structures.* Heriot-Watt University, Edinburgh.

Bachman, H. *et al.* (1995) *Vibration Problems in Structures.* Birkhäuser Verlag, Basel.

Bareš, R. and Massonnet, C. (1966) *Analysis of Beam Grids and Orthotropic Plates.* Crosby-Lockwood, London.

Bleich, F. and Melan, E. (1927) *Die Gewöhnlichen und Partiellen Differenzengleichungen der Baustatik.* (Ordinary and Partial Differential Equations of Structural Statics. In German.) Julius Springer, Berlin.

Bleich, F. (1952) *Buckling Strength of Metal Structures.* McGraw-Hill, New York.

Bolotin, V.V. (1964) *The Dynamic Stability of Elastic Systems.* Holden-Day, San Francisco.

Bolotin, V.V. (1969) *Statistical Methods in Structural Mechanics.* Holden-Day, San Francisco.

Boresi, A.P. and Chong, K.P. (1987) *Elasticity in Engineering Mechanics.* Elsevier, U.K.

Borrego, J. (1968) *Space Grid Structures,* The MIT Press, Cambridge, Massachusetts.

Britvec, S.P. (1973) *The Stability of Elastic Systems.* Pergamon Press, New York.

Brown, B.M. (1965) *The Mathematical Theory of Linear Systems.* Chapman & Hall, London.

Chilver, A.H. (*Ed.*) (1967) *Thin-Walled Structures.* Chatto & Windus, London.

Cox, H.L. (1965) *The Design of Structures of Least Weight.* Pergamon, Oxford.

Dean, D.L. (1976) *Discrete Field Analysis of Structural Systems.* Springer-Verlag, Vienna.

Dean, R.G. and Dalrymple R.A. (1991) *Water Wave Mechanics for Engineers and Scientists.* World Scientific, Singapore.

Den Hartog, J.P. (1956) *Mechanical Vibrations.* (4th ed.) McGraw-Hill, New York.

Donnell, L.H. (1976) *Beams, Plates, and Shells.* McGraw-Hill, New York.

Dugdale, D.S. and Ruiz, C. (1971) *Elasticity for Engineers.* McGraw-Hill, London.

Engel, H. (1968) *Structure Systems.* Frederick A. Praeger, New York.

Flügge, W. (1962) *Handbook of Engineering Mechanics.* McGraw-Hill, New York.

Ford, H. and Alexander, J.M. (1977) *Advanced Mechanics of Materials.* Ellis Horwood, Chichester, U.K.

Gelfland, I.M. and Fomin, S.V. (1963) *Calculus of Variations.* Prentiss-Hall, Englewood Cliffs.

Gere, J.M. and Timoshenko, S.P. (1984) *Mechanics of Materials.* (2nd ed.) Brooks/Cole Engineering Division, Wadsworth, California.

Goldberg, D.E. (1989) *Genetic Algorithms in Search, Optimisation and Machine Learning.* Addison-Wesley, New York.

Green, A.E. and Zerna, W. (1968) *Theoretical Elasticity.* (2nd ed.) Oxford University Press, London.

Haug, E.J. and Arora, J.S. (1979) *Applied Optimal Design.* Wiley, New York.

Hearmon, R.F.S. (1961) *An Introduction to Applied Anisotropic Elasticity.* Oxford University Press, London.

Hemp, W.S. (1973) *Optimum structures.* Clarendon Press, Oxford.

Horne, M.R. and Merchant, W. (1965) *The Stability of Frames.* Pergamon Press, Oxford.

Ishii, K. (1995) *Membrane Structures in Japan.* SPS Publishing Company, Tokyo.

Ishii, K. (1999) *Membrane Designs and Structures in the World.* Shinkenchiku-sha Co. Ltd., Tokyo.

Jordan, C. (1965) *Calculus of Finite Differences.* Chelsea Publishing Company, New York.

Kirby, P.A. and Nethercot, D.A. (1979) *Design for Structural Stability.* Collins, London.

Koiter, W.T. (1945) *Over de Stabiliteit Van Het Elastisch Evenwicht.* (On the Stability of Elastic Equilibrium. In Dutch.) Thesis, Delft. English translation (1967) N.A.S.A. TT F *10*, 833.

Kolář, V., Kratochvíl, J., Leitner, F. and Ženíšek, A. (1979) *Výpočet plošných a prostorových konstrukcí metodou konečných prvků.* (Design of Plane and Space Structures by the Finite Element Method. In Czech.) SNTL Prague.

Korn, G.A. and Korn, T.M. (1961) *Mathematical Handbook for Scientists and Engineers.* McGraw-Hill, New York.

Lazan, B.J. (1968) *Damping of Materials and Members in Structural Mechanics.* Pergamon Press, Oxford.

Lekhnitskii, S.G. (1981) *Theory of Elasticity of an Anisotropic Body.* Mir Publishers, Moscow.

Levy, M. (1992) *Why buildings fall down: how structures fail.* Norton, New York.

Lin, T.Y. and Stotesbury, S.D. (1981) *Structural Concepts and Systems for Architects and Engineers.* Wiley, New York.

Livesley, R.K. (1975) *Matrix Methods of Structural Analysis.* (2nd ed.) Pergamon Press, Oxford.

Love, A.E.H. (1952) *A Treatise on the Mathematical Theory of Elasticity.* University Press, Cambridge.

Makowski, Z.S. (1965) *Steel Space Structures.* Michael Joseph, London.

Makowski, Z.S. (1985) *Analysis, Design and Construction of Barrel Vaults.* Elsevier, U.K.

Margarit, J. and Buxadé (1972) *las mallas espaciales en arquitectura.* (Space grids in architecture. In Spanish.) Gustavo Gili, Barcelona.

McCallion, H. (1973) *Vibration of Linear Mechanical Systems.* Longmans, London.

Modi, J.J. (1988) *Parallel Algorithms and Matrix Computation.* Oxford University Press, Oxford.

Morgan, W. (1968) *The Elements of Structure.* Pitman Press, Bath.

Morris, A.J. (*Ed.*) (1982) *Structural Optimisation: A Unified Approach.* Wiley, Chichester.

Muskhelishvili, N.I. (1963) *Some Basic Problems of the Mathematical Theory of Elasticity.* P.Noordhoff, Groningen.

Neal, B.G. (1964) *Structural Theorems and their Applications.* Pergamon Press, Oxford.

Niven, W.D. (*Ed.*) (1890) *The Scientific Papers of James Clerk Maxwell.* Vols. 1 and 2, Cambridge University Press, Cambridge.

Oden, J.T. (1967) *Mechanics of Elastic Structures.* McGraw-Hill, New York.

Ogden, R.W. (1984) *Non-Linear Elastic Deformations.* Ellis Horwood, Chichester.

Parkes, E.W. (1965) *Braced Frameworks.* Pergamon, Oxford.

Pilkey, W.D. *et al.* (1994) *Stress, Strain and Structural Matrices.* Wiley Interscience, New York.

Prager, W. (1961) *Introduction to Mechanics of Continua.* Ginn & Co., Boston, U.S.A.

Rayleigh, Lord {Strutt, J.W.} (1894, 1896) *The Theory of Sound Vols. I &II* (2nd ed.) Macmillan, London. Reprinted by Dover Publications, New York in 1945.

Reddy, J.N. (1984) *Energy and Variational Methods in Applied Mechanics.* Wiley, New York.

Renton, J.D. (1987) *Applied Elasticity - matrix and tensor analysis of elastic continua.* Ellis Horwood, Chichester, U.K.

Rosenblueth, E. (*Ed.*) (1980) *Design of Earthquake Resistant Structures.* Pentech Press, London.

Scanlan, R.H. and Rosenbaum, R. (1951) *Introduction to the Study of Aircraft Vibration and Flutter.* Macmillan, New York.

Smith, J.W. (1988) *Vibration of Structures - Applications in civil engineering design.* Chapman and Hall, London.

Soare, M.V. (1986) *Structuri discrete şi structuri continue în mecanica construcţiilor. Probleme unidimensionale.* (Discrete structures and equivalent continua in the mechanics of solids. One-dimensional problems. In Romanian.) Editura Academiei, Bucharest.

Sokolnikoff, I.S. (1956) *Mathematical Theory of Elasticity.* (2nd ed.) McGraw-Hill, New York.

Southwell, R.V. (1936) *An Introduction to the Theory of Elasticity.* Oxford University Press.

Tauchert, T.R. (1974) *Energy Principles in Structural Mechanics.* McGraw-Hill, New York.

Thompson, J.M.T. and Hunt, G.W. (1984) *Elastic Instability Phenomena.* Wiley, New York.

Thomson, W.T. (1988) *Theory of Vibrations with Applications.* (3rd ed.) Allen & Unwin, London.

Timoshenko, S.P. and Gere, J.M. (1961) *Theory of Elastic Stability.* (2nd ed.) McGraw-Hill, New York.

Timoshenko, S.P. and Goodier, J.N. (1970) *Theory of Elasticity.* (3rd ed.) McGraw-Hill, New York.

Timoshenko, S.P. and Woinowsky-Krieger, S. (1959) *Theory of Plates and Shells.* (2nd ed.) McGraw-Hill, New York.

Timoshenko, S.P. and Young, D.H. (1965) *Theory of Structures.* (2nd ed.) McGraw-Hill, New York.

Timoshenko, S.P. Young, D.H. and Weaver, W. (1974) *Vibration Problems in Engineering.* (4th ed.) Wiley, New York.

Topping, B.H.V and Khan, A.I. (1993) *Neural Networks and Combinatorial Optimisation in Civil and Structural Engineering.* Civil-comp Press, Edinburgh.

Vlasov, V.Z. (1959) *Тонкостенные Упругие Стержни.* (Thin-Walled Elastic Beams. In Russian.) English translation (1961) National Science Foundation. Oldbourne Press, London.

Wah, T. and Calcote, L.R. (1970) *Structural Analysis by Finite Difference Calculus.* Van Nostrand Reinhold, New York.

Warburton, G.B. (1964) *The Dynamical Behaviour of Structures.* Pergamon Press, Oxford.

Washizu, K. (1968) *Variational Methods in Elasticity and Plasticity.* Pergamon Press, Oxford.

Weyl, H. (1952) *Symmetry.* Princeton University Press, Princeton.

Yang, Y.B. and Kuo, S.R. (1994) *Theory and Analysis of Nonlinear Framed Structures.* Prentiss-Hall, Singapore.

Young, W.C. (1989) *Roark's Formulas for Stress and Strain.* (6th ed.) McGraw-Hill, New York.

Zbirohowski-Kościa, K. (1967) *Thin Walled Beams.* Crosby Lockwood & Son, London.

Index

Printed and bound by CPI Group (UK) Ltd, Croydon, CR0 4YY
15/05/2025
01873299-0001